DESIGN AND OPERATION OF FARM IRRIGATION SYSTEMS

Revised Printing

Edited by

M. E. JENSEN

National Research Program Leader, Water Management
Science and Education Administration, Agricultural Research
U.S. Department of Agriculture, Beltsville, MD 20705

An ASAE Monograph
Number 3 in a series published by

American Society of Agricultural Engineers
2950 Niles Road
St. Joseph, Michigan 49085
(phone 616-429-0300)

ASAE Technical Editor: James A. Basselman
September 1983

PREFACE

Irrigation is one of the oldest agricultural practices that enabled civilizations to establish permanent settlements. Irrigation has been practiced for at least 4,000 years and some of the ancient techniques used to distribute water to croplands are still in use today. The history of irrigation in various regions of the world is currently being written by member countries of the International Commission on Irrigation and Drainage. These ICID publications are expected to reveal the sources of many current technologies.

We have not attempted to assemble a history of irrigation in this monograph, but rather we have tried to bring together in one volume the state-of-the-art of the many facets involved in designing and operating farm irrigation systems. We also tried to improve the state-of-the-art where significant advances have been made.

This monograph differs from many in that, rather than providing a typical indepth review of the literature in a subject area, we also tried to provide guidance to practicing engineers and engineering students in using the material presented for designing irrigation systems. For this reason, we included sample calculations.

We also tried to bring out the importance of managing irrigation systems so that the potential efficiencies of the various systems can be attained. System evaluation provides the necessary feedback to system innovators and designers so that new concepts and further refinements in design criteria can be developed to correct obvious gaps.

We also tried to assemble in one place recent advancements in irrigation theory and science so as to form a foundation on which to build future technology. We expect graduate students will use this monograph as a major reference in their research. Hopefully, it also will help future irrigation and engineering students to identify the original sources of basic engineering and scientific principles that have advanced irrigation to its present level of technology.

First copyright and printing dates are shown in 1980, but some chapters were completed and typeset early in 1979. As a result, the latest references cited by authors range from 1978 to 1980.

In this revised printing we have made needed corrections that were identified by the senior authors and users of this monograph. We also corrected typographic errors and updated the list of standards and specifications given on Page 441.

M. E. Jensen, Editor
Fort Collins, Colorado

September 1983

ACKNOWLEDGEMENTS

Many people contributed to the development of individual chapters. We have not attempted to recognize such contributions.

Norman R. Scott, chairman of the ASAE monographs committee, encouraged the development of this monograph following solicitation of subject areas in which ASAE members felt comprehensive publications were needed (1974). The ASAE monograph committee reviewed the preliminary proposal and suggested changes and authors.

The Soil and Water Steering Committee reviewed the original outline and list of tentative authors and provided comments and suggestions (1975). Coauthors for each chapter were selected cooperatively with the senior author of each respective chapter (1975-76). Final chapter outlines were developed cooperatively with the coauthors (1976).

Special credit is due James Basselman and the ASAE publication staff for their efforts in finalizing the monograph for distribution after the second national irrigation symposium held in Lincoln, Nebraska in October 1980. I also appreciate the patience of ASAE's budget and management committees as several target dates for printing slipped by. Many senior authors, after being contacted to prepare chapters, advanced to new positions with added responsibilities. In several cases, senior authors retired and no longer had access to stenographic services. These factors significantly affected our ability to meet the original target dates.

Special appreciation is due my former secretary at Kimberly, Idaho (Irma Boyd) and my current secretary at Beltsville, Maryland (Judith Rundle) for handling the heavy correspondence load, including the typing of several chapters or parts of chapters.

Finally, I would like to acknowledge my wife Doris and son Eric for their patience in tolerating my preoccupation with lengthy manuscripts on weekends.

M. E. Jensen

CONTRIBUTORS

J. W. Addink	President, Addink Engineering Company, Inc, Lincoln, NE
F. K. Aljibury	Irrigation Specialist, Cooperative Extension, Research Center, University of California, Parlier, CA
C. L. Anderson	Irrigation Engineer, Soil Conservation Service, U.S. Department of Agriculture (Retired), Columbia, MO
R. S. Ayers	Soil and Water Specialist, Agricultural Extension Service, University of California, Davis, CA
R. H. Brooks	Project Director, Egypt Water Use and Management Project, Agency for International Development, U.S. Department of State, Cairo, Egypt
D. L. Bassett	Associate Professor, Agricultural Engineering, Washington State University, Pullman, WA
R. D. Burman	Professor, Agricultural Engineering Division, University of Wyoming, Laramie, WY
C. M. Burt	Associate Professor, California Polytechnic State University, San Luis Obispo, CA
C. R. Camp, Jr.	Agricultural Engineer, Coastal Plains Soil and Water Conservation Research Center, Science and Education Administration, Agricultural Research, U.S. Department of Agriculture, Florence, SC
H. G. Collins	Soil Conservation Service, U.S. Department of Agriculture, Portland, OR (Retired)
E. J. Doering	Agricultural Engineer, Northern Great Plains Research Center, Science and Education Administration, Agricultural Research, U.S. Department of Agriculture, Beltsville, MD
W. W. Donnan	Consulting Engineer, Pasadena, CA
H. R. Duke	Agricultural Engineer, Science and Education Administration, Agricultural Research, U.S. Department of Agriculture, Fort Collins, CO
D. D. Fangmeier	Professor, Department of Soils, Water, and Engineering, University of Arizona, Tucson, AZ
H. M. Gitlin	Extension Specialist, Agricultural Engineering Department, University of Hawaii, Honolulu, Hawaii
A. D. Halderman	Extension Specialist, Cooperative Extension Service, University of Arizona, Tucson, AZ
D. S. Harrison	Professor, Extension Agricultural Engineer, Florida Cooperative Extension Service, University of Florida, Gainesville, FL
W. E. Hart	Wastequip, Inc., Davis, CA
D. F. Heermann	Agricultural Engineer, Science and Education Administration, Agricultural Research, U.S. Department of Agriculture, Fort Collins, CO
G. J. Hoffman	Agricultural Engineer, U.S. Salinity Laboratory, Science and Education Administration, Agricultural Research, U.S. Department of Agriculture, Riverside, CA
T. A. Howell	Agricultural Engineer, Water Management Laboratory, Science and Education Administration, Agricultural Research, U.S. Department of Agriculture, Fresno, CA
A. S. Humpherys	Agricultural Engineer, Snake River Conservation Research Center, Science and Education Administration, Agricultural Research, U.S. Department of Agriculture, Kimberly, ID
W. R. Johnston	Assistant Manager, Westlands Water District, Fresno, CA
J. Keller	Professor, Department of Agricultural and Irrigation Engineering, Utah State University, Logan, UT
R. A. Kohl	Associate Professor, Department of Plant Science, South Dakota State University, Brookings, SD
H. C. Korven	Contract Specialist, Western Regional Headquarters, Research Branch, Agriculture Canada, Saskatoon, SK, Canada
J. N. Krider	Water Management Engineer, Northeast Technical Service Center, Soil Conservation Service, U.S. Department of Agriculture, Broomall, PA
E. G. Kruse	Agricultural Engineer, Science and Education Administration, Agricultural Research, U.S. Department of Agriculture, Fort Collins, CO
R. A. Longenbaugh	Professor, Civil Engineering Department, Colorado State University, Fort Collins, CO
B. L. McNeal	Professor, Department of Agronomy and Soils, Washington State University, Pullman, WA

v

J. L. Merriam	Professor Emeritus, California Polytechnic State University, San Luis Obispo, CA
D. E. Miller	Research Soil Scientist, Irrigated Agriculture Research and Extension Center, Science and Education Administration, Agricultural Research, U.S. Department of Agriculture, Prosser, WA
J. T. Musick	Agricultural Engineer, Conservation and Production Research Laboratory, Science and Education Administration, Agricultural Research, U.S. Department of Agriculture, Bushland, TX
P. R. Nixon	Agricultural Engineer, Science and Education Administration, Agricultural Research, U.S. Department of Agriculture, Weslaco, TX
W. J. Ochs	Chief Drainage Engineer, Engineering Division, Soil Conservation Service, U.S. Department of Agriculture, Washington, DC
C. H. Pair	Soil Conservation Service, U.S. Department of Agriculture, Kimberly, ID (Retired)
H. A. Paul	California State University, Chico, CA
J. T. Phelan	Soil Conservation Service, U.S. Department of Agriculture, Washington, DC (Retired)
E. J. Pope	Soil Conservation Service, U.S. Department of Agriculture, Lincoln, NE
W. O. Pruitt	Irrigation Engineer, Water Science Department, University of California, Davis, CA
P. A. C. Raats	Soil Scientist, Institute for Soil Fertility, Haren, The Netherlands
E. Rapp	Professor, Department of Agricultural Engineering, University of Alberta, Edmonton, AB, Canada
J. A. Replogle	Hydraulic Engineer, U.S. Water Conservation Laboratory, Science and Education Administration, Agricultural Research, U.S. Department of Agriculture, Phoenix, AZ
F. E. Robinson	Water Scientist, Division of Agricultural Sciences, University of California, El Centro, CA
M. N. Shearer	Extension Irrigation Specialist, Oregon State University, Corvallis, OR
R. W. Skaggs	Professor, Department of Biological and Agricultural Engineering, North Carolina State University, Raleigh, NC
R. E. Sneed	Associate Professor, Irrigation and Drainage, North Carolina State University, Raleigh, NC
L. B. Spiess	Irrigation System Specialist, Irrigation Division, Alberta Agriculture, Lethbridge, AB, Canada
E. C. Stegman	Professor, Agricultural Engineering Department, North Dakota State University, Fargo, ND
D. S. Stevenson	Head, Soil Science and Agricultural Engineering Section, Canada Research Station, Sumerland, BC, Canada
J. I. Stewart	Agriculture and Forestry Research Organization, Agency for International Development, U.S. Department of State, Mugaga, Kenya
T. Strelkoff	Professor, Land, Air, and Water Resources Department, University of California, Davis, CA
L. R. Swarner	Chief, Maintenance Branch, Engineering Research Center, Water and Power Resources Service, U.S. Department of Interior, Denver Federal Center, Denver, CO
G. T. Thompson	Extension Irrigation Engineer, Irrigated Agriculture Research and Extension Center, Cooperative Extension Service, Washington State University, Prosser, WA
A. W. Warrick	Professor, Department of Soils, Water, and Engineering, University of Arizona, Tucson, AZ
L. S. Willardson	Professor, Department of Agricultural and Irrigation Engineering, Utah State University, Logan, UT
R. J. Winger, Jr.	Chief, Ground Water and Drainage Branch, Water and Power Resources Service, U.S. Department of Interior, Denver Federal Center, Denver, CO
J. W. Wolfe	Professor, Department of Agricultural Engineering, Oregon State University, Corvallis, OR
G. Woodward	Soil Conservation Service, U.S. Department of Agriculture, Lincoln, NE (Deceased)
J. L. Wright	Soil Scientist, Snake River Conservation Research Center, Science and Education Administration, Agricultural Research, U.S. Department of Agriculture, Kimberly, ID
I-pai Wu	Professor, Agricultural Engineering Department, University of Hawaii, Honolulu, Hawaii

CONTENTS

CHAPTER 5 — SALINITY IN IRRIGATED AGRICULTURE
G. J. Hoffman, R. S. Ayers, E. J. Doering and B. L. McNeal

CHAPTER 6 — WATER REQUIREMENTS
R. D. Burman, P. R. Nixon, J. L. Wright and W. O. Pruitt

CHAPTER 7 — DRAINAGE REQUIREMENTS AND SYSTEMS
W. J. Ochs, L. S. Willardson, C. R. Camp, Jr.,
W. W. Donnan, R. J. Winger, Jr. and W. R. Johnston

CHAPTER 8 — LAND SHAPING REQUIREMENTS
C. L. Anderson, A. D. Halderman, H. A. Paul and E. Rapp

CHAPTER 9 — FARM WATER DELIVERY SYSTEMS
J. A. Replogle, J. L. Merriam, L. R. Swarner and J. T. Phelan

CHAPTER 10 — FARM PUMPS
R. A. Longenbaugh and H. R. Duke

CHAPTER 11 — FARM WATER DISTRIBUTION SYSTEMS
E. G. Kruse, A. S. Humpherys and E. J. Pope

chapter 1

INTRODUCTION

1

INTRODUCTION

by   M. E. Jensen, USDA-SEA/AR, Beltsville, MD

In the past, irrigation enabled civilization to establish permanent home sites in semiarid and arid lands. Today, irrigated agriculture continues to make food and fiber supplies less dependent on fluctuations in climate.

Irrigation is one of the oldest known agricultural technologies, but improvements in irrigation methods and practices are still being made. The future will require even greater improvements as competition for limited water supplies continues to increase.

Fukuda (1976) recently summarized irrigation history. He emphasized the development of the technical aspects of irrigation and drainage. The oldest civilizations developed along the Nile, Tigris, Euphrates, Indus and Yellow Rivers and in Latin America. Cultivation along the Nile began about 6,000 B.C. Practices to keep canals free of sediment were in effect in Mesopotamia in 4,000 B.C. Shallow wells and flooding from the Indus River were used about 2,500 B.C. Records show that irrigation was practiced along the Yellow River in 2,627 B.C. and in Peru about 1,000 B.C.

A major project to develop the history of irrigation throughout the world has been initiated by the International Commission on Irrigation and Drainage (ICID). The first volume by Y. Yousry (1978) describes irrigation and drainage in the Arab Republic of Egypt. The history of irrigation can provide insights to some problems that irrigators must face in the near future. Other problems, like the rapid depletion of a major groundwater aquifer that is the main source of irrigation water, have not been encountered in the past. Similarly, the rapid rise in pumping costs because of escalating energy costs has created new challenges to irrigators, engineers and the irrigation industry. These problems were expected and this monograph was prepared to meet this emerging need.

1.1 DEVELOPMENT OF IRRIGATED LAND

Irrigated land development has kept pace with world population since about 1800 (Shmueli, 1973). In 1977 the Food and Agriculture Organization of the UN (FAO, 1977) estimated that the total global irrigated area was 223 million hectares (ha) and that this would increase to about 273 million ha by 1990. Buringh et al. (1975) estimated that, of the 3,419 million ha of potential agricultural land in the world, 470 million ha could be irrigated. A summary of irrigated area by regions and countries was presented by Zonn (1974) and reproduced by Fukuda (1976). A brief summary is presented in Table 1.1. The total of 233 million ha differs from the 1977 FAO estimate, although similar sources were used. Additional estimates and projections are presented in Section 2.2.

3

TABLE 1.1. MAJOR WORLD IRRIGATED AREAS*†

Continent and country	Agricultural land		
	Cultivated	Cultivated land irrigated	
	ha (1,000's)	ha (1,000's)	percent
Africa	214,000	8,929	4.2
Asia, excluding USSR	463,000	164,640	35.5
Australia and Oceania	47,000	1,701	3.6
Europe, excluding USSR	145,000	12,774	8.8
North and Central America	271,000	27,431	10.1
South America	84,000	6,662	7.9
USSR	233,000	11,500‡	4.9
Total	1,457,000	233,637	16.0

*Adapted from Zonn, ICID Bul., July 1974.
†FAO Production Yearbook, 1972, Vol. 26.
‡FAO data as of 1971.

The FAO (1979) estimated irrigated agriculture represented only 13 percent of the global arable land, but the value of crop production is 34 percent of the world total. The 1974 U.S. Census of Agriculture (USDC, 1978) indicated that 27 percent of the $81 billion of agricultural products, sold from farms with sales totaling $2,500 or more, was produced on 18 percent of the cropland. These data clearly indicate the role managing water resources (irrigation) plays in agricultural production today, as well as its future role.

Future development of irrigated land will be more difficult than it was in the past. New projects will be more expensive and escalating energy costs and limited energy supplies will affect project feasibility more than it did in the past (also see Sections 2.1 and 3.7).

Various groups are considering the future development of irrigated lands. In 1977, 15 national ICID committees presented reports on irrigation developments (ICID, 1977). The ICID currently is updating its 1969 publication, "Irrigation and Drainage in the World—A Global Review." In 1979, the FAO Committee on Agriculture emphasized the role of irrigation in food production.

1.2 IRRIGATION IN THE U.S.A.

1.2.1 Early Development

A continuous history of irrigation in the U.S. is not available, but evidence indicates that around 100 B.C. the Hohokam Indians built extensive canals in the Salt River Valley of Arizona. The Spanish settlers and missionaries established small irrigation projects in the Southwest during the 16th and 17th Centuries. Gulhati and Smith (1967) published a brief history of irrigation in the U.S.

In the mid-1800's, modern irrigation development in the U.S began along streams with the settlement of the West. Most of the earlier projects were developed by private enterprise. The Desert Land Act of 1877 and Carey Act of 1894 were created to stimulate private and state developments. The Reclamation Act of 1902 enabled the Federal Government to become involved in irrigation development. Currently, the Water and Power Resources Service (formerly the Bureau of Reclamation) projects deliver the total irrigation

TABLE 1.2. IRRIGATED AREA IN THE UNITED STATES

Year	U.S. Census data*			Irrigation Journal data†		
	Total irrigated area in the U.S.		Rate of growth per year	Total irrigated area in the U.S.		Rate of growth per year
	ha (1,000's)	ac (1,000's)	percent	ha (1,000's)	ac (1,000's)	percent
1939	7,278	17,983	—	—	—	—
1944	8,312	20,539	2.7	—	—	—
1949	10,484	25,905	4.8	—	—	—
1954	11,960	29,552	2.7	—	—	—
1959	13,421	33,164	2.3	—	—	—
1964	14,997	37,057	2.2	—	—	—
1969	15,832	39,122	1.1	—	—	—
1971	—	—	—	—	—	—
1972	—	—	—	20,215	49,951	—
1973	—	—	—	20,834	51,480	3.1
1974	16,691	41,243	1.1	21,461	53,029	3.0
1975	—	—	—	21,871	54,044	1.9
1976	—	—	—	23,032	56,911	5.3
1977	—	—	—	23,658	58,459	2.7
1978	20,700‡	51,000‡	3.0‡	23,834	58,893	0.7
1979	—	—	—	24,746	61,148	3.8

*U.S. Department of Commerce. 1978. 1974 Census of Agriculture, Vol. II, Part 9, Irrigation and Drainage on Farms.
†Irrigation Journal. 1972-1979. Irrigation Survey.
‡Estimate based on a compound growth rate of 3 percent since 1969.

water supply to 11 percent of the total irrigated area in the 17 Western States and a supplemental supply to an additional 9 percent.

1.2.2 Trends in Irrigation

The irrigated area in the U.S. has increased steadily since 1939 (Table 1.2). The largest recent percentage increases occurred in the subhumid and humid South and Southeast States (also see Table 2.1), but the largest increase in area occurred in the semiarid Central and Southern Great Plains (Table 1.3). The only apparent decrease in irrigated area was in the Central Mountain States. This was caused by an adjustment in the values for Utah in 1979 and by the decrease in area in Montana from 1974 to 1979.

The largest sprinkler irrigated areas are in the arid Pacific Northwest and in the semiarid Great Plains. In the subhumid Cornbelt and the arid Pacific Northwest, 84 and 53 percent respectively, of the irrigated land, are sprinkler irrigated. In 1979, 32 percent of the total irrigated area was sprinkler irrigated.

Irrigation trends also are related to major fluctuations in climate. For example, the rapid expansion of irrigation in the Texas High Plains in the 1950's was primarily caused by the severe drought in that region from 1953 to 1957 (Brown and Bark, 1971). More efficient pumps, better well-drilling techniques, and low cost and readily available fuels also stimulated the expansion of irrigation in areas where groundwater supplies were available. In

**TABLE 1.3. CHARACTERISTICS OF IRRIGATION DEVELOPMENT
IN THE UNITED STATES FROM 1974 TO 1979 BY REGIONS**

Region	Total area irrigated*			Sprinkler irrigated*			
	1974	1979	Change, percent	1974	1979	Change, percent	Percent of total†
	- - - - - - - - - - - - - - - - - - - hectares (1,000's) - - - - - - - - - - - - - - - - - - -						
1. Arid Southwest (AZ and CA)	4,010	4,470	+460 (+11)	651	835	+184 (+28)	19
2. Arid Pacific Northwest (ID, OR, and WA)	2,963	3,158	+195 (+7)	1,006	1,664	+658 (+65)	53
3. Semiarid Central Mountains (CO, MT, NV, UT, and WY)	4,587	4,280	-307‡ (-7)	529	559	+30 (+6)	13
4. Semiarid Central and Southern Great Plains (KS, NE, NM, OK, and TX)	7,343	8,987	+1,644 (+22)	1,766	2,884	+1,118 (+63)	32
5. Subhumid Cornbelt (IL, IN, MN, MO, and WI)	261	602	+341 (+131)	172	504	+332 (+193)	84
6. Subhumid and Humid South and Southeast (AR, FL, GA, LA, MS, NC, and SC)	1,993	2,511	+518 (+26)	504	799	+295 (+59)	32

*Irrigation Journal. 1979. Irrigation survey 1979. Survey Issue, p. 58A-58H.
†Percentage of the total 1979 area irrigated in the region.
‡Most of this decrease was caused by a downward adjustment of the amount of irrigated area in Utah in 1979.

1975, about 40 percent of the irrigation water supplies were from ground-water resources (Murray and Reeves, 1977) as compared with 11 percent in 1929 (Meinzer, 1939). The center pivot system also played a major role in the expansion of irrigation in the U.S. in the 1970's because it enabled farmers to commercially irrigate medium to coarse texture good agricultural land and other lands that previously were not suitable for irrigation. The center pivot system, which was patented in 1952, enabled automatic operation, good control of the amount of water applied per revolution and uniform water application (Splinter, 1976). The irrigation industry also became well organized to deliver, install and service center pivot systems.

1.2.3 Total Irrigated Areas

The "official" irrigated area in the U.S. is based on U.S. Census of Agriculture data. Only the land actually irrigated during the census year is counted. The last Census of Agriculture was conducted in 1978. Data from the 1978 Census of Agriculture were not available for inclusion in this monograph. The Water Resources Council estimated 18.2 million ha (45 million ac) were irrigated in 1975 for the Second National Water Assessment (U.S. WRC, 1978).

The Irrigation Journal (IJ) has been conducting annual surveys during the past decade because irrigation industries need current market data for planning product development and production levels (Morey, 1977). Census data were not available soon enough for this purpose. For example, the preliminary 1974 census data were released about 2 years after the survey and the official figures were released 4 years later in 1978 (USDC, 1978).

The IJ survey accounted for double cropping which would affect some estimates in Southern states and some alternate year irrigation, but these two items would not make up a very large part of the differences between the two surveys. Irrigated areas in some states, like Nebraska, that have been determined by other techniques including satellite methods to count center pivot systems, show large differences (Splinter, 1976).

The difference between the two 1969 irrigated area surveys was relatively small if we use the average growth rate from 1972 to 1976 in the IJ survey to extrapolate the IJ survey back to 1969. Using this procedure, I estimated that the IJ survey would have reported only about 18.5 million ha, which is only 17 percent greater than the 1969 census total. By 1974, the difference had increased to 29 percent. The IJ survey has been improved and the U.S. Census survey has been revised. Assuming that the compound rate of growth indicated by the IJ from 1972 to 1979 is valid (3 percent per year), I estimated that the 1978 census might show 20.7 million ha irrigated (51 million ac) in 1978.

1.3 COMPETITION FOR WATER

Competition for limited water supplies will significantly affect future irrigation development and practices. Depletion of groundwater reserves in the Great Plains will reduce the area of fully irrigated land. Currently, several alternatives to transfer water into this region are being evaluated, but these projects are not very encouraging because of the elevation of these lands relative to large fresh water supplies in the U.S. and because of competition for water in the areas of origin. A national public policy to use limited groundwater supplies mainly to offset droughts has not been established (NRC, 1976). Other similar water resources problems are discussed in Sections 2.2, 2.6, and 2.7.

1.4 SCOPE AND SIGNIFICANT ISSUES

This monograph brings together in 1 volume the many facets of technology required to design and operate farm irrigation systems. The design criteria are presented in SI metric units to encourage greater use of this monograph throughout the world and to speed the transition to the use of metric units in irrigation engineering in the U.S.

We tried to explain the current role of irrigation in food and fiber production because modern irrigation practice involves more than providing water for plant use and leaching. We included information on soil water characteristics, salinity, drainage, land shaping and new management practices so that practicing engineers, irrigation students and agronomists would have a ready reference to this information. This monograph also emphasizes the engineering aspects of irrigation. This is why we included sample calculations to assist the students and practicing engineers in using this book as a reference and a guide. We also tried to include the most recent advances in irrigation theory and science as a foundation for future technology.

1.4.1 State-of-the-Art

The current state-of-the-art in farm irrigation system technology by components of systems can be determined using the Table of Contents and Index. I have attemped to help identify some of these areas in the following paragraphs.

Resources. The resources that should be considered in planning farm systems are discussed in Sections 3.1 to 3.7. Section 3.7 briefly describes the economic and financial feasibility aspects and Section 3.8 covers environmental aspects that have recently become more important.

Soils. Important soil water parameters are described in Chapter 4 along with a summary of soil water dynamics. New concepts of describing infiltration capacities for design purposes are especially significant.

Salinity. Basic concepts of soil salinity, irrigation effects and salt tolerance of most common crops are presented in Chapter 5. Particularly important are the management practices for coping with salinity and sodium problems and toxic ion effects, and for reclaiming sodic soils. Also, criteria are presented to evaluate the suitability of waters for irrigation.

Water requirements. Alternative procedures for estimating water requirements are presented in Chapter 6. Recently developed crop curves for more accurately relating evapotranspiration (ET) to ET from standard reference crops like alfalfa and grass are presented. Procedures for estimating ET for trickle irrigated systems are presented in Section 16.5.

Drainage. Drainage is vital to irrigated agriculture. General drainage guidelines for irrigated land are presented in Chapter 7 along with design criteria.

Land shaping. Current practices in land shaping for surface irrigation are described in Chapter 8. Of particular interest is the recent development and use of laser-controlled equipment for final smoothing of level basins to tolerances that were not practical in the past.

Water delivery systems. Alternative schemes for delivering water to farms are described in Chapter 9. Future mechanization and automation probably will be needed to achieve high farm irrigation efficiencies along with improved water-delivery policies and practices.

Pumps and pump maintenance. Pump design is important if farmers are to obtain achievable pumping efficiencies. Of particular importance with escalating energy costs is pump maintenance. Guidelines for pump design and maintenance are presented in Chapter 10.

Farm distribution systems. Chapter 11 provides criteria for designing various farm distribution systems and structures. These include lined and unlined open channels and pipelines.

Surface irrigation systems. Chapter 13 summarizes the USDA Soil Conservation Service design criteria for the many types of surface irrigation systems. A detailed summary of current automation practices and structures is presented in Section 13.9. During the next decade I expect to see significant refinements in surface system design criteria using the surface hydraulics concepts presented in Chapter 12.

Sprinkler irrigation systems. Current sprinkler systems are described and design criteria are presented in Chapter 15.

Trickle irrigation systems. Trickle irrigation is the newest major irrigation method. Detailed design and operating criteria are presented in Chapter 16. Design and operating criteria for trickle irrigation systems differ significantly from those used to design other systems. In this report, we can consider trickle system design as being in a developmental period. Many challenges face engineers in designing systems that can achieve their potential efficiencies. Operation and maintenance criteria also are significantly different from those required by other irrigation systems. Water filtration, for example, is of special importance.

System evaluation. Changing to a new irrigation system is often not feasible for achieving higher irrigation efficiencies. There are many ways to improve existing systems, and system evaluation is needed to identify where changes can and should be made in both design and operation. System evaluation techniques are described in Chapter 17.

Irrigation water management. Basic water management concepts to meet various objectives are described in Chapter 18. Of particular importance are the yield-ET relationships and modern irrigation scheduling methods.

1.4.2 Recent Theoretical Advances and Future Design Needs

Soil water dynamics. Infiltration theory and flow of water in one, two and three dimensions are summarized in Chapter 4.

Hydraulics of surface irrigation. A comprehensive summary of new advances in surface irrigation hydraulics, developed during the past 2 decades, is presented in Chapter 12. The material in this chapter will be the building blocks of future surface irrigation design criteria. It will also enable developing better irrigation system management criteria.

Sprinkler irrigation flow dynamics. A summary of principles and current theory affecting sprinkler irrigation system design and operation is presented in Chapter 14.

Future design criteria. Some of the problems that are not adequately considered in designing surface irrigation systems concern infiltration uncertainties. Variation in infiltration rates from one irrigation to the next can be large and difficult to predict. Further refinement in design criteria may not be justified until soil parameters can be predicted more accurately. Future surface systems may require automatic feedback controls to regulate flow rates to compensate for unpredictable infiltration rates.

Surface irrigation design criteria need to make more effective use of computer solutions to surface hydraulics problems, like those presented in Fig. 12.9, 12.10, 12.12, and 12.13 and as described in Section 12.6.

More basic flow dynamics need to be incorporated into sprinkler system design criteria. Also, better criteria are needed to efficiently and effectively apply chemicals and plant nutrients in irrigation water.

1.5 FUTURE ENGINEERING CHALLENGES

Problems of water delivery to farms need to be addressed. Innovative new concepts will be needed to modernize many older irrigation projects so that the water delivery system, or delivery policies, do not limit irrigation efficiencies.

We still do not have economical irrigation systems that apply water with near perfect uniformity. Linear move lateral systems have this potential. When this goal has been achieved, we can develop methods to control the amount of water applied to only that essential for ET and leaching. Many existing surface systems cannot be operated efficiently without a large labor input.

Escalating energy costs discussed in Section 3.7 present new challenges to engineers attempting to develop optimum designs. New uses of alternative energy sources like wind will be receiving much attention during the 1980's.

Environmental concerns discussed in Section 3.8 and health hazards in developing countries (Sections 2.1 and 9.5) will require evaluating more project design alternatives and coordinating proposed designs with social and health scientists.

Selecting the most effective project improvement programs will require actual data on the design and operational constraints of existing irrigation systems. Too often assumptions are used to make management decisions in allocating limited resources for project improvements.

The irrigation challenges facing engineers will be greatest in developing countries where improved water management practices have a large potential to increase food and fiber production. A plan of action for an international program was discussed at a meeting of farm water management experts in Beltsville, Maryland, on May 13-15, 1980. This meeting was organized by FAO and hosted by USDA. The final plan of action, which will include significant irrigation system inputs, will involve programs at the farm and village level, the national government level and at the international level. This monograph is expected to significantly affect the designs of projects implemented under this United Nations program.

Dallaire (1977), in discussing the role of the engineer in the development of civilization, refered to the writer Peter F. Drucker, who argued that yesterday's technological changes were hardly greater than the first great revolution technology wrought on human history 7,000 years ago, when irrigation civilization established itself. Technology, the result of creative efforts of engineers and scientists, may be the major force shaping civilization. Drucker stated that technological changes create the need for social and political innovations and that new technology makes the existing institutional arrangements obsolete. Irrigation technology played a major role in bringing about social and political changes. We can expect new irrigation technology to play a similar future role.

The authors and I challenge engineers and scientists to use information presented in this monograph as building blocks for the future. The material presented should be considered as an interim state-of-the-art publication as bold new approaches are developed to use water resources for increasing and stabilizing food and fiber production.

References
1 Brown, M. J., and D. Bark. 1971. Drought in Kansas. Kansas State Agr. Exp. Sta. Bul. 547.

2 Buringh, P., H. D. J. van Heemst and G. J. Stering. 1975. Computation of the absolute maximum food production of the world. Publ. No. 598, Agr. Univ., Wageningen, The Netherlands.

3 Dallaire, G. 1977. The engineer: what role in the development of civilization? Civ. Eng., Oct., p. 64-70.

4 FAO. 1977. Water for agriculture. Food and Agr. Organ. Of the UN, UN Water Conf., Mar del Plata, March, 26 p.

5 FAO. 1979. The on-farm use of water. Com. on Agr., 5th Session, 22 p.

6 Fukuda, H. 1976. Irrigation in the world. Univ. of Tokyo Press, Tokyo, 341 p.

7 Gulhati, N. D., and W. C. Smith. 1967. Irrigated agriculture: an historical review. Chp. 1, p. 3-11. In: Irrigation of Agricultural Lands, R. M. Hagan, H. R. Haise, T. W. Edminster (Ed.) Am. Soc. Agron. Monog. 11.

8 ICID. 1977. Likely irrigated agriculture of 2,000 A.D.: role of ICID and its national committees to meet the situation. Intern'l. Comm. Irrig. and Drain. and Iranian Nat'l Com., 270 p.

9 Jensen, M. E. 1978. Irrigation water management for the next decade. Proc. New Zealand Irrig. Conf., Ashburton, April, p. 245-302.

10 Meinzer, O. E. 1939. Groundwater in the United States. p. 157-232. In: USGS Water Supply Paper 836D, U.S. Govt. Printing Office.

11 Morey, D. 1977. Crystal balling the irrigation industry. Irrig. J. 27(2):6-7.

12 Murray, R. C., and E. C. Reeves. 1977. Estimated use of water in the United States in 1975. U.S. Geo. Surv. Cir. 765, 45 p.

13 NRC. 1976. Climatic fluctuations and water resources. Chp. 2, p. 31-41. In: Climate and Food. Rpt. of the Committee on Climate and Weather Fluctuations and Agricultural Production, BARR, Comm. on Natural Resources, Nat'l. Res. Council, Nat'l. Acad. Sci.

14 Schmueli, E. 1973. Efficient utilization of water in irrigation. p. 411-423. In: Arid Zone

Irrigation, D. Yaron, E. Danfors and Y. Vaadia (Ed.), Springer-Verlag, New York.

15   Splinter, W. E. 1976. Center-pivot irrigation. Scientific Am. 234:90-96.

16   U.S. Department of Commerce. 1978. 1974 Census of Agriculture, Chp. I Irrigation of Agricultural Lands. p. I-1—I-63. In: Vol. II, part 9, Irrigation and Drainage on Farms.

17   U.S. Water Resources Council. 1978. The Nation's Water Resources 1975-2000, Vol. 1: Summary, 96 p.

18   Yousry, Y. 1978. Past and present irrigation and drainage in Arab Republic of Egypt. Min. of Agr., Mimeo., 268 p.

19   Zonn, I. 1974. Irrigation in the world. Intern'l. Comm. Irrig. and Drain. Bul., mimeo. July, p. 26-33.

chapter 2

THE ROLE OF IRRIGATION IN FOOD AND FIBER PRODUCTION

2

2

THE ROLE OF IRRIGATION IN FOOD AND FIBER PRODUCTION

by   M. E. Jensen, USDA-SEA-AR, Beltsville, MD; D.
S. Harrison, University of Florida, Gainesville, FL;
H. C. Korven Research Branch, Agriculture
Canada, Saskatoon, SK, Canada; and F. E.
Robinson, University of California, El Centro, CA

2.1 INTRODUCTION

Water is essential for plant growth. Water is needed for seeds to germinate, seedlings to emerge, and for the many plant growth functions. Water prevents the dehydration of plants, and water provides the transport mechanism for plant nutrients and the products of photosynthesis. When water for plant growth can be controlled by irrigation, average yields under comparable climatic conditions generally are higher than those obtained under rainfed conditions. The difference in yields between irrigated and nonirrigated lands is greatest during seasons that have periods of drought and above normal evaporative demands. Because yields on irrigated lands are higher and more consistent, irrigation plays a major role in stabilizing food and fiber production.

Irrigation also affects food and fiber production in other ways. For example, irrigation may prevent severe freeze or frost damage to orchards, citrus nurseries, strawberries, ferns, and subtropical fruits. If fruit trees are severely damaged, many years would be required to reestablish them. Irrigation may provide the soil moisture needed to prepare a seed bed after an extensive period of drought before a normal rainy season. Irrigation has enabled otherwise nonproductive lands to be reclaimed. Irrigation has enabled the productive use of waste effluent from various food processing and municipal sewage plants that would otherwise require expensive renovation before being returned to rivers or applied to land disposal sites. Irrigation also enables controlling high plant temperatures by wetting the foliage, and increasing soil moisture to permit harvesting root crops such as sugarbeets, potatoes and peanuts.

Irrigation technology has advanced significantly during the past two decades, but many existing projects and on-farm irrigation systems have not been improved significantly for decades. The disasterous weather of 1972 caused a wide spread anxiety about world food supplies. The resulting acute food crisis caused the United Nations General Assembly to convene a World Food Conference in November 1974. During this conference, the top priority for joint development of land and water resources was the improvement and rehabilitation of irrigation systems because systems that once were neglected were not being fully utilized for food production. Also, productive agricultural areas were being damaged by applying excess water to lands with inadequate drainage facilities.

Substantial improvements in the water storage and project distribution systems also are needed. However, improving the water storage and distribution systems without improving on-farm irrigation systems may have little effect on food production and farm water management problems. Likewise, substantial improvements of on-farm irrigation systems to improve water utilization efficiency may require corresponding improvements in the water storage and distribution systems. Adequate water must be available when needed by the crops. Future distribution systems will enable the farmer to obtain water in the amount and at the rate necessary to achieve efficient irrigations and to reject delivery when it is no longer needed to minimize waste.

Centuries ago, irrigation enabled civilization to establish permanent sites of residence in arid and semiarid lands. Nomad activity no longer was needed to secure food for people and their animals. Today, irrigated agriculture continues to make civilization less dependent on the vagaries of climate for food and fiber needed to sustain life.

But large irrigation schemes in developing countries have recently resulted in increased adverse publicity. This publicity emphasizes potential adverse effects that may accompany some irrigation developments while downplaying the main purpose of bringing water and a better standard of life to millions of people (Worthington, 1976). Of major concern is the incidence of diseases that may be transmitted by mosquitoes, Simulium fly, tse-tse fly, snails and freshwater crustacean (White, 1976). Public health considerations might be given proper attention where the potential for increased disease problems exist if irrigation is introduced into an arid area that normally has a dispersed population. These precautions involve assuring potable drinking water supplies and adequate sanitation and washing facilities in areas where population density increases as a result of irrigation development. Socioeconomic impacts also need to be considered.

2.2 INCREASING AND STABILIZING FOOD AND FIBER PRODUCTION

2.2.1 Population Growth and Food Production

O. W. Israelsen devoted his life to the development and improvement of irrigation principles and practices. He is best known throughout the world for his text book entitled "Irrigation Principles and Practices" which was first published in 1932. In his second edition (Israelsen, 1950), he stated, "If the population of the world continues to increase at its present rate, where is the food for these people to come from? The men and women with knowledge of irrigation will be called upon to assist in the solution of the world problem."

The development of irrigated lands has paralleled increases in world population. Shmueli (1973) estimated that less than 10 million hectares (ha) were irrigated in 1800 and about 40 million ha were irrigated in 1900. By 1950 the irrigated land had increased to about 160 million ha and in 1969 to about 200 million ha. The UN Food and Agricultural Organizaton (FAO, 1977) estimated that the total world irrigated area was 223 million ha in 1975 and that this will increase to about 273 million ha by 1990. The second high priority activity that was developed at the 1974 World Food Conference was a targeted increase of 25 percent in the irrigated area during the 10-yr period,

1975 to 1985. Buringh et al. (1975) estimated that of the 3,419 million ha of potential agricultural land in the world, 470 million ha could be irrigated.

In 1930 the world population was two billion and by 1960 it had reached three billion. In 1976, it had doubled the 1930 level, reaching four billion. World population is expected to reach at least six billion by 2000 and without extensive population control could easily reach eight billion (Frejka, 1973).

In estimated 86 million ha throughout the world now need improvement of both the main and on-farm systems for distributing and applying irrigation water. The FAO also estimates that by 1990, developing countries in Africa, Latin America, the Near East and Asia will need 22 million ha of new irrigated land and 45 million ha will need improvements. The costs of making these improvements and developing new irrigated lands will be high, but FAO suggests that one way to decrease these costs is by developing national facilities and skills to reduce the dependence on imported expertise, equipment and materials. This Monograph is expected to play a significant role in application of current technology for improving and modernizing on-farm irrigation systems to increase water use efficiency and crop production, not only in the United States but throughout the world.

2.2.2 Irrigation Efficiency and Food Production

Food production can be increased by improving water storage and conveyance and on-farm irrigation systems. One of the most important terms that is used extensively by irrigation specialists in designing and operating irrigation projects is irrigation efficiency. However, the same term is not well understood by many policy makers and others only casually acquainted with irrigated agriculture. For example, in a recent report to Congress by the General Accounting Office entitled "Better Federal Coordination Needed to Promote More Efficient Farm Irrigation," (GAO, 1976) the report basically concluded that 50 percent of the western irrigation water supplies are wasted. The report also indicates that savings in water supposedly obtained by increasing irrigation efficiency could be used to irrigate additional farm land. Undoubtedly, many irrigated projects could reduce the net consumption of water by substantial improvements in the distribution and on-farm systems, but the savings in water generally will not be proportional to the changes in irrigation efficiency as is so often erroneously assumed. This is a very common misconception that is expressed by the general public when evaluating or considering the use of water for crop production.

In response to the GAO report, the interagency task force report (ITFR, 1979) indicated that "If all the measures in the Soil Conservation Service survey were implemented under a 25-yr accelerated program, it is estimated that conveyance efficiency could be increased 10 percent, and on-farm efficiencies 13 percent. In the 17 Western States, with no increase in irrigated acreage and providing water to water-short areas, return flows could be reduced by 35.3 million acre-feet and net depletion by 3.3 million acre-feet. This results in a reduction in diversions of 38.6 million acre-feet. The only water that could be available for other consumptive use is 3.3 million acre-feet." The estimated cost of implementing all of the measures involved was $14.6 billion. One of the State's views indicated that "..., the reduction in gross diversion will only provide instream flows between the diversion point and where the water returns to the stream." These comments clearly indicate the need for a better understanding of irrigation efficiency terminology.

The concept of irrigation efficiency is so important that it must be reviewed in this section in relation to potential increases in food production. Israelsen (1950) stated that "As a general rule, reduction of wastes and consequent increase in effective use of natural resources, including irrigation water, results in *economical* use, provided the costs of reducing wastes are not excessive." The basic concepts defined by Israelsen have been accepted by irrigationists for many decades. The expression "irrigation efficiency" was defined as the ratio of water consumed by the crops of the irrigation farm or project to the water diverted from a river or natural source into the farm or project canal or canals.

$$E_i = \frac{V_c}{V_w} \dots\dots\dots\dots\dots\dots\dots\dots\dots\dots\dots\dots\dots\dots \quad [2.1]$$

where V_c is the irrigation water consumed (evapotranspiration) by the crops of an irrigation farm or project during their growth period, and V_w is the water diverted from a river or other natural source into the farm or project canals during the same period of time.

To further elaborate on the significance of this term and the implications resulting from changes in irrigation efficiency we can consider that the volume of water withdrawn from the stream or another natural source such as groundwater can be used consumptively (evaporated) or nonconsumptively, that is V_w is equal to V_c plus V_{nc}. Thus, the irrigation efficiency term defined by Israelsen can also be expressed as

$$E_i = \frac{V_c}{V_w} = 1 - \frac{V_{nc}}{V_w} \dots\dots\dots\dots\dots\dots\dots\dots\dots\dots\dots \quad [2.2]$$

The net depletion of water within a river or groundwater basin, V_{dep}, is

$$V_{dep} = V_c + (1 - E_r)V_{nc} \dots\dots\dots\dots\dots\dots\dots\dots\dots \quad [2.3]$$

E_r is the fraction of V_{nc} that is or can be recovered (Jensen, 1977). The value of E_r is very site specific. The net or effective irrigation efficiency, E_e, is

$$E_e = \frac{V_c}{V_w} + E_r \frac{V_{nc}}{V_w} \dots\dots\dots\dots\dots\dots\dots\dots\dots\dots \quad [2.4]$$

which also can be expressed as

$$E_e = E_i + E_r(1 - E_i) \dots\dots\dots\dots\dots\dots\dots\dots\dots\dots \quad [2.5]$$

The irrigation efficiency term is misused because the recovery of irrigation water that is not consumed, $E_r(1 - E_i)$, in equation [2.5] is often completely ignored. This may be justifiable where E_r is very small or negligible but this is often not the case in many mountain valleys of the irrigated west and in many river basins. The recovery efficiency in many of the older projects may be as high as 0.9 or more. However, where energy is required to lift water from a river or a groundwater aquifer as illustrated in the following equation, any improvement in irrigation efficiency will reduce the net energy required for an irrigation project or farm if there is no significant recovery of this energy (Jensen, 1977).

$$Q = Q_w + Q_p - Q_r \quad \dotfill \quad [2.6]$$

where Q_w = the energy required to withdraw water from a river or groundwater aquifer, $Q_w \sim V_c \, z/E_i E_p$ where z is the lift, and E_p the pumping plant efficiency; Q_p = the energy required to apply water under pressure; and Q_r = the energy that can be recovered within the distribution network or from the surface return flow to a stream. The quantity of energy required to lift 10 m^3 of water (equivalent to 1 mm depth on 1 ha) a height of 1 m is 0.0272 kW·h at 100 percent irrigation and pumping plant efficiencies.

Irrigation systems needed to achieve high irrigation efficiencies generally require increased capital and in some cases increased energy consumption. Escalating energy costs and limited energy supplies are now focusing attention on possible conflicting objectives. During the past two decades low energy, labor intensive irrigation systems have been replaced with more efficient, but also more energy intensive systems because of readily available, cheap energy supplies at a time when labor costs were increasing. Conversion to more efficient irrigation systems that have higher energy requirements is being advocated. A key question is emerging, however, and that is, can both water conservation and energy conservation goals be met by converting from low energy to high energy irrigation systems?

Increased food production per unit of water consumed can be achieved on existing irrigation projects by increasing the water use efficiency. The total water evaporated from soil and transpired (ET) by a given crop during the growing season is generally very similar in different areas that have similar climatic conditions. Therefore, the major emphasis in the past has been to increase the production per unit area of land, or to increase the numerator of the following equation.

$$U_e = \frac{DM}{ET} \quad \dotfill \quad [2.7]$$

where DM is equal to the dry matter produced per unit area and ET is the volume of water used in evapotranspiration (Viets, 1965). More recently Shmueli (1973) emphasized that the optimization approach is to maximize the ratio of yield to irrigation water applied by maximizing the yield per unit area at the same time minimizing the seasonal amount of irrigation water applied. He emphasizes that there are two hazards in attempting to minimize the denominator of equation [2.7]. First, crop yield and financial return are reduced from the investment in irrigation systems that apply inadequate water. Second, soil salinity is increased from continued partial wetting of the root zone that does not leach out the salts. He stressed that current information on the amounts of water used and the salt content of the soil must be measured as part of the routine farm operations so that the salts can be leached when necessary at the proper time and in the proper manner.

Water use efficiency is also concerned in transferring water between basins. This will receive increased attention as supplies become limited. There are many who advocate the transfer of water from cool northern to warm semiarid and arid climates in the United States to permit using longer growing seasons and double cropping. The primary justification for the proposed transfer is increased production per unit land area and increased value of crops in the warm arid southwest relative to the cooler northern climates. A crucial and unanswered question is, "What will this do to water use efficiency?" Policy makers should weigh the production per unit of irrigation water in the cooler northern climates, where land is available on which to apply easily available water, against the cost of transporting water over great distances to the warmer climates to achieve greater production per unit land area. A critical analysis of these concepts will indicate whether production of crops per unit of irrigation water is higher in the northern or southern climates. Also, any increased production in warmer climates should be weighed against the costs and associated losses of water encountered in the transfer over great distances (also see Section 2.6.3).

2.2.3 Indirect Effects of Irrigation on Food Production

There are indirect beneficial effects of irrigation on food production that are known to agriculturalists in irrigated areas, but not to the general public. For example, the frost protection provided by irrigation significantly stabilizes production of orchard and citrus crops. Ranchers that graze their livestock on extensive areas of dry rangelands in the western United states depend upon a supply of feed to sustain their breeding herds and to overwinter their livestock. The role of irrigated agriculture in sustaining the livestock base was clearly illustrated during the severe drought of 1977 in the United States. During this drought, alfalfa hay produced on irrigated lands in Montana and Idaho was shipped as far east as Minnesota and South Dakota and as far southwest as California to sustain dairy and beef herds. In many arid and semiarid regions of the world, the livestock base that can utilize the limited forage on nonproductive rangeland depends upon a reliable supply of feed to survive dry periods.

In most countries the lands most easily irrigated have already been developed. Reservoirs and canals to develop new agricultural lands could cost $5,000 to $6,000/ha or more. Planners need to consider the benefits and costs of developing new lands relative to renovating old projects.

Escalating energy costs and limited energy supplies are creating rapid changes in operating costs of irrigation systems. Energy is requiring a greater portion of the annual irrigation budget. Designers of future irrigation systems should use an energy escalating factor to determine realistic operating costs. This factor is the equivalent annualized cost of escalating energy, or a sinking fund deposit factor, which is applied to the first year energy cost to estimate the average annual energy costs for the life of the system. Farm managers must place more emphasis on maximizing net returns on a unit area and per unit of water when they are justifying capital investments in modern irrigation systems (also see Chapter 3).

The irrigated agricultural base in the United States increased rapidly during the past decade and is now estimated to be about 21 million ha (see Chapter 1). However, large areas of irrigated lands in the High Plains of Texas and western Kansas will convert back to nonirrigated lands unless new sources of water are obtained to replenish diminishing groundwater supplies. Can the United States afford to lose this stabilized agricultural base? Should groundwater reserves be maintained to stabilize production during periods of major droughts both in the United States and throughout the world? On the other hand, we also must consider whether to pump existing groundwater supplies when fossil fuels are readily available and relatively inexpensive.

Another problem in the Pacific Northwest is whether to remove water for crop produciton, or to leave the water in rivers for hydroelectric energy production. Whittlesey (1978) indicated that increasing irrigation to achieve economic development can impose significant social costs upon a region. Large capital investments are required to service increased population growth in arid areas and to supply the energy needed for irrigation. In addition, stream flow, diverted from hydroelectric production to the consumptive use of agriculture, must be replaced.

Project formulation has encountered new constraints that require increasing skills and time resources. The formulation process has become more complex and more concerned with social and environmental aspects. Cobb (1976) raised the question, "Will the additional time requirement seriously depreciate our ability to respond effectively to problems?"

Improved on-farm irrigation technology will be a vital component of increasing food production in the future. This Monograph is only a first step towards improving the application of current technology and stimulating the development of new on-farm irrigation technology. Improved technology is needed not only to increase food production, but to maximize net return from the investment of all available resources such as capital, labor and especially water resources.

2.3 CLIMATE AND IRRIGATION REQUIREMENTS

2.3.1 Climate and Crops

The major climatic element determining which crops can be grown in an area is temperature. One of the principle indices characterizing growing seasons for annual crops is the frost free period. Some annual crops, such as beans and tomatoes, are extremely sensitive to light frosts. These crops cannot be planted in an area until minimum air temperatures remain above 0 °C. These sensitive crops grow only until the first frost in the fall. Other annual crops can survive light freezes in the spring and fall. Perennial crops, however, begin growing when the average temperature in the spring reaches and remains above about 5 °C and continue growing until the average temperature in the fall drops below 5 °C. Deciduous orchards are most common in the northern climates. Citrus can be grown only in southern states where only mild freezes occur during the winter months.

Maximum and minimum temperatures during the growing season also influence which crops can be grown since some crops grow well under high temperatures, whereas other crops do better under mild or cool temperatures. Day length is also a factor since the development and maturation of some crops are day length dependent.

2.3.2 Irrigation Requirements

Irrigation supplements natural precipitation. In arid areas annual precipitation is generally less than 200 to 250 mm and irrigation is the primary source of water for crops. Normally farm crops cannot be grown in arid areas without irrigation.

Semiarid areas receive from 400 to 500 mm of annual precipitation. Crops can be grown in semiarid areas without irrigation but low yields are common and the risk of crop failure is fairly high. Summer fallow or keeping

the land barren in alternate summer seasons is a common practice on nonirrigated lands in semiarid areas to increase the amount of precipitation stored
in the soil for use during the next cropping season.

Subhumid areas receive from 600 to 700 mm of annual precipitation. Irrigation may be needed for short periods during the growing season in
subhumid areas depending on the available water storage capacity of soils
and the rooting depth of the crops grown.

Humid areas generally receive more than 800 mm of annual precipitation. Irrigation is normally not needed in these areas except on very sandy
shallow soils or for crops that have shallow root systems.

Irrigation requirements depend on both the precipitation during the
growing season and the soil type (water holding characteristics) and the
rooting depth of the crops. Precipitation during the crop growing season provides some of the water needed by crops. Also, it generally decreases daily
evapotranspiration (ET) by decreasing solar radiation, increasing humidity,
and reducing the transfer of sensible heat from adjacent nonirrigated areas
to well-watered crop areas. The latter process is commonly known as advection.

The distribution of precipitation during the growing season also influences irrigation needs. For example, in some areas with a continental type
climate such as the Great Plains, summer precipitation tends to coincide
with crop water requirements. In some of the intermountain states,
precipitation is uniformly distributed throughout the entire year, or most of
the precipitation may occur during winter. Some tropical climates have wet
and dry seasons with favorable temperatures for crop growth throughout the
entire year. Crops can be grown during the wet season without irrigation, but
irrigation is necessary during the dry season. Some countries have a mediterranean type climate where the winters are wet and cool which enable growth
of many crops with little or no irrigations, and the summers are warm and
very dry.

Subhumid climates can also be thought of as areas where precipitation
is about the same as ET. These climates are typically characterized by short
dry periods in some years. Humid area climates are areas where precipitation
normally is greater than or equal to ET most of the year.

As previously mentioned the water holding characteristics of the soil and
the normal rooting depth and the evaporative demand determine the need for
irrigation in semiarid or subhumid climatic zones. For example, vegetable
crops are generally very shallow rooted and are seldom grown in subhumid
climates without provisions for irrigation. Occasionally soil characteristics
severely limit the normal rooting depth for crops. For example, winter wheat
grown on deep medium textured soils in western Kansas and in the Texas
High Plains normally will extract water to a depth of 2 m if the crop is stressed for water. However, winter wheat grown in Pakistan on medium textured
soils that are not saline and with a deep water table do not extract water
much below 0.5 m. Although the soil textures are similar, root environment
involving such factors as soil compaction and aeration limit the extractable
soil water. Time of planting and the duration of the tillering and root
development period also may be involved.

Detailed information on crop water requirements for optimum production is presented in Chapter 6. Information on water management practices
is presented in Chapter 18. As water supplies become more limited, the trend

in irrigation water management will be to apply just enough water so that the next incremental increase in yield just returns the cost of providing this additional increment. As water supplies become scarce, even less water will be applied.

2.3.3 Determining Irrigation Requirements

Most of the water requirements have been determined by field and plot experiments. These were initiated in the United States beginning in the 1890's. The objectives of these experiments were twofold: (a) to determine the quantity of water required for ET when crop growth was not limited by soil water; and (b) to determine the amount of irrigation water needed, the frequency at which the water should be applied for optimum crop production and water use efficiency. The early crop water requirement studies were largely determined using gravimetric soil sampling techniques to determine the change in soil water that occurred between irrigations. In the early 1960's, the neutron soil water probe essentially replaced gravimetric procedures except in the surface layers where gravimetric soil samples are still used if a surface probe cannot be used.

During the past two decades, more emphasis has been placed on determining the optimum water requirements of crops, i.e., their total ET and distribution during the growing season when they are irrigated to produce near maximum yields. More recently the emphasis has shifted to determining the marginal response of crops to decreasing increments of water as water supplies have become limited and to alternative management strategies.

Also during the past two decades many weighing lysimeters have been installed. Major sites using lysimeters to determine irrigation water requirements are those at Brawley and Davis, CA; Kimberly, ID; Prosser, WA; and Temple, TX. One experimental technique that is being used more extensively because of its convenience is a continuously variable irrigation treatment. This technique involves a single sprinkler lateral with closely spaced sprinklers to give a triangular water application pattern with the maximum amount applied at the lateral. Irrigations may be applied daily or every few days to maintain an optimum or above optimum water level near the lateral. Other variables such as fertilizer rates are also often used with this system. The major disadvantages of this technique are that all treatments usually are irrigated at the same time, low wind periods must be used to reduce the wind effects on the distribution pattern and the limitation of nonrandomized plot design.

Other techniques are also being used to determine the optimum water requirements. Miller (1977) recently summarized results of experiments on sugarbeets, winter wheat and dry beans where daily irrigations were applied using a sprinkler system. The treatments involved applying a range of more to less than the normal ET rate as predicted from pan evaporation the previous day. His results indicate that the amount of water applied relative to potential ET after full crop cover has been established can be substantially reduced without significantly reducing the value of the numerator (marketable yield) of equation [2.7] from its maximum. The resulting decrease in the denominator of equation [2.7] increases the water use efficiency.

One of the principal uses of the yield-irrigation relationships being developed is to validate computer models that simulate plant growth and production along with ET and irrigation requirements. In the past, most yield-water applied relationships indicated a curvilinear yield-water trend where yields first tended to increase linearly with increasing amounts of water. Then, the yield response to increasing increments decreased until there was no response to additional water. Yields sometimes decreased when excessive water was applied. Unfortunately these curvilinear relationships generally have not been single-valued functions because of the differences in climate from year to year, the differences in the distribution of irrigation water being applied, and the yield responses to water deficits at different growth stages. Therefore, one of the more recent objectives of research has been to develop simulation models or techniques to approximate single valued functional relationships. These relationships can be used in computer programs to simulate the probable effects of managing irrigation water in different ways. They also can be used to evaluate the probable need for irrigation in subhumid and humid areas without running costly field experiments for many years to obtain adequate representation of climatic conditions. If a good computer model that has been validated in an area is used with perhaps 20 to 50 yr of climatic data, an economic assessment, including a probability factor, can be made as to whether or not the farmer should invest in an irrigation system.

Detailed studies of optimization techniques in Israel summarized by Shumeli (1973) clearly show that storage of water in the soil during the winter precipitation period can permit reduced seasonal water application on summer field crops such as cotton. Because of limited water supplies, Israel has been able to increase its irrigated area about 25 percent without increasing irrigation water supplies.

As new technology such as higher yielding genotypes is developed, some field experiments will still be needed to validate simulation models for predicting yield response to various degrees of plant water stress, soil moisture regimes, and irrigation practices. Generally, the water requirements of the new dwarf, high yielding wheat varieties, for example, are not significantly different from that required by the older varieties (Shimshi et al., 1973). These results are to be expected since there is no unique relationship between transpiration and the harvested portion of dry matter production for a given type of crop. Previous experiments with fertilizer rates illustrated that doubling yields by using fertilizer may increase ET no more than 10 percent (Viets, 1965).

2.4 SUPPLEMENTAL IRRIGATION IN SUBHUMID AND HUMID AREAS

2.4.1 Introduction

Israelsen (1950) in discussing irrigation in humid climates, emphasized the fact that many soils in humid areas are shallow and are capable of storing small amounts of water available to plants between rains. He also discussed the frequency of drought periods in Michigan, Minnesota, Illinois, Indiana, and Ohio. Experience indicated that although rainfall was sufficient for most of the ordinary crops in typical years, short periods without rainfall have nearly ruined crops that would have brought ample returns to an investment in an irrigation system. He also mentioned that a 77-yr study in Missouri from 1870 to 1947 showed that one-fourth of the years have been very dry and caused excessive crop losses. From 1870 to 1930 there were 15 dry years in

TABLE 2.1. INCREASE IN IRRIGATED CROPLAND IN
THE SUBHUMID AND HUMID AREAS OF THE
UNITED STATES DURING A 5-YEAR PERIOD
DURING THE 1970'S (IRRIG. J., 1979)

Rank (1979)	State	Area irrigated 1974	1979	Increase 1974-1979
		- - 1,000's of ha - -		percent
1	Florida	759	925	22
2	Arkansas	687	687	0
3	Georgia	78	339	335
4	Louisiana	274	268	- 2
5	Mississippi	134	203	51
6	Minnesota	48	182	279
7	Missouri	88	143	63
8	Wisconsin	66	102	55
9	Iowa	24	89	271
Subtotal		2,158	2,938	36
Other Eastern States		284	603	112
Total		2,442	3,541	45

1 hectare (ha) = 2.471 acres.

which the Missouri corn crop was much below average. Since that time, there have been numerous studies in variations in climate and irrigation experiments in humid and subhumid areas. Acid subsoils, aluminum toxicity, and a dense A_2 soil layer have been shown to be major factors limiting rooting depth.

Israelsen emphasized that farmers in humid climates can assure themselves of larger and more dependable crop yields by providing irrigation systems and developing water supplies. However, he stressed the fact that this does not necessarily mean the farmer's profit will be increased.

Many farmers in subhumid and humid areas of the U.S. consider irrigation as insurance. Irrigation has developed very rapidly during the past decade in these areas (see Table 2.1). Ross (1976), referring to the Eastern Cornbelt as the sleeping giant, summarized some economic data that indicated that unless the farmer could obtain $99 per 1,000 kg ($2.52/bu) for corn in Michigan, irrigation would not cover the cost of the sprinkler irrigation system. In addition, he would need an average increase of at least 1,120 kg/ha (44 bu/acre).

Because of the recent droughts in 1974 and 1977, coupled with substantial technological improvement in low labor sprinkler irrigation systems, there has been a tremendous increase in irrigated land in the subhumid and humid areas of the United States (see Table 2.1). An important question is whether this upward trend will continue or level off. Energy costs, an emerging key economic factor, are escalating and represent a major component of the annual operating costs for sprinkler systems. With the current low prices for farm products, the additional annual operating cost of running sprinkler systems as compared with surface irrigation systems, can make the difference between profit and loss. Industry effort in developing improved sprinkler irrigation technology has been large. There has not been a comparable industry research and development effort in improving low energy gravity or surface irrigation systems utilizing a closed pipe distribution system.

In spite of Florida's annual precipitation of 1,350 to 1,680 mm, irrigation had increased rapidly since 1959. In addition to having the largest irrigated area in the humid South, Florida ranks among the top ten U.S. states having increased from 167,000 ha in 1959 to over 0.9 million ha in 1979.

Currently the irrigated area in the six Southeastern states (Alabama, Florida, Georgia, Mississippi, North Carolina, and South Carolina) amounts to 1.6 million ha.

Limited water holding capacity of southeastern soils, better knowledge of crop water requirements and crop responses to irrigation, erratic rainfall or precipitation distribution and double cropping have made irrigation more appealing today than ever before. Some farmers can no longer justify the high investment in high value crop production without supplemental irrigation to minimize the uncertainties of rainfall. In addition, there are other potential advantages and benefits to be derived from irrigation, such as environmental control (freeze protection and cooling), fertilizer application, and application of herbicides.

Irrigation is not a cure-all for problems of crop production. It is a production component that requires careful management and planning as with any other input resource. Irrigation requires large capital expenditures. Costs of initial investments range from $600 to $3,700/ha depending on whether water application is for soil water control only or to provide soil water control and cold protection.

2.4.2 Common Systems for Supplemental Irrigation

Sprinkler systems are the most common types used in humid areas of Florida. Initially, portable systems which required vast amounts of labor for moving pipe were used. Recent advancements in mechanization of sprinkler systems have resulted in the installation of principally permanent set, solid set, or mechanical move systems. The permanent systems are used for high cash crops such as citrus, for orchard crops or ornamentals, and golf lawns or courses with initial costs ranging from $2,500 to $3,000/ha. The large investment costs can be justified where the systems are used for frost or freeze protection. There are about 65,000 ha of citrus in Florida with permanent overhead sprinkler systems.

Seepage systems account for 40 percent of Florida's irrigation. The seepage or subsurface irrigation system distributes water through a series of ditches to maintain a water table to irrigate the crop. Additional information on seepage systems is available from the Florida Extension Service.

Most of the irrigated systems in the other Southeastern states are also sprinkler systems. The two most common mechanical move systems used are the center pivot and the self-propelled, or traveling big guns. The center pivot systems have increased rapidly during the 1970's (see Chapters 14 and 15).

The self-propelled, traveling gun units were introduced in 1966. The "traveler," as it is called, is a large sprinkler mounted on a four-wheel transport and powered by an internal combustion engine, water turbine, water piston, or oil hydraulic motor. Travelers are adaptable to odd sized and shaped fields and to a wide range of plant and soil conditions. Their primary disadvantages are the friction head loss in the 200-m length of 100-mm-dia hose (140 to 200 kPa), the adverse effect of wind on the uniformity of water application, and a medium labor requirement. However, they are extremely flexible which makes the unit attractive for citrus, vegetable, and field crops grown on relatively small fields in the Southeast.

Trickle irrigation is relatively new in the Southeast and it still has potential problems and unanswered questions associated with it. There are about 5,000 ha of citrus and other orchard crops and 1,200 ha of vegetable crops under trickle irrigation in Florida and other Southeastern states. The greatest problem in Florida is clogging which is caused largely by the iron and sulfur in the irrigation water. More knowledge and experience are still needed for trickle irrigation design (see Chapter 16).

2.4.3 Optimizing Irrigation and Drainage Systems

In humid areas the principal water management problem is drainage. Properly managed systems improve plant growth by increasing soil temperatures early in the spring permitting more rapid germination and establishment of a crop, and by increasing the rate organic matter is mineralized to nitrate nitrogen. Drainage also indirectly affects plant growth and crop production by permitting more timely field operations. Typically, the earlier most crops can be planted, the greater the yield. Also, wet soils are difficult to till, especially those having high clay contents. Studies in Ohio have shown that if the soil is well-drained, twice as many suitable days can be utilized for tilling early in the spring. Without drainage, wet spots cannot be planted or harvested. Typically, drainage may enable planting 1 or 2 wk earlier which reduces farming risks and generally results in higher yields.

Excessive drainage can increase the risk of a water deficiency during short periods of drought. Thus, drainage systems should be managed so as not to overdrain subsoil moisture that should be retained for carrying a crop through short drought periods. When this cannot be done, supplemental irrigation should be considered along with a manageable drainage system. Drainage and irrigation systems need to be optimized for given climatic, soil, and cropping conditions. Additional information on drainage is provided in Chapter 7.

2.5 IRRIGATION IN ARID AREAS

2.5.1 Introduction

Crops cannot be grown in arid areas without irrigation. Irrigation must provide almost all of the crop water requirements to meet ET needs. Also, irrigation must be managed to remove salt that accumulates in the soils since nearly pure water is removed by ET. In addition to a very limited amount of salt free water from precipitation, arid zones typically have higher ET rates. These high ET rates occur because of higher solar radiation at a given latitude associated with a lack of cloud cover, larger vapor pressure deficits, and energy advection from surrounding nonirrigated areas.

2.5.2 Leaching—A Vital Component of Irrigation Water Management

In arid areas crops require adequate irrigation to meet ET needs, but soils must also receive additional water to leach accumulated salts from the root zone. All water obtained from surface streams or subsurface aquifers typically contains dissolved solids. Repeated irrigations lead to accumulation of salts in the root zone as plants extract nearly pure water thereby concentrating the salt solution in the soil. Leaching provides a fraction of the applied water in excess of that which can be retained within the root zone to flush out the concentrated salts that have accumulated.

Specific crops add constraints to leaching efforts. Alfalfa, for example, develops root rots and scalding if the crowns are covered with ponded water during the hot summer months. Allowing the salts to accumulate during summer and leaching during winter is one solution.

As the salt concentration in the soil increases where leaching is inadequate or absent, the crop yields that decrease first are those from the least salt tolerant species—citrus, grapes, and stone and pome fruits. The yields of the moderately tolerant crops—sorghum, sugarbeets, and tomatoes—are affected next, and finally, those of the most tolerant crops—barley, bermuda grass, cotton, and sugarbeets—begin to decrease as the salts accumulate to higher concentrations. Additional water above ET requirements must be applied to the soils for leaching to minimize the yield reductions and to sustain agriculture in the area (also see Chapter 5).

Soil texture influences hydraulic conductivity and thus the ability to leach arid soils. The natural aggregation of soils can be broken down or defloculated as the soil exchange complex approaches saturation with sodium. This is especially a problem on clay soils but is relatively unimportant on sandy soils. Chemical amendments containing calcium can be added to the soil or the water to counter the influence of sodium.

Other physical properties of soil also influence the leaching of salts. Soils that have laminar layers of different textures are often associated with salt precipitation where movement of water is restricted. In the Imperial Valley of California, slip plows and deep chiseling are used to perforate the layers and allow water movement through the less permeable strata.

Impermeable strata beneath the depth of tillage often cause the water table or capillary fringe to rise into the root zone. If the water table level is not controlled then water containing more concentrated salts moves into the root zone by capillarity as evaporation from the soil surface or removal of water by crops take place. Where natural drainage is not sufficient to remove the water that percolates through the root zone then artificial drainage systems must be installed. Detailed information on salinity and leaching requirements are presented in Chapter 5.

2.5.3 Drainage—An Essential System Component

In areas where natural drainage is very limited, the water table is often controlled by cement or clay tile lines, or by plastic drain tubing. In most irrigated projects, the return flow from the drainage system reenters the main rivers and mixes with the normal flows. In other situations, provisions are made to provide a sink where the concentrated salts (leachates) may be deposited. The Salton Sea in California was designated as an evaporation sink for drainage water by the developers of the agricultural area. Other arid areas allow the leachates to drain to the oceans by special waterways. Such a system is utilized in Yuma, AZ and Mexicalli, MX areas to transport saline drainage waters to the California Gulf. Recently developed irrigated land in the San Joaquin Valley of California is producing drainwater that is utilized as wildlife areas for fish, game and bird habitat. But, additional drainage will be needed to avoid degradation of the San Joaquin River that serves as a water source for its northern reaches. Where highly saline drain water is expected from irrigation projects, a suitable sink for the drain water should be located before the development of the project to avoid costly litigation and future modification of the irrigation system after it is in operation.

2.5.4 Water Costs and Systems Used

The cost of water development in arid areas together with the energy required are reflected in the price paid for water. This price in turn can influence the type of irrigation system used. For example, the Colorado River water delivered by gravity system to the arid Imperial Valley of California currently costs $5.07/1000 m³ ($6.25/acre-ft). At this price surface irrigation systems have come into common use in the area. These systems have significant runoff which normally is not recirculated. The same Colorado River water pumped over a mountain range and through a costly distribution network costs over $138/1000 m³ ($170/acre-ft) in some areas of San Diego County, CA. As the cost of water increased in San Diego, irrigation systems were designed to operate more efficiently. Trickle irrigation of fruits, vegetables, and floral crops became common with additional sprinkler facilities provided to leach surface salt accumulations when they are not removed by winter rainfall.

Recent price increases for water in the Imperial Valley and penalties for wasting excess water has spurred the introduction of side roll sprinklers for use on alfalfa. Tailwater systems are being used with furrow and border irrigation systems to reduce surface runoff from farms. Other methods being tried are level basins and trickle irrigation systems.

Another factor that is resulting in an increase in irrigation efficiency is the rapid urbanization that is occurring in arid areas. People are attracted to the warm, snow free winters and bountiful sunshine. As more and more people move into the arid areas, water needs for cities, suburbs, municipalities, and industries increase the competition for available water supplies. Increases in labor and material costs also have caused growers to improve their ability to control irrigation applications. Solid set sprinklers on time clocks are now appearing in the San Joaquin Valley of California. As the sophistication of the system increases, the need for more precise information on ET and leaching requirements has resulted in an increase in the sophistication of measuring techniques.

2.6 COMPETITION FOR AVAILABLE WATER SUPPLIES

2.6.1 Introduction

The growing impact of society attitudes towards water supplies and management is having a greater impact on water users, planners, legislators and managers (Hagan, 1977). There are many new emerging demands for water. Many citizens today seem more concerned about environmental protection than about agriculture and are opposed to expanding irrigation. Other foreseeable demands for water will necessitate development of new energy sources. Increased emphasis is being placed on preserving free flowing streams and preserving natural conditions for fish and wildlife.

Direct competition for water use from irrigation and hydroelectric power has become a major issue in the Pacific Northwest. The demands on water supplies are steadily increasing and in many areas the groundwater supplies are being depleted. Irrigation is the largest consumer of water and, hence, has received a great deal of attention during the past decade and will be receiving greater attention in the future. At the same time, there is a tremendous reluctance to construct additional storage facilities, mainly because of environmental concerns and current economic aspects. There is also greater opposition to interbasin transfer and increased pressures to limit water supplies so as to form a growth controller (Hagan, 1977).

2.6.2 Arid Southwest

The competition for declining water supplies in the arid Southwest has been increasing steadily for decades. The continued growth of population is paralleled by the increase in water needs of municipal, industrial, and urban interests. In Arizona one legal source of nonagricultural water is the purchase of agricultural land and water entitlement and the use of water for nonagricultural purposes after retiring the land (Miller, 1977). Agricultural interests in the lower Colorado basin supported the development of the California Water Plan. This Plan permitted importation of northern California water for many San Joaquin Valley needs in addition to those of the Metropolitan Water District of southern California (MWD). This was a realistic approach to replacing the Colorado River water presently being used in excess of the MWD entitlement. As the population in the coastal California regions continued to increase, agricultural interests could envision a loss of their water source by condemnation or price manipulation if the alternate California Water Plan had not been developed. Even with the additional state water available, water for irrigation is under competitive pressure in coastal California. The MWD until recently provided substantial agricultural discounts for water. Increasing energy costs to pump water across the Tehachapi and San Bernandino Mountain ranges forced an increase in water prices to all users, including agriculture.

The competition for Colorado River water is not just local, however. It extends to state, regional, and even international proportions. The latest international action with Mexico occurred in 1973 with the signing of a treaty to guarantee $1.85 \times 10^9 \text{m}^3$ (1.5 million acre-ft) of water at a total dissolved solids level of not greater than 115 ± 30 ppm higher than the average of that delivered at Imperial Dam (Minute 242, Aug. 30, 1973).

Regional competition between the "Upper Basin" (Colorado, New Mexico, Utah, and Wyoming) and "Lower Basin" (Nevada, California, and Arizona) led to the Colorado River Compact of 1922 (Oregon State, 1967, and Holburt and Pellegrin, 1969). A usage limit of $9.25 \times 10^9 \text{m}^3$ (7.5 million acre-feet) was placed on each basin. The Upper Colorado River Basin Compact of 1948 settled the division of water among the four states. Then the Colorado River Storage Act of 1956 provided major storage and obligated the Upper Basin to deliver the agreed upon amount to the Lower Basin.

In contrast to the placid Upper Basin, the competition between states in the Lower Basin has a history of stormy confrontation. At one time Arizona troops were lined up along the Colorado banks to physically prevent development. Arizona at first refused to ratify the Colorado River Compact of 1922. After 22 yr of legal and even military shows of force, Arizona finally signed the compact in 1944 and received entitlement to $3.45 \times 10^9 \text{m}^3$ (2.8 million acre-ft) of water for consumptive use per year. The Central Arizona Project (CAP), envisioned by Arizona to utilize her Colorado River entitlement, is still at this time suffering pangs of birth. The U.S. Supreme Court decision (Arizona vs. California, 373 U.S. 546) cleared the way for CAP by dividing between these states the water of the Lower Basins' entitlement (7.5 million acre-ft) plus any surpluses. This agreement also placed the responsibility for delivery of water during drought years with the Secretary of Interior (Scott and Johnston, 1968). Plans for construction of the CAP suffered a setback in early 1977 with President Carter's removal of federal funds from the project in the interest of national solvency. The local press reported that Arizona state officials are contemplating a continuation of the project with state funding, as was done in California (Wuertz, 1977). Sober reflection of the cost of CAP, particularly to cities such as Phoenix and Tucson, may further delay the construction of CAP (Martin, 1976).

2.6.3 Pacific Northwest

Large additional areas of agricultural soils could be irrigated in the Snake River Basin and the Columbia River Basin of the Pacific Northwest. However, for each new hectare of land brought under irrigation in Idaho there is a potential annual loss of 10,400 kW·h (4,200 kW·h per acre) of electricity that could be generated if the water used in ET were allowed to flow through each of the 21 power plants on the Snake and Columbia Rivers. In addition, depending on whether the water needs to be pumped or whether the water is delivered by gravity, a substantial increase in demand for electricity may be required with each new hectare developed. Similarly, Whittlesey (1978) indicated that each new hectare of land in the Columbia Basin in Washington will result in a combination of lost hydropower and increased annual demand for power, ranging from 10,400 to 13,300 kW·h/ha (4,200 to 5,400 kW·h/acre) depending on the site.

One of the major concerns in the Pacific Northwest is whether or not developers of new irrigated lands should pay for power at the replacement cost rather than at the current average cost for generating power. The present cost of generating hydroelectric power is about $0.007 per kW·h, whereas the replacement cost is about $0.03 per kW·h. Other arguments used against further development of irrigated land in the Pacific Northwest relate to the total social overhead cost when considering new roads, schools, and related services that must be provided. One of the arguments against this policy, however, is that these services and social costs may be required regardless of whether irrigation is developed or not.

The key issue in the future will probably be whether new irrigated lands are needed to increase food production. With substantial areas of good land still available for development in the Pacific Northwest, the production of food per unit of irrigation water might be higher with local usage than if this water were transferred out of the basin to the Pacific southwest as proposed. Transfer out of the basin to the southwest would result in a loss of hydroelectric power and large transportation costs for mass transfer of large volumes of water.

2.6.4 Southern Canada

A relatively small portion (less than 1 percent) of Canada is irrigated.

Only in the southcentral portion of British Columbia is irrigation essential for crop production (Canadian National Committee, ICID, 1977). the potential for expanding the irrigated area is large. In Alberta the irrigated area could be increased by a factor of 10 and Saskatchewan by a factor of 8.5. Changes in water use in Canada are currently being studied. Canada faces an emerging scarcity problem of water quantity in the southern prairies. Currently there is little conflict between industrial and recreational uses and withdrawals for irrigation. In 40 yr Canada has developed more than 400,000 ha of irrigated land and anticipates, with a probability of 50 percent, of developing an additional 800,000 ha in the next 25 yr. (Canadian National Committee, ICID, 1977).

2.6.5 Humid Southeast

Currently, the southeast has ample water for its urban, recreational, industrial, and irrigation needs. However, in certain states, particularly near urban areas, competition exists between agriculture and other uses for some of the groundwater supplies. Restrictions on the use of available groundwater supplies and allocation of groundwater to urban, industrial, and agricultural users will increase in the next decade.

Most of the present regulation of water use involves underground supplies. As urban population centers continue to grow, agriculture may be seeking other sources of water, such as surface waters, and municipal and industrial effluents, for some of its irrigation needs. Substantial investments will be required to fully develop and utilize surface water supplies and urban people must develop an understanding that some of the lakes, streams, and canals must be used for irrigation purposes. New regulations on water use is expected to present new challenges to irrigated agriculture during the coming decade. These include development of water use regulation to justify large uses and to demonstrate the benefits derived therefrom, development of surface water reservoirs and preservation of existing natural reservoirs. New regulations affecting the quality of water will also have an impact on agricultural practices. The objective of these regulations will be to reduce the rates at which nutrients are released into water supplies, as organic soils and mineral soils are used intensively for agricultural production.

It is anticipated that large users of water will require a use permit. Permit authorization will require agricultural producers to document their water needs when priorities for allocating water supplies are being established.

Many agricultural producers now use deep wells for irrigation while surface water supplies may be available in lakes and streams. Storage of rainfall in reservoirs and preservation of natural surface reservoirs for irrigation may enable reducing the competition for urban users for fresh groundwater supplies and reduce salt water intrusion into groundwater.

2.7 IRRIGATION WITH LIMITED WATER SUPPLIES

2.7.1 Practices in California

After many years of intense confrontation over water entitlements, the factions of California's water community have reached an admirable stage of maturity. In 1977, the second of two severe drought years in the Southwest, the Metropolitan Water District (MWD) of Southern California voluntarily agreed to forego over $370 \times 10^6 m^3$ (300,000 acre-ft) of its entitlement of state project water north of the Tehachapi range for the benefit of drought stricken San Joaquin farms and communities. A like draft of water was shifted to the Colorado River for pumping through the MWD Colorado River aquaduct. Those benefitting from the state water were required to pay the additional costs for pumping encountered by the MWD in the trade-off. United States Bureau of Reclamation (USBR) officials regulating the Shasta, Trinity, and Folsum Dams in California indicated that USBR contractors for water delivery in the San Joaquin Valley would have only a fraction of their previous deliveries available in 1977 because of the drought (Schuster, 1977). Local press reports indicated that municipalities received only 90 percent of the previous year's amount and agricultural contractors 20 percent, however, the actual amount was not set. Those with the most recent contracts received no water. Growers in the Sacramento delta area shifted from row crops, such as tomatoes, to crops, such as wheat, that required significantly less water. Farmers with established orchards and vineyards marshalled their allotments, supplemented them with local and pumped water, and stretched the water as far as possible with sprinkler and trickle irrigation. The USBR organized a water bank to get willing sellers of water together with owners of permanent crops and foundation cattle herds.

Rice growers in the Sacramento area reduced their rice crop from 170,000 ha to 81,000 ha to conserve water. The 2.44 m of water applied to rice was conserved and crops such as safflower, wheat, or barley were grown on the released area with less than 1.2 m of applied water. Where lower quality waters were available, sugar beets and corn replaced tomatoes.

Alfalfa growers were advised to keep alfalfa well watered in the spring and early summer months to strengthen healthy crowns that better prepared it to survive a late season drought. Also, they were advised to lengthen the time between cuttings to 50 percent bloom instead of the earlier 10 percent bloom that is customary to reduce the number of irrigations. This would lower hay quality but would further strengthen the alfalfa crowns toward a probable drought period in the late summer and fall months.

The Director of California's Drought Emergency Task Force indicated that over half of the water used by agriculture was from expanded use of groundwater supplies by new and refurbished wells. Pacific Gas and Electric in Northern California provided many kilometers (miles) of electric transmission lines and transformers to power pumps. Vineyard owners irrigated every other line alternately throughout the drought period.

The drought spurred investigation of the minimum water required for alfalfa and other crops. Robinson et al. (1976) showed that irrigating with an application to class A pan evaporation ratio of 0.75 would allow alfalfa to grow without reducing yields for 2 yr. The salts applied in the irrigation water accumulated in the lower horizons and were leached before reestablishing the next crop. Deep chiseling to 56 cm and leaching at one time more efficiently removed salt than applying a leaching quantity with each irrigation of the crop.

An evolution of irrigation systems was pressed forward by the drought. A grower in Cochise County, AZ, installed 10,900 ha (27,000 acres) of center pivots to obtain a uniform application at the same time that growers in Antelope Valley, CA, replaced 11 center pivots with 73 wheel-line systems. The wheel lines were 411 m long and were used for three 8-h sets per day.

Risers were 18.3 m apart. Pumps supplied 1.14 m³/s for the 1,200 ha.

2.7.2 Practices in Canada

The basic problem in irrigated agriculture, except in humid regions, is that of being short of water. Even in Canada, there is more land than there is water with which to irrigate the land. In Saskatchewan, for instance, less than 1 percent of the cultivated land is under irrigation and in most cases, the readily available water has been allocated (Korven and Randall, 1975).

The first criteria that must be determined are the total water requirements (consumptive use) of crops to be grown in the area and the portion of the total that has to be added by irrigation. A study was conducted at the Research Station, Lethbridge from 1950 to 1961 to measure water requirements of irrigated crops grown in southern Alberta (Sonmor, 1963). A similar study was conducted from 1961 to 1965 at the Research Station, Swift Current to measure the water requirements of selected crops for Southwestern Saskatchewan. Pohjakas et al. (1967) concluded that the consumptive use rates measured at Swift Current were somewhat higher than those reported by Sonmor for southern Alberta (Table 2.2).

The shortage of water and the trend to mechanical-move sprinkler irrigation systems, such as the wheel-move and center-pivot, prompted growers to irrigate more land than that for which the system was designed. The mechanical-move systems can be extended without increasing the labor cost unduly and thereby reduce the initial cost of equipment per unit area. The shortage of water produces a strong desire to use the scarce resource as efficiently as possible.

A study was initiated at the Swift Current Research Station in Saskatchewan in 1969 to determine the economic ramifications of producing less

TABLE 2.2 WATER REQUIREMENTS OF CROPS IN CANADA

| | Water use, mm | | | | | |
| | Consumptive use | | Irrigation requirement | | Seasonal U/E ratio* | |
Crop	S. Alta	S. Sask	S. Alta	S. Sask	S. Alta	S. Sask
Alfalfa	660	660	406	508	0.66	0.75
Alfalfa + bromegrass	—	610	—	457	—	0.67
Pasture	610	—	305	—	0.68	—
Sugar beets	559	610	330	356	0.54	0.69
Potatoes	508	533	356	381	0.56	0.61
Wheat	457	533	254	330	0.66	0.57
Oats	406	—	229	—	0.59	—
Barley	406	483	254	330	0.64	0.53
Flax	381	—	127	—	—	—
Corn	381	—	203	—	0.50	—
Tomatoes	356	—	178	—	—	—
Peas	330	483	203	254	—	0.53

*Mean seasonal ratio of water requirement to evaporation as measured by U.S.W.B. Class A pan

Consumptive use - rainfall + irrigation + soil moisture deficit

Southern Alberta - 11 years' data (1950-61); Sonmor (1963).
Southern Saskatchewan - 5 years' data (1961-65); Pohjakas et al. (1967).

than maximum yields of alfalfa, bromegrass and wheat by extending
sprinkler irrigation systems (Korven and Wiens, 1974). Each treatment, except for the nonirrigated check, was irrigated in six settings with a sprinkler
system of the same capacity. The three treatments were achieved by varying
the time required to complete the six settings. The basic treatment that met
the water requirement established in the previous study (Pohjakas et al.
1967) completed the six settings in 14 days. The second and third irrigation
treatments, in which the plots were irrigated in 21 days and 28 days,
simulated extending the system one and one-half and two times respectively.

The economic ramifications of extending the irrigation system were
studied by a theoretical exercise. Two irrigation systems were designed to
supply average peak water requirements, one to irrigate 32.5 ha (80 acres)
and the second to irrigate 65 ha (160 acres). The system designed for 32.5 ha
was extended to distribute the same quantity of water over a larger acreage,
i.e. 49 ha (120 acres) or 65 ha (160 acres) by adding main line. The four
systems are referred to as the 32.5- and 65-ha full irrigation systems, and the
49- and 65-ha partial systems.

The capital requirements per hectare were about the same for the
32.5 and 65-ha full irrigation systems, being about $410/ha ($165/acre).
The larger-sized pipe requirements for the latter offset the lower pump unit
investment cost per hectare. However, the capital requirements were much
lower for the 65-ha partial irrigation system, being about $250/ha
($100/acre).

Cost of hay production. The comparative costs of production and
returns are shown in Fig. 2.1. For some farmers, the value of alfalfa hay may
be greater than the cost of production. In such cases, the optimum system
may not be one whose cost of production is at a minimum (line ABCD in Fig.
2.1), but will be the one that maximizes the net return. Along the line EB,
the net return from the 65-ha partial irrigation system and 49-ha partial ir-

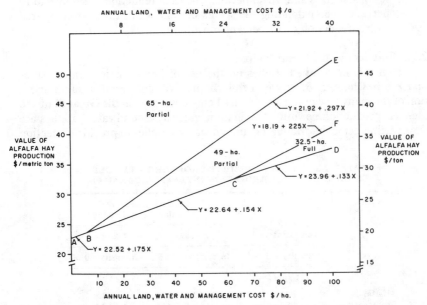

FIG. 2.1 Estimated cost of alfalfa hay production for three irrigation systems.

rigation system will be the same. Above and to the left of this line, the 65-ha partial irrigation system will maximize net return from alfalfa hay production. Along the line CF, the net return from the 49-ha partial irrigation system and 32.5-ha full irrigation system will be the same. Below and to the right of this line, the 32.5-ha full irrigation system will maximize net return from alfalfa hay production. In the area between these two lines, BE and CF, the 49-ha partial irrigation system will maximize net returns. The system that will maximize net returns will depend on the value of hay to the farmer and on the annual land and water costs per acre.

The yield reduction from using a full irrigation system over a larger area is much less for bromegrass and for wheat than for alfalfa (Table 2.3). When the farm value of bromegrass hay was over $57.30/metric ton ($52/ton) or wheat over 66.2¢/kg ($18/bu), a system designed for full irrigation should be considered.

Conclusions. These studies in Canada showed that:

1 The mean total yield of alfalfa, brome, and wheat was not significantly decreased when extending the sprinkler irrigation system for an area 1.5 times the design area, except in extremely dry years.
2 When annual land, water and management costs are under $62/ha ($25/acre), the 49-ha partial irrigation system will minimize cost of alfalfa hay production.
3 When the water supply is limited to a full irrigation on 32.5 ha (80 acres) and the value of alfalfa hay is high, the 65-ha (160-acre) partial irrigation system should be considered.
4 The full irrigation system should only be considered for alfalfa hay production when annual land, water and management costs are over $62/ha ($25/acre).
5 The 65-ha (160-acre) partial irrigation system should be considered for bromegrass and wheat because the yield reduction was only 8 to 10 percent when extending the sprinkler system two times as compared with a reduction of 24 percent for alfalfa.

2.7.3 Practices in the Great Plains

Some of the detailed studies on the use of limited irrigation are summarized in Chapter 18. Space in this chapter will not permit a summarization of the many new practices that are being developed in the Great Plains to increase production per unit of irrigation water as groundwater supplies are slowly being depleted. Managing limited water supplies to provide protection

TABLE 2.3. YIELDS FOR THREE CROPS GROWN UNDER THREE LEVELS OF IRRIGATION INTENSITY (CANADA)

Crop	Yield 14-day interval Intensity = 1.00	Yield reduction in % of 14-day interval yield	
		21-day Intensity = 0.75	28-day Intensity - 0.50
	kg/ha		
Alfalfa	9440	13.8	23.8
Bromegrass	8860	7.6	10.6
Wheat (grain)	2401	7.6	8.4

against severe drought effects is a common practice on crops like grain sorghum and cotton. A detailed summary of these practices is being prepared by J. T. Musick and B. A. Stewart for a chapter entitled "Use of Limited Water Supplies." This chapter is scheduled to be printed in 1981 in a monograph on dryland agriculture by the American Society of Agronomy (Personal Communication).

2.8 OTHER BENEFICIAL USES FOR IRRIGATION WATER

2.8.1 Germination

Irrigation can meet special needs of crops other than the essential evapotranspiration requirements. Where salts contained in irrigation water accumulate on the surface of furrow irrigated beds, sprinkler irrigation is applied during seed germination to provide a low salinity seed bed (Robinson et al. 1968; Robinson, 1969). A procedure used in California for several years is to first set up a furrow system and then utilize a portable solid set sprinkler system to germinate the seed. The sprinklers are then removed to germinate a second field while the first field is irrigated subsequently with furrow irrigation. This allows the more costly sprinkler system to be used over a large area while allowing the seedling roots to penetrate below the zone of salt accumulation on the surface of the furrow irrigated beds. Basically the same irrigation sequence is utilized on several crops. This consists of applying 12-h applications alternately on 8.1-ha (20-acre) blocks in a 16.2 ha (40-acre) field until the soil profile has been adequately irrigated. The alternate application reduces the pump size needed to cover the 16.2 ha (40 acres) and allows the application time to extend to that normally required for small seed to initiate radical development.

2.8.2 Crop Cooling

In some instances, as when lettuce is being grown before September 15 in the Imperial Valley, it is also common to apply daily 3-h applications to cool the germinating lettuce seedlings until they emerge. Cooling in vineyards is also practiced in California (Meyer and Gilbert, 1970). Sprinkler irrigation for 2.5 to 3.0 min with 12- to 15-min intervals between applications proved adequate to provide 8 to 14 °C (15 to 25 °F) cooling at relative humidities of 15 to 40 percent. Application rates used in the vineyards were 3.1 to 3.6 mm/h. Some of the crops cooled by sprinkler irrigation are almonds, apples, beans, cherries, cotton, cranberries, cucumbers, flowers, grapes, potatoes, prunes, strawberries, sugarbeets, tomatoes, and walnuts (Sneed, 1972; Gray, 1970; Kidder, 1970; Carolus, 1971; Unrath, 1972a; and Unrath, 1972b). Air conditioning of crops for higher yield and better quality also received much attention from researchers in California, Michigan, and Washington. Recent summaries of crop cooling principles were presented by Merva and Vandenbrink (1979) and Chessness et al. (1979).

Researchers in Louisiana and Georgia (Chessness et al., 1969) worked on the magnitudes of soil, plant and air temperatures, air speeds, and humidities present in a strawberry field during periods of atmospheric heat stress. In addition, other factors such as microclimate modification and intermittent sprinkling rate for reducing heat stress were determined.

An intermittent sprinkling rate of 2 mm/h with a 15-min on-15-min off period gave good temperature reduction with only a 45-percent recovery of presprinkling temperatures during the off period. Temperature reductions of 3, 8, and 16 °C for soil surface, ambient air, and leaves, respectively, were obtained.

Unrath and Sneed (1974) collected 4 yr of data on evaporative cooling of delicious apples in North Carolina and studied its economic feasibility. Evaporative cooling consistently resulted in fruit temperature reductions of 6 °C; increased total reddish color 8 to 20 percent, solid red color 13 percent and soluble solids 10 percent. Corking and bitter pit were reduced 8 and 7 percent respectively.

Ryan et al., (1973) used sprinkler irrigation in Florida to reduce heat stress in early Amber Peaches. Sprinkled trees yielded 47 percent fruit 5 cm and larger whereas the control yielded only 10 percent fruit 5 cm and larger. The sprinkling system was turned on for 15 s every 10 min between 8 a.m. and 6 p.m.

Griffin et al. (1974) reported that color of 'Red Delicious' apples can be enhanced by reducing fruit temperature by overhead sprinkling. This required 1 to 1.3 L/s of water per hectare, and 24-h continuous sprinkling. Temperature reduction in the fruit was about 11 °C (from 43 °C to 24 °C). The increase in income for these evaporative cooled apples was estimated to be $405/ha; however, available water supply and possible damage to the trees from this amount of water when ET rates are low is of concern.

Kemp et al. (1974) evaluated the effect of cooling snap beans with intermittent sprinkling at Lethbridge, Alberta, Canada. They concluded that the quality was not impaired when the crop was properly irrigated by conventional methods.

2.8.3 Delay of Fruit and Bud Development

Sprinkling to delay fruit bud development is new to the irrigation industry and workers at Utah State University are enthusiastic about this role (Alfaro et al., 1974; Anderson, 1974). Utah growers have suffered damaging temperatures during 9 out of 15 recent years and this new development may mean a constant market for them.

Evaporative cooling with sprinklers during early spring months, when frost danger is still prevalent, seems to be an effective way to delay bud development. In one experiment, bud growth was delayed 17 days for apples, 15 days for cherries, and 14 days for peaches. A 2-min wetting cycle followed by a 2-min drying cycle produced maximum cooling. The system was programmed to automatically activate when the air temperature reached 7 °C and to turn off at 6 °C. Design criteria for sprinkler systems to delay bud development was presented by Griffin and Richardson (1979). (Also see Section 18.4).

2.8.4 Frost and Freeze Protection

Frost protection is another special use for irrigation. By coating plants with water, the heat of fusion is released as the water freezes, maintaining plant temperatures that would otherwise drop well below freezing. The ice coating on the plant must be continually in contact with unfrozen water until the ice melts. Observations in California showed that where ice-coated alfalfa plants were continually sprinkled at 2.8 mm/h (0.11 in./h), the temperature did not drop below - 2.2 °C (28 °F), but where the sprinkling was stopped, the sublimation of the ice dropped the temperature to - 11 °C (12 °F), 2.8 °C below the 8.3 °C air temperature.

Much of the pioneer work in frost prevention was carried out in Michigan and California. This early work indicated that application rates somewhere between 2 and 3 mm/h would protect crops against certain radiation frosts. Subsequent work in Florida and Louisiana (Braud and Hawthorne, 1965; Braud et al., 1971 and 1973; Harrison et al., 1974; and Locascio et al., 1967) showed that plants could be protected against freezing temperatures as low as -9 °C with zero winds when application rates were increased to 6.5 mm/h. Also, at this same application rate, protection was obtained under winds of 5.4 m/s, down to - 1 °C air and - 13 °C dewpoint temperatures, respectively. Only 7 to 10 percent injury to mature strawberries was observed at these temperatures at which protection was obtained. The check plots contained 100 percent injured mature fruit.

Workers at N. C. State University showed yield increases in apples ranging from 18 to 205 percent when irrigated for freeze protection, in 1974.

Sprinkler frost protection has been successful for small fruit, potatoes, flowers, and grapes (Harrison and Gerber, 1965; Harrison et al., 1974; Kidder, 1970; Lamade, 1968). Sneed (1970) warned that advective frosts associated with cooling air movement require higher application rates than the radiative frosts on clear, calm, cloudless nights. Winds of over 4.5 m/s preclude advective frost protection and freeze protection. A series of recent articles in an ASAE Monograph entitled "Modification of the Aerial Environment of Crops" discussed various techniques for cold and frost protection (Barfield and Gerber, 1979). (Also see Section 18.4).

2.8.5 Application of Chemicals in Irrigation Water

Another special use of irrigation is the conjunctive application of chemicals with the water. The inclusion of fertilizers, pesticides, herbicides, desiccants, and defoliants is well documented. Various commercial pump injectors and flow regulators have been used to apply various chemicals and nitrate fertilizer (Carey, 1970; Fischbach, 1970; Robinson and Cudney, 1973; Sneed, 1972). (Also see Section 18.4).

References

1 Alfaro, J. F., R. E. Griffin, J. Keller, G. R. Hanson, J. L. Anderson, G. L. Ashcroft, and E. A. Richardson. 1974. Preventive freeze protection by preseason sprinkling to delay bud development. TRANSACTIONS of the ASAE 17:1025-1028.

2 Anderson, J. L. M. 1974. Effects of over tree sprinkling on fruit bud development and fruit quality. Plant Sci. Dep., Utah State Univ. Mimo. Rpt.

3 Barfield, B. J. and J. F. Gerber (Ed.). 1979. Modification of the Aerial Environment of Crops. ASAE Monog. No. 2, 436 p.

4 Braud, H. J., Jr. and P. L. Hawthorne. 1965. Cold protection for Louisiana strawberries. Bul. No. 591, Louisiana State Univ., March.

5 Braud, H. J. and M. Esphahani. 1971. Direct water spray for citrus freeze protection. ASAE Paper No. 71-234, ASAE, St. Joseph, MI 49085.

6 Braud, H. J., P. L. Hawthorne, and R. T. Brown. 1973. Direct water spray for citrus freeze protection annual report. Agr. Eng. Dep., Louisiana State Univ.

7 Buringh, P., H. D. J. van Heemst, and G. J. Stering. 1975. Computation of the absolute maximum food production of the world. Publ. No. 598, Agr. Univ., Wageningen, The Netherlands.

8 Canadian National Committee, ICID. 1977. Water development in Canada to meet agricultural production requirements for 2000 A.D. Proc. Spec. Sess., Int. Comm. Irrig. Drain., Tehran, May, Rpt. 1, p. 1-21.

9 Carey, P. J. 1970. Applying herbicide and other chemicals through sprinkler systems. Natl. Irrig. Symp., Lincoln, NE, p. X1-X6.

10 Carolus, R. L. 1971. Evaporative cooling techniques for regulating plant water stress. Hort Sci. 6(1):23-25.

11 Chessness, J. L. and H. J. Braud. 1969. Sprinkling to reduce heat stressing of strawberry plants. ASAE Paper No. 69-229, ASAE, St. Joseph, MI 49085.

12 Chessness, J. L., L. A. Harper, and T. A. Howell. 1979. Sprinkling for heat stress reduction. p. 388-393. In: Modification of the Aerial Environment of Crops, B. J. Barfield and J. F. Gerber (Ed.), ASAE Monog. No. 2, 436 p.

13 Cobb, G. D. 1976. Project formulation in today's world: Economic and financial aspects. Proc. Am. Soc. Civ. Eng. Spec. Conf., Ottawa, Canada, p. 660-672.

14 FAO. 1977. Water for agriculture. Food and Agric. Organ. of the UN, UN Water Conf., Mar del Plata, March, 26 p.

15 Fischbach, P. E. 1970. Applying fertilizer through the irrigation system. National Irrigation Symp. Lincoln, NE, p. W1-W6.

16 Frejka, T. 1973. Reference tables to the future population growth: Alternative paths to equilibrium. The Population Council, New York, p. 18.

17 General Accounting Office (GAO). 1976. Better Federal Coordination Needed to Promote More Efficient Farm Irrigation. Rep. to the Congress, June, 49 p.

18 Gray, A. S. 1970. Environmental control using sprinkler systems. Natl. Irrig. Symp., Lincoln, NE, p. RR1-RR10.

19 Griffin, R. E., J. F. Alfaro, J. Keller, G. R. Hanson, J. L. Anderson, G. L. Ashcroft, and E. A. Richardson. 1974. Enhancing anthocyamin in 'Red Delicious' apples by evaporative cooling. ASAE Paper No. 74-5033, ASAE, St. Joseph, MI 49085.

20 Griffin, R. E. and E. A. Richardson. 1979. Sprinklers for microclimate cooling of bud development. p. 441-455. In: Modification of the Aerial Environment of Crops, B. J. Barfield and J. F. Gerber (Ed.), ASAE Monog. No. 2, 436 p.

21 Hagan, R. M. 1977. Societal attitudes affecting water users. Proc. Am. Soc. Civ. Engr., Irrig. and Drain Div., Spec. Conf., Reno, NV, p. 108-113.

22 Harrison, D. S. and J. F. Gerber. 1965. Research with sprinkler irrigation for cold protection. TRANSACTIONS of the ASAE 7(4):464-468.

23 Harrison, D. S., J. F. Gerber, and R. E. Choate. 1974. Sprinkler irrigation for cold protection. Cir. 348 Tech., Florida Coop. Ext. Serv., Univ. of Florida. 19 p.

24 Holburt, M. B. and H. F. Pellegrin. 1969. California's stake in the Colorado River.Colo. River Bd. of Calif. Publ., Los Angeles, p. 13-15.

25 Irrigation Journal. 1979. Irrigation Survey, 1979. Irrig. J., Nov.-Dec. p. 58A-58H.

26 Israelsen, O. W. 1950. Irrigation principles and practices. (2nd Ed.) John Wiley and Sons, Inc., New York, 405 p.

27 Interagency Task Force Report. 1979. Irrigation water use and management. U.S. Dep. Int., U.S. Dep. Agr. and Environ. Prot. Agency. 143 p.

28 Jensen, M. E. 1977. Water conservation and irrigation systems. Proc. Climate—Technology Seminar, Columbia, MO, p. 208-250.

29 Kidder, E. H. 1970. Climate modification with sprinklers. Natl. Irrig. Symp. Lincoln, NE, p. V1-V6.

30 Korven, H. C. and J. K. Wiens. 1974. Evaluation of an over-extended sprinkler irrigation system. Can. Agr. Eng. 16:51-56.

31 Korven, H. C. and W. E. Randall. 1975. Irrigation on the Prairies, CDA Pub. 1488. (Rev.) 26 p.

32 Lamade, W. 1968. Frost control on potatoes. Veg. Crop Mgmt. 4(4):27-41.

33 Locascio, S. J., D. S. Harrison, and V. F. Nettles. 1967. Sprinkler irrigation of strawberries for freeze protection. Proc. Florida State Hort. Soc. 80:208-211.

34 Martin, W. E. 1976. Economic implications: Issues and alternatives in salt management. Proc. Conf. on Salt and Salinity Management. Santa Barbara, Calif. Water Res. Ctr. Rep. No. 38, p. 17-29.

35 Merva, G. E. 1979. Physical principles involved in alleviating heat stress. p. 373-387. In: Modification of the Aerial Environment of Crops, B. J. Barfield and J. F. Gerber (Ed.), ASAE Monog. No. 2, 436 p.

36 Meyer, J. L. and D. E. Gilbert. 1970. Solid sets in orchard and vineyards with consideration to frost control and climate modification. Proc. Sprinkler Irrig. Assoc. Ann. Tech. Conf., Denver, CO, p. 71-77.

37 Miller, C., Jr. 1977. Arizona Farm Bureau Fed. Conference, Phoenix, AZ. Quoted in Ariz. Farmer—Ranchman, 56:6-7.

38 Miller, D. E. 1977. Deficit high-frequency irrigation of sugarbeets, wheat, and beans. Proc. Am. Soc. Civ. Engr. Irrig. Drain. Div. Spec. Conf., Reno, NV, p. 269-282.

39 Orgeon State University Water Resources Research Institute. 1967. Northwest-Southwest Water Division Issues. Corvallis, OR, p. 1-58.

40 Pohjakas, K. D., W. L. Read, and H. C. Korven. 1967. Consumptive use of water by crops at Swift Current, Saskatchewan. Can. J. Soil Sci. 47:131-138.

41 Robinson, F. E. 1969. Stands for automation achieved by sprinkling. Proc. Am. Soc. Civ. Engr., J. Irrig. Drain. Div. 95:385-389.

42 Robinson, F. E. and D. Cudney. 1973. Use of sprinklers to study the influence of population density upon seed cotton production in arid areas. Agron. J. 65:266-268.

43 Robinson, F. E., J. N. Luthin, R. J. Schnagl, W. Padgett, K. K. Tanji, W. F. Lehman, and K. S. Mayberry. 1976. Adaptation to increasing salinity of the Colorado River. Contr. No. 160. Calif. Water Resour. Ctr., Univ. of Calif., Davis.

44 Robinson, F. E., O. D. McCoy, G. F. Worker, and W. F. Lehman. 1968. Sprinkler and surface irrigation of vegetable and field crops in an arid environment. Agron. J. 60:696-700.

45 Ross, R. 1976. Eastern Corn Belt: The Sleeping Giant? Irrig. Age. Part I, September, p. 6, 7, 10, and 12; Part II, October, p. 50-53; Part III, November-December, p. 6, 7, and 11.

46 Ryan, P. L., J. F. Bartholic, and D. W. Buchanan. 1973. Sprinkler and soaker irrigation of peach trees to reduce plant water stress and increase fruit size. Proc. Florida State Hort. Soc. 85:311-315.

47 Schuster, D. R. 1977. Chief Water Operations Branch, USBR, Sacramento. Central Valley Projects Operations. Calif. Irrig. Inst., Sacramento, CA. (Presentation).

48 Scott, V. H. and W. E. Johnston. 1968. Renewable resources in the development of a desert area of California. Task Force Study by Univ. of Calif. Desert Agr. In Calif., p. 7-28.

49 Shimshi, D., H. Bielorai, and A. Mantel. 1973. Irrigation of farm crops. p. 369-381. In: Arid Zone Irrig., D. Yaron, E. Danfors, and Y. Vaadia (Ed.). New York, Springer-Verlag.

50 Shmueli, E. 1973. Efficient utilization of water in irrigation. p. 411-423. In: Arid Zone Irrig., D. Yaron, E. Danfors, and Y. Vaadia (Ed.). New York, Springer-Verlag.

51 Sneed, R. E. 1970. Using sprinkler irrigation for crop cooling. Sprinkler Irrig. Assoc., May, p. 5-7.

52 Sneed, R. E. 1972. Frost control. Sprinkler Irrig. Assoc. Proc. Annual Tech. Conf. Denver, CO, p. 41-53.

53 Sonmor, L. G. 1963. Seasonal consumptive use of water by crops grown in southern Alberta and its relationship to evaporation. Can. J. Soil Sci. 43:287-297.

54 Unrath, C. R. 1972a. The evaporative cooling effects of over tree sprinkler irrigation on 'Red Delicious' apples. J. Am. Soc. Hort. Sci. 97(1):55-58.

55 Unrath, C. R. 1972b. The quality of 'Red Delicious' apples as affected by over tree sprinkler irrigation. J. Am. Soc. Hort. Sci. 97(1):58-61.

56 Unrath, C. R. and R. E. Sneed. 1974. Evaporative cooling of 'Delicious Apples'—the economic feasibility of reducing environmental heat stress. J. Am. Soc. Hort. Sci. 99(4):372-375.

57 Viets, F. G., Jr. 1965. Increasing water use efficiency by soil management. p. 259-274. In: Plant Environment and Efficient Water Use. W. H. Pierce, D. Kirkham, J. Pesek, and R. Shaw (Ed.). Am. Soc. Agron., Madison, WI.

58 Wuertz, W. 1977. State must help build CAP. Ariz. Farmer-Ranchman. 56:75.

59 White, G. 1976. The main effects and problems of irrigation. p. 1-72. In: Arid Land in Developing Countries: Environmental Problems and Effects. E. B. Worthington (Ed.). Pergamon Press, New York.

60 Wittlesey, N. K., K. C. Gibbs, and W. R. Butcher. 1978. Social overhead capital costs of irrigation development in Washington State. Water Resour. Bul. 14(3):663-678.

61 Worthington, E. B. (Ed.). 1976. Arid Land Irrigation in Developing Countries: Environmental Problems and Effects. Pergamon Press, New York. 463 p.

chapter 3

FARM RESOURCES AND SYSTEM SELECTION

3

FARM RESOURCES AND SYSTEM SELECTION

3

FARM RESOURCES AND SYSTEM SELECTION

by   G. T. Thompson, Washington State University, Prosser, WA; L. B. Spiess, Alberta Agriculture, Irrigaton Division, Lethbridge, Canada; J. N. Krider, Soil Conservation Service, USDA, Broomall, PA

This chapter is presented to provide guidance for planning and selecting a farm irrigation system. The methodology for obtaining the data necessary to carry out specific evaluations and analyses, and for preparing the designs required for planning and selecting a farm irrigation system are either presented in other chapters of this monograph or in the listed references. The acquisition and evaluation of specific information to establish the requirements for planning and selecting a farm irrigation system may require legal assistance and the services of different disciplines such as agronomists, horticulturalists, and soil scientists.

The planning process for selecting a farm irrigation system requires an inventory of the resources available to the farmer. The evaluation of these resources is necessary to identify the production potentials, and the physical and operational constraints which affect the selection of viable alternative farm irrigation systems. The analysis and comparison of these alternatives provides a basis for selecting the farm irrigation system.

Regardless of the size of the farm to be irrigated, the physical resources of water supply, soils, topography and climate will affect the farm cropping potential. These resources along with the crops to be grown, must be evaluated in conjunction with other resources such as available labor and energy, marketing potentials, economic conditions, the farmers' preferences, and his financial situation as all will influence the selection of the farm irrigation system. Therefore, the initial step in the planning process is to identify the parameters which are needed to determine the viable irrigation methods for a given farm. After assessing the applicable irrigation methods for a farm, alternative farm irrigation systems can be designed and analyzed, and the appropriate system selected.

3.1 FARM WATER SUPPLY

The characteristics of the farm water supply can have a significant effect on the selection of a farm irrigation system. Such factors as location, available quantity, time distribution of the quantity, and the quality of the water supply can affect the considerations for defining and evaluating applicable farm irrigation systems.

The analysis of water supply factors is required to answer questions such as:

1 What is the cost of delivering the necessary flows from the water source to the farm?
2 Is the flow large enough for border irrigation?
3 Is water available in the necessary quantities during the growing season?
4 Will the salinity of the water affect the irrigation method?

3.1.1 Sources

The water supply for farm irrigation use may be obtained either from surface or subsurface sources or both. Surface sources usually are obtained either by direct diversions from streams and rivers, or pumped from lakes or ponds, or by deliveries from irrigation project distribution facilities. Surface water supplies may be obtained by gravity diversions from streams or rivers in a limited number of cases, although most new farm diversions from surface sources are now accomplished by pumping. Water deliveries to farms from irrigation project distribution facilities may be by gravity flow from canals or as pressure deliveries from closed pipe systems. When farm deliveries are made through project type closed pipe distribution systems, the pressure at which the farm deliveries are made can influence the selection of the farm irrigation system. For example, if water is delivered at pressures adequate for operating sprinkler or trickle irrigation systems, this may dictate the selection of the farm irrigation system.

Subsurface waters are obtained by pumping from ground water supplies using wells. Some irrigation projects use wells to obtain part or all of their water supply, but most of the ground water used for irrigation is from wells installed and operated by individual farmers. When wells are installed by individual farmers the cost of the installation, as well as the quantity and quality of the water supply become important factors in evaluating and selecting a farm irrigation system.

When wells are used; the well yield, height of lift and friction loss to the irrigated area, and the long-term reliability of the ground water source must be carefully evaluated before selecting the irrigation system.

The optimal well yield will determine whether limitations on quantities of available water will affect either the size and type of farm irrigation system that can be selected for direct water delivery or require on-farm storage to provide desired flow quantities. The long-term reliability of a well to produce a desired quantity of water is an important feature to assure full utilization of capital investments in the irrigation system. If the long-term reliability is somewhat uncertain the farmer should be aware of the risks involved for the development.

The height of lifting the water from the operating elevation to the ground surface is an important factor in determining the total dynamic head (TDH) of the pumping system. The TDH of the system includes the height of lift from the well to the surface, friction losses in the pump column, fittings, pipelines, surface topographical elevation differences, and required operating pressures for farm irrigation systems such as sprinklers.

The techniques of well design, development, and testing are beyond the scope of this discussion. Excellent treatises on these subjects such as those by the Bureau of Reclamation (1977), Campbell and Lehr (1973), Davis and

DeWiest (1966), and Johnson, Inc. (1966), are available to the planner and engineer. Further details of farm water delivery systems are presented in Chapter 10.

3.1.2 Water Quantity

The quantity and timing of potential water deliveries can affect the planning and selection of a farm irrigation system. The identification of viable farm irrigation system alternatives requires knowledge of the available flow rates during the irrigation season, and the seasonal quantity of water allocated to the farm. Limiting physical conditions such as well discharge capacity, low stream flow periods, limited volumes stored in ponds, and/or any legal restrictions on withdrawals may constrain either the flow rates, or the seasonal volume of water available for irrigation.

The available flow rate **from the source** may be inadequate to directly supply the water needed for the proposed crops during their peak consumptive use period, or be inadequate to provide the flows desired to efficiently operate surface irrigation systems. The provision of facilities such as on-farm storage to provide adequate flows rates, or modifications in design discharges may be involved in a design of the farm system.

The timing of available flows may also affect the selection of the system. Intermittent large flows such as may occur from an irrigation project using a rotational delivery method may result in selection of a surface method such as border irrigation rather than providing storage or enlarging a sprinkler system to use the flow when it is available.

The seasonal quantity of available water if limited may dictate limitations on cropping plans, and/or water management strategies to optimize the economic returns. These limitations are then used in design and selection of the irrigation system.

The legal process of granting water rights may result in the allocation of both the flow rate and the annual quantity of water to be supplied for a farm. Many states in the United States, and other countries, when granting water rights specify maximum flow rates and annual quantities of water to both individual farmers and to irrigation projects. The conditions included in these rights must be evaluated to determine their effects on water use and management strategies which may affect the design and costs of owning and operating the alternative farm irrigation systems being considered.

3.1.3 Water Quality

The quality of the available irrigation water can influence the selection of the farm irrigation system. The chemical constituents of the water may limit the crops which can be grown and affect the drainability of the soils. The suitability of water for irrigation with respect to salinity, and the effects of irrigation methods on salinity profiles in the soil, is given in Chapter 5.

Since water quality controls the leaching requirments and may affect crop selections, it can influence farm irrigation system selection. Some chemical constitutents in the water can result in corrosion problems when using various types of metals in the irrigation system. The sediment load carried by the water can also influence the design for adequate performance of a farm irrigation system. Where heavy sediment loads are encountered, trickle irrigation system filtration requirements are increased to provide reliable performance. Sediment laden water will increase wear and reduce the life of irrigation system components such as pump impellers and sprinkler nozzles.

3.2 SOILS

Soils are one of the basic input resources for agricultural production. The selection of a farm irrigation system is influenced by the soil factors of texture, structure, depth, salinity, infiltration or intake rates, and the water holding characteristics. The factors of topography and soil erodibility by wind and water can also influence the farm irrigation system design and selection.

Soil survey and land classification maps are available for many of the lands being considered for irrigation. In the United States, the Soil Conservation Service, U. S. Bureau of Reclamation and many land grant universities have conducted soil surveys and land classifications of much of the arable lands. In countries other than the U. S., a large number of arable land areas have also been surveyed and classified. Maps and details of the surveys are generally available from government agencies in these countries. These soil maps and accompanying descriptions provide the basis for assessing the production capabilities and special management requirements when planning irrigation activities.

When areas of non-irrigated lands are considered for irrigation development the irrigability of the soils should be determined by soil scientists and agricultural engineers. The soil and land factors affecting farm irrigation system selection are those related to physical characteristics which affect irrigation water management. These factors are discussed in general terms in the following sections of this chapter, while detailed discussions covering soil-water properties and dynamics are presented in Chapter 4.

3.2.1 Soil Factors Affecting System Selection

A number of soil related factors affect farm irrigation system selection. These soil factors can be listed as texture, structure, profiles and depth, salinity, subsurface drainage, macro- and micro- relief (topography), erodibility, and the climatic considerations which exert an indirect soil related impact. These items are discussed individually to identify possible impacts on system selection.

Soil texture: The proportion of the different sizes of particles making up a soil determines its texture. The primary soil particle sizes, as measured by effective particle diameter, are designated by textural groups as gravel, sand, silt, and clay. The textural classification system commonly used for soil evaluations associated with selecting farm irrigation systems is the one developed by the U. S. Department of Agriculture (1964). The size ranges of the particles associated with textural groups are specified as follows:

Textural group	Particle size, mm
Gravel	> 2.0
Sand	
very coarse sand	1.0 - 2.0
coarse sand	0.5 - 1.0
medium sand	0.25 - 0.5
fine sand	0.10 - 0.25
very fine sand	0.05 - 0.10
Silt	0.002 - 0.05
Clay	< 0.002

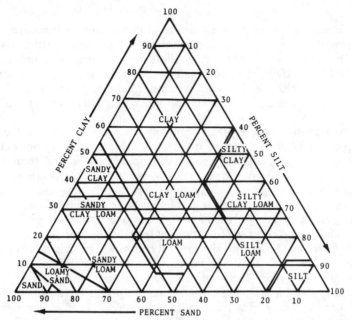

FIG. 3.1 Proportions of sand, silt and clay for different soil textures.

A textural classification chart for 12 classes, as shown by Fig. 3.1, has also been developed by the U. S. Department of Agriculture (1951). In general, sandy soils are classed as coarse-textured, loam soils as medium-textured, and clay soils as fine-textured. The textural characteristics affect the selection of a farm irrigation system by the influence it exerts on the water-holding capacity of a given depth of soil, and the effect it exerts on the infiltration, or water intake characteristics of the soil. Textural characteristics may also influence traction ability of heavy irrigation systems such as center pivot sprinkler systems.

Soil structure: Soil structure refers to the degree which individual soil particles aggregate into groups. The particles of coarse-grained soils tend to function as individuals, while the aggregated particles of fine-textured soils tend to form granules. The size and shape of these particle groups, and their stability is defined as the soil structure. Descriptions of various soil structural forms are provided in a number of references such as the U. S. Department of Agriculture (1951).

Fundamentally soil structure is developed by wetting and drying, freezing and thawing, and combinations of these conditions. The primary function of organic matter in the soil is to add stability to soil aggregates and serve as a cushion against the effects of tillage. Excessive irrigations, plowing, or other tilling of fine-textured soils, when either too wet or too dry, will tend to break down the groups of soil particles forming the soil structure. For good crop production, the maintenance of a good soil structure is recognized as being equally important as maintaining a good chemical balance. This is because the soil structure influences the ease of root penetration, the infiltration rate, aeration, and the movement of water in the soil, all of which are essential to good crop production.

Since soil structure can influence infiltration, or the intake rate of a soil it thereby affects the design and selection of the farm irrigation system. The intake rate of a soil can be a controlling factor in length of runs and size of borders for surface irrigation methods, as well as permissible application rates for sprinkler and trickle systems. Intake is an especially important design parameter for center pivot sprinkle systems to prevent runoff and erosion.

Soil profiles and depth: When soils are examined below the surface, distinct layers or horizons of different soil materials may be found. In a number of cases a depth of good soil material may be underlain with hard pans, dense clays, coarse sands, gravels, or rock which provides little, if any, support for plant growth. Deep homogeneous medium-textured soils with good structure will provide for storage of large volumes of water to sustain plant growth for relatively long periods between irrigation applications. But there are many soils irrigated at present with depths as shallow as 30 cm, and likely additional lands with shallow soil depths will be irrigated in the future. Since shallow soils limit the effective rooting depth of the crop and the amount of moisture which can be stored they require more frequent irrigation to supply the crop water needs than deep soils of equivalent texture.

The layering which exists in nonuniform soil profiles can significantly affect the water-holding capacity of the soil moisture reservoir. The effect of layering and stratification on soil moisture capacities has been examined by a number of investigators such as Miller and Bunger (1963). Further details are given in Chapter 4.

Soil salinity: Where saline or alkaline soils occur they must first be reclaimed before undertaking irrigated crop production. The farm irrigation system must provide the water for both leaching and the production of any salt tolerant crops grown during the reclamation period. For additional information on the sources of the soluble salts and procedures for reclaiming soils with undesirable levels of salinity or alkalinity see Chapter 5.

Drainage: Subsurface drainage of irrigated soils, whether it is accomplished naturally or by provision of drainage facilities, is essential and complementary to irrigation of any soil or farm. Adequate subsurface drainage is necessary to prevent an undesirable rise of the water table and increases in soil salinity. The surface and subsurface drainage characteristics of the soils on an irrigated farm or project may affect the design and analysis of alternative farm irrigation systems during the selection process. The provision of irrigation systems with adequate capability to accurately control well-managed water applications may significantly reduce the requirements for drainage facilities. The need to provide irrigation control may influence the selection of the farm irrigation system. Details for determining drainage requirements and designing subsurface drainage systems are given in Chapter 8.

Topography: The topography of the farm lands to be irrigated significantly affects farm irrigation system selection. Land slopes may limit the adaptability of surface irrigation methods by affecting the length of run, labor required for operations, or the particular type of surface system which may be appropriate. Micro-relief, in contrast to general land slope, can significantly affect the land grading required for installation of surface irrigation systems, and in some cases affect the surface drainage provisions required for specific types of sprinkler systems. Where lands with shallow soil depths occur the alternative farm irrigation systems will be limited to those systems requiring minimal land grading.

Erodibility factors: Sandy soils with limited structural development may be highly susceptible to wind erosion during periods when they are without crop cover. Mech and Woodruff (1967) reported in an evaluation of wind erosion control methods, that maintenance of a high moisture content in the surface soils will assist in reducing wind erosion. Selection of a farm irrigation system that will provide for applying light frequent irrigations to maintain a high moisture level in the surface soils may be desirable in such cases.

Soils with little structural development and low organic matter content tend to be susceptible to water erosion. This, combined with sloping lands and soils with high intake rates, can affect the design requirements for furrow-type gravity irrigation systems, and thereby influence the selection of the farm system. The erosion of soils by irrigation practices for various irrigation methods are discussed by Bondurant (1977) and Mech and Smith (1967).

3.3 CLIMATE—CROP—SYSTEM INTERACTIONS

The crops to be grown on any farm will be selected from those that are adaptable to the water supply, soils, and the climatic conditions. The crops selected must also be marketable with the prospect of an attractive economic return. The array of crops which are adaptable to a given farm may be obtained in part from those which are currently, or have been, successfully grown in the region, but if the farmer desires to fully examine the irrigated cropping potentials, agronomic, horticultural, and economic assistance may be required. When planning and evaluating alternative farm irrigation systems, an understanding of the cultural practices used to grow each potential crop is particularly important. The farm irrigation system selected must be applicable to the crops to be grown, as well as the physical constraints of the farm.

The irrigation systems which are adaptable to specific crops may be somewhat limited by either the cultural practices employed for production, or by the growth characteristics of the crop itself. For example, the production of paddy rice requires controlled partial submergence of the rice plant for most of its growth period. This results in requiring a farm irrigation system which uses carefully graded basins, or flat (level) border-type surface irrigation systems. Another example is the production of corn which limits the types of sprinkler systems which, in a practical sense, can be efficiently used for irrigation. The desire to produce shallow rooted crops, requiring light frequent irrigations, may also influence the type of irrigation system which is selected. Another consideration when selecting an irrigation system may be the need to provide efficient light irrigations for germination of seeds to establish new crop plantings.

3.4 CAPITAL AND LABOR

In addition to the water, land, and climatic resources the availability of capital for development of an irrigated farm and labor for operating the system exerts a major influence on the selection of the initial irrigation system.

3.4.1 Available Capital

The capital investment required to install a farm irrigation system can vary considerably among applicable systems. Generally costs increase with the sophistication of the water control capability that is provided by the system, and the components included to reduce the labor required for operation. Therefore, in planning and selecting a farm irrigation system, it is important to evaluate available capital and labor so that possible trade-offs can be identified. This is especially important when initial investment capital is limited.

3.4.2 Available Labor and Technical Skill

The amount of labor and technical skill available for operating the farm irrigation system also influences the type of system to be selected. When labor is either difficult to obtain, relatively unskilled, and/or expensive the present trend is to select systems which use lesser amounts of labor. Many systems which require lesser amounts of operating labor also require greater maintenance and repair skills. Consideration must be given to the level of technical skill available to maintain and operate labor reducing irrigation facilities.

With the present emphasis on improving the efficiency of irrigation water use, there is a need to develop technical skills necessary to apply water efficiently.

Considerable effort has been expended by researchers in developing labor saving devices and procedures to control the irrigation flows used for surface irrigation methods. Some of these developments are discussed by Humpherys (1967 and 1969), Haise and Kruse (1969), and Bowman (1969) (also see Chapter 13). In addition, many sprinkler and trickle irrigation systems are being designed with features that permit them to be operated with low labor requirements and at the same time provide the means to manage irrigation water efficiently (see Chapters 15 and 16).

3.5 ENERGY

Energy is required for pumping and for operating the mechanized facilities of many farm irrigation systems. The energy used for pumping varies with static lift; the pressure supplied to operate gated pipe, sprinklers and trickle emitters; friction losses in the pipelines; and the efficiency of the pump and motor. These all must be carefully evaluated when designing and selecting a farm irrigation system. Efficient energy use requires correct scheduling of the time and amount of irrigation applications, and maintaining an efficiently operating pump and power unit.

Farm irrigation systems are usually powered by electric motors or internal combustion engines. The source of electricity is primarily from hydropower generating plants, or by electrical generating plants which use fossil fuels. Initial efforts are underway to examine and develop ways to utilize solar radiation (Alexander, 1978; and Fischbach and Matlin, 1978) to generate electricity for operating farm irrigation systems. In addition, efforts are underway to develop and evaluate wind generation which will assist with supplying electrical power for pumping (Clark and Schneider, 1978). They report that their wind-assist method provides 65 percent of the energy used to pump water from a deep well. The energy source for internal combustion engines is the petroleum fuels of gasoline, diesel, LPG and natural gas.

The current realization that energy resources are finite and depletable along with its continually increasing cost has resulted in increasing concern about efficient energy use. Barnes et al. (1973) commented that the energy used for pumping water in the United States may be several times greater than for all other agricultural field operations. Sloggett (1979) estimates that 23 percent of the on-farm energy use for crop production in the U. S. is for on-farm pumping.

The procedures for evaluating the energy use by farm irrigation systems is a subject which is being addressed by a number of concerned individuals and groups associated with farm irrigation system manufacture, selection, installation and operation.

Chen et al. (1976) summarized the energy requirements for several selected types of farm irrigation systems. Seven representative farm irrigation systems in Oregon were used which included five types of sprinkler systems, a trickle system, and a surface system. The sprinkle irrigation systems selected were; (1) portable hand-move system, (b) mechanically moved lateral system (termed side-roll), (c) continuous move center-pivot system, (d) solid-set system using buried pipe. For the trickle irrigation method, a system with multiple polyethylene (PE) micro-tube emitters and polyvinyl chloride (PVC) surface pipe was selected, while for the surface irrigation method a system designed for use of small furrows (rills) was selected.

Sample energy use comparisons for these systems were provided for specific parameters involving the size of farm, design flow rates, annual crop consumptive use, irrigation farm efficiencies, sprinkler and trickle system operating pressures, and land grading for surface irrigation. Annual operational energy requirements, including the energy used to manufacture, transport and install the systems, were calculated. It was assumed, for these evaluations, that each system was applicable for the resources of soils and water, and the cropping selections, except for the trickle system which was applied only to orchard irrigation.

Chen et al. (1976) used four configurations of the hand-move sprinkler systems, three configurations of mechanical side-roll sprinkler systems, and one configuration of each of the other systems. Farm areas of 8.1, 32.4, and 64.8 hectares were selected for study with design flow capacities based on supplying 2.5, 5.1, and 7.6 mm/day, while providing a constant annual consumptive use of 508 mm.

Based on their design parameters they found that energy requirements increased for the different systems examined in the following order; surface, trickle, side-roll, hand-move, center pivot, permanent, and portable solid set.

The assessment of the energy use by viable alternative farm irrigation systems is an important aspect in selection because of the increasing cost and need to conserve energy. The cost of energy is projected to increase in importance as an economic factor in the selection and operation of farm irrigation systems. The designer therefore, must not only apply current energy costs when evaluating alternative systems, but also examine the economic sensitivity of the systems to energy cost increases over their expected life.

Energy availablity is also an important factor which must be assessed when carrying out evaluations to select a farm irrigation system. This is important whether energy is supplied by petroleum products or electricity. The projected availability of specific petroleum fuels may affect the selection of internal combustion engines for an irrigation system.

The interruption of electrical energy supplied for irrigation during periods of peak demand by other users, may either be required by the power supplier, or may provide an opportunity for significant savings in energy cost. Electrical load and water management relationships have been examined by Heermann and Duke (1978). They propose procedures which integrate irrigation scheduling with electrical load management to reduce peak power demands on the electrical system. Heermann and Duke estimate that such load management can reduce peak demands by as much as 20 percent.

Stetson et al. (1977), has demonstrated a reduction in peak demand of 15 percent. This was accomplished by dividing individual pumping systems into 11 different groups which could be selected for shut off. These groups were divided into six weekly and five daily groups. The six weekly groups were assigned a given day of the week on which they were subject to control, with Sunday being a noncontrolled day. When power demands reached a pre-selected demand level, power was shut off to a selected group of irrigation systems. Control was applied first to the weekly groups. Then, if required, the daily groups were shut off on a rotation basis until the pre-selected demand level was obtained. Actual control was accomplished from a main office control console. The "ripple control" system used coded, high frequency signals imposed on the transmission line to activate individual pump switches by use of a decoder.

When operating under electrical load management where energy will be supplied only part of a given day or period farm irrigation system designs must assure adequate system flow capacity to provide the required water during the critical water use periods. Evaluations of seasonal crop water use requirements may indicate that adjustments in farm cropping to use a shorter period of operation during the critical use period will be more economical than providing larger irrigation system capacity.

The selection of a farm system from the viable alternative irrigation methods for a given farm requires that an analysis of energy requirements be carried out to determine energy availability and costs. Additional discussions on minimizing farm irrigation energy requirements are presented in Chapter 18.

3.6 FARM IRRIGATION SYSTEMS

Farm irrigation systems can be classified into irrigation methods by the manner in which water is applied to the soil. A common classification used is to divide the systems into (a) surface irrigation methods where small open channels or overland flow is used to distribute the water over a cropped field, (b) sprinkler irrigation methods where water is distributed aerially to the cropped field, and (c) trickle irrigation methods where small point applicators are used to apply the water. A fourth method, subsurface irrigation is accomplished by controlling the level of the water table or by applicators which apply the water below the soil surface which results in the water being distributed through the soil by capillary forces.

Detailed descriptions of the irrigation systems employed to apply these methods of irrigation are given in Chapters 13, 15, and 16. These chapters provide the requirements for designing and installing each irrigation system along with discussions of usage and operational characteristics. Specific details for required investigations and acquisition of data, along with details of how to design the components of each specific system is also provided in other chapters of this monograph.

3.7 ECONOMIC AND FINANCIAL FEASIBILITY

Farm irrigation development is occurring around the world on areas ranging from less than a hectare of land to areas in excess of several thousand hectares. The complexity of the economic and financial analyses required increases with the size of the area being developed, the type of cropping system which is to be applied and the amount of funds involved. The procedures discussed in this section provide a basis for carrying out economic and financial analyses when developing an irrigated farm enterprise. The procedures can be simplified for small developments involving limited inputs of capital and operational funds but may require further elaboration for large complex farm developments which involve large amounts of capital to develop and operate.

When planning an irrigation system both an economic and a financial feasibility evaluation should be carried out. The economic feasibility evaluation assesses the economic viability of the planned development and assists in selecting the farm irrigation system from among adaptable alternatives, while the financial feasibility evaluation assesses the financial conditions that will be encountered in developing and operating the irrigated farm. The financial feasibility analysis is prepared to assist the investor in determining the cash flow characteristics under various types of financing. Therefore, the financial analysis should evaluate possible financing alternatives with respect to interest rates, repayment schedules and the length of the loan period. In summary, the economic feasibility studies establish the justification for an irrigation development while the financial feasibility studies evaluate the capability for the development to repay the funding costs associated with the capital investment and the continuing operational costs of the irrigation enterprise.

Farm irrigation system selection from viable alternatives can be carried out on the basis of an economic and financial comparisons of the costs to own and/or rent, operate, and maintain the various systems. The analysis may result in an individual farm using only one type of irrigation method, or it may conclude that a large complex farm should use several methods for the different land and crop conditions.

3.7.1 Economic Feasibility

Evaluation of the economic feasibility of an irrigation system requires estimating all of the costs and returns expected from the development. These economic studies should include comparisons of the costs and returns for the viable alternative farm irrigation and cropping systems, and be a part of the feasibility report for a farm or project development.

The results of an economic feasibility study will provide the farmer with the necessary information as to the attractiveness of proceeding with an irrigation development and for selecting the farm irrigation and cropping system.

Financing for farm irrigation developments is provided by banks, life insurance companies, government agencies, and others with investment capital. The managers of these agencies normally require a detailed evaluation of the proposed agricultural development to assist them in determining whether it is economically sound. Therefore, a thorough economic study not only assists the developer in determining whether a decision to develop is sound, but provides the lending agency with necessary information to determine the type of financial arrangements that will be required for the development venture to be successful.

For determination of the economic feasibility, the development of the costs and returns as annual values on a common time basis is suggested. To develop annual cost values for investment in facilities requires estimating the depreciation of the components of the irrigation system based on their expected life, and the cost of the interest on the capital invested in these items.

In some areas of the world irrigation equipment can be rented rather than purchased. This can reduce the actual initial investment costs for part or all of the irrigation equipment. The costs for renting the equipment then becomes part of the annual cost. Obtaining irrigation equipment by renting may be attractive where limited investment capital is available for development. But it should be recognized that rental fees must include equipment depreciation, interest on the investment, and a reasonable profit to the equipment owner.

Annual irrigation costs must include all costs associated with owning and/or renting, operating and maintaining the irrigation system. In addition to the irrigation system costs must be added the other costs associated with the production of irrigated crops. Data for production costs, other than the irrigation costs, must be obtained to develop a budget for each crop and farm production enterprise to be included in the development. The preparation of cost budgets for farm crop and livestock enterprise may require assistance from various agricultural specialists such as agronomists, livestock specialists, economists, etc. Many locations have available from universities, lending institutions, and others, detailed production cost budgets for many agricultural crops and enterprises.

The following is a list of annual cost items which should be developed to assess the annual cost of an irrigation enterprise:

   1 Water costs, which may include costs of obtaining water rights and permits, or annual assessments for water deliveries from a project entity such as irrigation districts.

   2 Annual fixed costs of owning and/or renting the irrigation system, which will include depreciation of the investment in all irrigation facilities and interest on this investment.

   3 Energy costs for operating the irrigation system.

   4 Costs for repair, operation, and maintenance of the system including labor.

   5 Taxes and insurance.

   6 Other irrigated crop and agricultural enterprise production costs.

If actual price quotations are obtained at a given date for all elements of the irrigation enterprise further adjustment is not required, but many times costs from similar developments carried out at an earlier date can be obtained and adjusted. These adjustments to a common times basis can be made by use of cost trend indexes such as those published by Engineering News

Record, or the Water and Power Resources Service (formerly the U. S. Bureau of Reclamation). Future expenditures can also be adjusted to the given date by use of the present worth procedures. The actual investment and operating costs are then reduced to annual values by application of the amortization concept using capital recovery procedures. These procedures, when applied for the expected useful life of a component are sometimes referred to as "life cycle-costing" (Keller, 1976). Details for carrying out the procedures discussed above are presented later in this section.

The general approach described can be used to evaluate the fixed and variable costs associated with a proposed irrigation enterprise and to carry out value engineering when selecting irrigation system components that will minimize the annual costs over the expected life of the system (or selected period of analysis). The approach can be used to evaluate selection of individual system components, or for selection of alternatives among large complex systems. Procedures (to be discussed later) are available to evaluate the effect or sensitivity of costs during the period of analysis to escalation of individual inputs such as labor, fuel, electrical energy, and other cost items, as well as revenues from sales of the farm production.

3.7.2 Fixed Costs

Depreciation: The depreciation of the components of an irrigation system is based on the expected life of the element. Table 3.1, which gives the expected life of a number of irrigation system components, has been prepared from numerous sources as a guideline for estimating depreciation. It is recognized that variability in expected life can occur for many of these components due to different physical conditions, the level of the repair, operation and maintenance practiced, and the length of time the system is used each year. The guideline for the expected life of a number of the listed items is given as a range. The lower expected life values are generally used for smaller units and are based on normal operation and maintenance practices which have generally occurred with their usage. The upper ranges of depreciation life are suggested as a guideline for rigorously engineered, carefully constructed and installed items for which a high level of maintenance is to be practiced. Engineering judgment should be carefully exercised when applying depreciation life values during the economic analysis. It should be emphasized that the depreciation period in years suggested by the guidelines is based on an average of 2,000 hours of use each year. Such factors as less usage, off-season handling and storage, and maintenance will all affect the life of a particular item included in the irrigation system.

Analysis period: A period of analysis must be selected for the studies to determine the economic feasibility of the development. Although approaches such as suggested by Pair et al. (1975), have been used where the life of each item is assummed and an annual cost value determined, an approach of selecting a much longer life for complex projects is usually used. For large projects, periods of 40, 50, and 100 years are used, while for analyses of on-farm irrigation systems periods of 20, 25, or 30 years are more common.

Present worth: To carry out an annual cost calculation where individual items in the system are depreciated in less time than the period of analysis being used requires some means of paying for replacement. This is accomplished by determining the time at which the replacement would occur, then

TABLE 3.1 DEPRECIATION GUIDELINES FOR IRRIGATION SYSTEMS COMPONENTS

Component	Depreciation, h	Period, yr	Annual maintenance and repairs, %*
Wells and casings	—	20-30	0.5 - 1.5
Pumping plant structure	—	20-40	0.5 - 1.5
Pump, vertical turbine			
bowls	16,000-20,000	8-10	5 - 7
column, etc.	32,000-40,000	16-20	3 - 5
Pump, centrifugal	32,000-50,000	16-25	3 - 5
Power transmission			
gear head	30,000-36,000		5 - 7
V-belt	6,000	3	5 - 7
flat belt, rubber and fabric	10,000	5	5 - 7
flat belt, leather	20,000	10	5 - 7
Prime movers			
electric motor	50,000-70,000	25-35	1.5 - 2.5
diesel engine	28,000	14	5 - 8
gasoline engine			
air cooled	8,000	4	6 - 9
water cooled	18,000	9	5 - 8
propane engine	28,000	14	4 - 7
Open farm ditches (permanent)		20-25	1 - 2
Concrete structures		20-40	0.5 - 1.0
Pipe, asbestos - cement and PVC buried		40	0.25 - 0.75
Pipe, aluminum, gated, surface		10-12	1.5 - 2.5
Pipe, steel, waterworks class, buried		40	0.25 - 0.50
Pipe, steel, coated and lined, buried		40	0.25 - 0.50
Pipe, steel, coated, buried		20-25	0.50 - 0.75
Pipe, steel, coated, surface		10-12	1.5 - 2.5
Pipe, steel, galvanized, surface		15	1.0 - 2.0
Pipe, steel, coated and lined, surface		20-25	1.0 - 2.0
Pipe, wood, buried		20	0.75 - 1.25
Pipe, aluminum, sprinkler use, surface		15	1.5 - 2.5
Pipe, reinforced plastic mortar, buried		40	0.25 - 0.50
Pipe, plastic, trickle, surface		10	1.5 - 2.5
Sprinkler heads		8	5 - 8
Trickle emitters		8	5 - 8
Trickle filters		12-15	6 - 9
Land grading†		none	1.5 - 2.5
Reservoirs†		none	1.0 - 2.0
Mechanical move sprinklers		12-16	5 - 8
Continuously moving sprinklers		10-15	5 - 8

*Annual maintenance and repair costs are expressed as a percentage of the initial cost.
†Various ranges of expected life, from 7-50 yr, have been applied to land grading and reservoir costs. If adequate maintenance is practiced, these items will remain unaffected by depreciation. For economic analysis interest on the investment will cover the costs involved. Life may be limited for reservoirs if watershed sedimentation will reduce its usefulness. Costs associated with water rights can also be handled by an interest charge.

calculating the present worth value of the replacement that will occur at the beginning of the project. The present worth value is calculated by applying the interest rate being used for the analysis to the replacement cost of the item. The present worth factor (PWF) is calculated by:

$$PWF = (1 + i)^{-n} \dots\dots\dots\dots\dots\dots\dots\dots\dots\dots\dots [3.1a]$$

and the present worth value (PW) by:

$$PW = S (1 + i)^{-n} = (S) (PWF) \quad \dotfill \quad [3.1b]$$

where S is the replacement cost, i is the interest rate and n is the number of years the cost will be incurred in the future. This value is added to the initial investment value.

Similarly, there will be items which still have a useful life and have not been completely depreciated at the end of the period of analysis. This undepreciated value is commonly termed the salvage value. The present worth of the salvage value remaining at the end of the analysis period should be calculated and subtracted from the initial capital and present worth costs. This set of calculations results in a single present worth value for the proposed development. Whenever a single or series of values are reduced to a present worth value, the values are said to be discounted.

Amortization: The annual cost of capital invested in the irrigation system can be determined from the present worth value of the investments plus the interest during the period of analysis. A commonly used approach for determining annual costs is to calculate a uniform series of annual values for depreciation and interest over the analysis period that is equivalent to the single present worth value. The value of this uniform series of annual costs is determined by application of an amortization factor, generally referred to as the capital recovery factor (CRF). This factor is calculated by:

$$CRF = i (1 + i)^n ((1 + i)^n - 1)^{-1} \quad \dotfill \quad [3.2a]$$

The annual amortization value (AV) is then calculated by:

$$AV = (PW) (CRF) \quad \dotfill \quad [3.2b]$$

The present worth and amortization values can be easily calculated with the above equations with small calculators, or can be obtained from financial tables available in many books.

3.7.3 Variable Costs

Energy costs: Energy costs are estimated by first calculating the quantity of energy (electricity and/or fuel) to be used annually for irrigation and applying the appropriate prices to determine the cost. Annual pumping energy use can be estimated using the irrigation requirements, irrigation application efficiency, pumping plant efficiencies including the power unit energy conversion efficiency, and the total dynamic pumping lift. This is determined by:

$$PE = (K A D H)/(E_i E_p) \quad \dotfill \quad [3.3]$$

where PE is the pumping energy in the desired units; K is the conversion factor depending on the input measurement and desired output measurement units; A is the irrigated area; D is the depth of the net irrigation requirement; H is the total dynamic pumping head which includes elevation differences, friction and minor losses, and operating pressures; E_i is irrigation application efficiency as a ratio of irrigation water used for evapotranspiration to the amount of water pumped; and E_p is the pumping plant efficiency as a ratio of

TABLE 3.2 ENERGY VALUES FOR PUMPING

Energy source	Measurement units	Heat energy	
		kcal	(Btu)
Electricity	kWh	860	(3,410)
Diesel fuel			
No. 1	liter (gal)	9,017	(135,250)
No. 2	liter (gal)	9,250	(138,750)
Gasoline, regular	liter (gal)	8,274	(124,100)
Tractor fuel	liter (gal)	9,043	(135,630)
Butane	liter (gal)	6,814	(102,200)
Propane	liter (gal)	6,143	(92,140)
Natural gas	m^3 (ft^3, Therms)	8,896	(1,000, 100,000)

water energy output to energy input. Measurement units commonly used for irrigation activities are area in hectares (acres); depth of net irrigation requirement in mm (inches); and total dynamic head in meters (feet). Energy may be calculated for these measurement dimensions in kilocalories (kcal) by a K value of 23.42 (or British Thermal Units (BTU) by $K = 291.0$). (Note: the preferred metric energy unit is the joule and 1 kcal = 4.1868 kJ).

These units can readily be converted to standard units of energy purchase by using the appropriate conversions such as: kilowatt hours for electricity, liters (gallons) for liquid fuels, and cubic meters (cubic feet or therms) for natural gas. Table 3.2 provides the average energy contained in various sources of energy used for irrigation pumping.

Pricing schedules for energy must be obtained for each specific location due to cost variations between regions. Appropriate energy conversion efficiency values for pumping can be obtained from Chapter 10 of this monograph while applicable irrigation efficiencies are presented in other chapters describing the systems being evaluated.

A concern of engineers when planning, evaluating, and selecting an irrigation system is the effect of varied costs over the period of analysis. Continued trends of increasing costs have been encountered for a long period of time with projections that escalation of costs will continue in the future. This is especially true in the area of energy prices. A procedure for evaluating potential effects of cost escalations is presented later in this section.

Operating costs: The costs for evaluating the day-to-day operation of the irrigation facilities should assess the cost of labor and other items involved in operating the system, such as transport, energy for mechanical moves, irrigation scheduling services which are purchased, etc.

The amount of labor required for system operation should be estimated for the particular irrigation enterprise being evaluated. A number of variables are involved such as the charactristics of the particular irrigation system used, amount of automation and other controls installed, type of crop grown, frequency of irrigations, number of irrigations to be applied, and the duration of water applications. Some limited guidelines from specific evaluations of labor requirements for different irrigation systems are available, such as those summarized by Pair et al. (1975) and Fischbach (1975). Estimates can be prepared by careful analysis of operations, or obtained from actual irrigators with similar conditions and systems.

Maintenance and repair costs: The annual cost of maintenance and repairs of irrigation system is another item which will vary depending on the actual operational hours, the individual irrigation system component involved, and the manner in which the operations are carried out. For estimating annual maintenance and repair costs a normal practice is to utilize a percentage of the initial investment cost in the various components of the irrigation system. Table 3.1 lists some ranges of percent of initial costs which have been used for estimating maintenance and repairs of irrigation facilities and may be useful as guidelines during evaluation.

Taxes and insurance: The annual cost of taxes can usually be obtained from the taxing entity in the particular location the irrigation development is occurring, and insurance costs obtained similarly from insurance companies. The combined cost for taxes and insurance normally runs in the range of 1.5 to 2.5 percent of the initial investment value of the irrigation facilities.

Other crop production costs: The annual costs for other items required for crop and other agricultural production such as livestock have been developed for both irrigated and nonirrigated conditions by local agricultural personnel and economists, lending institutions, and others in many of the areas where irrigation development will occur. Where these costs are not available, assistance will need to be obtained from personnel knowledgeable in the recommended practices for production and their costs for both irrigated and nonirrigated crops. In addition to providing the cost data for the cultural practices required, they wll need to provide information on yields being obtained under nonirrigated conditions and projections of the yields which can be expected through the use of irrigation.

3.7.4 Cost Escalation

As noted previously, the fixed and variable costs associated with owning and operating an irrigation system have increased for a number of years and are projected to continue to increase in the future. If the costs of all the items required to develop and operate an irrigation enterprise increase uniformly with time an initial evaluation using current input costs to own and operate an irrigated farm along with the assessment of current market values of produce will adequately describe the attractiveness of the enterprise. Unfortunately, some individual inputs, such as energy or labor, may increase more rapidly than the costs for other inputs. This may affect either the initial design selection of a particular irrigation system component, such as size of the pipeline and pumping plant, and can be assessed by a sensitivity analysis using the escalated annual variable component costs.

An approach to carry out such an analysis for a projected rate of escalating costs during the period of analysis provides a method for adjusting the present worth and annual cost values.

The present worth factor escalating a fixed cost component is obtained by:

$$PWF(r) = \frac{(1+r)^n}{(1+i)^n} \quad \dots\dots\dots\dots\dots\dots\dots\dots\dots\dots\dots\dots \quad [3.4]$$

where PWF (r) is the escalated present worth value; r is the annual rate of escalation; i is the annual interest rate; and n is the length of life of the component in years. This can be used to evaluate the effect of escalation on fixed cost items such as a pump replacement. A similar analysis should be carried out to determine the escalation effect on the present worth of any salvage values.

Given an annual cost value (AV) based on non-escalated costs an equivalent annual cost factor for a uniform series can be obtained (Pearson, 1974) by:

$$EAF(r) = \left[\frac{(1 + r)^n - (1 + i)^n}{(1 + r) - (1 + i)} \right] \cdot \left[\frac{i}{(1 + i)^n - 1} \right], \text{ For } r \neq i \dots\dots\dots \quad [3.5a]$$

and:

$$EAC(r) = (S) (EAF(n)) \dots\dots\dots\dots\dots\dots\dots\dots\dots\dots\dots\dots\dots \quad [3.5b]$$

where EAF (r) is the uniform equivalent annual cost factor for a series of costs which escalate over a period of time, and EAC (r) is the uniform equivalent annual cost.

Example 1: If a set of pump bowls which initially costs $5,000 is expected to be replaced in 10 years and the period of analysis for the total project was 20 years, what would be the present worth value of the replacement if an interest rate of 12 percent is the value of the money but the projected escalation rate of costs is 9 percent?

The present worth value is then:

$$PW(r) = S \cdot PWF(r) = \$5,000 \times \left[\frac{(1.09)^{10}}{(1.12)^{10}} \right]$$

$$= \$3,811$$

Example 2: Suppose the energy cost for a farm is estimated to cost $5,000 per year based on present energy cost but these costs are expected to escalate at a rate of 9 percent per year for the 20 year period of analysis. What is the equivalent annual energy cost that should be used for engineering value judgments for the selection of irrigation systems? When the cost of investment capital is 12 percent?

$$EAF(r) = \left[\frac{(1 + 0.09)^{20} - (1 + 0.12)^{20}}{(1 + 0.09) - (1 + 0.12)} \right] \cdot \left[\frac{0.12}{(1 + 0.12)^{20} - 1} \right]$$

$$= 1.8699$$

Therefore, the equivalent uniform series which would represent the cost of energy becomes:

$$EAC(r) = \text{Energy Cost} \cdot EAF(r) = \$5,000 \times 1.8699 = \$9,349$$

It should be noted that all costs and income can be reduced to equivalent present worth values for comparison purposes. This would require that the uniform series of equivalent annual costs be reduced to a corresponding present worth value for each item.

3.7.5 Economic Evaluation

Benefit-costs: After value engineering is used to minimize the cost of each alternative system and comparison data between systems is assembled, the irrigation system can be selected and the economic feasibility of the development assessed. This assessment is made by comparing the estimated annual costs of owning and operating the farm to the projected increase in the returns, or benefits, the irrigation enterprise will generate by increasing the yield of crops and other products sold from the farm. This comparison provides a benefit-cost assessment of the proposed development and provides the means to assess its attractiveness. The annual benefits when divided by the annual costs are termed the benefit-cost ratio which is used by many lending institutions to evaluate the attractiveness of the enterprise. This method of approach is also used to compare the attractiveness of the selection of viable alternative irrigation systems and methods, as well as different types of irrigated agricultural enterprises which may be feasible for a given development.

3.7.6 Application of Economic Procedures

The analytical concepts and procedures presented in this chapter provide a basis for carrying out the economic and financial evaluations for irrigation development. The following example of "value engineering" illustrates application of the analytical procedures.

Example: Select the minimal cost diameter for an irrigation mainline using PVC pipe. Carry out the analysis for (a) a 20-year period such as might be requird by a lending agency, and (b) a 40-year period which is the expected useful life of the pipe. The annual interest rate on capital invested is 9 percent. For both analysis periods test the sensitivity of the annual costs to own and operate the pipeline to projected escalation of costs for replacements, operation, maintenance and repair. Use a projected annual escalation rate of 7 percent for capital invested in replacement and all annual cost items except for energy which is projected to increase at 12 percent annually. The requirements for the pipeline are:

```
1. Service Area              =  50.6 ha (125 acre)
2. Volume of Water Conveyed  =  5.55 X 10⁵ m³ /yr (450 acre ft/yr)
3. Flow Rate                 =  63.1 L/s (1000 gal/min)
4. Pipeline Length           =  607 m (1850 ft)
```

Table 3.3 provides initial unit investment costs for the pipe and associated components, their estimated life and values for calculating the annual operation, maintenance and repair costs. The initial energy cost is $0.02/kWh, the annual cost of taxes and insurance is 2.5 percent of the initial investment costs.

TABLE 3.3 ASSOCIATED PIPELINE COSTS

Item	Investment cost $/unit	Estimated life yr	Operation, maintenance, and repair %/yr*
Pumping			
Bowls	40.23/kW (30/bhp)	10	6.0
Column	147.51/kW (110/bhp)	20	4.0
Motors, control, etc	80.46/kW (60/bhp)	25	2.0
Pipe (PVC installed)			
Diameter			
203 mm (8 in.)	11.06/m (3.63/ft)	40	0.5
254 mm (10 in.)	16.67/m (5.47/ft)	40	0.5
305 mm (12 in.)	22.74/m (7.46/ft)	40	0.5

*Percent of initial investment cost.

Solution: The selection of the pipeline diameter which minimizes the ownership and operational costs is accomplished by comparing the effect of changing pipeline diameters on the annual cost. This includes the investment cost for the pipeline and the associated pumping station, including replacement; plus the annual costs for operation, maintenance, repair, taxes and insurance, and energy.

The first step is to calculate the amount of friction loss for the 63.1 L/s (1000 gal/min) flow rate for the different pipe diameters selected for comparison in this analysis. Then the size of pumping equipment and the annual energy used for pumping is calculated. The friction losses, pump sizes and the annual friction energy use calculated for the pipeline are given in Table 3.4.

The second step is to calculate the present worth of the capital to be invested during the period of analysis. This present worth value includes the initial investment costs and the cost of the replacements required during the analysis period. The value of each undepreciated component at the end of the analysis period is determined and reduced to a present worth value. This present worth value (termed salvage value) is subtracted from the present worth values for investment.

TABLE 3.4 PIPELINE, PUMPING, AND ENERGY REQUIREMENTS

Pipe dia. mm	(in.)	Flow velocity m/s	(ft/s)	Friction loss* m	(ft)	Power unit size† bkW	(bhp)	Friction energy use‡ kWh/yr
203	(8)	2.09	(6.38)	11.1	(34.2)	8.05	(10.8)	21,600
254	(10)	1.34	(4.08)	3.77	(11.5)	2.68	(3.6)	7,260
305	(12)	0.93	(2.84)	1.56	(4.75)	1.12	(1.5)	3,000

*Based on pipe length of 607 m.
†Based on an assumed pump efficiency of 80 percent.
‡Based on a pumping plant efficiency of 73 percent.

TABLE 3.5 20-YR ANALYSIS OF PIPELINE COSTS

Item	mm (in.)	mm (in.)	mm (in.)
	203 (8) $	254 (10) $	305 (12) $
Present worth values			
Pumping			
Pump bowls - initial cost	324	108	45
Replacement - 10 yr	136	46	19
Pump columns - initial cost	1,188	396	165
Motors, controls, etc — initial cost	648	216	90
Salvage - 20 yr	(−) 23	(−) 8	(−) 3
Pipeline - 607 m (1,850 ft)	6,716	10,120	13,801
Salvage - 20 yr	(−) 599	(−) 903	(−)1,231
Total	8,390	9,975	12,886
Annual cost			
Depreciation and interest	919	1,093	1,412
Maintenance and repair			
Pumping			
Pump bowls	19	6	3
Pump columns	48	16	7
Motors, controls, etc	13	4	2
Pipeline	34	51	69
Energy	432	145	60
Taxes and insurance	222	271	353
Total	1,687	1,586	1,906

TABLE 3.6 40-YR ANALYSIS OF PIPELINE COSTS

Item	mm (in.)	mm (in.)	mm (in.)
	203 (8) $	254 (10) $	305 (12) $
Present worth values			
Pumping			
Pump bowls - initial cost	324	108	45
Replacement - 10 yr	136	46	19
Replacement - 20 yr	58	19	8
Replacement - 30 yr	24	8	3
Pump column - initial cost	1,188	396	165
Replacement - 20 yr	212	79	33
Motors, controls, etc - initial cost	648	216	90
Replacement - 25 yr	75	25	10
Salvage	(−) 8	(−) 3	(−) 2
Pipeline - 607 m (1,850 ft)	6,716	10,120	13,801
Total	9,373	11,014	14,172
Annual cost			
Depreciation and interest	871	1,024	1,317
Maintenance and repair			
Pumping			
Pump bowls	19	6	3
Pump columns	48	16	7
Motors, controls, etc.	13	4	2
Pipeline	34	51	69
Energy	432	145	60
Taxes and insurance	222	271	353
Total	1,639	1,517	1,811

TABLE 3.7 20-YEAR ANALYSIS OF A PIPELINE WITH
ESCALATING COSTS

Item	Pipe diameters					
	mm	(in.)	mm	(in.)	mm	(in.)
	203	(8)	254	(10)	305	(12)
	$		$		$	
Present worth values						
Pumping						
Pump bowls - initial cost	324		108		45	
Replacement - 10 yr	269		90		37	
Pump columns - initial cost	1,188		396		165	
Motors, controls, etc - initial cost	648		216		90	
Salvage - 20 yr	(-) 89		(-) 30		(-) 13	
Pipeline - 607 m (1,850 ft)	6,716		10,120		13,801	
Salvage - 20 yr	(-) 2,318		(-) 3,494		(-) 4,765	
Total	6,738		7,406		9,360	
Annual cost						
Depreciation and interest	738		811		1,025	
Maintenance and repair						
Pumping						
Pump bowls	33		11		5	
Pump columns	81		27		11	
Motors, controls, etc	22		7		3	
Pipeline	57		86		117	
Energy	1,138		382		158	
Taxes and insurance	376		459		597	
Total	2,445		1,783		1,916	

The third step is to calculate the annual cost to own, operate, and maintain the pipeline and associated system elements. This is accomplished by use of the capital recovery approach.

Tables 3.5 and 3.6 show the present worth and annual cost comparisons for use of the three pipe diameters tested without consideration of escalating costs for the 20 and a 40-year period of analysis, respectively. In both cases, the analysis results in selecting a 254-mm (10-in. diameter) pipe. These results show that for non-escalated costs the result is relatively insensitive to length of analysis period.

To examine the sensitivity of the costs to differential price escalation, the effect of escalating the costs of pumping equipment replacement such as bowls, etc., as well as all annual elements such as maintenance and repairs, and energy should be included in the analysis.

Tables 3.7 and 3.8 show the cost comparisons resulting from projections of continuing cost increases throughout the period of analysis. For a 20-year period of analysis, the 254-mm diameter pipe is the minimal cost selection, but for the 40-year period, the 305 mm diameter pipe results in the minimum annual cost. This shows the selection of pipe diameters to be sensitive to escalating cost values as well as for different analysis periods.

Sample calculations for values in the tables are given below to enable the reader to reproduce individual values in the table.

1 Initial costs are obtained by direct application of investment cost data.

**TABLE 3.8 40-YEAR ANALYSIS OF A PIPELINE WITH
ESCALATING COSTS**

Item	Pipe diameters					
	mm	(in.)	mm	(in.)	mm	(in.)
	203 $	(8)	254 $	(10)	305 $	(12)
Present worth values						
Pumping						
Pump bowls - initial cost	324		108		45	
Replacement - 10 yr	269		90		37	
Replacement - 20 yr	224		75		31	
Replacement - 30 yr	186		62		26	
Pump columns - initial cost	1,188		273		165	
Replacement - 20 yr	820		273		114	
Motors, controls, etc - initial cost	648		216		90	
Replacement - 25 yr	408		136		57	
Salvage - 40 yr	(-) 185		(-) 62		(-) 26	
Pipeline - 607 m (1,850 ft)	6,716		10,120		13,801	
Total	10,598		11,414		14,340	
Annual cost						
Depreciation and interest	985		1,061		1,333	
Maintenance and repair						
Pumping						
Pump bowls	46		15		7	
Pump columns	117		39		17	
Motors, controls, ect	32		10		5	
Pipeline	83		124		167	
Energy	2,627		881		365	
Taxes and insurance	540		659		857	
Total	4,430		2,789		2,751	

2 The present worth value of non-escalated costs for replacements are obtained by use of equation [3.1b]. For example, the present worth value for replacement for pump bowls used with the 203 mm (8 inch) diameter pipe is:

$$PW = S \cdot \left[\frac{1}{(1+i)^n} \right] = \$324 \times \left[1/(1+0.09)^{10} \right] = \underline{\$136}$$

The equivalent present worth value for replacement of the same item using given escalated costs is obtained by equation [3.4], this results in:

$$PW(r) = S \left[\frac{(1+r)^n}{(1+i)^n} \right] = \$324 \times \left[(1+0.07)^{10}/(1+0.09)^{10} \right] = \underline{\$269}$$

3 Salvage values are obtained by using a straight-line depreciation to determine the values which are not fully depreciated at the end of the analysis period. The calculations are similar to those used in 2 above to determine either non-escalated or escalated present worth values. For example, when using escalated values, the salvage value for pipeline at the end of the 20-year analysis is:

Salvage value = [20 yr/40 yr (life)] $\times$ \$6716 = $\underline{\underline{\$3358}}$

PWC = \$3358 $\times$ $[1/(1 + 0.09)^{20}]$ = \$599

4 The annual value of depreciation and interest is obtained by applying the capital recovery factor to the present worth value of the investment. For example the present worth value of the components associated with the 203-mm pipe is \$8,390. Therefore, the annual uniform series over the period of analysis which will pay the interest and recover the depreciation for the 20-year period is:

AV = PW $\cdot$ CRF = \$8390 $\times$ $[0.09(1 + 0.09)^{20}]/[(1 + 0.09)^{20} - 1]$ = \$919

This approach is applied to both non-escalated and escalated values since escalation, when used, is included in the present worth values.

5 The annual operation, maintenance and repair costs are obtained by direct application of the appropriate percentage values to the initial cost of each component, or the unit cost for energy used each year. For example, the annual non-escalated energy cost (S_e) for the 203-mm diameter pipe is:

S_e = 21,600 kWh $\times$ \$0.02/kWh = \$432

This cost is used directly if non-escalated costs are to be used for evaluation. To obtain an equivalent uniform annual series of costs for evaluation of the effects of increasing costs the annual value (S_e) obtained above is adjusted by use of equation [3.5a] and [3.5b]. For example, the equivalent annual cost of energy for the 203-mm (8 in.) diameter pipe in Table 3.7 is obtained by:

EAC (r) = $S_e \cdot$ EAF(r)

EAC(r) = \$432 $\times \left[\dfrac{(1 + 0.12)^{20} - (1 + 0.09)^{20}}{(1 + 0.12) - (1 + 0.09)} \right] \cdot \left[\dfrac{0.09}{(1 + 0.09)^{20} - 1} \right]$ = \$432 $\times$

2.633 = \$1,138

Similarly by applying the appropriate projected escalation percentage for each of the cost items a total equivalent annual cost is developed. Use of total equivalent annual cost estimates to determine the cost-benefit ratio requires that projections for annual escalation of revenues be made to provide the equivaluent annual benefits for the comparison.

3.7.7 Financial Feasibility

If the economic feasibility as measured by the cost-benefit ratio is attractive, the next step is to carry out a financial evaluation to assess the financial feasibility. The purpose of the financial feasibility study is to identify the actual year-by-year costs and revenues which can be expected after starting the development. The annual repayments for loans obtained to carry out the development plus all annual variable costs expected for operating the enterprise must be identified. These annual costs are then compared with the projected revenue from the enterprise to assess whether particular cash-flow problems will occur. In the study, cash outlays which result from actual expenditures for loan repayments, energy purchase, hiring of labor, payment of taxes and all other expenditures to operate the enterprise are tabulated on an annual basis.

Likewise the expected revenue from farm sales are projected to provide a year-by-year comparison of expenses and income. This type of study can be used to identify problems which may be encountered from lending periods that are too short or other loan stipulations. This study also provides the means for assessing the effect of the installation period and the lower than normal yields which may be encountered during the first several years of development.

With respect to the installation period, many larger developments may use several years to install a total system. When a system is developed in this manner it is likely some components of the system which will service the total system will be installed during initial construction. This results in expenditures for a particular component which may not be fully utilized until development is completed

After installation of irrigation facilities to either change an existing system or to start a new development several years may occur before the full potential of the development is realized with respect to production and consequently, revenue. There are a number of causes for this period of time (development period). Some of these causes are inherent in the land resources, such as the recovery period for graded lands, and changes resulting from converting from a rainfed to fully irrigated cropping environment. Other causes may be managerial in nature due to the requirement for the operator to learn new skills, and timing of cultural practices (including water applications). Although variations may occur, a period of three to five years is normally expected before the projected yields are obtained.

In summary, a financial feasibility analysis is essential for an operator to assess the financial attractiveness of the development.

3.8 ENVIRONMENTAL CONSIDERATIONS

Many irrigation developments, especially on newly irrigated lands, require an environmental assessment to identify changes in the environment that can be expected to occur as a result of the development. These assessments require an inventory of the factors associated with the environmental changes that will occur during and after the development, and evaluation of the impacts caused by the changes. A thorough environmental assessment normally requires the services of several specialists to cover the particular areas of concern.

Whether or not an environmental assessment is required, for an irrigation development will need to be determined by the planner. When planning and selecting farm irrigation systems, which are part of an overall irrigation project, it is likely that environmental concerns have been addressed as part of the overall planning effort. But for individual irrigation developments on new lands an environmental assessment may be necessary to comply with current development regulations.

In the United States, the purpose of the environmental assessment is to provide concerned government agencies with information with which they can evaluate the environmental effect of the project, and as may be required by law, issue environmental impact statements. In countries other than the United States, an environmental assessment may be required by either the government or by the funding agency.

This discussion is not intended to be complete, but is presented to highlight some of the factors which need to be addressed during the assessment. Since before and after development examination of the various factors involved is required to assess changes that will occur as a result of the development, the initial step is to inventory the existing environmental conponents during the planning phase of project development.

Various approaches are used to identify the items which need to be addressed during the environmental assessment. The preparation of a matrix such as suggested by Leopold et al. (1971), may assist the planner with identification of the items to be considered.

Guidelines and engineering criteria which are applicable to environmental concerns when developing an irrigation enterprise should be obtained from the responsible entity prior to carrying out the designs or the environmental assessment. This will assist in identifying environmental items which should be addressed by the assessment and provide design criteria for elements of the project which may be specified due to a particular environmental factor.

3.8.1 Environmental Inventory
The environmental inventory for irrigation development assesses the characteristics of the environment which exist prior to development. Such an inventory will include the baseline data for assessing changes that will result from the development. A complete environmental inventory could be identified under four major areas of concern and can be obtained as follows:

A. Physical and Chemical Characteristics
  1 Earth: (a) soils, (b) land form, and (c) unique physical features
  2 Water: (a) surface, (b) groundwater, (c) quality, (d) temperatures, and (e) recharge
  3 Atmosphere: (a) climate (macro, micro), (b) temperature, and (c) quality
  4 Processes: (a) flooding, (b) erosion (wind, water), (c) air movement, (d) soil stability (slides, slumps, settling), (e) depositions (sedimentation), and (f) salinization
B. Biological Conditions
  1 Flora: (a) trees and shrubs, (b) grasses, (c) crops, (d) microflora, and (e) endangered species

    2 Fauna: (a) birds, (b) land animals including reptiles, (c) fish and shellfish, (d) benthic organisns, (e) insects, (f) microfauna, and (g) endangered species

C. Cultural Factors
    1 Land Use: (a) grazing and forestry, (b) agriculture, and (c) wetlands
    2 Recreation: (a) fishing and hunting, (b) camping and hiking, and (c) picknicking
    3 Aesthetics and Human Interest: (a) open space qualities, (b) scenic views and vistas, (c) rare and unique species or ecosystems, (d) unique physical features, and (e) historical or archaeological sites
    4 Cultural Status: (a) cultural patterns, (b) health and safety, (c) employment, and (d) population density
    5 Man-Made Facilities: (a) structures, (b) transportation network, and (c) utility networks

D. Ecological Relationships such as: (a) salinization of water, (b) salinization of soils, (c) food chains, and (d) eutrophication

E. Other Items: (a) site specific items

The inventory, when completed, provides a basis for identifying changes that will occur as land is developed for irrigated agriculture. The complexity of the changes which can be expected is a function of the size and complexity of the development.

3.8.2 Environmental Impacts

    Irrigation development will result in environmental changes and impacts. A partical list of possible changes are categorized in the following check list.

A. Modification of Regimes
    (a) Alteration of flora
    (b) Alteration of fauna
    (c) Alteration of streamflows
    (d) Alteration of groundwater
    (e) Alteration of crops

B. Land Transformation and Construction
    (a) Land grading
    (b) Irrigation-system (pipelines, pumps, canals, etc.)
    (c) Well drilling
    (d) Fencing
    (e) Water impoundments and diversions

C. Traffic Changes
    (a) Automobile
    (b) Trucking

D. Chemical Treatment
    (a) Fertilizers
    (b) Weed control
    (c) Insect control

E. Recreational Changes
    (a) Sports opportunities (fishing, hunting)
    (b) Others

F. Other Changes
    (a) Unique, site specific

Some of the environmental concerns which may be encountered in irrigation developments are discussed below.

Water diversions: Numerous irrigation farm developments in the western and other areas of the United States are obtaining the water supply by pumping from streams, rivers, and lakes. These sources may support fish which are used for recreational and commercial purposes. Because of fish populations, specific design criteria for water diversions may be required by the water regulating agency. This criteria may specify the minimum depth below the normal water surface and the location with respect to the shoreline where intake of water can occur. Specifications for the maximum dimensions of screen openings and allowable velocities of flow to and through screens may also be a requirement for design.

Stream flow considerations: Irrigation diversions, especially from small streams, will cause a change in the flow pattern. This change will correspond to the seasonal patterns of irrigation diversions. The hydrologic changes will need to be identified during the environmental assessment to assure that minimum flows as may be required by governing regulations are provided.

Another aspect of operating an irrigation enterprise is that return flows which eventually discharge into streams or other water bodies usually occur. Both the quality and quantity of these flows will need to be assessed to evaluate not only the expected volume but the sediments, salts, and other materials which may be returned. Applicable environmental controls may result in requiring pump-back systems to reclaim the surface runoff from irrigation, or installation of appropriate facilities to control the sediments being returned to waterways. These requirements can affect the selection of the farm irrigation systems, as well as the design of a system.

Salinization: Water quality with respect to salt loads is another important environmental concern. As discussed in Chapter 5 the prevention of the build up of an undesirable salt level in the soil is a necessary component of maintaining a viable irrigation enterprise. The portion of the water which is leached through the soil to prevent harmful salt concentrations increases the concentration of salts in the return flow waters. The salt concentrations which can be expected to occur in this portion of return flow waters as well as their location of reentry to waterways needs to be determined during the environmental assessment. If environmental guidelines or regulations as to the maximum allowable content of salts in return flows are applicable, this may require development of approriate water management strategies to assure these levels are not exceeded. Such management strategies may affect farm irrigation system design and selection.

References

1 Alexander, G. H. 1978. A solar-powered pump for irrigation. Proc. Ann. Tech. Conf. Irrig. Assoc., Silver Springs, MD, p. 1-19.

2 Barnes, K. K. et al. 1973. Energy in Agriculture. A task force report of the Council for Agricultural Science and Technology, Ames, IA.

3 Bondurant, J. A. 1977. Erosion and sedimentation on irrigated lands. ASAE Publ. H-77, 146-151, St. Joseph, MI.

4 Bowman, C. C. 1969. Semi-automatic control of irrigation. VII Cong. Internl. Comm. Irrig. & Drain., 24, 0. 271-275. New Delhi, India.

5 Campbell, M. D. and J. H. Lehr. 1973. Water Well Technology. McGraw-Hill, New York, NY, 681 p.

6 Chen, K., J. W. Wolfe, R. B. Wensink, and M. A. Kizer. 1976. Minimum energy designs for selected irrigation systems. Paper No. 76-2037, ASAE, St. Joseph, MI.

7 Clark, R. N., and A. D. Schneider. 1978. Irrigation pumping with wind energy. Paper No. 78-2549, ASAE, St. Joseph, MI.

8 Davis, S. N. and E. J. M. DeWiest. 1966. Hydrogeology. J. Wiley and Son, New York, NY, 463 p.

9 Fischbach, P. E. 1975. Mechanical moved sprinkler systems and gated pipe with a reuse system. Proc. Ann. Tech. Conf. Sprinkler Irrig. Assoc., Silver Springs, MD, p. 63-70.

10 Fischbach, P. E. 1978. Solar Cell Irrigation. Proc. Ann. Tech. Conf. Irrig. Assoc., Silver Springs, MD, p. 213-217.

11 Haise, H. R. and E. G. Kruse. 1969. Automation of surface irrigation systems. Proc. Am. Soc. Civ. Engr., J. Irrig. & Drain. Div., 95(IR-4):503-516.

12 Heermann, D. F. and H. R. Duke. 1978. Electical load and water management. Proc. Ann. Tech. Conf. Irrig. Assoc., Silver Springs, MD, p. 60-67.

13 Humpherys, A. S. 1967. Control Structures for Automatic Surface Irrigation Systems. TRANSACTIONS of the ASAE 10(1):21-23, 27.

14 Humpherys, A. S. 1969. Mechanical Structures for Farm Irrigation. Proc. Am. Soc. Civ. Engr., J. Irrig. Drain. Div., ASCE 95(IR-4):463-479.

15 Johnson, E. E. 1966. Ground Water and Wells. E. E. Johnson, Inc., St. Paul, MN.

16 Keller, J. 1976. Energy economics in pipe selection. Proc. Ann. Tech. Conf. Sprinkler Irri. Assoc., 134-145, Silver Springs, MD, p. 134-145.

17 Leopold, L. B., F. E. Clarke, B. B. Hanshaw, and J. R. Balsley. 1971. A Procedure for Evaluating Environmental Impact. Circ. 645. U. S. Geolo. Survey, Washington, DC.

18 Mech, S. T., and D. D. Smith. 1967. Water Erosion Under Irrigation. p. 950-963. In: Irrigation of Agricultural Lands. R. M. Hagan, H. R. Haise, and T. W. Edminister (Ed.), Am. Soc. Agron., No. 11, Madison, WI.

19 Mech, S. T., and N. P. Woodruff. 1967. Wind erosion on irrigated lands. p. 964-973. In: Irrigation of Agricultural Lands, R. M. Hagan, H. R. Haise, and T. W. Edminister (Ed.), Am. Soc. Agron. No. 11, Madison, WI.

20 Miller, D. E., and W. C. Bunger. 1963. Moisture retenetion by soils with coarse layers in the profile. Soil Sci. Soc. Am. Proc. 27:586-589.

21 Pair, C. H., W. W. Hinz, C. Reid, and K. R. Frost. 1975. Sprinkler Irrigation. Chapter XVIII, Silver Springs, MD.

22 Pearson, G. F. 1974. Life-cycle costing in an energy crisis era. Professional Engineer, Vol. 44, No. 7, 22-29, Washington, DC.

23 Sloggett, G. 1979. Energy and U. S. Agriculture: Irrigation Pumping, 1974-1977. Agri. Econ. Report No. 436, USDA, Washington, DC.

24 Stetson, L. E., D. G. Husted, and F. J. Potthoff. 1977. Ripple control of irrigation loads at Southwest Public Power District. Proc. Univ. of Nebr. Short Course, Lincoln, NB.

25 U. S. Department of Agriculture. 1951. Soil Survey Manual. USDA Handbook 18, Washington, DC, 503 p.

26 U. S. Department of Agriculture, Soil Conservation Service. 1964. Soil-Plant-Water Relationships. Nat'l. Eng. Handb., Irrigation, Sec. 15, Chapter 1, Washington, DC.

27 U. S. Department of Interior, Bureau of Reclamation. 1977. Ground Water Manual. Washington, DC, 480 p.

chapter 4 ▬▬▬▬▬

SOIL WATER
PART I — Properties
PART II — Dynamics

4 chapter

4

SOIL WATER
PART I — Properties

by R. W. Skaggs, North Carolina State University,
Raleigh, North Carolina; D. E. Miller, Irrigated
Agriculture Research and Extension Center,
USDA-SEA/AR, Prosser, Washington; and R. H.
Brooks, AID Egypt Water Use and Management
Project, Cairo, Egypt, Oregon State University,
Corvallis, Oregon

The design and operation of efficient irrigation systems require
knowledge of the factors and processes controlling movement and storage
of water in soil. This chapter begins with a discussion of the nature of soil
water and the interactive forces that affect the distribution and movement of
water in the soil-water-plant system. Methods for measuring the state of soil
water and the soil properties that describe the water holding characteristics
of soils are presented and discussed. The rate that water moves in soil is
directly proportional to its hydraulic conductivity. Methods to measure hydraulic conductivity in both saturated and unsaturated soils are presented
and techniques for predicting the unsaturated hydraulic conductivity function from the soil water characteristic are discussed. The rate that irrigation
water can be applied as well as the methods of application are often dictated
by the infiltration characteristics of the soil. Factors controlling infiltration
and procedures for measuring infiltration characteristics are also presented
and discussed in Part I of this chapter.

4.1 WATER HOLDING CHARACTERISTICS OF SOILS

4.1.1 Introduction
Soil water is of interest largely because of its influence on plant growth
and crop production. A growing plant must be able to balance the atmospheric demand for water with the amount it can extract from the soil. The
soil water supply is alternately depleted through evapotranspiration and replenished by irrigation or precipitation. The importance of soil water storage
and availability in crop production has long been recognized and much research has been done to characterize the soil properties responsible for water
absorption and retention.
Soil water content alone is not a satisfactory criterion for describing the
availability of water to plants and attempts have been made to describe the
energy state of water. Initially, empirical measurements and relationships
were developed, but these gave way to consideration of fundamental mechanisms and expressions. The soil-water-plant system is now treated as a continuous dynamic system where water moves through the soil to the plant root

surfaces, into the root, through the plant and into the atmosphere along a path of continuously decreasing potential energy. The removal of soil water depends not only upon its amount and energy state but also upon the plants ability to absorb water and the atmospheric demand for water from the plant. The amount of water that is sufficient for satisfactory plant production depends upon the crop species and variety, stage of growth, and the marketable product. This entire process, called "SPAC" by Philip (1966) for Soil-Plant-Atmosphere Continuum, was discussed in detail by Hillel (1971).

In light of the importance of the energy state of soil water it is proper to begin this discussion with soil water potential. A more detailed discussion of the various potential components was given by Rawlins (1976).

4.1.2 Soil Water Potential

About the turn of the century, soil water was arbitrarily classified into different forms such as gravitational water, capillary water, hygroscopic water, etc. (Briggs, 1897, cited by Richards and Wadleigh, 1952). These early groupings have been replaced by a fundamental concept referred to as soil water potential. Soil water does not vary in form, within the range of our interest, but it does vary in the energy with which it is retained in the soil. The work required to move an incremental volume of water from some reference state to the soil water is known as the total soil water potential, ψ. The usual reference state, arbitrarily defined as having zero potential, is pure water with a flat air-water interface at some specified elevation, temperature, and air pressure. In an unsaturated soil energy must be expended to remove water, so the soil water potential has a negative sign. The potential gradient, or rate of decrease of potential energy with distance, is the driving force causing soil water flow (Section 4.2). Thus, water will move from a wet soil where the potential is near zero, to a dry soil where the potential is lower (a larger negative value).

The total soil water potential may be expressed as the sum of four component potentials.

$$\psi = \psi_g + \psi_p + \psi_o + \psi_n \quad \dots\dots\dots\dots\dots\dots\dots\dots\dots\dots\dots\dots\dots\dots \quad [4.1]$$

where ψ is the total soil water potential and ψ_g, ψ_p, ψ_o, and ψ_n are gravitational, matric or pressure, osmotic, and pneumatic potentials, respectively. Rawlins (1976) lists other potentials, but they are not of interest for this discussion.

Gravitational potential. Gravitational potential is determined by position in the gravitational field and may be expressed in units of potential energy per unit mass as $\psi_g = zg$ where z is the distance above an arbitrary reference plane and g is the acceleration due to gravity. The reference elevation may be set at any convenient height, such as the soil surface.

Matric or pressure potential. Matric or pressure potential is that due to the attraction of soil surfaces for water as well as to the influence of soil pores and the curvature of the soil-water interface. It may be expressed in terms of soil water pressure, P, as $\psi_p = Pg$, also in units of energy per unit mass. It is convenient to consider pressure potential as a continuous function of water content so that it is positive in a saturated soil below the water table and negative in unsaturated soil.

Osmotic potential. The decrease in potential energy of soil water due to solutes relative to pure water is called osmotic potential. This effect does not influence soil water flow appreciably, because the solutes can move with the water. It is very important, however, in the uptake of water by plants through root cell membranes. Solutes also lower the vapor pressure of water so that osmotic potential may be important in vapor diffusion.

Pneumatic potential. This potential (energy per unit mass) may be expressed as $\psi_n = P_{sa}\, g$, where P_{sa} is the soil air pressure. Usually the air pressure is considered to be uniform throughout the soil profile and the pneumatic potential is ignored in characterizing soil water flow. Such assumptions are not always justified, however (Section 4.3).

Expression of soil water potential. Soil water potential may be expressed in either of three ways:

Energy per unit mass, ψ: Potential energy is defined in terms of the energy required to remove a unit mass of water from the soil under specified conditions. This is considered to be the fundamental expression. The dimensions are L^2T^{-2} and the usual units are ergs per gram or joules per kilogram.

Energy per unit volume, ψ_v: Multiplying energy per unit mass by density gives energy per unit volume. The density of water is about 1 g/cm^3, so that energy per unit mass and per unit volume are numerically equal in the cgs system of units. Energy per unit volume has the dimensions of pressure, $ML^{-1}T^{-2}$, and may be expressed as millibars, bars, or pascals.

Energy per unit weight (hydraulic head, H): Energy per unit weight can be expressed in terms of the height of a liquid column that exerts the given pressure at the bottom of the column. If the liquid is water, this energy is called hydraulic head, H, the dimensions are L, and a common unit is cm of water. This is a convenient way to express soil water potential and is often used in characterizing soil water movement in both saturated and unsaturated soil. When osmotic and pneumatic potentials are ignored, the potential energy per unit weight may be written as $H = h + z$, where h is the matric or pressure potential expressed as a head and z is the gravitational head, which is the vertical distance of the point of measurement above the reference plane. If the point of measurement is below the reference plane, z is negative.

The various methods of expressing soil water potential can be easily converted from one to another, i.e., $\psi_v = \psi\varrho$ and

$$h = \frac{\psi_v}{\rho g} = \frac{\psi}{g},$$

where ρ is the density of water and g is the acceleration of gravity. A useful conversion table taken from Hillel (1971) is given as Table 4.1.

To avoid confusion among the various expressions for soil water energy status, one must keep in mind that a high matric or pressure potential refers to a wet soil and means approaching zero from the negative side, while a low potential refers to dry soil and is a large negative number. A high potential would be -0.10 bar while a low potential would be -15 bars. On the other

TABLE 4.1 ENERGY LEVELS OF SOIL WATER EXPRESSED IN VARIOUS UNITS
(Adapted from Hillel, 1971, Table 3.1)

Soil-water potential				Soil water suction		Relative humidity at 20 °C
per unit mass		per unit volume	per unit weight			
ergs/g	joules/kg	bars	cm H_2O	bars	cm H_2O	percent
-1×10^4	-1	-0.01	- 10.2	0.01	10.2	100.00
-1×10^5	-10	-0.1	-102.0	0.1	102.0	99.993
-5×10^5	-50	-0.5	-510.0	0.5	510.0	99.964
-1×10^6	-100	-1.0	-1020	1.0	1020	99.927
-5×10^6	-500	-5	-5100	5	5100	99.637

hand, low suction or tension refers to wet soil but means approaching zero from the positive side. High suction means a dry soil and is a large positive number; i.e., a low suction is +0.10 bars and a high suction is +15 bars.

4.1.3 Soil Water Characteristic

As water is removed from a soil, the matric or pressure potential of the water remaining is decreased (a larger negative number). If water is added to the soil, the matric potential is increased (a smaller negative number). A curve showing the functional relationship between matric potential and soil water content is known as the soil water characteristic. Potential and water content can be expressed in any convenient units. When the relationship is determined as the soil dries, the curve is known as the desorption, water retention, or water release curve. When the relationship is determined as a dry soil wets, it is called the sorption or imbibition curve.

Water is retained in the soil by a combination of the attraction of particle surfaces for water and the capillary action of water in the soil pores. The matric potential is related to the curvatures of the air-water interfaces, which in turn are affected by the soil pore geometry as well as the wetness. At high matric potentials (near zero), most of the soil pores are filled with water and the total porosity and pore size distribution greatly influence the water retained. Inasmuch as soil texture dominates the total porosity and pore size distribution, it has a marked effect on the soil water characteristic. In general, the higher the clay content of a soil, the higher will be the water content at any given potential. Soil aggregation, especially of the fine-textured soils, tends to increase the number of large pores. Thus, soil structure is important in the amount of water retained at high potentials. When the large pores empty, the water remaining in the soil is held in the smaller interaggregate pores and on the particle surfaces. As the soil dries, the amount of particle surface area determines the water retained, and this is strongly influenced by soil texture. Soil compaction also influences the water characteristic because compaction results in smaller pores, reduced total porosity, and increased surface area in a given soil volume. It is usually the larger pores that are reduced most by compaction, so that the influence is greater at high potentials. Examples of soil water characteristics for three soils of different textures are given in Fig. 4.1.

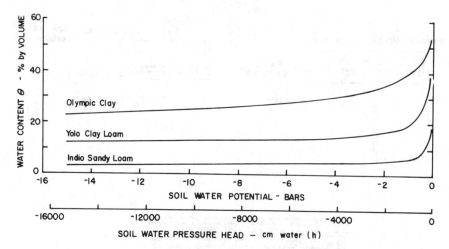

FIG. 4.1 Soil water characteristic curve as influenced by soil texture [Adapted from Richards and Wadleigh, 1952].

The term "soil water capacity" is given to the slope ($d\theta/dh$ in Fig. 4.1) of the soil water characteristic at any point on the curve. This value represents the change in water content per unit change in matric potential and is an important property in relation to soil water storage and release.

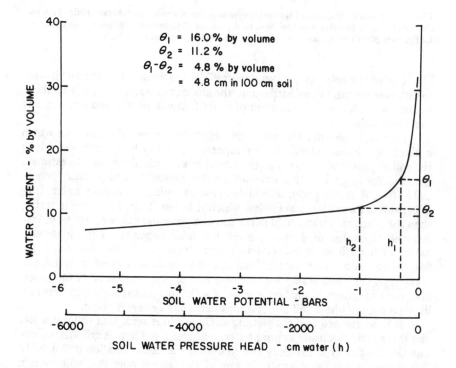

FIG. 4.2 Soil water characteristic for Ritzville loam, and estimation of water released between two potentials h_1 and h_2.

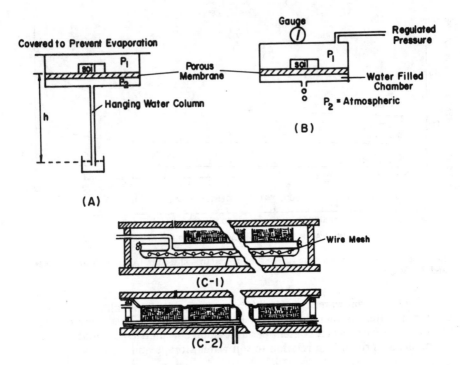

FIG. 4.3 Schematic diagrams of apparatus used to measure a soil water characteristic. A. A suction table using a hanging water column. B. A pressure chamber used for lower potentials, and C. Pressure plate and pressure membrane units used by Richards [1965].

Use of the soil water characteristic. The soil water characteristic is used to estimate the pore size distribution, the amount of water a soil will retain at a given potential, and the amount of water that will be released between any two potentials.

Although the soil water potential largely determines the ease with which a plant can obtain water, it is also important to know how much water is in the soil at potentials above a given critical level. This allows one to estimate the need for irrigation. In a soil with a water content θ_1 and pressure potential h_1 (Fig. 4.2), the potential will decrease as a plant withdraws water. If h_2 is considered a critical level below which it is not desired to deplete water, then the amount of water held in the soil at a potential above h_2 will be $\theta_1 - \theta_2$. Because the value of θ at any given potential depends on the hysteretic nature of the soil water characteristic (see "Hysteresis"—this Section, page 84), the values of water release should be considered only as estimates. Also, many soils swell and shrink with wetting and drying, so that all of the water does not come from a constant volume of soil. This is especially important at high potentials where soil structure influences the characteristic.

If it is assumed that the wetting angle of soil is zero, that soil water has the same surface tension as pure water, and that a soil pore is cylindrical, one can use the theory of capillary rise to obtain the useful relation $d \cong 3/h$. In this equation d is the diameter in mm of the largest pore that will contain water at the pressure potential h expressed in cm of water. A decrease in matric potential from h_1 to h_2 will result in a release of a volume of water.

This water can be considered to have come from pores having an effective size range of d_1 to d_2.

Methods of determining the soil water characteristic. The soil water characteristic is usually determined in the laboratory using tension tables or pressure plates (Fig. 4.3). In all of the techniques used, a porous membrane or plate hydraulically connects the soil water with water in the lower chamber. The pores in the membrane are small enough that water but no air can pass through under the imposed pressure. In all cases $P_1 > P_2$, so that water is forced from the soil into the lower chamber. At equilibrium, the imposed pressure (expressed in suitable terms) can be considered as the potential of the water remaining in the soil.

For high potentials the membrane may be blotter paper, fine sand, sintered glass, porous steel, or similar materials. In this case P_1 is often atmospheric and P_2 is obtained with a hanging water column (Fig. 4.3a) or with regulated vacuum. At lower potentials of about -1 bar or less, the pores of these materials are too large to remain water filled and air will pass through the membrane. A fine-pored ceramic is then used as the membrane, P_2 is atmospheric, and P_1 is obtained with compressed gas — usually air or nitrogen. Ceramic membranes are available with bubbling pressures of up to 20 bars. For still higher pressures of up to 100 bars a cellulose acetate film is used as the membrane.

In practice a sample of soil is placed in the pressure chamber in a retaining ring and saturated overnight. The desired pressure is then applied until outflow ceases and the soil water is considered to be in equilibrium with the applied pressure. The amount of water in the soil is then determined, usually by oven drying. The process is repeated, with a second sample being subjected to a different pressure. The resulting soil water contents are plotted against applied pressure or vacuum, expressed as potential units, usually bars or millibars, to form the water characteristic. Details of apparatus and procedure are given by Richards (1965), Tanner and Elrick (1958) and others.

Because the pore size distribution has such influence on water retention at high potentials, disturbed samples (dried and sieved) often give erroneous results. Use of so-called undisturbed soil cores is preferable, but even with these, some error is inevitable because of the swelling and shrinking that accompanies wetting and drying of many soils. At low potentials (approaching -15 bars) the soil specific surface dominates water retention and the error introduced by using disturbed soil samples is quite small.

The water characteristics of soils in the field may be obtained by use of water potential measuring devices in combination with soil water measurement. Tensiometers are the instruments most commonly used to measure potential and are discussed in Section 4.1.5. Their main disadvantage is that they are reliable only at potentials above about -0.8 bar. However, in most soils, much of the useable water is held in this range. Tensiometers are installed in the field at desired depths and soil water potentials are measured. Then the soil is sampled gravimetrically to determine water content. If water content on a volume basis is needed, the soil bulk density must also be measured or estimated. Because of natural soil variability and the variation in soil water content with wetting history (hysteresis), the field-determined water characteristic curve is not precise and is difficult to duplicate.

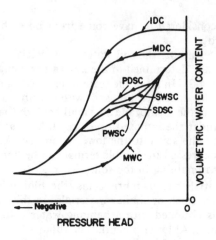

FIG. 4.4 Diagram of typical hysteresis curves, where IDC is the initial drainage curve, MWC and MDC are main wetting and drainage curves respectively, PWSC and PDSC are primary wetting and drainage scanning curves, and SWSC and SDSC are secondary wetting and drainage scanning curves [From Gillham, 1972, cited by Rawlins, 1976].

Sorption curves are more difficult and tedious to determine then desorption curves because equilibrium is reached very slowly. At low potentials (in the wilting range and drier) the soil is allowed to become wet by absorbing water vapor from an atmosphere of known potential which is controlled by a specified concentration of sulfuric acid or salts. At higher potentials, the process is like desorption in reverse. The soil absorbs water through a membrane, against an imposed pressure or vacuum. In the ranges of from about -0.5 to - 15 bars a ceramic membrane is used and the imposed pressure is compressed air. At higher potentials, the membrane may be coarser material like sintered glass and the soil absorbs water against controlled vacuum or from a hanging water column. The apparatus described by Tanner and Elrick (1958) was designed for both sorption and desorption measurements.

Hysteresis. The desorption and sorption soil water characteristics will differ because the water content in a soil at a given potential depends upon the wetting and drying history of the soil. This history dependence in the relationship between potential and water content is called hysteresis. An example of the desorption and sorption curves for a soil is given in Fig. 4.4. When the desorption curve is obtained with an initially saturated sample and the sorption curve is obtained with an initially dry sample, the two moisture characteristics are known as the primary hysteresis loops or main branches. If the soil is not completely dry before rewetting or not completely wet before drying, the resulting curves will be between the two primary curves and they are known as scanning curves. Soils are rarely saturated in nature, even when the water table rises to the surface, because of air that is entrapped during wetting. Thus, curve MDC in Fig. 4.4 is more realistic than curve IDC.

At any given potential the water content will be greater in a drying soil (desorption) than in a wetting soil (sorption). The effect of hysteresis is most marked in wet soils where the potential is determined by the capillary effect, i.e., the curvature of the air-water interfaces and the surface tension of water. Excellent discussions of hysteresis are given by Baver et al. (1972), Hillel (1971), and Nielsen et al. (1972).

Field soil is almost always neither completely dry before rewetting nor completely wet before drying. During infiltration only the surface few centimeters are near saturation, with deeper layers being less wet, depending upon rates of flow through the soil, time of rewetting, etc. When infiltration stops, redistribution begins and the soil loses water by drainage and plant use at water contents ranging from near saturation at the surface, to considerably less than saturation at the wetting front. Thus, the water characteristic can be used only with reservation in interpreting soil water status. The water content or potential can only be estimated from measurement of the other, unless the wetting history is accurately known. However, the amount of error involved is relatively small, compared with other errors involved such as soil variability, climatic changes, and plant variabilities.

4.1.4 Soil Water "Constants"

Field capacity and permanent wilting point once were considered to be soil water constants. They are now recognized as very imprecise but qualitatively useful terms.

Field capacity. After infiltration ceases, water within the wetted portion of the profile will redistribute under the influence of potential gradients. Downward movement is relatively rapid at first but decreases rapidly with time. Field capacity refers to the water content in a field soil after the drainage rate has become small and it is an estimate of the amount of water that may be temporarily stored in the soil profile for plant use. For many years after its introduction by Veihmeyer and Hendrickson (1927), field capacity was accepted as a physical property characteristic of each soil and it was assumed that an application of a given amount of water would wet a soil to field capacity to a definite depth, depending upon the soils deficit in relation to field capacity. Field capacity now is used only as a very rough term that refers in general to the water content of a soil a few days after it has been wetted. For most soils this is a near optimum condition for growing plants. However, soil water may continue to drain for many days after irrigation (Robins et al., 1954; Ogata and Richards, 1957; Miller and Aarstad, 1974; and others).

The field capacity concept is more applicable to coarse- than to fine-textured soils because in coarse soils most of the pores empty soon after irrigation and the capillary conductivity becomes very small at relatively high potentials. In fine-textured soils, with a narrower range of soil pore sizes, the hydraulic conductivity will not change so rapidly with time and the drainage can continue for weeks or months. Soil characteristics that have greatest influence on field capacity are soil texture and layers within the profile that impede water flow. Fine soils retain more water than coarse soils as well as drain longer at significant rates. Any layer interface will inhibit water movement across the interface and thus restrict redistribution and increase field capacity (see Section 4.1.6). Other factors that may influence field capacity are organic matter content, depth of wetting, wetting history, and plant water use.

In general, field capacity cannot be determined on a single soil sample in the laboratory, because it is influenced by the water transmission characteristics of the whole soil profile as well as the properties of a soil at a given point. Such laboratory procedures as the moisture equivalent and

1/3-bar percentage can be taken only as rough estimates and apply only to uniform soil. Field capacity, to be useful, should be determined in the field under conditions that will normally exist during the growing season, except for water removal by a growing crop. The plot of soil must be large enough to avoid significant errors due to lateral flow (Miller, 1963). The time of sampling to characterize field capacity is greater for fine- than for coarse-textured soils and for a deep root zone than for a shallow one (Miller and Aarstad, 1973; also see equation 18.14).

Wilting point. The permanent wilting point or percentage is the soil water content below which plants growing in that soil remain wilted even when transpiration is nearly eliminated. It represents a condition where the rate of water supply to the plant roots is very low. The water content corresponding to the wilting point applies to the average water content of the bulk soil and not to the soil adjacent to the root surfaces. The soil next to the root surfaces will usually be drier than the bulk soil (Gardner, 1960), because water cannot move toward the root surfaces fast enough to supply plant demands and a water content gradient develops near the root.

Like field capacity, permanent wilting is not a soil constant or a unique soil property. There is no single soil water content at which plants cease to withdraw water. Even though wilted, plants will absorb water, but not at rates sufficient to regain turgor. Plants growing under low atmospheric demand can dry soil to lower water contents than if the demand is high, because more time is allowed for water to move through the soil to the roots. Also, when atmospheric demands are high, plants may temporarily wilt even though soil water contents are considered adequate. Sugarbeet wilting in midday during the summer is an example. The wilting is simply the result of water not moving to and through the root surfaces as fast as the plant demands it.

In the wilting range, almost all soil pores are empty of water and the water content is determined largely by the specific surface area and the thickness of water films on these surfaces. The thickness of these films is related to potential, and Richards and Weaver (1944) found the water content in soil subjected to a pressure potential of – 15 bars was closely correlated with the permanent wilting percentage, as determined with sunflowers, for a wide range of soils. Because of its simplicity and the availability of reliable equipment, the 15-bar percentage is now commonly used to estimate the permanent wilting percentage. Sunflowers are the standard test plant used for determining permanent wilting percentage (Peters, 1965).

Available water. The amount of water released by a soil between field capacity and permanent wilting is called the available water. The term implies that the available water can be used by plants, but this is misleading. If the soil water content approaches the wilting range, especially during periods of high atmospheric demands for water or flowering and pollination, yield or quality of most crops will be greatly decreased. Some crops, however, can extract soil water to potentials considerably below – 15 bars. Wheat is a notable example, as are many drought- or salt-tolerant species. Plants may remove water from a wet soil before it drains to field capacity. Wilcox (1962) and Miller and Aarstad (1971, 1973) showed that this water partially compensates for drainage that occurs after the field capacity range is attained.

Inasmuch as the difference between field capacity and available water can be no more meaningful than either of the terms, available water itself is only an estimate of the amount of water a crop can use from a soil. It was once generally accepted that soil water was equally available to plants from field capacity to wilting point (Veihmeyer and Hendrickson, 1927, 1949, 1950). This was disputed by Richards and Wadleigh (1952), and it is now recognized that more energy must be exerted to extract water as the soil dries, in accord with the potential theory of water movement.

Many farmers irrigate when the available water has been depleted a certain amount, depending upon the crop. For high-water-requiring crops such as potatoes, irrigation may be scheduled at 15 to 25 percent depletion (85 to 75 percent available water remaining in the soil); for many other crops, the depletion may go to 50 to 75 percent before irrigation. This bank account type of irrigation ignores any relation between depletion and water potential. For a given soil, the degree of depletion allowed before irrigation may be roughly related to potential through the characteristic curve.

As with field capacity, available water is a useful concept, providing that one recognizes its limitations and that it varies with soil depth, climatic factors that influence evapotranspiration, and the soil profile characteristics.

4.1.5 Field Methods For Characterizing Soil Water

Soil water content. It is often necessary to be able to determine the amount of water in a soil. Gardner (1965) and Rawlins (1976) have discussed the various methods and associated errors for measuring soil wetness. The discussion here will be limited to those techniques considered most useful in the field.

Gravimetric: The accepted standard for soil water measurement is the gravimetric method, which involves weighing a sample of moist soil, drying it to a constant weight at a temperature of 105 to 110 °C, and reweighing to determine the amount of water lost on drying. The results are commonly expressed as the ratio of mass of water lost to mass of dry soil. The required drying time depends upon the soil texture, soil wetness, loading of the oven, sample size, whether the oven is a forced draft or convection type, and other factors. Usually 24 h is sufficient but the required time is obtained by repeatedly weighing a sample after various periods of drying. Microwave ovens have been used to reduce drying time, but they need more evaluation before they can be generally accepted (Rawlins, 1976).

The water content is calculated as:

$$\theta_{dw} = \frac{\text{Wt. of Moist Soil} - \text{Wt. Oven Dry Soil}}{\text{Wt. of Oven Dry Soil}}$$

or

$$\theta_{dw} = \frac{\text{Wt. of Moist Soil}}{\text{Wt. of Oven Dry Soil}} - 1$$

where θ_{dw} refers to water content expressed on the basis of the dry weight of soil.

It is often convenient to express soil water contents on a volumetric basis, i.e., the ratio of the soil water volume to the total soil volume. This is done by multiplying the water content on a dry weight basis by the ratio of the soil bulk density, ρ_b, and water density, ρ, as follows:

$$\theta_v = \theta_{dw} \frac{\rho_b}{\rho}$$

where

θ_v = water content on a volume basis.

The soil bulk density (or volume weight) is defined as the oven dry weight of soil in a given volume, as it occurs in the field. It may be measured by drying and weighing a known volume of soil, or by using the clod, core, or excavation method (Blake, 1965). The core method is the most commonly used. A cylindrical metal sampler of a known volume is forced into the soil at the desired depth. The resulting soil core is removed and dried at 105 to 110 °C and weighed. The bulk density is the weight of the oven-dry soil divided by volume of the cylinder. Samples may be taken at successive depths from the surface by cleaning out the sample hole to the desired depths with an auger and then forcing the sampler into the soil at the bottom of the hole. Alternatively, a trench may be dug and samples taken by forcing the sampler into the soil horizontally at the desired depth. Obviously, this latter method involves more labor but the sampling zones can be better observed. Excellent core samplers are available commercially.

Soil bulk density can also be determined by radiation methods using either a double- or single-probe instrument as discussed by Blake (1965). The equipment is expensive and the soil water content must be known when density is measured.

Neutron scattering: The neutron scattering procedure to estimate soil water content has gained wide acceptance in recent years. It has some advantages over the gravimetric method because repeated measurements may be made at the same location and depth, thus minimizing the effect of soil variability on successive measurements. It also determines water content on a volume basis, with the volume of soil involved being influenced by the particular instrument used and on the soil type and wetness. Disadvantages are the initial high investment in equipment and the time required per site because of the need to install access tubes.

A source of high energy or fast neutrons is lowered to the desired depth into a previously installed access tube. Radium-beryllium and americium have been used, but all current equipment uses a americium. The fast neutrons are emitted into the soil and gradually lose energy by collision with various atomic nuclei. Hydrogen, present almost entirely in soil water, is the most effective element in the soil in slowing down the neutrons. Thus, the degree of the slowing down of neutrons is a measure of the soil water content. The slowed or thermalized neutrons form a cloud around the source and some of these randomly return to the detector, causing an electrical pulse on a charged wire. The number of such pulses is measured over a given interval of time with a scalar or the rate of pulsation can be measured with a ratemeter. The count rate is almost linearly related to the water content.

When not in use the radiation source is housed in a shield that contains a high hydrogen material such as paraffin wax. This material serves as a standard by which proper operation of the instrument can be verified. Inasmuch as instrument variations and source decay occur, it is more satisfactory to use the count ratio method rather than just a count. The ratio of sample count/standard count is plotted versus water content. This eliminates any systematic errors due to instrumentation that may vary from day to day. The volume of soil measured depends upon the energy of the initial fast neutrons and upon the wetness of the soil. With the radium-beryllium source the volume of soil measured is a sphere of about 15 cm dia in a wet soil and up to 50 cm or more in a dry soil (de Vries and King, 1961; Van Bavel, 1961). Measurements near the surface are not accurate because neutrons are lost through the surface. It is difficult to accurately detect any sharp delineation such as a wetting front or soil layering effect.

The manufacturer usually supplies a calibration curve but one should verify whether it can be used for a given soil. If changes in water content are desired rather than absolute values, a single curve is more widely applicable because the same error will be in successive readings and will cancel out.

Two calibration procedures have been used — field calibration in natural soil profiles and laboratory calibration in large prepared soil standards. Calibration should be done with the same type of access tubes as will be used in the field. For details of calibration, see Douglas (1962).

Gamma ray attenuation: The mathematics of attenuation of monochromatic radiation have been discussed by Gardner (1965) and Rawlins (1976). A beam of known intensity is passed through a soil column and the decrease in intensity is measured. The attenuation is related to the mass of material through which the beam passes. If the soil bulk density remains constant or varies in a known way, the water content may be inferred or if the water content is known, the soil bulk density may be inferred. This technique has been widely used in recent years in laboratory studies.

A double-tube attenuation unit has been used in the field (Reginato and Van Bavel, 1964) and commercial equipment is available to measure water contents by this method. The main advantage of this system is that it allows nondestructive measurement of water within a narrow depth range, as opposed to the large volume measured with the neutron probe. Reginato (1974) has developed a dual energy gamma ray technique by which changes in both soil water and bulk density can be measured.

Other methods: The water content of porous blocks in equilibrium with the soil water has been used as a measure of soil water content. This technique is more widely used to estimate soil water potential than soil water content and it is discussed in the section on measuring potential.

If the relation between soil water content and matric potential is known, any method of measuring potential can be used to estimate soil water content. The effect of hysteresis must be kept in mind. LaRue et al. (1968) indicated that when field variability was considered, water contents estimated from tensiometer readings were as accurate as gravimetric samples. They used moisture characteristics derived from soil cores.

Rapid methods have been proposed for measuring soil water, including burning with alcohol to generate the heat of drying (Gardner, 1965), evaporation from a mixture of soil and nonvolatile oil (Thijssen et al., 1954; cited by Rawlins, 1976), extraction of water with methyl alcohol (Bouyoucos,

1931; Bowers and Smith, 1972), and generation of acetylene gas by adding carbide to wet soil (Blystone et al., 1961; Ely, 1965). Additional ones were discussed by Cope and Trickett (1965). Although the accuracy of most of these methods is said to be about the same as routine oven-drying, they have not been widely used. The carbide technique has been used to a limited degree for irrigation scheduling* and in road construction (Blystone et al., 1961).

Soil water potential. It is often desirable to measure soil water potential in addition to, or instead of, soil water content. The estimation of soil water potential from water content data via the characteristic curve may not be sufficiently accurate. Matric (pressure) potential, due only to the surface and capillary forces, can be easily measured in wet soil with instruments called tensiometers. The name is derived from the term "tension" that was initially applied to the energy of retention of soil water. The tensiometer does not measure osmotic potential. The sum of the matric and osmotic potentials can be measured by freezing point depression or with thermocouple psychrometers. Osmotic potential only can be estimated with salinity sensors.

Tensiometer: The tensiometer is widely used for measuring soil water potential in the field and laboratory. Many commercial models are available, or the necessary parts can be purchased and assembled at a significant savings. The theory and use of tensiometers have been discussed by Richards (1949) and Richards (1965).

Schematic diagrams of tensiometers are shown in Fig. 4.5. A tensiometer consists of a porous ceramic cup filled with water and connected through a water-filled tube to a suitable vacuum measuring device. The cup, when saturated with water, must be capable of withstanding air pressures of about 1 bar without leaking air. For normal field applications, the vacuum is measured with a reliable vacuum gauge (Bourdon vacuum gauges). For more precise work, a mercury manometer is used. In reacting to soil water changes, water moves into or out of the tensiometer cup and this movement requires finite time. For special conditions where rapid response time is needed, the vacuum measurement is made with pressure transducers and almost no water flow through the cup is required (Klute and Peters, 1962). The pressure transducers can also be used to continuously record the vacuum in the tensiometer, but the associated electronic read-out equipment is expensive. Temperature-compensated pressure transducers have been developed (Rawlins, 1976). Other null-type tensiometers have been developed for laboratory use (Miller, 1951; Leonard and Low, 1962) but these are not generally suited for field use. A commercial portable null-type tensiometer with a Bourdon vacuum gauge is available that allows measurements to be made in the field in about 5 min. Tensiometers have been used as sensors for automating irrigation systems to maintain a desired soil water range.

The major criticism of the tensiometer is that it functions reliably only in the wet soil range at potentials of about - 0.8 bar or higher. At lower poten-

---

*Unpublished information on file at the Engineering and Research Center, Bureau of Reclamation, U.S. Department of Interior, Denver, CO 80225.

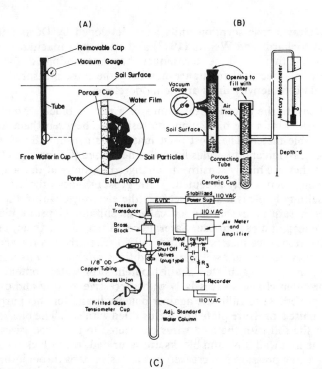

FIG. 4.5 Schematic diagram of tensiometers. A. With a vacuum guage showing the cup-soil water contact. B. With a mercury manometer, and C. A laboratory null-type model equipped with a recording pressure transducer [A is from a brochure of Soil Moisture Equipment Co., B is from Richards 1965, and C is from Klute and Peters, 1962].

tials air is drawn out of the water inside the tensiometer and readings are not reliable. This is not as serious as it may seem, because about 75 percent or more of the available water in coarse-textured soils and 25 to 50 percent in fine-textured soils is in this range (Richards and Marsh, 1961).

In sandy soils, where hydraulic conductivities are very low, tensiometers may not function properly, unless a null-type measurement is used. Soil water flow away from the cup as the soil dries may be so slow that hydraulic equilibrium with the bulk soil will not be achieved. Under such conditions the bulk soil water potential may be much lower than the tensiometer indicates.

Inasmuch as the water column in the tensiometer becomes unreliable at about - 0.8 bar, the type of tensiometer shown in Fig. 4.5 is useful in the root zone but not for deep installations. For example if the tensiometer is installed at a depth of 300 cm, the potential that can be reliably measured at that depth is about - 0.5 bar because the water column within the tensiometer exerts a potential of about - 0.3 bar at the gauge. If it is necessary to measure water potentials at great depths, tensiometers equipped with integral pressure transducers can be used and only electrical leads need come to the soil surface (Watson, 1967a).

Porous blocks: Porous blocks in contact with moist soil will equilibrate with the soil water. The water content of a given block will depend largely upon the matric potential of the soil water with which it is in contact and to a lesser degree upon its wetting history. Various properties of a porous block can be used to indicate its water content and thus indirectly the soil water potential. Gravimetric sorption units were developed by Davis and Slater (1942) and Richards and Weaver (1943) and electrical capacitance has been used also. Neither of these techniques has been adopted for routine measurements. Electrical resistance has been the most measured property and thermal conductivity units are being developed.

To measure the electrical resistance of a porous block, two electrodes are imbedded within the block and the resistance between them measured with a suitable meter. The most used material is gypsum (Bouyoucos and Mick, 1940). Nylon and fiberglas have also been used (Colman and Hendrix, 1949), but they are highly sensitive to changes in salinity of the soil solution and have not been widely accepted. With gypsum blocks the effect of soil solution salinity levels is masked because the electrolyte within the block is essentially a saturated solution of calcium sulfate. Gypsum blocks are relatively cheap and easy to use. Several companies supply them, as well as inexpensive resistance meters (Gardner, 1965). Even though the accuracy is not good, they do indicate soil water conditions qualitatively as wet or dry. The blocks must be calibrated against potential or water content and the calibration often changes excessively with successive wetting and drying.

It is preferable to calibrate against potential, and for this purpose it is common to use a pressure plate apparatus filled with soil. The blocks are imbedded in the soil with the lead wires connected to insulated wires leading through the unit wall. Soil and blocks are saturated, then block resistance is measured as the pressure is increased in successive steps through the desired range. Block resistance is plotted against applied pressure to get a working curve (Taylor et al., 1961). As discussed in previous sections, the applied pressure can be taken as a measure of the soil water potential. Calibration curves are usually supplied with commercial blocks and it is not necessary to calibrate individual blocks if they are made nearly identical. However, uniformity of readings can be improved by immersing a number of blocks in water to saturate them, and then selecting a group with similar resistances at saturation. Blocks differing markedly should be discarded.

When installed in the field, wires from many blocks can be led to a single convenient location, thereby eliminating the need to walk out into a field. Resistances can easily be continually recorded with suitable strip chart recorders. Gypsum blocks are not very accurate in the wet soil range. However, this is the range where tensiometers work well. Thus, tensiometers and gypsum blocks are often used together — tensiometers to characterize wet soil and gypsum blocks to characterize potentials less than about - 1 bar.

Shaw and Baver (1939) suggested that heat conductivity could be used as an index to water content, using the principle that heat is conducted much faster in water than in dry soil. A constant current may be passed through a heating element imbedded in a porous block for a given time.The resulting heat is conducted away from the element, and the element temperature can be related to the water content of the porous block. Phene et al. (1971 a,b) measured temperatures with a linear diode temperature sensor, eliminating the need to correct for ambient temperature changes. Phene et al. (1973) subsequently used this system to control irrigation to maintain soil matric potential within narrow range. Rawlins (1976) indicates that these instruments are next to tensiometers in accuracy. They are still essentially a research tool, although they do show promise for field applications. Units are available commercially and have been incorporated into irrigation controllers.

Psychrometric methods: Detailed discussions of the use of psychrometry to measure water potential have been given by Wiebe et al. (1971), Brown and Van Haveren (1972) and Rawlins (1976). The technique measures the sum of the matric and osmotic potentials. The method most widely adopted for *in situ* measurement of soil water potential is the measurement of wet-bulb temperature, with a small thermocouple serving as the wet-bulb. Water is condensed on the thermocouple by passing a current through it to cool it below the dew-point by the Peltier effect. The current is then removed and the wet-bulb temperature is measured. Ambient temperature is also measured with the same thermocouple to allow corrections to be made for temperature effects on the calibration curve. Units are calibrated against potentials of standard solutions.

Thermal gradients in the soil can cause measurement errors (Rawlins, 1976; Wiebe et al., 1977). This error can be reduced by installing the units with axes of symmetry parallel to the soil surface, or can be avoided by use of a double-junction unit in which the reference and sensing junctions are placed close to each other within the psychrometer chamber. Problems associated with temperature have also been reduced by using dew-point hygrometers, also made with thermocouples, rather than the wet-bulb type. Although both wet-bulb and dew-point instruments are commercially available, the methods are still essentially research tools.

Osmotic Potential: Where salinity is a possible problem, it is desirable to measure osmotic as well as matric potential. Vapor pressure or freezing point depression techniques can be used to measure osmotic potential of solutions, but they are not suitable for routine measurements in the field. Electrical conductivity is utilized in the salinity sensor and 4-electrode probe.

Salinity sensors were developed at the USDA Salinity Laboratory at Riverside, CA and they have been used extensively. Oster and Willardson (1971) estimated that commercial sensors available are accurate to about ± 0.25 bar. The sensor consists of a thin ceramic disk that has very small pores (bubbling pressure of about 15 bars) that remain water-filled through the matric potential range involved. When buried in the soil, the solution in the ceramic equilibrates with the soil solution and any change in electrical conductivity can be attributed to salinity. The units also include a thermistor to allow measurement of soil temperature, which is used to correct for temperature effects on electrical conductivity.

The 4-electrode probe has been used to measure soil solution conductivity directly in the field (Rhoades and Ingvalson, 1971). Although not so accurate in measuring osmotic potential as the salinity sensor, the technique is useful in evaluating salinity problems in the field.

4.1.6 Effect of Soil Layers on Water-Holding Characteristics

In comparison with a uniform soil, any profile discontinuity that affects pore size distribution, such as a textural or structural change, will decrease water movement. Soil water content above the boundary will be increased (Robins, 1959) if the soil has received sufficient water to allow drainage below the depth in question. The effect of soil layers on water retention and movement was reviewed recently by Miller (1973).

When the soil is wet, if the discontinuity is a layer of finer texture than the soil above, water will drain through the upper soil faster than through the layer. Water will accumulate above the layer and positive hydraulic pressures may develop. Unless an appreciable slope exists, the water will remain in the soil until it is lost by evapotranspiration and slow drainage through the substrata. Inasmuch as the water would drain away if the layer were not present, the layer greatly increases water storage capacity of the soil above. These soils must be irrigated carefully to prevent adverse effects of excess water and poor aeration. A slowly permeable layer will transmit water slowly, but for a long period after irrigation. Thus, the effective available water will be much higher with high evapotranspiration rates than with lower ones because the plant can extract water while the soil is draining — water that would be lost as drainage if plants were not present.

If the discontinuity is a layer that is coarser than the soil above, the layer will not conduct significant amounts of water until many of the pores are filled with water. This will occur only at potentials much higher than those at which the pores in the soil above are filled. As the profile drains, the layer will stop transmitting water at relatively high potentials, and the water content in the soil above the layer will remain higher than if the layer were not present.

The major items to consider in evaluating the influence of coarse layers on water retention are (i) the coarseness of the layer (including the amount of fine materials), (ii) depth of the layer, and (iii) desorption characteristics of the soil (Miller, 1973).

In a wet soil with plant use of water prevented, the coarseness of the layer determines the matric potential at the soil:layer interface. Water cannot drain from the soil faster than the coarse layer can transmit it and movement through a coarse layer will be negligible at some high potential, which is determined by its unsaturated conductivity characteristics. Inasmuch as the matric potential decreases about 1 mb/cm increase in elevation above a coarse layer, the depth to the layer determines the potential distribution within the soil above. The water content distribution is then determined by the desorption characteristics of the soil above the layer. A coarse layer will have the greatest influence in increasing water retention, compared to a uniform soil, when the soil loses appreciable water at some potential lower than what it will drain to over the coarse layer.

Field profiles range from those that contain sharply defined layers with a pronounced influence on water retention to those where the layering has negligible effect. Miller (1969) measured the unsaturated conductivities of the coarse layers of several profiles and related them to the amount of < 0.25-mm material in the samples. He made the following generalizations:

1   If the coarse material contains less than about 1 percent of < 0.1 mm material, soil over such material will not drain to potentials less than about -30 to - 50 mbars at the interface. The expected potentials decrease with an increase of < 0.1 mm material, and at about 7 percent will be about -70 to -90 mbars.

2   If more than about 15 percent of the < 0.1 mm material is present, the potential distribution probably will be little affected by the layer and, thus the water retention will be little affected.

4.2 HYDRAULIC CONDUCTIVITY

The basic relationship for describing soil water movement was derived from experiments by Darcy in 1856 who found that the flow rate in porous materials is directly proportional to the hydraulic gradient. Darcy's law may be written as:

$$q_s = -K \, \Delta H / \Delta s \quad \dotfill \quad [4.2]$$

where q_s is the volume of water moving through the soil in the s-direction per unit area per unit time and $\Delta H / \Delta s$ is the hydraulic gradient in the same direction. The proportionality factor, K, is the hydraulic conductivity, which depends on both properties of the fluid and the porous medium. The quantity H is the hydraulic head which for practical purposes may be considered the sum of the pressure head, h, and the elevation-head, z (Section 4.1.2). The negative sign in equation [4.2] indicates flow is in the direction of decreasing H.

As long as the hydraulic head is measured below the water table or where the pressures of the soil-water are positive, the hydraulic conductivity, K is constant with respect to H. In the limit as Δs approaches zero, the hydraulic conductivity is assumed to be a soil property at a point in the soil mass and equation [1] may be written as,

$$q_s = -K \, \partial H / \partial s \quad \dotfill \quad [4.3]$$

Whenever the hydraulic conductivity varies from point to point in the soil mass, the soil is said to be heterogeneous. If the hydraulic conductivity is dependent on the flow direction, the soil is said to be anisotropic. An excellent discussion of Darcy's law for anisotropic soils is given by Childs (1969).

For regions of the soil mass which are only partially saturated and in which the water content varies with both time and position the equation for flux may be written as,

$$q_s = -K(\theta) \frac{\partial H}{\partial s} \quad \dotfill \quad [4.4]$$

where the hydraulic conductivity is a function of water content, θ. Since the relationship $\theta = \theta(h)$ is a property of the soil (albiet also dependent on the wetting and drying history due to hystersis — Section 4.1.3), we may write K = K(h), and

$$q_s = -K(h) \frac{\partial H}{\partial s} \quad \dotfill \quad [4.5]$$

Recall that H = h + z (Section 4.1.2) where z is the vertical distance above the datum. Then for flow in the horizontal direction, x,

$$q_x = -K(h) \frac{\partial H}{\partial x} \quad \dotfill \quad [4.6]$$

and for vertical flow

$$q_z = -K(h) \frac{\partial}{\partial z} (h + z),$$

$$q_z = -K(h) \frac{\partial h}{\partial z} - K(h) \quad \dotfill \quad [4.7]$$

We may also write

$$\frac{\partial h}{\partial s} = \frac{\partial \theta}{\partial s} \frac{dh}{d\theta}$$

where $dh/d\theta$ can be evaluated from the soil water characteristic and is itself a function of θ. Then we can write equation [4.6] as:

$$q_x = -K(\theta) \frac{dh}{d\theta} \frac{\partial \theta}{\partial x},$$

or

$$q_x = -D(\theta) \frac{\partial \theta}{\partial x} \quad \dotfill \quad [4.8]$$

where $D(\theta) = K(\theta)dh/d\theta$ is defined as the soil water diffusivity having units of L^2/T (e.g. cm^2/h). Likewise equation [4.7] may be written:

$$q_z = -D(\theta) \frac{\partial \theta}{\partial z} - K(\theta) \quad \dotfill \quad [4.9]$$

For unsaturated soils the water moves primarily in small pores and through films located around and between solid particles. As the water content decreases, the cross-sectional area of the films also decreases and the flow paths become more limited. The result is a hydraulic conductivity function that decreases rapidly with water content as shown schematically in Fig. 4.6a. In most cases hysteresis in the $K(\theta)$ relationship is small. However, when $K = K(h)$ is used as in equation [4.5], hysteresis may be quite pronounced due to hysteresis in the $h(\theta)$ relationship (Fig. 4.6b). We noted earlier (Section 4.1.3) that soils are usually not completely saturated in nature because of air entrappment during the wetting process. Thus, even for apparently saturated regions below the water table the volumetric water content may not be equal to total porosity, but to θ_s, the water content at residual air saturation. The corresponding hydraulic conductivity is K_s (Fig. 4.6a), which may still be considered constant in regions below the water table and is sometimes referred to as the apparent saturated conductivity.

There is some advantage in expressing the flux in terms of the water content gradient (equations [4.8] and [4.9]) in that $D(\theta)$ varies over a smaller

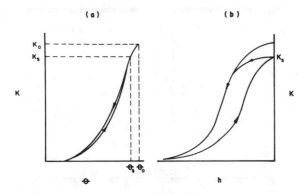

FIG. 4.6 Hydraulic conductivity versus water content [a] and soil water pressure head [b].

range than the corresponding $K(\theta)$. Note, however, that θ is constant for $h > 0$, so $d\theta/dh = 0$ and $D(\theta)$ is not defined for positive pressure heads. Therefore, these equations cannot be used for saturated flow or for combined saturated-unsaturated flow.

Swartzendruber (1969) noted that the essence of the development of equations for unsaturated flow (equations [4.4] through [4.9] was originally set forth by Buckingham (1907) and suggested that they be called the Buckingham-Darcy equations to reflect their historical origins.

4.2.1 Saturated Hydraulic Conductivity

Numerous methods have been developed for measuring saturated conductivity in the field and on soil cores in the laboratory (Bouwer and Jackson, 1974). When a water table is present the most commonly used methods are the pumped-hole techniques. These include, in order of increasing size of the region sampled for K, the piezometer method (Kirkham, 1946); the tube method (Frevert and Kirkham, 1948); the augor hole method (Van Bavel and Kirkham, 1948; Boast and Kirkham, 1971), the two-well method (Childs et al., 1953); and the four-well method (Kirkham, 1954; Snell and van Schilfgaarde, 1964). While all of these methods have been rigorously developed and thoroughly tested, conductivity often varies widely from point to point in the field and numerous measurements are usually required to obtain field effective values.

Auger hole method. This is probably the most widely used technique for measuring hydraulic conductivity in the field. As indicated by the schematic in Fig. 4.7, an auger hole is drilled to some distance below the water table. After the water level in the hole has reached equilibrium with the field water table, the water level is lowered by quickly pumping or bailing water from the hole. The saturated hydraulic conductivity is then determined from the rate that the water level rises in the hole as,

$$K = -C \, dy/dt \quad\quad\quad\quad\quad\quad\quad\quad\quad\quad\quad\quad\quad\quad\quad\quad\quad\quad\quad\quad\quad\quad\quad\quad [4.10]$$

Where C is a constant that depends on the radius of the hole, a; the depth of the hole below the water table, d, and the distance from the bottom of the

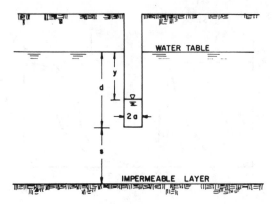

FIG. 4.7 Schematic of auger hole method for determining saturated hydraulic conductivity.

hole to the impermeable layer.

Diserens (1934) is credited with being the first to develop and apply an equation to calculate hydraulic conductivity from auger hole measurements. An exact solution of the form of equation [4.10] was obtained by Kirkham and van Bavel (1948) for a hole penetrating to the impermeable layer (s = 0). Numerical solutions for the general case shown in Fig. 4.7 were obtained by Ernst and are presented in nomograph form by van Beers (1958). The value of C in equation [4.10] can be obtained from these nomographs for given values of a, d, s and y.

An exact mathematical solution for flow into an auger hole was developed by Boast and Kirkham (1971). They considered a lower boundary condition of either an impermeable layer as shown in Fig. 4.7 or an infinitely permeable layer with a pressure head equal to the water table height. They presented their solutions in the form of equation [4.10] with C values tabulated for various ratios of d/a and s/d (Table 4.2). Values of C are given for holes that are initially pumped empty, 1/4 full and 1/2 full (y/d values of 1, 0.75 and 0.5, respectively). The rate of rise of water in the auger hole, dy/dt, , is measured in cm/s and this value is multiplied by C to find K in m/d. C-values corresponding to s/d and d/a values between those given in Table 4.2 can be obtained by interpolation. Boast and Kirkham (1971) suggest that for accurate work logarithms of C, d/a and y/d should be taken before doing an interpolation.

The auger hole method is based on the assumptions that there is uniform soil, negligible water table drawdown around the auger hole and no water movement in the unsaturated zone above the water table. For layered soils an estimate of K for each layer can be obtained by using auger holes of different depths. In order to prevent excessive drawdown near the hole the water table should be lowered quickly by bailing or pumping. Continuous pumping to clean mud or sediment from the auger hole should be avoided or, if used, should be followed by a 'resting period' to allow equilibrium with the surrounding water table before conducting the test.

Although the diameter of the auger hole is not restricted, sizes between the range of 5 to 15 cm are commonly used. Care must be taken in heavy soils to prevent compacting and smearing the sides of the hole as it is being dug. Shady et al. (1972) found that the use of helical power driven augers remold-

TABLE 4.2 VALUES OF C FOR EQUATION 4.10 FOR AUGER HOLE UNDERLAIN BY BY IMPERMEABLE OR INFINITELY PERMEABLE MATERIAL. THE RATE OF RISE OF WATER IN THE AUGER HOLE IS MEASURED IN cm/s AND THIS VALUE IS MULTIPLIED BY C TO FIND K IN m/d (BOAST AND KIRKHAM, 1971)

d/a	y/d	s/d for impermeable layer								s/d ∞	s/H for infinitely permeable layer			
		0	0.05	0.1	0.2	0.5	1	2	3		5	2	1	0.5
1	1	447	423	404	375	323	286	264	255	254	252	241	213	166
	0.75	469	450	434	408	360	324	303	292	291	289	278	248	198
	0.5	555	537	522	397	449	411	386	380	379	377	359	324	264
2	1	186	176	167	154	134	123	118	116	115	115	113	106	91
	0.75	196	187	180	168	149	138	133	131	131	130	128	121	106
	0.5	234	225	218	207	188	175	169	167	167	166	164	156	139
5	1	51.9	48.6	46.2	42.8	38.7	36.9	36.1		35.8		35.5	34.6	32.4
	0.75	54.8	52.0	49.9	46.8	42.8	41.0	40.2		40.0		39.6	38.6	36.3
	0.5	66.1	63.4	61.3	58.1	53.9	51.9	51.0		50.7		50.3	49.2	46.6
10	1	18.1	16.9	16.1	15.1	14.1	13.6	13.4		13.4		13.3	13.1	12.6
	0.75	19.1	18.1	17.4	16.5	15.5	15.0	14.8		14.8		14.7	14.5	14.0
	0.5	23.3	22.3	21.3	21.5	20.6	19.0	18.3		18.7		18.6	18.4	17.8
20	1	5.91	5.53	5.30	5.06	4.81	4.70	4.66		4.64		4.62	4.58	4.46
	0.75	6.27	5.94	5.73	5.50	5.25	5.15	5.10		5.08		5.07	5.02	4.89
	0.5	7.67	7.34	7.12	6.88	6.60	6.48	6.43		6.41		6.39	6.34	6.19
50	1	1.25	1.18	1.14	1.11	1.07	1.05			1.04			1.03	1.02
	0.75	1.33	1.27	1.23	1.20	1.16	1.14			1.13			1.12	1.11
	0.5	1.64	1.57	1.54	1.50	1.46	1.44			1.43			1.42	1.39
100	1	0.37	0.35	0.34	0.34	0.33	0.32			0.32			0.32	0.31
	0.75	0.40	0.38	0.37	0.36	0.35	0.35			0.35			0.34	0.34
	0.5	0.49	0.47	0.46	0.45	0.44	0.44			0.44			0.43	0.43

ed the soil around the hole and resulted in much lower K-values than when hand augers were used. They suggested that different types of augers be used for different soil types.

Hoffman and Schwab (1964) presented a method for determining hydraulic conductivity based on drain outflow. Although somewhat difficult to employ, this method allowed measurement of the effective K value for a field-size soil mass. Skaggs (1976) presented a method to determine the hydraulic conductivity-drainable porosity ratio from water table drawdown measurements. The method can be used for water table measurements at points between parallel drains or at an arbitrary distance from a single drain.

Field methods for measuring the apparent saturated hydraulic conductivity when a water table is not present were reviewed in some detail by Bouwer and Jackson (1974). These methods include the shallow well pump-in technique, and the cylinder permeameter method (Winger, 1960); the infiltration gradient technique (Bouwer, 1964); the air-entry permeameter technique (Bouwer, 1966); and the double-tube method (Bouwer, 1961). The reader is referred to Bouwer and Jackson (1974) and to the individual references given above for details concerning these methods. Again all of these methods provide measurement of K at a point, so, due to field variability, numerous measurements may be required to obtain a field effective K value.

4.2.2 Unsaturated Hydraulic Conductivity

The measurement of unsaturated hydraulic conductivity is considerably more difficult than measuring saturated hydraulic conductivity. Since the K value is dependent on water content, both the hydraulic gradient and water content must be determined for a range of water contents to adequately define the hydraulic conductivity function. Most of the reported measurements for unsaturated soils have been conducted in the laboratory where boundary conditions can be carefully controlled and soil water content and flow rates precisely measured. Field measurements have also been reported but are much more difficult because of the number of variables that must be measured and soil variability in the field.

Laboratory methods. Klute (1965) and Anat et al. (1965) described similar steady state methods for measuring $K(\theta)$ based on the defining relationship given in equation [4.4]. Essentially the method consists of setting up boundary conditions to obtain steady, one directional flow for adjustable pressure heads. A soil sample is placed in an airtight cavity between two horizontal porous plates through which water flows into and out of the sample. The average pressure head in the sample is controlled by the air pressure in the cavity. The mean hydraulic gradient between two points in the sample is determined by using tensiometers to measure the difference in pressure head. By measuring the steady state flow rate, q, the conductivity may be calculated directly from equation [4.4]. Then the air pressure is changed and the procedure repeated for another value of pressure head.

Although simple in concept, this method has several disadvantages. Since soil conductivities are small in general, particularly at the lower water contents, long times are required to reach steady state. This is especially true for imbibition. Also the conductivity function obtained using the above method represents a "point" determination, or at most a determination for a short soil section. In order to incorporate some of the heterogeneities of natural soils in the conductivity function, it is desirable to make conductivity determinations on rather large soil samples. This can be done by increasing the length of the soil sample but the time required to reach steady state would also increase.

Youngs (1964) determined $K(\theta)$ by measuring the rate of infiltration into soil columns from water supplied at the surface at a negative pressure. Although the effective conductivity function for a rather large sample was determined and the experimental measurements were simple in nature, the method required a separate soil column for each water content. Also several days were required to determine the conductivity for low water contents.

A zone of entrapped air was used by Watson (1967b) to speed up the formation of steady state conditions for infiltration into a soil column initially drained to a water table at its base. A large number of tensiometers were used to determine the pressure head and hydraulic gradient distribution at steady state. Then, by measuring the steady state flow rate, q, the conductivity at each point in the column was calculated. While this method is attractive from the standpoint of required time, it is restricted to coarse grained soils and to relatively low pressure heads.

Field methods. Hillel and Gardner (1970) and Bouma et al. (1971) developed an *in situ* method to measure $K(\theta)$ by infiltrating water through artificial crusts of different hydraulic resistances. The crust may be a porous plate or it may be formed in place by puddling a thin layer at the soil surface. After the crust is formed, water is applied to the surface using a cylindrical infiltrometer (Section 4.3.11) and a small constant head is maintained for the duration of the measurement. The effect of the crust resistance is to prevent saturation immediately below the crust. Small tensiometers are used to measure the pressure head at a point just below the crust and at a second point about 3 cm below the first. Flow is continued until the pressure heads at the two points approach the same value giving a unit hydraulic gradient and a constant infiltration rate. The hydraulic conductivity corresponding to the measured pressure head is then equal to the steady infiltration rate (equation [4.4] with $\partial H / \partial s = -1$).

A series of crusts of progressively lower resistance can be used to measure progressively higher K values corresponding to higher water contents. Such a series of tests can be carried out either successively in the same location or concurrently on adjacent sites. In order to insure one-dimensional flow, Bouma et al. (1971) found that it was necessary to carve an in-place soil column about 30 cm deep and wrap it with an impervious foil material. Pits may be dug to measure conductivities for profile layers below the surface.

A method originally described by Nielsen et al. (1964) has been frequently used in recent years (e.g. Nielsen et al., 1973; Cassel, 1974) to determine $K(\theta)$ for soils. Water is applied to the surface of a field plot approximately 2 to 4 m square. After a predetermined amount of water has infiltrated, usually 5 to 15 cm, application is ceased and the plot covered with plastic film to prevent evaporation from the surface. Changes in soil water pressure head during redistribution of the infiltrated water are measured using tensiometers located at 15 cm depth increments to a total depth of about 1.5 m below the surface. Changes in the soil water content with time are inferred from tensiometer readings and the soil water characteristics. The soil water characteristics are obtained from cores taken at each tensiometer depth. The flux at a given depth is calculated from changes in the soil water contents above that depth. Then the hydraulic gradient is determined directly from tensiometer readings and the conductivity calculated from equation [4.4].

A major problem with both field and laboratory methods for determining $K(\theta)$ is the time and expense required. Measurement of a single $K(\theta)$ function by a well trained technician may require several days. Furthermore, several measurements may be needed to adequately characterize $K(\theta)$ for a given soil type because of field variability of the soil properties.

Calculation of $K(\theta)$ from the soil water characteristic. To evade the difficulty of directly measuring $K(\theta)$, numerous attempts have been made to formulate some sort of computational scheme so that the partial or unsaturated hydraulic conductivity may be acquired through the knowledge of other soil properties which are easier to measure. Such properties should be representative of the geometry of pores and their distribution in space. Since the microscopic structure of a porous medium is too complicated to deal with in exact mathematical terms, simplifying assumptions are necessary.

Childs and Collis-George (1950) adopted an approach for finding the relation between hydraulic conductivity and the pore-size distribution of a porous medium by assuming that pores of various sizes were randomly distributed in the medium. Similar statistical approaches were taken by

Marshall (1958) and Millington and Quirk (1959, 1961) who modified the
original concepts of Childs and Collis-George to include pore interaction. In
both cases the pore-size distribution was quantified by use of the soil water
characteristic. The conductivity is calculated by dividing the soil water
characteristic into m equal water content increments and obtaining the
pressure head, h, at the midpoint of each increment. Jackson (1972) showed
that both methods could be reduced to the same general form as,

$$K(\theta_i) = \frac{30 \, \gamma^2 \, \theta_i^P}{\rho g \, \eta \, n^2} \sum_{j=i}^{m} (2j + 1 - 2i)h_j^{-2} \quad\dotfill \quad [4.11]$$

Where

$K(\theta_i)$ = hydraulic conductivity in cm/min corresponding to the water
content, θ_i

γ = surface tension of water

ρ = density of water

η = viscosity and

g = gravitational constant.

For the Marshall method, p = 2 and n is the number of water content increments from zero to the water content in question. For the Millington and
Quirk method, p = 4/3 and n = m = the total number of water content increments.

Experimental evaluations of the prediction methods have shown that
best results are obtained when a matching factor is used to force the
calculated and measured conductivities to agree at a given water content,
usually saturation. When the matching factor is based on the saturated conductivity, both the Millington and Quirk and Marshall equations can be
written in the following form (Jackson, 1972).

$$K(\theta_i) = K_s \left(\frac{\theta_i}{\theta_s}\right)^P \frac{\sum_{j=i}^{m} (2j + 1 - 2i)/h_j^2}{\sum_{j=1}^{m} (2j - 1)/h_j^2} \quad\dotfill \quad [4.12]$$

Where

p = 4/3 for Millington and Quirk

p = 0 for the Marshall method.

Kunze et al. (1968) reported that Millington and Quirk's method with a
matching factor did not produce the best fit with their experimental data.
They claimed a better fit could be obtained if the volumetric water content in
equation [4.11] was not raised to 4/3 power but to the 1.0 power. This result
was in agreement with later work by Jackson (1972). Green and Corey (1971)
also tested the methods of Marshall and Millington and Quirk as well as their
own modified version of the Marshall method. They discovered all three
methods gave reliable predictions of measured hydraulic conductivity, and
suggested they be used routinely for field applications. They pointed out that
to characterize the variation in the water retention curves from many sites in
a field might be more important than to accurately measure the $K(\theta)$ values
on a very limited number of cores or field sites.

Brooks and Corey (1964) derived a relatively simple expression for unsaturated conductivity based on methods originally developed by Burdine (1953).

$$K = K_s \, S_e^{\,(2 + 3\lambda)/\lambda} \qquad \dots\dots\dots\dots\dots\dots\dots\dots\dots\dots\dots\dots\dots \quad [4.13]$$

$$S_c = (\theta - \theta_r)/(\theta_s - \theta_r), \qquad \dots\dots\dots\dots\dots\dots\dots\dots\dots\dots\dots\dots \quad [4.14]$$

where
θ_r = water content at residual saturation at which point the conductivity is assumed zero
θ_s = water content at saturation
K_s = saturated hydraulic conductivity.
The Brooks and Corey equation may also be written as,

$$K = K_s(h_b/h)^{2 + 3\lambda}, \, h < h_b \qquad \dots\dots\dots\dots\dots\dots\dots\dots\dots\dots\dots \quad [4.15]$$

$$K = K_s, \, h > h_b \qquad \dots\dots\dots\dots\dots\dots\dots\dots\dots\dots\dots\dots\dots \quad [4.16]$$

where
h_b = bubbling pressure
λ = pore size distribution index.
h_b, λ and θ_r may be obtained from the soil water characteristic by methods given by Brooks and Corey (1964) and also discussed by Bouwer and Jackson (1974).

Brust et al. (1968) compared Brooks and Corey's method with that of Millington and Quirk, and concluded that the former gave better results than the latter when compared with hydraulic conductivity measured in the field. Nielsen et al. (1964) pointed out that in general, the computational methods for obtaining the partial hydraulic conductivity or permeability appeared most successful for soils with narrow ranges of pore-size distribution. In a highly aggregated soil, a considerable portion of the total water content is retained as relatively immobile water. Under such circumstances better results were obtained if the concept of effective saturation as used in equation [4.14] was adopted. Bouwer and Jackson (1974) stated that although the computational procedure of Brooks and Corey's method was relatively simple, care ought to be exercised to obtain the best value of residual saturation. They concluded the calculated hydraulic conductivity compared favorably with other methods and with measured data.

4.3 INFILTRATION

Infiltration, usually defined as the entry of water into the soil profile, is a process of great practical importance to irrigation design. It is the infiltration capacity of the soil that determines the rate that water can be applied to the surface without runoff. For surface irrigation, the most efficient furrow or border length depends on the infiltration capacity. Failure to adequately consider the infiltration process may result in nonuniform distribution of water in the field as well as excessive water loss due to deep percolation and runoff. Many of the soil related factors that control infiltration also govern soil water movement and distribution during and after the infiltration process. Hence an understanding of infiltration and the factors affecting it is important to the design and operation of efficient irrigation systems. Many aspects of infiltration are treated in detail in a compilation by Hadas, et al. (1973). Philip (1969), Hillel (1971), and Morel-Seytoux (1973) have presented excellent reviews of the infiltration processes.

Infiltration may involve one-dimensional, vertical soil water movement such as occurs during sprinkler or flood irrigation; water movement in two dimensions such as flow from an irrigation furrow; or in three dimensions, such as flow from a drip irrigation emitter. Our discussion will begin with a general description of the vertical infiltration process. This will be followed by a review of a theoretical method that has been proposed for characterizing infiltration and subsequent soil water movement under various initial and boundary conditions. Solutions to the governing equations are used to demonstrate the effects of factors such as initial soil water content and application rates on infiltration. The influence of other factors such as surface sealing and resistance to air movement are also considered. Finally methods for measuring infiltration capacities of field soils applicable to both surface and sprinkler irrigation methods are presented and discussed.

4.3.1 General
Consider infiltration into a deep, homogeneous soil column with a uniform initial water content. At time, $t = 0$, water is ponded at a shallow depth on the soil surface and is continually added at a rate necessary to keep the ponded depth constant. The flux or the rate water enters the soil surface is called the infiltration rate, f. If the rate is measured we will find that it decreases with time as shown schematically in Fig. 4.8. The decrease is primarily due to reduction in the hydraulic gradients at the surface but may also be affected by other factors such as surface sealing and crusting. If the experiment is continued for a sufficiently long time the infiltration rate will approach a constant rate, f_c. The constant f_c is generally assumed to be equal to the saturated hydraulic conductivity, K_0, but will actually be somewhat less than K_0 due to entrapped air. In most cases f_c is more accurately approximated by K_s, the hydraulic conductivity at residual air saturation.

Since water is always ponded on the surface in our hypothetical experiment, the infiltration rate is limited only by soil-related factors. The rate that water will infiltrate as limited by soil factors is often called the infiltration capacity of the soil. Hillel (1971) noted that the term capacity, is generally used to denote an amount or volume and can be misleading when applied to a time-rate process. He proposed the term soil infiltrability rather than infiltration capacity.

Now consider the same soil column as described above with irrigation water or rainfall applied at a constant rate, R, to the surface. For this case the initial infiltration rate will be equal to R and is limited by the application rate rather than the soil profile. As long as the application rate is less than the infiltration capacity, water will infiltrate as fast as it is supplied and the

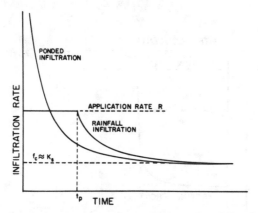

FIG. 4.8 Schematic of infiltration rate-time relationships for a ponded surface and for an application rate, R.

infiltration rate will be controlled by the application rate. However, the infiltration capacity may become less than R after a period of time (Fig. 4.8). Then the surface will become ponded and the infiltration rate again controlled by the soil profile. Water supplied in excess of the infiltration capacity will become available for surface storage and/or runoff.

Infiltration rate is normally expressed in units of length per unit time (or volume per unit area per unit time), e.g., cm/h, mm/h. Cumulative infiltration, $F = F(t)$, is the total amount of water infiltrated at any time t and may be expressed as,

$$F(t) = \int_{o}^{t} f \, dt \quad \ldots\ldots\ldots\ldots\ldots\ldots\ldots\ldots\ldots\ldots\ldots\ldots\ldots\ldots\ldots\ldots\ldots\ldots \quad [4.17]$$

where f is the infiltration rate which may or may not be equal to infiltration capacity as discussed above.

The soil water distribution during infiltration into a uniform, relatively dry soil was first presented by Bodman and Coleman (1943). They showed that the profile could be divided into the four zones shown schematically in Fig. 4.9. The saturated zone extended from the surface to a maximum depth of approximately 1.5 cm. The transition zone, a region of rapid decrease of soil water content, extended from the zone of saturation to the transmission zone, a zone of nearly constant water content which lengthens as infiltration proceeds. The wetting zone maintains a nearly constant shape during infiltration and culminates in the wetting front which is the visible limit of water penetration into the soil. Except for the saturation and transition zones, the results of Bodman and Coleman have been generally confirmed by other investigators. While there has been considerable disagreement in the literature it is generally agreed that, in most cases, the soil will not be completely saturated at the surface due to air entrapment and possible counterflow of the air phase. Most theories of soil water movement do not predict the transition zone. However, McWhorter (1976) showed that an abrupt steepening of the profile near the surface would be predicted for rainfall infiltration if the resistance to air movement is considered.

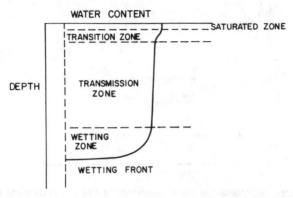

FIG. 4.9 Infiltration Zones of Bodman and Colman [1943].

One of the basic relationships used in characterizing water movement in soil is Darcy's law discussed in Section. 4.2 The other basic relationship is the principle of conservation of mass, which, assuming the density of the water to be constant, may be written for a point in the soil medium as,

$$\frac{\partial \theta}{\partial t} = -\nabla \cdot q \quad \dots \quad [4.18]$$

where

    q = flux
    t = time
    θ = volumetric soil water content.

For flow in the vertical z direction only, equation [4.18] may be written as,

$$\frac{\partial \theta}{\partial t} = \frac{\partial q}{\partial z} \quad \dots \quad [4.19]$$

Combining equations [4.19] and [4.9] and substituting H = h + z from Section 4.1.2 yields the Richards equation for the vertical direction:

$$C(h)\frac{\partial h}{\partial t} = \frac{\partial}{\partial z}\left[K(h)\frac{\partial h}{\partial z}\right] + \frac{\partial K}{\partial z} \quad \dots \quad [4.20]$$

The soil water capacity C(h) is defined in Section 4.1.3 and may be obtained from the soil water characteristic. Note that use of equation [4.20] assumes that there is no resistance to soil air movement and the air pressure remains constant throughout the profile. It is also usually assumed that the soil matrix is rigid and does not change with time, so that the soil water characteristic and hydraulic conductivity relationships are not time variant. These assumptions do not always hold and may cause large errors in predicted results as will be discussed in subsequent sections.

Equation [4.20] can be expanded to include two-dimensional flow by adding the term $\partial / \partial y (K(h) \partial h / \partial y)$ to the right side of the equation. It can also be written with the water content, θ, as the dependent variable by defining soil water diffusivity as $D(\theta) = K(h) \, dh/d\theta$:

$$\frac{\partial \theta}{\partial t} = \frac{\partial}{\partial z}\left(D(\theta)\frac{\partial \theta}{\partial z}\right) + \frac{\partial K}{\partial z} \quad \dotfill \quad [4.21]$$

Equations [4.20] and [4.21] were first derived by Richards (1931) and are referred to as forms of the Richards equation (Swartzendruber, 1969).

Both forms of the Richards equation for flow in the vertical direction contain two soil parameters; the θ-based equation contains $D(\theta)$ and $K(h)$, and the h-based equation contains $C(h)$ and $K(h)$. These parameters are related for unsaturated soil by $D = K/C$. For most soils, all three parameters vary markedly with water content or pressure head. The pronounced nonlinearity of these parameters is the prime source of difficulty in solving the Richards equation subject to boundary conditions pertinent to infiltration.

Advantages can be stated for both the h-based and θ-based equations in describing the movement of water in unsaturated soil. When saturated conditions are reached, the h-based equation is reduced to the familiar Laplace equation describing saturated flow. For saturated flow $K(h)$ reaches a constant value, $C(h) = 0$, and the pressure head, h, changes from a negative to a positive quantity. For cases where both saturated and unsaturated flow conditions exist the solution to the h-based equation will be valid; however, the θ-based equation "blows up" for saturated conditions as $D(\theta)$ tends to infinity. On the other hand there are advantages to the θ-based equation for describing completely unsaturated flow as the changes in both θ and D are often 1 to 3 orders of magnitude less than corresponding changes in h and C. In general, round off errors in numerical solutions to the θ-based equation are of lesser consequence than for the h-based equation.

4.3.3 Solutions to Richards Equation

In order to characterize the infiltration process using the Richards equation it is first necessary to solve the equation for existing boundary conditions. As an example, the boundary conditions for the ponded infiltration case in Section 4.3.1 may be written as

$$h = \delta, \, z = 0, \, t > 0 \quad \dotfill \quad [4.22a]$$

$$h = h_i, \, z \to \infty, \, t \geq 0 \quad \dotfill \quad [4.22b]$$

$$h = h_i, \, z \geq 0, \, t = 0 \quad \dotfill \quad [4.22c]$$

where
    δ = ponded water depth
    h_i = soil water pressure head corresponding to the initial water content.

The solution obtained will yield h = h(z,t) either in functional form, if obtained analytically, or in tabular form if obtained numerically. Then the infiltration rate can be determined for any time by simply applying Darcy's law at the surface. Thus, both the pressure head distribution at all times during the process and the infiltration rate may be determined by solving the Richard equation.

The nonlinearity of the soil properties K, D and C has prevented exact analytical solutions of equations [4.20] and [4.21] except for a few limited cases. However, many numerical techniques have been developed for solution of the equations subject to various boundary conditions of interest. Philip (1957a) presented a rapidly converging numerical procedure for infiltration into a deep homogeneous soil with uniform initial water content. This method was the basis of a series of papers in which Philip (1957b-1957e) analyzed the effects of various factors on infiltration. Other numerical procedures which were more general in application but which required the use of the digital computer followed. Whisler and Klute (1965) used an iterative procedure to solve equation [4.20] subject to a nonuniform initial water content. Their technique took into account hysteresis, a factor that was also included in the numerical technique of Staple (1966). Whisler and Klute (1966) presented a method to solve equation [4.20] for infiltration into a layered soil of arbitrary depth and initial head distribution. The method is also valid for rainfall surface conditions. Techniques for solving equations [4.20] and [4.21] subject to rainfall boundary conditions were presented by Rubin (1966) and Rubin and Steinhardt (1963). The technique given by Rubin (1966) can also be used to account for the effects of non-zero air entry suction. Smith and Woolhiser (1971) solved the Richards equation for rainfall-ponding surface conditions. The solutions were used in combination with solutions for unsteady overland flow to predict infiltration and runoff from sloping soils. Theoretical predictions were in good agreement with the results of both laboratory and field experiments.

Numerical procedures for solving the Richards equation for two and three dimensional flow have been developed by Rubin (1968), Amerman (1969) and Freeze (1971). Green et al. (1970) considered the movement of both air and water in soil and, assuming two-phase flow, presented numerical techniques for both one and two dimensional flow by which infiltration, groundwater recharge, water table depth, and stream base flow could be quantified for a groundwater basin. More recently J. Y. Parlange (1971, 1972a-c) has published a series of papers in which approximate but analytical techniques are used to characterize infiltration and the factors affecting it.

We will not dwell on methods used to solve the governing equations here. The reader is referred to references given above for full treatment of the solution methods. In some cases we will use solutions to the Richards equation to examine effects of soil properties and boundary conditions on the infiltration process. In others the approach will be more general and a qualitative discussion of the effects will be given. In addition certain approximate methods for quantifying infiltration have been found sufficiently flexible for general use and these will be discussed in some detail.

4.3.4 Effect of Soil Properties

Infiltration rate-time relationships as predicted from numerical solu-

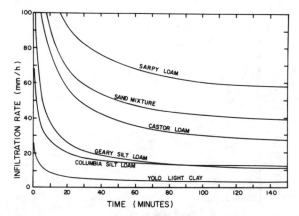

FIG. 4.10 Predicted infiltration rates from numerical solutions to the Richards equation for deep soils with a shallow ponded surface. Soil properties were obtained from the literature as follows: Sarpy 1. and Geary s.1.—Hanks and Bowers [1962]; Castor 1.—Staple and Gupta [1966]: Yolo 1.c.—Philip [1957a]; Sand m.—Skaggs et al. [1970]; Columbia s.1—Kirkham and Powers [1972]. Initial water contents corresponded to a uniform initial pressure head of h_i = -2400 cm.

tions of equation [4.20] for vertical infiltration from a ponded surface into deep homogeneous soil profiles are plotted in Fig. 4.10 for six soils. Corresponding plots for cumulative infiltration are given in Fig. 4.11. As expected, infiltration rates tend to increase with coarser soil texture. However, soil structure is also important so it is not always possible to relate infiltration capacities to texture alone just as the hydraulic conductivity does not always increase with coarser texture soils. The influence of the shapes of the soil water characteristic and the hydraulic conductivity-water content relationships on infiltration was studied by Hanks and Bowers (1963). They showed that variations in the soil-water diffusivity at low water contents had negligible effect on infiltration from a ponded water surface. However, variations in either the diffusivity or soil-water characteristic at water contents near

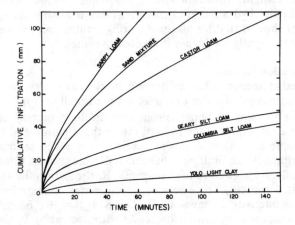

FIG. 4.11 Predicted cumulative infiltration relationships for the soils in Fig. 4.10.

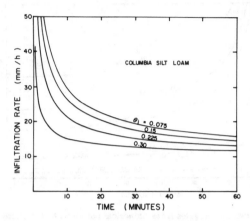

FIG. 4.12 Predicted infiltration rates for a deep Columbia silt loam with different initial water contents. Saturated volumetric water content for this soil is $\Theta_s = 0.34$.

saturation have a very strong influence on predicted infiltration. Therefore, errors in measuring the hydraulic soil properties are of far greater consequence for water contents near saturation than for drier conditions so far as infiltration is concerned.

4.3.5 Initial Water Content

One of the important factors influencing infiltration of water into the soil profile is the initial water content. Fig. 4.12 shows predicted flux curves for infiltration from a shallow ponded surface into a deep Colombia silt loam. The three solutions were obtained by numerically solving Richards equation for the uniform initial water contents given. Soil properties used were those reported by Kirkham and Powers (1972). Note that infiltration rates are high for drier initial conditions but that the dependence on initial water content decreases with time. If infiltration is allowed to continue indefinitely, the infiltration rate will eventually approach K_S regardless of the initial water content. Infiltration rates are higher at low initial contents because of higher hydraulic gradients and more available storage volume. Philip (1958) showed that for all times during infiltration, the wetting front advances more rapidly for higher initial water contents.

4.3.6 Application Rates

As noted in Section 4.3.1, infiltration depends on the rate of application as well as soil conditions. For example, if the rainfall or application rate is less than K_S for a deep homogeneous soil, infiltration may continue indefinitely at a rate equal to the rainfall rate without ponding at the surface. The water content of the soil in this case does not reach saturation at any point but approaches a limiting value which depends on the rainfall intensity. Specifically, for a given rainfall intensity, R, the soil profile approaches a uniform water content, θ_L; where θ_L is the water content for which the hydraulic conductivity is equal to R, i.e. $K(\theta_L) = R$. Because the unsaturated hydraulic conductivity decreases with decreasing θ, the lower the rain intensity, the lower the value of θ_L. Detailed investigations of rainfall in-

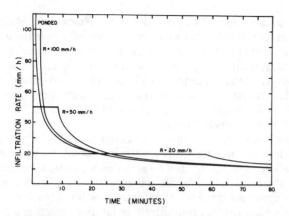

FIG. 4.13 Predicted infiltration rates for a deep, homogeneous Geary silt loam profile for constant surface application rates and for a shallow ponded surface. The initial water content was uniform at $\Theta_i = 0.26$ which corresponds to $h_i = -750$ cm of water.

filtration have been conducted by Rubin and his co-workers (Rubin and Steinhardt, 1963, 1964; Rubin et al., 1964; and Rubin, 1966) and the reader is referred to their work for more details on this subject. The effect of application intensity R, on infiltration rate as predicted by solutions to the Richards equation is shown in Fig. 4.13 for Geary silt loam. Initially, water infiltrates at the application rate. If $R > K_S$, the infiltration capacity falls below R after some period of time, surface ponding begins and water becomes available for runoff. The time that surface ponding is initiated decreases with increasing R and the infiltration rate-time relationships are clearly dependent on the application intensity (Fig. 4.13).

If the infiltration rate is plotted versus cumulative infiltration, F, (Fig. 4.14) rather than time, the resulting relationships are much less dependent on R. In fact, for approximate purposes, the relationships may be assumed to be independent of R and a single curve used for all application rates greater than K_S. This assumption is inherent in the Green-Ampt methods used to predict time of ponding by Mein and Larson (1973) and in the

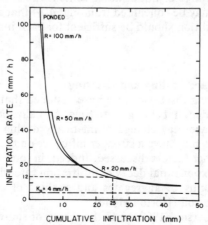

FIG. 4.14 Infiltration rates of Fig. 4.13 versus cumulative infiltration.

parametric model presented by Smith (1972). Reeves and Miller (1975) extended this assumption to the case of erratic rainfall where the unsteady rainfall or application rate dropped below the infiltration capacity for period of time followed by a high intensity application. They obtained numerical solutions to the Richards equation which considered hysteresis and surface crusting. Their investigations showed that the infiltration capacity could be approximated as a simple function of cumulative infiltration, regardless of the application rate versus time history.

For sprinkler irrigation, application rates should be selected so that surface runoff does not occur. This is usually insured by determining the steady state infiltration capacity, f_c, and by choosing an application rate that is less than or equal to this value. While this procedure will work for some cases, there are many situations in which infiltration will not reach a steady rate, or in which the steady rate is prohibitively small. For example, a soil that has a restricting layer a meter or so below the surface may have a steady state infiltration rate of nearly zero yet be capable of large intake rates during the first stages of infiltration. An alternative method originally proposed by Rubin (1966) for choosing the application rate is to make use of the relationship between infiltration rate and cumulative infiltration. The total amount of water to be applied at one irrigation is usually fixed by the crop and soil water holding capacity. Thus the maximum application intensity can be obtained from plots such as Fig. 4.14.

For example, lets assume that 25 mm of water are to be applied in a single irrigation. Using the relationship for a ponded surface in Fig. 4.14, we can see that the application rate should not exceed 12 mm/h in order to prevent surface ponding and runoff. The saturated hydraulic conductivity for the Geary soil is about 4 mm/h so the application rate chosen is greater than $3f_c$. Notice that use of the 'ponded' curve tends to give a safe-side estimate of the permissible application rate. That is, based on the other relationships plotted, we would not expect surface ponding to occur for $R = 12$ mm/h until cumulative infiltration is somewhat greater than 25 mm. The relationships given in Fig. 4.14 were predicted for a uniform initial water content of 0.26 (volumetric) which corresponds to approximately 50 percent of the available water holding capacity for this soil. For irrigation design, such relationships are needed for the range of initial water contents at which irrigation might begin. However, it may be concluded from Fig. 4.14 that a single curve for a ponded surface condition should be sufficient for each initial water content.

4.3.7 Effect of Surface Sealing and Crusting

In the proceeding discussions we have assumed that the soil matrix or skelton is rigid and does not change with time. Actually the hydraulic properties at the soil surface may change dramatically during the application of water. Such changes may have a stronger influence on the rate of infiltration than some of the other factors discussed. Indeed, in some of the early studies of infiltration, the exponential decay of infiltration rate with time was entirely attributed to slaking of aggregates and swelling of colloids which progressively sealed the soil (Horton, 1940).

Edwards and Larson (1969) used the theory of soil-water movement to investigate the influence of surface seal development on infiltration of water into a tilled soil. They determined the saturated hydraulic conductivity of a surface layer 5 mm thick as a function of time of exposure to simulated rainfall. In one example, the saturated hydraulic conductivity of the surface layer decreased from 19 to 2 mm/h after 2 h of simulated rainfall applied at a rate of 70 mm/h. The predicted cumulative infiltration for the two hour period was reduced by about 50 percent and the infiltration rate from 25 to 10 mm/h because of surface sealing. Greater reduction in infiltration rate did not occur because of an increase of the suction gradient at the surface as the seal was formed. Hillel and Gardner (1969, 1970) used approximate analytical methods to evaluate the effect of surface sealing on steady state and transient infiltration processes. They showed that infiltration into a crusted profile may be approximated by assuming that water enters the soil layer below the crust at a nearly constant suction, the magnitude of which depends on the conductivity and thickness of the crust and the hydraulic properties of the underlying soil.

In a recent study Morin and Benjamini (1977) concluded that the crust formed by raindrop impact is the dominant factor influencing the infiltration capacity of a bare Hamra soil. In order to characterize the infiltration process they considered three different soil conditions, uncrusted dry soil, crusted wet soil and crusted dry soil. For each different initial condition the infiltration capacity could be described in terms of the cumulative rainfall striking the surface rather than cumulative infiltration as in Fig. 4.14. When treated in this way, infiltration capacity was found to be independent of the rainfall intensity for field tests on each of the three initial conditions. When simulated rainfall was applied at a rate of 130 mm/h, the infiltration capacity of a bare Hamra soil decreased to 8 mm/h after 60 min as opposed to a continuous rate of 130 mm/h when the surface was protected by a straw mulch.

Because of the complex nature of the sealing process and the difficulty of describing the manner in which hydraulic properties of the surface layer change with time, there have been few attempts to use the theory of soil-water movement to analyze the phenonemena. However, the results of numerous experimental investigations indicating the importance of surface sealing have been reported. For example, Mannering and Meyer (1963) showed that straw mulch applied at densities of greater than 2,200 kg/ha prevented surface sealing due to rainfall impact. For mulch densities of 1,100 kg/ha or less, surface sealing occurred and cumulative infiltration was reduced from approximately 32 mm to 12 mm for a 30 min test on Wea silt loam. Corresponding final infiltration rates were reduced from about 6 to 2 mm/h. Other studies of the influence of rainfall energy and various soil factors on surface sealing and infiltration have been reported by Moldenhauer and Kemper (1969), Moldenhauer and Long (1964), Burwell and Larson (1969), Koon et al. (1970), Horton (1940), and Duley (1939). Clearly, surface sealing and crusting has a significant, perhaps dominant effect on infiltration in bare or unprotected soils. This factor is of less importance when the surface is protected by a crop canopy or a mulch. However it must be considered in the design of irrigation systems and in selecting methods for measuring infiltration capacities.

4.3.8 Layered Soils

The Richards equation was solved by Hank and Bowers (1962) for infiltration from a ponded surface into a two layered soil. A continuous pressure head distribution was predicted at the layer boundaries, but the predicted water content distributions were discontinuous because of the difference in the soil-water characteristics of the two soils. For a coarse over a fine soil, infiltration proceeded exactly as for a coarse soil alone until the wetting front arrived at the boundary between the two layers. Then the progress of the wetting front was slowed, a positive pressure head developed in the top layer and the infiltration rate approached that predicted for the fine soil alone. The infiltration rate-time relationship predicted for fine over coarse soils was nearly the same as that predicted for a uniform fine-textured soil except for a decrease in the infiltration rate when the wetting front reached the coarse layer. Whisler and Klute (1966) obtained similar solutions for coarse textured soils over fine soils, fine texture lenses in coarse soils and coarse textured lenses in fine soils. Their solutions for a coarse textured lens in fine soil predicted that the wetting front would be held up for a period before entering the coarse layer. The pressure head at the wetting front is nearly always negative. Therefore, the larger pores in a coarse textured layer do not fill with water resulting in only partial saturation and low hydraulic conductivity in the layer (see Section 4.1.6). The result is a sharp decrease in the infiltration rate followed by a partial recovery when the wetting front moves into a coarse layer. This effect was observed by Miller and Gardner (1962) for infiltration into air dry stratified soils. Thus a thin layer of either coarser or finer texture than the otherwise uniform profile serves to cause a decrease in infiltration rate when the wetting front reaches that layer.

4.3.9 Effects of Air Entrapment

We noted in Section 4.3.2 that the derivation of the Richards equation assumes that displaced soil air moves through the profile with negligible resistance and that the air pressure remains constant throughout. This assumption is usually justified by the small viscosity of air relative to that of water and by assuming that air can escape through large pores that remain partially open during infiltration. While these assumptions may hold in some instances, there are numerous cases where air is trapped by infiltrating water causing an air pressure buildup in advance of the wetting front and a reduction of the infiltration rate. Even for deep profiles, pressure buildup and subsequent counterflow of escaping air causes infiltration rates to be less than would occur if resistance to air movement was truly negligible.

The fact that air movement may significantly affect infiltration has been recognized for many years (e.g., Free and Palmer, 1940). Studies showing the effects of air pressure buildup and the flow of air on the infiltration process have been conducted by Wilson and Luthin (1963), Peck (1965) and Adrian and Franzini (1966). A particularly detailed study of this phenomena with oil as the infiltrating liquid was reported by McWhorter (1971). He conducted infiltration experiments for conditions representing both semi-infinite and finite profile depths as well as presented analytical methods for predicting the effects of two-phase flow. Methods for characterizing the effects of air movement on infiltration in terms of soil properties and boundary conditions have been presented by Morel-Seytoux and his colleagues. These methods

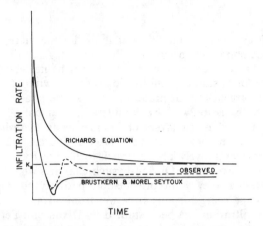

FIG. 4.15 Schematic of the effect of the air phase on infiltration as predicted by solution of the Richards equation by the methods of Brustkern and Morel-Seytoux [1970] and as observed by McWhorter [1971].

are summarized in a detailed treatment of two-phase flow in porous media by Morel-Seytoux (1973).

The effect of air movement on infiltration as predicted by methods presented by Brustkern and Morel-Seytoux (1970) is shown schematically in Fig. 4.15 for a soil with a water table or restrictive layer at a relatively shallow depth. When the air phase is neglected, the infiltration rate predicted by solution of the Richards equation asymptotically approaches K_S, the hydraulic conductivity at residual saturation, as discussed in Section 4.3.1. However, when air is entrapped between the wetting front and the water table or a restrictive layer, air pressure increases causing a rapid reduction in infiltration rate. As air pressure increases, upward flow of the air phase begins, followed by escape of air from the surface and an increase in the infiltration rate. This predicted infiltration rate also asymptotically approaches a constant value but this value is significantly less than K_S. The predicted 'dips' in the infiltration rate curve occur sooner and are more pronounced for smaller depths to the water table or restricting layers which is consistent with the observations of McWhorter (1971). However, McWhorter observed that when air began to escape from the surface, the infiltration rate increased to values higher than the final rate (shown schematically by the broken curve in Fig. 4.15), then decreased. A disturbance of the surface zone was observed at the time air began to escape and the increase was attributed to deformation of the soil matrix resulting in higher hydraulic conductivity of the surface layers. As an example of the effect of the air phase, McWhorter's data for a 1.85-m long, closed column of Poudre sand shows that the long-term infiltration rate asymptotically approaches a value that is approximately 60 percent of K_S. Examples presented by Morel-Seytoux (1973) predicted steady state infiltration rates of approximately $0.3\ K_S$ for soils with shallow water tables and between $0.8\ K_S$ and $0.9\ K_S$ for infinitely deep profiles.

McWhorter (1976) presented approximate methods for characterizing two-phase flow during steady infiltration as occurs during rainfall or sprinkler irrigation. His analysis, which assumed that air is incompressible and thus tends to predict the maximum effect of the air phase on infiltration, showed that the time required for surface ponding may be much smaller than that predicted from a single phase flow analysis. Furthermore, resistance to air flow will cause ponding to occur at infiltration rates that would persist indefinitely if resistance to air movement could truly be neglected. In one example the maximum sustainable infiltration rate (i.e. without surface ponding) was only 32 percent of that predicted from single phase flow theory.

In the above discussion we have considered movement of water and air in a homogeneous soil or in layers of homogeneous materials. However natural soils are seldom homogeneous and often are permeated, especially in the surface layers, by relatively large channels formed by roots, cracking due to shrinkage, and worm holes. Obviously such channels would have a great effect on infiltration as they would provide both pathways for rapid inflow of water and an escape route for air as the wet front advances. The effect of macropores on infiltration has been studied by Dixon and Peterson (1971) and by Dixon and Linden (1972). They found that large pores, open to the soil surface, can contribute greatly to infiltration, in some cases raising total infiltration by a factor of 10 or more. However, small soil air pressures can block this contribution so large pores close to the surface, but not open, have a much smaller effect on infiltration. Thus infiltration can be increased by using cultural practices designed to prevent the sealing off of large pores near the surface.

4.3.10 Approximate Equations

Infiltration can be characterized by theoretical methods for most boundary and initial conditions of interest. The methods, some discussed in detail, others by reference only, provide physically consistent means of quantifying infiltration in terms of the soil properties governing movement of water and air. However these elaborate procedures are rarely used in practice to describe the infiltration process. One reason is that the large computers required to implement many of the calculation methods are usually not available to the irrigation specialist at the time he wishes to use them. A more severe limitation, however, has been the difficulty of obtaining the necessary soil property data. Variation of the soil properties, both with depth and from point to point in the field, causes numerous measurements to be required to adequately describe field conditions. Present methods of determining the properties are difficult and such data are only available for a limited number of soils. Thus, while the prediction methods discussed in this chapter are extremely valuable in analyzing the effects of various factors on the infiltration process, they have so far been of limited value for direct application to design.

Attempts to characterize infiltration for field applications have usually involved simplified concepts which permit the infiltration rate or cumulative infiltration volume to be expressed algebraically in terms of time and certain soil parameters. The most obvious characteristic of the infiltration process is that the rate decreases rapidly with time during the early part of an infiltration event. Although attributed to different physical phenomena, this characteristic is reflected by all of the approximate infiltration equations.

One of the simplest infiltration equations was proposed by Kostiakov (1932),

$$f = K_k \, t^{-\alpha} \quad \dotfill \quad [4.23]$$

where

f = infiltration rate

t = time after infiltration starts

K_k, α = constants which depend on the soil and initial conditions.

The parameters in this equation have no physical interpretation and must be evaluated from experimental data.

Horton (1939, 1940) presented a three parameter infiltration equation which may be written as,

$$f = f_c + (f_o - f_c)\, \bar{e}^{\beta t} \quad\dots\dots\dots\dots\dots\dots\dots\dots\dots\dots\dots\dots\dots\dots\dots\dots\dots \quad [4.24]$$

where

f_c = final constant infiltration rate

f_o = infiltration capacity at $t = 0$

β = soil parameter which controls the rate of decrease of infiltration rate.

From our previous discussion we know that f_o and β must depend on the initial water content as well as the application rate, and that for homogeneous profiles, f_c will be somewhat smaller than the saturated hydraulic conductivity. Again the equation parameters are usually evaluated from experimental infiltration data.

Philip (1957d) proposed that the first two terms of his series solution for infiltration from a ponded surface into a deep homogeneous soil be used as a concise infiltration equation. The equation may be written for infiltration rate as,

$$f = \frac{S}{2}\, t^{-1/2} + C \quad\dots\dots\dots\dots\dots\dots\dots\dots\dots\dots\dots\dots\dots\dots\dots\dots \quad [4.25]$$

where S is a parameter which Philip calls sorptivity and, along with C, can be evaluated analytically if the soil properties $D(\theta)$ and $h(\theta)$ are known. The resulting value for C will be approximately $K_s/3$ (Youngs, 1968) so the equation will not be physically consistent for large times if Philips methods are used to define C. A regression fit to experimental data will tend to give C = f_c, however.

An empirical equation based on a storage concept was described by Holtan et al. (1967). The infiltration capacity is expressed in terms of cumulative infiltration, initial water content and other soil parameters as,

$$f = a\,(S_t - F)^n + f_c \quad\dots\dots\dots\dots\dots\dots\dots\dots\dots\dots\dots\dots\dots\dots\dots\dots \quad [4.26]$$

where

F = cumulative infiltration

S_t = storage potential of the soil above the impending strata

a, n = constants dependent on soil type, surface and cropping conditions.

A major difficulty with the Holtan equation is the evaluation of the control depth on which to base S_t. Holtan and Creitz (1967) suggested using the depth to the first impeding layer; however, Huggins and Monke (1966) found that the effective depth was highly dependent on both the surface condition and cultural practices used in preparing the seed bed. Smith (1976) argued that the infiltration curves are physically related to gradients and hydraulic conductivity far more than to soil porosity and that we should not expect the Holtan equation to adequately describe the process.

The approximate method that has received the most attention during recent years is the infiltration equation originally derived by Green and Ampt (1911). The equation may be derived by assuming that water enters soil as slug flow resulting in a sharply defined wetting front which separates a zone that has been wetted from a totally uninfiltrated zone. Direct application of Darcy's law yields the following form of the Green-Ampt equation.

$$f = K_s (H_o + S_f + Z_f)/Z_f \dotfill [4.27]$$

where

K_s = hydraulic conductivity of the transmission zone
H_o = pressure head at the surface,
S_f = effective suction at the wetting front
Z_f = distance from the surface to the wetting front.

Expressing the cumulative infiltration as $F = (\theta_o - \theta_i)Z_f = M_d Z_f$ and assuming the depth of ponding at the surface is shallow so that $H_o \approx 0$, equation [4.27] may be written as,

$$f = K_s + K_s M_d S_f/F \dotfill [4.28]$$

where M_d is the difference between initial and final volumetric water contents $(\theta_o - \theta_i)$.

In addition to uniform profiles for which it was originally derived, the Green-Ampt equation has been used with good results for profiles that become denser with depth (Childs and Bybordi, 1969) and for soils with partially sealed surfaces (Hillel and Gardner, 1970). Bouwer (1969) showed that is may also be used for nonuniform initial water contents.

Mein and Larson (1973) used the Green-Ampt equation to predict infiltration from steady rainfall. Their results were in good agreement with rates calculated from solutions to the Richards equation for a wide variety of soil types and application rates. Such results are consistent with Fig. 4.15 which shows that the relationship between infiltration capacity and cumulative infiltration is approximately independent of the application rate. Thus, the Green-Ampt equation has an inherent advantage over the approximate equations that specify infiltration rate in terms of time, in that a single curve can be used for all application rates. Whereas a family of curves (or set of equation parameters) would be required if time is used as the dependent variable.

Another advantage of the Green-Ampt approach is that the equation parameters have physical significance, although the definition of S_f has caused some difficulty. Bouwer (1966) suggests that S_f is approximately one-half of the air entry suction which can be measured in the field using methods he

presented. Mein and Larson (1973) used an average suction at the wetting
front, defined as,

$$S_{av} = \int_0^1 S \, dK_r \quad \dotfill \quad [4.29]$$

where
   $K_r =$ relative hydraulic conductivity, $K_r = K/K_0$
   $K_0 =$ saturated conductivity.
   While the Green-Ampt equation is attractive from several points of
view, the forms presented above still neglect the effect of air movement which
may lead to significant errors as discussed in Section 4.3.9. Morel-Seytoux
and Kanji (1974) showed that the form of equation [4.27] remains the same
when the simultaneous movement of both water and air are considered. They
substituted H_c, the effective matric drive, for S_{av}, although they noted that
S_{av} is a reasonable approximation of H_c. The resistance to air movement was
accounted for by the introduction of a viscous resistance correction factor, β,
which was defined as a function of the soil and fluid properties. Then the
Green-Ampt equation as modified by Morel-Seytoux and Kanji (1974) may
be written as,

$$f = K_s/\beta + K_s M_d H_c/\beta F \quad \dotfill \quad [4.30]$$

Calculated values of β for five soils showed a range from 1.1 to 1.7 as opposed
to an assumed value of $\beta = 1$ when the air phase is neglected. Neglecting the
air phase for the five soils considered would have resulted in overpredicting
the infiltration rates by 10 to 40 percent. However this only holds if the equation parameters are defined from basic soil properties rather than from direct
infiltration measurements. Morel-Seytoux and Kanji (1974) pointed out that
if K_s is determined in the field using an apparatus such as the one described
by Bouwer (1966), the effect of air resistance will be lumped into the K_s value
obtained and equations [4.27] and [4.28] will give reliable predictions for
large times.
   Which approximate infiltration equation is the best? Numerous comparisons of the equations and prediction methods have been reported in the
literature. Skaggs et al. (1969) used regression analyses on large plot rainfall
simulator data to determine equation parameters for the Green-Ampt,
Horton, Philip and Holtan equations. Based on the results of 52 field tests,
they found that all of the equations would adequately fit the measured infiltration data. The two-parameter Green-Ampt and Philip equations gave
fits nearly equal to those of the three-parameter Horton and Holtan equations. However, the parameters for all four equations varied widely due to
field variation and crusting effects.
   Swartzendruber and Youngs (1974) and Smith (1976) compared the
Green-Ampt and Philip equations and found they predicted very similar
results when the parameter C is set equal to f_c. Whisler and Bouwer (1970)
compared the Green-Ampt, Philip and numerical solutions to the Richards
equation with each other and with experimental data from laboratory soil
columns. They found that solutions to the Richards equation were in best

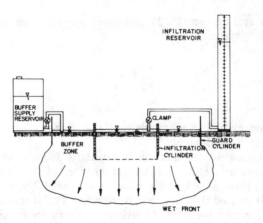

FIG. 4.16 Schematic of cylindrical infiltrometer. Ponded depths in the infiltration cylinder and buffer zone are held constant by manual adjustment of inflow rates.

agreement with observed results but that the Green-Ampt equation was the easiest to use and give predictions sufficiently accurate for most field problems. The flexibility of the Green-Ampt equations for describing infiltration under varied initial, boundary and soil profile conditions makes it an attractive method for field applications. The fact that the equation parameters have physical significance and can be computed from soil properties is an added advantage. In practice, however, it will usually prove advantageous to determine the equation parameters from field measurements such as those proposed by Bouwer (1966) or by fitting measured infiltration data. Such measurements are usually easier than standard soil property determination and also tend to lump the effects of such factors as heterogeneities, worm holes and crusting in the equation parameters. This results in more reliable infiltration predictions than if the parameters are determined from basic soil property measurements.

4.3.11 Infiltration Measurements

Parr and Bertrand (1960) published a thorough review of field methods for measuring infiltration capacity. Basically two types of devices have been used — sprinkling infiltrometers and flooding infiltrometers. From the application point of view, it would be advantageous to use a sprinkling infiltrometer if sprinkler irrigation is to be used; while flooding infiltrometers would be more appropriate for soils that are to be furrow or flood irrigated. However, the flooding devices are far more frequently used because they require less equipment and are easier to install and operate than the sprinkling type.

The most commonly used infiltrometer is probably the ring or cylindrical infiltrometer which was described in detail by Haise et al. (1956) and is shown schematically in Fig. 4.16. It consists of a metal cylinder 20 - 45 cm in dia which is pressed or driven into the soil. Infiltration is measured by ponding water inside the cylinder and measuring the rate that the free surface falls, or by measuring the rate that water must be added to maintain a constant ponded depth. Once the wetting front exceeds the buried cylindrical depth, lateral flow may cause the measured infiltration rates to be higher than would occur during irrigation. Lateral flow is especially troublesome if restrictive layers such as plow pans exist or if the hydraulic conductivity decreases with depth. When restrictive layers are at a shallow depth, it is recommended that the infiltration cylinder should be driven into or through the layer if possible. Another means of preventing erroneous measurements due to lateral flow is to use a guarded ring or buffer area around the outside of the infiltration cylinder as shown in Fig. 4.16. Water is ponded between the two cylinders at all times to prevent edge effects and to maintain vertical flow below the central infiltration cylinder.

Wooding (1968) presented analytical solutions for steady state infiltration from a circular ponded surface into a deep, homogeneous soil. His solutions can be used to estimate the guard ring diameter needed for deep soils. It may be concluded from Wooding's solutions (his Fig. 5) that a guard ring diameter of 1.25 times that of the infiltration cylinder would be sufficient for coarse soils. For general purposes, a ratio of the cylinder diameters of two or greater (D/d≥2) is recommended, however. Bouwer (1963) showed that differences in the water level between the infiltration cylinder and the buffer area can cause errors in the infiltration measurements. To eliminate this source of error it is recommended that water be ponded at a shallow depth, say 1 to 2 cm, in both the buffered area and the infiltration cylinder. The water level can be maintained by manually controlling the water supply from separate reservoirs or by using a mariotte syphon or float valves to automatically control the inflow. If a tall cylinder of uniform cross-section is used as the supply reservoir, the infiltration volume can be determined easily and accurately by simply recording the water level in the reservoir.

Infiltration under furrow irrigation involves soil water movement in both vertical and lateral directions. Because the rate of infiltration depends on the size and shape of the furrow, the rate water moves into the soil is often called intake rate rather than infiltration rate. Regardless of the term used, the determination of the intake rate is important to the design of an efficient furrow irrigation system. Infiltration rates determined by sprinkler or cylindrical infiltrometers represent primarily vertical flow, so it is difficult to apply these results directly. One method frequently used to determine intake rates is to make inflow and outflow measurements in an irrigation furrow. Measuring flumes are used to make flow measurements at two points in an irrigation furrow located 30 to 90 m apart. Intake rates are computed from the difference of inflow and outflow for various times after water application begins. A detailed discussion of the use of the inflow-outflow method is given by Merriam (1968) (also see Chapter 17). While this method provides a good means of evaluating furrow irrigation systems that are already installed, it is often not convenient to use this method to determine intake rates for the design of a new system. Bondurant (1957) developed a furrow infiltrometer to measure intake rates in a short section of an irrigation furrow. The furrow is blocked off by metal plates and water is applied at a rate sufficient to maintain a constant depth. The intake rate is then determined in a manner similar to that described for the cylinderical infiltrometer.

There are many types of sprinkling infiltrometers as discussed by Parr and Bertrand (1960). Construction and operation of one such infiltrometer was presented in detail by Dixon and Peterson (1964). Sprinkling or spray infiltrometers usually consist of a plot surrounded by partially buried sheet metal barriers with facilities for measuring the rate of surface runoff. Water

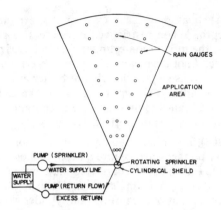

FIG. 4.17 Schematic of apparatus for measuring infiltration capacities for sprinkler irrigation design [after Tovey and Pair, 1966].

is sprinkled onto the plot surface at a constant intensity and the infiltration rate is determined from recorded runoff measurements. In most cases the infiltration rate is determined by simply subtracting the runoff rate from the application intensity. However the rate of surface storage during the initial stages of runoff should also be considered as shown by Skaggs et al. (1969). Failure to correct for surface storage may result in significant errors in the early part of the measured infiltration relationships (Smith, 1976).

A relatively simple method for measuring infiltration capacities for sprinkler irrigation design was developed by Tovey and Pair (1966). A shielded rotating sprinkler head is used to apply water to a circular section of soil as shown in Fig. 4.17. Water is supplied to the sprinkler from a trailer-mounted recirculating system or directly from a field supply such as an irrigation canal. Three rows of rain gauges set at about 0.75 m intervals outward from the sprinkler are used to measure the application rates at each location. The sprinkler is started and run for a selected period of time; Tovey and Pair used a 1-h period. Water application to the soil is observed and notes made for each row of gauges as to whether water is applied too fast, too slow or equal to the infiltration capacity. The design infiltration capacity is taken as the rate at the gauge position where the applied water just disappears from the surface as the sprinkler jet returns to apply more water to the same spot. Application rates at each location are determined at the end of the test by measuring the amount of water caught in the various rain gauges.

Tovey and Pair (1966) suggested that the soil should be at field capacity when their test is initiated so that the infiltration capacity measured is the long term or constant rate, f_c, as defined in Section 4.3.1. If this value is used for the design application rate, any desired amount of irrigation water can be applied without danger of surface runoff. However, if f_c is prohibitively small it may be possible to take advantage of the higher infiltration capacities for drier conditions and apply water at a faster rate as discussed in Section 4.3.6. The methods given by Tovey and Pair could still be used to determine the design application rate by running the tests for soil water contents representative of those normally found when irrigation begins. The design rate would be chosen for these drier initial conditions such that the desired amount of irrigation water could be applied before continuous surface ponding takes place.

One advantage of the Tovey and Pair (1966) method is that it is relatively easy to set up and make the required measurements. Perhaps a more important advantage is that the infiltration capacity measurements are made for conditions very similar to those that will exist during actual irrigation. Thus, the effects of surface sealing, air entrapment, soil layers, etc., will be reflected in the infiltration capacity measurement in the same way that they will affect the irrigation process.

4

SOIL WATER
PART II—DYNAMICS

by   P. A. C. Raats, Institute for Soil Fertility, Haren,
The Netherlands; A. W. Warrick, University of
Arizona, Tucson, AZ

4.4 PLANT-SOIL-WATER INTERACTIONS

Along with solar radiation and nutrients, water is a crucial input for successful crop production. The amount of water directly involved in the photosynthetic reaction is very small. However, the biochemical reactions and transport processes in plants require an aqueous environment. During the day CO_2 diffuses via the stomata from the atmosphere to the inside of the plant leaves. Water is then lost through these same stomata to the drier atmosphere with the losses replenished from uptake by the plant roots. If the soil is depleted or cannot supply water rapidly enough, the stomata close and photosynthesis comes to a halt. Before the transpiration rate drops appreciably, desiccation affects other processes. Generally dry matter production is the largest if the water potential is kept high, but dry matter production is not always equivalent to marketable yield. Sometimes water stress is needed to induce differentiation or maturation of the portion of the plants to be harvested. Even if the soil is kept wet, there will often be a period of high water stress around midday, when the evaporative demand is highest. Then growth may decrease, while photosynthesis continues. Assimilates can be stored for several hours before photosynthesis is decreased. They will then be used in accelerated growth when stress is relieved.

The soil should be sufficiently wet to satisfy the evaporative demand although not waterlogged. From the surface downward there should be a continuous air filled pore space, allowing rapid release of CO_2 and entrance of O_2, that is, the root zone should be sufficiently aerated. In order to maximize the yield, one must also keep the osmotic pressure of the soil solution low. The tolerance for salts varies widely among species.

4.5 EQUATIONS OF MASS BALANCE AND FLOW

At any point in the soil the time rate of change in water content is equal to the net flux towards that point minus the rate of uptake by plant roots. Mathematically, this balance of mass is expressed by

$$\partial\theta/\partial t = - \nabla \cdot q + \lambda T \dotfill [4.31]$$

where
  t = time
  θ = volumetric water content
  ∇ = vector differential operator
  q = volumetric flux
  T = rate of transpiration
  λ = distribution function for the uptake of water by plant roots.
The dimension of λ is reciprocal length and at any instant the integral of λ over the entire root zone is unity. The flux of water, q, is assumed to be the sum of a matric component - K ∇h and a gravitational component K ∇z:

$$q = -K \, \nabla h + K \, \nabla z \quad \dots\dots\dots\dots\dots\dots\dots\dots\dots\dots\dots \quad [4.32]$$

where
  K = hydraulic conductivity
  h = pressure head of the water
  z = vertical coordinate taken positive downward
The sum of pressure and gravitational heads is equal to the total head H:

$$H = h + z \quad \dots\dots\dots\dots\dots\dots\dots\dots\dots\dots\dots\dots\dots \quad [4.33]$$

Nonlinear, hysteretic relationships among the variables describing the flow make the mathematical analysis of movement of water in the root zone interesting but difficult. The pressure head, h, is generally a function of the history of the water content, θ. The plot of h versus θ is called the water retentivity curve (Also see Section 4.1.3). The differential water capacity, dθ/dh, is a measure of the capacity of the soil to store water at a given pressure head. The hydraulic conductivity, K, is a function of the water content and hence a function of the history of the pressure head.

For the root zone as a whole, the balance of mass for the water requires

$$dS/dt = (R + U + I) - (E + T + D) \quad \dots\dots\dots\dots\dots\dots\dots \quad [4.34]$$

where
  S = total amount of water stored in the root zone at any instant
  R = rate of supply from rainfall
  U = rate of upward flow into the root zone
  I = rate of irrigation
  E = rate of evaporation from the soil surface
  T = rate of transpiration
  D = drainage out of the root zone.
Fig. 4.18 illustrates several of these factors for a growing sorghum crop after Ritchie (1972). Water added in the form of irrigation is shown along the top. Also included are the evaporation from the soil surface, E, and transpiration by the plants, T. The evaporation is a function of the surface wetness as well as the plant cover. As the surface wets up after each irrigation, the evaporation tends to increase. The subsequent decrease of the evaporation rate is

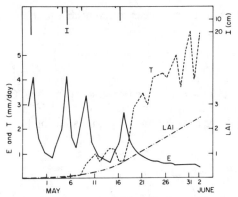

FIG. 4.18 Evaporation, transpiration, water added and leaf area index for grain sorghum (after Ritchie, 1972).

roughly proportional to the square root of time. The transpiration rate is closely related to the area of the leaves, as indicated by the figure. Note, as leaf area index increases, so does the transpiration.

Let us now examine the supplies of water in equation [4.34]. The rainfall R is highly variable in most climates. As a result, the capacity of the soil to absorb and retain water in the root zone is often crucial. An excess supply can cause loss of water and nutrients out of the root zone as well as poor aeration and trafficability. Drainage of excess water out of the root zone at one time of the year can provide a store of water of good quality available for capillary rise at another time of the year. Such supplies are very important in humid areas. In arid regions capillary rise usually leads to undesirable accumulation of salts in the root zone.

Irrigation is needed whenever rainfall and capillary rise do not suffice. In the following, we take a close look at the movement of water in the soil profile. First, we will look at examples for which the movement is primarily one-dimensional, that is, only in the vertical direction. Such is the case for most gravity and sprinkler irrigation systems. In other words, the water is distributed over the entire soil surface and the movement through the soil itself is primarily in the vertical. The accumulation of salts in the profile will also be considered briefly. Later we will look at two and three-dimensional problems. These relate to trickle and subsurface irrigation, in which case the water is delivered in close proximity to where it will be taken up by the plants rather than distributed over a larger area of the land.

4.6 ONE-DIMENSIONAL FLOWS

Suppose we have a steady flow without supplies from rainfall, in other words, in equation [4.34], $dS/dt = O$, $R = O$, and I, E, T, and U or D are constants. Integration of the steady balance of mass gives

$$q(z) = I - E - T_o \int^z \lambda \, dz \quad\dots\dots\dots\dots\dots\dots\dots\dots \quad [4.35]$$

If $K = K_o \exp(\alpha h)$ then the flux q (z) also satisfies

$$q(z) = -\alpha^{-1} \, dK/dz + K \quad\dots\dots\dots\dots\dots\dots\dots\dots \quad [4.36]$$

The solution of this ordinary, linear differential equation for K is:

$$K = \left\{ K' - \alpha_0 \int^z q\,(z)\,\exp(-\alpha z)\,(dz) \right\}\,\exp\,(\alpha z) \quad \dots\dots\dots\dots\dots \quad [4.37]$$

where K' is the hydraulic conductivity at the soil surface. Generally K' is not known a priori. More likely the hydraulic conductivity at some depth $Z = \varsigma$ will be known. Setting $K = K_\zeta$ and $z = \zeta$ in equation [4.37] and solving for K' gives

$$K' = K_\zeta \exp(-\alpha\zeta) + \alpha_0 \int^\zeta q\,(z)\,\exp(-\alpha z)\,dz \quad \dots\dots\dots\dots \quad [4.38]$$

Introducing equation [4.35] and [4.38] into equation [4.37] gives:

$$K = (I - E) + \left\{ K_\zeta - (I - E) \right\}\,\exp[-\alpha(\zeta - z)]$$
$$- \alpha T \exp\alpha z \, _z\!\int^\zeta (_0\!\int^z \lambda\,dz)\,\exp(-\alpha z)\,dz \quad \dots\dots\dots\dots \quad [4.39]$$

Three special cases are of particular interest
1. If $T = O$, then equation [4.39] reduces to

$$K = (I - E) + \left\{ K_\zeta - (I - E) \right\}\,\exp[-\alpha\,(\zeta - z)] \quad \dots\dots\dots\dots \quad [4.40]$$

If $I - E > O$, then the flow is downward. At a large distance above the water table, K will approach the uniform value $(I - E)$. If $I - E < O$, then equation [4.40] describes steady upward flow. For a given value of $I - E$, the maximum possible depth of the water table ζ_{max} is found by setting $K = O$ at $z = O$ and $K = K_{sat}$ at $z = \zeta_{max}$

$$\zeta_{max} = \alpha^{-1} \ln \left\{ 1 - K_{sat}/(I - E) \right\} \quad \dots\dots\dots\dots\dots \quad [4.41]$$

If $\alpha = 0.05\ cm^{-1}$, $K_{sat} = 10\ cm/d$, and $I - E = -0.1\ cm/d$, then $\zeta_{max} = 92.3$ cm.
2. If $\zeta \to \infty$, then equation [4.39] reduces to

$$K = (I - E) - \alpha T \exp(\alpha z) \, _z\!\int^\infty (_0\!\int^z \lambda\,dz)\,\exp(-\alpha z)\,dz \quad \dots\dots\dots \quad [4.42]$$

Equation 4.42] is equivalent to equation [8] of Warrick (1974). If $\lambda = \delta^{-1} \exp(-z/\delta)$, then equation [4.42] reduces to

$$K = \left[L + \frac{T}{E + T}\,(1 - L)\,\frac{\alpha\delta}{1 + \alpha\delta}\,\exp(-z/\delta) \right]\,I \quad \dots\dots\dots\dots \quad [4.43]$$

where $L = D/I$ is the leaching fraction.
3. If $\lambda = \delta^{-1} \exp(-z/\delta)$, then equation [4.39] reduces to

$$
\begin{aligned}
K = &(I - E - T) \\
&+ (K_\zeta - q^*) \exp[-\alpha(\zeta - z)] \\
&+ [\alpha\delta/(1 + \alpha\delta)] \, T \exp(-z/\delta) \\
&- \left\{ [\alpha\delta/(1 + \alpha\delta)] \, T \exp(-z/\delta) \right\} \exp[-(1 + \alpha\delta)(\zeta/\delta - z/\delta)]
\end{aligned}
$$

$$\dots\dots\dots\dots\dots\dots\dots\dots\dots\dots\dots\dots\dots\dots\dots\dots \quad [4.44]$$

where
$$q^* = (I - E - T)/(E + T).$$

In equation [4.44], the first term represents the net upward or downward flow at large depth; the second term represents the primary perturbation due to the imposed condition at $z = \zeta$; the third term represents the primary perturbation due to the distributed uptake in the root zone; and the fourth term represents the interaction between the imposed condition at $z = \zeta$ and the distributed uptake in the root zone. Fig. 4.19 shows calculated distributions of K for $T = 0.8$ cm/d and surface flux values of - 0.2 to 2.

When water evaporates from the soil surface, all the salts are left behind. Plants also selectively take up water, excluding virtually all the salts. The transport of the salt is mainly by convection with the soil water. Managing saline waters for irrigation requires finding the space-time trajectories of parcels of water and determining the changes in quality of these parcels.

The progress of a parcel along its path can be calculated from

$$
t - t_o = \int_{z_o}^z v^{-1} \, dz \dots\dots\dots\dots\dots\dots\dots\dots\dots\dots\dots\dots\dots\dots \quad [4.45]
$$

where
$t - t_o$ = time it takes for a parcel to travel from z_o to z
v = speed along its path.

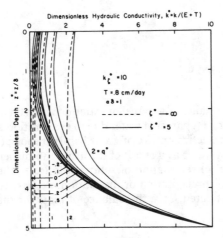

FIG. 4.19 Hydraulic conductivity in the root zone as a function of depth for several values of deep seepage (after Raats, 1977a).

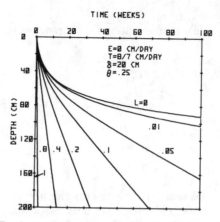

FIG. 4.20 Depth-time trajectories of parcels of water (after Raats, 1977a).

Within the root zone, the velocity at depth z is on the average given by the right hand size of equation [4.35] divided by θ. Fig. 4.20 shows depth time trajectories of parcels of water for different leaching fractions when the rate of uptake decreases exponentially with depth according to $\lambda = \delta^{-1}\exp(-z/\delta)$. For large z/δ these trajectories approach straight line asymptotes with a slope of (1 − L)/L and an intercept of d/δ with the z/δ-axis. The slope (1 −L)/L is the dimensionless velocity at large depth; with the aid of tracers this slope could be used to determine L. The intercept d/δ represents the dimensionless cumulative displacement induced by water absorbed at some distance below the soil surface. The cumulative displacement d increases as δ increases and as L decreases. A large rooting depth induces relatively rapid leaching of the soil solution over a large depth even if the leaching fraction is very small. For any uptake distribution, it follows from equation [4.45] that

$$\lim_{z \to \infty} (t - t_0) = z/v_\infty - d/v_\infty \dots\dots\dots\dots\dots\dots\dots\dots\dots\dots\dots [4.46]$$

where

$$d = {_0\!\int^\infty} (1 - v_\infty/v)\, dz \dots\dots\dots\dots\dots\dots\dots\dots\dots\dots\dots [4.47]$$

The change in quality of a parcel of water depends on (a) increase of the solute concentration of a parcel of water due to evaporation at or near the soil surface and due to selective uptake of water by plant roots; (b) gain or loss of solutes by parcels of water as a result of diffusive and dispersive mixing with their surroundings; (c) retardation of solutes relative to the water resulting from adsorption; (d) precipitation or dissolution of salts. If only the first of these factors is considered, then the balance of mass for the solutes reduces to

$$\partial(\theta c)/\partial t = -\,\partial(\theta v c)/\partial z, \dots\dots\dots\dots\dots\dots\dots\dots\dots\dots\dots [4.48]$$

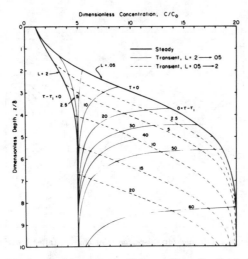

FIG. 4.21 Steady and transient salinity profiles (after Raats, 1975).

where

   c = concentration of salt in the soil water

   v = conductive velocity as before.

Combining equation [4.48] with equation [4.31], the balance of mass for the water gives:

$$\partial c/\partial t + v\,\partial c/\partial z = \frac{dc}{dt}\;\bigg|\;\text{parcel} = (\frac{\beta t}{\theta})\,c \quad \dots\dots\dots\dots\dots\dots\dots \quad [4.49]$$

Thus, the rate of increase of the concentration due to selective uptake is proportional to the rate of uptake βT, the current concentration c of the parcel of water, and the dryness θ^{-1}. The dependence on θ^{-1} is important in comparing sands, loams, and clays. Integration of equation [4.49] gives

$$c_{parcel} = c_I \exp {}_{t_o}\!\!\int^t \beta T/\theta\; dt = c_I \exp {}_o\!\!\int^x \beta T/(\theta\, v)\; dz \quad \dots\dots\dots \quad [4.50]$$

where c_I is the salt concentration at the soil surface (corrected for the increase due to evaporation from the soil surface). This equation applies always, but is easiest to implement if βT and v are fixed functions of position and time, i.e., if the flow is steady. Following the change in concentration of a parcel of water along its depth-time trajectory allows one to determine the resulting steady (Raats, 1974a, b) and unsteady (Raats, 1975) salinity profiles. Examples of such profiles are shown in Fig. 4.21.

4.7 TWO AND THREE-DIMENSIONAL FLOWS

In several types of irrigation, water is delivered to a point, a line or a confined region representing only a small fraction of the total surface area. Examples include trickle irrigation, subsurface irrigation and bubbler systems with a separate outlet for individual trees or shrubs. Once the water enters the soil, the flow is governed by gravity and capillarity, just as for the one-dimensional cases. However, the dynamics of the flow will be more com-

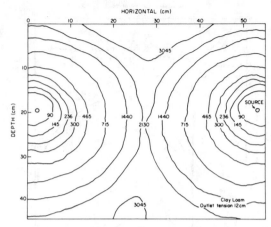

FIG. 4-22 Infiltration advance for subsurface line sources (after Ferreira, 1977).

plex, with the water moving not only in the upward and downward directions, but rather in a two or three-dimensional pattern.

Fig. 4.22 shows experimental results for a subsurface line source. These results were obtained by Ferreira (1977) using a Pima clay loam in a 45 × 55 × 30 cm tank. Water was added at two locations, each at a depth of 22 cm under a constant 12 cm suction inside a ceramic tube. The soil was unsaturated. The numbers on the curves show the infiltration time in minutes. For small times such as t = 90 min, the movement is approximately symmetric about each source and also each side is acting independently. However, at larger times the effect of gravity is such that the wetting fronts have advanced farthest in the downward direction. For the largest time t = 3,045 min, the wetting fronts of the two sources have intersected. Also, note that for the larger times the wetting fronts approach the surface as well as the lower boundary of the system. At the start, the intake occurs relatively fast and then tapers off to a constant value of about 0.1 cm³/cm·min. Such a constant tension source was used by Ferreira to grow sorghum and lettuce. For the lettuce, he used a narrower row spacing of 30 cm, however.

Several attempts have been made to model infiltration from point and line sources. Three different approaches have been used: (a) Numerical techniques, (b) Semi-analytical techniques, and (c) Analytical techniques. Examples of the first include the papers by Brandt et al. (1971) and by Neuman et al. (1975). Brandt et al. (1971) used a finite difference method to analyze flow from line and point sources. They compared the results with field and laboratory experiments. In the work of Neuman et al. the finite element technique was used and results were compared with field measurements from a stratified profile with subsurface irrigation. Examples of the second technique, semi-analytical solutions, are given by Philip (1969b) and Parlange (1972b). Their examples are primarily for infiltration only, that is, the techniques are not adapted to drainage or to boundary conditions where water is added and then shut off as in a cyclic system. Also, many of these solutions do not include the influence of gravity.

Analytical solutions have been found for a variety of point and line source geometries. The main advantage is that the solutions are exact and calculations can be made with negligible roundoff. The disadvantage, for the

TABLE 4.3 CHRONOLOGICAL LIST OF TWO AND THREE-DIMENSIONAL
SOLUTIONS OF THE MOISTURE FLOW EQUATION USING THE LINEARIZING
ASSUMPTION OF $K = K_0 \exp(\alpha h)$

1. Philip (1968) Single, buried point sources and spherical cavities
2. Wooding (1968) Single disk source
3. Philip (1969b) Single buried line source
4. Raats (1970) Array of surface line sources
5. Philip (1971) Surface sources
6. Raats (1971) Single surface point source
7. Philip (1972) Buried, surface, and perched sources in heterogeneous soils
8. Raats (1972) Sources buried at some distance below the soil surface
9. Maaledj and Malavard (1973) Influence of impermeable layers
10. Zachmann and Thomas (1973) Array of buried line sources
11. Gilley and Allred (1974) Array of buried line sources, including influence of uptake
    of water by plants
12. Lomen and Warrick (1974) Line sources (Includes time dependence)
13. Thomas et al. (1974) Array of buried line sources
14. Warrick (1974) Point source (Includes time dependence)
15. Philip and Forrester (1975) Buried, surface and perched sources in heterogeneous soils
16. Lomen and Warrick (1976) Summary paper; includes ring sources
17. Warrick and Lomen (1976) Strip and disk sources
18. Raats (1976) Summary paper; includes flow to line sinks
19. Thomas et al. (1976) Array of buried line sources; includes comparison with experi-
    ment
20. Batu (1977a, b) Line and strip sources
21. Warrick and Lomen (1977) Line above a shallow water table
22. Lomen and Warrick (1978) Surface flux proportional to K
23. Raats (1977b) Laterally confined flows from sources and to sinks
24. Warrick and Amoozegar-Fard (1977) Water regimes for porous cup samplers
25. Thomas et al. (1977) Line sources, plant uptake measurement
26. Merrill, Raats, and Kirksen (1978) Point source on a heterogeneous column, including
    cyclic applications
27. Dirksen (1978) Basically confined, buried line sources in homogeneous and layered
    soils
28. Warrick et al. (1979) Line sources with extraction and design nomographs

most part, is that special forms for the soil-water hydraulic properties must be chosen. For example, in order to linearize the flow equation, the hydraulic conductivity is often assumed to be an exponential function of the pressure head, just as we did earlier for the one-dimensional case. Table 4.3 summarizes solutions of line and point source problems using this assumption. If the sources are sufficiently far removed from both the soil surface and any sort of impeding layer below, they may be regarded as isolated sources embedded in an infinite medium. Flows from sources located at the soil surface or at a finite depth below the soil surface have also been analyzed. Solutions for confined systems both laterally and with depth are also available. For the time-dependent solutions, it must be assumed in addition that the derivative of the hydraulic conductivity with respect to the water content, $dK/d\theta$, is a constant. Often this assumption is not realistic.

If the point source is left on continuously, the distribution of the water content eventually approaches a steady-state. This is in contrast to the one-dimensional case in which the wetted part of the profile continues to expand. For a one-dimensional case the water content increases in a limited volume of soil near the wetting front, whereas in the multi-dimensional cases the added water is distributed over a continually increasing volume. Physically, the water added on the surface is balanced by flow at regions very remote from the source. The solid lines of Fig. 4.23 denote constant pressure head h = −23, −59 and −144 cm for such a system when $K = (1.23)(10)^{-5} \exp(0.019$

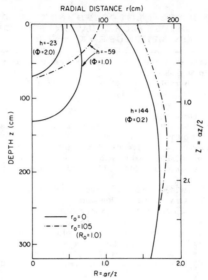

FIG. 4.23 Steady-state pressure head distribution for a point and disc surface source (after Warrick and Lomen, 1976).

h) cm/s and a discharge of 1000 cm³/h. This was the conductivity best fitted to Yolo light clay by Gilley and Allred (1974). The region close to the source is wet, further away it is drier. The dashed lines show constant pressure head surfaces for a disc source of radius $r_o = 105$. Such a source results when the discharge from a point emitter leads to ponding on the soil surface. The dashed and dotted lines represent different patterns, especially for regions close to the source. The lines of constant h close to the disk source tend to be flatter and broader than for the point source. At large distances from the source the curves coalesce. In other words, at points removed from the source one cannot tell whether the water has entered through a very restricted region close to the source or whether it spread over the soil surface. An alternative approach to modeling such a system is to take the soil surface as saturated over whatever region is necessary to allow the given discharge rate to enter through the soil surface. This was done by Brandt et al. (1971). Also shown on Fig. 4.23 are the values of the dimensionless matric flux potential $\Phi = 8\pi K/(\alpha^2$ Discharge Rate) and the dimensionless coordinates $R = \alpha r/2$ and $Z = \alpha z/2$ after Warrick and Lomen (1976). These values may be compared for any soil for which the conductivity is exponentially related to h and for any discharge rate q either at a point or over a dimensionless radius $R_o = 1.0$.

Philip (1972) and Philip and Forrester (1975) looked at steady infiltration from point sources into heterogeneous soils. Their analysis is based on a conductivity of the form $K = K_o \exp(\alpha h + \beta z)$. Physically this corresponds to a profile for which the conductivity varies exponentially with the pressure head h as well as exponentially with the depth z. The value of β tells how the variation with depth occurs. If $\beta > 0$ the conductivity at a given pressure head increases with depth; if $\beta < 0$ it decreases with depth. If $|\beta|$ is large, the conductivity changes very rapidly with depth. If $|\beta|$ is small, the change with depth is more gradual. The lines of constant h or the lines of equal pressure head for a given discharge rate will tend to be flatter and wider for the larger values of β.

4.8 NOTATION AND DEFINITIONS FOR PART II

c, C_l, c_{parcel}	Concentration of soil water solution
d	Rooting depth — units L.
D	Rate of drainage from the root zone — units L/T.
E	Rate of evaporation from the soil surface — units L/T.
h	Pressure head (soil water matric potential) — units L.
I	Rate of irrigation — units L/T.
K_o, K', K_ζ, K_{sat}	Hydraulic conductivities — units L/T.
L	Leaching fraction — dimensionless.
q	Soil water flux density — units L/T.
R	Rainfall rate — units L/T; also used as dimensionless radial coordinate.
S	Amount of water stored in the soil profile — units L.
t, t_o	Time — units T.
T	Rate of transpiration — units L/T.
U	Rate of upward flow — units L/T.
v, v_∞	Convective velocities — units L/T.
z, z_o	Depth — units L.
Z	Dimensionless depth
α, α_o	Empirical slope of logarithm of conductivity versus pressure head — units 1/L.
β	Empirical factor used to describe hydraulic conductivity varying with depth — units 1/L. Also, empirical factor for selective uptake of salts by roots.
δ	Root distribution factor — units L.
λ	Distribution function of root distribution function — units 1/L.
ϕ, Φ	Matrix flux potential — units L^2/T and dimensionless flux potential, respectively.
θ	Volumetric water content — dimensionless
ζ	Depths used for the one-dimensional analysis — units L.
∇	Vector gradient operator — units 1/L.

4.9 References

1   Adrian, D. D. and J. B. Franzini. 1966. Impedance to infiltration by pressure build-up ahead of the wetting front. J. Geophys. Res. 71(24):5857-5863.

2   Amerman, R. C. 1969. Finite difference solutions of unsteady, two-dimensional, partially saturated porous media flow. Ph.D. Thesis, Purdue University, Lafayette, IN.

3   Anat, A., H. R. Duke, and A. T. Corey. 1965 Steady upward flow from water tables. Hydrol. Paper No. 7. Colorado State Univ., Fort Collins.

4   Batu, V. 1977a. Steady infiltration from a ditch. Theory and experiment. Soil Sci. Soc. Am. J. 41:677-682.

5   Batu, V. 1977b. Steady infiltration from single and periodic strip sources. Soil Sci. Soc. Am. J. 42:544-549.

6   Baver, L. D., W. H. Gardner, and W. R. Gardner. 1972. Soil Physics, 4th ed., John Wiley & Sons, New York, 498 p.

7   Blake, G. R. 1965. Bulk density. p. 374-390. In: Methods of Soil Analysis I. C. A. Black, D. D. Evans, J. L. White, L. E. Ensminger and F. E. Clark (Ed.), Am. Soc. Agron. Monog. 9.

8   Blystone, J. R., A. Pelzer, and G. P. Steffens. 1961. Moisture content determination by the calcium carbide gas pressure method. Public Roads J. of Highway Res. 31:177-181.

9   Boast, C. W. and Don Kirkham. 1971. Auger hole seepage theory. Soil Sci. Soc. Am. Proc. 35:365-374.

10   Bodman, G. B. and E. A. Colman. 1943. Moisture and energy conditions during downward entry of water into soils. Soil Sci. Soc. Am. Proc. 7:116-122.

11   Bondurant, J. A. 1957. Developing a furrow infiltrometer. AGRICULTURAL ENGIENERING 38:602-604.

12   Bouma, J. D., I. Hillel, F. D. Hole, and C. R. Amerman. 1971. Field measurement of unsaturated hydraulic conductivity by infiltration through artificial crusts. Soil Sci. Soc. Am. Proc. 35:362-364.

13 Bouwer, H. 1961. A double tube method for measuring hydraulic conductivity of soil *in situ* above a water table. Soil Sci. Soc. Am. Proc. 25:334-339.

14 Bouwer, H. 1963. Theoretical effect of unequal water levels on the infiltration rate determined with buffered cylindrical infiltrometers. J. Hydrol. 1:29-34.

15 Bouwer, H. 1964. Measuring horizontal and vertical hydraulic conductivity of soil with the double-tube method. Soil Sci. Soc. Am. Proc. 28:19-23.

16 Bouwer, H. 1966. Rapid field measurement of air-entry value and hydraulic conductivity of soil as significant parameters in flow system analysis. Water Resour. Res. 2:729-738.

17 Bouwer, H. 1969. Infiltration of water into nonuniform soil. Proc. Am. Soc. Civ. Engr., J. Irrig. and Drain. Div. 95(IR4):451-462.

18 Bouwer, H. and R. D. Jackson. 1974. Determining soil properties. p. 611-672. In: Drainage for Agriculture, Am. Soc. of Agron., van Schilfgaarde, J. (Ed.), Madison, WI.

19 Bouyoucos, G. 1931. The alcohol method for determining moisture content of soils. Soil Sci. 32:173-178.

20 Bouyoucos, G. J. and A. H. Mick. 1940. An electrical resistance method for continuous measurement of soil moisture under field conditions. Mich. Agr. Exp. Sta. Tech. Bul. 172, 38 p.

21 Bowers, S. A. and S. J. Smith. 1972. Spectrophotometric determination of soil water content. Soil Sci. Soc. Am. Proc. 36:978-980.

22 Brandt, A., E. Bresler, N. Diner, J. Heller, and D. Goldberg. 1971. Infiltration from a trickle source: I. Mathematical models. Soil Sci. Am. Proc. 35:675-682.

23 Brooks, R. H. and A. T. Corey. 1964. Hydraulic properties of porous media. Hydrol. Paper No. 3. Colorado State Univ., Fort Collins, 27 p.

24 Brown, R. W. and B. P. Van Haveren (Eds.). 1972. Psychrometry in water relations research. Utah Agr. Exp. Sta., Logan, UT.

25 Brust, K. J., C. H. M. van Bavel and C. B. Stirk. 1968. Hydraulic properties of a clay loam soil and the field measurement of water uptake by roots: III. Comparison of field and laboratory data on retention and of measured and calculated conductivities. Soil Sci. Soc. Am. Proc. 32:322-326.

26 Brustkern, A. and H. J. Morel-Seytoux. 1970. Analytical treatment of two-phase infiltration. Am. Soc. Civ. Engr. J. Hydr. Div. 96(HY 12):2535-2548.

27 Buckingham, E. R. 1907. Studies on the movement of soil moisture. U.S. Dept. Agr. Bur. of Soils Bul. 38, 61 p.

28 Burdine, N. T. 1953. Relative permeability calculations from pore size distribution data. Petroleum Trans., Am. Inst. Mining and Metallurg. Engr. 198:71-78.

29 Burwell, R. E. and W. E. Larson. 1969. Infiltration as influenced by tillage-induced random roughness and pore space. Soil Sci. Soc. Am. Proc. 33:449-452.

30 Cassel, D. K. 1974. *In situ* unsaturated hydraulic conductivity for selected North Dakota soils. Bul. No 494, Agr. Exp. Sta., ND State Univ., Fargo. 20 p.

21 Childs, E. C. 1969. An introduction to the physical basis of soil water phenomena. John Wiley & Sons, NY. 493 p.

32 Childs, E. C. and M. Bybordi. 1969. The vertical movement of water in stratified porous material — 1. Infiltration. Water Resour. Res. 5(2):446-459.

33 Childs, E. C., A. H. Cole and D. H. Edwards. 1953. The measurement of the hydraulic permeability of saturated soil *in situ*. II. Roy Soc. (London) Proc., 216:72-89.

34 Childs, E. C., A. H. Cole, and N. Collis-George. 1950. The permeability of porous materials. Roy. Soc. (London), Proc., A.201:392-405.

35 Colman, E. Z. and T. M. Hendrix. 1949. Fiberglass electrical soil-moisture instrument. Soil Sci. 67:425-438.

36 Cope, F. and E. S. Trickett. 1965. Measuring soil moisture. Soils Fert. 28:201-208.

37 Davis, W. E. and C. S. Slater. 1942. A direct weighing method for sequential measurements of soil moisture under field conditions. J. Am. Soc. Agron. 34:285-287.

38 De Vries, J. and K. M. King. 1961. Note on the volume of influence of a neutron surface moisture probe. Can. J. Soil Sci. 41:253-257.

39 Dirksen, C. 1978. Transient and steady flow from subsurface line sources at constant hydraulic head in anisotropic soil. TRANSACTIONS of the ASAE 21:913-919.

40 Diserens, E. 1934. Beitrag zur Bestimmung der Durchlassigkeit des Bodens in naturlicher Bodenlagerung. Schweizerische Landw. Monatscefte XII: 1 and following (as referenced by Bouwer and Jackson, 1974).

41 Dixon, R. M. and D. R. Linden. 1972. Soil-air pressure and water infiltration under border irrigation. Soil Sci. Soc. Am. Proc. 36(5):948-953.

42 Dixon, R. M. and A. E. Peterson. 1971. Water infiltration control: a channel system concept. Soil Sci. Soc. Am. Proc. 35(6):968-973.

43   Dixon, R. M. and A. E. Peterson. 1964. Construction and operation of a modified spray infiltrometer and a flood infiltrometer. Res. Rep. 15, Exp. Sta., Univ. of Wisconsin, Madison.

44   Douglas, J. 1962. Handbook on measurement of soil moisture by the neutron scattering method. USDA Forest Service, M-4084, 34 p.

45   Duley, F. L. 1939. Surface factors affecting the rate of intake of water by soils. Soil Sci. Soc. Am.Proc. 4:60-64.

46   Edwards, W. M. and W. E. Larson. 1969. Infiltration of water into soils as influenced by surface seal development. TRANSACTIONS of the ASAE 12:463-465, 470.

47   Ely, G. W. 1965. Irrigation water management by the carbide method of water determination. Irrig. Drain. and Maint. 15:16-17, 26.

48   Ferreira, P. A. 1977. Evapotranspiration and soil matric potentials using tension irrigation. Unpublished Ph.D. Dissertation, The University of Arizona, Tucson.

49   Free, J. R. and V. J. Palmer. 1940. Relationship of infiltration air movement and pore size in graded silica sand. Soil Sci. Soc. Am. Proc. 5:390-398.

50   Freeze, R. A. 1971. Three-dimensional, transient, saturated-unsaturated flow in a groundwater basin. Water Resour. Res. 7:347-366.

51   Frevert, R. K. and Don Kirkham. 1948. A field method for measuring the permeability of soil below the water table. Highw. Res. Bd., Proc. 28:433-442.

52   Gardner, W. R. 1960. Dynamic aspects of water availability to plants. Soil Sci. 89:63-73.

53   Gardner, W. H. 1965. Water content. In: Methods of Soil Analysis I. C. A. Black (Ed.-in chief), Am. Soc. Agron. Monog. 9. pp. 82-127.

54   Gilley, J. R. and E. R. Allred. 1974. Irrigation and root extraction from sursurface irrigation laterals. TRANSACTIONS of the ASAE 17:927-933.

55   Green, W. H. and G. Ampt. 1911. Studies of soil physics, part I. — the flow of air and water through soils. J. Agr. Sci. 4:1-24.

56   Green, D. W., H. Dabiri, C. F. Weinaug and R. Prill. 1970. Numerical modeling of unsaturated groundwater flow and comparison of the model to field experiment. Water Resour. Res. 6:862-874.

57   Green, R. E. and J. C. Corey. 1971. Calculation of hydraulic conductivity: a further evaluation of some predictive methods. Soil Sci. Soc.Am. Proc. 35:3-8.

58   Hadas, A., D. Swartzendruber, P. E. Rijtema, M. Fuchs and B. Yaron. Chp. A1, A3, A5, A9 and A10, Physical aspects of soil water and salts in ecosystems. Springer-Verlag, NY.

59   Haise, H. R., W. W. Donnan, J. T. Phelan, L. F. Lawhon and D. G. Shockley. 1956. The use of cylinder infiltrometers to determine the intake characteristics of irrigated soils. USDA Publ. ARS 41-7, 10 p.

60   Hanks, R. J. and S. A. Bowers. 1963. Influence in variation in the diffusivity-water content relation on infiltration. Soil Sci. Soc. Am. Proc. 27:263-265.

61   Hanks, R. J. and S. A. Bowers. 1962. A numerical solution of the moisture flow equation for infiltration into layered soils. Soil Sci. Am. Proc. 26:530-534.

62   Hillel, D. 1971. Soil and water — physical principles and processes. Academic Press, NY, 299 p.

63   Hillel, D. and W. R. Gardner. 1969. Steady infiltration into crust topped profiles. Soil Sci. 108:137-142.

64   Hillel, D. and W. R. Gardner. 1970. Transient infiltration into crust topped profiles. Soil Sci. 109:69-76.

65   Hoffman, G. J. and G. O. Schwab. 1964. Tile spacing prediction based on drain outflow. TRANSACTIONS of the ASAE 7:444-447.

66   Holton, H. N. and N. R. Creitz. 1967. Influence of soils, vegetation and geomorpholgy on elements of the flood hydrograph. Proc. Symp. on floods and their computation, Leningrad, Russia, I.A.S.H. Bul. 85(II, Part III: 755-767.

67   Holton, H. N., C. B. England and V. O. Shanholtz. 1967. Concepts in hydrologic soil grouping. TRANSACTIONS of the ASAE 10(3):407-410.

68   Horton, R. E. 1939. Analysis of runoff plot experiments with varying infiltration capacity. Trans. Am. Geophys. Union, Part IV: 693-694.

69   Horton, R. E. 1940. An approach toward a physical interpretation of infiltration-capacity. Soil Sci. Soc. Am. Proc. 5:399-417.

70   Huggins, L. F. and E. J. Monke. 1966. The mathematical simulation of the hydrology of small watersheds. TR1, Purdue Water Resour. Res. Center, Lafayette, IN. 129 p.

71   Jackson, R. D. 1972. On the calculation of hydraulic conductivity. Soil Sci. Soc. Am. Proc. 36:372-380.

72   Kirkham, Don. 1946. Proposed method for field measurement of permeability of soil below the water table. Soil Sci. Soc. Am. Proc. 10:58-68.

73   Kirkham, Don. 1954. Measurement of the hydraulic conductivity in place. Symp. on permeability of soil. Amer. Soc. Test. Mater. Spec. Tech. Pub. 163:80-97.

74   Kirkham, Don and W. L. Powers. 1972. Advanced soil physics. Wiley-Interscience, NY, 534 p.

75   Kirkham, Don and C. H. M. van Bavel. 1948. Theory of seepage into auger holes. Soil Sci. Soc. Am. Proc. 13:75-81.

76   Klute, A. 1965a. Laboratory measurement of hydraulic conductivity of unsaturated soil. p. 253-261. In: Methods of soil analysis, Part 1. C A. Black (Ed.) Am. Soc. of Agron., Madison, WI

77   Klute, A. and D. B. Peters. 1962. A recording tensiometer with a short response time. Soil Sci. Soc. Am. Proc. 26:87-88.

78   Koon, J. L., J. G. Hendrick and R. E. Hermanson. 1970. Some effects of surface cover geometry on infiltration rate. Water Resour. Res. 6:246-253.

79   Kostiakov, A. N. 1932. On the dynamics of the coefficient of water-percolation in soils and on the necessity for studying it from a dynamic point of view for purposes of ameliora-tion. Trans. 6th Comm. Internl. Soil. Sci. Soc., Russian Part A: 17-21.

80   Kunze, R. J., G. Uehara, and K. Graham. 1968. Factors important in the calculation of hydraulic conductivity. Soil Sci. Soc. Am. Proc. 32:760-765.

81   LaRue, M. E., D. R. Nielsen, and R. M. Hagan. 1968. Soil-water flux below a rye grass root zone. Agron. J. 60:625-629.

87   Leonard, R. A. and P. F. Low. 1962. A self-adjusting, null-point tensiometer. Soil Sci. Soc. Am. Proc. 26:123-125.

3   Lomen, D. O. and A. W. Warrick. 1974. Time-dependent linearized infiltration: II. Line sources. Soil Sci. Soc. Am. Proc. 38:568-572.

84   Lomen, D. O. and A. W. Warrick. 1976. Time-dependent linearized moisture flow solutions for surface sources. p. 169-177. In: System simulation in water resources. G. C. Van Steenkiste (Ed.), North Holland.

85   Lomen, D. O. and A. W. Warrick. 1978. Linearized moisture flow with loss at the soil surface. Soil Sci. Soc. Am. J. 42:396-400.

86   Maaledj, M. and L. Malavard. 1973. Resolutions analogiques et numeriques de prob-lems d'irrigation des sols par canaux equidistants. C. R. Acad. Sci. Paris Ser. A 276:1433-1436.

87   Mannering, J. V. and L. D. Meyer. 1963. The effects of various rates of surface mulch on infiltration and erosion. Soil Sci. Soc. Am. Proc. 27:84-86.

88   Marshall, T. J. 1958. A relation between permeability and size distribution of pores. J. Soil Sci. 9:1-8.

89   McWhorter, D. B. 1971. Infiltration affected by flow of air. Hydrol. Paper No. 49. Colorado State Univ., Fort Collins, 41 p.

90   McWhorter, D. B. 1976. Vertical flow of air and water with a flux boundary condi-tion. TRANSACTIONS of the ASAE 19(2):259-261, 265.

91   Mein, R. G. and C. L. Larson. 1973. Modeling infiltration during a steady rain. Water Resour. Res. 9(2):384-394.

92   Merriam, J. L. 1968. Irrigation system evaluation and improvement. Blake Printing, San Luis Obispo, CA.

93   Merrill, S. D., P. A. C. Raats and C. Dirksen. 1978. Laterally confined flow from a point source at the surface of a inhomogeneous soil column. Soil Sci. Am. J. 42:851-857.

94   Miller, D. E. 1963. Lateral moisture flow as a source of error in moisture retention studies. Soil Sci. Soc. Am. Proc. 27:716-717.

95   Miller, D. E. 1969. Flow and retention of water in layered soils. USDA Cons. Res. Rep. No. 13, 28 p.

96   Miller, D. E. 1973. Water retention and flow in layered soil profiles. pp. 107-117. In: Field Soil Water Regime. R.R. Bruce et al. (Ed.), Soil Sci. Soc. Am., Madison.

97   Miller, D. E. and W. H. Gardner. 1962. Water infiltration into stratified soil. Soil Sci. Soc. Am. Proc. 26:115-118.

98   Miller, D. E. and J. S. Aarstad. 1971. Available water as related to evapotranspira-tion rates and deep drainage. Soil Sci. Soc. Am. Proc. 35:131-134.

99   Miller, D. E. and J. S. Aarsted. 1973. Effective available water and its relation to evapotranspiration rate, depth of wetting, and soil texture. Soil Sci. Soc. Am. Proc. 37:763-766.

100   Miller, D. E. and J. S. Aarstad. 1974. Calculation of the drainage component of soil water depletion. Soil Sci. 118:11-15.

101   Miller, R. D. 1951. A technique for measuring tensions in rapidly changing systems. Soil Sci. 72:291-301.

102   Millington, R. J. and J. P. Quirk. 1959. Permeability of porous media. Nature 183:387-388.

103   Millington, R. J. and J. P. Quirk. 1961. Permeability of porous solids. Faraday Soc., Trans. 57:1200-1206.

104   Moldenhauer, W. C. and W. D. Kemper. 1969. Interdependence of water drop energy and cold size on infiltration and clod stability. Soil Sci. Soc. Am. Proc. 33:297-301.

105   Moldenhauer, W. C. and D. C. Long. 1964. Influence of rainfall energy on soil loss and infiltration rates: I. Effect over range of texture. Soil Sci. Soc. Am. Proc. 28:813-817.

106   Morel-Seytoux, H. J. 1973. Two phase flows in porous media. Adv. in Hydroscience 9:119-202.

107   Morel-Seytoux, H. J. and J. Khanji. 1974. Derivation of an equation of infiltration. Water Resour. Res. 10(4):795-800.

108   Morin, J. and Y. Benjamini. 1977. Rainfall infiltration into bare soils. Water Resour. Res. 13(5):813-817.

109   Neuman, S. P., R. A. Feddes, and E. Bresler. 1975. Finite element analysis of two-dimensional flow in soils considering water uptake by roots: I. Theory. Soil Sci. Soc. Am. Proc. 39:224-230.

110   Nielsen, D. R., J. M. Davidson, J. W. Biggar, and R. J. Miller. 1964. Water movement through Panoche clay loam soil. Hilgardia 35:491-506.

111   Nielsen, D. R., R. D. Jackson, J. W. Cary, and D. D. Evans. (Ed.). 1972. Soil Water Am. Soc. Agron., Madison. 175 p.

112   Nielsen, D. R., J. W. Bigger, and K. T. Erh. 1973. Spatial variability of field measured soil water properties. Hilgardia 42(7):215-260.

113   Ogata, G. and L. A. Richards. 1957. Water content changes following irrigation of bare-field soil that is protected from evaporation. Soil Sci. Soc. Am. Proc. 21:355-356.

114   Oster, J. D. and L. S. Willardson. 1971. Reliability of salinity sensors for the management of soil salinity. Agron. J. 63:695-698.

115   Parlange, J. Y. 1971. Theory of water movement in soils: 2, one-dimensional infiltration. Soil Sci. 111(3):170-174.

116   Parlange, J. Y. 1972a. Theory of water movement in soils; 3, two and three dimensional adsorption. Soil Sci. 112(5):313-317.

117   Parlange, J. Y. 1972b. Theory of soil water movement in soils 5. Multidimensional unsteady infiltration. Soil Sci. 113:156-161.

118   Parlange, J. Y. 1972c. Theory of water movement in soils: 8, one-dimensional infiltration with constant flux at the surface. Soil Sci. 114(1):1-4.

119   Parr, J. F. and A. R. Bertrand. 1960. Water infiltration into soils. Adv. in Agron. 12:311-363.

120   Peck, A. J. 1965. Moisture profile development and air compression during water uptake by bounded porous bodies. 3: vertical columns. Soil Sci. 100(1):44-51.

121   Peters, D. B. 1965. Water availability. pp. 279-285. In: Methods of soil analysis I. C. A. Black (Ed.-in-chief), Am. Soc. Agron. Monog. 9.

122   Phene, C. J., G. J. Hoffman, and S. L. Rawlins. 1971a. Measuring soil matric potential *in situ* by sensing heat dissipation within a porous body. I. Theory and sensor construction. Soil Sci. Soc. Am. Proc. 35:27-33.

123   Phene, C. J., S. L. Rawlins, and G. J. Hoffman. 1971b. Measuring soil matric potential *in situ* by sensing heat dissipation within a porous body. II. Experimental results. Soil Sci. Soc. Am. Proc. 35:225-229.

124   Phene, C. J., G. J. Hoffman, and R. S. Austin. 1973. Controlling automated irrigation with a soil matric potential sensor. TRANSACTIONS of the ASAE 16(4):773-776.

125   Philip, J. R. 1957a. Numerical solution of equations of the diffusion type with diffusivity concentration dependent. II. Austral. J. Physics. 10:29-42.

126   Philip, J. R. 1957b. The theory of infiltration: 1. The infiltration equation and its solution. Soil Sci. 83:435-448.

127   Philip, J. R. 1957c. The theory of infiltration: 3. Moisture profiles and relation to experiment. Soil Sci. 84:163-178.

128   Philip, J. R. 1957d. The theory of infiltration: 4. Sorptivity and algebraic infiltration equations. Soil Sci. 84:257-264.

129   Philip, J. R. 1957e. The theory of infiltration: 5. The influence of initial water content. Soil Sci. 84:329-339.

130   Philip, J. R. 1958. The theory of infiltration: 6. Effect of water depth over soil. Soil Sci. 85:278-283.

131   Philip, J. R. 1966. Plant water relations: some physical aspects. Ann. Rev. Plant Physiol. 17:245-268.

132 Philip, J. R. 1968. Steady infiltration from buried point sources and spherical cavities. Water Resour. Res. 4:1039-1047.

133 Philip, J. R. 1969a. Theory of infiltration. Advances in Hydroscience, 5:215-296.

134 Philip, J. R. 1969b. Theory of infiltration. 5:215-296. In Ven Te Chow (Ed). Adv. in Hydrosci. Academic Press, New York.

135 Philip, J. R. 1971. General theorem on steady infiltration from surface sources with application to point and line sources. Soil Sci. Soc. Am. Proc. 35:867-871.

136 Philip, J. R. 1972. Steady infiltration from buried surface, and perched point and line sources in heterogeneous soils. I. Analysis. Soil Sci. Soc. Am. Proc. 36:268-273.

127 Philip, J. R. and R. I. Forrester. 1975. Steady infiltration from buried surface, and perched points and line sources in heterogeneous soils: II. Flow details and discussion. Soil Sci. Soc. Am. Proc. 39:408-414.

138 Raats, P. A. C. 1970. Steady infiltration from line sources and furrows. Soil Sci. Soc. Am. Proc. 34:709-714.

139 Raats, P. A. C. 1971. Steady infiltration from point sources, cavities and basins. Soil Sci. Soc. Am. Proc. 35:689-694.

140 Raats, P. A. C. 1972. Steady infiltration from sources at arbitrary depth. Soil Sci. Soc. Am. Proc. 36:399-401.

141 Raats, P. A. C. 1974a. Steady flow of water and salts in uniform profiles with plant roots. Soil Sci. Soc. Am. Proc. 38:717-722.

142 Raats, P. A. C. 1974b. Movement of water and salt under high frequency irrigation. p. 222-227. Proc. 2nd International Drip Irrigation Congress, San Diego, CA. USA, (Available from Dept. of Soil Sci., and Agr. Eng., Univ. of Calif., Riverside, CA 92502, USA).

143 Raats, P. A. C. 1975. Distribution of salts in the root zone. J. of Hydrol. 27:237-248.

144 Raats, P. A. C. 1976. Analytical solutions of a simplified flow equation. TRANSACTIONS of the ASAE 19:683-689.

145 Raats, P. A. C. 1977a. Convective transport of solutes in and below the root zone. Managing saline water for irrigation. Proc. Internl. Salinity Conf. Texas Tech. Univ., Lubbock, August, 1976.

146 Raats, P. A. C. 1977b. Laterally confined, steady flows of water from sources and to sinks in soils. Soil Sci. Soc. Am. J. 41:294-304.

147 Rawlins, S. L. 1976. Measurement of water content and the state of water in soils. p. 1-55. In: Water Deficits and Plant Growth, Vol. IV. T. T. Koslowski (Ed.). Academic Press, NY.

148 Reeves, M. and E. E. Miller. 1975. Estimating infiltration for erratic rainfall. Water Resour. Res. 11(1):102-110.

149 Reginato, R. J. 1974. Gamma radiation measurements of bulk density changes in a soil pedon following irrigation. Soil Sci. Soc. Am. Proc. 38:24-39.

150 Reginato, R. J. and C. H. M. Van Bavel. 1964. Soil water measurement with gamma attenuation. Soil Sci. Soc. Am. Proc. 28:721-724.

151 Rhoades, J. D. and R. D. Ingvalson. 1971. Determining salinity in field soils with soil resistance measurements. Soil Sci. Soc. Am. Proc. 35:54-60.

152 Richards, L. A. 1949. Methods of measuring soil moisture tension. Soil Sci. 68:95-112.

154 Richards, L. A. 1965. Physical conditions of water in soil. pp. 128-151. In: Methods of Soil Analysis I. Am. Soc. Agron. Monogr. 9.

155 Richards, L. A. and C. H. Wadleigh. 1952. Soil water and plant growth. pp. 73-251. In: Soil Physical Conditions and Plant Growth. B. T. Shaw (Ed.), Am. Soc. Agron. Monog. 2.

156 Richards, L. A. and L. R. Weaver. 1943. The sorption-block soil moisture meter and hysteresis effects related to its operation. J. Am. Soc. Agron. 35:1002-1011.

157 Richards, L. A. and L. R. Weaver. 1944. Fifteen atmosphere percentage as related to the permanent wilting percentage. Soil Sci. 56:331-339.

158 Richards, S. J. 1965. Soil suction measurements with tensiometers. p. 153-163. In: Methods of Soil Analysis I. C. A. Black (Ed.-in-chief). Am. Soc. Agron. Monog. 9.

159 Richards, S. J. and A. W. Marsh. 1961. Irrigation based on soil suction measurements. Soil Sci. Soc. Am. Proc. 25:65-69.

160 Ritchie, J. T. 1972. Model for predicting evaporation from a row crop with incomplete cover. Water Resources. Res. 8:1204-1213.

161 Robins, J. S. 1959. Moisture movement and profile characteristics in relation to field capacity. Internl. Comm. Irrig. Drain. 8:509-521.

162 Robins, J. S., W. O. Pruitt, and W. H. Gardner. 1954. Unsaturated flow of water in field soils and its effect on soil moisture investigations. Soil Sci. Soc. Am. Proc. 18:344-347.

163 Rubin, J. 1966. Theory of rainfall uptake by soils initially drier than their field

capacity and its applications. Water Resour. Res. 2(4):739-749.

164   Rubin, J. 1968. Theoretical analysis of two-dimensional, transient flow of water in unsaturated and partly unsaturated soils. Soil Sci. Soc. Am. Proc. 32:607-615.

165   Rubin, J. and R. Steinhardt. 1963. Soil water relations during rain infiltration: I. Theory. Soil Sci. Soc. Am. Proc. 27:246-251.

166   Rubin, J. and R. Steinhardt. 1964. Soil water relations during rain infiltration: III. Water uptake at incipient ponding. Soil Sci. Soc. Am. Proc. 28:614-619.

167   Rubin, J., R. Steinhardt and P. Reiniger. 1964. Soil water relations during rain infiltration: II. Moisture content profiles during rains of low intensities. Soil Sci. Soc. Am. Proc. 28:1-5.

168   Shady, A. M., R. S. Broughton, Y. Bernier, U. Delisle and G. Sylvestre. 1977. Experience with production scale field measurement of hydraulic conductivity. Proc. Third National Drain. Symp., ASAE Publ. 1-77. pp. 112-116. ASAE, St. Joseph, MI. 49085.

169   Shaw, B. and L. D. Baver. 1939. An electrothermal method for following moisture changes of the soil in situ. Soil Sci. Soc. Am. Proc. 4:78-83.

170   Skaggs, R. W. 1976. Determination of the hydraulic conductivity-drainable porosity ratio from water table measurements. TRANSACTIONS of the ASAE 19(1):73-80, 84.

171   Skaggs, R. W., E. J. Monke and L. F. Huggins. 1970. An approximate method for determining the hydraulic conductivity function of unsaturated soil. TR5, Purdue Water Resour. Res. Center, Lafayette, IN. 124 p.

172   Skaggs, R. W., E. J. Monke, G. H. Foster and L. F. Huggins. 1969. Experimental evaluation of infiltration equations. TRANSACTIONS of the ASAE 12(6):822-828.

173   Smith, R. E. 1976. Approximations for vertical infiltration rate patterns. TRANSACTIONS of the ASAE 19(3):505-509.

174   Smith, R. E. 1972. The infiltration envelope: Results from a theoretical infiltrometer, J. of Hydrol. 17:1-21.

175   Smith, R. E. and D. A. Woolhiser. 1971. Overland flow on an infiltrating surface. Water Resour. Res. 7(4):899-913.

176   Snell, A. W. and J. van Schilfgaarde. 1964. Four-well method of measuring hydraulic conductivity in saturated soils. TRANSACTIONS of the ASAE 7:83-87, 91.

177   Staple, R. J. 1966. Infiltration and redistribution of water in vertical columns of loam soil. Soil Sci. Soc. Am. Proc. 30:533-558.

178   Staple, W. J. and R. P. Gupta. 1966. Infiltration jinto homogeneous and layered columns of aggregated loam silt loam, and clay soil Can. J. Soil Sci. 46:293-305.

179   Swartzendruber, D. 1969. The flow of water in unsaturated soils. p. 215-292. In: Flow through porous media. R. M. DeWiest (Ed.). Academic Press, New York.

180   Swartzendruber, D. and E. G. Youngs. 1974. A comparison of physically based infiltration equations. Soil Science 117(3):165-167.

187   Tanner, C. B. and D. E. Elrick. 1958. Volumetric porous (pressure) plate apparatus for moisture hysteresis measurements. Soil Sci. Soc. Am. Proc. 22:575-576.

182   Taylor, S. A., D. D. Evans, and W. D. Kemper. 1961. Evaluating soil water. Utah State Agric. Exp. Sta. Bull. 426, 67 p.

183   Thomas, A. W., E. G. Kruse, and H. R. Duke. 1974. Steady infiltration from line sources buried in soil. TRANSACTIONS of the ASAE 17:115-128, 133.

184   Thomas, A. W., H. R. Duke, D. W. Zachman, and E. G. Kruse. 1976. Comparisons of calculated and measured capillary potentials from line sources. Soil Sci. Soc. Am. J. 40:10-14.

185   Thomas, A. W., H. R. Duke, and E. G. Kruse. 1977. Capillary potential distribution in root zones using subsurface irrigation. TRANSACTIONS of the ASAE 20(1):62-67.

186   Tovey, R. and C. H. Pair. 1966. Measurement of intake rate for sprinkler irrigation design. TRANSACTIONS of the ASAE 9:359-363.

187   Van Bavel, C. H. M. 1961. Neutron measurement of surface soil moisture. J. Geophys. Res. 66:4193-4198.

188   Van Bavel, C. H. M. and Don Kirkham. 1948. Field measurement of soil permeability using auger holes. Soil Sci. Soc. Am. Proc. 13:90-96.

189   van Beers. 1958. The auger-hole method. Internl. Inst. Land Reclam. and Improve. Bul. No. 1. Wageningen, The Netherlands. 32 p.

190   Viehmeyer, F. J. and A. H. Hendrickson. 1927. Soil moisture conditions in relation to plant growth. Plant Physiol. 2:71-78.

191   Viehmeyer, F. J. and A. H. Hendrickson. 1949. Methods of measuring field capacity and wilting percentages of soils. Soil Sci. 68:75-94.

192   Viehmeyer, F. J. and A. H. Hendrickson. 1950. Soil moisture in relation to plant growth. Ann. Rev. Plant Physiol. 1:285-304.

193   Warrick, A. W. 1974. Solution to the one-dimensional linear moisture flow equation with water extraction. Soil Sci. Soc. Am. Proc. 38:573-576.

194   Warrick, A. W. and A. Amoozergar-Fard. 1977. Soil water regimes near porous cup water samplers. Water Resour. Res. 131:203-207.

195   Warrick, A. W. and D. O. Lomen. 1976. Time-dependent infiltration: III. Strip and disc sources. Soil Sci. Soc. Am. Proc. 40:639-643.

196   Warrick, A. W. and D. O. Lomen. 1977. Flow from a line source above a shallow water table. Soil Sci. Soc. Am. J. 41:849-852.

197   Warrick, A. W., A. Amoozegar-Gard, and D. O. Lomen. 1979. Linearized moisture flow from line sources with water extraction. TRANSACTIONS of the ASAE 22(3):549-553, 559.

198   Watson, K. K. 1967a. A recording field tensiometer with rapid response characteristics. J. Hydrol. 5:33-39.

199   Watson, K. K. 1967b. The measurement of the hydraulic conductivity of unsaturated porous materials utilizing a zone of entrapped air. Soil Sci. Soc. Am. Proc. 31:716-720.

200   Whisler, F. D. and H. Bouwer. 1970. Comparison of methods for calculating vertical drainage and infiltration for soils. J. of Hydrol. 10:1-19.

201   Whisler, F. D. and A. Klute. 1965. The numerical analysis of infiltration, considering hysteresis, into a vertical soil column at equilibrium under gravity. Soil Sci. Soc. Am. Proc. 29:489-494.

202   Whisler, F. D. and A. Klute. 1966. Analysis of infiltration into stratified soil columns. Symp. on water in the unsaturated zone, Wageningen, The Netherlands. p. 451-461.

203   Wiebe, H. H., R. W. Brown, and J. Barker. 1977. Temperature gradient effects on *in situ* hygrometer measurements of water potential. Agron. J. 69:933-939.

204   Wiebe, H. H., G. S. Campbell, W. H. Gardner, S. L. Rawlins, J. W. Cary, and R. W. Brown. 1971. Measurement of plant in soil water status. Utah Ag. Exp. Sta. Bul. 484, 71 p.

205   Wilcox, J. C. 1962. Rate of soil drainage following an irrigation. III. A new concept of the upper limit of available water. Can. J. Soil Sci. 42:122-128.

206   Wilson, L. G. and J. N. Luthin. 1963. Effect of air flow ahead of the wetting front on infiltration. Soil. Sci. 96(2):136-143.

207   Winger, R. J., Jr. 1960. In-place permeability tests and their use in subsurface drainage. Internl. Comm. Irrig. and Drainage, 4th Cong., Madird, Spain. 48 p.

208   Wooding, R. 1968. Steady infiltration from a shallow circular pond. Water Resour. Res. 4:1259-1273.

209   Youngs, E. G. 1968. An estimation of sorptivity for infiltration studies from moisture movement considerations. Soil Sci. 106:157-163.

210   Youngs, E. G. 1964. An infiltration method of measuring the hydraulic conductivity of unsaturated porous materials. Soil Sci. 97:307-311.

211   Zachman, D. W. and A. W. Thomas. 1973. A mathematical investigation of steady infiltration from line sources. Soil Sci. Soc. Am. Proc. 37:495-500.

chapter 5

SALINITY IN
IRRIGATED AGRICULTURE

5

5

SALINITY IN IRRIGATED AGRICULTURE

by G. J. Hoffman, U.S. Salinity Laboratory, SEA-AR,
USDA, Riverside, CA; R. S. Ayers, University of
California, Davis, CA; E. J. Doering, Northern
Great Plains Research Center, SEA-AR, USDA,
Mandan, ND; B. L. McNeal, Washington State
University, Pullman, WA

5.1 SCOPE AND SIGNIFICANCE OF SALINITY

Salinity control is frequently a major objective of irrigation management, even though the primary objective of irrigation is to maintain soil matric potential in a range suitable for optimum crop yield. These objectives are often closely related because salinity is almost a universal threat in arid areas where irrigation waters typically contain significant amounts of dissolved salts. In humid regions, salinity is of limited concern because rain water is almost free of dissolved salts; thus, salinity problems may occur in humid areas only in coastal regions subject to sea water intrusion or from occurrences of brine spillage, waste disposal, or excess fertilization. As drainage problems or supplemental irrigation increases in humid regions, salinity problems analogous to those in arid regions will no doubt become more prevalent.

All soils and irrigation waters contain a mixture of soluble salts, not all of which are essential for plant growth. The salt concentration of the soil solution is usually higher than that of the applied water. This increase in salinity results from plant transpiration and soil surface evaporation, which selectively remove water, concentrating the salts in the remaining soil water. To prevent soil salinity from reaching the harmful levels illustrated in Fig. 5.1, it is necessary to remove a portion of the concentrated soil solution from the crop root zone by leaching. Salts will be leached whenever water applications exceed evapotranspiration, providing that soil infiltration and drainage rates are adequate. The key to salinity control is a net downward movement of soil water in the root zone. Even in well-managed, high-yielding fields, however, the soil water may be several times more saline than the irrigation water. With insufficient leaching, this ratio can easily increase to tenfold or more, with resultant damage to crops. How much leaching is required depends on the quality of the irrigation water, the crop grown, and the frequency and uniformity of irrigation.

This chapter outlines the influence of soil salinity on crop production, introduces management techniques to cope with saline soils and waters, including choice of an irrigation system, and evaluates methods of reclaiming salt-affected soils.

Salt-affected field in California resulting from insufficient leaching.

Severe soil salinity problem in a Colorado corn fie

Sodic soil in Idaho.

Effect of salinity on cotton in Arizona.

FIG. 5.1 Examples of salinity damage to agricultural crops.

5.1.1 Source and Accumulation of Soluble Salts

The initial source of salts in soils is the weathering of rocks and minerals in the earth's crust. During weathering, bonds between elements of primary minerals are broken and new combinations are formed, including both secondary silicates and soluble salts. Salts weathered from acid magmatic rocks (e.g., granite, porphyry, gneiss) are principally chlorides, sulfates, and carbonates of sodium and potassium, whereas those from basic magmatic rocks (basalt and diabase) typically are sulfates and carbonates of calcium and magnesium. Although weathering of primary minerals is the initial source of all soluble salts, seldom have sufficient quantities of salt accumulated from such weathering to form a saline soil. Saline soils usually develop in areas that receive and accumulate salts carried from other locations by water.

Throughout geological time, soluble salts have been exchanged between land and ocean through the water cycle. During such exchanges, salts have tended to separate because of their different solubilities, with the less soluble sulfates and carbonates predominating on land and the more soluble chlorides accumulating in the seas. Rainfall contains relatively little salt, but an annual rainfall of 250 mm can still deposit from 50 to 1250 kg/ha of salt on the landscape, depending on the distance from the sea coast and prevailing wind direction (Junge and Werby, 1958). With time, such air-borne salts can contribute substantial amounts of salinity to infrequently leached, poorly drained, or land-locked areas.

The most significant transfer mechanism for salts from oceans to the continents is by the accumulation of salts in marine sediments, followed by periods of uplift and exposure of the sediments at the earth's surface. The Mancos shales of Colorado and Wyoming, the Permian Red Beds of Kansas and Oklahoma, the Green River formation of Utah, and the marine sediments along the west side of the San Joaquin Valley in California are prominent examples of exposed marine deposits that contain large quantities of salts. The ocean is also a direct source of salts in low-lying soils along sea coasts because of sea water intrusion. Completing the water cycle, soluble salts are transported from the continents to the oceans by water moving over and under the land to water courses that terminate in the sea.

Relatively small salt inputs from rainfall and mineral weathering can lead to substantial soil salinity problems as salts are redistributed and accumulated with drainage waters. An example is southwestern Australia where salts have been deposited by rainfall for centuries on relatively old (and hence already well-weathered) soils. Such salts have been concentrated in low-lying areas during excessive deep percolation that has accompanied the transition from deep-rooted shrubs to shallow-rooted grasses in the area. Another example is the "saline seep" areas of eastern Montana and the western Dakotas, where percolating soil water becomes saline, is intercepted by impermeable horizontal layers, and conducted laterally on top of these layers to hillside surfaces or landscape depressions (Doering and Sandoval, 1976). The opportunity for soil water to percolate past the root zone is enhanced by cultural practices that increase infiltration and reduce evapotranspiration (e.g., summer fallow).

Other examples where long-term salt accumulation has led to substantial salinity problems are the Columbia Basin area of Washington and the Red River Valley of North Dakota. Although the Columbia River has unusually high-quality irrigation waters, salt problems have developed wherever inadequate drainage has resulted in shallow ground waters that contain high salt concentrations. Some of the salts in this ground water are associated with former lacustrine (lake-laid) sediments, although additional salinity has accumulated in the slowly permeable subsoils from rainfall and mineral weathering in overlying and surrounding soils with inadequate deep percolation. In the Red River Valley of North Dakota, the cause of the salinity problem is a shallow water table that is sustained by the migration of dissolved salts with upward-flowing artesian water (Benz et al., 1976).

In addition to natural salt sources, salinity can sometimes arise from excessive applications of soluble potassium and nitrogen fertilizers. Soluble nitrate also can contribute to the degradation of nearby municipal water supplies.

Irrigation with municipal, industrial, or other waste waters is yet another source of soluble salts. In addition to the salinity hazard, certain waste waters may contain small quantities of trace elements, such as cadmium, selenium, lithium, lead, and other metals, which may be toxic to plants at relatively low concentrations.

5.1.2 Extent and Distribution of Salt-Affected Soils

Salinity is estimated to be a potential threat to about half of the 20 million hectares of irrigated land in the Western United States, with crop production being limited by salinity on about 25 percent of this land (Wadleigh,

1968). The total area of salt-affected soils throughout the world is not precisely known. A recent survey indicated that of the 756 million ha cultivated and the 91 million irrigated in 24 countries (Shalhevet and Kamburov, 1976), over 50 million are affected by salinity.

Attempts have been made to assemble maps delineating the extent and distribution of salt-affected soils. However, the detail of such maps is usually inadequate to aid in individual farm management decisions. The maps will be very helpful, however, for watershed management decisions and pointing out areas of potential concern. Salinity surveys should be used to determine whether salinity is a problem for a particular management unit.

5.2 SALINITY MANAGEMENT

5.2.1 Leaching

How much leaching is required to maintain a viable irrigated agriculture depends upon (a) the salt content of the irrigation water, soil, and groundwater; (b) the salt tolerance of the crop; (c) climatic conditions; and (d) soil and water management. If leaching is not adequate, deleterious salt accumulations can develop within a few cropping seasons.

The only economical means of controlling soil salinity is to ensure a net downward flow of water through the crop root zone. The ratio of the additional increment of water, beyond that required for crop evapotranspiration, that must pass through the root zone to prevent harmful salt accumulations in the soil to the total irrigation water applied has been termed the leaching requirement (U.S. Salinity Laboratory Staff, 1954). Once salts have accumulated to the maximum tolerable limit for the crop under a given set of conditions, any salt added with subsequent irrigations must be balanced by a similar amount removed by leaching to prevent a loss in yield. This is termed maintaining a salt balance.

One method of estimating the leaching requirement uses a steady-state salt-balance model. This model applies to a soil profile that has been irrigated long enouth to achieve steady-state with respect to salt accumulation and distribution. Such a model can be expressed as

$$\Delta S_{SW} = V_I\, C_I + V_R\, C_R + V_G\, C_G + S_M + S_A - V_D\, C_D - S_P - S_C \quad \dots \quad [5.1]$$

where V_I, V_R, V_G, V_D, and C_I, C_R, C_G, C_D are the respective volumes and salt concentrations of the irrigation water, rainfall, groundwater, and drainage waters. V_G refers to that volume of water that moves by capillarity into the root zone from an underlying water table; and S_M, S_A, S_P, and S_C are the quantities of salt added to or removed from the soil solution by mineral weathering, amendments (fertilizer, manure, chemicals, etc.), precipitation as insoluble minerals, or the harvested crop, respectively. The net difference between these inputs and outputs is the resultant change in soil-water salinity (ΔS_{SW}). Assuming that ($S_M + S_A$) is essentially equal to ($S_P + S_C$) and that V_G and V_R are small compared to V_I, the steady-state ($\Delta SW = 0$) form of equation [5.1] reduces to

$$V_D/V_I = C_I/C_D. \quad \dots\dots\dots\dots\dots\dots\dots\dots\dots\dots\dots\dots\dots \quad [5.2]$$

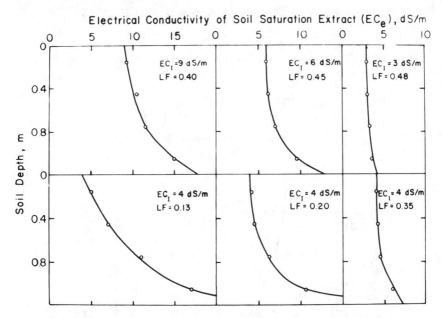

FIG. 5.2 Steady-state soil salinity profiles from an alfalfa lysimeter study with different irrigation water qualities and leaching fractions (Bower et al., 1969).

With uniform areal infiltration of water and uniform evapotranspiration, the volume terms in equation [5.2] can be reduced to equivalent depths of water (D). Furthermore, because the electrical conductivity (EC) of a water is generally a reliable index of total salt concentration (U.S. Salinity Laboratory Staff, 1954), EC can be substituted for C in equation [5.2]. The resulting equation,

$$LF = D_D/D_I = EC_I/EC_D, \dotfill [5.3]$$

shows the approximate equality, under the stated conditions, between leaching fraction ($LF = D_D/D_I$) and the ratio of salinity in the irrigation and drainage waters (EC_I/EC_D). Thus, by varying the fraction of applied water of concentration EC_I that percolates through the root zone, it is possible to control the concentration of salts in the drainage water (EC_D), and hence, to control either the average or the maximum salinity of soil water in the crop root zone at some desired level.

The most common method of measuring soil salinity is to determine the electrical conductivity of saturation extracts, EC_e. The use of EC_e is currently recommended because the saturation percentage is easily and reproducibly determined in the laboratory and is related to the field-moisture range of soils varying widely in texture (U.S. Salinity Laboratory Staff, 1954). For many soils, the soluble salt concentration of the soil solution at field capacity is about twice that at saturation. Some recent developments in instrumentation now permit direct determinations of electrical conductivity of soil water, EC_{SW}. Two such devices are salinity sensors (Oster and Ingvalson, 1967) and four-electrode probes (Rhoades and Ingvalson, 1971). Obviously, salinity measurements are more reliable if made on soil solutions in the field-moisture range.

Electrical Conductivity of Soil Saturation Extract (EC$_e$), dS/m

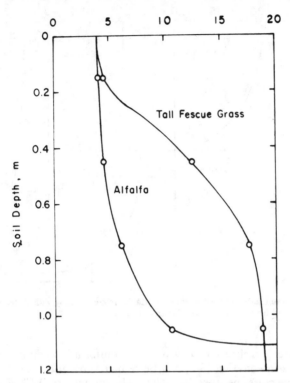

FIG. 5.3 Steady-state salinity profiles in lysimeters for tall fescue grass and alfalfa (Bower et al., 1969, 1970).

5.2.2 Soil Salinity Profiles

Based on equation [5.3], EC$_I$ and LF control the salt concentration of the drainage water (EC$_D$). Experimental evidence of the validity of this relationship is given in Fig. 5,2, where EC$_D$ and the distribution of soluble salts in the root zone are shown for alfalfa grown in a controlled lysimeter study (Bower et al., 1969). Irrigation water quality influenced the salinity profile at shallow depths, whereas the leaching fraction controlled the soil salinity level near the bottom of the root zone. Other factors, such as the crop's water extraction pattern, the influence of groundwater, and the irrigation frequency and method, can also significantly alter soil salinity profiles.

The effect of crop water extraction on soil salinity can be seen from Fig. 5.3. Both soil (Pachappa sandy loam) profiles were treated identically, with the same irrigation water quality (EC$_I$ = 4 dS/m)* and leaching fraction (0.2), except that deep-rooted alfalfa was grown in one and shallow-rooted tall fescue grass in the other. For the grass, most of the water was extracted above the 0.5-m depth, with almost no water extracted below 1.0 m. Thus, because plant roots extract relatively pure water, the salt concentrates where water is extracted. Alfalfa extracted water from much deeper, with considerable quantities of salt being accumulated below the 1.0-m depth.

Groundwater can influence the salinity profile in the crop root zone only if the net flow of water is upward for significant periods of time. Thus,

*1 dS/m = 1 millimho/cm.

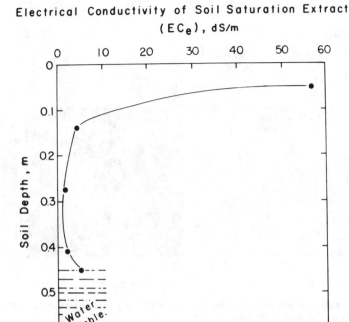

Electrical Conductivity of Soil Saturation Extract
(EC_e), dS/m

FIG. 5.4 Example of the influence of a shallow, saline water table on a soil salinity profile (adapted from Mohamed and Amer, 1972).

the high salt concentration near the soil surface depicted in Fig. 5.4 occurs in the absence of sufficient irrigation or rainfall to maintain downward water flow.

One further example of how leaching influences soil salinity is given in Fig. 5.5 from Hoffman et al. (1979). Here, water was applied daily by drip irrigation along every other plant row. The soil salinity distributions, as measured by soil-water chloride concentration, are given for 17 and 2 percent leaching. The chloride concentration of the irrigation water was 7.5 mol/m^3 and EC_I was 2.2 dS/m. Thus, assuming no significant amount of chloride uptake by the crop, a soil-water chloride concentration of 15 mol/m^3 would be associated with locations at which half of the irrigation water had been evapotranspired. A soil-water chloride concentration of 75 mol/m^3 would indicate the points at which 90 percent of the water had been evapotranspired. While salt concentrations were generally much higher for 2 than for 17 percent leaching, the salinity patterns for the two leaching treatments were similar, and typical of profiles under line sources of irrigation. Yield responses to such salinity profiles are highly crop dependent, as will be discussed in the next section. Beginning with the second year of treatment, yields of grain sorghum, wheat, and lettuce were reduced approximately 10, 30, and 70 percent, respectively, by the reduction in leaching from 17 to 2 percent.

5.2.3 Influence of Irrigation Method on Soil Salinity

Soil salinity profiles differ markedly among various irrigation methods because of differences in water distribution. Irrigation water may be applied

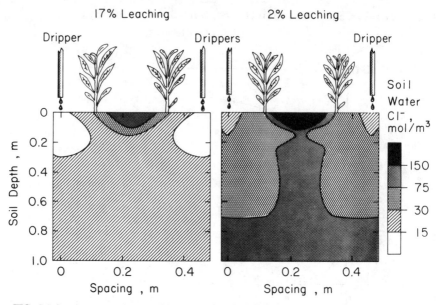

FIG. 5.5 Steady-state soil-water chloride profiles for 17 and 2 percent leaching in a field plot study with trickle irrigation (Hoffman et al., 1979).

over an entire area, as line sources, or as point sources. Irrigation methods that apply water over the entire area are flooding and sprinkling. Furrows, porous or multi-emitter drip, and subsurface irrigation systems apply water along lines. Point sources of irrigation include microbasins and drip or sub-surface systems with widely spaced emitters. In addition, soil salinity profiles within each method of irrigation may differ from point to point because of application nonuniformities in both time and space.

Irrigation systems that apply water over the entire area typically cause a relatively uniform increase in soil salinity with depth to the bottom of the root zone, providing that leaching is not excessive and application is relatively uniform. If the irrigation frequency is low or if evaporation from the soil surface is high, soil salinity, particularly near the soil surface, increases with time between irrigations. Salt accumulation can also vary widely within a given field if soil hydraulic conductivity, uniformity of water application, or crop water extraction differs.

The distribution of salts in a soil profile under line water sources has both lateral and vertical components. A typical cross section of the profile, see Fig. 5.5 for example, shows both an isolated pocket of accumulated salts at the soil surface midway between line sources and a second, deep zone of accumulation, whose location depends on the degree and efficiency of leaching. Directly beneath the line source is a leached zone, with size dependent on the rate and frequency of irrigation and the water extraction pattern of the crop. This type of soil salinity profile is typical of most furrow- and drip-irrigation systems for row crops.

Whereas the salt distribution from line sources generally increases both laterally and with depth, the distribution from point irrigation sources increases radially in all directions below the soil surface. As the rate of water application increases, the shape of the salinity distribution changes. Based on the mathematical model of Bresler (1975), the salinity distribution in

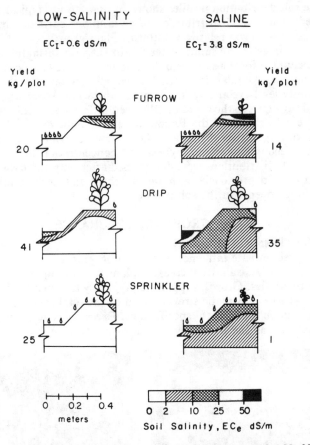

FIG. 5.6 Influence of the irrigation system on the soil salinity pattern and yield of bell pepper at two levels of irrigation water quality (Bernstein and Francois, 1973a).

sand changes from elliptical (with the maximum axis vertical) to more circular as the rate of water application increases. In fine-textured soils, and particularly in layered soils, considerably more water can move horizontally than vertically as the rate of application increases, resulting in relatively shallow depths of salt accumulation (Bresler, 1975). For tree crops irrigated with several drip emitters per tree, the wetting patterns may overlap, thus reducing the level of salt accumulation midway between emitters under a tree.

Irrigation systems are difficult to compare because of subtle, but often significant, differences in cultural practices associated with each system. For example, the addition of fertilizer with the water in one irrigation method and not another can cause unassessable effects on crop yields. Differences in water application efficiency, even though equal amounts of water are applied, can cause significant differences in salt distribution. The experiment of Bernstein and Francois (1973a), as summarized in Fig. 5.6, shows that pepper plants, when drip irrigated with good quality water (EC_I = 0.6 dS/m), outyielded by about 50 percent sprinkler- and furrow-irrigated plants receiving the same amount of water. With saline water (EC_I = 3.8 dS/m), the yield benefits of drip irrigation were even larger. Part of the yield differences can be explained by more frequent irrigation with the drip system.

The salt distribution profiles shown in Fig. 5.6 are typical of those just described. The lateral distribution of salts under sprinkler irrigation (an areal-type system) was relatively uniform. Plant growth was severely limited by foliar damage in the saline water treatment with sprinkling, and evaporation accounted for the salt accumulation near the soil surface. The salinity distribution was similar for both line source systems (furrow and drip), with salinity levels relatively low beneath the water sources and relatively high midway between the sources. Observed differences in salt accumulation below the side slopes of the furrow were no doubt caused by additional leaching during furrow irrigation. Of course, the more saline irrigation treatment accounted for the highest salt concentration in each case. Because the salt distributions were determined after only one irrigation season for soil that had previously been well leached, salt accumulation had not yet developed deeper in the soil profile.

5.3 CROP SALT TOLERANCE

5.3.1 Plant Response to Salinity

The predominate influence of salinity on plants is growth suppression (e.g., Fig. 5.7). As salinity increases, plant growth suppression increases, until the plant dies. Growth suppression typically is a nonspecific salt effect, depending more on osmotic stress created by the total concentration of soluble salts than on the levels of specific ions. Excessive concentrations of some individual ions, as discussed later, can also be detrimental to plants. Typical-

Growth reduction of cotton for salinity levels of 1, 14, 28, and 42 dS/m (left to right)

Growth reduction of pea caused by salinity levels of 1, 7, 11, and 14 dS/m (left to right)

Yield reduction of sweet corn caused by salinity levels 1, 6, 10, and 14 dS/m (left to right)

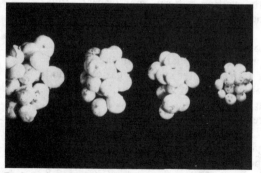

Decline in onion yields due to salinity levels of 1, 6, 11, and 17 dS/m (left to right)

FIG. 5.7 Influence of root-medium salinity [EC_e] on growth or yield of several crops.

FIG. 5.8 Influence of soil salinity on oat seedlings. Mean root zone salinity (EC_e) is 5 dS/m on the top and 16 dS/m on the bottom.

ly, growth decreases linearly as salinity increases beyond a threshold salinity level, and the effect is similar whether salinity is increased by raising the concentration of nutrient ions or by adding non-nutrient salts, such as sodium chloride, sodium sulfate, or calcium chloride, which are common in saline soils (Hayward and Long, 1941; Gauch and Eaton, 1942).

Salinity retards cell enlargement and cell division, the production of proteins and nucleic acids, and the rate of increase in plant mass (Maas and Nieman, 1978). Although salinity affects plants in many ways physiologically, visible injury symptoms, such as leaf marginal burn or necrosis, seldom occur except under extreme salination. An example of visible salinity damage to oats is shown in Fig. 5.8. Crop yields have been reduced drastically if symptoms are visible. Without visible symptoms, salinity may still cause morphological and anatomical changes, some of which undoubtedly help sustain favorable plant-water relations and improve the chances of plant survival. These changes include fewer and smaller leaves, fewer stomata per unit leaf area, increased succulence, thickening of both leaf cuticles and surface layers of wax, reduced differentiation and development of vascular tissues, and earlier lignification of roots (Poljakoff-Mayber, 1975 and Strogonov, 1962). Unfortunately, experimental evidence for these structural responses is inadequate, but it indicates that responses vary with plant species and with the type of salinity present.

5.3.2 Factors Influencing Salt Tolerance

Growth Stage. Many crops seem to tolerate salinity equally well during seed germination and later growth stages (Bernstein and Hayward, 1958). Germination failures that occur frequently on saline soils are not normally due to crops being especially sensitive during germination, but rather to exceptionally high concentrations of salt where the seeds are planted. Salt can accumulate at the soil surface as a consequence of upward water movement and evaporation.

The salt tolerance of some crops, however, does change with growth stage (Maas and Hoffman, 1977). For example, barley, wheat, and corn are more sensitive to salinity during early seedling growth than during germination or larger growth stages, while sugarbeet and safflower are relatively sensitive during germination. The tolerance of soybean may either increase or decrease from germination to maturity, depending on the variety.

Varieties and Rootstocks. Varietal differences, while not common, must also be considered. Although most known varietal differences are among grass species (i.e., bermudagrass, bromegrass, creeping bentgrass, barley, rice, and wheat), some variation has been noted among legumes (birdsfoot trefoil, soybean, and berseem clover) (Maas and Hoffman, 1977). Increased varietal differences in salt tolerance are anticipated in the future as many new varieties are being developed from a more diverse genetic base.

Rootstock differences are an important factor affecting the salt tolerance of tree and vine crops. Varieties and rootstocks of avocado, grapefruit, and orange that differ in ability to absorb and transport sodium and chloride have different salinity tolerances (Cooper, 1951, 1961). Similar effects of rootstocks on salt accumulation and tolerance have been reported for stonefruit trees and grape.

Nutrition. Apparent salt tolerance may also vary with soil fertility. Crops grown on infertile soils generally have abnormally high apparent salt tolerances when compared with crops grown on fertile soils, because yields on nonsaline soil are severely limited by inadequate fertility. Obviously, proper fertilization would increase absolute crop yields even though apparent salt tolerance (i.e., relative yield) would be decreased.

Unless salinity causes specific nutritional imbalances, fertilization in excess of nutritional need generally has little effect on, or may even reduce, relative salt tolerance. Apparent decreases in salt tolerance accompanying excessive nitrogen fertilization have been reported for corn, cotton, rice, wheat, and spinach (Maas and Hoffman, 1977). Bernstein et al. (1974) concluded from sand culture studies that high nitrogen or potassium levels did not improve the salt tolerance of wheat, barley, corn, or six vegetable crops (garden beet, broccoli, cabbage, carrot, lettuce, and onion).

Irrigation Management. Irrigation practices can influence salinity damage. Assuming that plants respond primarily to the soil water salinity in that part of the root zone with the highest total water potential, then time-integrated salinity measured in the zone of maximum water uptake should correlate best with crop response. When irrigating frequently, this zone corresponds primarily to the upper part of the root zone where soil salinity is influenced primarily by the salinity of the irrigation water (Bernstein and François, 1973b). With infrequent irrigation, the zone of maximum water uptake becomes larger as the plant extracts water from increasingly saline solutions at greater depths. The decreasing soil matric potential during a typical irrigation cycle is an additional inhibitory factor, the effect of which appears additive to that of a decreased soil water osmotic potential (Wadleigh and Ayers, 1945, and Wadleigh et al., 1946).

Environment. Environmental factors, such as temperature, atmospheric humidity, and air pollution, markedly influence crop salt tolerance. Many crops seem less salt tolerant when grown in hot, dry environments than in cool, humid ones (Magistad et al., 1943). High atmospheric humidity alone tends to increase the salt tolerance of some crops, with high humidity generally benefiting salt-sensitive more than salt-tolerant crops (Hoffman and Rawlins, 1971; Hoffman and Jobes, 1978).

A strong interaction between the effects of ozone, a major air pollutant, and salinity has been found for pinto bean, garden beet, and alfalfa. At ozone concentrations often prevalent in agricultural settings near metropolitan areas, alfalfa yields may be increased by maintaining moderate, but not detrimental, salinity levels (Hoffman et al., 1975). Salinity also reduced ozone damage to pinto bean and garden beet, but effects were beneficial only at salinity and ozone levels too high for economical crop production. These results, however, indicate that salinity-ozone interactions may be commercially important for leafy vegetable and forage crops.

5.3.3 Salt Tolerance Evaluations

The relative salt tolerances of selected agricultural crops are given in Table 5.1, as taken from Maas and Hoffman (1977). The alphabetical crop list provides two essential parameters sufficient to evaluate salt tolerance: (a) the threshold salinity level (the maximum allowable salinity that does reduce yield measurably below that of a nonsaline control treatment) and (b) the percent yield decrease per unit of salinity increase beyond the threshold. All salinity values are reported as EC_e [the electrical conductivity of soil saturation extracts reported in units of dS/m and corrected for temperature to 25 °C] and rounded to two significant digits. A qualitative salt-tolerance rating is also given for quick, relative comparisons among crops. These ratings are defined by the boundaries shown in Fig. 5.9.

The data presented in Table 5.1 were normally obtained by artificially salinizing field plots and using cultural practices that closely simulated field conditions. Thus, they indicate the salt tolerance to be expected under normal growing conditions for that particular crop. Typically, salinity treatments were imposed at the seedling stage, so the data do not represent salt tolerance for the germination and early seedling growth stages. Soil salinity was maintained relatively uniform by irrigating frequently at a leaching fraction of about 0.5 with waters of a given salt concentration. Crop yields were correlated with EC_e of soil samples from the major part of the crop root zone.

In general, decreases in yield with increasing salinity depart from linearity only in the lower part of the relative yield curves where yields are commercially unacceptable anyway. With some crops, e.g., bean, onion, clover, and pepper, yield approaches zero asymptotically; with a few others, yields

TABLE 5.1. SALT TOLERANCE OF AGRICULTURAL CROPS AS A FUNCTION
OF SOIL SATURATION EXTRACT SALINITY (EC_e) WHERE RELATIVE YIELD (Y)
IN PERCENT = $100 - b(EC_e - a)$ (FROM MAAS AND HOFFMAN, 1977)

Crop	Salinity* at initial yield decline (threshold) (A)	Percent yield decrease per unit increase in salinity beyond threshold (B)	Qualitative salt tolerance rating†
	dS/m	%/(dS/m)	
Alfalfa			
Medicago sativa	2.0	7.3	MS
Almond			
Prunus dulcis	1.5	19	S
Apple			
Malus sylvestris	—	—	S
Apricot‡			
Prunus armeniaca	1.6	24	S
Avocado‡			
Persea americana	—	—	S
Barley (forage) §			
Hordeum vulgare	6.0	7.1	MT
Barley (grain) §			
Hordeum vulgare	8.0	5	T
Bean			
Phaseolus vulgaris	1.0	19	S
Beet, garden‖			
Beta vulgaris	4.0	9	MT
Bentgrass			
Agrostis palustris	—	—	MS
Bermudagrass#			
Cynodon Dactylon	6.9	6.4	T
Blackberry			
Rubus spp.	1.5	22	S
Boysenberry			
Rubus ursinus	1.5	22	S
Broadbean			
Vicia Faba	1.6	9.6	MS
Broccoli			
Brassica oleracea botrytis	2.8	9.2	MS
Bromegrass			
Bromus inermis	—	—	MT
Cabbage			
Brassica oleracea capitata	1.8	9.7	MS
Canary grass, reed			
Phalaris arundinacea	—	—	MT
Carrot			
Daucus Carota	1.0	14	S
Clover, alsike, ladino, red, strawberry			
Trifolium spp.	1.5	12	MS
Clover, berseem			
T. alexandrinum	1.5	5.7	MS
Corn (forage)			
Zea Mays	1.8	7.4	MS
Corn (grain)			
Zea Mays	1.7	12	MS
Corn, sweet			
Zea Mays	1.7	12	MS
Cotton			
Gossypium hirsutum	7.7	5.2	T
Cowpea			
Vigna unguiculata	1.3	14	MS
Cucumber			
Cucumis sativus	2.5	13	MS
Date palm			
Phoenix dactylifera	4.0	3.6	T

TABLE 5.1. SALT TOLERANCE OF AGRICULTURAL CROPS AS A FUNCTION
OF SOIL SATURATION EXTRACT SALINITY (EC_e) WHERE RELATIVE YIELD (Y)
IN PERCENT = $100 - b(EC_e - a)$ (FROM MAAS AND HOFFMAN, 1977)

Crop	Salinity* at initial yield decline (threshold) (A)	Percent yield decrease per unit increase in salinity beyond threshold (B)	Qualitative salt tolerance rating†
	dS/m	%/(dS/m)	
Fescue, tall			
Festuca elatior	3.9	5.3	MT
Flax			
Linum usitatissimum	1.7	12	MS
Grape ‡			
Vitis spp.	1.5	9.6	MS
Grapefruit ‡			
Citrus x paradisi	1.8	16	S
Hardinggrass			
Phalaris tuberosa	4.6	7.6	MT
Lemon ‡			
Citrus limon	—	—	S
Lettuce			
Lactuca sativa	1.3	13	MS
Lovegrass**			
Eragrostis spp.	2.0	8.4	MS
Meadow Foxtail			
Alopecurus pratensis	1.5	9.6	MS
Millet, Foxtail			
Setaria italica	—	—	MS
Okra			
Abelmoschus esculentus	—	—	S
Olive			
Olea europaea	—	—	MT
Onion			
Allium Cepa	1.2	16	S
Orange			
Citrus sinensis	1.7	16	S
Orchardgrass			
Dactylis glomerata	1.5	6.2	MS
Peach			
Prunus Persica	1.7	21	S
Peanut			
Arachis hypogaea	3.2	29	MS
Pepper			
Capsicum annum	1.5	14	MS
Plum ‡			
Prunus domestica	1.5	18	S
Potato			
Solanum tuberosum	1.7	12	MS
Radish			
Raphanus sativus	1.2	13	MS
Raspberry			
Rubus idaeus	—	—	S
Rhodesgrass			
Chloris Gayana	—	—	MS
Rice, paddy §			
Oryza sativa	3.0	12	MS
Ryegrass, perennial			
Lolium perenne	5.6	7.6	MT
Safflower ‡			
Carthamus tinctorius	—	—	MT
Sesbania §			
Sesbania exaltata	2.3	7	MS
Sorghum			
Sorghum bicolor	—	—	MS

TABLE 5.1. SALT TOLERANCE OF AGRICULTURAL CROPS AS A FUNCTION
OF SOIL SATURATION EXTRACT SALINITY (EC_e) WHERE RELATIVE YIELD (Y)
IN PERCENT = 100 – b(EC_e – a) (FROM MAAS AND HOFFMAN, 1977)

Crop	Salinity* at initial yield decline (threshold) (A)	Percent yield decrease per unit increase in salinity beyond threshold (B)	Qualitative salt tolerance rating†
	dS/m	%/(dS/m)	
Soybean			
Glycine Max	5.0	20	MT
Spinach			
Spinacia oleracea	2.0	7.6	MS
Strawberry			
Fragaria spp.	1.0	33	S
Sudangrass			
Sorghum sudanense	2.8	4.3	MT
Sugarbeet‖			
Beta vulgaris	7.0	5.9	T
Sugarcane			
Saccharum officinarum	1.7	5.9	MS
Sweet potato			
Ipomoea Batatas	1.5	11	MS
Timothy			
Phleum pratense	—	—	MS
Tomato			
Lycopersicon, Lycopersicum	2.5	9.9	MS
Trefoil, Big			
Lotus uliginosus	2.3	19	MS
Trefoil, Birdsfoot narrowleaf‡‡			
L. corniculatus tenuifolium	5.0	10	MT
Vetch			
Vicia sativa	3.0	11	MS
Wheat §, ††			
Triticum aestivum	6.0	7.1	MT
Wheatgrass, crested			
Agropyron desertorum	3.5	4	MT
Wheatgrass, fairway			
A. cristatum	7.5	6.9	T
Wheatgrass, slender			
A. trachycaulum	—	—	MT
Wheatgrass, tall			
A. elongatum	7.5	4.2	T
Wildrye, Altai			
Elymus angustus	—	—	T
Wildrye, beardless			
E. triticoides	2.7	6	MT
Wildrye, Russian			
E. junceus	—	—	T

*   Salinity expressed as EC_e (dS/m = decisiemens per meter = 1 millimho/cm, referenced to 25 °C).

†   Ratings are defined by the boundaries in Fig. 5.9.

‡   Tolerance is based on growth rather than yield.

§   Less tolerant during emergence and seedling stages. EC_e should not exceed 4 to 5 dS/m at these times.

‖   Sensitive during germination. EC_e should not exceed 3 dS/m at this time for garden beet and sugarbeet.

\#   Average of several varieties. Suwannee and Coastal are about 20 percent more tolerant, and Common and Greenfield are about 20 percent less tolerant than the average.

**   Average for Boer, Wilman, Sand, and Weeping varieties. Lehmann seems about 50 percent more tolerant than the other varieties tested.

††   The salt tolerance of some semidwarf varieties may be higher.

‡‡   Broadleaf birdsfoot trefoil seems less tolerant than narrowleaf trefoil.

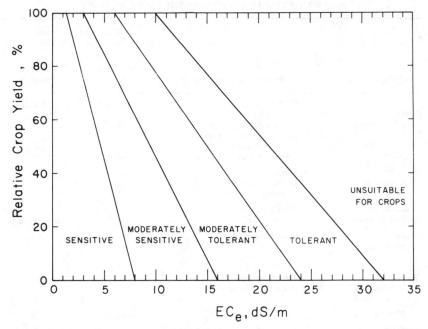

FIG. 5.9 Division boundaries for qualitative salt tolerance ratings of agricultural crops. (Maas and Hoffman, 1977).

decrease linearly with increasing salinity to a point beyond which yields drop sharply to zero as the plants die. Salinity at zero yield is of interest because it may be used to estimate the maximum salinity that a crop can tolerate (Bernstein and Francois, 1973b; van Schilfgaarde et al., 1974).

Relative yield (Y) in percent at any given soil salinity (EC_e) can be calculated by the equation

$$Y = 100 - b(EC_e - a), \dotfill [5.4]$$

where a is the salinity threshold value and b is the yield decrease per unit salinity increase, as given in Table 5.1. For example, alfalfa yields decrease approximately 7.3 percent per dS/m when the soil salinity exceeds 2.0 dS/m. Therefore, at a soil salinity of 5.4 dS/m, the relative yield, $Y = 100 - 7.3$ $(5.4 - 2.0) = 75$ percent.

Division boundaries for the salt tolerance ratings defined in Fig. 5.9 were chosen to approximate the family of linear curves that represent most of the crops reported. Four divisions were labeled to correspond with previously published terminology ranging from sensitive to tolerant. With few exceptions, the linear salt tolerance curves for each crop remained within a single division. Where the linear salt tolerance curve for a crop happened to cross division boundaries, the crop was rated based on its relative tolerance at the lower salinity levels.

5.3.4 Specific Ion Effects

Toxicity. Unlike most annual crops, trees and other woody perennials may be specifically sensitive to ions, such as chloride and sodium, which are taken up with soil water, move with the plant transpiration stream, and accumulate in the leaves. Crop, varietal, and rootstock differences in tol-

TABLE 5.2. CHLORIDE TOLERANCES OF FRUIT CROP ROOTSTOCKS
AND VARIETIES IF LEAF INJURY IS TO BE AVOIDED

Crop	Rootstock or variety	Maximum permissible chloride concentration	
		Soil saturation extract*	Plant leaf analysist†
	Rootstocks	mol/m^3	%(g/g)
Citrus	Rangpur lime,		
Citrus spp.	mandarin	25	0.7
	Rough lemon, tangelo		
	sour orange	15	
	Sweet orange, citrange	10	
Stone fruit	Marianna	25	0.3
Prunus spp.	Lovell, Shalil	10	
	Yunnan	7	
Avocado	West Indian		
Persea americana	Mexican	8	0.25 to 0.50
Grape	Salt creek, 1613-3	40	0.5
Vitis spp.	Dog ridge	30	
	Varieties		
Grape	Thompson seedless, Perlette	25	0.5
Vitis spp.	Cardinal, Black rose	10	
Olive			
Olea europaea			0.5
Berries‡			
Rubus spp.	Olallie blackberry	10	
	Indian summer raspberry	5	
Strawberry	Lassen	8	
Fragaria spp.	Shasta	5	

*From Bernstein (1965).
†From Reisenauer (1976).
‡Data available for a single variety of each crop only.

erances to chloride and sodium depend largely upon the rate of transport of these ions from the soil to the leaves. In general, the slower the chloride absorption, the more tolerant the plant is to this solute.

Leaf injury symptoms appear in chloride-sensitive crops when leaves accumulate about 0.3 to 0.5 percent chloride on a dry-weight basis. Symptoms develop as leaf burn or drying of leaf tissue, typically occurring first at the extreme tip of older leaves and progressing back along the leaf edges. Excessive leaf burn is often accompanied by defoliation. Chemical analysis of soil or leaves can be used to confirm probable chloride toxicity. The maximum permissible concentrations of chloride in the soil saturation extract or in plant leaves for several sensitive crops are given in Table 5.2.

Symptoms of sodium toxicity occur first on older leaves as a burning or drying of tissue at the outer edges of the leaf. As severity increases, the affected zone progresses toward the center of the leaf between the veins. For many tree crops, sodium concentrations in leaf tissue of 0.2 to 0.5 percent (dry-weight basis) are the maximum allowable without injury symptoms. Sodium toxicity is often modified and reduced if calcium is present. Because of this interaction, a reasonable evaluation of the potential toxicity is given by the exchangeable-sodium-percentage (ESP) of the soil or the sodium-adsorption-ratio (SAR) of soil extracts or irrigation waters (U.S. Salinity Laboratory Staff, 1954). The tolerance of representative crops to sodium is given as a function of soil ESP in Table 5.3.

TABLE 5.3. TOLERANCE OF VARIOUS CROPS TO EXCHANGEABLE-SODIUM-
PERCENTAGE (ESP) UNDER NONSALINE CONDITIONS (PEARSON, 1960)

Tolerance to ESP and range at which affected	Crop	Growth response under field conditions
Extremely sensitive (ESP = 2-10)	Deciduous fruits Nuts Citrus *Citrus* spp. Avocado *Persea americana*	Sodium toxicity symptoms even at low ESP values
Sensitive (ESP = 10-20)	Bean *Phaseolus vulgaris*	Stunted growth at these ESP values even though soil physical conditions may be good
Moderately tolerant (ESP = 20-40)	Clover *Trifolium* spp. Oat *Avena sativa* Tall fescue *Festuca elatior* Rice *Oryza sativa* Dallisgrass *Paspalum dilatatum*	Stunted growth due to nutritional factors and adverse soil conditions
Tolerant (ESP = 40-60)	Wheat *Triticum aestivum* Cotton *Gossypium hirsutum* Alfalfa *Medicago sativa* Barley *Hordeum vulgare* Tomato *Lycopersicon* Beet *Beta vulgaris*	Stunted growth usually due to adverse soil physical conditions
Most Tolerant (ESP = more than 60)	Crested and Fairway wheatgrass *Agropyron* spp. Tall wheatgrass, *Agropyron elongatum* Rhodesgrass *Chloris gayana*	Stunted growth usually due to adverse soil physical conditions

Irrigation methods that wet plant leaves, such as overhead sprinkling, may cause specific ion toxicity problems at sodium or chloride concentrations lower than those that cause problems with surface irrigation methods. This occurs primarily during periods of high temperature and low humidity. Excess chloride and sodium can accumulate in leaves by foliar absorption, and the more frequent the wetting and drying cycles, the greater the leaf damage. Citrus, sprinkler irrigated in several California valleys, has been damaged with chloride and sodium concentrations as low as 3 mol/m^3. These same concentrations caused no toxic effects with furrow or flood irrigation (Harding et al., 1958). Slight damage has been reported for more tolerant crops, such as alfalfa, when sprinkled under extremely high evaporative conditions with water containing chloride and sodium concentrations in excess of 6 mol/m^3. In contrast, waters with sodium and chloride levels of 24 and 37 mol/m^3, respectively, have caused little or no damage when evaporation rates are low (Nielson and Cannon, 1975).

Boron, although an essential minor element, is phytotoxic if present in excess. Most boron toxicity problems arise from high concentrations in well waters or springs located near geothermal areas or geological faults. Few surface waters contain sufficient boron to cause toxicity. Sensitivity to boron is not limited to woody perennials, but affects a wide variety of crops. Boron toxicity symptoms, illustrated in Fig. 5.10 for grape, typically appear at the tip and along the edges of older leaves as yellowing, spotting, and/or drying of leaf tissue. The damage gradually progresses interveinally toward midleaf. A gummosis or exudate on limbs or trunks is sometimes noticeable on boron-affected trees, such as almond. Many sensitive crops show toxicity symptoms when boron concentrations in leaf blades exceed 250 mg/kg, but not all sensitive crops accumulate boron in their leaves.

FIG. 5.10 Comparisons between normal grape leaves (top) and those showing boron toxicity (Eaton, 1935).

Stonefruits (e.g., peach, plum, almond) and pome fruits (pear, apple, and others) may not accumulate boron in leaf tissue to the extent that leaf analysis is a reliable toxicity indicator.

A wide range of crops have been tested for boron tolerance in sand cultures. The results of these tests are summarized in Table 5.4. The crops have been grouped according to their relative tolerance to boron in the irrigation water. These data were based on the boron level at which toxicity symptoms were observed, and do not necessarily indicate corresponding reductions in yield.

Many trace elements are toxic to plants at extremely low concentrations. Fortunately, irrigation waters derived from surface and ground waters do not normally contain harmful levels of these elements. An exception is the occurrence of selenium in some well waters in Wyoming (Miller, 1956). Municipal and industrial waste waters are more likely to contain toxic concentrations of trace elements. Little information is available on plant tolerances to trace elements. Only suggested maximum concentrations for irrigation waters have been reported. The concentrations given in Table 5.5 are based on the indefinite protection of soils for crop production under con-

TABLE 5.4. RELATIVE TOLERANCE OF CROPS TO BORON*
(ADAPTED FROM WILCOX, 1960)

Tolerant† 4.0 mg/L of boron	Semitolerant 2.0 mg/L of boron	Sensitive 1.0 mg/L of boron
Asparagus *Asparagus officinalis*	Sunflower, native *Helianthus annus*	Pecan *Carya illinoinensis*
Date palm *Phoenix dactylifera*	Potato *Solanum tuberosum*	Walnut, black and Persian or English *Juglans spp.*
Sugarbeet *Beta vulgaris*	Cotton, Acala and Pima *Gossypium spp.*	Jerusalem artichoke *Helianthus tuberosus*
Garden beet *Beta vulgaris*	Tomato *Lycopersicon, Lycopersicum*	Navy bean *Phaseolus vulgaris*
Alfalfa *Medicago sativa*	Radish *Raphanus sativus*	Plum *Prunus domestica*
Broadbean *Vicia Faba*	Field pea *Pisum sativum*	Pear *Pyrus communis*
Onion *Allium Cepa*	Olive *Olea europaea*	Apple *Malus sylvestris*
Turnip *Brassica Rapa*	Barley *Hordeum vulgare*	Grape *Vitis spp.*
Cabbage *Brassica oleracea capitata*	Wheat *Triticum aestivum*	Kadota fig *Ficus carica*
Lettuce *Lactuca sativa*	Corn *Zea Mays*	Persimmon *Diospyros virginiana*
Carrot *Daucus Carota*	Sorghum *Sorghum bicolor*	Cherry *Prunus spp.*
	Oat *Avena sativa*	Peach *Prunus Persica*
	Pumpkin *Cucurbita spp.*	Apricot *Prunus Armeniaca*
	Bell pepper *Capsicum annum*	Thornless blackberry *Rubus spp.*
	Sweetpotato *Ipomoea Batatas*	Orange *Citrus sinensis*
	Lima bean *Phaseolus lunatus*	Avocado *Persea americana*
		Grapefruit
		Lemon *Citrus limon*
2.0 mg/L of boron	1.0 mg/L of boron	0.3 mg/L of boron

*Relative tolerance is based on the boron concentration in irrigation water at which boron toxicity symptoms were observed when plants were grown in sand culture. It does not necessarily indicate a reduction in crop yield.
†Tolerance decreases in descending order in each column between the stated limits (adapted from Wilcox, 1960).

tinuous use of degraded water. Slightly higher values may be used for short times, particularly for soils having the capability to inactivate the trace element. For an in-depth discussion of the effects of such trace element concentrations, the reader is referred to other publications (National Academy of Sciences, 1972; Pratt, 1972).

Nutritional Disturbance. The concentrations of some ions in saline soils may be as much as three orders of magnitude greater than those of several of the essential nutrients. With differences of this magnitude, it is surprising that plant nutritional disturbances are not more common in salt-

TABLE 5.5. RECOMMENDED MAXIMUM CONCENTRATIONS OF TRACE ELEMENTS
IN IRRIGATION WATERS THAT ARE TO BE USED FOR SENSITIVE CROPS
ON SOILS WITH LOW CAPACITIES TO RETAIN THE ELEMENTS
IN UNAVAILABLE FORMS (FROM PRATT, 1972)

Element*	Concentration	Element	Concentration
	mg/L		mg/L
Aluminum	5.0	Lead	5.0
Arsenic	0.1	Lithium†	2.5
Beryllium	0.1	Manganese	0.2
Cadmium	0.01	Molybdenum	0.01
Chromium	0.1	Nickel	0.2
Cobalt	0.05	Selenium	0.02
Copper	0.2	Vanadium	0.1
Fluoride	1.0	Zinc	2.0
Iron	5.0		

*These levels normally will not adversely affect either plants or soils. No data were available for mercury, silver, tin, titanium, or tungsten.
†Recommended maximum concentration for irrigated citrus is 0.075 mg/L (Bingham et al., 1964).

affected soils. However, with the exception of the disturbance of orthophosphate utilization (Nieman and Clark, 1976), salinity has not been found to influence the utilization of essential nutrients by plants (Bernstein, 1964; Bernstein et al., 1974).

In some instances, however, if the proportion of calcium to other cations becomes either extremely high or low, nutritional imbalances can occur that reduce crop yield below that expected from osmotic effects alone (Bernstein, 1964). Bean plants, which avidly accumulate calcium, cannot tolerate excess calcium. On the other hand, calcium deficiency under some saline conditions results in blossom-end rot of tomatoes, internal browning of lettuce, and reduced corn growth. Because soil salinity in the field normally involves a mixture of salts, the effects of specific ions on crop nutrition tend to be minimized so that the osmotic effect usually predominates.

5.3.5 Effects of Exchangeable Sodium

Although many tree crops are sensitive to sodium, the major concern from high concentrations of sodium in soils and irrigation waters is the eventual deterioration of soil structure, resulting in decreased water infiltration and hydraulic conductivity. When calcium and/or magnesium are the predominant cations adsorbed on the soil exchange complex, the soil tends to have a granular structure that is easily tilled and readily permeable. When the amount of adsorbed sodium exceeds 10 to 15 percent of the total cation-exchange-capacity (CEC), however, soil mineral particles tend to disperse and hydraulic conductivity decreases. Because drainage is essential for salinity management and reclamation, good soil structure must be maintained.

Permeability becomes a problem when the rate of soil water infiltration is reduced to the point that the crop is not adequately supplied with water. Sodium may also add to cropping difficulties through crusting of seed beds, temporary saturation of surface soil, and/or possible disease, weed, oxygen, nutritional, and salinity problems. Water of very low salt content can aggravate a sodium problem because it allows a maximal swelling and dispersion of soil minerals and organic matter, and also has a tremendous capacity to dissolve and remove calcium. The sodium-adsorption-ratio (SAR) of soil extracts is generally a good indicator of the exchangeable sodium status of the soil and it can be determined more readily than ESP. The SAR is defined as

$$SAR = Na/\sqrt{(Ca + Mg)}, \dotfill \quad [5.5]$$

where all ion concentrations are in mol/m^3. For many soils, the SAR of the soil saturation extract is approximately equal to the ESP below ESP values of 25 or 30. This covers the range of greatest concern for irrigated agriculture. In some cases, SAR of the irrigation water has not proved a satisfactory guide to potential soil permeability problems because of the influence of carbonates and bicarbonates on the precipitation of calcium and magnesium. Thus the relationship,

$$adj\ SAR = \frac{Na}{\sqrt{(Ca + Mg)}} [1 + (8.4 - pH_c)] \dotfill \quad [5.6]$$

was developed (Rhoades, 1972) to estimate the permeability hazard of irrigation waters, taking into account the relative concentrations of sodium, calcium, magnesium, carbonate, and bicarbonate. All ion concentrations for the relationship are in mol/m^3. Values for pH_c, a theoretical pH for the water if in contact with solid-phase lime at atmospheric levels of carbon dioxide, can be obtained from Table 5.6.

Most crops show small or only moderate losses in yield when favorable soil structure is maintained in the presence of high levels (15 to 30 percent) of exchangeable sodium (Bernstein and Pearson, 1956; Pearson and Bernstein, 1958). In general, higher ESP levels can be tolerated in coarse-textured than in fine-textured soils. Unfortunately, no economically feasible amendment or conditioner is available to maintain favorable structure in fine-textured sodic soils containing swelling clays.

5.4 IRRIGATION WATER QUALITY

5.4.1 Composition of Irrigation Waters

The quality of an irrigation water is judged not only by the total concentration of ions, but also by the individual ions present. The most common cations in irrigation water are calcium, magnesium, sodium, and potassium, and the most common anions are bicarbonate, sulfate, and chloride. Other solutes, such as nitrate, carbonate, and trace elements, may also affect water quality in some instances. Irrigation water may contain significant concentrations of nitrate in localized areas. High nitrate levels are beneficial to crop production in most instances, but can potentially degrade ground-water quality. Appreciable amounts of carbonate are present only at pH values above about 8.5. Trace elements, such as boron, selenium, or lithium, are not common in irrigation waters. However, minute concentrations can severely limit production of certain crops.

Ion concentrations are typically reported as moles per cubic meter of solution (mol/m^3) or as milligrams per liter (mg/L). The latter is numerically equivalent to parts per million (ppm), an outmoded unit. To convert from mol/m^3 to mg/L, multiply the mol/m^3 by the atomic weight of the

TABLE 5.6. VALUES FOR CALCULATING pH_c, THE
THEORETICAL pH OF WATER IN CONTACT WITH LIME AT
ATMOSPHERIC LEVELS OF CARBON DIOXIDE (FROM
AYERS AND WESCOT, 1976). THE VALUE OF pH_c IS
EQUIVALENT TO THE SUM OF $(pk_2' - pk_c')$, p(Ca + Mg),
AND p(Alk)

Concentration sum	$pk_2' - pk_c'$*	p(Ca + Mg)†	p(Alk)‡
mol/m^3			
0.05	2.0	4.6	4.3
0.10	2.0	4.3	4.0
0.15	2.0	4.1	3.8
0.20	2.0	4.0	3.7
0.30	2.0	3.8	3.5
0.40	2.0	3.7	3.4
0.50	2.1	3.6	3.3
0.75	2.1	3.4	3.1
1.5	2.1	3.1	2.8
2.0	2.2	3.0	2.7
4.0	2.2	2.7	2.4
6.0	2.2	2.5	2.2
8.0	2.3	2.4	2.1
10	2.3	2.3	2.0
15	2.3	2.1	1.8
20	2.4	2.0	1.7
30	2.4	1.8	1.5
50	2.4	1.6	1.3
80	2.5	1.4	1.1

*$(pk_2' - pk_c')$ is evaluated from the concentration sum of Ca + Mg
+ Na in mol/m^3.
†p(Ca + Mg) is obtained from using the concentration sum of Ca
+ Mg in mol/m^3.
‡p(Alk) is evaluated from the concentration sum of CO_3 + HCO_3
in mol/m^3.

ion (given for convenience in Table 5.7). Total salt concentrations are often
expressed in terms of the electrical conductivity (EC) of the solution in units
of millimhos/cm (mmho/cm) or deciSiemens per meter (dS/m). In addition,
the sodium-adsorption-ratio (SAR) or an appropriately adjusted SAR (adj.
SAR) may be calculated to assess the sodium hazard of the water (Rhoades,
1972). The determinations required to assess a water's suitability for irrigation are summarized in Table 5.7, along with their symbols and units
of measure.

The quantities and types of salts present in waters proposed for irrigation use vary widely. Water sources include rainfall, rivers, groundwaters,
lakes, brackish waters, and waste waters. Rainfall, rivers, and groundwaters are the major sources of irrigation water; but, as supplemental irrigation and the disposal of municipal and industrial waste waters on land
increase, the other sources will become significant in selected areas.

Even though rainwaters have the lowest salt content of any group of
natural waters, they are not completely free of dissolved salts. Salts present
in minute amounts in the atmosphere from terrestrial and marine sources
are removed and concentrated by rainfall. As oceanic air masses move inland, atmospheric salinity decreases rapidly, with typical salt concentrations of rainfall decreasing from 40 mg/L along the coast to only a few
mg/L in the continental interiors (Junge and Werby, 1958).

TABLE 5.7. DETERMINATIONS REQUIRED TO EVALUATE
WATER QUALITY FOR IRRIGATION

Determination	Symbol	Unit of measure	Atomic weight
Total Salt Content			
(1) electrical conductivity	EC	dS/m*	—
or (2) concentration	C	mg/L or mol/m^3	—
Sodium Hazard			
(1) sodium adsorption ratio†	SAR		
(2) adjusted sodium			
adsorption ratio‡	adj SAR	—	
Constituents			
(1) cations			
calcium	Ca	mol/m^3	40.1
magnesium	Mg	mol/m^3	24.3
sodium	Na	mol/m^3	23.0
potassium §	K	mol/m^3	39.1
(2) anions			
bicarbonate	HCO$_3$	mol/m^3	61.0
sulphate	SO$_4$	mol/m^3	96.1
chloride	Cl	mol/m^3	35.5
carbonate §	CO$_3$	mol/m^3	60.0
nitrate §‖	NO$_3$	mg/L	62.0
Trace elements §‖			
boron	B	mg/L	10.8
Acidity or Alkalinity	pH	—	—

*1 millimho/cm, referenced to 25 °C.

†SAR = $Na/\sqrt{(Ca + Mg)}$.

‡Adjusted SAR = $\left[Na/\sqrt{(Ca + Mg)}\right] \left[1 + (8.4 - pH_c)\right]$.

§ Constituent that may be important in special situations only.

‖ Because of its low concentration, this constituent is usually expressed on a mg/L basis.

The average salt content of rivers, worldwide, is estimated at about 120 mg/L (Livingstone, 1963), but the range is wide. Rivers used for irrigation generally have above-average salt contents because they are in arid areas, although some in the Pacific Northwest and California remain low in salinity for much of their lengths. Compositions of representative river waters are given in Table 5.8. Specific ions in river waters also vary widely, but trends in composition are evident. As an example, the effect of irrigation diversions and return flows on river water composition is illustrated for the Rio Grande River in Table 5.9. The data are for the stretch of river from near Santa Fe, New Mexico to south of El Paso, Texas. Salinity, boron, and the proportions of sodium and chloride increase markedly with distance down the river, whereas the proportions of calcium and bicarbonate concurrently decrease. The proportions of magnesium and sulfate remain nearly constant. Consequently, the adjusted SAR tends to increase as total salt concentration of river water increases. Similar trends are observed for many rivers in arid regions, although the magnitudes of the effects vary from river to river.

Typical water analyses of wells used for irrigation are also given in Table 5.8. Groundwaters are generally more saline than surface waters, and commonly contain higher proportions of sodium, boron, and nitrate. Changes in water quality with pumping duration are not common, but quality frequently differs significantly among relatively closely spaced wells. Such

TABLE 5.8. WATER QUALITY OF REPRESENTATIVE IRRIGATION WATERS

Water Source and location	Total Dissolved Salts EC (mS/m*)	Total concentration (mg/L)	Total concentration (mol_c/m³†)	B (mg/L)	Constituents (mol_c/m³) Ca^{2+}	Mg^{2+}	Na^+	K^+	CO_3^{2-} & HCO_3^-	SO_4^{2-}	Cl^-	adj SAR $(mol/m^3)^{1/2}$
Individual Rivers‡												
San Joaquin, Biola, CA	6	53	0.6	0.06	0.2	0.1	0.2	0.1	0.4	0.1	0.1	0.1
Feather, Nicolaus, CA	9	70	0.9	0.01	0.4	0.3	0.1	0.1	0.7	0.1	0.1	0.1
Columbia, Canadian Border	15	87	1.6	—	1.1	0.4	0.1	0.0	1.3	0.3	0.0	0.2
Sacramento, Knights Landing, CA	15	111	1.6	0.05	0.7	0.5	0.4	0.0	1.2	0.2	0.2	0.6
Snake, King Hill, ID	50	312	5.3	0.04	2.3	1.6	1.3	0.1	3.5	1.1	0.7	2.0
Missouri, Williston, ND	65	426	6.9	0.11	2.9	1.6	2.3	0.1	2.8	3.8	0.3	2.8
Rio Grande, Falcon Dam, TX	67	419	6.7	0.15	3.0	1.0	2.6	0.1	2.4	2.4	1.9	3.5
South Platte, Julesburg, CO	120	910	14.2	0.23	6.6	2.8	4.5	0.3	3.8	9.4	1.0	5.0
Colorado, Yuma, AZ	140	914	14.3	0.18	5.2	2.6	6.4	0.2	2.8	7.6	4.0	7.1
Salt, Stewart Mtn Dam, AZ	140	755	12.9	0.14	1.2	2.6	8.9	0.2	2.8	1.0	9.1	12.0
Arkansas, John Martain Dam, CO	140	1130	16.6	0.15	7.1	4.3	5.1	0.1	2.9	12.8	0.9	4.9
Pecos, Artesia, NM	340	2530	39.9	—	20.4	6.2	13.3	0.0	2.3	23.8	13.8	9.1
Gila, Gillespie Dam, AZ	740	5120	83.3	0.28	17.0	12.0	53.1	1.2	5.5	28.1	49.7	40.0
Well Water§												
Indio, CA	30	205	3.0	0.02	1.4	0.3	1.3	0.0	2.2	0.6	0.2	2.0
Bakersfield, CA	79	500	7.5	6.93	0.2	0.0	7.3	0.0	2.5	2.5	2.5	12.0
Scottsdale, AZ	120	718	12.7	—	3.2	2.7	6.8	—	3.2	3.1	6.2	8.3
Tolleson, AZ	400	2878	43.8	—	6.4	4.1	33.3	—	10.8	8.1	24.9	38.0
Pecos, TX	440	2793	47.9	—	9.3	16.1	21.8	0.7	1.9	16.9	29.1	14.0
Roll, AZ	720	4535	74.8	1.28	14.8	11.7	48.0	0.3	6.3	13.1	55.4	36.9

*mS/m = millisiemens per meter = 0.01 millimho/cm, referenced to 25°C.
†mol_c/m^3 = moles of charge per cubic meter = milliequivalents per liter.
‡Data adapted from U.S. Geological Survey (1967 and 1974).
§Data taken from Wilcox (1948), Longenecker and Lyerly (1959), and Smith et al. (1964).

TABLE 5.9. CHANGE IN WATER QUALITY OF THE RIO GRANDE RIVER WITH DISTANCE, 1963 ANNUAL MEANS (ADAPTED FROM WILCOX, 1968)

Location	Distance downstream	Total Dissolved Salts			B	Constituents							adj SAR
		EC	Total concentration			Ca^{2+}	Mg^{2+}	Na^+	K^+	CO_3^{2-} & HCO_3^-	SO_4^{2-}	Cl^-	
	km	mS/m*	mg/L	mol_c/m^3†	mg/L	mol_c/m^3							$(mol/m^3)^{1/2}$
Otowi Bridge, NM	0	37	151	3.8	0.03	2.2	0.6	1.0	0.0	2.3	1.3	0.2	1.5
Elephant Butte, NM	386	63	360	6.2	0.11	2.7	0.9	2.5	0.1	2.6	2.6	1.0	3.4
Caballo Dam, NM	431	69	398	6.8	0.13	2.9	0.9	2.9	0.1	2.8	2.6	1.5	4.2
Leesburg Dam, NM	512	80	471	8.0	0.12	3.4	1.1	3.4	0.1	3.0	3.2	1.9	4.5
El Paso, TX	604	130	812	13.4	0.23	4.6	1.5	7.2	0.1	3.7	5.9	4.0	9.1
Fort Quitman, TX	734	580	3734	62.5	0.62	15.6	7.0	39.7	0.2	4.8	18.5	39.2	30.7

*mS/m = millisiemens per meter = 0.01 millimho/cm, referenced to 25 °C.
†mol_c/m^3 = moles of charge per cubic meter = milliequivalents per liter.

changes reflect the composition of different strata from which water is being pumped. For example, wells within 1 km of each other in the Coachella Valley of California have been reported to have salt concentrations varying from 0.6 to 12 dS/m.

5.4.2 Suitability of Waters for Irrigation

Both the quantity and quality of water supply are important for irrigation. A water supply must be adequate to fulfill anticipated irrigation needs. If poor quality water is applied, however, special management practices may be required to maintain full crop productivity. The problems that result from using poor quality waters will vary in type and degree of severity. Osmotic effects on crop yield, effects on soil permeability, and specific ion toxicities have been discussed. Other problems can arise, such as excessive vegetative growth, lodging, or delayed crop maturity resulting from excessive nutrients (usually nitrogen) in the water supply; white deposits on fruit or leaves due to sprinkling with waters high in bicarbonate; and nutritional disorders caused by an unusual pH of the water.

Guidelines for evaluating water quality for irrigation are given in Table 5.10. These guidelines are limited to the aspects of irrigation water quality that are normally encountered and that materially affect crop production. Emphasis is on long-term, dominating influences of water quality on soil water-plant systems, especially as related to crop production. Such water quality guidelines are intended as a management tool only, and the user should guard against drawing unwarranted conclusions based strictly on generalizations.

The water quality guidelines of Table 5.10 cover a wide range of conditions common to irrigated agriculture and incorporate some of the newer concepts in soil-water-plant relationships. Several basic assumptions are made to better define the range of suitability of the guidelines. The guidelines may need adjustment if the water is to be used under conditions where these assumptions are not valid. The basic assumptions for these guidelines are:

Soil and Climate. The soil texture is sandy loam to clay loam, with good internal drainage. The climate is assumed to be semi-arid to arid, with low effective annual rainfall. Drainage is assumed to be adequate. Full crop production, including necessary management inputs, is assumed when the guidelines indicate that water quality does not constitute a problem.

TABLE 5.10. GUIDELINES FOR INTERPRETATION OF WATER QUALITY
FOR IRRIGATION (FROM AYERS AND WESTCOT, 1976)

Water Quality Criterion	Degree of Problem		
	None	Increasing	Severe
Salinity			
EC (dS/m)	< 0.75	0.75-3	> 3
Permeability			
1. Low salt water			
EC (dS/m)	> 0.5	0.5-0.2	< 0.2
2. Sodium hazard			
adj. SAR*			
Montmorillonitic soils	< 6	6-9	> 9
Illite-Vermiculitic soils	< 8	8-16	>16
Kaolinite-Sesquioxidic soils	<16	16-24	>24
Specific Ion Toxicity			
Sodium† (adj SAR)	< 3	3-9	> 9
Chloride† (mol/m^3)	< 4	4-10	>10
Boron (mg/L)	< 0.75	0.75-2.0	> 2.0
Miscellaneous Effects			
Nitrogen (mg/L)	< 5	5-30	>30
Bicarbonate (mol/m^3)‡	< 1.5	1.5-8.5	> 8.5
pH	[Normal range 6.5 to 8.4]		

*adj. SAR (adjusted Sodium-Adsorption-Ratio) can be calculated using equation [5.6].
  Values presented are for the dominant types of minerals in the clay fraction of a given soil.
  The higher the salinity of the water, the less likely that permeability problems will occur.
†Values are for sensitive crops such as trees and woody plants. Most annual crops are less
  sensitive.
‡Overhead sprinkler irrigation with high bicarbonate waters during periods of extremely low
  humidity may cause white deposits on fruit or leaves.

The existence of a potential problem indicates that certain tolerant crops
may have to be grown to maintain full productivity. It does not indicate that
the water is unsuitable for use on any crop.

Methods and Timing of Irrigations. Surface and sprinkler methods
of irrigation are assumed, including flood, basin, strip-check, furrow, corrugation and sprinkler. It is assumed that the crop utilizes a considerable
portion of the stored soil water between irrigations. With these irrigation
methods, about 15 percent of the applied water is assumed to percolate
below the rooting depth. The guidelines are believed to be too restrictive
for high-frequency irrigation. They may also need modification when applied
to subsurface irrigation.

Uptake of Water by Crops. Water is assumed to be taken up by the
crop from within the root zone where it is most readily available. This is
normally about 40 percent from the upper one-quarter, 30 percent from the
second quarter, 20 percent from the third quarter, and 10 percent from the
lowest quarter of the root zone. Each irrigation is assumed to leach the upper
root zone and to maintain it at a relatively low salinity. Salinity is assumed
to increase with depth to the bottom of the root zone. The average salt concentration of the soil solution in the root zone is assumed to be three times
the salinity of the applied water. This average is assumed to be the salinity
to which the crop responds.

Degree of Problem. The division of Table 5.10 into "No Problem,"
"Increasing Problem," and "Severe Problem" is somewhat arbitrary, because changes are gradual. Changes of 10 to 20 percent above or below the
guideline values may have little significance if considered along with other
factors affecting yield. Many field studies and observations, as well as
controlled research experiments, were used as a basis for the listed divisions,
and they have proved practical under a range of field conditions.

5.5 SALINITY MANAGEMENT TECHNIQUES

5.5.1 Salinity Control

The major objective in selecting management procedures to control salinity is to improve soil water availability to the crop. Procedures that require relatively minor changes in management are more frequent irrigations, selection of more salt-tolerant crops, additional leaching, preplant irrigation, and seed placement. Alternatives that require significant changes in management are changing the irrigation method, altering the water supply, land-grading, modifying the soil profile, and installing artificial drainage.

More Frequent Irrigations. Salts concentrate in the soil solution as water is extracted by the crop. Hence, concentrations are lowest following an irrigation and highest just before the next irrigation. Increasing irrigation frequency increases the average soil water content (Rawlins, 1973). Particularly the upper portion of the root zone is maintained low in salinity if each irrigation is adequate. Frequent irrigations also permit small water applications that minimize surface runoff. Simply applying more water less frequently often will not be beneficial, because the extra water is lost to surface runoff or lowers application efficiency.

Crop Selection. When using saline irrigation water, selection of a salt-tolerant crop may be required to avoid yield reductions. There is an approximate tenfold range in salt tolerance of agricultural crops (Table 5.1). The selection of a more salt-tolerant crop, however, will not eliminate the need for leaching and for better management practices. Planting earlier in the spring or growing cool-season crops where salinity problems are marginal may reduce the water requirement sufficiently to attain full production even with rather salt-sensitive crops.

Additional Leaching. Soluble salts that accumulate in soils must be leached below the crop root zone. The amount of irrigation water that must drain below the crop root zone to maintain full crop production is referred to as the leaching requirement and was discussed in Section 5.2.1. The time interval between leachings does not appear to be critical, provided that crop tolerances are not exceeded. Hence, leaching can be accomplished with each irrigation, every few irrigations, once yearly, or after even longer intervals, depending on the severity of the salinity problem and crop salt tolerance. Often, an annual leaching during the noncrop or dormant period is sufficient. Rainfall is inadequate in many irrigated areas to meet the leaching requirement.

Preplant Irrigation. Salts often accumulate near the soil surface during fallow periods, particularly when water tables are high or winter rainfall is below normal. Under such conditions, both seed germination and seedling growth can be seriously reduced unless the soil is leached before planting. A preplant irrigation must be applied before most cultivation and seedbed preparation operations.

Seed Placement. Obtaining a satisfactory stand of furrow-irrigated crops on saline soils or when using saline water is often a problem. Growers sometimes compensate for poor germination by planting two or three times as much seed as would normally be required. In other instances, planting procedures are adjusted to ensure that the soil around the germinating seeds is low in salinity. This can be done by selecting suitable practices, bed shapes, or irrigation management.

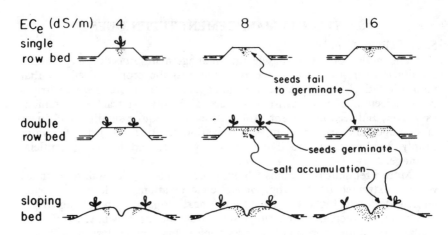

FIG. 5.11 Pattern of salt build-up as a function of seed placement, bed-shape, and irrigation water quality (Bernstein et al., 1955).

In furrow-irrigated soils, planting seeds in the center of a single-row, raised bed will place the seeds exactly in the area where salts are expected to concentrate (Fig. 5.11). With a double-row raised planting bed, the seed is placed near the shoulder of the bed and away from the area of greatest salt accumulation. Thus, higher soil salinities can be tolerated at germination than with the single-row plantings because the water moves the salts through and away from the seed area toward the center of the ridge.

There are other alternatives. Alternate-furrow irrigation may help in some cases. If beds are wetted from both sides, the salt accumulates in the top and center of the bed. If alternate furrows are irrigated, however, salts often can be moved beyond the single seed row to the nonirrigated side of the planting bed. Salts may still accumulate, but accumulation at the center of the bed will be reduced. The longer the water is held in the furrow, the lower will be the salt accumulation at the mid-bed seed area. Off-center, single-row plantings on the shoulder of the bed, close to the water furrow, have also been used as aids to germination under saline conditions. Double-row planting under alternate-furrow irrigation is not recommended because salt will accumulate on the edge of the bed away from the irrigated furrow.

With either single- or double-row plantings, increasing the depth of water in the furrow can also improve germination in salt-affected soils. Salinity can be controlled even better by using sloping beds, with the seeds planted on the sloping side just above the water line. Irrigation is continued until the wetting front has moved well past the seed row. During the first cultivation after planting, the sloped bed can be converted to a conventional raised bed.

Changing Irrigation Method. Gravity irrigation methods, such as flood, basin, furrow, or border methods, are usually not sufficiently flexible to permit changes in frequency of irrigation or depth of water applied per irrigation and still maintain efficiency. For example, with furrow irrigation it may not be possible to reduce the depth of water applied below 80 to 100 mm per irrigation. As a result, irrigating more frequently might improve water availability to the crop, but might also waste water. If a change to more frequent irrigations is advisable, a sprinkler or drip irrigation system may be required.

With adequate system design and management, sprinklers can apply water with good uniformity at application rates low enough to prevent surface runoff. The depth of water applied to supply the crop's water requirement plus leaching can readily be controlled by adjustments in the duration and frequency of application. Sprinklers are sometimes used during germination and early seedling growth when some crops may be particularly sensitive to salinity, high temperatures, and/or soil crusting.

With good-quality water, yields with drip irrigation should be equal to, or perhaps slightly better than, those obtained with other methods under comparable conditions. With poor-quality water, yields may be better with drip irrigation because of the continuously high soil water content and daily replenishment of water lost by evapotranspiration.

Changing Water Supply. Changing to a water supply of better quality is a simple solution to a salinity problem, but alternative supplies are not always available. If waters of two different qualities are available, a blend may be feasible to reduce the salinity hazard of the more saline one. Any change in quality due to blending may be evaluated by use of the Guidelines in Table 5.10.

Land Grading. In some instances fields are not graded accurately enough to permit satisfactory water distribution by surface irrigation. High spots in the field reduce water-intake by the soil and may lead to salinity problems. As an alternative, sprinkler or drip irrigation can be used without precise grading.

Soil Profile Modification. If soils have layers that impede or inhibit root and water penetration, water management and salinity control can be greatly simplified if these layers are fractured, destroyed, or at least rendered more permeable to roots and water. Subsoiling and chiseling may improve internal drainage of the soil profile, but results are often short lived. Deep plowing, however, often results in permanent improvement. It is usually performed after land grading but before leaching. This is a drastic treatment and often necessitates growing an annual crop such as barley the first year after deep plowing, and then regrading.

Drainage. Lack of adequate surface or subsurface drainage greatly complicates water management for salinity control. Land grading and improved surface drainage systems may be required to alleviate poor surface drainage due to flat or uneven slopes that cause ponding and waterlogging. Subsurface drainage may be impeded by a layer that is slowly permeable to water. Subsurface drainage problems may also arise because of overirrigation, seepage of water from higher elevations, or leakage from canals. Water tables less than 1.5 to 2 m below the soil surface may cause salts to accumulate in the root zone if net downward water movement is not maintained. Salt moves with the water to the soil surface and is deposited when the water evaporates. This can cause salinity problems even with good quality irrigation water. The salinity problem is solved by first improving drainage, then leaching.

5.5.2 Improving Water Penetration

Both chemical and physical methods can be used to improve soil permeability lost due to excess sodium in the soil. Beneficial chemical methods include using soil or water amendments and blending or changing the irrigation water supply. Physical methods that may increase the amount of water penetration are increasing the irrigation frequency, cultivating or deep tilling, extending the duration of each irrigation, changing the grade for surface irrigations, collecting and recirculating surface runoff waters, using sprinklers to match the rate of water application to the soil infiltration rate, and using organic residues.

Chemical Methods. Amendments may be effective where soil permeability has been decreased by the use of irrigation waters low in salinity ($EC_W < 0.5$ dS/m) or by the presence in the soil or water of excessive amounts of sodium, carbonate, or bicarbonate (a high adj. SAR). Amendments will not be useful if low permeability is caused by soil texture, compaction, water-restricting layers or high groundwater. When low infiltration rates are due to a high soil ESP, improved permeability should result if either the concentration of sodium in the irrigation water is decreased or the concentrations of calcium and magnesium are increased. No inexpensive process or chemical is available for removing sodium from irrigation waters. Calcium, however, can be added to the soil or the water to decrease the sodium to calcium ratio. The source of calcium may be direct (gypsum) or indirect from acid or acid-forming substances (i.e., sulfuric acid or sulfur) which dissolve calcium from lime in the soil. Field trials should always be conducted to determine if results are sufficiently beneficial to justify the expense.

Granular gypsum has been broadcast on soils at rates ranging from 2 to 20 Mg/ha. Where the permeability problem is primarily due to low water infiltration rates, granular gypsum may be more effective if left on the soil surface or mixed to shallow soil depths, rather than worked deeper into the soil. No more than about 700 kg of gypsum per 1000 m^3 of applied irrigation water can be dissolved from gypsum applied directly to the soil.

Applying gypsum in the irrigation water usually requires less gypsum per unit area than for soil applications. Water applications of gypsum are particularly effective for restoring lost permeability caused by low-salinity waters ($EC_I < 0.5$ dS/m) but the gypsum becomes less effective as the salinity of the irrigation water increases.

Sulfur may also be effective as a soil amendment for correcting a sodium problem (high ESP) if the soil contains lime. The sulfur must first be oxidized to sulfuric acid by soil bacteria, which in turn reacts with soil lime to produce gypsum. The oxidation process is slow and requires a warm, well-aerated, moist soil. Because sulfur is not water soluble and must react with soil lime, it is not normally effective as an amendment for improving water infiltration. Sulfur has been used successfully on calcareous soils having extremely high ESP levels.

Sulfuric acid is used occasionally as a amendment and can be applied either to the soil or to the irrigation water. It reacts rapidly with soil lime because oxidation is not required. However, it is highly corrosive and dangerous to handle. It may damage concrete pipes, steel culverts, checkgates, and aluminum pipes if not handled properly.

Other amendments may also be effective but they are not extensively used because of their relatively high costs. Comparative data for some of these amendments are given in Table 5.11. In some cases, an alternative source of high-quality water, although of limited volume, may be available for blending (Section 5.5.1).

Physical Methods. Only the physical methods for improving soil permeability not presented in Section 5.5.1 will be described here.

Cultivation and deep tillage may increase water penetration, although they are usually only temporary solutions. Deep tillage (chiseling, subsoiling) can also be expected to improve water penetration, but because many permeability problems are at or near the soil surface, the shallow soil soon reverts to its previous condition. Where slow infiltration is caused by a sur-

TABLE 5.11. WATER AND SOIL AMENDMENTS AND THEIR
RELATIVE EFFECTIVENESS IN SUPPLYING CALCIUM

Amendment	Amount equivalent to 100% gupsum
Gypsum (Ca $SO_4 \cdot 2H_2O$)*	1.00
Sulfur (S)†	0.19
Sulfuric acid (H_2SO_4)*	0.61
Ferric Sulfate [$Fe_2(SO_4)_3 \cdot 9H_2O$]†	0.61
Lime Sulfur (9% Ca + 24% S)*	0.78
Calcium chloride ($CaCl_2 \cdot 2H_2$)*	0.86
Calcium nitrate [$Ca(NO_3)_2 \cdot 2H_2O$]*	1.06

*Suitable for use as a water or soil amendment.
†Suitable only for soil application

face crust or a nearly impermeable soil surface, cultivation can roughen the
soil and open cracks and air spaces that will slow the surface flow of water
and, for a time, greatly increase infiltration.

Extending the duration of each irrigation may increase the amount of
irrigation water infiltrating, but aeration, waterlogging, excessive surface
runoff, and surface drainage problems may result. The duration of the preplant irrigation can safely be extended to allow the soil profile to fill. This
irrigation may provide the only opportunity to fill the deeper part of the crop
root zone without secondary effects on the growing crop.

Crop residues left on the soil or cultivated into the surface will often
improve water penetration. For significant improvement in water penetration, relatively large quantities of crop or other organic residues are usually
required. From 40 to 400 Mg/ha of animal wastes may be required to improve water penetration significantly. Rice hulls, sawdust, shredded bark
and many other waste products have been tried with various degrees of success at rates equal to 10 to 20 percent of the soil by volume. Nutritional imbalances and nitrogen shortages may develop after the use of sawdust, and
chloride or potassium toxicities have been noted from the use of rice hulls.

5.6 RECLAMATION OF SALT-AFFECTED SOILS

Reclamation is discussed separately from other management techniques
to emphasize the differences between the relatively continual management
procedures required to control salinity, as described in Section 5.5, and
reclamation procedures required to restore productivity lost because of severe
soil salinity or sodicity. Reclamation may require the removal of excess soluble salts as well as the reduction of soil ESP. The only proven way to reduce
the soluble salt concentration in the root zone is leaching. The ESP is more
difficult to reduce because sodium ions adsorbed on soil-exchange sites must
first be replaced with divalent cations from the soil solution, through a chemical reaction, and then be leached from the root zone. Hence, the reclamation of a sodic soil is a combination of chemical and mass-transfer processes.

5.6.1 Removal of Soluble Salts

Reclamation by leaching requires adequate drainage. Natural internal
drainage is normally adequate if the soil profile below the crop root zone is
permeable and provides sufficient internal storage capacity or if permeable
layers are present to provide drainage to a suitable outlet. Where such
natural drainage is lacking, artificial systems must be constructed. van
Schilfgaarde (1974) gives information on drainage requirements for agriculture.

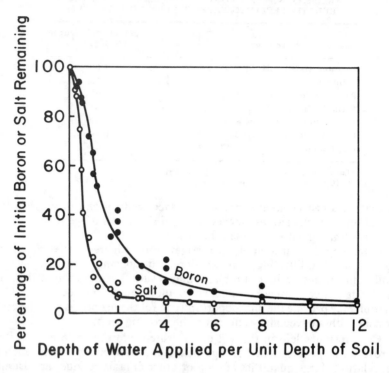

FIG. 5.12 Percentage of initial boron or salt remaining in a soil profile as related to the depth of leaching water applied per unit depth of soil (Reeve et al., 1955).

The amount of water that must be applied for reclamation depends largely on the initial soil salinity. Generally, about 80 percent of the soluble salts initially present in a soil profile will be removed by leaching with a depth of water equivalent to the soil depth to be reclaimed. This is shown in Fig. 5.12 (Reeve et al., 1955). For crops sensitive to salinity, further leaching may be required, while somewhat less leaching may be required for salt-tolerant crops. Achieving maximum crop production and minimizing the recurrence of salinity problems depend upon uniform infiltration and uniform leaching. Thus, any differences in leaching uniformity caused by differences in soil properties or by inefficiencies in the irrigation system should be carefully evaluated.

Leaching efficiency depends on the irrigation system and its management. Reclamation is normally achieved by flooding, either continuously or intermittently, or by sprinkling. The principal difference among the methods is the soil-water content maintained during reclamation. Less water is required to remove salt from a given volume of soil when the soil-water content is low, but more time is required because of the reduced hydraulic conductivity associated with the larger unsaturated pore volume. Intermittent leaching, particularly when a crop is grown, results in a lower average soil-water content than does continuous flooding. In the Imperial Valley of California on a silty-loam soil, a depth of 1100 mm of water was required for continuous flooding to achieve the same degree of reclamation as did 720 mm applied intermittently (Oster et al., 1972). In Israel and sometimes in the U.S., leaching is done by sprinkling. The advantage of sprinkling over flooding is that water can be applied at a rate less than the soil infiltration rate, thereby avoiding ponding. In a field experiment on a silty clay soil classified as moderately alkali and high in salts, Nielsen et al. (1965) found that 260 mm of water applied intermittently by sprinkling reduced the salt content of the upper 0.6 m of the profile to the same degree as 750 mm applied by continuous flooding.

Reclamation of salt-affected soils can be enhanced by the presence of plants. If the initial salinity is very high, the topsoil must be leached before even salt-tolerant plants can be grown. The beneficial effects of plants are not well understood, but they are probably due to the physical action of plant roots; the increased dissolution of lime in the presence of carbon dioxide evolved from plant roots; or the addition of organic matter (Shalhevet and Kamburov, 1976).

Boron can also be removed by leaching, but the leaching efficiency is about one-third that for soluble salts (Fig. 5.12). Therefore, about 3 m of leaching water must be applied per meter of soil depth reclaimed to remove about 80 percent of the boron.

5.6.2 Reclamation of Sodic Soil

To reclaim sodic soil, one must apply water relatively high in calcium and magnesium ions, [i.e., the SAR of the percolating solution must be less than the SAR of the equilibrium soil solution (SAR_e) at the initial ESP of the soil] and leaching water must percolate through the soil profile to transport the divalent cations to the cation-exchange sites for exchange with the adsorbed sodium. Chemical exchange will occur until the SAR of the percolating solution equals SAR_e.

If sufficient gypsum is not naturally present, any soluble calcium salt can be applied as an amendment to reclaim sodic soils, but the application of gypsum, calcium carbonate, or calcium chloride is most common. Sulfur and sulfuric acid are sometimes used to enhance conversion of naturally occurring calcium carbonate to gypsum which is more soluble than calcium carbonate. Calcium chloride is much more soluble than either gypsum or calcium carbonate. When sufficient gypsum is naturally present in the upper part of the soil profile (i.e., no less than chemically equivalent to the exchangeable sodium present) and when the clay-sized minerals in the soil are of the nonexpanding type (illite, kaolin, vermiculite), chemical reclamation can often be achieved simply by leaching, providing that hydraulic conductivity is adequate. It may be advantageous in some cases, particularly if soil permeability has been reduced drastically by exchangeable sodium, to superimpose a wetting and drying, freezing and thawing, or crop growth cycle on the chemical reclamation process before the soil is fully reclaimed.

The flow of leaching solution through the profile is essential to the reclamation process. Hydraulic conductivity of a sodic soil is a function of both ESP of the soil and electrolyte concentration of the percolating solution. Hydraulic conductivity decreases as ESP increases when electrolyte concentration remains constant, and increases as electrolyte concentration increases when ESP remains constant. The functional relationships vary with soil texture and mineralogy.

Leaching solutions having low-electrolyte concentration cause sodic soils to disperse, and thus, cause hydraulic conductivity to be low. Leaching solutions having high-electrolyte concentrations have a flocculating effect on soil particles and cause clay packets to contract (Norris and Quirk, 1954; Aylmore and Quirk, 1959). As a result the higher the salt concentration of the leaching solution the higher hydraulic conductivity will be (Quirk and

TABLE 5.12. AMOUNTS OF GYPSUM, CALCIUM CHLORIDE,
OR SULFUR NEEDED TO REPLACE THE INDICATED
AMOUNTS OF EXCHANGEABLE SODIUM IN A 0.3-m DEPTH
OF SODIC SOIL HAVING A BULK DENSITY OF 1.47 Mg/m^3

Exchangeable sodium	Gypsum	Calcium chloride	Sulfur*
mol/kg of soil	---------------- Mg/ha ----------------		
10	3.8	2.4	0.7
20	7.6	4.9	1.4
30	11.0	7.3	2.1
40	15.0	9.8	2.8
50	19.0	12.0	3.5
60	23.0	15.0	4.2
70	27.0	17.0	5.0
80	30.0	20.0	5.7
90	34.0	22.0	6.4
100	38.0	24.0	7.1

*Applicable only if the soil contains sufficient lime with which the sulfur can react.

Schofield, 1955; Quirk, 1957; Gardner et al., 1959; Reeve and Bower, 1960; Reeve 'and Tamaddoni, 1965; Doering and Reeve, 1965; Reeve and Doering, 1966a and 1966b; and Doering and Willis, 1975). Clay minerals having expanding-type lattices (smectite) influence hydraulic conductivity more than do minerals of the non-expanding type (illite, kaolin, vermiculite).

The amount of amendment required to reclaim a sodic soil is a function of the soil cation-exchange capacity (CEC), the desired change in ESP, the soil bulk density, and soil depth. When the CEC is expressed in mol/kg of soil, the needed calcium (expressed as mol/kg of soil) is simply the product of CEC and the desired change in ESP divided by 100. The amounts of gypsum, calcium chloride, or sulfur needed to exchange various amounts of exchangeable sodium from a 0.3-m depth of soil having a bulk density of 1.47 Mg/m^3 are given in Table 5.12. The amount of amendment required for other soil bulk densities can be obtained by multiplying the values given in Table 5.12 by the ratio of the new bulk density to 1.47.

The amount of water that must pass through the profile for chemical reclamation with gypsum depends on the amount of gypsum needed for chemical exchange. For example, assume that the soil has a CEC of 150 mol/kg, a bulk density of 1.47 Mg/m^3, an ESP of 50, and that the average ESP is to be reduced to 12 (ΔESP = 38) in the 1-m soil depth. Therefore, the desired exchange is [150 mol/kg x (50 - 12)]/100 or 57 mol/kg of soil. From Table 5.12, 21.6 Mg/ha of gypsum are required for a soil depth of 0.3 m or 72 Mg/ha for a soil depth of 1 m. Typically, a 10-mm depth of applied water will dissolve about 250 kg of gypsum per hectare. For this example, 2.9 m of water would be required. Although, the average ESP after reclamation was assumed to be 12, the final ESP will be lower near the soil surface than at the bottom of the profile. Complete exchange by all the dissolved gypsum was also assumed for the calculations. In practice, complete exchange should not be expected. If a chemical amendment is required, it is generally recommended that the amount be chemically equivalent to about 1.25 times the amount of exchangeable sodium that is to be removed (U.S. Salinity Laboratory Staff, 1954).

Gypsum, sulfur, and limestone amendments are usually broadcast and then cultivated into the soil. When sulfur is used, leaching should be delayed until the sulfur has oxidized and gypsum has been formed. McGeorge and Greene (1935) showed that about 2.2 Mg/ha of sulfur would oxidize in 2 or 3 weeks under favorable temperature and moisture conditions in Arizona. When acids or acid-formers are being used alkaline-earth carbonates must be either in or above the sodic layer to ensure that downward percolating water will carry dissolved calcium to the exchange sites. Because of its limited solubility, gypsum does not create an additional salinity problem when it is used. Hence, if hydraulic conductivity is acceptable and sufficient leaching takes place, crops that are not sensitive to sodium can often be grown during reclamation.

When soil physical conditions have deteriorated and hydraulic conductivity is so low that the time required for chemical reclamation is excessive, the high-electrolyte method for sodic-soil reclamation may be warranted. The high-electrolyte method consists of applying successive dilutions of a high-salt water containing divalent cations (Reeve and Doering, 1966a). Exchangeable sodium is replaced by divalent cations from the leaching solution, while water penetration is maintained by the flocculating effect of the high-salt water. Soil hydraulic conductivities often are extremely low when clay minerals of the expanding-lattice type (smectite) are present in the soil. For example, in a field study with highly sodic Waukena clay loam soil, the infiltration rate for saturated gypsum solution was only about 1.6 mm/day (Reeve and Doering, 1966b). With such a low infiltration rate, it was estimated that 7 yr of continuous leaching would be required to reclaim the profile to a depth of 0.9 m. However, the profile was chemically reclaimed in 3 days with a series of $CaCl_2$ solutions, starting with a concentration of 300 mol/m^3 and following with solutions containing 150, 75, 38, 19, and 6 mol/m^3. The average infiltration rate for the $CaCl_2$ solutions was 230 mm/day, or about 150 times the infiltration rate for the gypsum solution. The high-electrolyte method has also been used to reclaim a slowly permeable, mildly sodic, low-electrolyte soil in a humid environment, where hydraulic conductivity and infiltration were increased by 30 to over 100 percent (Rahman et al., 1974).

The effectiveness of applied amendments decrease as the electrolyte concentration increases (Doering and Willis, 1975). Therefore, highly soluble salts should be applied at controlled concentrations. Successful use of the high-electrolyte method requires a balance between keeping the electrolyte concentration high to reduce reclamation time and keeping the electrolyte concentration low to reduce the amount of amendment required. A practical technique for utilizing the high-electrolyte method is to satisfy about three-fourths of the exchange requirement with three $CaCl_2$ solutions, starting with a solution containing about 250 mol/m^3 and following with solutions containing 100 and 50 mol/m^3. This is followed by a gypsum application to satisfy the remaining exchange requirement, and completed by leaching with about a 0.3-m depth of water per meter depth of soil to be reclaimed. This final step is essential to leach the high-salt solutions below the root zone.

High-salt solutions containing sodium as well as calcium ions, including even sea water and appropriate dilutions thereof, can also be used to reclaim sodic soils (Doering and Reeve, 1965; Reeve and Doering, 1966b). Hydraulic conductivity is maintained at acceptably high levels if the waters are not diluted too rapidly, but the water requirement increases significantly as the sodium fraction of the high-salt water increases.

5.7 SUMMARY

Salinity is a major threat to irrigated agriculture because many of the soils and irrigation waters contain significant amounts of dissolved salts. These salts limit crop production on about 25 percent of the irrigated land in the western United States because of the total osmotic effects, individual ion toxicities, or reduced soil permeability caused by excess sodium.

Originating from geological weathering, the salts are transported and accumulated in the soil profile as a result of water movement. Some of these accumulated soluble salts must be removed by leaching to maintain a viable irrigated agriculture. The actual amount of water that must leach below the crop root zone depends upon the salt contributions of the irrigation water, soil, and groundwater; the salt tolerance of the crop; climatic conditions; and soil and water management.

The predominate influence of salinity on plants is growth suppression. Typically, growth decreases linearly as salinity increases beyond a threshold salinity level. The maximum allowable salinity that does not result in a measurable yield reduction (threshold level) and the rate of relative yield reduction per unit increase in salinity are presented as relative salt tolerance parameters for more than 60 crops. These parameters must be taken as tentative, however, because salt tolerance can be influenced by irrigation management, climatic conditions, stage of plant growth, crop variety and rootstock, and plant nutrition. Individual ions, such as chloride, sodium, and boron, and trace elements, may be toxic to certain crops at concentrations too low to cause total osmotic effects. Lists are given of crops particularly sensitive to each of these ions or elements. Also presented are techniques for evaluating sodium levels that may be detrimental to soil structure.

Water-quality guidelines are presented to aid in evaluating water suitability for irrigation. For waters of marginal quality, various management options are discussed to improve crop production. These are changes in the frequency, amount, or system of irrigation, more suitable seed placement, proper crop selection, and soil profile modification. Likewise, both chemical and physical methods are given for improving water penetration in sodic soils.

Reclamation is required to restore soil productivity where salinity is excessive. About 80 percent of the soluble salts can be removed by applying a depth of water equivalent to the soil depth to be reclaimed, providing that soil permeability and drainage are adequate. For reclamation of sodic soils, the exchangeable sodium must be replaced, normally by calcium, and the displaced sodium must be leached from the root zone. Amendments may have to be applied for successful reclamation of sodic soils with the choice of amendment depending on the soil's physical and chemical properties.

References

1 Ayers, R. S., and D. W. Westcot. 1976. Water quality for agriculture. FAO Irri. and Drain. Paper 29, 97 p.

2 Aylmore, L. A. G., and J. P. Quirk. 1959. Swelling of clay-water systems. Nature 183:1752-1753.

3 Benz, L. C., F. M. Sandoval, E. J. Doering, and W. O. Willis. 1976. Managing saline soils in the Red River Valley of the North. USDA-ARS-NC-42.

4 Bernstein, L. 1964. Effects of salinity on mineral composition and growth of plants. Proc. 4th Intern. Colloquium Plant Anal. and Fert. Problems (Brussels) 4:25-45.

5 Bernstein, L. 1965. Salt tolerance of fruit crops. USDA Inform. Bull. 292, 8 p.

6 Bernstein, L., M. Fireman, and R. C. Reeve. 1955. Control of salinity in the Imperial Valley, CA. USDA-ARS-41-4, 16 p.

7   Bernstein, L., and L. E. Francois. 1973 a. Comparisons of drip, furrow, and sprinkler irrigation. Soil Sci. 115:73-86.
8   Bernstein, L., and L. E. Francois. 1973b. Leaching requirement studies: Sensitivity of alfalfa to salinity of irrigation and drainage waters. Soil Sci. Soc. Am. Proc. 37:931-943.
9   Bernstein, L., L. E. Francois, and R. A. Clark. 1974. Interactive effects of salinity and fertility on yields of grains and vegetables. Agron. J. 66:412-421.
10   Bernstein, L., and H. E. Hayward. 1958. Physiology of salt tolerance. Ann. Rev. Plant Physiol. 9:25-46.
11   Bernstein, L., and G. A. Pearson. 1956. Influence of exchangeable sodium on the yield and chemical composition of plants. I. Green beans, garden beets, clover, and alfalfa. Soil Sci. 82:247-258.
12   Bingham, F. T., A. L. Page, and G. R. Bradford. 1964. Tolerance of plants to lithium. Soil Sci. 98:4-8.
13   Bower, C. A., G. Ogata, and J. M. Tucker. 1969. Rootzone salt profiles and alfalfa growth as influenced by irrigation water salinity and leaching fraction. Agron. J. 61:783-785.
14   Bower, C. A., G. Ogata, and J. M. Tucker. 1970. Growth of sudan and tall fescue grasses as influenced by irrigation water salinity and leaching fraction. Agron. J. 62:793-794.
15   Bresler, E. 1975. Two-dimensional transport of solutes during nonsteady infiltration from a trickle source. Soil Sci. Soc. Am. Proc. 39:604-613.
16   Cooper, W. C. 1951. Salt tolerance of avocados on various rootstocks. Texas Avocado Soc. Year. 1951:24-28.
17   Cooper, W. C. 1961. Toxicity and accumulation of salts in citrus trees on various rootstocks in Texas. Florida State Hort. Soc. Proc. 74:95-104.
18   Doering, E. J., and R. C. Reeve. 1965. Engineering aspects of reclamation of sodic soils with high-salt waters. Proc. Am. Soc. Civ. Engr., J. Irrig. and Drain. Div. 91:59-72.
19   Doering, E. J., and F. M. Sandoval. 1976. Hydrology of saline seeps in the Northern Great Plains. TRANSACTIONS of the ASAE 19(5):856-861, 865.
20   Doering, E. J., and W. O. Willis. 1975. Chemical reclamation for sodic strip-mine spoils. USDA-ARS-NC-20.
21   Eaton, F. M. 1935. Boron in soils and irrigation waters and its effect on plants, with particular reference to the San Joaquin Valley of California, USDA Tech. Bul.448.
22   Gardner, W. R., M. S. Mayhugh, J. O. Goertzen, and C. A. Bower. 1959. Effect of electrolyte concentration and exchangeable-sodium-percentage on diffusivity of water in soils. Soil Sci. 88:270-274.
23   Gauch, H. G., and F. M. Eaton. 1942. Effect of saline substrate on hourly levels of carbohydrates and inorganic constituents of barley plants. Plant Physiol. 17:347-365.
24   Harding, R. B., M. P. Miller, and M. Fireman. 1958. Absorption of salts by citrus leaves during sprinkling with water suitable for surface irrigation. Proc. Am. Soc. Hort. Sci. 71:248-256.
25   Hayward, H. E., and E. M. Long. 1941. Anatomical and physiological responses of the tomato to varying concentration of sodium chloride, sodium sulphate, and nutrient solutions. Bot. Gaz. 102:437-462.
26   Hoffman, G. J., and J. A. Jobes. 1978. Growth and water relations of cereal crops as influenced by salinity and relative humidity. Agron. J. 70:765-769.
27   Hoffman, G. J., S. L. Rawlins, J. D. Oster, J. A. Jobes, and S. D. Merrill. 1979. Leaching requirement for salinity control. I. Wheat, sorghum, and lettuce. Agr. Water Mgmt. (in press)
28   Hoffman, G. J., E. V. Maas, and S. L. Rawlins. 1975. Salinity-ozone interactive effects on alfalfa yield and water relations. J. Environ. Qual. 4:326-331.
29   Hoffman, G. J., and S. L. Rawlins. 1971. Growth and water potential of root crops as influenced by salinity and relative humidity. Agron. J. 68:877-880.
30   Junge, C. E., and R. T. Werby. 1958. Concentration of chloride, sodium, potassium, calcium, and sulfate in rain water over the United States. J. Meteorol. 151:417-425.
31   Livingstone, D. A. 1963. Chemical composition of rivers and lakes. In Data of Geochemistry. M. Fleischer. (Ed.), 6th ed. Geol. Survey. Prof. Paper 440-6, 64 p.
32   Longenecker, D. E., and P. J. Lyerly. 1959. Some relations among irrigation water quality, soil characteristics and management practices in the Trans-Pecos area. Texas Agr. Exp. Sta. Bul. MP-373, 17 p.
33   Maas, E. V., and G. J. Hoffman. 1977. Crop salt tolerance — Current assessment. Am. Soc. Civ. Engr. Proc. J. Irrig. & Drain. 103(IR2):115-134.
34   Maas, E. V., and R. H. Nieman. 1978. Physiology of plant tolerance to salinity. In Crop tolerance to suboptimal land conditions. G. E. Jung (Ed.) Chap. 13, ASA Spec. Publ.: 277-299.

35 Magistad, O. C., A. D. Ayers, D. H. Wadleigh, and H. G. Gauch. 1943. Effect of salt concentration, kind of salt, and climate on plant growth in sand cultures. Plant Physiol. 18:151-166.
36 McGeorge, W. T., and R. A. Greene. 1935. Oxidation of sulphur in Arizona soils and its effect on soil properties. Ariz. Agr. Exp. Sta. Tech. Bul. 59, pp. 297-325.
37 Miller, W. M. 1956. Summary of partial analyses of Wyoming waters. Salinity and selenium. Wyo. Agr. Exp. Sta. Mines, Cir. 64.
38 Mohamed, N. A., and F. Amer. 1972. Sodium carbonate formation in Ferhash area and possibility of biological dealkalization. Proc. Internatl. Symp. on New Developments in the Field of Salt Affected Soils, Dec. 4-9, 1972. Min. of Agr. ARE, Cairo, 346 p.
39 National Academy of Sciences and National Academy of Engineering. 1972. Water Quality Criteria. EPA-R3-73-033, U.S. Gov. Printing Off., Washington, DC, 592 p.
40 Nielsen, D. R., J. W. Biggar, and J. N. Luthin. 1965. Desalinization of soils under controlled unsaturated flow conditions. Internl. Comn. Irrig. and Drain. 6th Cong., New Delhi, India. Q. 19, p. 15-24.
41 Nielson, R. F., and O. S. Cannon. 1975. Sprinkling with salt well water can cause problems. Utah Sci., Agr. Exp. Sta. 36(2):61-63.
42 Nieman, R. H., and R. A. Clark. 1976. Interactive effects of salinity and phosphorous nutrition on the concentrations of phosphate and phosphate esters in mature photosynthesizing corn leaves. Plant Physiol. 57:157-161.
43 Norris, K., and J. P. Quirk. 1954. Crystalline swelling of montmorillonite. Nature 173:255.
44 Oster, J. D., and R. D. Ingvalson. 1967. In situ measurement of soil salinity with a sensor. Soil Sci. Soc. Am. Proc. 31:572-574.
45 Oster, J. D., L. S. Willardson, and G. J. Hoffman. 1972. Sprinkling and ponding techniques for reclaiming saline soils. TRANSACTIONS of the ASAE 15(6):115-117.
46 Pearson, G. A. 1960. Tolerance of crops to exchangeable sodium. USDA Inform. Bul. 216, 4 p.
47 Pearson, G. A., and L. Bernstein. 1958. Influence of exchangeable sodium on yield and chemical composition of plants: II. Wheat, barley, oats, rice, tall fescue, and tall wheatgrass. Soil Sci. 86:254-261.
48 Poljakoff-Mayber, A. 1975. Morphological and anatomical changes in plants as a response to salinity stress. In Plants in saline environments. A. Poljakoff-Mayber and J. Gale (Eds.), Ecolog. Studies 15:97-117. Springer-Verlag, New York.
49 Pratt, P. F. 1972. Quality criteria for trace elements in irrigation waters. Calif. Agr. Exp. Sta., 46 p.
50 Quirk, J. P. 1957. Effect of electrolyte concentration on soil permeability and water entry in irrigation soils. 3rd Cong. Internat. Comn. Irrig. and Drain., San Francisco, Calif. Q. 8, R6:115-123.
51 Quirk, J. P., and R. K. Schofield. 1955. The effect of electrolyte concentration on soil permeability. J. Soil Sci. 6:163-178.
52 Rahman, M. A., E. A. Hiler, and J. R. Runkles. 1974. High electrolyte water for reclaiming slowly permeable soils. TRANSACTIONS of the ASAE 17:129-133.
53 Rawlins, S. L. 1973. Principles of managing high frequency irrigation. Soil Sci. Soc. Am. Proc. 37:626-629.
54 Reeve, R. C., and C. A. Bower. 1960. Use of high-salt waters as a flocculant and source of divalent cations for reclaiming sodic soils. Soil Sci. 90:139-144.
55 Reeve, R. C., and E. J. Doering. 1966a. The high-salt-water dilution method for reclaiming sodic soils. Soil Sci. Soc. Am. Proc. 30:498-504.
56 Reeve, R. C., and E. J. Doering. 1966b. Field comparison of the high-salt water dilution method and conventional methods for reclaiming sodic soils. Internl. Comn. Irrig. and Drain., 6th Cong. (New Delhi, India) Q. 19R.1:19.1-19.14.
57 Reeve, R. C., A. F. Pillsbury, and L. V. Wilcox. 1955. Reclamation of a saline and high boron soil in the Coachella Valley of California. Hilgardia 24(4):69-91.
58 Reeve, R. C., and Gh. Tamaddoni. 1965. Effect of electrolyte on laboratory permeability and field intake rate of sodic soil. Soil Sci. 99:261-266.
59 Reisenauer, H. M. 1976. Soil and plant-tissue testing in California. Div. of Agr. Sci., Univ. of Calif. Bul. 1879, 54 p.
60 Rhoades, J. D. 1972. Quality of water for irrigation. Soil Sci. 113:277-284.
61 Rhoades, J. D., and R. D. Ingvalson. 1971. Determining salinity in field soils with soil resistance measurements. Soil Sci. Soc. Am. Proc. 35:54-60.
62 Shalhevet, J., and J. Kamburov. 1976. Irrigation and salinity—a world-wide survey. Internl. Comn. Irrig. and Drain., New Delhi, India. 106 p.

63   Smith, H. V., G. E. Draper, and W. H. Fuller. 1964. The quality of Arizona irrigation waters. Ariz. Agr. Exp. Sta. Rep. 223, Tucson, 96 p.

64   Strogonov, B. P. 1962. Physiological basis of salt tolerance of plants. Trans. A. Poljakoff-Mayber and A. M. Mayer, Israel Program Sci. Trans., Jerusalem, 1964, 279 p.

65   U. S. Geological Survey. 1967. Quality of surface waters of the United States, 1961. U. S. Geol. Survey Water-Supply Papers Nos. 1883, 1884, and 1885.

66   U. S. Geological Survey. 1974. Quality of surface waters of the United States, 1969. U. S. Geol. Survey Water-Supply Papers Nos. 2145, 2146, 2147, 2148, 2149, and 2150.

67   U. S. Salinity Laboratory Staff. 1954. Diagnosis and improvement of saline and alkali soils. USDA Agr. Handb. 60, 160 p.

68   Van Schilfgaarde, J. (Ed.). 1974. Drainage for Agriculture. Agronomy 17. Am. Soc. Agron., Madison, WI, 700 p.

69   van Schilfgaarde, J., L. Bernstein, J. D. Rhoades, and S. L. Rawlins. 1974. Irrigation management for salt control. Proc. Am. Soc. Civ. Engr., J. Irrig. & Drain. Div. 100(IR3): 321-338.

70   Wadleigh, C. H. 1968. Wastes in relation to agriculture and forestry. U.S. Dep. Agr. Misc. Publ. 1065, 112 p.

71   Wadleigh, C. H., and A. D. Ayers. 1945. Growth and biochemical composition of bean plants as conditioned by soil moisture tension and salt concentration. Plant Physiol. 20:106-132.

72   Wadleigh, C. H., H. G. Gauch, and O. C. Magistad. 1946. Growth and rubber accumulation in guayule as conditioned by soil salinity and irrigation regime. USDA Tech. Bul. 925, 34 p.

73   Wilcox, L. V. 1948. The quality of water for irrigation use. USDA Tech. Bul. 962, 40 p.

74   Wilcox, L. V. 1960. Boron injury to plants. USDA Bul. 211, 7 p.

75   Wilcox, L. V. 1968. Discharge and salt burden of the Rio Grande above Fort Quitman, Texas, and salt-balance conditions on the Rio Grande Project. U.S. Salinity Lab. Res. Rpt. 113, USDA, ARS, Riverside, CA, 200 p.

chapter 6

WATER REQUIREMENTS

6

6

WATER REQUIREMENTS

by   R. D. Burman, University of Wyoming, Laramie,
WY; P. R. Nixon, USDA-SEA/AR, Weslaco, TX;
J. L. Wright, USDA-SEA/AR, Kimberly, ID; and
W. O. Pruitt, University of California, Davis, CA

6.1 INTRODUCTION

The main objective of irrigation is to provide plants with sufficient water to prevent stress that may cause reduced yield or poor quality of harvest (Haise and Hagan, 1967; Taylor, 1965). The required timing and amount of applied water is governed by the prevailing climatic conditions, crop and stage of growth, soil moisture holding capacity, and the extent of root development as determined by type of crop, stage of growth, and soil.

Need for irrigation can be determined in several ways that do not require knowledge of evapotranspiration (ET) rates. One way is to observe crop indicators such as change of color or leaf angle, but this information may appear too late to avoid reduction in crop yield or quality. This method has been used successfully with some crops like beans (Haise and Hagan, 1967). Other similar methods of scheduling, which involve determining the plant water stress, soil moisture status, or soil water potential are described in Chapter 18.

This chapter describes methods of estimating crop water requirements expressed as equivalent depth of water over the horizontal projection of the crop growing area. This information, when combined with soil water holding characteristics, has the advantage of not only being useful in determining when to irrigate, but also enables specifying how much water to apply. ET information is also needed in determining the volume of water required to satisfy short-term and seasonal water requirements for fields, farms and irrigation projects, and in designing water storage and distribution systems. In addition, this information is essential for most water right transfers from agriculture to other uses because most such transfers are limited to historic crop water use amounts.

Water use measurements have been made in many field experiments and at many locations. The data available from various sources are of varying quality depending upon the conditions and techniques that were used. The material presented in this chapter emphasizes methods of estimating ET rates and provides guidelines for estimating irrigation water requirements.

6.2. IMPORTANT DEFINITIONS

Several important quantities are defined before measurement or estimation methods are described. Most of these definitions are commonly used in agricultural literature.

6.2.1 Evapotranspiration and Potential Evapotranspiration

The definition of evapotranspiration, abbreviated ET or symbolically E_t, presented in this chapter is in widespread use. The definition of potential ET (E_{tp}) is controversial and may have different meanings in various parts of the world and to different people in the same country. The following definitions include several variations of potential ET.

Evaptranspiration. The combined process by which water is transferred from the earth's surface to the atmosphere. It includes evaporation of liquid or solid water from soil and plant surfaces plus transpiration of liquid water through plant tissues expressed as the latent heat transfer per unit area or its equivalent depth of water per unit area.

Potential evapotranspiration. The rate at which water, if available, would be removed from the soil and plant surface expressed as the latent heat transfer per unit area or its equivalent depth of water per unit area.

Other definitions of potential evaptranspiration. Mathematically, in the common derivation of the combination equation, potential ET is the ET that occurs when the vapor pressure at the evaporating surface is at the saturation point (van Bavel, 1966). This definition is not limited to any particular degree of vegetation or growth stage of a crop. Since this definition is not restricted to a standard surface, it has had limited direct use by the designer or operator of an irrigation system.

Some investigators in the Western United States have used the ET from a well-watered crop like alfalfa with 30 to 50 cm of top growth and at least 100 m of fetch as representing potential ET (Jensen, 1974). Others have used ET from well-watered clipped grass as a potential ET. The height of the grass has been historically uncertain. Penman (1948) used clipped grass similar to a lawn to develop his version of the combination equation. Recently, this has been defined as "the rate of evapotranspiration from an extensive surface of 8- to 15-cm, green grass cover of uniform height, actively growing, completely shading the ground, and not short of water" (Doorenbos and Pruitt, 1977). In testing the Penman formula, Makkink (1957) found that the height of the grass did have an influence on the ET rate.

Crop versus potential ET. The realtionship between the ET of a specific crop (E_t) at a specific time in its growth stage and potential ET is of practical interest to the designer or operator of an irrigation system because ET estimates are often made from potential ET (E_{tp}). The relationship has lead to crop coefficients:

$$K_c = \frac{E_t}{E_{tp}} \quad \dots\dots\dots\dots\dots\dots\dots\dots\dots\dots\dots\dots\dots\dots\dots \quad [6.1]$$

where K_c is referred to as a crop coefficient incorporating the effects of crop growth stage, crop density, and other cultural factors affecting ET. Crop coefficients are discussed in more detail in Section 6.5. The crop coefficient defined in equation [6.1] is not the K factor used in the original Blaney-Criddle method.

6.2.2 Reference Crop Evapotranspiration

Because of the ambiguities involved in the interpretation of potential evapotranspiration, the term "Reference Crop Evapotranspiration," or E_{tr}, is frequently being used. Doorenbos and Pruitt (1977) use ET_o, hereafter denoted as E_{to}, to replace E_{tp}. They define E_{to} as "the rate of evapotranspiration from an extensive surface of 8 to 15 cm, green grass cover of uniform height, actively growing, completely shading the ground, and not short of water."

An alternate definition of E_{tr} which is widely used in the Western United States was presented by Jensen et al. (1970); E_{tr} "represents the upper limit or maximum evapotranspiration that occurs under given climatic conditions with a field having a well-watered agricultural crop with an aerodynamically rough surface, such as alfalfa with 12 in. to 18 in. of top growth."

The irrigation engineer or scientist should make sure that the definition of E_{tr} being used is completely understood and that written documentation carefully identifies the basic definitions used in calculations, designs, or reports. Actual E_t is estimated using equation [6.2].

$$E_t = K_c \, E_{tr} \text{ or } E_t = K_c \, E_{to} \quad \dots\dots\dots\dots\dots\dots\dots\dots\dots\dots\dots\dots \quad [6.2]$$

E_{tr} refers to reference crop ET based on alfalfa and E_{to} refers to reference crop ET based on grass.

The definition of K_c used in equation [6.2] is essentially the same as that used in equation [6.1] except that the use of E_{tr} or E_{to} requires identifying the reference base. E_{tr} or E_{to} can either be based on direct measurements or estimates. The use of equation [6.2] is greatly expanded in Section 6.5.

6.2.3 Effective Precipitation

Effective rainfall or precipitation (P_e) is more difficult to define than potential ET. At this point it is sufficient to define P_e according to Dastane (1974) as "that which is useful or usable in any phase of crop production." The definition of P_e is expanded and several methods for estimating P_e are presented in Section 6.8.

6.2.4 Other Factors

Irrigation water requirements may be influenced by salt management, seed germination, crop establishment, climate control, frost protection, fertilizer or chemical application, and soil temperature control. Leaching requirements are discussed in Section 5.2, salt management in Section 5.5 and reclamation of salt affected soils in Section 5.6 Other beneficial uses of water connected with irrigation water requirements are discussed in Section 6.6 (also see Sections 2.8, 14.8 and 18.4).

6.2.5 Irrigation Water Requirements

The designer or operator of an irrigation system must determine irrigation water requirements, R, for both short periods and on a seasonal basis. The units of R usually are volume per unit area or depth. The irrigation water requirement was defined by Doorenbos and Pruitt (1977) as "the depth of water needed to meet the water loss through ET of a disease-free crop, growing in large fields under non-restricting soil conditions including soil water and fertility and achieving full production potential under the given growing environment." R also can be stated as:

$$R = E_t - P_e + (\text{other beneficial uses}) \quad \dots\dots\dots\dots\dots\dots\dots\dots \quad [6.3]$$

Other beneficial uses are enumerated in detail in Section 6.6.

6.3 DETERMINING EVAPOTRANSPIRATION

The designer or operator obtains ET data from direct field measurements or from estimates based on climatological and crop data. Direct field measurements are very expensive and are mainly used to provide data to calibrate methods for estimating ET from climatic data. Real time field measurements are being used for water administration in some areas such as Colorado. The main thrust of research has been to determine the amounts of water used for crop production and to develop methods of predicting ET from climatic data.

6.3.1 Direct Measurements

Water balance field measurement. The water balance approach to measuring ET involves periodic determinations of root zone soil moisture and recording intervening rainfall, irrigation, or drainage. Soil tanks in which crops are grown, known as lysimeters, have been used to facilitate accurate water accounting. Weighing-type lysimeters, operated in a representative field environment, provide the most accurate ET information. In western areas of the United States the water balance method has also involved stream inflow-outflow measurements. Average ET for the land area involved is equal to inflow, including ground water, surface water and rainfall, minus outflow after taking into account changes in soil moisture storage.

Other methods of field measurement. Short-period ET (i.e. hourly or less) can be determined by applying meteorolgical equations that require involved meteorological measurements. These approaches, based on mass transfer and related concepts, usually require very accurate vapor pressure and wind speed measurements at two or more heights above the crop, and other measurements that may be necessary.

Essentially instantaneous ET can be determined with measurements that enable solving the energy balance equation. This approach is based on the fact that most of the transformed radiant energy (measured net radiation) goes into latent heat (evaporation or dew), and the balance goes into soil heat (measured soil heat flux), and sensible heat (heating or cooling of air). The partitioning between latent and sensible heat is obtained by using vapor pressure and temperature gradient measurements to calculate Bowen's ratio (Fritschen, 1965).

ET for periods of a day or longer can be determined by summing the short-period data obtained with the above methods. The calculations are voluminous, and data uncertainties may occur. These methods are useful for research and currently are seldom used in irrigation scheduling or water resource calculations.

6.3.2 Estimation from Climatic Data

Confidence is developing in the practical utility of ET equations that require weather records. This confidence comes from comparisons of calculated daily and longer-period ET values with water balance measurements, especially those from weighing lysimeters.

Numerous equations that require meteorological data have been proposed, and several are commonly used to estimate ET for periods of a day or more. These equations are all empirical to various extents; the simplest requiring only average air temperature, daylength, and a crop factor. The generally better performing equations require daily radiation, temperature, vapor pressure and wind data.

A method of estimating ET should not be automatically rejected because of the lack of available climatic data. It is often possible to estimate unavailable data; for example, several methods of estimating net radiation exist, (see Subsections 6.4.2 and 6.5.3) and dew point data can be estimated from minimum temperature data (Pochop et al., 1973).

A comprehensive evaluation of common evapotranspiration equations was made by the Technical Committee on Irrigation Water Requirements, American Society of Civil Engineers (Jensen, 1974) using data from 10 world wide locations. They concluded "that no single existing method using meteorlogical data is universally adequate under all climatic regimes, especially for tropical areas and for high elevations, without some local or regional calibration." Local calibration is discussed in Subsection 6.4.6.

The calculation of ET estimates from weather records is appealing because the approach is relatively simple compared with on-site ET measurements. The calculated reference crop ET can be used to estimate actual ET by using coefficients to account for the effect of soil moisture status, stage of growth and maturity of a crop. Coefficients for many crops have been developed from field experiments and are discussed in Section 6.5.

Estimates of actual ET for fields with incomplete cover also can be made using models that separate ET into evaporation and transpiration components (Ritchie, 1972; Tanner and Jury, 1976). The models attempt to account for reduction of evaporation with surface drying.

Crop ET can also be estimated using coefficients which relate crop ET to evaporation as measured with pans (Pruitt, 1966; Doorenbos and Pruitt, 1977). The 1.2-m (4-ft) diameter U.S. Weather Service Class A evaporation pan has been used successfully for this purpose. The evaporation pan provides a measurement of evaporation from an open water surface integrating the effects of radiation, wind, temperature, and humidity. While plants respond to the same climatic variables, pans and plants respond differently on a daily basis. Pan coefficients therefore are better suited for longer time periods. Pans are also very sensitive to the wetness of the immediate surroundings.

A flow chart is presented in Fig. 6.1 outlining the sequential steps for estimating irrigation water requirements from climatic data. These steps are intended to apply to the information presented in this chapter. A similar sequence would be valid for any other source of data.

Important considerations. Observed ET rates for a given crop and growth stage depend on climatic conditions. Water use rates observed at one location may not apply elsewhere. For example, the peak monthly ET rate at Brawley, California, an arid inland location is 2.5 times that at a coastal location at Lompoc, California (Jensen, 1974). In a California coastal valley the summertime ET from alfalfa 37 km (23 mi) inland was found to be more than 1.5 times that 29 km (13 mi) nearer the ocean (Nixon et al., 1963). Conversely, measured or calculated ET values might properly be transferred considerable distance where rather uniform conditions of climate and cropping practices exist on relatively flat terrain.

Obviously, weather records that are used to calculate ET should be representative of the area in question. Thus, weather data should not be used indiscriminately without knowledge of the weather station, site exposure and the care with which the station was maintained.

FLOW CHART

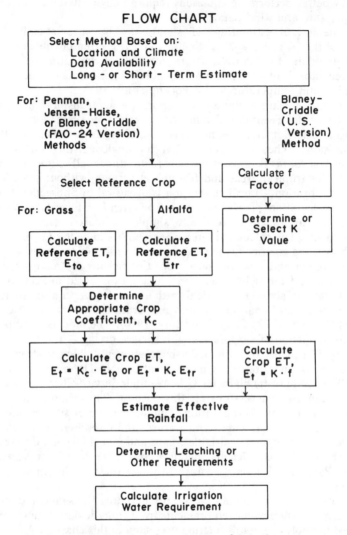

FIG. 6.1 Typical flow chart for estimation of irrigation water requirements from climatic data.

Factors contributing to water requirements. ET is the principal factor in determining irrigation water requirements, but losses in storage, conveyance and applying water, the inability to apply water uniformly, and the need for soil leaching are additional factors. The planning and operation of irrigation systems must take all these factors into consideration in determining water requirements. Other possible requirements and uses for water not directly required for ET are discussed under other beneficial uses in Section 6.6, and in Chapters 2, 14 and 18.

6.4 SELECTED METHODS OF ESTIMATING REFERENCE CROP ET

Many methods of estimating ET have been proposed. The methods may be broadly classified as those based on combination theory, humidity data, radiation data, temperature data, and miscellaneous methods which usually involve multiple correlations of ET and various climatic data. The design engineer or hydrologist unfamiliar with methods is often faced with a bewildering choice. Several publications discuss the choice of methods for various climatic conditions and for various amounts of input climatic data. Among these are a United Nations Food and Agriculture Organization publication, (FAO-ID 24), (Doorenbos and Pruitt, 1977) and a report of the ASCE Irrigation Water Requirements Committee (ASCE-CU Report) (Jensen, 1974).

Recent research by micrometerologists and soil scientists has separated ET calculations into evaporation from the soil and transpiration components (Ritchie, 1974). The transpiration rate has been successfully related to the leaf area index of the plants, the soil moisture status and potential transpiration rate. These have not been used in engineering calculations and have not been refined for a wide range of conditions and therefore, are not presented here. The reader should be aware that these methods may come into wider use in the future.

This chapter presents detailed step-by-step instructions for three of the most commonly used methods of estimating ET for a reference crop plus the use of evaporation from pans as an index of E_{rr}. The reader is referred to other sources for other methods such as Doorenbos and Pruitt (1977) and Jensen (1974).

6.4.1 Basis for Reference ET

Reference crop ET selected must be compatible with the crop coefficients (K_c) that are to be used. For example, K_c used to calculate ET based on alfalfa reference ET must not be used with an E_{rr} intended to simulate grass. The reverse is equally illogical. Engineers also must be certain that the method of estimating E_{rr} is related to the same base as was used for the development of the crop curves that they are using. The Penman and Jensen-Haise methods cited in this chapter both estimate E_{rr} based on alfalfa because these are compatible with recently developed crop coefficients for the Western United States (Wright, 1979). The Blaney-Criddle and pan evaporation methods described in this section are recent FAO modifications which estimate grass based reference ET.

Doorenbos and Pruitt (1977) also present modifications of the Penman method and radiation methods in the FAO publication, which as the first step requires estimates of grass based reference ET. The FAO procedures also require using grass based crop coefficients. The FAO procedures cover a very broad range of wind, sunshine, and humidity conditions because they are based on a world-wide data set. The Penman method presented in this chapter is particularly suited to irrigated areas in the Western United States because of recently developed alfalfa based crop coefficients (Wright, 1979).

6.4.2 Penman Method

The Penman method, first introduced in 1948 (Penman, 1948) and later simplified (Penman, 1963) was the first of several combination equations. Combination equations are derived from a combination of energy balance and a mass transport or aerodynamic term. The ASCE-CU Report shows that the combination methods are the most accurate methods for a very wide range of climatic conditions. The accuracy of combination methods results from the theoretical basis of the methods. Estimates obtained with a combination equation are reliable for periods of from 1 day to 1 month. With modifications, reliable hourly estimates are possible.

The Penman equation, modified for estimating alfalfa based reference ET in cal/cm²·d is :

$$E_{tr} = \frac{\Delta}{\Delta + \gamma} (R_n + G) + \frac{\gamma}{\Delta + \gamma} \, 15.36 \, W_f \, (e_a - e_d) \cdots\cdots\cdots\cdots \quad [6.4]$$

where E_{tr} = reference crop ET in cal/cm²·d; Δ is the slope of the vapor pressure-temperature curve in mb/°C; γ is the psychrometer constant in mb/°C; R_n is net radiation in cal/cm²·d; G is soil heat flux to the surface in cal/cm²·d; W_f is the wind function(dimensionless); ($e_a - e_d$) is the mean daily vapor pressure deficit in mb; and 15.36 is a constant of proportionality in cal/cm²·d·mb. An expression adapted from Bosen (1960) can be used to approximate Δ:

$$\Delta = 2.00(0.00738 \, T + 0.8072)^7 - 0.00116 \cdots\cdots\cdots\cdots\cdots \quad [6.5]$$

where T is mean daily temperature (°C). An expression by Brunt (1952) can be used to find γ:

$$\gamma = \frac{0.386 \, P}{L} \cdots\cdots\cdots\cdots\cdots\cdots\cdots\cdots\cdots\cdots\cdots\cdots \quad [6.6]$$

where P is average station barometric pressure (mb) and L is the latent heat of vaporization (cal/g). P is usually assumed to be a constant for a given location and may be calculated using a straight line approximation of the U.S. standard atmosphere;

$$P = 1013 - 0.1055 \, E \cdots\cdots\cdots\cdots\cdots\cdots\cdots\cdots\cdots \quad [6.7]$$

where E is sea level elevation (meters). L may be calculated as follows (Brunt, 1952):

$$L = 595 - 0.51 \, T \cdots\cdots\cdots\cdots\cdots\cdots\cdots\cdots\cdots\cdots \quad [6.8]$$

where T is °C. The variations of $\Delta/(\Delta + \gamma)$ with elevation and temperature are given in Table 6.1.

The W_f term is usually determined by regression techniques where W_f has the form:

$$W_f = a_w + b_w U_z \cdots\cdots\cdots\cdots\cdots\cdots\cdots\cdots\cdots\cdots \quad [6.9]$$

where a_w and b_w are regression coefficients and U_z is the daily wind travel (km/d) at z m above the ground. Many investigators recommend that a_w and b_w be determined for a location if the necessary data are available. Some values of a_w and b_w previously determined are listed in Table 6.2 for z = 2 m. Wright (1981) has developed functional relationships for a_w and b_w which vary with the season and are discussed later in this chapter. Wind travel, U_z, is frequently obtained at an elevation of 2 m above the ground for use in

TABLE 6.1. VARIATION OF $\Delta/(\Delta + \gamma)$ WITH
ELEVATION AND TEMPERATURE*

°C	Elev., m					
	0	500	1000	1500	2000	2500
0.0	0.401	0.414	0.428	0.443	0.458	0.475
5.0	0.477	0.491	0.505	0.520	0.536	0.552
10.0	0.551	0.564	0.578	0.593	0.608	0.624
15.0	0.620	0.632	0.645	0.659	0.673	0.688
20.0	0.681	0.693	0.705	0.717	0.730	0.743
25.0	0.735	0.745	0.756	0.767	0.778	0.790
30.0	0.781	0.790	0.799	0.809	0.818	0.828
35.0	0.820	0.828	0.835	0.844	0.852	0.860
40.0	0.852	0.858	0.867	0.872	0.879	0.886
45.0	0.878	0.884	0.889	0.895	0.901	0.907
50.0	0.900	0.904	0.909	0.914	0.919	0.924

* $\dfrac{\gamma}{\Delta + \gamma} = 1 - \dfrac{\Delta}{\Delta + \gamma}$, based on the U.S. standard atmosphere.

developing the wind functions for the Penman equation. Wind data collected at another elevation can be extrapolated to the 2-m elevation by the following expression which approximates a logrithmic velocity profile and is based on an aerodynamically "rough" crop surface such as alfalfa:

$$U_2 = U_z \left(\frac{2}{z}\right)^{0.2} \quad \dots\dots\dots\dots\dots\dots\dots\dots\dots\dots\dots\dots\dots\dots\dots \quad [6.10]$$

where z (m) is the elevation of the wind measurement and U_2 is the estimated wind travel at 2 m.

Various procedures have been used to calculate the saturation vapor pressure deficit term $(e_a - e_d)$ of equation [6.4] and sometimes the method used has not been clearly identified. Two possible methods are described here. Method 1 uses the saturation vapor pressure at mean air temperature as e_a and the saturation vapor pressure at the mean daily dew point temperature as e_d. This method is described in more detail by Doorenbos and Pruitt. Method 2 is more applicable in arid areas and high elevations where large diurnal temperature changes occur:

$$e_a = \frac{1}{2} (e_a \max + e_a \min) \quad \dots\dots\dots\dots\dots\dots\dots\dots\dots\dots\dots\dots\dots \quad [6.11]$$

TABLE 6.2. SELECTED VALUES OF a_w AND b_w FOR
VARIOUS WIND FUNCTIONS FOR THE PENMAN METHOD

No.	Author(s)	Reference crop	a_w	b_w	Method of calculating $(e_a - e_d)$
1	Penman (1963)	Clipped grass	1.0	0.00621	1
2	Wright and Jensen (1972)	Alfalfa	0.75	0.0115	2
3	Doorenbos and Pruitt (1977)	Grass	1.0	0.01	1
4	Wright (1981)	Alfalfa	(varies with time)		2

where e_a max is the saturation vapor pressure at maximum daily air temperature, e_a min is the saturation vapor pressure at minimum daily air temperature, and the saturation vapor pressure at the mean daily dew point temperature is used for e_d. Procedures for calculating the mean daily dew point temperature or mean daily vapor pressure are sometimes not clear or consistent. Future studies and publications are expected to establish a standard procedure for this.

It is extremely important to make certain that the crop coefficients to be used are based on the same W_f that was used to estimate reference crop ET. For example, use the W_f by Wright and Jensen (1972) or Wright (1981) for crop coefficients presented in Subsection 6.5.3. If the grass based E_{ro} as defined by Doorenbos and Pruitt (1977) is used, use K_c values from Subsection 6.5.4 or the crop coefficient procedures presented in FAO-ID 24. They emphasize that the wind function used must also be compatible with the method used to calculate the vapor pressure deficit term $(e_a - e_d)$ and the crop coefficients used must have been developed using the same procedure for calculating $(e_a - e_d)$ and the wind function W_f.

The absence of humidity data is often cited as a reason for not using combination equations in engineering calculations of ET. There are alternatives for estimating average daily dew point temperature. For example, Pochop et al. (1973) presented empirical relationships between average daily dew point temperature and daily minimum temperature for Wyoming. Saturation vapor pressure (mb) for any temperature T (°C) may be determined from the following approximation of Bosen (1960):

$$e_s \simeq 33.8639 \left[(0.00738\ T + 0.8072)^8 - 0.000019\ |1.8\ T + 48| \right.$$

$$\left. + 0.001316 \right] \quad \dots\dots\dots\dots\dots\dots\dots\dots\dots\dots\dots\dots\dots\dots\dots\dots [6.12]$$

Net radiation (R_n) in langleys per day (ly/d) can be calculated from solar radiation data. A langley is a cal/cm². The signs of R_n and G (equation [6.4]) assume that heat movement toward the soil surface is positive. In practice, G is often assumed to be zero for daily E_{tr} calculations. To estimate R_n:

$$R_n = (1 - \alpha)\ R_s - R_b \quad \dots\dots\dots\dots\dots\dots\dots\dots\dots\dots\dots\dots\dots\dots\dots [6.13]$$

where α is reflected short wave radiation, called albedo, expressed as a decimal. Albedo is often taken to be 0.23 for commercial irrigated crops. Merva (1975) presented an extensive table of α values. However, albedo is known to change with sun angle and can be estimated with an equation such as equation [6.36] for alfalfa at Kimberly, Idaho (Wright, 1981), if sufficient data are available. R_s is incoming short wave solar radiation. R_b is net outgoing long wave radiation and may be estimated as follows:

$$R_b = \left[a\ \frac{R_s}{R_{so}} + b \right] R_{bo} \quad \dots\dots\dots\dots\dots\dots\dots\dots\dots\dots\dots\dots\dots [6.14]$$

where R_{so} is clear day solar radiation, i.e. the solar radiation expected on a day without clouds. A clear day radiation curve can be plotted from several years of solar radiation data with the upper envelope forming the clear day radiation curves. Some experimentally determined coefficients, a and b, are

**TABLE 6.3. EXPERIMENTAL COEFFICIENTS FOR
NET RADIATION EQUATIONS [6.14] AND
[6.16] (from Jensen, 1974)**

Region	(a b)	$(a_1$ $b_1)$
Davis, California	(1.35, – 0.35)	(0.35, – 0.046)
Southern Idaho	(1.22, – 0.18)	(0.325, – 0.044)
England	(not available)	(0.47, – 0.065)
England	(not available)	(0.44, – 0.080)
Australia	(not available)	(0.35, – 0.042)
General	(1.2, – 0.2)	(0.39, – 0.05)
General	(1.0, 0)	

shown in Table 6.3. R_{bo} is net outgoing long wave radiation on a clear day
and may be estimated as follows:

$$R_{bo} = \epsilon\, 11.71 \times 10^{-8}\, T_k^4 \quad\dots\dots\dots\dots\dots\dots\dots\dots\dots \quad [6.15]$$

$$= (a_1 + b_1 \sqrt{e_d})\, 11.71 \times 10^{-8}\, T_k^4 \quad\dots\dots\dots\dots\dots\dots \quad [6.16]$$

where e_d has previously been defined in this chapter, T_k is average daily air
temperature in °K and some values for a_1 and b_1 can be found in Table 6.3.
If humidity data are not available, the following expression developed by
Idso and Jackson (1969) may be used to calculate ϵ:

$$\epsilon = -0.02 + 0.261\, \exp[-7.77 \times 10^{-4}\, (273 - T_k)^2] \quad\dots\dots\dots\dots \quad [6.17]$$

where T_k is in °K.

R_n can also be calculated from the following simplified procedure:

$$R_n = a_3 R_s + b_3 \quad\dots\dots\dots\dots\dots\dots\dots\dots\dots\dots\dots\dots\dots\dots\dots \quad [6.18]$$

An extensive table of values of a_3 and b_3 was presented in the ASCE-CU
Report (Jensen, 1974).

Penman's original method (Penman, 1948) called for an initial estimate
of evaporation from a hypothetical open water surface and then its conversion to potential ET by an empirical coefficient which varied with the season.
Doorenbos and Pruitt (1977) developed a somewhat similar approach, but
their corrections are related to maximum humidity, the ratio of daytime to
night-time winds and wind velocity; their procedures are recommended for
E_{to} estimates of periods from 10 days to 1 month.

6.4.3 Jensen-Haise Method

The Jensen-Haise method (Jensen and Haise, 1963) is another procedure
for estimating ET from climatic data. Though the method is often classified
as a solar radiation method, air temperature is also used and the coefficients
are based on other input parameters such as elevation and long term mean
temperature. The method produces an estimate of an alfalfa E_{tr} as defined by
Jensen et al., (1970). Doorenbos and Pruitt (1977) also presented a solar
radiation method for estimating E_{to} for grass. The reader is again cautioned
that both the method of estimating E_{tr} and the crop coefficients must be based on the same reference crop.

The Jensen-Haise method is the result of a review of about 3000 measurements of ET that were made in the Western United States over about a 35 year period. The method presented in this chapter is known as the "Modified Jensen-Haise" method. The ASCE Irrigaton Water Requirements Committee recommended that estimates using the Jensen-Haise method be made for periods of 5 days to a month.

The Jensen-Haise method is as follows:

$$E_{tr} = C_T (T - T_x) R_s \qquad\qquad\qquad\qquad\qquad\qquad [6.19]$$

where E_{tr} has the same units as R_s and is compatible with alfalfa based crop coefficients.

$$C_T = \frac{1}{C_1 + 7.3\, C_H} \qquad\qquad\qquad\qquad\qquad\qquad [6.20]$$

$$C_H = \frac{50\ mb}{e_2 - e_1} \qquad\qquad\qquad\qquad\qquad\qquad\qquad [6.21]$$

where e_2 is the saturation vapor pressure of water in mb at the mean monthly maximum air temperature of the warmest month in the year (long term climatic data), and e_1 is the saturation vapor pressure of water in mb at the mean monthly minimum air temperature of the warmest month in the year.

$$C_1 = 38 - \frac{2\,E}{305} \qquad\qquad\qquad\qquad\qquad\qquad\qquad [6.22]$$

where $E = $ the site elevation in m.

$$T_x = -2.5 - 0.14(e_2 - e_1) - \frac{E}{550} \qquad\qquad\qquad\qquad [6.23]$$

Solar radiation may be measured or estimated.

6.4.4 Blaney-Criddle Method

The Blaney-Criddle method was first proposed in 1945 by H. F. Blaney and W. D. Criddle (Blaney and Criddle, 1945) and was based on Western USA field measurements of ET. The method has been revised many times and there are so many variations that when the method is used the authors must be very careful and complete in their identification of the exact variation used. Perhaps the best known variation in the United States is that found in Technical Release No. 21 of the USDA Soil Conservation Service (USDA SCS, 1970). The method has been used on a world-wide basis but local calibration has been considered highly desirable.

The Blaney-Criddle method is based on the principle that ET is proportional to the product of daylength percentage and mean air temperature. The monthly constant of proportionality has been called the crop growth stage coefficient. This coefficient is not the same as the crop coefficient defined by equation [6.1] and [6.2]. Estimating ET by the early versions of the Blaney-Criddle method is a single stage process which does not involve the intermediate step of estimating reference crop ET. Estimates have been considered to be valid for monthly periods (Jensen, 1974). The one stage Blaney-Criddle method is widely used in the intermountain region of the United States, with local calibration, for water right deliberations (Kruse and Haise,

1974; Burman, 1979).

A recent major revision of the Blaney-Criddle method was published by FAO (Doorenbos and Pruitt, 1977). The FAO Blaney-Criddle method first produces a reference crop ET estimate for grass (see Subsection 6.2.2). The FAO modifications were based on data from 20 locations representing a very wide range of climatic conditions.

The FAO variation uses air temperature measurements for the site in question. The need for local calibration is minimized by the classification of climate at a site based on daytime wind, humidity and sunshine. For these classifications general estimates of wind, humidity or sunshine from sources such as a climatic atlas or more exact data may be used.

The FAO variation of the Blaney-Criddle method is as follows:

$$E_{to} = a_4 + b_4\, f \quad \dotfill \quad [6.24]$$

$$f = p(0.46\,T + 8) \quad \dotfill \quad [6.25]$$

where E_{to} is in mm/d, p is the percentage of daytime hours of a day compared to the entire year (see Table 6.4), and T is the average monthly air temperatures, °C.

The numbers a_4 and b_4 represent the intercept and slope of a straight line relationship between E_{to} and f. E_{to} may be determined directly from f using Fig. 6.2 and classifications of daytime wind, minimum humidity and percent sunshine.

Daytime wind may be estimated from daily wind by using the ratio of day to night winds.

U day/U night ratio	1.0	1.5	2.0	3.0	3.5	4.0
Correction for U day	1.0	1.5	1.33	1.5	1.56	1.6

TABLE 6.4. MEAN DAILY PERCENTAGE (p) OF ANNUAL
DAYTIME HOURS FOR DIFFERENT LATITUDES

Latitude North / South*	Jan / July	Feb / Aug	Mar / Sept	Apr / Oct	May / Nov	June / Dec	July / Jan	Aug / Feb	Sept / Mar	Oct / Apr	Nov / May	Dec / June
60 deg	0.15	0.20	0.26	0.32	0.38	0.41	0.40	0.34	0.28	0.22	0.17	0.13
58	0.16	0.21	0.26	0.32	0.37	0.40	0.39	0.34	0.28	0.23	0.18	0.15
56	0.17	0.21	0.26	0.32	0.36	0.39	0.38	0.33	0.28	0.23	0.18	0.16
54	0.18	0.22	0.26	0.31	0.36	0.38	0.37	0.33	0.28	0.23	0.19	0.17
52	0.19	0.22	0.27	0.31	0.35	0.37	0.36	0.33	0.28	0.24	0.20	0.17
50	0.19	0.23	0.27	0.31	0.34	0.36	0.35	0.32	0.28	0.24	0.20	0.18
48	0.20	0.23	0.27	0.31	0.34	0.36	0.35	0.32	0.28	0.24	0.21	0.19
46	0.20	0.23	0.27	0.30	0.34	0.35	0.34	0.32	0.28	0.24	0.21	0.20
44	0.21	0.24	0.27	0.30	0.33	0.35	0.34	0.31	0.28	0.25	0.22	0.20
42	0.21	0.24	0.27	0.30	0.33	0.34	0.33	0.31	0.28	0.25	0.22	0.21
40	0.22	0.24	0.27	0.30	0.32	0.34	0.33	0.31	0.28	0.25	0.22	0.21
35	0.23	0.25	0.27	0.29	0.31	0.32	0.32	0.30	0.28	0.25	0.23	0.22
30	0.24	0.25	0.27	0.29	0.31	0.32	0.31	0.30	0.28	0.26	0.24	0.23
25	0.24	0.26	0.27	0.29	0.30	0.31	0.31	0.29	0.28	0.26	0.25	0.24
20	0.25	0.26	0.27	0.28	0.29	0.30	0.30	0.29	0.28	0.26	0.25	0.25
15	0.26	0.27	0.27	0.28	0.29	0.29	0.29	0.28	0.28	0.27	0.26	0.25
10	0.26	0.27	0.27	0.28	0.28	0.29	0.29	0.28	0.28	0.27	0.26	0.26
5	0.27	0.27	0.27	0.28	0.28	0.28	0.28	0.28	0.28	0.27	0.27	0.27
0	0.27	0.27	0.27	0.27	0.27	0.27	0.27	0.27	0.27	0.27	0.27	0.27

*Southern latitudes: apply 6 mo difference as shown.

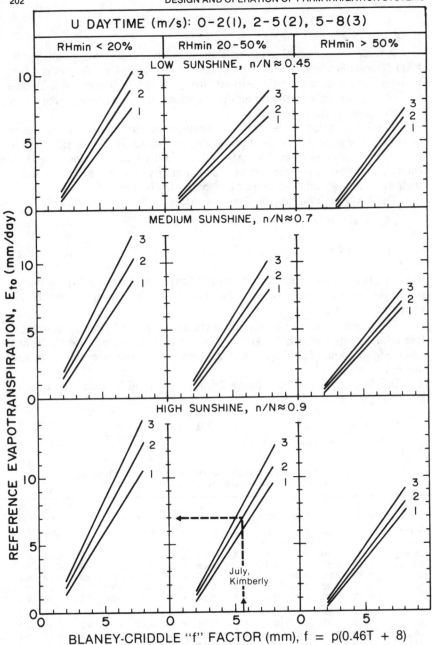

FIG. 6.2 Prediction of reference ET for grass (E$_{ro}$) from Blaney-Criddle f factor for different conditions of minimum relative humidity, sunshine duration and day-time wind (from Doorenbos and Pruitt, 1977).

The minimum relative humidity is the ratio of saturation vapor pressure at average dew point temperature to that at maximum air temperature.

Doorenbos and Pruitt (1977) recommend that individual calculations be made for each month of record and that values of E$_{ro}$ may need to be increased for higher elevations or latitudes. They recommend estimation periods of from 10 days to one month. For computerized applications, Doorenbos and Pruitt (1977) recommend interpolation of the slope of the line from an extensive table and the intercept from humidity and sunshine inputs.

6.4.5 Pan Evaporation Method

Evaporation pans are an integral part of most agricultural weather stations. If the stations are visited weekly or more often and the operator is diligent, excellent data may be collected. Reference crop ET may be estimated by the following relationship.

$$E_{to} = K_p E_p \quad \dots\dots\dots\dots\dots\dots\dots\dots\dots\dots\dots\dots\dots\dots\dots\dots \quad [6.26]$$

where E_p = pan evaporation in any desired units, for example mm/d, K_p = dimensionless pan coefficient, and E_{to} = reference crop ET (grass) in the same units as E_p.

Since E_{to} represents grass ET (see Subsection 6.2.2) it is therefore mandatory that crop coefficients (K_c) used to convert E_{to} to ET for a specific crop and time be taken from Subsection 6.5.4 or from FAO-ID 24. The information in this Subsection, while useful in interpreting data from existing pans, is intended more as guidelines for locating evaporation pans specifically intended for estimating ET.

Data from evaporation pans have been correlated with ET for many years because pan evaporation integrates many of the factors involved in ET; these include wind, radiation, humidity and air temperature. The evaporation pan however is inanimate and does not reflect heat storage and transfer characteristics of a crop. For literature review the reader is referred to Doorenbos and Pruitt (1977) and Jensen (1974).

Types of pans. Discussion in this Subsection is limited to the *U.S. Class A Pan*. This pan is 121 cm in diameter and 25.5 cm deep. The pan is usually constructed of galvanized steel or Monel metal. The pan is placed on a wooden platform and leveled. The bottom of the pan is usually about 15 cm above ground level. The water level is maintained within a range of from 5 to 7.5 cm below the rim by careful water additions, or by a float system and a supply tank. Changes in water level are measured by a vernier hook gage placed in a stilling well. Many other types of evaporation pans have been used; these include different sizes, depths, screens and many are buried below the ground surface (also see Subsection 16.5.3). Doorenbos and Pruitt (1977) present a table of factors plus narrative discussion relating various sizes of pans to the *Colorado Sunken Pan*. Hounam (1973) also discusses various sizes, types of pans, and their relative performance.

Selection of K_p values. The pan coefficient varies with pan exposure, wind velocity, humidity, and distance of homogeneous material to the windward side (fetch). Values of K_p for periods of 10 days to a month may be selected from Table 6.5. Additional factors are discussed later. Table 6.5 is self explanatory except Cases A and B need further elaboration. Case A defines the condition where air moves across at least 50 m of dry surface and then across from 1 to 1000 m of a green crop. The situation is reversed in Case B; see the sketch below for a visual interpretation. Doorenbos and Pruitt (1977) also present a similar table for use with the Colorado sunken pan.

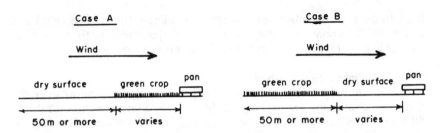

Additional factors. Many additional factors can modify the pan coefficients found in Table 6.5. For example E_p may be increased by 10 percent if the pan is painted black. If pans are placed in a small enclosure surrounded by tall crops, K_p may need to be increased by up to 30 percent for dry windy climates, and only from 5 to 10 percent for calm humid climates. The coefficients presented in Table 6.5 assume no screen is present, that no crops taller than 1 m are within 50 m and that the area within 10 m of the pan is covered by a frequently mowed green grass cover or by bare soils. Doorenbos and Pruitt (1977), Jensen (1974), and Hounam (1973) discuss additional factors that influence pan evaporation.

Location and operation of pans. A weather station which includes an evaporation pan should be located so that its surrounding conditions are easy to classify and maintain in as constant a condition as possible. The tempta-

TABLE 6.5. PAN COEFFICIENT K_p FOR CLASS A PAN FOR DIFFERENT GROUND COVER AND LEVELS OF MEAN RELATIVE HUMIDITY AND 24 h WIND
(For use in equation [6.26] to estimate E_{to})

Class A Pan	Case A Pan surrounded by short green crop			Case B† Pan surrounded by dry-fallow land		
RH mean %	low < 40	medium 40-70	high > 70	low < 40	medium 40-70	high > 70
Wind‡ km/day	Upwind distance of green crop m			Upwind distance of dry fallow m		
Light	0	0.55	0.65	0.75	0	0.7	0.8	0.85
< 175	10	0.65	0.75	0.85	10	0.6	0.7	0.8
	100	0.7	0.8	0.85	100	0.55	0.65	0.75
	1 000	0.75	0.85	0.85	1 000	0.5	0.6	0.7
Moderate	0	0.5	0.6	0.65	0	0.65	0.75	0.8
175-425	10	0.6	0.7	0.75	10	0.55	0.65*	0.7
	100	0.65	0.75	0.8	100	0.5	0.6	0.65
	1 000	0.7	0.8	0.8	1 000	0.45	0.55	0.6
Strong	0	0.45	0.5	0.60	0	0.6	0.65	0.7
425-700	10	0.55	0.6	0.65	10	0.5	0.55	0.65
	100	0.6	0.65	0.7	100	0.45	0.45	0.6
	1 000	0.65	0.7	0.75	1 000	0.4	0.45	0.55
Very strong	0	0.4	0.45	0.5	0	0.5	0.6	0.65
> 700	10	0.45	0.55	0.6	10	0.45	0.5	0.55
	100	0.5	0.6	0.65	100	0.4	0.45	0.5
	1 000	0.55	0.6	0.65	1 000	0.35	0.4	0.45

†For extensive areas of bare-fallow soils and not agricultural development, reduce Kpan values by 20 percent under hot windy conditions, by 5 to 10 percent for moderate wind, temperature and humidity conditions.
‡Total wind movement km/d.

tion to place the station in an unused or otherwise convenient but unrepresentative location should be resisted. The pan's location should be dictated by the intended purposes. With proper location and care in use, reference crop ET estimates to ± 10 percent accuracy should be possible.

6.4.6 Local Calibration

All methods of estimating ET from climatic data involve empirical relationships to some extent. Even the combination equation, the Penman method for example, utilizes an empirical wind function. The empirical relationships account for many local conditions. The ASCE Irrigation Water Requirements Committee stated that ". . . no single existing method using meteorological data is universally adequate for all climatic regimes, especially for tropical areas and for high elevations, without some local or regional calibration" (Jensen, 1974). If the crop economic importance is high, local calibration is needed to at least give confidence to irrigation water requirement estimates. Doorenbos and Pruitt (1977) present a detailed description of a world wide calibration of the Blaney-Criddle, radiation, and Penman methods. The principles can be applied to a local or regional calibration.

Calibration involves the simultaneous collection of field E_t data and the corresponding climatic data. The time interval for ET estimates has an influence on the methods that are used for field measurements. Preferably, if the method is to be used for short period estimates, comparable data should be used in calibration.

Blaney-Criddle method. The Blaney-Criddle method is suited for monthly estimates of ET, (Jensen, 1974). Therefore, field measurements of ET can be made using careful soil moisture measurements, water table lysimeters, drainage lysimeters, weighing lysimeters or inflow-outflow techniques. Only air temperature and rainfall data are needed to complete the calibration by determining the appropriate monthly crop coefficient.

Jensen-Haise method. The Jensen-Haise method is recommended for 5-day to 1-month periods (Jensen, 1974). Drainage lysimeters are only suitable for 10-day or longer periods (Doorenbos and Pruitt, 1977), and can be eliminated if short period calibration is desired. ET measured by soil moisture change can also be eliminated for short period calibrations. Therefore, if 5-day periods are desired, weighing lysimeters or Bowen ratio techniques should be used to collect the necessary field ET data for local calibration. For monthly calibration, ET may be determined by properly performed measurements of soil moisture depletion, inflow-outflow, lysimeters or other techniques. Climatic data should include solar radiation, air temperature and rainfall data on at least a daily basis.

Local calibration of both C_T and T_x can be obtained by regression of measured E_{tr}/R_{so} against mean air temperature if data are available from about 5 to 30 °C, or higher. If only a few data points are available over a narrow temperature range, then these data should be used to adjust the T_x value, but not the C_T value.

Penman method. The Penman method can provide accurate estimates of ET for periods of 1 month to 1 hour depending on the method of calibration. For short periods only weighing lysimeters can provide the necessary E_t data. Climatic data must include, solar radiation, net radiation if possible, wind movement, air temperature, vapor pressure and precipitation all collected on intervals suitable for the desired prediction periods. Usually local calibration is accomplished by calibrating the transfer coefficient identification of the variables.

$$h = 15.36 \, W_f \, (e_a - e_d) \quad \dots\dots\dots\dots\dots\dots\dots\dots\dots\dots\dots\dots \quad [6.27]$$

Whenever local calibration is made, consistency between any reference crop used, crop coefficients, and calculation method used to obtain terms as $(e_a - e_d)$ must be followed. If consistency is not followed ET estimates will be illogical and may not represent the crop grown. For daily calibration of the Penman method see Wright (1981) and Subsection 6.5.3.

6.5 ESTIMATING ET FOR CROPS

Estimating ET for a specific crop can be a very complex matter depending on the degree of refinement desired. To obtain the most accurate estimates, all of the major contributing crop and environmental conditions need to be taken into account. These involve climate, soil moisture, the type of crop, stage of growth and the extent to which the plants cover the soil. This section is intended to provide the means for the practicing engineer or irrigation scientist to integrate these inter-related factors into the best possible ET estimates. The procedures primarily involve the use of an estimated reference ET and experimentally developed ET crop coefficients. Such procedures are now extensively used in irrigation scheduling methods and in estimating crop water requirements and have been described in detail in previous publications. For purposes of this section, the most salient principles and information are provided. Those desiring more information should consult the listed references.

The common Blaney-Criddle method does not use ET crop coefficients. Rather, the estimations of crop ET are made in one step. The method was revised by Doorenbos and Pruitt (1977) to provide an estimate of E_{to} for grass so that appropriate crop coefficients could be used to estimate ET for a specific crop. Such procedures produce estimates with accuracies suitable for 10-day to monthly periods.

Detailed and specific procedures and guidelines were summarized by Doorenbos and Pruitt (1977) for predicting crop water requirements for a wide range of crops and conditions and availability of associated information. They outlined a three-stage procedure involving (a) a reference crop ET, (b) a crop coefficient, and (c) the effects of local conditions and agricultural practices. They chose ET for 8- to 15-cm tall, green, well-watered grass as the reference ET and selected or adapted crop coefficients accordingly. Four methods of estimating this reference ET were presented, namely: (a) Blaney-Criddle, (b) radiation, (c) Penman, and (d) pan evaporation. In this section, we present crop coefficients for E_{tr} based on alfalfa, as defined by Jensen et al. (1970) suitable for daily estimates of ET when E_{tr} is determined by the Penman method described in this chapter. These alfalfa based coefficients are also suitable for the Jensen-Haise method as presented in Subsection 6.4.3. We also present a limited set of crop coefficients based on grass E_{to} which are intended for use with the FAO Blaney-Criddle and pan evaporation methods described in Subsections 6.4.4 and 6.4.5.

6.5.1 Crop Coefficients

Experimentally developed crop coefficients reflect the physiology of the crop, the degree of crop cover, and the reference ET. In applying the coeffi-

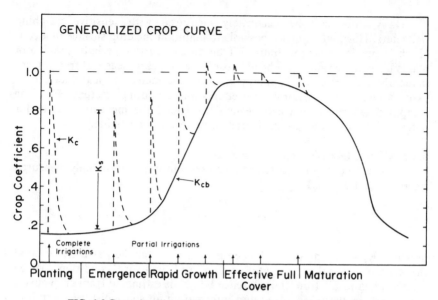

FIG. 6.3 Generalized basal ET crop coefficient curve (K_{cb}) with adjustment for increased evaporation due to surface soil wetness (K_s) to determine the over-all crop coefficient (K_c).

cients, it is important to know how they were derived since they are empirical ratios of crop ET to the reference ET, as shown in equation [6.1]. The combined crop coefficient includes evaporation from both the soil and plant surfaces. The contribution of soil evaporation is strongly dependent upon the surface soil wetness and exposure. Transpiration is primarily dependent upon the amount and nature of plant leaf area, and the availability of water within the root-zone. Crop coefficients can be adjusted for soil moisture availability and surface evaporation. The distribution of crop coefficients with time is known as a crop curve. See Fig. 6.3 and 18.1 for examples of crop curves. Other time-related crop parameters may also be used as a base.

In the experimental determination of crop coefficients, ideally both crop ET and reference ET are measured concurrently. The crop coefficient is then calculated as the dimensionless ratio of the two measurements. Well sited, sensitive weighing lysimeters provide ideal daily measurements and problems with soil-water drainage are avoided. Care must be taken to insure that border effects are minimized, that fetch is adequate, and that crop and soil moisture conditions are similar in the lysimeter and the field.

6.5.2 Reference ET

Alfalfa has frequently been selected as a reference crop because it has relatively high ET rates in arid areas where there is considerable advective sensible heat input from the air (Wright and Jensen, 1972; and Wright, 1979, 1981). In such cases, reference ET (E_{rr}) is equal to daily alfalfa ET when the crop occupies an extensive surface, is actively growing, standing erect and at least 20-cm tall, and is well watered so that soil water availability does not limit ET. Reference ET obtained with such an alfalfa surface will usually be greater than that for a clipped grass surface, particularly in windy arid areas.

Daily rates can be accurately measured with sensitive weighing lysimeters. However, it is not possible to maintain the crop surface in a condition to provide near maximum ET because of cutting periods, lodging of plants by wind or rain, and the effects of late and early seasonal frosts. Consequently, daily alfalfa ET, energy balance, and meteorological data can be used to develop and calibrate procedures for computing reference ET. The computed reference then can be used to extend the measured values for periods or locations where measured values are not available.

6.5.3 Alfalfa Related Crop Coefficients

An overall daily crop coefficient can be determined from daily measured reference and crop ET by:

$$K_c = \frac{E_t}{E_{tr}} \quad\quad\quad [6.28]$$

in which K_c = the dimensionless crop coefficient for the particular crop at the existing growth stage and surface soil moisture condition. When estimating crop ET from the reference ET, K_c is estimated from crop curves for the day or period involved and information on soil moisture conditions by:

$$K_c = K_{cb} K_a + K_s \quad\quad\quad [6.29]$$

in which K_c = daily crop coefficient, K_{cb} = daily basal ET crop coefficient, K_a = a coefficient dependent upon available soil moisture, and K_s = a coefficient to allow for increased evaporation from the soil surface occurring after rain or irrigation. These procedures are described in greater detail by Jensen (1974), and Jensen et al. (1971). The generalized basal crop coefficient, K_{cb}, was defined by Wright (1979) to represent conditions when the soil surface was dry so that evaporation from the soil was minimal but soil-water availability did not limit plant growth or transpiration, i.e. $K_c = K_{cb}$ with K_a = 1 and K_s = 0. He determined daily values of K_{cb} by manually fitting a basal crop curve to overall crop curves obtained with equation [6.28]. This specific designation also distinguished the K_{cb} values obtained with lysimeter ET data from mean crop coefficients previously developed from soil-water-balance data.

When available water within the root zone limits growth and ET, K_a of equation [6.29] will be less than 1.0 and can be approximated by relationships similar to:

$$K_a = [\ln(A_w + 1)]/[\ln(101)] \quad\quad\quad [6.30]$$

in which A_w = the percentage of available water (100 when the soil is at field capacity), and K_a = 1 when A_w = 100, and K_a goes to zero as A_w goes to 0. This algorithm was developed from published ET-soil water data (Jensen et al., 1971). Other relationships for K_a were reviewed by Howell (1979).

Increased soil evaporation due to rainfall or irrigation, can be estimated by:

$$K_s = (K_1 - K_{ci})\exp(-\Lambda t), \quad K_1 > K_{ci} \quad\quad\quad [6.31]$$

in which t = the number of days after the rain or irrigation; Λ = the combined effects of soil characteristics, evaporative demand, etc; and K_{ci} = the value of K_{cb} at the time the rain or irrigation occurred. This algorithm will also vary for various soils and locations. At Kimberly, Idaho K_s was approximated by: $(0.9 - K_c)0.8$; $(0.9 - K_c)0.5$; and $(0.9 - K_c)0.3$; for the first, second, and third days after a rain or irrigation, respectively (Jensen et al., 1971). When K_c exceeds 0.9 no adjustment is needed for rain or irrigation. A diagramatic representation of the expected changes in the crop coefficient as affected by stage of growth and wet surface soil, is presented in Fig. 6.3.

A summary of basal crop coefficients for several crops is presented in Table 6.6 for arid areas. These were derived for use with estimated ET for a reference crop of actively growing, well watered alfalfa at least 20-cm tall. Dates typical of Kimberly, Idaho for planting, emergence, effective cover, and harvest for the various crops are presented in Table 6.7.

Values of K_{cb} are listed on a normalized time scale, instead of actual dates, with time from planting until full cover on a percentage basis, PCT, and time after as elapsed days, DT. Coefficient relationships of this type have been used extensively in irrigation scheduling (Jensen, 1974). The normalized time scale helps account for the effects of seasonal differences on crop development. Alfalfa cuttings are listed individually because of major differences in climate for each of the growth periods.

The alfalfa related crop coefficients described in this section were computed using the Penman method discussed in Subsection 6.4.2 with some modifications. Suitable procedures have been described in many publica-

TABLE 6.6. DAILY BASAL ET CROP COEFFICIENTS (K_{cb}) FOR DRY SURFACE SOIL CONDITIONS
for use with a reference ET representative of alfalfa for irrigated crops grown in an arid region with a temperate inter-mountain climate. Coefficients were determined experimentally using ET data obtained with sensitive weighing lysimeters at Kimberly, Idaho, from 1968 through 1978, (from Wright, 1979)

Crop	Basal ET crop coefficients, K_{cb}									
	PCT, time from planting to effective cover (%)									
	10	20	30	40	50	60	70	80	90	100
Small grains	0.15	0.16	0.20	0.28	0.55	0.75	0.90	0.98	1.00	1.02
Beans	0.15	0.17	0.18	0.22	0.38	0.48	0.65	0.78	0.93	0.95
Peas	0.20	0.17	0.16	0.18	0.20	0.28	0.48	0.67	0.86	0.95
Potatoes	0.15	0.15	0.15	0.21	0.35	0.45	0.60	0.72	0.78	0.80
Sugar beets	0.20	0.17	0.15	0.15	0.16	0.20	0.30	0.50	0.80	1.00
Corn	0.15	0.15	0.16	0.17	0.18	0.25	0.40	0.62	0.80	0.95
Alfalfa (1st)	0.50	0.58	0.67	0.75	0.80	0.85	0.90	0.95	0.98	1.00
(2nd & 3rd)	0.50	0.25	0.25	0.40	0.55	0.79	0.80	0.90	0.98	1.00
Winter wheat	0.65	0.70	0.75	0.80	0.85	0.90	0.95	0.98	1.00	1.02
	DT, days after effective cover									
	10	20	30	40	50	60	70	80	90	100
Small grains	1.02	1.00	0.80	0.50	0.25	0.10	0.10	—	—	—
Beans	0.95	0.94	0.65	0.36	0.18	0.15	0.10	—	—	—
Peas	0.93	0.82	0.50	0.37	0.20	0.10	0.10	—	—	—
Potatoes	0.80	0.80	0.75	0.74	0.73	0.72	0.70	0.50	0.25	0.20
Sugar beets	1.00	1.00	1.00	0.96	0.93	0.89	0.86	0.83	0.80	0.75
Corn	0.95	0.95	0.93	0.91	0.89	0.83	0.76	0.30	0.20	0.15
Alfalfa (1 & 2)	1.00	1.00	1.00	0.25	—	—	—	—	—	—
(3rd)*	1.00	1.00	0.52	0.30	—	—	—	—	—	—
Winter wheat	1.02	1.00	0.96	0.50	0.20	0.10	0.10	—	—	—

*Final cutting.

TABLE 6.7. DATE OF VARIOUS CROP GROWTH STAGES IDENTIFIABLE
IN THE FIELD FOR CROPS STUDIED AT KIMBERLY, IDAHO, 1968-1978
(from Wright, 1979)

| Crop | Date of occurrence | | | | | | | Time (days) | |
	Plant-ing	Emer-gence	Rapid growth	Full cover	Heading or bloom	Ripen-ing	Harvest	Planting to full cover	Full cover to harvest
Small grains	4/1	4/15	5/10	6/20	6/15	7/20	8/15	80	55
Beans	5/22	6/5	6/15	7/15	7/5	8/10	8/30	55	45
Peas	4/10	4/25	5/10	6/5	6/15	7/5	7/25	55	50
Potatoes	4/25	5/25	6/10	7/10	7/1	—	10/10	75	90
Sugar beets	4/15	5/15	6/10	7/15	—	—	10/15	91	100
Corn	5/5	5/25	6/10	7/15	7/30	9/10	9/20	79	70
Alfalfa 1st	4/1	—	4/20	5/15	—	—	6/15	45	35
2nd	6/15	—	6/25	7/5	—	—	8/1	20	35
3rd	8/1	—	8/10	8/25	—	—	9/20	25	25
Winter wheat	10/1	10/15	3/20	4/25	6/5	7/15	8/10	205	60

tions, such as those of Jensen (1974), Jensen et al. (1971), Wright and Jensen (1972), Wright and Jensen (1978), and Wright (1981). Other methods can also be adapted, but as mentioned earlier in this chapter, the combination equation seems to give the most consistent results, particularly in arid irrigated regions subject to considerable sensible heat advection. To adequately account for advection, even the combination equation should be calibrated or verified for local conditions.

The changes necessary to permit estimating reference ET for a crop of well watered, actively growing alfalfa, at least 20-cm tall, are presented here for convenience of the reader. This follows procedures developed earlier with recent refinements by Wright (1981). Measurements or estimates of the following daily meteorological parameters are required: (1) solar radiation, (2) maximum and minimum air temperature, (3) average humidity, or at least an 0800-h dew-point temperature, and (4) wind travel.

A combination equation similar to that in Subsection 6.4.2 was used to estimate a reference ET for the development of the basal crop coefficients by:

$$E_{tr} = 10 \frac{E_t}{L} \qquad \qquad \qquad [6.32]$$

where E_{tr} is on a water depth equivalent basis (mm/d), E_t is the latent heat flux computed with the calibrated equation (cal/cm^2·d), L is the latent heat of vaporization (cal/cm^3), and 10 is for unit conversion (mm/cm). A wind function with time dependent coefficients was used.

$$W_f = a_w(t) + b_w(t) U_2 \qquad \qquad \qquad [6.33]$$

where W_f is the wind function and $a_w(t)$ and $b_w(t)$ are variable coefficients to adapt the function to the location or time of year. Varying the wind function permits adapting W_f to changing conditions of the surrounding area which influence sensible heat advection. The following empirical relationships were derived for Kimberly, Idaho.

$$a_w(t) = 23.8 - 0.7865D + (9.7182E\text{-}03)D^2 - (5.4589E\text{-}05)D^3$$

$$+ (1.42529E\text{-}07)D^4 - (1.41018E\text{-}10)D^5 \qquad \qquad [6.34]$$

$$b_w(t) = -0.0122 + (5.2956E\text{-}04)D - (5.9923E\text{-}06)D^2$$
$$+ (3.4002E\text{-}08)D^3 - (9.00872E\text{-}11)D^4 + (8.79179E\text{-}14)D^5$$

$$\dotfill [6.35]$$

where D is the day of the year and the polynomial coefficients are for wind travel measured at 2 m in km/d. Respective values for 4/15, 6/15, 8/15, 10/15, and seasonal mean for a_w are: 0.74, 1.83, 1.01, 0.55, and 1.06; and for b_w: 0.0069, 0.0088, 0.0107, 0.0099, and 0.0091. These mean values compare with the seasonal Penman coefficients of 1.0 and 0.0062 and 0.75 and 0.0115 of Wright and Jensen (1972, 1978) (also see Table 6.2).

The net radiation term, R_n, of equation [6.4] was estimated from daily solar radiation, temperature, and humidity data by equations [6.13] to [6.16] using values and functions as developed by Wright (1981) for Kimberly, Idaho. The albedo (α) was computed by:

$$\alpha = 0.29 + 0.06 \ \text{SIN} \left\{ 30[M+(N/30) + 2.25] \right\} \quad \dotfill [6.36]$$

where M is the number of the month and N is the number of the day. The season long regression coefficients for Kimberly, Idaho are: a_1 is 0.325 and b_1 is -0.044 (Wright and Jensen, 1972). The coefficient a_1 of equation [6.16] was computed with a "normal" distribution equation:

$$a_1 = 0.26 + 0.1 \ \text{exp} \left\{ -[30(M+N/30)-207)/65]^2 \right\} \quad \dotfill [6.37]$$

A constant value of b_1 of -0.044 was used with the variable a_1. Coefficients for equation [6.14] were: for R_s/R_{so} greater than 0.7; a = 1.054 and b = 0; and for R_s/R_{so} less than or equal to 0.7, a = 1.0 and b = 0.

TABLE 6.8. DAILY BASAL ET CROP COEFFICIENTS (K_{cb}) FOR USE WITH GRASS REFERENCE ET (E_{to})
for irrigated crops grown in an arid Mediterranean climate. Coefficients are for dry soil surface conditions and were determined experimentally with ET data obtained with sensitive weighing lysimeters at Davis, CA, 1965-1975. Days from planting to effective full cover and from then to harvest or maturity are listed

Crop	Planting date	Days to peak K_c	Time from planting to peak K_c, %									
			10	20	30	40	50	60	70	80	90	100
Sorghum	5/17	45	0.12	0.13	0.14	0.16	0.22	0.33	0.50	0.75	1.00	1.07
Beans	6/21	43	0.10	0.12	0.16	0.21	0.28	0.39	0.53	0.75	0.98	1.08
Tomatoes	4/29	80	0.14	0.15	0.17	0.19	0.22	0.33	0.48	0.71	1.04	1.18
Barley	10/31	100	0.18	0.20	0.22	0.24	0.28	0.34	0.47	0.66	0.90	1.07
Corn	5/14	52	0.12	0.13	0.15	0.20	0.29	0.45	0.81	0.99	1.08	1.13
Sugar beets (late)	6/16	55	0.12	0.13	0.16	0.20	0.29	0.45	0.65	0.87	1.04	1.10
Sugar beets (early)	3/25	90	0.14	0.16	0.18	0.22	0.27	0.37	0.53	0.77	1.04	1.10

Crop	Harvest date	Days to harvest	Days after peak K_c									
			10	20	30	40	50	60	70	80	90	100
Sorghum	9/13	74	1.08	1.06	1.03	0.99	0.94	0.88	0.79	0.65	—	—
Beans	9/18	46	1.12	1.12	1.10	0.71	0.15	—	—	—	—	—
Tomatoes	9/24	68	1.24	1.21	1.12	1.03	0.90	0.75	0.58	—	—	—
Barley	5/19	100	1.15	1.17	1.19	1.21	1.19	1.12	0.98	0.75	0.50	0.24
Corn	9/20	77	1.17	1.17	1.17	1.14	1.03	0.87	0.67	—	—	—
Sugar beets (late)	11/18	100	1.15	1.16	1.16	1.16	1.15	1.14	1.13	1.12	1.10	1.08
Sugar beets (early)	9/20	90	1.13	1.15	1.15	1.14	1.13	1.11	1.08	1.05	1.01	—

6.5.4 Grass Related Crop Coefficients

Crop coefficients derived for use with a reference ET for grass (Doorenbos and Pruitt, 1977) are discussed in this section. A summary of basal crop coefficients for several crops is presented in Table 6.8 similarly to those in Table 6.6 except that E_{ro} was used as a base in their development. These coefficients were obtained at Davis, California and are therefore representative of an arid, Mediterranean-type climate. Data for many additional crops are presented in FAO-ID 24 (Doorenbos and Pruitt, 1977).

The adjustments to the Blaney-Criddle and evaporation pan methods of Subsections 6.4.4 and 6.4.5 may be used to estimate E_{ro} for use with the grass-based crop coefficients. Compatible Penman and radiation methods may also be used. (Doorenbos and Pruitt, 1977). However, the grass-based crop coefficients should not be used with the Penman and Jensen-Haise methods as presented in Subsections 6.4.2 and 6.4.3.

Field and vegetable crops. The growing season may be divided into four stages:

(1)	Initial stage	: germination and early growth when the soil surface is mostly bare, crop ground cover < 10 percent.
(2)	Crop development stage	: from the initial stage to effective full crop ground cover (70 to 80 percent).
(3)	Mid-season stage	: from effective full crop ground cover to the start of maturation as indicated by changes in leaf color or dropping of leaves.
(4)	Late season stage	: from the end of the mid season stage until full maturity or harvest.

Curves for other crops may be constructed in the following manner for a given location.

1   Establish planting date from local information or practices in similar climatic zones.

2   Determine total growing season and length of crop development stages from local information. Guidelines to crop development stages are presented in Table 6.9.

3   Initial stage: predict irrigation and/or rainfall frequency, then select K_c value and plot as shown in Fig. 6.3 or 6.8. This is an alternate approach to correcting K_c for rain or irrigation (Wright, 1981).

4   Mid-season stage: based on local climate (humidity and wind), select K_c from Table 6.10 and plot as a straight line.

5   Late-season stage: for time of full maturity select a K_c value from Table 6.10. Assume a straight line between the end of the mid-season stage to the full maturity date.

6   Development stage: assume a straight line between the end of the initial stage and the start of the mid-season stage.

The curve may be refined by sketching a smooth curve, but this may only make a small difference in results. The construction of such a curve for field corn at Kimberly, Idaho is shown in the example calculations, in Section 6.14.

Forage crops comprise millions of hectares of irrigated land in the world. K_c values for these crops reach a high value just prior to cutting and a

TABLE 6.9. LENGTH OF GROWING SEASON AND CROP
DEVELOPMENT STAGES OF SELECTED FIELD CROPS:
SOME INDICATIONS
(from Doorenbos and Pruitt, 1977)

Beans (dry) Pulses	Continental climates late spring planting 20/30/40/20 and (110); June planting Central California and West Pakistan 15/25/35/20 and (95); longer season varieties 15/25/50/20 and (110).*
Corn (grain) (sweet)	Spring planting East African highlands 30/50/60/40 and (180); late cool season planting, warm desert climates 25/40/45/30 and (140); June planting sub-humid Nigeria, early October India 20/35/40/30 and (125); early April planting Southern Spain 30/40/50/30 and (150).
Grain, small	Spring planting Mediterranean 20/30/60/40 and (150); October-November planting warm winter climates; Pakistan and low deserts 25/35/65/40 and (165).
Potato (Irish)	Full planting warm winter desert climates 25/30/30/20 and (105); late winter planting arid and semi-arid climates and late spring-early summer planting continental climate 25/30/45/30 and (130); early-mid spring planting central Europe 30/35/50/30 and (145); slow emergence may increase length of initial period by 15 days during cold spring.
Sugarbeet	Coastal Lebanon, mid-November planting 45/75/80/30 and (230); early summer planting 25/35/50/50 and (160); early spring planting Uruguay 30/45/60/45 and (180); late winter planting warm winter desert 35/60/70/40 and (205).

*15/25/50/20 and (110) stand respectively for initial, crop develop-
ment, mid-season and late season crop development stages in days
and (110) for total growing period from planting to harvest in days.

low value just after cutting. It is essential that local harvest dates be considered in making ET estimates for forage crops. Table 6.11 gives high,
average, and low values for alfalfa, grass hay, legumes, and pasture. For
seasonal estimates average K_c values may be used. For irrigation timing and
depth, the variation due to cutting also must be considered. More detail and
a graphical presentation of the seasonal variation in K_c for alfalfa is
presented in FAO-IR 24 (Doorenbos and Pruitt, 1977).

TABLE 6.10. SELECTED CROP COEFFICIENTS BASED
ON GRASS E_{to} FOR FIELD CROPS FOR DIFFERENT
GROWTH STAGES AND CLIMATIC CONDITIONS
(from Doorenbos and Pruitt, 1977)

Crop	Crop stage	Humidity: RH min 0-5	> 70% 5-8	RH min 0-5	< 20% 5-8
		Wind m/s:			
Beans (dry)	3	1.05	1.1	1.15	1.2
	4	0.3	0.3	0.25	0.25
Corn (field)	3	1.05	1.1	1.15	1.2
	4	0.55	0.55	0.6	0.6
Grain	3	1.05	1.1	1.15	1.2
	4	0.3	0.3	0.25	0.25
Potato	3	1.05	1.1	1.15	1.2
	4	0.7	0.7	0.75	0.75
Sugarbeet	3	1.05	1.1	1.15	1.2
	4	0.9	0.95	1.0	1.0

TABLE 6.11. CROP COEFFICIENTS (K_c) FOR ALFALFA,
CLOVER, GRASS-LEGUMES AND PASTURE
with mean values for between cuttings, low values for just after
cuttings with dry soil conditions, and peak values for just
before harvest. For wet soil conditions increase low values
by 30% (adapted from Doorenbos and Pruitt, 1977)

Climatic conditions	Period	K_c			
		Alfalfa	Grass hay	Clover, grass-legumes	Pasture
Humid with light	mean	0.85	0.80	1.00	0.95
to moderate	peak	1.05	1.05	1.05	1.05
winds	low	0.50	0.60	0.55	0.55
Dry with light	mean	0.95	0.90	1.05	1.00
to moderate	peak	1.15	1.10	1.15	1.10
winds	low	0.40	0.55	0.55	0.50
Strong winds	mean	1.05	1.0	1.10	1.05
	peak	1.25	1.15	1.20	1.15
	low	0.30	0.50	0.55	0.50

6.5.5 Effect of Irrigation Method on Evapotranspiration

The method of irrigation may affect ET rates while water is being applied and possibly for several days following irrigation. During irrigation, the ET rate may be highest with sprinklers because of the added evaporation opportunity provided by the increased availability of a vapor sink and the sensible energy supplied by the air layer through which the water drops travel. During windy conditions these effects are especially important due to the transport of droplets outside of the area being irrigated.

Wetting of a crop surface by irrigation (or precipitation) does not necessarily result in greater ET than otherwise. A number of studies have shown that surface evaporation replaces vegetative transpiration in equal amounts (Christiansen and Davis, 1967). In such cases ET is already at the potential rate and the site of the evaporative process is merely changed from plant stoma to the wet vegetative surface. Wetting the crop increases ET where ET has been restricted by such factors as low vegetative density and a dry soil surface, limited soil moisture available for plants, high stomatal resistance, or xerophytic plant adaptation.

At low vegetative densities evaporation from wet soil can be an important factor in contributing to ET (Ritchie, 1971). Thus, an irrigation method that does not wet the entire bare soil area can result in less ET than one that does. An advantage of drip irrigation is that it does not wet the entire soil area. However the saving of evaporation is less than the ratio of unwetted area to total bare soil area would suggest because of advective influences (also see Section 16.5).

The effect of irrigation method on ET, while of some consequence during and immediately following irrigation, may be small on a seasonal basis. For example, Bucks et al. (1974) found that the seasonal ET for high production of cabbage in Arizona was about the same with drip, modified furrows and furrow irrigation. Lysimeter studies of grain sorghum in Texas showed no significant differences in yield or water use efficiency (ratio of grain yield to total crop water use) between drip and sprinkler irrigation with three irrigations per week (Ravelo et al., 1977).

6.6 OTHER BENEFICIAL USES

Water applied at appropriate times can sometimes make additional contributions to improved crop production besides the replenishment of soil moisture. While meeting the ET need of crops is the primary purpose of irrigation, conditions may require providing water for additional beneficial uses as discussed in Chapters 2 and 18 and briefly described in this Section.

6.6.1 Germination of Seeds

Germination of seeds may be enhanced by irrigation at planting, and sometimes irrigation is essential for seed germination. Subsequent crop development and harvest are aided by the uniform seed germination and plant emergence. Sprinkler irrigation is especially suited to this application because the amount of water applied can be limited to the amount necessary; this is especially important where water supplies are limited. Soil wetting for germination by furrow irrigation is successfully practiced in many areas, but more water is required than with sprinklers when "subbing" from furrow to ridge planted seed is involved. Furthermore, salinity tends to be concentrated in the ridge by evaporation.

6.6.2 Climate Modification

Climate modification may be possible using water. A large-scale effect is apparent as one drives from the desert into an irrigated area on a hot summer day and feels the effect of evaporative cooling on the atmosphere. This lowering of dry bulb temperature is accompanied by an increase in vapor pressure and may be accompanied by a reduction in wind speeds (Burman et al., 1975). Experiments using sprinkler or mist applications at field sites within irrigated areas have typically decreased crop temperatures 4 to 12 °C. Increases in yield of 10 to 70 percent with such crops as peas, tomatoes, cucumbers, muskmelons and strawberries are reported, and improved quality of apples and grapes have been observed (Westerman et al., 1976). However, crop response to lowered temperature stress may sometimes be less beneficial than judged from the amount of air temperature suppression. Design procedures for climate-control sprinkling and misting systems are not well developed. Misting to improve greenhouse environments is a common practice.

Evaporative cooling experiments to delay bloom of fruit trees, with attendant reduced danger from freeze damage, were reported by Wolfe et al. (1976). They found that with application rates of 3 L/s·ha misting systems did better than low-pressure sprinklers in keeping daytime orchard temperatures down, and thus more successfully delayed bud development until the danger of frost had passed. The mist system required only about 60 percent as much water per day of bloom delay as did the sprinklers.

6.6.3 Freeze Protection

Freeze protection can result from water applied to the soil to increase soil heat conduction and soil heat storage capacity. Significant protection may be achieved by continuous wetting of plant parts by sprinkler water during critical hours.

In general, oil releases much more heat to a crop if it is used to pump water instead of being burned. A more complete discussion of freeze protection methods can be found in Chapters 2 and 18.

Application rates during a freeze period may be dictated by the available irrigation system (2.5 to 6.4 mm/h for sprinklers). Under-tree sprinklers designed specifically for freeze protection may have rates as low as 1.3 mm/h.

Blanc et al. (1963) stated that protection down to -6 °C can be achieved by overhead sprinkler rates of 1.5 to 2.0 mm/h for low growing plants; 2.0 mm/h for fruit trees; and 2.0 to 2.5 mm/h for vines. These rates must be increased when atmospheric dew points are low. The application of water by overhead sprinklers should begin when falling air temperatures reach 1 °C, or when wet bulb temperatures reach freezing. Sprinkling should be continued until ice is melting on its own and air temperature remains above freezing.

6.6.4 Fertilizer Application

Fertilizer application by irrigation water is often the cheapest way, and may be the only way of applying it (except by air) to a crop that runs out of N in mid or late season. Anhydrous and aqua NH_3 and solutions made from dry fertilizers are commonly used as sources of N. Liquid H_3PO_4 and solutions of K are also applied by irrigation water. In some areas the harmful effects of high Na water on infiltration rates are counteracted by the addition of gypsum to the irrigation water.

The amount of water applied during fertilization is usually governed by the ET needs of the crop. Nutrients that do not move rapidly in the soil are applied during the beginning of the irrigation period, whereas nitrate is applied late in the period to prevent penetration to excessive depths.

Fertilizers can be applied by surface or sprinkler systems. All components of the system must be corrosion resistant, and the system should be thoroughly flushed with water at the end of the irrigation period. Further details of fertilizer application techniques and precautions are given by Viets et al. (1967) and in Section 16.9.

6.6.5 Soil Temperatures

Soil temperatures can be markedly affected by irrigation water. Low water temperatures may depress soil temperatures and impede plant development. The literature tends to support the generalization that vegetative growth is largely correlated with root temperature, reproductive events, and with shoot temperature (Raney and Mihara, 1967). Soil cooling may be desirable under certain circumstances, such as establishing seedling stands of head lettuce.

6.6.6 Dust Suppression

Dust suppression, though not related to irrigation, can be achieved by using sprinkler systems. The feedlot dust generated in hot, dry climates when cattle become active in the early evening can be suppressed with sprinkling. Carroll et al. (1974) report applying just enough water in two increments to suppress dust while avoiding problems of odor and pests associated with excessive wetness of pens.

6.7 LEACHING REQUIREMENTS

The amount of water required to maintain a favorable salt balance depends upon local conditions. These include the amount of soluable salts present in the soil, soil type (texture), quality of irrigation water, ET rates, rainfall amounts and distribution, and depth of groundwater (drainage practices). Guidance as to the amount of leaching required for specific situations is available from several sources, especially the U.S. Salinity Laboratory, Riverside, California (U.S. Salinity Lab. Staff, 1954) and the Hebrew University of Jerusalem, Israel (Yaron et al., 1974). Salinity problems and control are discussed in detail in Chapter 5. Procedures for estimating leaching requirements are presented in FAO-ID 29 and the ASCE-CU Report (Ayres and Wescot, 1976; and Jensen, 1974).

6.8 ESTIMATING EFFECTIVE RAINFALL

Effective rainfall is that portion of rainfall that contributes to meeting the ET requirement of a crop (Hershfield, 1964). This differs diametrically from the hydrologic definition which describes effective rainfall as that portion of the total rain that produces runoff. Thus, rain water that neither leaves as surface runoff nor contributes to excess subsurface drainage may be effective precipitation in the context of irrigation water management. An extensive review of models for estimating effective rainfall from measured rainfall has been published by the Food and Agriculture Organization of the United Nations (Dastane, 1974).

Rain water retained by the plant canopy contributes to the satisfaction of the meteorological evaporative demand. This results in a consequent reduction in use of soil moisture. However, some engineers discount each rainfall event by a small amount, say 2 mm (0.08 in.), in situations where vegetative cover is incomplete or where prevailing ET rates are otherwise less than potential.

Estimates of effective precipitation should take local conditions into account. Rainfall of high intensity or large amounts that produce runoff should be considered to be of reduced value. Similarly, rainfall on an already wet soil profile is ineffective to the extent that subsurface drainage exceeds leaching requirements. Soil moisture accretion after the crop reaches physiological maturity is nonbeneficial unless it is stored in the soil for use by a crop during the next growing season.

Heermann and Shull (1976) upon analyzing seasonal, monthly, daily and hourly occurrence and dissipation of different rainfall amounts concluded that daily ET is increased after a rainfall during the early development of the crop (alfalfa). Frequent irrigations and rainfall increased the total seasonal ET as compared with infrequent rainfall and irrigation. Small rainfall amounts are important, not only in the amount of water received, but because of the associated decrease in potential ET due to cloudy, humid conditions. Techniques are available (Jensen, 1974; Jensen et al., 1971; Ritchie, 1972) to account for increased evaporation immediately after an irrigation or rainfall.

Two of the simple models of estimating effective rainfall from measured rainfall are presented here. The first method is very simple and was apparently developed by the U.S. Bureau of Reclamation for monthly water resource calculations. Stamm (1967) makes the following comments about its use. The method is intended for the arid and semi-arid areas of the Western United States. To be conservative the method should be applied to the driest 5 consecutive years in the growing season only. The latter requirement has

TABLE 6.12. EFFECTIVE PRECIPITATION BASED ON
INCREMENTS OF MONTHLY RAINFALL (U.S. BUREAU
OF RECLAMATION METHOD)

Precipitation increment range			Effective precipitation accumulated - range	
mm	in.	Percent	mm	in.
0.0- 25.4	0-1	90-100	22.9- 25.4	0.90-1.00
25.4- 50.8	1-2	85- 95	44.4- 49.5	1.75-1.95
50.8- 76.2	2-3	75- 90	63.5- 72.4	2.50-2.85
76.2-101.6	3-4	50- 80	76.2- 92.7	3.00-3.65
101.6-127.0	4-5	30- 60	83.8-107.9	3.30-4.25
127.0-152.4	5-6	10- 40	86.4-118.1	3.40-4.65
Over -152.4	Over 6	0- 10	86.4-120.6	3.40-4.75

often been ignored. Table 6.12 shows factors used to estimate monthly effective rainfall from measured rainfall.

A second commonly used method in the United States of estimating effective rainfall from field measurements was developed by the Soil Conservation Service. The method, which is described in more detail by Dastane (1974), is based on a soil moisture balance performed for 22 stations using 50 years of data. The method recognizes both monthly ET estimates and monthly precipitation measurements. In addition the method indicates that effective rainfall defined for irrigation purposes by the depth of irrigation water applied is directly related to irrigation frequency. The monthly effctive rainfall may be estimated for a 75-mm irrigation application using Table 6.13. If the irrigation application differs from 75 mm the effective rainfall may be corrected by an appropriate factor selected from Table 6.14.

6.9 IRRIGATION EFFICIENCY

6.9.1 Estimating Expected Irrigation Efficiency

After determining net irrigation water requirements, an estimate of the expected irrigation efficiency is needed to determine gross irrigation water requirements. No irrigation system is capable of applying an exact amount of water with perfect uniformity. In addition, some water will be lost by evaporation during application, especially with sprinkler systems. Loss of water by evaporation during sprinkling may reduce the rate at which soil water normally would be extracted when not being irrigated so that this may not be a total loss. The effectiveness of evaporation in reducing soil water extraction is expected to vary from near 100 percent when evaporation occurs from water ponded on the surface of an actively growing crop to near zero for evaporation from sprinkler spray discharging several meters above the crop under windy conditions (also see Section 14.5).

Surface runoff, water spillage and leakage from the on-farm water distribution system also affect the expected farm irrigation efficiency. A major part of surface runoff and spillage may be recovered for use on a given farm if an effective reuse system is used.

Seepage from unlined farm ditches and deep percolation through the soil profile due to nonuniform and excessive water applications usually cannot be recovered for use on a given farm so as to affect the design irrigation efficiency. However, from a water supply viewpoint, water returning to the groundwater below a farm reduces net depletion of the water supply. Likewise, recovery of surface runoff, and in some cases deep percolation

TABLE 6.13. AVERAGE MONTHLY EFFECTIVE RAINFALL AS RELATED TO MEAN MONTHLY RAINFALL AND MEAN MONTHLY CONSUMPTIVE USE (USDA, SCS)

Monthly mean rainfall mm	Mean monthly consumptive use mm													
	25	50	75	100	125	150	175	200	225	250	275	300	325	350
	Mean monthly effective rainfall mm													
12.5	7.5	8.0	8.7	9.0	9.2	10.0	10.5	11.2	11.7	12.5	12.5	12.5	12.5	12.5
25.0	15.0	16.2	17.5	18.0	18.5	19.7	20.5	22.0	24.5	25.0	25.0	25.0	25.0	25.0
37.5	22.5	24.0	26.2	27.5	28.2	29.2	30.5	33.0	36.2	37.5	37.5	37.5	37.5	37.5
50.0	25	32.2	34.5	35.7	36.7	39.0	40.5	43.7	47.0	50.0	50.0	50.0	50.0	50.0
62.5	at 41.7	39.7	42.5	44.5	46.0	48.5	50.5	53.7	57.5	62.5	62.5	62.5	62.5	62.5
75.0		46.2	49.7	52.7	55.0	57.5	60.2	63.7	67.5	73.7	75.0	75.0	75.0	75.0
87.5		50.0	56.7	60.2	63.7	66.0	69.7	73.7	77.7	84.5	87.5	87.5	87.5	87.5
100.0		at 60.7	63.7	67.7	72.0	74.2	78.7	83.0	87.7	95.0	100	100	100	100
112.5			70.5	75.0	80.2	82.5	87.2	92.7	98.0	105	111	112	112	112
125.0			75.0	81.5	87.7	90.5	95.7	102	108	115	121	125	125	125
137.5			at 122	88.7	95.2	98.7	104	111	118	126	132	137	137	137
150.0				95.2	102	106	112	120	127	136	143	150	150	150
162.5				100	109	113	120	128	135	145	153	160	162	162
175.0				at 160	115	120	127	135	143	154	164	170	175	175
187.5					121	126	134	142	151	161	170	179	185	187
200.0					125	133	140	145	158	168	178	188	196	200
225					at 197	144	151	160	171	182				
250						150	161	170	183	194				
275						at 240	171	181	194	205				
300							175	190	203	215				
325							at 287	198	213	224				
350								200	220	232				
375								at 331	225	240				
400									at 372	247				
425										250				
										at 412				
450	25	50	75	100	125	150	175	200	225	250				

within a project, reduces the net depletion of water in a river-groundwater system. Recovery of return flows, both surface and subsurface, for use on downstream projects affects the net depletion of water in river basins. The reuse of return flow is one of the main foundations of Western water right management, and its importance is impossible to overestimate.

The overall farm irrigation efficiency to be used in design should be estimated by considering all components that affect irrigation efficiency.

TABLE 6.14. MULTIPLICATION FACTORS TO RELATE MONTHLY EFFECTIVE RAINFALL VALUE OBTAINED FROM TABLE 6.13 TO NET DEPTH OF IRRIGATION APPLICATION (d)

d mm	factor	d mm	factor	d mm	factor
10.0	0.620	31.25	0.818	70.0	0.990
12.5	0.650	32.5	0.826	75.0	1.000
15.0	0.676	35.0	0.842	80.0	1.004
17.5	0.703	37.5	0.860	85.0	1.008
18.75	0.780	40.0	0.876	90.0	1.012
20.0	0.728	45.0	0.905	95.0	1.016
22.5	0.749	50.0	0.930	100.0	1.020
25.0	0.770	55.0	0.947	125.0	1.040
27.5	0.790	60.0	0.963	150.0	1.060
30.0	0.808	65.0	0.977	175.0	1.070

Identifying the magnitude of the various components will assist in determining the alternative design or types of systems that should be considered.

6.9.2 Irrigation Efficiency Definitions

The following terms proposed by the Irrigation Water Requirements Committee of the American Society of Civil Engineers (Jensen, 1974) are applicable to on-farm systems as well as projects. They are similar to those proposed by the International Commission of Irrigation and Drainage (Bos and Nugteren, 1974).

Reservoir storage efficiency, E_s, is the ratio of the volume of water available from the reservoir for irrigation, to the volume of water delivered to the storage reservoir—surface or underground—for irrigation.

Water conveyance efficiency, E_c, is the ratio of the volume of water delivered to the point of use by an open or closed conveyance system to the volume of water introduced into the conveyance system at the supply source or sources.

Unit irrigation efficiency, E_u, is the ratio of the volume of irrigation water required for beneficial use in the specified irrigated area to the volume of water delivered to this area.

Farm irrigation efficiency, E_i, is the product of the component terms, expressed as ratios.

$$E_i = E_s E_c E_u \dots\dots\dots\dots\dots\dots\dots\dots\dots\dots\dots\dots\dots\dots \quad [6.38]$$

The overall irrigation efficiency for a project or a river basin can be expressed in a similar manner. For clarity and comparative purposes, all efficiency estimates or evaluations should be identified as to the size of unit, the period of time or number of irrigations involved, the adequacy of irrigations in meeting net irrigation requirements, and computational procedures used.

Effective irrigation efficiency, E_e, of a farm, project, or river basin is necessary to estimate or evaluate the net depletion of water within a river basin or groundwater system (Jensen, 1977). It is based on the assumption that irrigation efficiency ($E_i = V_c/V_w$) as defined by Israelsen (1950) is the ratio of water consumed (V_c) by the agricultural crops on a farm project to the water diverted (V_w) from a natural source into the farm or project canals and laterals. The net depletion of water, V_{dep}, specifically for irrigation is

$$V_{dep} = V_c + (1 - E_r) V_{nc} \dots\dots\dots\dots\dots\dots\dots\dots\dots\dots\dots\dots \quad [6.39]$$

where V_c is the volume consumed by agricultural crops; V_{nc} is the volume diverted to a farm or project that is not consumed by the crops; and E_r is the fraction of E_{nc} that is recovered (or could be when evaluating the potential efficiency) for agriculture or other uses. The effective efficiency is

$$E_e = \frac{V_c}{V_w} + E_r \frac{V_{nc}}{V_w} \dots\dots\dots\dots\dots\dots\dots\dots\dots\dots\dots\dots \quad [6.40]$$

which also can be expressed as

$$E_e = E_i + E_r (1 - E_i) \dots\dots\dots\dots\dots\dots\dots\dots\dots\dots\dots\dots \quad [6.41]$$

Additional discussions and definitions of similar irrigation efficiency terms can be found in articles by Bos and Nugteren (1974), Jensen et al. (1967), Kruse and Heermann (1977) and Schmueli (1973). A summary of observed and attainable field and farm irrigation efficiencies was presented by Jensen (1978).

Irrigation water use efficiency, E_{iw}, is a measure of the increase in the production of the marketable crop component relative to the increase in water consumed when irrigated, over the consumption under nonirrigated conditions. The Committee on Irrigation Efficiencies of the International Commission on Irrigation and Drainage (Bos, 1980) recently defined this efficiency as the yield/ET ratio, Rye

$$Rye = \frac{V_i - V_o}{ET_i - ET_o} \quad \dots\dots\dots\dots\dots\dots\dots\dots\dots\dots\dots\dots\dots\dots \quad [6.42]$$

where V_i is the mass of marketable crop produced with irrigation; V_o is the mass of marketable crop (that could be) produced without irrigation: ET_i is the mass of water used in ET by the irrigation crop; and ET_o is the mass of water (that could be) used in ET by the same crop if not irrigated. Rye as defined is dimensionless, but in practice irrigation water use efficiency would be more conveniently expressed as mass of marketable crop per unit volume of water (kg/m^3) as has been done by many others over the past two decades. Typical maximum values to be expected for grain crops like corn and wheat are 1.5 to 2.0 kg/m^3.

6.10 DESIGN REQUIREMENTS

For many years it has been traditional to base the design capacity of sprinkler or other irrigation systems on what is called the peak ET rate. The peak ET rate is for the irrigation interval (I) and is higher for a one or two day period than for a week or more as the irrigation interval. Several recent studies have shown that the design ET rate (E_{td}) should be based on a probability level of expected ET which changes throughout the growing season. The system designer must make a choice of E_{td} based on soil moisture holding capacity, climatic probability, and the crop grown. The variables involved are: E_{td} is the peak ET rate for the irrigation interval used for design purposes, in depth per time, commonly mm/d (in./d); I is the irrigation interval in days; and D_n is the net depth of water to be applied during the design period in mm (in.). D_n is a function of soil characteristics, plant growth stage, and may include an allowance for leaching. See Chapters 4 and 18 for more information on the determination of D_n.

Two methods of estimating E_{td} are presented. The first involves the use of historic climatic data to estimate the expected ET on a probability basis and the second uses empirical relationships between estimated average monthly ET and E_{td}. The second approach does not involve probability.

6.10.1 Estimating E_{td} Using Climatic Records

An array of daily estimates of ET can be generated by using a long term climatic data set and a method of estimating ET suitable for daily values. One of the combination equations, such as the Penman, should be used and a frequency analysis made. E_{td} can then be selected on a probability basis for any desired interval during the growing season.

A series of recent papers show the statistical variation of E_{td} for selected

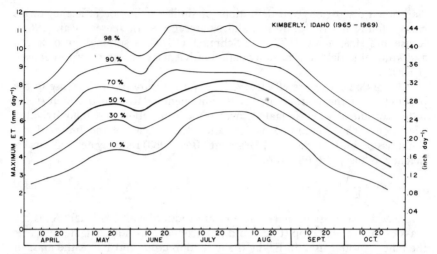

FIG. 6.4 Frequency distributions for reference ET (E_{tr}) for well watered alfalfa with full cover as calculated from 5 years of climatic data for Kimberly, Idaho (from Wright and Jensen, 1972).

locations in California, Idaho, and Nebraska (Pruitt et al., 1972; Wright and Jensen, 1972; Rosenberg, 1972; and Nixon et al, 1972). Typical results for Kimberly, Idaho (Wright and Jensen, 1972) are shown in Fig. 6.4 and 6.5.

6.10.2 E_{td} Based on Monthly Estimates

Engineers often do not have the time and the data needed to perform a statistical analysis to evaluate E_{td} requirements for design purposes. For many years the Soil Conservation Service has used an empirical method of

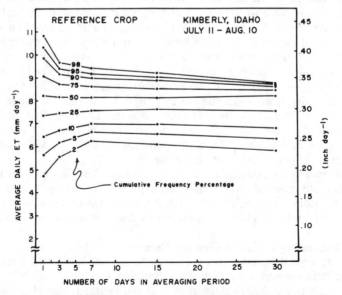

FIG. 6.5 Cumulative frequency percentages of average daily ET, estimated from data in Fig. 6.4 for 1-day, 3-day, 7-day, 15-day, and 30-day averaging periods for the peak 30-day period at Kimberly, Idaho (from Wright and Jensen, 1972).

estimating peak ET based on mean monthly values of ET as follows:

$$E_{td} = 0.034 \ E_{tm}^{1.09} \ I^{-0.09} \quad\dots\dots\dots\dots\dots\dots\dots\dots\dots\dots\dots\dots\dots\dots \quad [6.43]$$

where E_{tm} = mean ET for the month in mm and I = the net irrigation application in mm. For example, if the mean monthly ET is 200 mm (or about 6.7 mm/d) and the net irrigation is 100 mm, the E_{td} will be 7.2 mm/d. This procedure does not involve climatic probability but does consider the time period between irrigations by accounting for the depth of irrigation water applied. A soil with low water holding capacity would have a short irrigation interval because of the small amount of water retained for plant use. The method does not give the designer the opportunity of selecting a probability level for use in claculating the peak ET rate.

The relationship between monthly ET and peak ET for design purposes is very dependent on climatic conditions. These climatic differences are considered in a method recommended by the FAO (Doorenbos and Pruitt, 1977). The designer can utilize a simple graphical procedure for estimating peak E_t from monthly estimates (Fig. 6.6). The method also does not involve a probability level.

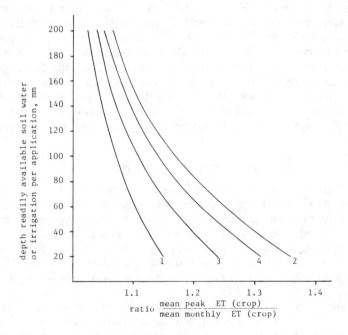

1. Arid and semi-arid climates and those with predominantly clear weather conditions during month of peak ET crop.

2. Mid-continental climates and sub-humid to humid climates with highly variable cloudiness in month of peak ET crop.

3. and 4. Mid-continental climates with variable cloudiness and mean ET crop of 5 and 10 mm/day respectively.

FIG. 6.6 FAO procedure for estimating peak ET from monthly estimates (from Doorenbos and Pruitt, 1977).

TABLE 6.15. APPROXIMATE RANGES OF SEASONAL CROP
ET FOR VARIOUS CROPS (from Dorenbos and Pruitt, 1977)

Crop	Seasonal ET, mm	Crop	Seasonal ET, mm
Alfalfa	600-1500	Onions	350- 600
Avocado	650-1000	Orange	650- 950
Bananas	700-1700	Potatoes	350- 625
Beans	250- 500	Rice	500- 950
Cocoa	800-1200	Sisal	550- 800
Coffee	800-1200	Sorghum	300- 650
Cotton	550- 950	Soybeans	450- 825
Dates	900-1300	Sugarbeets	450- 850
Deciduous trees	700-1050	Sugar cane	1000-1500
Flax	450- 900	Sweet potatoes	400- 675
Grains, small	300- 450	Tobacco	300- 500
Grapefruit	650-1000	Tomatoes	300- 600
Maize	400- 750	Vegetables	250- 500
Oil seed	300- 600	Vineyards	450- 900
—		Walnuts	700-1000

6.11 ANNUAL REQUIREMENTS

Seasonal ET estimates are often needed for a variety of water resource deliberations. The Irrigation Water Requirements Technical Committee, American Society of Civil Engineers, published an extensive table of seasonal ET measurements for a wide variety of crops at several locations (Jensen, 1974). Table 6.15 presents a summary of the approximate range in seasonal ET to be expected for various crops (Doorenbos and Pruitt, 1977). Seasonal ET is dependent on climate, time of planting, crop conditions, length of growing season, and other factors, such as the soil water level that is maintained. If ET estimates are greater or less than those shown in Table 6.15, calculations should be reviewed carefully and efforts should be made to verify that conditions are sufficiently different to account for differences in the estimates.

6.12 DESIGN CAPACITY

The design capacity of irrigation systems should meet peak evapotranspiration requirements. The delivery volume is determined by the expected cropping pattern serviced by the system. This involves considering the area devoted to each type of crop and its expected ET rate.

System design also involves the frequency with which each field must be irrigated. This is a function of the soil moisture holding capacity, effective depth of crop rooting, and the rate at which soil water is depleted as governed by the ET rate (Stamm, 1967).

Theoretically, an irrigation system can be designed for less than the peak daily ET rate as long as it can provide the peak average rate during the period between irrigations. The design capacity must allow for conveyance losses in the system and inefficiencies of applying water to the land. Also, the actual delivery rate of a system may be less than the design rate because of such factors as misaligned joints, dented pipe, or changed friction coefficients of channels, etc.

It may be prudent to include a flexibility or safety factor to allow for breakdowns, holidays, requirements for faster coverage for insect or disease control or other agrotechnical reasons, changes from the assumed cropping pattern, and occasional very windy days in the case of sprinkler (Zimmerman, 1966).

The design capacity should provide flow rates that are sufficient for the method of irrigation employed. A parallel consideration is that the design be compatible with the infiltration rate of the soil.

6.13 ESTIMATED RETURN FLOW AND QUALITY

Irrigation water applied in excess of crop requirements will result in surface runoff from the lowest point on the field and/or will percolate beyond the root zone. The surface run-off and deep percolation, moving under the influence of gravity and eventually re-entering streams or lakes, is referred to as "return flow". Return flow quality and quantity is of very great hydrologic importance.

Return flow becomes divertable water for downstream water users and therefore changes in return flow may disrupt the management of water resources. Western water right laws require that changes in water rights must not harm vested water rights. This means that when irrigation water rights are converted to municipal or industrial uses stream flow may have to be augmented by releases from reservoirs to make up lost return flow. Excess soil water which reappears as return flow is water in temporary storage and tends to stabilize Western stream flow.

Irrigation in excess of crop water requirements may create drainage problems. Some excess water is needed to maintain an acceptable salt balance in the soil (see Section 6.7 and Chapter 5 for a detailed discussion of leaching for salt management, and Chapter 7 for details concerning drainage).

Return flows contain more dissolved solids than the irrigation water because ET removes pure water. In addition flow through or over the soil and geologic formations in their path may cause further changes in water quality both chemically and biologically. These changes may be environmentally desirable or undesirable. An example of a desirable change results from the application of wastewater using irrigation methods as a means of renovating the wastewater. This method is now receiving increasing attention.

Irrigation management practices which assure high quality return flow are also receiving widespread attention. Sufficient research has been completed to permit intelligent decision making processes to proceed in solving many water quality problems in irrigated agriculture. Results of irrigation return flow research and development programs were summarized in the proceedings of a national conference on irrigation return flow quality management sponsored by the U.S. Environmental Protection Agency (Law and Skogerboe, 1977).

6.14 EXAMPLE CALCULATIONS

These example calculations are intended for the trained engineer or irrigation scientist with access to a scientific electronic calculator or to a computer. Most of the procedures followed are easy to adapt to a modern computer.

Daily Estimates, Penman method. A calibrated version of a combination equation such as Penman's is probably the most suitable method of accurately estimating daily ET. These calculations refer to the Penman method described in Subsections 6.4.2 and 6.5.3. The data used represent a typical summer day, at Kimberly, Idaho.

Day Number 200, July 19

Elevation	1195 m
Maximum air temperature	32.2 °C
Minimum air temperature	12.2 °C
Average air temperature	22.2 °C
Average dew point temperature	10.0 °C
Average air temperature for the previous 3 days	20.9 °C
Clear day solar radiation	747 ly
Measured solar radiation	686 ly
Measured net radiation	350 ly
Wind velocity at 3.66-m elevation	164 km/day
Estimated daytime wind/nighttime wind	4.0
Measured ET for alfalfa	8.5 mm

Step 1. Estimate E_{tr}, using constant albedo and W_f

$$E_{tr} = \frac{\Delta}{\Delta + \gamma} (Rn + G) + \frac{\gamma}{\Delta + \gamma} 15.36 (W_f)(e_a - e_d) \quad \ldots\ldots\ldots\ldots \quad [6.4]$$

$$\Delta = 2.00 (0.00738 \times 22.2 + 0.8072)^7 - 0.00116 \quad \ldots\ldots\ldots\ldots \quad [6.5]$$

$$= 1.627 \text{ mb/°C}$$

$$P = 1013 - 0.1055 \times 1195 = 887 \text{ mb} \quad \ldots\ldots\ldots\ldots\ldots\ldots \quad [6.7]$$

$$L = 595 - 0.51 \times 22.2 = 584 \text{ cal/g} \quad \ldots\ldots\ldots\ldots\ldots\ldots \quad [6.8]$$

$$\gamma = \frac{0.386 \times 887}{584} = 0.586 \text{ mb/°C} \ldots\ldots\ldots\ldots\ldots\ldots\ldots \quad [6.6]$$

$$\frac{\Delta}{\Delta + \gamma} = \frac{1.627}{1.627 + 0.586} = 0.735$$

$$\frac{\gamma}{\Delta + \gamma} = 1.000 - 0.735 = 0.265$$

$\frac{\Delta}{\Delta + \gamma}$ and $\frac{\gamma}{\Delta + \gamma}$ also can be interpolated from Table 6.1.

$$U_2 = 164 \left(\frac{2}{3.66} \right)^{0.2} = 145 \text{ km/d} \quad \ldots\ldots\ldots\ldots\ldots\ldots \quad [6.10]$$

$$e_d = 33.8639 [(0.00738 \times 10.0 + 0.8072)^8 - 0.000019 |1.8 \times 10.0 + 48|$$

$$+ 0.001316]$$

$$= 12.3 \text{ mb} \quad \ldots\ldots\ldots\ldots\ldots\ldots\ldots\ldots \quad [6.12]$$

$$e_a = \frac{1}{2} (48.1 + 14.2) = 31.1 \text{ mb} \quad \ldots\ldots\ldots\ldots\ldots\ldots \quad [6.11]$$

Meteorological tables also can be used for vapor pressures.

$$R_{bo} = (0.325 - 0.044 \sqrt{12.3}) \; \frac{11.71(273 + 22.2)^4}{10^8} \quad\dots\dots\dots\dots \quad [6.15]$$
and Table 6.3

$$R_{bo} = 152 \, ly$$

Alternate R_{bo}

$$\epsilon = -0.02 + 0.261 \exp \left[\frac{-7.77}{10^4} (22.2)^2 \right] \quad \dots\dots\dots\dots \quad [6.17]$$

$$= 0.158$$

$$R_{bo} = \frac{0.158 \times 11.71}{10^8} (273 + 22.2)^4 \quad \dots\dots\dots\dots\dots \quad [6.15]$$

$$= 140 \, ly$$

$$R_b = \left[1.22 \times \frac{686}{747} - 0.18 \right] 152 = 143 \, ly \quad \dots\dots\dots\dots \quad [6.14]$$

$$R_n = (1 - 0.23) \, 686 - 143 = 385 \, ly \quad \dots\dots\dots\dots\dots \quad [6.13]$$

Assume $G = 0$

$$E_{tr} = 0.735 \, (385 + 0) + 0.265(15.36)(0.75 + 0.0115 \times 145)(31.1$$

$$- 12.3) \quad \dots\dots\dots\dots\dots\dots\dots\dots\dots\dots\dots \quad [6.4]$$

$$= 468 \, ly/d$$

$$= \frac{468 \; \dfrac{cal}{cm^2 \; day}}{584 \; \dfrac{cal}{cm^3}} \times 10 \; \frac{mm}{cm}$$

$$= 8.01 \, mm/day$$

Monthly ET estimates. Data used for these estimates represent average July conditions for Kimberly, Idaho. Estimates of E_{tr} are based on procedures found in Subsections 6.4.3, 6.4.5, 6.5.3, and 6.5.4:

Mean maximum air temperature	30.0 °C
Mean minimum air temperature	11.7 °C
Mean air temperature	20.8 °C
Mean dew-point air temperature	9.4 °C
Mean vapor pressure	11.8 mb
Mean wind travel at 3.66 m	206 km/day
U-day/U-night (assumed)	3.0
Mean percent sunshine (estimated from radiation data)	84%
Mean day length	14.8 h
Mean pan (Class A) evaporation	8.9 mm/day
Mean measured alfalfa ET	8.1 mm/day
Latitude	42.2 deg N
Mean solar radiation	640 ly/day
Crop (assume)	field corn

ET Estimated by Jensen-Haise Method:

$$E_{tr} = C_T (T - T_x)^P \cdot s \quad \dots\dots\dots\dots\dots\dots\dots\dots\dots\dots\dots \quad [6.19]$$

e_2 = 42.4 mb, for 30 $^\circ$C .... [6.12]

e_1 = 13.8 mb, for 11.7 $^\circ$C .... [6.12]

C_H = 50/(42.4 – 13.8) = 1.75 .... [6.21]

C_1 = 38 – (2 $\times$ 1195)/305 = 30.2 .... [6.22]

C_T = 1/(30.2 + 7.3 $\times$ 1.75) = 0.0233 .... [6.20]

T_x = – 2.5 – 0.14(42.4 – 13.8) – 1195/550 = – 8.7 $^\circ$C .... [6.23]

E_{tr} = 0.0233(20.8 – (–8.7))640 = 440 ly/day = 7.5 mm/d ..... [6.19]

ET Estimated by Blaney-Criddle (FAO Method):

E_{to} = a_4 + b_4f (a regression relationship) .... [6.24]

f   = p(0.46T + 8) .... [6.25]

p   = 0.33 (Table 6.4, July at Lat. 42.2 $^\circ$N)

(0.46T + 8) = 0.46 $\times$ 20.8 + 8 = 17.57

f = P(0.46T + 8) = 5.80 .... [6.25]

From Fig. 6.2 for f = 5.80, n $\simeq$ 0.9,

U Daytime $\simeq$ 2 – 5 m/s, and

RH min = 100 $\times$ 11.8/42.4 = 28% $\simeq$ 20 – 50 range

E_{to} = 7.1 mm/day (ET for grass)

Since E_{tr} $\simeq$ 1.15 E_{to} (for light to moderate winds in arid climates)

E_{tr} $\simeq$ 1.15 $\times$ 7.0 = 8.2 mm/day

ET Estimated by Pan Evaporation, FAO:

E_{to} = $K_p E_p$ .... [6.26]

K_p for case A, with 100 m fetch,

RH_{mean} = $\dfrac{100}{2}$ [11.8/13.8 + 11.8/42.4] = 57%, and

U_2 = 183 km/day (Light to Moderate, extrapolated to 2 meters)

K_p = (0.8 + 0.75)/2 $\simeq$ 0.78 (Table 6.5)

E_p = 8.9 mm/day for July mean

E_{to} = $K_p E_p$ = 0.78 $\times$ 8.9 = 6.9 mm/day .... [6.26]

Since $E_{tr} \cong 1.15\, E_{to}$ (for light-moderate winds in arid climates)

$E_{tr} = 1.15 \times 6.9 = 7.9$ mm/day

Crop Curve Development, FAO Method. An example of the construction of a grass related crop curve, using the procedure of Doorenbos and Pruitt (1977), is presented for field corn at Kimberly, Idaho. The necessary dates pertaining to crop development from Table 6.7 are planting, 5/5; emergence, 5/25; rapid growth, 6/10; full cover, 7/15; tasselling, 7/30; ripening, 9/10; harvest (silage), 9/20; 70 days. Assuming an E_{to} for May of 6.5 mm/d and irrigation on 7-day intervals; an initial K_c of 0.45, as determined from Fig. 6.7; and the beginning of the mid-season stage of growth on 7/1; the constructed crop curve would be as shown in Fig. 6.8. The maximum K_c for mid-season of 1.05 was determined using a maximum e_s of 42.4 mb (30 °C) and a minimum e_s of 11.8 mb (9.4 °C), giving a minimum Relative Humidity of 28 percent; and a U_2 for daytime wind of 3.2 m/s. A K_c of 0.55 was assumed for stage of maturity for silage harvest.

References

1 Ayers, R. S., and D. W. Westcot. 1976. Water quality for agriculture. FAO Irrig. and Drain. Paper 29, 97 p.

2 Blanc, M. L., H. Geslin, I. A. Holzberg, and B. Mason. 1963. Protection against frost damage. Tech. Note No. 51. World Meteorol. Org., Geneva, 62 p.

3 Blaney, H. F. and W. D. Criddle. 1945. Determining water requirements in irrigated areas from climatological data. (processed) 17 p.

4 Bos, M. G. 1980. Irrigation efficiencies at crop production level. Intern'l Comm. on Irrig. and Drain. Bul. (In print).

5 Bos, M. G. and J. Nugteren. 1974. On irrigation efficiencies. Intern'l Inst. for Land Reclam. and Improve. Publ. 19, Wageningen, Neth., 95 p. (Note: 2nd edition 1978, 142 p.).

6 Bosen, J. F. 1960. A formula for approximation of the saturation vapor pressure over water. Monthly Weather Rev. 88(8):275-276.

7 Brunt, D. 1952. Physical and dynamical meteorology. 2nd ed. University Press, Cambridge, 428 p.

8 Bucks, D. A., L. J. Erie, and O. F. French. 1974. Quantity and frequency of trickle and furrow irrigation for efficient cabbage production. Agron. J. 66:53-57.

9 Burman, R. D. 1979. Estimation of mountain meadow water requirements. Symp.

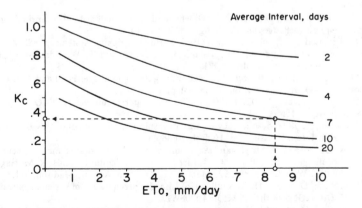

FIG. 6.7 Average crop coefficients (K_c) for grass reference ET (E_{to}) for the initial crop development stage as related to the average recurrence interval of irrigation and/or significant rains from an example for Cairo, Egypt. (adapted from Doorenbos and Pruitt, 1977).

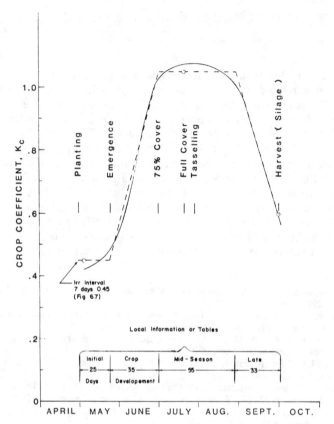

FIG. 6.8 Example of crop coefficient curve constructed for a crop of field corn using grass reference ET (E_{ro}).

Proc., Management of Intermountain Meadows, RJ 141, Wyo., 11-23.

10   Burman, R. D., J. L. Wright and M. E. Jensen. 1975. Changes in climate and estimated evaporation across a large irrigated area in Idaho. TRANSACTIONS of the ASAE 18(6):1089-1093.

11   Carroll, J. J., J. R. Dunbar, R. L. Gibens, and W. B. Goddard. 1974. Sprinkling for dust suppression in a cattle feedlot. Calif. Agric. 28(3):12-14.

12   Christiansen, J. E., and J. R. Davis. 1967. Sprinkler irrigation systems. p. 885-904. In: Irrigation of agricultural lands, R. M. Hagan, H. R. Haise, and T. W. Edminster (Ed.). Monog. 11, Am. Soc. Agron., Madison, WI.

13   Dastane, N. G. 1974. Effective rainfall in irrigated agriculture. Food and Agr. Org., United Nations, FAO Irrig. and Drain. Paper, 61 p.

14   Doorenbos, J. and W. O. Pruitt. 1977. Crop water requirements. FAO Irrig. and Drain. Paper 24 (rev.), 156 p.

15   Fritschen, L. J. 1965. Accuracy of evapotranspiration determinations by the Bowen ratio method. Bul. Intern'l. Assoc. Sci. Hydrol. 10:38-48.

16   Haise, H. A. and R. M. Hagan. 1967. Soil, plant, and evaporative measurements as criteria for scheduling irrigation. P 577-604. In: Irrigation of Agricultural Lands, R. M. Hagan, H. R. Haise, and T. W. Edminster (Ed.), Monog. 11, Am. Soc. Agron., Madison, WI.

17   Heermann, D. F. and H. H. Shull. 1976. Effective precipitation of various application depths. TRANSACTIONS of the ASAE 19(4):708-712.

18   Hershfield, D. M. 1964. Effective rainfall and irrigation water requirement. Proc. Am. Soc. Civil Eng., J. Irrig. and Drain. Div. 90:(IR2) 33-37.

19   Hounam, C. E. 1973. Comparison between pan and lake evaporation. World Meteorol. Org. Tech. Note No. 126, WMO-354, 52 p.

20   Howell, T. A. 1979. Evaporative demand as a plant stress. p. 97-113, In: Modification

of the Aerial Environment of Crops. B. J. Bardiels and J. F. Gerber (Ed.), ASAE monog. No. 2.

21 Idso, S. B. and R. D. Jackson. 1969. Thermal radiation from the atmosphere. J. Geophys. Res., 74:5397-5403.

22 Israelsen, O. W. 1950. Irrigation Principles and Practices. John Wiley and Sons., Inc. (2nd ed.), New York. 405 p.

23 Jensen, M. E. (Ed.). 1974. Consumptive use of water and irrigation water requirements. Rep. Tech. Com. on Irrig. Water Requirements, Am. Soc. Civ. Eng., Irrig. Drain. Div., 227 p.

24 Jensen, M. E. 1977. Water conservation and irrigation systems. Climate-Tech. Sem. Proc., Columbia, MO, p. 208-250.

25 Jensen, M. E. 1978. Irrigation water management for the next decade. Proc. New Zealand Irrig. Conf., Asburton, p. 245-302.

26 Jensen, M. E. and H. R. Haise. 1963. Estimating evapotranspiration from solar radiation. Proc. Am. Soc. Civ. Engr., J. Irrig. and Drain. Div. 89:15-41.

27 Jensen, M. E., D. C. N. Robb and C. E. Franzoy. 1970. Scheduling irrigations using climate-crop-soil data. Proc. Am. Soc. Civ. Engr., J. Irrig. and Drain. Div. 96(IR1):25-38.

28 Jensen, M. E., L. Swarner, and J. T. Phelan. 1967. Improving irrigation efficiencies, p. 1120-1142. In: Irrigation of agricultural lands. R. M. Hagan, H. R. Haise and T. W. Edminster (Ed.) Monog. 11, Am. Soc. Agron., Madison, WI.

29 Jensen, M. E., J. L. Wright, and B. J. Pratt. 1971. Estimating soil moisture depletion from climate, crop and soil data. TRANSACTIONS of the ASAE 14(5):954-959.

30 Kruse, E. G. and H. R. Haise. 1974. Water use by native grasses in high altitude Colorado meadows. USDA-SEA-AR, ARS-W-6, 60 p.

31 Kruse, E. G. and D. P. Heermann. 1977. Implications of irrigation system efficiencies. J. Soil and Water Conserv. 32(6):265-270.

32 Law, J. P., Jr., and G. V. Skogerboe (Ed.). 1977. Irrigation return flow quality management. Proc. Nat. Conf. sponsored by U.S. Environmental Protection Agency and Colorado State Univ. Colo. State Univ., Fort Collins, Colo. 451 p.

33 Makkink, G. F. 1957. Testing the Penman formula by means of lysimeters. J. Inst. Water Engr. 11(3):277-288.

34 Merva, G. E. 1975. Physioengineering principals. AVI Publishing Co., 353 p.

35 Nixon, P. R., G. P. Lawless, and G. V. Richardson. 1972. Coastal California evapotranspiration frequencies. Proc. Am. Soc. Civ. Engr., J. Irrig. and Drain. Div. (IR2):185-191.

36 Nixon, P. R., N. A. McGillivray, and G. P. Lawless. 1963. Evapotranspiration—climate comparisons in coastal fogbelt, coastal valley, and interior valley locations in California. Publ. No. 62, Internat'l Assoc. Sci. Hydrol. Com. for Evaporation, p. 221-231.

37 Penman, H. L. 1948. Natural evaporation from open water, bare soil, and grass. Proc. Roy. Soc. London. A 193:120-145.

38 Penman, H. L. 1963. Vegetation and hydrology. Tech. Communication No. 53. Commonwealth Bur. of Soils, Harpenden, England, 125 p.

39 Pochop, L. O., et al. 1973. Psychrometric data patterns and prediction models. Wyo. Water Resour. Ser. No. 48, Univ. of Wyo.

40 Pruitt, W. O. 1966. Empirical method of estimating evaportranspiration using primarily evaporation pans. p. 57-61. In: Proc., Evapotranspiration and its role in water resources management. M. E. Jensen (Ed.), ASAE, St. Joseph, MI 49085.

41 Pruitt, W. O., S. von Oettigen, and D. L. Morgan. 1972. Central California evapotranspiration frequencies. Proc. Am. Soc. Civ. Engr., J. Irrig. and Drain. Div. (IR2):203-206.

42 Raney, F. E. and Yoshiaki Milhara. 1967. Water and soil temperature. p. 1024-1036. In: Irrigation of Agricultural Lands. R. M. Hagan, H. R. Haise, and T. W. Edminster (Ed.). Monog. 11, Am. Soc. Agron., Madison, WI.

43 Ravelo, C. J., E. A. Hiler, and T. A. Howell. 1977. Trickle and sprinkler irrigation of grain sorghum. TRANSACTIONS of the ASAE 20(1):96-99, 104.

44 Ritchie, J. T. 1971. Dryland evaporative flux in a subhumid climate: I Micrometeorological influences. Agron. J. 63:51-55.

45 Ritchie, J. T. 1972. Model for predicting evaporation from a row crop with incomplete cover. Water Resour. Res. 8:1204-1213.

46 Ritchie, J. T. 1974. Evaluating irrigation needs for Southeastern U.S.A. Proc., Am. Soc. Civ. Engr., Irrig. and Drain. Div., Spec. Conf., Biloxi, MS.

47 Rosenberg, N. J. 1972. Frequency of potential evapotranspiration rates in central Great Plains. Proc. Am. Soc. Civ. Engr., J. Irrig. and Drain. Div. 98(IR2):203-206.

48   Schmueli, E. 1973. Efficient utilization of water in irrigation. p. 411-423. In: Arid Zone Irrigation, D. Yaron, E. Danfors, and Y. Vaadia (Ed.), Springer-Verlag, New York.

49   Stamm, G. G. 1967. Problems and procedures in determining water supply requirements for irrigation projects. p. 771-784, In: Irrigation of agricultural lands, R. M. Hagan, Monog. No. 11, Am. Soc. Agron., Madison, WI.

50   Tanner, C. B., and W. A. Jury. 1976. Estimating evaporation and transpiration from a row crop during incomplete cover. Agron. J. 68:239-243.

51   Taylor, S. A. 1965. Managing irrigation water on the farm. TRANSACTIONS of the ASAE 8:433-436.

52   U.S. Salinity Laboratory Staff. 1954. Diagnosis and improvement of saline and alkali soils. L. A. Richard (Ed.)., U.S. Dept. Agr. Handb. 60, 160 pp.

53   USDA, Soil Conservation Service. 1970. Irrigation water requirements. Eng. Div. Tech. Rel. No. 21, U.S. Gov't. Printing Office, 88 p.

54   van Bavel, C. H. M. 1966. Potential evapotranspiration: the combination concept and its experimental verification. Water Resour. Res. 2(3):455-467.

55   Viets, F. G., Jr., R. P. Humbert, and C. E. Nelson. 1967. Fertilizers in relation to irrigation practice. p. 1009-1023, In: Irrigation of Agricultural Lands. R. M. Hagan, H. R. Haise, and T. W. Edminster (Ed.), Monog. 11, Am. Soc. Agronomy, Madison, WI.

56   Westerman, P. W., B. J. Barfield, O. J. Loewer, and J. N. Walker. 1976. Evaporative cooling of a partially-wet and transpiring leaf—I. Computer model and its evaluation using wind-tunnel experiments. TRANSACTIONS of the ASAE 19(5):881-888.

57   Wolfe, J. W., P. B. Lombard, and M. Tabor. 1976. The effectiveness of a mist versus a low pressure sprinkler system for bloom delay. TRANSACTIONS of the ASAE 19(3):510-513.

58   Wright, J. L. 1979. Recent developments in determining crop coefficient values. (Abst) Proc., Am. Soc. Civ. Engr. Irrig. and Drain Div. Spec. Conf. July, p. 161-162.

59   Wright, J. L. 1981. New evapotranspiration crop coefficients. (In process).

60   Wright, J. L. and M. E. Jensen. 1972. Peak water requirements in Southern Idaho. Proc. Am. Soc. Civ. Engr., J. Irrig. and Drain. Div. 98(IR2):193-201.

61   Wright, J. L. and M. E. Jensen. 1978. Development and evaluation of evapotranspiration models for irrigation scheduling. TRANSACTIONS of the ASAE 21(1):88-96.

62   Yaron, D., J. Shalhevet, and E. Bresler. 1974. Economic evaluation of water salinity in irrigation. Res. Rep. to Resources for the Future, Inc. Dep. of Agr. Econ. and Manage. Hebrew Univ. of Jerusalem.

63   Zimmerman, J. D. 1966. Irrigation. John Wiley and Sons, New York, NY.

chapter 7

DRAINAGE REQUIREMENTS AND SYSTEMS

7

7

DRAINAGE REQUIREMENTS AND SYSTEMS

by   W. J. Ochs, USDA-SCS, Washington, D.C.; L. S.
Willardson, Utah State University, Logan, UT; C.
R. Camp, Jr., USDA-SEA/AR, Florence, SC; W.
W. Donnan, Consulting Engineer, Pasadena, CA;
R. J. Winger, Jr., U.S. Bureau of Reclamation,
Denver, CO; W. R. Johnston, Westlands Water
District, Fresno, CA

7.1 DRAINAGE REQUIREMENTS

Most agricultural crops require a well aerated root zone free of saturation by water. Excess water can be removed naturally or artificially by installing a drainage system. Soils that are considered prime agricultural land are
either naturally well drained or the water table is controlled by artificial
drainage. Poorly drained soils are less desirable. Artificial drainage of poorly
drained soils can change their suitability for agriculture from lower to higher
levels. In many cases, good drainage is the difference between a highly productive farm and a marginal farm.

7.1.1 Introduction

Agricultural drainage can be defined as the removal and disposal of excess water and salt from agricultural land to provide a good environment in
the soil for plant growth. The sources of excess water can be precipitation,
snowmelt, irrigation facilities, overland flow or underground seepage from
adjacent areas, artesian flow from deep aquifers, floodwaters, or water applied for such special purposes as leaching salts from the soil or for
temperature control.

An ideal condition in an irrigated area would be one where the water
table is deep and the soil is so constituted that there is unrestricted downward
movement of any excess water and salt from the root zone. Surface drainage
is unrestricted and runoff occurs quickly without causing erosion. In humid
nonsaline areas, a moderately permeable soil with a stable, relatively shallow
water table is a desirable drainage situation. However, where possible, a deep
soil without an encroaching water table that permeates the root zone is the
most desirable condition in humid or arid regions.

In arid areas, artificial drainage systems are installed to control water-
table level and salinity in the root zone. The drains are installed at a depth of
2 to 3 m for the purpose of lowering the water table. Proper irrigation water
management then insures a continual downward movement of the excess
water and salts brought to the soil by the irrigation water and the salts that
are concentrated there by the extraction of water by crop plants. Through
many years of experience, a 2-m depth of drains has been found to be

necessary in order to prevent excessive upward movement of salts by capillary rise of water during the period when crops are not growing. Experiments have been conducted and shallow water tables investigated in arid areas, but the conclusions have not justified installation of shallow drains in these areas. Deep drains are also advantageous from an economic point of view. Studies by Christopher and Winger (1975) and Willardson and Donnan (1978) have indicated that a 2.5-m drain depth is the approximate depth for maximum economy of installation. Deeper drains require higher excavation costs; shallower drains require closer spacing and a larger number of drains per unit area.

In humid irrigated areas, drains are usually installed at a depth of 1.0-1.5-m. Shallow drains mean higher water table and, possibly, a supplementary water supply for plants between intermittent rainstorms. A second, more important reason for shallow drains in humid areas is the tendency for decreasing permeability with depth in the soils. In humid areas where water table control is extensively practiced, the ideal constant depth of water table for maximum crop production is approximately 1 m.

The benefits of drainage may be difficult to determine on some soils because of other resources involved in land development. Soils that are marginally productive without drainage can be highly productive with drainage but may also require additional inputs to obtain that production. Irrigation, fertilizers, mechanization, weed control, and other inputs such as additional harvest costs are all involved in the increased production. The increased production would be impossible without drainage but is not entirely attributable to drainage. When land is drained, the farmer may change crop species also resulting in increased income. For example, land that is not properly drained may produce only poor pasture but may be converted to growing more intensive crops following drainage. Data from Spain reported by Willardson (1974) indicate an increasing annual rate of production from drained land that was converted from salt-tolerant vegetation to wheat farming.

Control of root-zone salinity and root-zone aeration are the main objectives of agricultural subsurface drainage. Salinity control is imperative for the establishment of a permanent irrigated agriculture. Good aeration of the root zone prevents denitrification and loss of valuable plant nutrients but, at the same time, allows the plant roots to carry on respiration in an unrestricted manner. Maximum production is obtained from fertile, well-drained soils.

7.1.2 Factors Affecting Drainage

Subsurface drainage of agricultural land is the removal of excess water and salt from the soil. Since water moves through the soil, the soil may be a limiting constraint. The kind of soil, therefore, plays an important role in drainage. Some soils have characteristics that make them easy to drain, whereas other soils are difficult, if not impossible, to drain under present economic conditions. Knowledge of the kind of soil in a proposed drainage project is essential to the planning of the drainage system. Soil surveys and soil classifications are valuable tools that need to be used during all phases of the drainage system planning, design, and operations.

Residual soils are typical in most of the agricultural areas now being drained throughout the world. These soils are formed by natural weathering processes from the parent material existing in that particular location. In residual soils formed directly from weathering of rocks, the subsoil material eventually grades into the soil parent rock. For residual soils that have been formed from materials moved by wind or water, the soils have different characteristics but also grade into the normal parent material deeper in the profile. Some of these parent materials are well drained such as loess or wind-deposited materials. Other soils are very impermeable, originating from glacial parent materials that are very fine textured. Since soil structure and soil porosity develop from the top of the profile through the activity of roots and micro-organisms, soil permeability tends to decrease with depth. The older the soil, the deeper the profile through the activity of roots and micro-organisms, soil permeability tends to decrease with depth. The older the soil, the deeper the profile development and higher the permeability deep in the profile.

Alluvial soils are typically found in most irrigated areas. These soils consist of mineral particles deposited primarily by the action of water. They are characterized by stratification of coarse and fine layers and a general heterogeneity. The soil may consist of layers of gravel, sand, silt, and clay, which may persist over large areas or may be relatively unconnected. Average values of soil properties required for drainage design are usually difficult to obtain for these soils. However, alluvial soils can be very productive and need to be drained where natural drainage is not adequate.

Lacustrine soils formed in lake bottoms are particularly difficult to drain. They often have undeveloped profiles in arid regions and fully developed profiles in humid regions. Since the mineral materials were deposited in lake waters, the soils tend to be finely stratified and have very low vertical permeability that interferes with adequate drainage. Lacustrine soils require careful study to determine whether or not they are drainable.

Soil Investigation: For good drainage design, adequate information must be obtained on the characteristics of the soils to be drained. This information is obtained by examining the physical characteristics of the soil that relate to drainage design. The soil can best be examined by excavation or boring. Soil surveys may be available for the proposed project area. If not, efforts should be made to obtain standardized soil surveys since valuable information on hydraulic conductivity, water table, salinity, and productivity is provided in these reports. Additional soil investigation is usually needed, however, and these investigations can best be made by borings or excavations.

The most common tool used for examining a soil profile ia a soil auger. Different types are available, the most common being a spiral auger. Spiral augers disturb the natural structure of the soil, so that only soil texture can be observed effectively. Other types of augers suitable for examining the soil profile include a barrel-type auger for sandy soils and an open "dutch" auger for clay soils. A posthole auger is suitable for examining soil profiles where other types of augers are not available. Typical soil augers are shown in Fig. 7.1.

Tube type probes are also a good tool for examining soil profiles and in many cases are considered better than an auger. The probe is a volumetric soil sampler and is used either in place of, or in conjunction with, a soil auger. The advantage of a soil sampler is that the undisturbed structure of

FIG. 7.1 Typical augers used for examining a soil profile.

the soil can be observed. The volumetric sampler provides a relatively un-disturbed sample of soil.

Notes should be kept that identify and locate each hole and describe the condition and depth of the various strata. Depth of strata should be referenc-ed to the soil surface. Depth at which the water appears during boring and depth at which the water is observed in the open auger hole the following day should be recorded. Auger holes, if protected, will remain open for periods as long as 1 year in stable soils and can be used as temporary observation wells for observing the fluctuation of the water table. If the auger holes close for some reason, they can be easily cleaned with the original auger to restore their effectiveness as water-table observation wells. Subsurface profile infor-mation can also be obtained from well drilling logs.

Jetting equipment can be used to obtain limited information on soil properties. A high-pressure water jet on the end of a rigid pipe can be used to erode a hole through the soil giving evidence of stratification and some evidence of soil texture. However, the information obtained is usually not very precise with respect to depth or soil texture since material from all the strata is mixed by the jetting.

Bore-hole grids are needed for design investigation of a soil profile for drainage purposes in most arid and semiarid areas. The grid should be coor-dinated and planned on the basis of a soil survey map. Locations of borings and observation holes should be recorded to provide the maximum amount of information for each kind of soil in the area. If it is desirable to know the direction of ground-water movement, holes should be bored on a regular grid and should be carefully noted on a topographic map. On very uniform land, one hole per hectare may be sufficient. In other conditions, the initial grid spacing might be as close as 100 m between auger holes. If additional detail is needed, the spacing can be halved or quartered to increase the intensity of the investigation in areas of specific interest. If direction of ground-water movement is important, elevations of the ground surface at each auger hole will be required so that the direction of ground-water movement can be plotted. Depth to water table can then be plotted from the auger-hole data without reference to elevations. Auger holes should penetrate to at least the estimated depths at which the drains should be installed. Some holes should be made 1.5 to 2.0 times the depth of the proposed drains in order to determine the characteristics of the deeper soils. Water moving toward subsurface drains also travels in the soil below the bottom of the drains unless that travel is restricted by impermeable layers.

Logging observation-hole data requires that a careful, written record be made of the soil properties in every observation hole. Any observation about soil characteristics, water-table position, and unusual soil horizons should be noted. The observation hole should be given a number so that it can be located on a map. Notes should be made of every significant change in the soil. The texture and thickness of each soil horizon should be recorded, and observations regarding the continuity of soil horizons in adjacent holes should be noted. For example, sand at 1.5 m in every hole is a good indication that the sand layer is continuous and the information can be used to advantage in the design of the drainage system. If similar layers are found at different depths in every hole, reference should be made to the differences in elevation to determine whether the soil strata surface is undulating or the subsurface layers are also undulating or are discontinuous.

Soil quality: Soil quality for drainage purposes can be classified in two general categories, physical and chemical. Physical properties include texture, density, and structure. The texture class of a soil is determined by the size of soil particles. The texture classes are identified by proportions of sand, silt, and clay. Soil structure indicates the arrangement of the soil particles. Even though porosity of the soil is unrelated to permeability, soils with a high permeability or hydraulic conductivity are usually referred to as being porous. For soils having the same porosity, the soil with the greater number of large pores has a higher hydraulic conductivity. Soils such as clays, which have a high porosity but very small pores, have low permeability. (See Chapter 4 for detailed discussions on soil water, hydraulic conductivity, etc.)

Soil chemical properties as well as physical properties must be considered in evaluating the soil quality for drainage purposes. If a soil has a high sodium content, application of irrigation water may cause deterioration of the soil structure and decrease the permeability to the point that both leaching and drainage will be affected. If evidence of salinity, such as poor or sparse plant growth, accumulation of salt on the surface, or a very soft, dry surface condition, is observed, soil samples should be taken to determine the chemical properties of the soil. Factors to be considered are electrical conductivity, sodium content, calcium content, magnesium and sodium content, and pH of the soil. Most problems occur in sodic soils. These soils have an exchangeable sodium content greater than 15 percent, a conductivity of the saturation extract less than 4 dS/m, and a pH greater than 8.5. (1 dS/m = 1 mmho/cm). Sodic soils require chemical amendments during irrigation and reclamation to prevent permanent destruction of the soil structure. Saline soils are defined as those having a conductivity of the soil structure. Saline soils are defined as those having a conductivity of the saturation extract greater than 4 dS/m, an exchangeable sodium content less than 15 percent, and a pH less than 8.5. Saline soils can be reclaimed easily be providing adequate drainage and by leaching with ordinary water. Saline-sodic soils, which are between these two extremes, are those having a conductivity of the saturation extract greater than 4 dS/m and an exchangeable sodium content greater than 15 percent. Saline-sodic soils must be treated with chemical amendments to preserve their structure.

Salinity problems in soils are relatively rare in humid areas because adequate leaching takes place from normal rainfall. In arid areas where a high water table may have caused salt accumulation in the surface soils, careful analysis should be made of the soil chemical condition to prevent damage to the soil by uncontrolled application of irrigation water. (See chapter 5 for detailed discussions on salinity.)

Kinds of crops: Although some crops are more adapted to shallow water tables than others, the ideal drainage for most economic crop plants is unrestricted downward movement of excess water. If the water table is being controlled at a specific depth for the purpose of providing water to the plants, then variables such as rainfall, plant rooting depth, plant variety, and soil texture must be considered. In high rainfall areas, the water table must be low enough to maintain soil aeration. Shallow-rooted plants require a relatively high water table if they are subirrigated by a controlled water table. Coarse-textured soils require a relatively high water table to provide adequate subirrigation water to plants. For fine-textured soils, the water table must be deeper for proper subirrigation. Since crop rotations are practiced on many drained soils, the depth of the water table in an artificially drained field should usually be designed to provide a water-table depth satisfactory for the deepest rooted plants.

Climate: The principal influence of climate on drainage requirements is reflected in whether the climate is humid or arid. If the climate is humid, removal of excess surface and subsurface water originating from rainfall is the principal purpose for agricultural drainage. If the climate is arid, controlling the water table and preventing an accumulation of salt in the soil's root zone resulting from irrigation water applications are the important factors. In semiarid areas or subhumid areas, drains may be required for both purposes. During certain periods of the year, excess precipitation may occur that requires rapid removal of excess water from the soil. During other periods of the year, upward movement of saline ground waters into the root zone should be minimized. The amount of water to be removed is a function of the infiltration characteristics of the soil. Factors related to climate that control the amount of water to be removed by drainage are the soil surface condition, which controls the part of the rain falling that penetrates into the deep soil layers; rainfall frequency; evapotranspiration; and irrigation application in excess of evapotranspiration needs.

For drainage of irrigated lands the method of determining the drainage requirement used by the U.S. Bureau of Reclamation takes into account the transient regimes of the ground-water recharge and discharge caused by rainfall and the application of irrigated water. Use of the Transient Flow Theory results in a drain spacing for which dynamic equilibrium is reached at a specific root-zone depth required by crops grown in the area.

A careful analysis of the water balance in the soil should be made to determine the drainage coefficient. In some areas of the country, the Soil Conservation Service or local drainage organizations have established drainage coefficients for use in ordinary drainage design. In humid areas, a drainage coefficient of approximately 10 mm/day is appropriate; in arid reas, it is approximately 3 mm/day. A specific water balance study should be conducted to determine the true drainage coefficient for the area of interest.

Topography: Water movement toward subsurface drains occurs primarily beneath the soil surface. Water always moves from high to lower potential energy levels. Darcy's law states that the quantity of water moving through the soil is a function of the gradient or slope, the hydraulic conductivity of the soil, and the cross-sectional area of soil through which the water moves. If the cross-sectional area is large and the hydraulic conductivity and the gradient are small, a large amount of water can be moved through a soil. If the hydraulic conductivity is high, the gradient is high, and the cross-sectional area is small, a large amount of water can also be moved through a soil. The combination of the three factors in Darcy's law determines the flow quantity.

Some drainage problems arise if the subsurface stratification does not correspond to the surface topography. For example, in an area with a rolling or undulating soil surface, stratification may all be horizontal. Water moving down through the soil may encounter an impermeable horizontal layer, move laterally, and create local side hill seeps and small drainage problems. If the slope of the soil surface is different from the slope of the subsurface layers, the layers may intersect the soil surface at some point, causing seeps over a large area. In soils without contrasting layers that have abrupt changes in the gradient of flow may cause the water table to rise and intersect the soil surface. Changes in slope are logical locations for the placement of interceptor drains. Interceptor drains must be placed in or above an area of local water concentration to remove the excess water appearing at that point. Since water moving in the subsoil is most difficult to control, a careful study should be made of both surface and subsurface conditions to determine the effect of topography on drainage requirements.

Depth of water table: Optimum depth to the water table cannot be precisely determined with criteria now available. The generally accepted practice of placing drains 2.5 m deep in arid areas and 1.5 m deep in humid areas is the general guide now. The crops to be grown and the salinity of soil and water should be considered when determining the correct water-table level for an area. Subsurface drains should be installed in the most permeable strata near these depths to obtain maximum drain effectiveness. As indicated earlier, the economical drain depth from an energy and installation cost viewpoint is approximately 2.5 m in arid areas. However, soil physical conditions may require drains to be placed at some other depth. Course-textured soils have a smaller capillary rise. Therefore, the water table can generally be shallower in coarse-textured soils without causing a problem of soil surface salinization.

The water table is a dynamic rather than a static condition in artificially drained soils. The water table rises during periods of recharge and falls during periods of drainage between recharge events. Recharge events are either irrigation or natural rainfall. In both humid and arid areas, the net annual movement of water should be downward through the soil but the quantity of water moving depends on the excess available. The rate of water movement depends on the soil and the ability of the drainage system to remove the excess water applied. If a water table rises temporarily into the root zone, then it must recede within 1 or 2 days to prevent damage to most plant roots due to lack of soil aeration. Specifications used by the U.S. Bureau of Reclamation for arid and semiarid areas do not allow a water table at a point midway between drains to rise closer to the soil surface than 1.3 m at any time during the year, particularly during the irrigation season. Since the drains are generally installed at a depth of 2.5 m, the water table should not rise closer to the soil surface than 0.5 m for most crops and should fall to the design water-table depth within 1 or 2 days.

7.2 NATURAL SUBSURFACE DRAINAGE

All soils drain naturally if an outlet is available. The rate of drainage depends on the physical charateristics of the soil and the condition of the outlet. The capacity of soils to drain naturally depends on site and soil physical factors and on the rate of recharge. If the rate of recharge exceeds the natural drainage capacity, which is controlled by permeability, slope and cross-sectional area of flow, then the excess water will cause drainage problems. Some soils in areas have very restricted natural drainage capacity, but this drainage problem is not apparent before irrigation since the natural rate of recharge is low. In humid areas on the other hand, swampy areas develop where natural drainage capacity is limited. The natural drainage capacity in arid areas may be sufficient to take care of normal precipitation because it is less than evapotranspiration and only occasional recharge events occur. With the relatively steady application of water by irrigation, however, the natural drainage capacity is often exceeded and drainage problems develop quickly. During the planning stage for a large irrigation project, the natural drainage capacity of the soil should be evaluated to determine the rate at which drainage problems may develop. The position of water table, the expected rate of recharge, and the possible water movement toward natural drainage outlets should be determined.

7.2.1. Internal Soil Water Movement

The basic soil physical property that must be determined for subsurface drainage design is the saturated hydraulic conductivity. Hydraulic conductivity is a measure of the soil's ability to transmit water and should be measured in the field. Laboratory measurements of disturbed or undisturbed samples are helpful in identifying a soil but should not be used for design calculations. Careful field measurements of hydraulic conductivity are preferable because they measure the soil properties in their natural condition and are representative of a large soil volume. The high variability of hydraulic conductivity for a given soil requires that a large number of measurements be made to define sufficiently a value to be used in designing dependable drainage. Fewer measurements are required for uniform soil, but soil permeability tests should be planned on the basis of at least one per 5 ha and they should be closely correlated with the kine of soil. Both auger-hole permeability tests and shallow well pump-in tests can be used according to Luthin (1973) to determine in situ hydraulic conductivity satisfactorily.

Determinations of hydraulic conductivity can be made by analyzing ground-water movement adjacent to subsurface drains or open drains. Water table drawdown measurements are required for most methods and volume discharged from a drain is required for some methods. Using these measurements, pertinent information describing the drainage system where measurements, and an appropriate water flow equations are used, the hydraulic conductivity or hydraulic conductivity-drainable porosity ratio can be calculated. With this method there is some difficiulty in obtaining accurate estimates of seepage.

Other soil parameters of interest in drainage design are saturation percentage, field capacity, permanent wilting percentage, and drainable porosity. The saturation percentage of the soil is used to determine the interconnected void space in the soil. Field capacity is the water content of a soil that has had an opportunity to drain freely. The difference between saturation and field capacity is an indication of the storage space in the soil that must be filled before the water table rises.

The difference between field capacity and permanent wilting percentage is the soil water that is available to plants and is managed for irrigated agriculture. Knowledge of the water extraction volume and the available storage space is important in determining the expected ground-water recharge or drainage coefficient. Specific yield, sometimes called drainable porosity, is a soil parameter used in transient drain-spacing equations and is the fraction of soil volume drained when the water table drops a given distance. New techniques have recently been developed by Skaggs (1976) and Skaggs et al. (1978) for determining the ratio of hydraulic conductivity to specific yield. If only hydraulic conductivity measurements are made, then an estimate of specific yield can be made by using a graph developed by the U.S. Bureau of Reclamation and published by Luthin (1973).

This graph was developed from hundreds of field hydraulic conductivity tests and specific yield determinations and from undisturbed core samples from the same strata on many soil types. Many field spot checks were made to check correlation between laboratory and field specific yield. The accuracy of the relationship has proven reliable for the design of thousands of kilometers of drains and is an important alternative to making difficult direct measurements of specific yield in the field.

The natural water-table depth in an area is an indication of the hydrologic balance between the natural drainage capacity and the natural recharge. The water-table position depends on the surface topography, subsurface soil strata configuration, and condition of the natural drainage outlet for the area. With the introduction of irrigation, the natural water-table depth changes because of the changed hydrologic conditions caused by irrigation. If the natural drainage capacity is high, the water table remains relatively unchanged after irrigation. If the natural drainage capacity of the area is restricted the water table may rise rapidly with the introduction of irrigation until a new equilibrium level is established. In some low areas, the water table may even rise above the soil surface.

7.2.2 Natural Outflow, Artesian Flow, and Seepage Sources

Natural outflow: The natural outflow of drainage water from an area is a function of the hydrologic balance and is the difference between the water applied and the water consumed or used. Water can be applied either through irrigation or precipitation; water can be consumed through evapotranspiration, evaporation, seepage, or other consumptive uses in the area. If a location is available where all outflow from a drainage basin is concentated, direct measurements provide an estimate of the natural drainage capacity of an area. Precipitation and the slope of the ground-water table are reasonably easy to measure, but the flow cross-sectional area, water-discharge volume, and evapotranspiration are difficult to measure.

Artesian water: Artesian pressures in the subsoil are caused by the interspersion of permeable and impermeable layers connected to supplies of water at higher elevations. Artesian pressure is an indication that a restricting layer is present in the soil. When this layer is penetrated, water discharges from the confined layer under pressure and, in some cases, the hydraulic pressure is sufficient to cause water to rise above the soil surface. If the confining layer is slightly permeable, water flow to subsurface drains may be considerable and will require closer drain spacing than normal. Artesian pressure is determined by installing piezometer to measure the vertical soil water pressure or potential gradient in the soil. Piezometers can be installed by jetting, by driving, and in some places, by augering. To measure the pressure at the point of the piezometer, it is very important to have a tight seal between the outside of the piezometer and the soil in which the piezometer is installed. At least two piezometers are necessary at a location for determining a vertical gradient and three piezometers provide more information. Thre piezometers and an auger hole provide even more information about the vertical direction of ground-water movement in a soil. In large areas, piezometer batteries are normally installed on a 1-km grid. Artesian pressure below a confining layer that is both impermeable and continuous does not interfere with drainage in the root zone. However, the confining layer is a barrier to downward water movement, thus restricting natural drainage.

Seepage sources: It is important in a drainage investigation to determine the sources of the excess water that must be removed from the soil so that, if possible, the problem can be corrected at the source. If the water causing the drainage problem is coming from leaking canals, canal linings should be considered as an alternative to an intensive drainage system. In some places, prevention of canal seepage remove's the excess water that exceeds the capacity of the natural drainage system and artificial drains are not required. Seepage from topographically higher areas in an irrigation project or in a humid area sometimes can be identified, and the drainage problems can be solved by a single interceptor drain. To use these alternative solutions to drainage problems, the source and magnitude of the seepage water supply must be determined. Piezometers and water-table observations wells can be used to determine the direction of ground-water movement and identify the seepage sources.

7.3 OPEN DRAINS

Most open drains (ditches) are used to provide needed surface drainage. Surface drainage is described in ASAE Engineering Practice EP-302.2 (ASAE, 1979b) as the removal of excess water from the soil surface in time to prevent damage to crops and keep water from ponding on the soil surface or, in surface drains that are crossed by farm equipment, without causing excess soil erosion. Surface drainage for farmland can be provided by improving natural channels, constructing new drains, or smoothing and grading the land surface. These techniques can be used alone or in combination to provide a suitable system for a given drainage situation. Surface drainage problems result where the excess water from either precipitation or irrigation cannot infiltrate into the soil and move through the soil to a suitable outlet or cannot move freely over the soil surface to a surface outlet.

Although a closed-conduit subsurface drainage system is the primary method used to control water tables and provide soil drainage, open drains of adequate depth can also provide subsurface drainage and serve as outlets for subsurface drains. The chief disadvantages of using open drains for subsurface drainage are the high maintenance costs and the removal of large land areas from production. Since closed subsurface drains are discussed elsewhere in this chapter, this discussion is directed primarily toward the use of open drains for surface drainage.

7.3.1 Kinds of Open Drain Systems

Open drains function both as collection drains and disposal drains. A collection system consists of all drains that collect water in the field. Their alignment, shape, and size must fit the topography, soils, drainage area, and land use of the field. They should be located and shaped so that water flows directly into them with minimal soil erosion. This is usually accomplished by making them shallow with flat side slopes and protecting the channel with stabilizing vegetation or structural devices.

The location of disposal drains is not as retrictive as the location of collection drains. The outlet is the primary influence on disposal drain location. A disposal-drain system consists of the larger drains that receive water from the collection system and transport it to a suitable outlet. Lateral and main drains can both be included in this system. The drains are usually trapezoidal in shape except for the smaller field laterals that frequently have a V-shaped channel. The size and shape of an outlet drain is determined by the volume of water discharged into it by the collection system; and it must be sufficient depth to drain all collection drains.

Types of open drain systems: Surface drainage systems can be classified in three broad categories: (a) random systems, (b) parallel systems, and (c) diversion on interceptor systems.

Random systems provide drainage of low areas by a series of drains connecting the depressions through which the excess water is delivered to a suitable lateral outlet drain. These systems are relatively low in cost but require frequent maintenance and are obstacles to farm machinery, particularly if they are not constructed with low side slopes and are more than 0.3 m deep.

Parallel systems consist of field drains established in a parallel but not necessarily equidistant pattern. A parallel design facilitates mechanization by eliminating point and short rows. The field drains, which can be crossed by farm machinery, are constructed perpendicular to the direction of the rows and the field laterals, permitting farm machinery to travel greater distances without turning. The field surface between field laterals should be improved by land smoothing or land grading for maximum benefits. In areas where the field drains do not dry in about the same time as the adjoining land surface, rows can be established parallel to field drains. Temporary row drains must be provided in this situation to reduce row length and to drain rows that pass through depressions.

A diversion or interceptor system consists of one or more diversions, terraces, interceptors, or field drains constructed across the slope. Diversions intercept upland runoff and prevent it from overflowing flatland at lower elevations. This simplifies the installation of drainage systems on the lower lands and aids in the control of sediment transport and deposition. Interceptor drains collect and remove surface runoff as well as subsurface water on sloping lands having shallow, permeable surface soils overlying relatively impermeable subsoils. This subsurface interception technique is directed toward removing the water before it resurfaces and becomes a surface drainage problem.

Types of drains in a collection system: Several types of drains can be used in a collection system, either singly or in combination, depending on the particular drainage requirement. A collection drain should be constructed with sufficient width, depth, and flatness of slope to allow tillage equipment to open furrows that will drain freely into it.

The furrow between planted rows is the initial water collector in a surface drainage system on row-cropped land. The rows should be directed so that water can move freely without ponding or causing excessive erosion. Field or row drains must be installed to provide free drainage of row furrows. If the row furrow is used for furrow irrigation, the location of field drains can be determined by irrigation requirements.

The row drain is a small temporary drain that cuts across rows or through depressions to provide drainage into field drains or field laterals; it is constructd with a plow or similar tillage implement. These drains are used primarily in conjunction with parallel rows and bedding systems in areas of high precipitation or for crops having a low water tolerance. They are short, normally less than 100 m in length, and are opwn at both ends to discharge into field drains or field laterals. Row drains are also known locally as cross drains, quarter drains, plow drains, annual drains, or header drains.

A field drain is a shallow graded drain for collecting water within a field, usually constructed with flat side slopes for ease of crossing with farm equipment. It can drain depressions and collect or intercept flow from the land surface or channel flow from natural depressions, furrows, row drains, and bedding systems.

Land forming: Although not a specific kind of open drain or type of surface drainage system, land forming is an important component of several drainage systems. Land forming can be described as mechanical alteration of the soil surface to improve surface water drainage. The various methods used to improve drainage are smoothing, grading, bedding, and leveling. In irrigated land, where water is applied to the surface, the land forming method selected is determined primarily by irrigation requirements instead of surface drainage requirements, particularly in the arid areas. In humid areas, the drainage requirements usually dominate the method selected because of the possibility of high precipitation and runoff, even during periods when irrigation may be required. Soil properties must be carefully considered to avoid exposure of unfavorable soil horizons as a result of soil removal. Land forming for irrigation is discussed in Chapter 8.

Land smoothing, according to ASAE Engineering Practice EP-302.2 (ASAE, 1979b), is smoothing the land surface with a land plane or land leveler to eliminate minor depressions and irregularities without changing the general topography. This procedure improves the surface drainage and increases farm machinery efficiency. In some places, land smoothing is an interim measure that will be replaced by land grading when economic conditions permit. In other places, where the soil surface has sufficient natural slope towards an outlet, land smoothing may be the only land forming method required. If smoothing is used, the soil must be such that small cuts do not expose soil layers that would hinder machinery operations or plant growth or create severe erosion problems.

Lnad grading according to ASAE Engineering Practice EP-302.2 (ASAE, 1979b) is the process of forming the soil surface to predetermined grades so that each row or the surface slopes to a drain. This is accomplished by cutting, filling, and smoothing to planned continuous surface grades with uniform slopes but not necessarily plane surfaces. In developing a land grading plan, filling depressions with soil from adjoining mounds and ridges and establishing grade in the direction of row grade, tillage, or predominant natural slope is emphasized to minimize cuts and fills.

Land leveling according to ASAE Engineering Practice EP-302.2 (ASAE, 1979b) is a precise operation that modifies the land surface to planned grades to provide for more efficient surface irrigation. Irrigated land also is frequently leveled to obtain good drainage. In humid areas, collection drains or tailwater-runoff drains at the ends of furrows and along the borders of leveled land must be sized to handle storm runoff that normally requires a greater capacity than that required for irrigation runoff water alone. In some places in arid areas, the land surface is shaped to provide a "dead-level" condition for irrigation. Also, water leveling is used in fields that are to be flooded and provides adequate drainage for rice and other similar close-growing, water-tolerant crops. In this method, the fields are divided into segments, separated by permanent contour levees, and leveled to a series of planes with zero grade. Leveling is accomplished by flooding the segments scraping the high areas between levees, and allowing the suspended soil to settle in the low areas.

Bedding, as outlined in ASAE EP-302.2 (ASAE, 1979b), is accomplished by plowing land to form a series of low, narrow ridges separated by parallel dead furrows having ridges oriented in the direction of greatest land slope. In areas with little or no slope, wider ridges can be established by shaping and smoothing the land surface and the drain spoil material between closely spaced and graded field drains. Crowning is a bedding technique where a series of broad, low ridges, approximately parabolic convex in shape, are developed with the ridge midway between the drains. The bed width depends primarily on the type of parallel drain provided between beds or lands. Dead furrows connected to field drains and laterals that collect and remove drainage water normally provide adequate surface drainage for land areas 8 to 25 m wide. For widths greater than 18 m, the crowning method is recommended with the surface slopes across the crown and with parallel field drains replacing the dead furrows.

7.3.2 Design of Open Drains

This section deals with the design of open drains used primarily as outlets for drainage systems where the general topography is flat to mildly sloping and where surface waters are diffused as is true for most irrigated land. The depth of small field drains is determined by the depth or row furrows and upstream field drains; the side slopes are primarily determined by the need for farm machinery to cross the drain and are usually very flat. This section is not intended to be a comprehensive treatment of design of open channels for general drainage. It concentrates on the unique design requirements for open drains used in irrigated areas, as well as presenting a general design procedure for open drains.

Design procedure: The basic procedure for design of open drains includes the following: (a) obtain field information such as field elevation, location and size of existing drains, control points, soil borings, and elevation of water in the outlet with stage-frequency information if available; (b) establish control points and set hydraulic gradeline; (c) determine watershed areas; (d) compute design discharge for the lower end of each channel reach; (e) select and record appropriate design criteria including values for channel roughness (n), side slopes, minimum bottom width, and minimum depth below hydraulic gradeline; and (f) design the drain cross section below the established hydraulic gradeline. Several problems may arise in applying this procedure, including the combining of flow at junctions of drains, bridges, and culverts and from different types of watersheds. The reader is referred to other more extensive treatments of design of open drains such as, ASAE EP-302.2 (ASAE, 1979b), USDA, Soil Conservation Service, National Engineering Handbook 16 (1971) and Schwab et al. (1966) for solutions to these problems.

Open drains should be designed to pass the required drainage flow throughout the length of the drain with the hydraulic gradeline sufficiently below the land elevation to provide good drainage. The hydraulic gradeline represents the surface of the water when the drain is operating at design flow. The slope(s) of the hydraulic gradeline is used in the Manning formula to determine velocity. The grade or slope of the drain bottom may have a different value because the drain bottom is not always parallel to the hydraulic gradeline.

Uniform flow is ordinarily assumed in the design of drainage channels except above culverts and at locations where the design requires backwater computations. With these exceptions the drain bottom normally can be established parallel to the hydraulic gradeline and a uniform channel section used. Even though nonuniform flow results where minor obstructions occur or where minor local drainage enters, it is of little practical significance and the general efficiency of the system is not impaired. The rate of water movement in open drains is influenced by: (a) the size of drainage area; (b) the irrigation or rainfall rate; (c) type of crop; (d) watershed runoff characteristics such as soil, vegetation, and slope; (e) height and frequency of flood waters, particularly as they affect the outlet; and (f) the degree of protection desired. The degree of protection desired is often difficult to evaluate because economic losses due to flood damage must be estimated and crops respond differently to given periods of flooding. Several methods used to determine the runoff or drain capacity are discussed in a later section.

Once the rate of water removal or drain capacity is known, the drain channel can be designed. The required channel dimension for a given flow rate (Q), hydraulic gradient (s), and channel roughness (n) is usually determined by solving the Manning equation to determine mean velocity (v) and by use of the relation:

$$Q = Av \quad\quad\quad\quad\quad\quad\quad\quad\quad\quad\quad\quad\quad\quad\quad [7.1]$$

where Q = rate of flow in m^3/s; A = cross-sectional area of the channel in m^2, and v = mean velocity in m/s.

In many places the drain grade is determined by either the outlet elevation, the lowest area in the watershed, or the depth of other drains in the system. In flat areas the drain grade should be the steepest possible, provided the maximum permissible velocity is not exceeded. The reader is referred to drainage design manuals such as ASAE Engineering Practice EP-302.2 (ASAE, 1979b) and Schwab et al. (1966) for additional detail regarding the determination of mean velocity, selection of channel roughness values, and channel shape.

In addition to the above, the following factors regarding channel shape and cross-sectional area should be considered: (a) in irrigated areas depth of drain rather than capacity may be the dominant consideration; (b) a deep drain gives a higher velocity than a shallow one; (c) a deep drain requires less area than a shallow one; (d) a deep drain may expose unstable soil layers that a shallow one would not expose, and (e) a deep drain may provide a better opportunity for future subsurface drainage in the area.

It is good practice to allow for sedimentation in a drain during the first 2 or 3 years after construction. This practice allows the drain to maintain the design capacity after it stabilizes and is provided for by increasing the design size. The amount of this allowance depends on the expected sediment transported from the watershed, erodibility of soiils exposed in the drain, and erosion from adjoining lands. After the sides of the drains are stabilized by vegetation, sedimentation normally decreases.

Plans and specifications: Construction plans and specifications for drainage work normally become a part of a contract, and the plans are often used by the landowners, engineers, and government agencies long after construction has been completed. Plans should be neat, clear, complete, and specific and reflect sound design. All separate items of the plan should include the names of the job, the landowner, and the designing engineer; location of the farm; and the scale used in all drawings. Construction plans for individual farm drains usually include a drainage plan map, profiles, cross-sections, ditch designs, structural details, soil profiles, and specifications. Cost estimates are usually provided only when required. Maintenance plans, to be discussed in detail later, should be provided, especially on large projects involving a group or several landowners.

The drainage plan map should show the location of all proposed drains, bridges, culverts, watershed boundaries, farm boundaries and ownership information, and other cultural features that would be helpful or needed during construction or for orientation purposes. If plan-profile sheets are used for detailed plans, an index of the plan profile should be included in the plan map. Plans for all drains down to the farm lateral size should include profiles. Completed profiles should include normal ground line and elevation of critical low points in the field to be drained, existing and proposed drain bottom, hydraulic gradeline, existing and proposed control structures, elevation of high water for the design storm at the outlet, width of drain, and elevation of the water table. Enough cross sections should be prepared for drains to allow accurate computation of the excavation volume required. This depends primarily on the uniformity of topography and the variations in existing ditches. Calculations made for designing all drains of the drainage system should be included as a part of the plans. Plans for all structures that are a part of the project should be included. Structures for small- to moderate-size projects are normally of standard size and design and a copy of a standard design can be used.

Written specifications are required for every item in a construction proposal. Specifications should be detailed enough to prevent any misunderstanding about the type of job desired but should not include unnecessary detail. In most cases, standard specifications are available for the more common types of work. Some of the items that should be included are clearing, channel excavation, spreading of spoil bank, structure excavation, material type and quality, and seeding of the drain sideslopes, berm and spoil bank.

Location: Drains should be located to provide the most effective drainage of the area. Topography, soils, existing drains, property lines, and cultural features all influence drain location. Natural outlets usually fix the location and elevation of a drain, but the alignment and efficiency of the channel can be improved by using curves and cutoffs. An open drain must terminate in an adequate outlet of sufficient capacity to carry the design discharge without causing stage increases and significant damage downstream.

In some places, open drains are needed where site conditions are likely to cause soil erosion problems. Channel erosion is primarily influenced by flow velocity, position of the water table, soil texture, soil structure, and vegetation. After careful study of these factors and the protection needed, alternative drain locations should be considered if significant erosion is probable. If alternative drain locations are not feasible, grade-control structures or bank protection may be required.

Open drains in humid areas provide outlets for farm drains, subsurface drains, interception drains, and irrigation return flow. In most cases, each farm unit should have a dependable outlet. For small farm units, it may not be feasible to provide a lateral for each farm. In these places, small groups of landowners may need to construct a group lateral. Most outlet drains are constructed in or near natural drainage channels.

In irrigated arid areas, open drains serve primarily to remove surface and subsurface irrigation waste water and irrigation canal seepage water. Drains can be located either parallel or perpendicular to the direction of ground-water flow, generally depending on whether the drain is intended to control the water table or intercept subsurface flow. In most places, location of the drains is affected by the irrigation or canal system and the depth and location of permeable aquifers.

Capacity: Drain capacity is determined by the size of the area to be drained, the intensity of drainage required, and the amount of precipitation, irrigation waste water, and natural runoff that could occur. In humid areas and in other irrigated areas having high intensity rainfall, the drain must have sufficient capacity to remove storm runoff as well as the irrigation waste water and natural runoff that could occur. For these areas, the drainage requirement or drainage intensity is normally expressed as a drainage coefficient that describes the rate of removal of excess water necessary to provide a specified level of protection. Drainage coefficients, expressed as depth of water per unit area for a given period of time, are available in USDA, Soil Conservation Service, National Engineering Handbook 16 (USDA, SCS, 1971) and in Schwab et al. (1966) for various geographic regions and crops. Drainage coefficients are specified for surface drainage, for subsurface drainage, and for a combination of the two. If the drain serves as both a surface drain and an outlet for subsurface drains, care must be used to select the correct drainage coefficient. In irrigated areas where subsurface drain flow is uniform for extended periods, it should be considered as a base flow in computing the drain capacity. In other areas where subsurface flow results from precipitation and is intermittent, the required drain capacity should be determined by surface drainage flow.

In arid and semiarid areas, the drainage requirement is usually expressed in terms of the amount of irrigation water supplied. Values of 10 to 40 percent of the applied irrigation water have been reported by Luthin (1957) as drainage requirements for different areas in these irrigated areas. Because of the wide variation in these values with location, the reader is directed to local sources for known values or estimated values based on experience. For large areas where irrigation water is applied to only a part of the total area, the drainage requirments can be expressed as a volume of water per unit area per unit of time on the basis of local experience.

Another method of determining the required rate of water removal, primarily for large watersheds, is by the empirical relationship:

$$Q = k\ M^x \quad \dotfill \quad [7.2]$$

where Q = runoff in m^3/s; k = a constant; M = watershed area in m^2, and x = an exponent.

The constants k and x vary with the drainage requirement and the location. Values for these constants, as well as curves for equation [7.2] and various combinations of values for k and x, are available in ASAE Engineering Practice EP-302.2 (ASAE, 1979b) and USDA, Soil Conservation Service, National Engineering Handbook 16 (USDA, SCS, 1971).

Auxiliary conservation structures and practices: Many conservation structures such as pipe drops, drop spillways, grade control structures, and chutes can be installed in open drains where required to control serious soil erosion and to stabilize the bottom grade of the drain (USDA, SCS, 1977). All spoil material from construction should be either uniformly spread over the adjacent fields or used for field roads. Vegetation should be established in drains and on spoil areas, if appropriate, as soon as possible after completing construction of the drain. Other measures for controlling erosion of the drain side slopes such as riprap or revetment are required in extreme cases. Ramps may be required to protect side slopes if the drain is accessible to livestock, and watergates may be required where fences cross drains.

In many cases, the designer has a choice of several acceptable methods of solving specific drainage-associated problems. The final method selected is determine by factors such as initial cost, longevity, hydraulic efficiency, and maintenance requirements and costs. Detailed descriptions of the many types of structures and practices and the hydraulic design of structures are not included here but are available in various hydraulic and drainage references such as USDA, Soil Conservation Service, National Engineering Handbook 16 (USDA, SCS, 1971) and Schwab et al. (1966).

7.3.3 Construction of Open Drains

Good quality construction is required for effective drainage and can best be insured by having qualified engineers responsible for the site survey; design, and layout of the system and by checking to insure that construction meets specifications. Once a drainage system has been designed and approved for construction, the location of all drains and auxiliary structures included in the system must be marked in the field with center-line and slope stakes. The capabilities and limitations of the construction equipment to be used should also be considered in the design of open drains.

Site conditions such as kinds of soil, soil stability, soil wetness, size of drains to be constructed, and location of the spoil banks, affect the type of equipment to be used. Other considerations in the selection of construction machinery include the volume of soil to be moved, degree of accuracy required, and time allowed to complete the project. Various types of equipment for construction of open drains are available including dragline, scraper, backhoe, grader, bulldozer, wheel-type excavator, shovel, and plow. For further details regarding types, description, operation, and selection of this equipment, consult manufacturer's brochures, construction equipment handbooks, or other drainage references such as Schwab et al. (1966) and Luthin (1957). The quality of the completed drainage system also depends on the use of the proper construction equipment and the skill of the equipment operator.

Berms are required to prevent the sloughing of drain side slopes caused by heavy loads too neat the edge of the drain, to provide roadways for maintenance equipment, to prevent excavated materials from eroding or falling back into the drain, to provide work areas, to facilitate spoil-bank spreading, and to eliminate the need for moving spoil banks in future operations. If spoil banks are to be spread, the berm required during construction and the method of spreading the spoil should be specified in the construction contract. The best use of the spoil is determined by the kind of soil excavated use of adjacent land, need for roads, and method of maintenance employed. In some soils, the spoil can be spread or used for farm roads. A road or travelway should be established on the berm or the spread spoil that will be adequate for movement and operation of the equipment to be used for future maintenance of the drain. Normally, in humid areas the spoil banks should be spread on the adjacent land. The spoil should slope away from the drain and should be left in a condition that allows the use of conventional farm equipment. If the spoil bank is not spread, it should be confined to as small as area as possible consistent with berm requirements. If infertile soils occur at the same depth in the excavated material, the fertile soil should be separated during construction and spread over the infertile soil or used to other good advantage. Provisions must be made for surface water entry into the drain through the spoil bank.

To prevent serious erosion and to stabilize the drain bottom and side walls, berm, and spoil area, vegetation should be established on these areas as soon as practical, soil and climatic conditions permitting. Additional details regarding berms and spoil banks are available in various technical documents provided by public agencies such as USDA, Soil Conservation Service, National Handbook of Conservation Practices (USDA, SCS, 1977).

7.3.4 Maintenance of Open Drains

Maintenance requirements: A good maintenance program is an necessary as proper design and construction and is always required, even though the system is of the best design and construction. Maintenance is a continuing requirement if the drainage channel is expected to continue to perform as designed. Preventive maintenance usually begins immediately following construction, and corrective maintenance follows any subsequent failure of the drainage system or some part of it. The primary causes of open-drain failure or deterioration are channels clogged by vegetation, sediment in the drain, control structures that restricts flow, improper location alignment of drains, erosion of drain channel and side slopes, high sediment load in drainage water, and failure to provide adequate preventive maintenance.

In most places, particularly in humid areas, vegetation (usually adapted grasses) should be established on the drain channel and side slopes and on the spoil bank immediately after construction is completed. This vegetation must be maintained to prevent excessive growth of weeds and grasses and to control the growth of undesirable woody vegetation, particularly during the first 2 years. Timeliness of maintenance is very important to prevent increases in both difficulty cost.

After construction, channel side slopes may slough and partially fill the channel or sedimentation may occur because of excessive erosion in the spoil and berm areas or other areas in the watershed. In some cases, the channel alignment, grade, and width and the side slopes may even need to be changed. Even where all possible measures have been taken to prevent erosion and sedimentation, the channel must be cleaned out periodically and returned to its original condition.

Adequate smoothing and establishment of vegetation on the spoil and berm areas is an important aid to reduce sedimentation in the channel. Maintenance of the roadway in the vegetated spoil and berm areas facilitates all maintenance operations. Maintenance of land surfaces, row drains, and field drains are also necessary if the drainage system is expected to operate adequately. Where land grading has been used to develop surface drainage, maintenance is critical during the first 2 years after construction because of soil settlement, especially where deep fills were made. For areas with large cuts and fills, several subsequent annual land-smoothing operations may be necessary.

The entire drainage system should be inspected during or immediately following a heavy runoff event soon after completion of construction and at least annually thereafter. Structures should be checked frequently for obstructions, excessive erosion due to overtopping, and piping along conduits.

Maintenance methods: Regular channel maintenance should be performed using various methods that can be broadly classified as mechanical or chemical. The same equipment used in construction is often used for the removal of sediment and reshaping of the channel, spoil area, field drains, and land surface at intervals after construction.

Rotary mowers mounted on booms extending from tractors and highway-type side mowers that allow movement of the blade above and below the horizontal plane are often used to control grass, weeds, and brush along drain side slopes and in the channel. Pasturing can be used effectively to control grass, but it must be controlled so that livestock are kept out of channels and off the side slopes while the soils are extremely wet. Controlled burning, particularly during winter, is useful in removing dead weeds, tall grass, and small brush if local regulations and site conditions permit it. If woody vegetation is not controlled early, more expensive measure such as special equipment or hand labor must be used to clean large open drains.

Excellent control of undesirable vegetation is possible with chemicals. Caution must be exercised in their application to prevent damage to adjoining crops from chemical drift and to avoid pollution of drainage water. Specific instructions regarding the application and effectiveness of a chemical are included on the product label. All Federal, State, and local laws and regulations governing the use of herbicides must be followed.

Responsibility for maintenance: Landowners, members of drainage or irrigation groups, and elected public officials of drainage or irrigation projects should accept responsibility for planning, financing, and carrying out the maintenance required for drainage systems. The full benefits of investment in construction will be realized only if the drainage system is maintained in the condition that allows it to function as designed. It is helpful, especially in a project including several landowners, to provide a written plan of maintenance that covers maintenance requirements, methods of maintenance to be used, an estimate of maintenance costs, and the method of financing the program. Particular attention should be given to practices that reduce the need for maintenance such as developing farm conservation plans for all landowners in the project, installing erosion control measures in the watershed, and early establishment of vegetation on the drain right-of-way. After construction, good drainage depends on good maintenance.

7.4 CLOSED GRAVITY DRAINS

Closed gravity drains are pipe systems that are installed underground to collect and carry away excess ground water. They consist of open-joint tiles, pipes, porous materials, or perforated tubes laid horizontally at depths of 1 to 3 m. The drains can be installed in a trench or flexible tubing can be plowed in directly. If a trench is excavated, it is backfilled after the drain is installed, and the area over the trench is farmed in the usual manner. Excess ground water enters the drain by seepage and flows by gravity to the disposal outlet.

Closed gravity drains lower a high water table caused by excess ground water. The major advantage of closed drains is the saving of the cropland area required for open drains and the increased efficiency of mechanical tillage, planting, and harvesting operations. Open drains occupy a considerable area of cropland and divide fields into small areas that are difficult to farm with large, multirow equipment. With closed drains, an entire field can be farmed with only a few open drains to serve as barriers. A properly designed closed-drain system, installed with a good protective envelope is easy to maintain, whereas sediment must be removed periodically and vegetation maintained in open drains. The primary disadvantage of a closed-drain system is the difficulty in diagnosing maintenance problems that may develop. Since the entire system except the outlet is buried, it is sometimes difficult to determine which part of the system is malfunctioning. Additionally, in areas where deposition of manganese and iron oxides is a problem, closed drains could clog. This requires costly cleaning operations to return the drain to efficient operating condition. An open drain, under these conditions, is rejuvenated each time it is cleaned.

7.4.1 Kinds of Closed Drains

Interceptor drains: The most familiar type of closed drain according to Donnan (1959) is the interceptor drain. It consists of a single drain installed to intercept the lateral flow of ground water coming from a known seepage source upslope to keep it from reaching the area to be protected. Interceptor drains are used to collect seepage from canals, reservoirs, and other man-made facilities. They are used to intercept lateral ground-water flow from higher lands or from entering levee-protected lowlands. They have also been used to check the lateral flow of ground water from irrigated areas at higher elevations.

Relief drains: The most common type of closed drain is the relief system. It consists of a system of parallel lateral drains that ae connected to a collector or main drain or that discharge directly into an open drain along one side of the field. The relief system is designed to lower a water table in the plant's root zone over the entire drainage area or field, thus providing relief from excessive water logging.

Relief mole drains: A third type of closed-drain system, used only occasionally in the United States, consists of a combination of deep lateral drains and a series of shallow unlined mole drains. The mole drains are installed perpendicular to the lateral drains and are designed to facilitate the movement of ground water to the laterals, increasing the drawdown of the water table between the lateral drains.

7.4.2 Materials for Closed Drains

Moles: Mole drains are constructed by pulling a smooth, bullet-shaped object through the soil at a depth of about 0.7 m. If the soil is plastic and resistant to erosion, the cavity or conduit created by the mole remains open and provides a conduit for removing drain water. Mole drains normally discharge into an open drain or a newly excavated and backfilled trench of a closed drain at a lower elevation. Mole drains have a half-life of about 5 years or less, depending on the kind of soil. Mole drains are rarely used on irrigated lands except in some humid areas.

Clay tile: Both clay and concrete drain tile are manufactured in short lengths of 0.3, 0.6, or 0.9 m. The tile is laid end to end with open butt joints or tongue and grove joints in the open trench and is often surrounded with a gravel envelope material before the trench is backfilled. The first clay tile installed for agricultural drainage in the United States was installed about

1850. Until 1965, this was by far the most widely used material for closed gravity drains. ASTM Specification C4-72 (ASTM, 1978) currently provides for three grades of clay drain tile. Clay tile is susceptible to freezing and thawing action where frost penetrates the ground to the depth of the tile. The water absorption characteristic of the tile is a valuable index of the resistance to freezing and thawing.

Concrete tile: Although portland cement was used to make agricultural drain tile before the advent of clay tile, it did not come into prominent use until the 1900's. ASTM Specification C-412-60 (ASTM, 1978) provides for three classes of concrete drain tile. The classs are standard quality, extra quality, and special quality. Special-quality concrete drain tile is very dense tile designed for use in soils in which there may be corrosive acids or sulfates in the soil or soil water. It is manufactured by using sulfate-resistant cement.

Plastic tubing: Plastic tubing was first used for agricultural drainage in the 1940's. However, this tubing had smooth, thick walls to provide bearing strength and the cost was such that it was not competitive with clay and concrete tile. In the early 1960's, corrugated plastic tubing began to be manufactured from bóth polyvinyl chloride and polyethylene. The corrugations provided sufficient strength to allow a thin-wall configuration and, therefore, a considerable savings in cost. An advantage of corrugated plastic drain tubing is that the material is inert and, when buried, is not subject to deterioration caused by ultraviolet rays, acids, or alakali solutions. It can be extruded in a continuous tube and packaged in a long coil. Continuous lengths of tubing can be laid in the trench with little hazard of misalignment. ASTM Specification F-405-77 (ASTM, 1978) covers the requirements for manufacturing corrugated polyethylene tubing. ASTM Standard Recommended Practice (F-449-76) and ASAE Engineering Practice 260.3 (ASAE, 1979a) are also related to installation. Corrugated plastic tubing for agricultural drainage appears to be rapidly replacing clay and concrete tile in all but the larger sizes. In 1975, more than 150 million meters of corrugated plastic tubing was manufactured in the United States and Canada.

7.4.3 Design of Closed Drains

The design of a closed gravity drain system involves a determination of required depth, spacing, and size of drain together with provision of an adequate outlet and appurtenant works. Depth and spacing are roughly proportional, depending on the permeability of the soil strata. Generally, the greater the depth of drain the wider the spacing between drains with the choice of depth and spacing often being an economic consideration (van Schilfgaarde et al., 1956).

Depth: The depth of the drain is often controlled by the depth of the outlet system. Where the outlet is not limiting, depth is determined by such factors as crop type, soild, drainable strata, and drain spacing. In arid areas with high water tables or where salts occur in the soil and the ground water, drains should be relatively deep. For irrigated areas, the drains should be placed approximately 2.5 m deep. In stratified soils, it is advisable to place the drain in the most permeable layer, provided it is below the depth to which the water table should be lowered, and at a depth which can be economically reached. Where soil layers are undulating or discontinuous, it may not be possible to place the drain in the most permeable layer since the tubing must be on grade and continue on grade to a point of discharge.

TABLE 7.1. AVERAGE FLOW RATE FROM CLOSED
GRAVITY DRAINS (WEEKS, 1959)

Size of drained area	Average yield from area
hectares	liters per second
0 - 16	11
16 - 32	20
32 - 400	20 plus 6 L/s for each added 16 ha over the original 32
400 - 1200	140 plus 3 L/s for each added 16 ha over the original 400

Spacing: Spacing between drains depends upon many factors but the texture and permeability of the soil, and the depth of the drain below soil surface are significant items (Donnan, 1946). Ground water usually moves through coarse textured soil more rapidly than through fine textured soil. Therefore, drains can be spaced farther apart in coarse textured soils. Several theories covering the flow of ground water to subsurface drains and consequent drain spacing equations have been proposed during the past 30 years. Most of the applicable equations are outlined in the USBR-Drainage Manual (USDAI, BR, 1978) and USDA-SCS NEH-16 (USDA, SCS, 1971). These theories have modified the approach the drainage engineer takes in obtaining a solution to the many drainage problems. Considerable progress has been made toward the rational design of drainage systems using such theories. In designing a closed gravity drain system, it is recommended that the engineer review the various drainage design equations, along with the conditions for which each equation applies. Then, depending on the drainage situation, design parameters available to him and the sophistication desired in the system design, the most appropriate of the several formulas can be used to arrive at a suitable design. Steady-state and transient equations are used in both humid and arid areas.

Grade: Sediment is likely to accumulate in buried subsurface drains laid with little or no grade. On flatlands, a minimum grade for drain lines should be established according to site conditions. For example, the drain line may be 2 m deep at the outlet but, to obtain a proper grade for the line, the upslope end is only 1.5 m deep. There must be a compromise between the depth of the line and the grade of the line to achieve an optimum balance. The USBR minimum grade criteria are established by maintaining a velocity of 0.3 m/s when the conduit is flowing fully and 0.001 grade is considered the minimum. USDA-SCS minimum grade criteria indicate that velocity cannot fall below 0.15 m/s if there is no sedimentation hazard and 0.45 m/s if there is a sedimentation hazard.

Yield from drain lines: The average discharge from closed gravity drains in an irrigated area varies according to the spacing, soil type, irrigation practice, crop type, consumptive use, and other related factors. According to Weeks (1959), the average flow rate from a closed gravity drain system can be estimated from the data presented in Table 7.1. This applies primarily to a portion of Southern California, however, and should be used with care. The

TABLE 7.2. VALUES OF MANNING'S n FOR SUBSURFACE DRAIN MATERIALS

Conduit Material	Minumum	Recommended
Clay and concrete tile	0.011	0.012 to 0.014
Corrugated plastic tubing		
75 to 200 mm diameter	0.015	0.015 to 0.016
200 to 380 mm diameter	0.017	0.017 to 0.018

maximum flow yield is recommended for use in design by the U.S. Bureau of Reclamation (USDI, BR, 1978, pages 163-166).

Drain size: Seventy-five millimeters is the smallest diameter of drain tubing in general use in the United States. The problems in maintaining a precise grade and alignment and the possibility of some sediment intake encourage the use of larger sizes even though the larger size is not for hydraulic-capacity. In designing drainage systems for hydraulic capacity, the size of the conduit is determined from the design flow, slope, and roughness coefficient. Mannings formula is commonly used to determine required size.

$$Q = CR^{2/3} S^{1/2} A \dots \dots \dots \dots \dots \dots \dots \dots \dots \dots [7.3]$$

Where Q = rate of flow in m³/s; C = 1/n; R = Hydraulic radius in m; S = slope in m/m; A = cross-sectional area of flow in m²; and n = roughness coefficient. Minimum and normally recommended values for "n" roughness coefficient are provided in Table 7.2.

There are certain "rules-of-thumb" with regard to drain-tube sizes that, if followed, generally prevent overloading the hydraulic capacity of a drain line and insure a trouble-free system. They are:

1 Lateral lines should be at least 25 mm in diameter.
2 The downstream diameter of any lateral line longer than 180 m should be 100 mm in diameter.
3 The downstream diameter of any lateral longer than 400 m should be 125 mm in diameter.
4 Collector lines should be at least 150 mm in diameter.
5 The downstream position of any collector line at the point where the aggregate length of upstream laterals and the collector line exceeds 4,600 m should be at least 200 mm in diameter.

7.4.4 Installation of Closed Drains

The installation of main and lateral drains in a closed gravity system should follow specific criteria. ASAE Engineering Practice ASAE EP 260.3 (ASAE, 1979a) provides some guidance. The system should provide a minimum number of outlets and, whenever practical, the system should be planned to have a short drain and long laterals. For economical installation, the laterals should be oriented to use the available field slope and the main or collector line should be oriented to parallel the natural waterways. Often, however, the most effictive drainage is provided if laterals are parallel with water-table contours. The trench should be excavated, starting at the outlet and proceeding upgrade. The collector or main drain should be installed first and then the laterals, again proceeding upgrade. The alignment of the drain should be such that the tubing can be laid in a straight line or with smooth curves. The use of manufactured connections is recommended. Furthermore, junction boxes or manholes are recommended wherever two main or collector lines meet or where two or more laterals connect to a collector line.

Closed drains must be installed in such a way that the tubing is not deformed excessively by the soil loads placed on them. The load that usually governs the strength required is the weight of the earth covering the drain. The magnitude of the load that the drain can safely support depends on the unit weight of the soil, the width and depth of the trench, and the method of bedding and installation of the drain. If the drain is less than 1 m deep, such as in humid area irrigated sites, there is a danger from impact loads imposed by heavy farm vehicles. Rigid conduits such as clay and concrete tile fail by rupture of the pipe walls. Their principal load-supporting ability lies in the inherent strength of the pipe. Flexible conduits such as corrugated plastic tubing fail by deflection. They rely only partly on their inherent wall strength to resist external loads. For flexible tubing, deflection causess the horizontal diameter of the corrugated plastic tubing to increase the compression of the soil at the sides of the tube. This circumstance builds up resistance that, in turn, helps support the vertically applied load. Flexible tubing should be installed in the trench in a manner that insures goof soil support from all sides. Trenching machines in use today dig a trench with a modified "V" bedding configuration to promote optimum support for the tube.

Envelope to protect subsurface drains: Closed gravity drains installed in irrigated areas are usually provided with a well graded envelope material or a synthetic-fabric envelope material to prevent sediment entering the drain. Envelope materials such as hay, straw, sawdust, wood chips, peat moss, and corncobs have been used primarily in humid areas; however, in the mechanical installation process, a more permanent installation can be achieved if a 75-mm or more layer of well graded aggregate envelope is laid directly behind the digging wheel. The tube is laid on the gravel and it is surrounded with at least 75 mm of the aggregate envelope material. This envelope material serves a threefold function. It serves as an ideal stabilized bed for the tubing and insures good transfer of deflected load. It also screens out soil particles that might otherwise migrate into the conduit, and it provides an improved zone of permeability around the conduit. The aggregate envelope is regarded as the most efficient and trouble-free protective envelope to use on closed drain systems. The unavailability of sand and gravel in many places has led to the use of cinders, slag, and crushed stone in some areas. Envelope materials of synthetic fabrics such as nylon, polyester, and polyethylene are being used in some locations instead of the conventional aggregate envelopes. Current practices are in an evolutionary stage and criteria are being developed for correct manufacturing control and for specification of these materials for use in soils where it can be assured that successful drainage systems will result and the useful life of the installation predicted.

Outlets: An outlet for a closed subsurface drain system should provide for the free flow of drain water to an open drain or other disposal point. A good practice is to have a free fall of at least 0.15 m in humid areas and 0.45 m in arid areas between the conduit invert and the level of water in an outlet drain. In some places, the design of the drain system and depth requirements of the lateral lines require an outlet deeper than the available disposal drain. In these places, the procedure is to install a sump with a pump to discharge the drain water into the outlet system at a higher level. Sumps are constructed by excavating a pit near the subsurface drain outlet and adjacent to the disposal drain. The excavated pit is lined with concrete, and the outlet of the drain enters the sump near the bottom. The sump usually extends about 0.3 to 0.6 m above the ground surface. The water is pumped from the sump into the outlet ditch with an automatic-float or electrode-activated pump.

Moles: Mole drains were discussed briefly in a previous section. In the arid irrigated areas, the moel technique of drainage is used very little because of plowpans and traffic compaction. This circumstance promotes the need for and widespread use of rippers, subsoilers, and other massive tillage practices. Farmers are more interested inproviding good penetration of irrigation water and would rather perform this ripping operation when the soil is relatively dry than install a mole drain under the moist conditions needed for this practice. In arid areas, the soils are also somewhat unstable and mole drains have a very short useful life.

7.4.5 Maintenance of Closed Drains
A properly designed and installed closed-drain system is not difficult to maintain. This is not to say that failure does not occur in an existing drain system. First of all, it is imperative to have a free flow at the outlet, which prevents physical clogging of the drain at that point. Frequent observation of the drain flow will indicate whether the system is functioning properly. If there is a marked decrease in flow, the conduit could be physically clogged by silt, plant roots, or chemicals. It could also be damaged by animals or it could collapse. Both physical high-pressure water-jet cleaning techniques and chemical treatment by introducing sulfur dioxide to dissolve the oxide deposits have been developed to restore flow in drain lines clogged by chemicals, roots, or sediment.

7.5 DRAINAGE PUMPING

7.5.1 Pumped Outlets for Drainage Systems
Need for pumping: Providing sumps and pumps for the disposal of water from both surface and subsurface drainage systems if discharge by gravity flow is not practical requires detailed investigation and a careful survey of site conditions for planning, design, and construction. Pumps can be used for disposal of drainage water from open drains where gravity outlets are not practical because of inadequate outlet conditions. For example, they can be used where backwater from storms or tidal flooding prevents the discharge of the drainage water into a natural outlet at frequent intervals or where the area is very flat and relift pumps and developed grades are required to keep the water moving.

Pumps are also used as outlets for subsurface drainage systems where it is not possible to maintain adequate grade and economical excavation depth without the use of relift pumps or where the subsurface-drain effluent must be relifted for discharge into a gravity outlet.

In some areas, the need for pumping develops when orgainc soils are drained and land subsidence makes pumping more practical than continuous improvement of an existing gravity outlet. Also, pumping may be the most practical way of controlling the subsidence rate by regulating the water level at the outlet. Control of the water-table level can provide subsurface irrigation as well as drainage for some site conditions.

In addition to agricultural drainage, the need for small individual sump pumps to control the water in basements must be considered by homeowners, developers, and irrigation and drainage districts as urban development extends into irrigated areas.

Location of the pumping plant: The location of the pumping plant is determined chiefly by topography and groundwater conditions. Normally, a pumping site is located at the point of lowest elevation of the area covered and is close to a suitable outlet. If pumping plants are located primarily to provide adequate grade in long pipe systems, they are usually constructed where the installed pipe drain reaches a depth of about 3 m.

Subsurface drains normally can be installed 3 m deep at an economical cost, but the cost increases rapidly for greater depths. Other factors to consider are: (a) Accessibility of electricity and fuel supply roads; (b) Adequate room for forebay storage; (c) Adequate foundation conditions; (d) Protection against vandalism; and (e) Ground-water conditions and fluctuations.

Incorporating the pumping plant into the drainage system: The drainage system served by a pumping plant should be designed to have adequate grade lines and nonerosive velocity between the sump and the most remote parts of the system during pumping. The collector drain should have ample depth and cross-sectional area so that flow to the pump is as uniform as possible between the high and low water stages at the sump.

Runoff from areas that can be drained by gravity should be diverted from the area served by the pumps. In some surface drainage systems, runoff can be diverted by gravity into an open outlet system until water in the outlet rises too high and closes a flapgate. The pump then takes over until the level of the outlet water drops below the elevation of the flapgate.

If pumping plants are used to provide outlets for pipe drain systems, open gravity collection drains should be provided as often as possible. In some places, it may be economical to line the collector drains with concrete through areas where seepage from shallow unlined gravity drains returns part of the water being carried to the ground water.

Static lift: Static lift is the difference in elevation between the water level in the sump and the water level in the discharge by or outlet if the discharge pipe is submerged. If the discharge pipe is not submerged, static lift is the difference between the water level in the sump and the centerline of the water in the discharge pipe at the high point of discharge.

The static lift is determined after the drainage system to be served by the pumping plant has been designed. The operating level at the sump is determined by the elevation of the land to be drained, by the hydraulic grade line and operating level of water in the collection drains feeding the sump, and by the elevation of the outlet.

Optimum stage: Optimum stage is the sump-water elevation that is desired for proper functioning of the drainage system. If pumping for subsurface drainage, the optimum stage should be the elevation that provides free drainage for the lowest area to be protected. If surface drainage is the primary consideration, the optimum stage is the forebay elevation that is consistent with or integrated with the hydraulic grade line of the complete drainage system served by the pumping plant.

The optimum stage for subsurface drainage is 1.2 m or more below the land surface in humid irrigated areas and 1.8 to 2.8 m in arid irrigated areas. For organic soils, the water table should be maintained at a shallow depth to reduce the oxidation of organic soil materials.

If the amount of storage of temporary runoff is significant, pumps should be designed and operated in a manner that provides adequate storage for all occasions. If both surface and subsurface water must be handled at the same pump station, a low-volume pump is usually used for the subsurface flow and a pump of larger capacity is used for the surface flow. If only one pump is required, the optimum stage for surface drainage governs the pump selection and the requirements for subsurface drainage determine the pump elevation.

Maximum, minimum, and average static lift: The maximum static lift is the difference in elevation between the maximum drawdown in the sump or forebay and the highest water level in the discharge bay or the centerline of the discharge line of a free-flowing pump.

The minimum static lift is the difference in elevation between the water level in the sump or forebay at the time the pump cycle starts and the water level in the discharge bay or the centerline of the discharge line of a free-flow pump.

The average static lift is simply the sum of the maximum and minimum lifts divided by two if the sump has straight side walls. For irregular-shaped sumps, the average static lift is the time-weighted average lift calculated between the maximum and minimum static lifts.

Pumping plant capacity: Inflow rates to the sump or forebay, sump storage, and allowable pump cycle frequency determine the pumping plant capacity. The maximum inflow rates from all sources should be estimated by using standard hydrologic methods for surface flow and steady or transient-state methods for ground-water flows. If the pumping plant includes storage in a forebay or sump, the storage capacity, inflow rate, and allowable cycles per hour are used to determine plant capacity. If storage is unavailable, then continuous running pumps ar maximum discharge should be installed.

Determining pumping plant capacity for drainage of small surface areas, those no more than 2.6 km², usually reqires simple design procedures. The maximum inflow to the sump or forebay from the entire area should be used to determine capacity. Pumping plants often include a sump with one or two pumps that cycle periodically. Typically, sumps should be designed and pump capacity selected so as to limit the number of cycles during maximum flows to no more than five per hour. Running time and idle time should be equal during peak flows. Running time should not be less than 3 min. In places where the rate of flow varies greatly over the year, variable speed or multiple pumps can be used.

For sump pump installations, the sump storage must equal the volume of inflow for one-half the cycle time (t_c). The pumping rate can be calculated from the following equation:

$$Q_p = \frac{S + Q_s t_c}{t_c} \quad \cdots\cdots\cdots\cdots\cdots\cdots\cdots\cdots\cdots\cdots\cdots\cdots\cdots\cdots \quad [7.4]$$

where Q_p is the pumping rate at maximum flow in m³/s; t_c is the running time in seconds; S is sump storage volume in m³; and Q_s is inflow rate in m³/s.

Procedures for determining plant capacity for large areas are basically the same as those described for small areas. Generally, more sophisticated hydrologic methods are used to determine the inflow rate. The methods account for time of travel, surface and subsurface storage, allowable periods of inundation, etc. Return flows from large areas are generally more uniform and constant than those for small areas. Constantly running pumps are usually used in large areas. Multiple pumps are often used in areas where storm flows cause large changes in the inflow rate.

7.5.2 Design of Pumping Plant

Chapter 10, also provides details that will be valuable when designing or evaluating drainage pumping plants.

Kinds of pumps: The head, capacity, number of pumps, and plant operation all influence pump selection. Pumps suited to most agricultural drainage conditions must be designed to operate efficiently at a low head. Where surface drainage waters are dominant, the pumps must have a large capacity and he capable of handling sediment and debris. For these reaseon, various types of centrifugal pumps are used.

Axial-flow propeller pumps are most commonly used in surface drainage systems. They are designed for a total head ranging from 0.9 m to 7.6 m and may exceed 5.7 m³/s in capacity. The pumps do not require priming, are low in cost, and can be used in small places. Their efficiency drops rapidly when the head exceeds design capacity. Deep sumps are required to minimize vortex action and loss of suction.

Radial-flow pumps (volute or turbine) are used for pumping heads ranging from 6 to 60 m. They handle large amounts of sediment. However, small amounts of trash clog turbine pumps. The units are more expensive than axial-flow pumps but are highly useful for deep-well drainage.

Mixed-flow pumps combine features of axial-flow and radial-flow pumps. They operate over a wider range of head and generally at a higher head than a propeller-type pump. They handle sediment and small trash without clogging.

Number of pumps: The drainage rate for a system determines the number of pumps installed. In most places, a single pump will handle subsurface flows from a 250-ha area or the flow from a small watershed. For large watersheds or where operations demand pumping at all times, two or more pumps may be required. Multiple pumps have the advantage of insuring efficient operation under a wide range of flow and head conditions. Also, continuous service during maintenance periods is insured if two or more pumps are installed.

Generally, capacity in multiple-pump installations should be distributed equally among the pumps. Where ground-water drainage (low volume) and surface drainage (high-volume storms) are handled in the same plant, one pump should be selected to handle the ground-water flow and one to handle the occasional storm. Optimum efficiency should be the controlling factor in selecting pumps that will operate continuously for long periods of time.

Pumping requirements: Pump performance data are normally supplied by the pump manufacturer. Pump selection depends on head, speed, and capacity, and on the power available.

The total or dynamic head used in selecting pumps is the sum of the static head, head losses in the suction and discharge lines, and the pressure head at the discharge.

Under certain conditions, specific speed as defined by the Hydraulic Institute is a factor in determining pumping requirements if the installation has a high suction lift (greater than 4.5 m). The suction lift should be as small as practical to reduce costs.

Pump size: Pump size is selected for maximum efficiency at the speed and head condition under which the pump will operate. Discharge velocity for optimum conditions generally ranges from 2.8 to 4 m/s. For preliminiary design, pump size is computed by dividing the square root of the required pumping capacity by the average discharge velocity. The final pump size can be determined from performance curves supplied by pump manufacturers or by direct testing.

Power sources and drive mechanism: Cost, reliability, and economy of operation, availability of power, and duration of pumping determines the selection of both the power sources and the drive mechanisms. Electricity, gasoline, gas, and diesel fuel are the energy sources generally used in drainage applications. Synchronous or squirrel-cage electric motors can be used. Where starting torque is high, capacitors should be installed to adjust for line voltage drops. Internal-combustion engines can be used if electrical power is not available. They offer the advantage of operation at variable speeds and are not subject to power outages.

Power requirements: Power requirement is a function of discharge (Q), weight of water (w), total dynamic head (H_t), and energy efficiency of the components of the system.

The theoretical power output, on net power required of the pumps, is:

$$Pnet = K \ Q \ H_t \quad \cdots \cdots \cdots \cdots \cdots \cdots \cdots \cdots \cdots \cdots \cdots \cdots \quad [7.5]$$

where Pnet = net power in kW; Q = pump discharge in m³/s; ^Ht = total dynamic head in m; and K = 9.81

The power output required of the engine or motor (gross power) is:

$$P_g = \frac{Pnet}{E_p \ E_t \ E_m} \quad \cdots \cdots \cdots \cdots \cdots \cdots \cdots \cdots \cdots \cdots \cdots \cdots \quad [7.6]$$

where P_g is gross power in kW, E_p is the efficiency of the pump, E_t is the efficiency of the transmission line, and E_m is the efficiency of the motor. Pump efficiency ranges from 65 to 80 percent. Electric motors are generally 90 percent efficient; diesel engines, 80 percent; water-cooled gas engines, 70 percent; and air-cooled engines 60 percent. Direct-drive transmissions are 100 percent efficient; gearbox transmission, 95 percent; V-belts, 90 percent; and flat belts, 80 percent.

Operating controls: Controls for operating the pumping plant can be manual or automatic. Since most drainage installations are comparatively remote, automatic controls are desirable, particularly where short cycling periods are common. However, alternate manual controls should always be provided in case the automatic system malfunctions. A wide variety of on-off switching devices are used in drainage pumping installations. The most common are float switches activated by the water level in a sump or stilling well. Electrodes, bubblers, and pressure systems are also used.

In large installations using internal-combustion engines, manual controls are generally installed. This allows the operator to inspect the equipment and service the pumps and engines before each startup.

Safety controls: Safety controls should be installed in every plant to protect the pumps and motors. A time-delay relay should be installed to prevent short cycling in installations using cyclic operations. Also, in multiple-pump installations, a relay should be installed to prevent simultaneous starting of all units. Other considerations should include low water level cutoffs, overload protection, governors on enginers, oil level and temperature sensors, and high voltage protection.

Sump capacity and dimensions: Sumps range from large open pits to precast concrete pipes of large diameter set on end. Sump design should not allow the water level to exceed the optimum operating level. For closed pipe drains, the water level should never exceed the top of the entrance pipe by more than one-half of the pipe diameter. The drawdown level in the sump should allow a minimum clearance of 0.6 m to the bottom or to a water depth limited by vortex action.

The storage capacity must equal the volume of inflow for one-half the cycle time. The equipment flood level should be high enough to prevent flooding of the equipment by surging water if pumps are suddenly stopped.

In multiple-pump installations, the units must be properly spaced (usually four times the diameter of the pump bowl) to avoid large vortex development and insure distribution of flow evenly to each unit. Velocity to all units should be low and flow should be directed in a straight line. Rectangular sumps are recommended. Although small installations can use circular sumps, vortex development may cause problems.

Provisions should be made to dewater sumps for maintenance. In large systems, stop-log retainers or a water-tight gate should be included in the inlet structure. In small systems, the pump and motors can simply be removed from the plant for repair and cleaning.

Trash racks and sediment ponds: Trash and debris should be excluded from the pump. Trash racks located at the intake should be inclined toward the plant. Openings in the trash rack should range from 40 to 80 mm. The rack should be hinged or be removable for maintenance. If large sediment loads are expected, stilling pools should be incorporated in the design.

Discharge pipes: Steel pipes protected against corrosion are most commonly used for discharge lines. Flexible couplings should be used if the line passes through the concrete sump walls. Sharp bends should be avoided and thrust blocks should be installed at changes of alignment in large systems. If the line discharges below a water surface, flapgates must be provided to prevent backflow. Unless a headwall is used, the discharge line must project well beyond the dike face or channel bank to prevent circulation erosion. Riprap protection is recommended for large discharge volumes. In some systems, siphons, air valves, backflow protection, etc., may be desirable.

Housing: Weather-proof housing should be provided for all installations using unsealed motors, engines, and transmission units. In large structures, special maintenance openings should be provided in the roof, walls, and floors. Gantry cranes may be necessary for large pumping units. Fencing should be provided to protect the plant against vandalism and to protect the public against hazards of pits, high voltage, and drive mechanisms. Adequate ventilation and fire protection must be provided for internal-combusion engines. Electric lines, switches, fuel lines, and tools should be adequately protected against vandalism.

7.5.3 Field Tests of Pumping Plants

Field tests of new drainage pumping plants are recommended to check their performance for comparison with the design specifications. Periodic tests of operating pumping plants are desirable to determine and maintain operating efficiency.

Procedure for field tests: Temporary staff gages in the suction and discharge bays should be established by using an assumed or sea level datum on gages. Gages already established can be used, provided elevations are checked to the nearest 3 mm.

Elevations using the same datum as the staff gages should be obtained for the following:

1 Floor of bay or sump.
2 Entrance lip of suction pipe.
3 Centerline of pump, motor, and engine shafts.
4 Each pump, engine, motor, suction, and discharge pipe for an accurate plan and the profile layout of each pumping unit. Manufacturers' catalogs should be consulted to obtain equipment dimensions.
5 The centerline of each hole tapped in the suction pipe, discharge pipe, and pump to the nearest 3 mm.
6 Water surface in sump and discharge bay and both channels before, during, and after the test.

Total head on pump: The total head equals the static lift plus head losses in the suction pipe and discharge pipes and the velocity head.

If the pump is submerged, it may not be feasible to measure the suction-pipe pressure head. The total head can be estimated by measuring the discharge pressure and the intake water level and then estimating the suction pressure head by taking into account the estimated entrance loss and the friction losses of the suction pipe.

Measurements of the discharge and suction heads in the field can be made by using manometers or piezometers. A hole is tapped in the suction or discharge pipe, 0.1 to 0.5 m away from the pump flange, and a manometer tube installed. If flow is unusuallly turbulent, additional holes around the circumference of the pipe may be necessary to obtain average readings. To obtain accurate results, straight lengths of pipe should be used. The pipe should be running full and the piezometer taps should be at least five times the pipe diameter away from any bends. If manometer or piezometer tubes are impractical, weir, orifice, flume, or current mater measurements can be used.

Pump efficiency: After the proper measurements have been made, the efficiency "E" of pumps driven by electric motors is computed by equation 7.7.

$$E = \frac{Pnet}{Pm} \dots\dots\dots\dots\dots\dots\dots\dots\dots\dots\dots\dots\dots\dots [7.7]$$

where Pm is the power input to the electric motor in kW.

7.5.4 Operation and Maintenance

An operations and maintenance manual should be prepared for the pumping plant and include repair manuals, shop drawings, wiring diagrams, plumbing diagrams, periodic inspection sheets, and directions for operation and "troubleshooting."

The manual should contain methods for testing the pump operation, pump controls, and safety switches. Operators should be trained and instructed on the operation of pumps, motors, engines, and control devices.

Pumps that require priming must be filled completely to eliminate pockets of air in the casing around the shaft. If engines that permit substantial variation in speed are used, pumps should be regulated to operate at the most efficient speed as determined from tests or characteristic curves. If several units are included in the facility, the most efficient unit or combination of units should be used at all times. Each unit should be operated periodically to insure reliable operation when needed. Equipment should be kept in good repair. Equipment, plant, and grounds should be kept clean and orderly to minimize fire hazard, insure ready access, and prevent accidents.

The facility should be thoroughly inspected periodically during operation and at least monthly during periods of nonoperation. Immediately after a major operation, the engines, motors, and pumps should be inspected, cleaned, and lubricated, the sumps flushed, and fuel and lubricants replenished.

Occasional performance and efficiency tests are desirable, particularly for larger facilities. The tests detect low operating efficiency resulting from wear and other causes and indicate the need for repairs or replacement.

Inspections should include observations on the condition of trash racks, plant forebay, and discharge bay. Arrangements should be made for disposal of debris, drift, and trash accumulations that might interfere with plant operation. Inspection should include checks for erosion, leaks, displacement of riprap, and foundations that should be repaired. Seasonally, the hinges and seats of flap gates and the slide controls of valve gates should be lubricated. Stop logs and other emergency equipment should be checked also.

Inspection, testing, and maintenance are important to the continued safe and efficient operation of drainage pumping plants.

7.6 WELL DRAINAGE

Drainage wells are ground-water wells designed, constructed, and spaced to relieve or control an existing or potential high water table or artesian condition. A discharging well creates a cone of depression in the water table or piezometric surface, the shape and dimensions of which depend primarily on the aquifer characteristics, rate of discharge, and boundary conditions. Drainage occurs only within the influence of this cone of depression (Donnan, 1959; Weeks, 1959). Most drainage-well installations consist of a group of wells so spaced that their individual cones of depression overlap sufficiently to lower the water-table level at all points in the well field.

7.6.1 Conditions Needed for Well Drainage

Two conditions must be present to permit drainage by wells: (1) an aquifer of adequate transmissivity and (2) adequate vertical permeability between the root zone and the aquifer.

Other conditions that favor well drainage are:

1 Extremely flat surface terrains that preclude gravity outletting of drains and the costs for installing and operating wells are comparable to or lower than those for horizontal drains plus relift pumping facilities as suggested by Ham (1974).
2 Uneven topography that would require excessively deep excavation for drains.
3 Subsurface conditions unsuitable for installing conventional drains such as hard rock or unstable soils.
4 Highly developed land where high rights-of-way and/or construction costs of drains might exceed costs of well installation and operation.
5 Areas where materials, equipment, and labor for constructing drains are scarce or more costly than the costs for installing and operating wells.
6 Areas where drain construction may result in environmental damage.
7 Low-cost energy available for pumping from wells.
8 Local conditions such that the pumped drainage water can be used as a supplemental irrigation supply.
9 The lack of uniform subsurface conditions make it extremely difficult to determine drain spacing.

7.6.2 Design and Construction of Drainage Wells

The types of drainage wells relate directly to the type of aquifer to be drained. For unconfined aquifers or water-table conditions, pumping of the well is necessary. For leaky confined aquifers or artesian conditions, free discharge may relieve the hydrostatic pressure sufficiently to retard upward flow of water. A third condition is an artesian aquifer where open wells are not free flowing and pumping is required.

Drilling methods, materials, and specifications: Typical drilling methods used to install drainage wells are the direct rotary, reverse circulation rotary, and cable tool methods. Representative samples of subsurface materials in test holes and pilot holes should be taken, and similar data from production wells in the area should be considered in selecting appropriate drilling methods.

Materials used for drainage wells are similar to those used for water supply wells, and the same consideration must be given to selecting materials to permit practical construction operations, provide adequate strength, and resist corrosion and other deterioration.

Specifications for individual drainage wells are similar to those for water supply wells. Specifications for a drainage well field differ from those for a water-supply well field only in one important aspect. In a water-supply well field, wells are spaced to minimize well interference. In a drainage well field, wells are spaced so as to cause intentional interference in order to lower the water level in a controlled manner. If at all possible, construction should be staged so that data obtained from testing one well can be used to locate subsequent wells.

Results of a standard ground-water survey and other investigations should be used to verify that the subsurface conditions are satisfactory for well drainage. These data should be used to locate the site of the first pilot hole if pilot wells or test holes are needed to investigate subsurface conditions.

Pilot holes may be needed to provide the additional data necessary for proper well design. Detailed logging and collection and analysis of representative samples are necessary to determine properly the subsurface characteristics that determine well depth, screen length, screen setting depth and slot size, gravel pack gradation, if used, and to allow preliminary estimates of well capacity and drawdown.

The depth of a drainage well depends on the subsurface conditions at the site and can be modified somewhat by economic considerations. The greater the percentage of aquifer screened, for example, the more productive the well but the higher the construction costs. The costs of making a single well more productive should be balanced against the costs of adding wells that would accomplish the same result.

In general, drainage wells are relatively shallow and, as a result, are subject to low confining pressures. Therefore, lightweight casings are usually adequate. However, consideration must be given to water quality in choosing the casing and screen weights and materials. Slot types and sizes should be determined from the aquifer grain-size gradations. The selection of slot types and sizes should not be separated from the gravel-pack design but should be considered at the same time.

Gravel packs increase the efficiency of wells and must be designed to provide positive mechanical stabilization of the aquifer to prevent sand entering the well. The gravel-pack design should be based on gradation data from the analysis of the test-hole and pilot-hole samples and should be completed along with the well screen design as part of a total well design.

The gravel pack should be placed in a manner that prevents or minimizes segregation of sizes. The wider the gradation of the designed gravel pack, the greater the importance of preventing segregation. Tremie pipes are usually the most satisfactory means of placing gravel packs (Johnson, 1966).

Selection of energy source, pump, and motor: Because drainage wells are long-term installations, selection of the energy source for pumping may become as critical as the initial installation costs. Items to consider include operational costs such as costs for energy and maintenance. Consideration should be given during the initial well design and pump selection to the possibility that changes in the availability or cost of a particular energy source may make it necessary to change the energy source at a later date. Gasoline and diesel engines, electric motors, and wind-driven pumps can all be considered.

Pumps must be selected to fit the physical requirements of the well, including yeild, drawdown, head, casing diameter, pump depth setting, etc. Motors (or engines) must be selected to provide adequate power to operate the pump at the proper rotation speed on a continuous basis.

Monitoring equipment and construction inspection: Two types of monitoring equipment need to be considered, permanent and portable or temporary. Permanent monitoring equipment includes an air line to measure the water level in the well, a fuel or energy use meter, and a discharge measuring device such as a propeller meter, venturi, or weir. Portable equipment includes a weighted tape or an electrical water-depth measuring device and a portable discharge measuring device such as an orifice meter.

Construction should be inspected at all stages of well construction, and the inspections should be documented to insure proper well installation and completion and to help pinpoint the cause of problems during subsequent operation. Inspection should include insuring the proper bit size and control of any drilling fluid additives, plumbness of hole and casing, proper welds or pipe and acreen connections, proper gravel-pack gradation and placement methods; proper well development; and proper installation and testing of production pumps, motor, controls, and discharge facilities. Also, accurate and reliable well logs, including a description of subsurface materials and an as-built drawing of all well design and construction details, should be kept as permanent records.

Testing drainage wells: Although pilot and test-hole drilling and sampling should provide sufficient data to design the first well, detailed aquifer testing of the first production well is necessary to develop information on aquifer characteristics proper well spacing and pumping rates. Spacing and pumping rates in turn determine the number of wells needed, pump capacity, bowl depth setting, and size and layout of discharge collection system.

Observation wells and piezometers: Although transmissibility can be computed from the results of one properly placed observation well, two or more observation wells allow the use of several interpretative techniques as checks on the results. Since drainage wells normally are completed in shallow aquifers, the costs of installing observation wells and piezometers is generally low compared with other costs and the data are usually well worth the cost.

The closest observation well is usually placed away from the pumped well at a distance of one to two times the thickness of the saturated thickness of the saturated aquifer; additional wells are spaced at multiples of this distance. For example, for an aquifer less than 15 m deep, observation wells might be placed at 15, 30, and 45 m.

Where appropriate, several shallow observation wells should be installed in the saturated materials of lower permeability being drained. These shallow wells will show the rate of vertical drainage in the upper soil, which is the critical area in drainage of agricultural lands.

In locating observation wells, the expected location of the second drainage well to be installed should be considered as well as the presence of suspected presence of any impermeable or recharge boundaries. Since the completed well field will operate on the concept of well interference, the line of observation wells for any one well should extend into the expected area of interference of any future surrounding wells.

Testing procedures: Testing procedures are outlined by many water resource agencies and in many publications. To the extent practical, the duration of a pump test should not be fixed, and the decision to terminate the test should be determined by a continuing preliminary field analysis of the collected data. Adequate and reliable data on the vertical drainage rates and volumes are needed as the test progresses.

Well drawdown recovery data should also be recorded because they permit further verification of the calculated transmissibility by specific capacity methods.

Data computation and use: Many published interpretative methods are available for analyzing the test data and, in general, it is worthwhile to use several methods (ASTM, 1978). Initial plots should include log-log plots and semilog plots of time-drawdown data for each observation well. Other plots could include a family of distance-drawdown plots for various time periods ans demilog plots of the recovery data.

Analysis of the aquifer test performed in the first well should give good data on the transmissibility, permeability, coefficient of storage and boundary conditions, all of which can be used for laying out the remainder of the drainage well field. As the well field is developed, adjustments for the heterogeneity of the area can be made.

Spacing of drainage wells: Vertical recharge from deep percolation and the size of the cone of influence of the well are interdependent. Because the size of the cone of influence directly affects well spacing, determining the deep percolation rate is a major requirement in the design of a drainage well field.

Transmissibility is another factor influencing the spacing of drainage wells. It determines the rate of flow to the well and, therefore, has a major influence on the shape of the cone of influence of the well (Brown et al., 1963).

Two general types of boundaries, impermeable and recharge, influence spacing. The presence of either type of boundary influences the shape and extent of the cone of influence.

Storativity is a boundary related factor that determines the rate of expansion of the cone of influence. Storativity, which affects the rate of expansion of the cone of influence, is affected by the type of aquifer present. Long-term pumping could convert a confined aquifer to a water-table aquifer.

The pumping rate also affects the drawdown everywhere within the cone of influence of the well and, therefore, the well spacing. Although a wide range in well capacity theoretically could be used by adjusting the well spacing, experience has shown that 30 L/s wells are about the smallest economically feasible under most field conditions.

Well fields generally cannot be laid out to follow precise theoretical pumping schedules. A network of observation wells generally must be installed to monitor the water level. Monitoring the ground-water level permits selective pumping from individual wells as necessary. Salinity control can be a critical factor in determining the permissible pumping schedule for a combined irrigation and drainage well field.

Spacing of drainage wells must be determined by the aquifer characteristics, well yields, recharge rates, etc., modified by geographic and other limitations. Several references are available on the theoretical spacing of drainage wells and their effects—Peterson (1957), Hantush (1964), and Glover (1967), to name a few. The formulas and graphical methods given are more accurate than the field data values necessary to apply them. These formulas are valid for the internal wells of a system, but the peripheral wells may require more specialized methods of analysis because of the uncontrolled hydrologic boundaries and the lateral as well as the vertical recharge that may exist.

In locating wells in a field, consideration must be given to the degree of land development, interference with farming or other activities, energy source, environmental concerns, need for access for service and maintenance, disposal of the well discharge, location of surface drainage, and presence of hydrologic boundaries. If the wells are to supplement existing field drains, the location of these drains should also be considered.

If water quality is such that some or all of the pumped water can be reused, consideration should be given to locating the wells to discharge directly into existing irrigation systems.

7.6.3 Disposal of Drainage Water

Direct application locally is the most economical way to dispose of the pumped water, but water-quality considerations may rule out this alternative. If so, imported surface water can be mixed with the pumped water to meet volume and water-quality standards.

Poor water quality may require that the pumped water be completely removed from the area. This requires a suitable collection system, a conveyance system, and an acceptable point of ultimate disposal. In some cases, poor quality water can be used to irrigate salt-tolerant crops to reduce the volume of drainage water (See Section 7.8.2). The lack of suitable disposal points may require other methods such as disposal in evaporation ponds or injection into aquifer zones where the water quality would not have an adverse environmental effect. These alternative methods of disposal are usually very expensive.

7.6.4. Cost Comparisons

In general, drainage wells are less costly to construct than a subsurface pipe drainage system but are more costly to operate and maintain because of the continuing energy and maintenance requirements. In certain situations, however, subsurface-pipe drainage is not feasible from an engineering, economic, or environmental point of view. In these situations, drainage wells may be an attractive alternative, regardless of the operation and maintenance costs. For artesian conditions, pump drainage usually provides the more reliable and economical drainage system.

Beneficial use of the water produced may also make well drainage feasible. A dependable supply of low-cost energy would also favor well drainage.

7.6.5. Operation and Maintenance of Drainage Wells

Many drainage wells are costly to operate and maintain (aside from their energy use) because of their intrinsic nature and susceptibility to deterioration by corrosion, incrustation, and plugging. Such adverse conditions can often be minimized, however, by use of proper design criteria and construction practices, use of suitable materials, and thorough sterilization of the well and pump on completion of the well construction and pump installation. In addition, major well problems can oftern be avoided by a program of systematic monitoring of well yield and drawdown, static water levels, and water quality. Pumps should be tested periodically to detect any loss of efficiency that may precede serious mechanical problems.

7.7 ECONOMIC AND LEGAL ASPECTS OF DRAINAGE

Drainage improvements, like irrigation improvements, should be designed so that the costs do not exceed the benefits. For large complex projects, an economic analysis should be included in the feasibility determination. For small jobs, comparison with similar work in the area, where the benefits are well known, is usually enough to determine feasibility. Complex projects usually require the assistance of an economist. There are various stages between these two extremes and the principles of good economic analysis found in engineering economics books should be followed.

Drainage on an individual farm is usually the responsibility of the landowner. Drainage works needed beyond a farm are usually obtained by landowners cooperating under State or local water laws or under informal group arrangements. State laws for organizing and operating special-purpose water districts vary considerably, and the terminology is often quite different, even in adjoining states.

Several legal permits are usually required for work related to installation of drainage systems. These also vary considerably by locality and the type of drainage activity. Individuals engaged in drainage work must therefore become familiar with the applicable drainage laws, the procedures used in special-purpose water districts, and the permits necessary for performance of drainage work in the area they are servicing.

7.8 RETURN FLOW AND DRAINAGE EFFLUENT

7.8.1 Types of Return Flow

In general, the volume of irrigation return flow produced on a farm is the difference between the amount of water delivered from the water supply for irrigation and the amount of water held by the soil and consumed by the irrigated crop The return flow can consist of two parts: (a) surface runoff (tailwater), which is produced during irrigation, and (b) subsurface drainage effluent resulting from deep seepage. The part of the applied water that percolates through the crop's root zone is called the leaching fraction and flows into the subsurface drainage system or continues to percolate vertically and adds to the ground water. Another type of surface return flow from farmland is stormwater runoff. The quality of these types of return flow can vary significantly. The quality of stormwater runoff is generally very good with regard to salinity, but it may contain significant amounts of sediment plus fertilizer or pesticide residues that could be picked up as the water flows across the fields. Surface runoff from irrigation applications (tailwater) usually has the same good quality as storm runoff, but it also reflects the quality of the applied water. Stormwater originates as precipitation, which is essentially pure.

Subsurface drainage effluent, on the other hand, has moved through the soil rather than over it. In semiarid and arid areas, subsurface drainage effluent usually contains total dissolved solids in concentrations much higher than those in surface return flows. Subsurface return flows in humid areas are generally of good quality and contain a minimum quantity of total dissolved solids.

Irrigation water in arid and semiarid areas is a precious natural resource and must be used a efficiently as possible. Even though the production of some surface runoff is generally unavoidable in any gravity irrigation system, all irrigators should minimize the production of return flow. The principal management option pertaining to the production of return flow is that of better irrigation management. Higher irrigation application efficiency through irrigation scheduling, a change in irrigation methods, or reduced leaching can help reduce the production of irrigation return flow. Once return flows are produced, however, they should be reused if the quality of the water permits. The quality factors that must be considered in making this decision are described below.

Salinity: All waters contain some dissolved solids, but the salinity level of surface-runoff return flows usually is only slightly higher than that of the water applied for irrigation. The salinity level of subsurface return flows usually is considerably higher than that of the applied water, which limits the reuse potential of this water, at least as an irrigation supply.

Suspended sediment: Surface return flows (tailwater) and storm-water runoff may contain large amounts of sediment, depending on local conditions. The sediment may limit the options for disposal. Subsurface return flows are generally free of suspended sediment.

Nutrients: Nitrogen and phosphorus are the two nutrients generally applied as fertilizer that can cause water-quality problems in return flows. Phosphorus is readily adsorbed on soil particles, which limits its presence in subsurface waters. Nitrogen is very soluble and can be expected in both surface and subsurface drainage waters. Nurtrients are beneficial in return flows if the water is used to irrigate a crop but are detrimental if discharged into a stream, thus promoting undesirable algae and aquatic weed growth.

Pesticides: Significant levels of pesticides are usually found only in surface return flows. Subsurface return flows usually do not contain pesticides because the materials applied to the irrigated fields are readily adsorbed on the soil particles and usually move only short distances in the soil. In addition, most pesticides currently used degrade rapidly to less toxic compounds that cause few problems in return flows.

7.8.2 Management Options

There are generally two management options in handling agricultural return flows, reuse and disposal. If an irrigation return flow is of good quality, it should be reused. If the irrigation return flow is of poor quality because of some constituent that limits its use for another beneficial purpose, then the water must be disposed of in some other way. Surface irrigation return flows are generally reused in the same field through the construction of a tailwater return system or are reused in a lower field through a system of ditches that direct the runoff to the lower elevation. Tailwater return systems generally consist of a sump or collection basin at the lower end of the field, which permits the sediment in the return flow to settle out, and a pump and the pipeline needed to carry the tailwater back to the head of the field. The tailwater is then remixed with the irrigation supply water and reapplied to the field. It is important that the sediment in the water is not recirculated through the irrigation system.

Saline subsurface return flows, and some surface return flows of poor quality should not be reused as irrigation water but should be used in some other beneficial manner, if possible, before final disposal. Other possible beneficial uses could be (a) development of marshland for wildlife habitat, (b) power plant cooling, (c) aquaculture, or (d) discharge into a stream to increase the fish population. If no beneficial use is possible, then it is necessary to dispose of the return flow at a carefully selected site where the discharge will not cause a pollution problem.

Several other factors should be considered in analyzing the potential reuse of irrigation return flows. They include the potential for energy conservation, the economics of reusing return flows, the need for water conservation, and the potential pollution problem that might be created through the discharge of the return flows.

7.8.3 Concern for the Environment

The United States Congress showed its concern for improving the quality of water supplies by passing the Federal Water Pollution Control Act Amendments of 1972 (Public Law 92-500, Fed. WPAA, 1972). This act expanded the role of the Federal Government in water-pollution control and called for increased public participation. It also authorized a regulatory mechanism requiring uniform, technology-based effluent standards together with a national permit system for all point-source discharges. The objective of the act is to "restore and maintain the chemical, physical, and biological integrity of the Nation's waters." Two national goals were proclaimed: (a) "that the discharge of pollutants into the navigable waters be eliminated by 1985" and (b) "that, whenever attainable, an interim goal of water quality which provides for the protection and propagation of fish, shellfish, and wildlife and provides for the recreation in and on water be achieved by July 1, 1983." It is clear that Congress intended this law to apply to the discharge of all agricultural pollutants into the Nation's waters. This, of course, means that is applies to the discharge of agricultural return flows. As defined in the act, the term "point source" means any discernible, confined, and discrete conveyance, including, but not limited to, any pipe, ditch, channel, tunnel, conduit, well, discrete fissure, container, rolling stock, concentrated animal feeding operation, or vessel, or other floating craft, from which pollutants are or may be discharged. Nonpoint discharges are discharges other than those defined as point-source discharges. As defined in the Environmental Protection Agency (EPA) regulations, the term "irrigation return flow" means surface water, other than navigable waters, containing pollutants that result from the controlled application of water by any person to land used primarily for crops, forage growth, or nursery operations. The term "surface water" means water that flows exclusively across the surface of the land from the point of application to the point of discharge. This definition seems to eliminate subsurface water or drainage effluent return flows from consideration under point-source discharge. Even though this may be the policy of EPA, such discharges probably will be controlled under the requirements of Public Law 92-500 or some State regulations if the subsurface drainage effluent is collected and discharged into public waters through any kind of a pipe or ditch. The authority to administer and issue National Pollutant Discharge Elimination (NPDES) permits for all discharges to public waters has been grated to many individual States by EPA (Federal Register, 1976).

State and local regulations: States that have been authorized to implement the NPEDS permit system must develop a program that meets Federal requirements. Although the requirements of an individual State can be more stringent than Federal requirements, the minimum Federal requirements must be met. It has been recognized that agricultural return flows vary considerably from area to area and that uniform Federal standards may not be the best way to handle the permits for agricultural discharges. Considerable work must be done to determine the "best available technology, economically achievable" that will result in reasonable progress toward the national goal of eliminating the discharge of pollutants in agricultural return flows. In contrast, the State of Washington has established local basin water quality operation committees to protect and maintain water quality in the local areas. Each committee develops farming practices that will reduce soil erosion and enhance the quality of water. The committees locate the problems within hydrologic basins and work with local farmers to design water-quality plans for individual farms. A completed plan then becomes an agreement between the dischargers and the State, which indicates that the best practical management method is being used to comply with water-pollution control laws.

Regardless of the approach taken by a particular environment agency, it is obvious that there is a growing concern for protecting the environment. The design and operation of all irrigation and drainage systems must reflect the need to maximize reuse of and minimize pollution from agricultural return flows.

References

1   Am. Soc. Agr. Engr., Subsurface Drainage in Humid Areas Committee (SW 232). 1979a. Design and construction of subsurface drains in humid areas. Engineering Practice; Am. Soc. Agr. Engr. EP 260.3. 9 p.

2   Am. Soc. Agr. Engr., Surface Drainage Committee (SW 233). 1979b. Design and construction of surface drainage systems on farms in humid areas. Engineering Practice; Am. Soc. Agr. Engr. EP 302.2. 8 p.

3   American Society for Testing and Materials. 1978. Annual Book of ASTM Standards, 48 parts. ASTM, Phila., Pa.

4   Brown, R. H., J. G. Ferris, C. E. Jocab, D. B. Knowles, R. R. Meyer, H. E. Skibitzke, and C. V. Theis. 1963. Methods of determining permeability, transmissibility and drawdown. U.S. Geol. Survey, Water Supply Paper 1536-I, 341 p.

5   Christopher, J. N., and R. J. Winger. 1975. Economical drain depth for irrigated areas. Proc. Am. Soc. Civ. Engr., Irrig. & Drain. Spec. Conf., Logan, Utah, p. 264-272.

6   Donnan, W. W. 1946. Model tests of a tile spacing formula. Soil Sci. Soc. of Am., Proc. 11:131-136.

7   Donnan, W. W. 1959. Drainage of agricultural lands using interceptor lines. Proc. Am. Soc. Civ. Engr., J. Irrig. & Drain. Div. 85(IRI):13-23.

8   Federal Register, 1976. Vol. 41, No. 134, July 12. p. 28493-28497.

9   Federal Water Pollution Control Act Amendments of 1972—Public Law 92-500. Oct. 18, 1972.

10   Glover, R. E. 1967. The effect of pumping over an area. Proc. Natl. Symp. on Ground-Water Hydrol., Am. Water Resources Assoc. Urbana, Ill. p. 149-156.

11   Ham, H. H 1974. Use of drainage wells. Am. Soc. Agr. Engr. Paper No. 74-2518, 10 p.

12   Hantush, N. S. 1964. Supplement to Peterson's "Design of replenishment wells Proc. Am. Soc. Civ. Engr. 90(IRI):67-76.

13 Johnson, E. E., Inc. 1966. Ground Water and Wells. Edward E. Johnson, Inc., St. Paul, Minn. 440 p.

14 Luthin, J. N., (Ed.). 1957. Drainage of agricultural lands. Monog. 7, Am. Soc. of Agron., Madison, Wisconsin. 620 pp.

15 Luthin, J. N. 1973. Drainage engineering. R. E. Krieger Publ. Co., Huntington, N.Y., 250 p.

16 Peterson, D. F., Jr. 1957. Hydraulics of wells. Trans. Am. Soc. Civ. Engr., 122:502-517.

17 Schwab, G. O., R. K. Frevert, T. W. Edminster, and K. K. Barnes. 1966. Soil and Water Conservation Engineering. 2nd ed., John Wiley and Sons, Inc., New York, N.Y. 683 pp.

18 Skaggs, R. W. 1976. Determination of the hydraulic conductivity—drainable porosity ratios from water-table measurements. TRANSACTIONS of the ASAE 19(1):73-80, 84.

19 Skaggs, R. W., L. G. Wells, and S. R. Ghate. 1978. Predicted and measured drainable porosities for field soils. TRANSACTIONS of the ASAE 21(3):522-528.

20 U.S. Dept. of Agr., Soil Conserv. Serv. 1971. Nat. Eng. Handb., Sect. 16, Drainage of Agricultural Land. 423 p.

21 U.S. Dept. of Agr., Soil Conserv. Serv. 1977. Nat. Handb. of Conserv. Practices. 450 p.

22 U.S. Dept. of Int., Bur. of Reclamation. 1978. Drainage Manual. 286 p.

23 van Schilfgaarde, J., D. Kirkham, and R. K. Frevert. 1956. Physical and mathematical theories of tile and ditch drainage and their usefullness in design. Iowa State Univ. Res. Bul. 436, p. 667-706.

24 Weeks, L. O. 1959. Drainage in the Coachella Valley of California. Proc. Am. Soc. Civ. Engr., J. Irrig. & Drain. Div. 85(1R3):83-88.

25 Willardson, L. S. 1974. Drainage for world crop production efficiency. Proc. Am. Soc. Civ. Engr. Irrig. & Drain. Spec. Conf., Biloxi, Miss., p. 9-19.

26 Willardson, L. S., and W. W. Donnan. 1978. Drain depth and spacing criteria. Trans. 10th Cong. on Irrig. & Drain, Internl. Comm. on Irrig. & Drain., Athens, Greece, 1978, p. 34.1.241-.246.

chapter 8

LAND SHAPING REQUIREMENTS

8

Chapter 8

LAND SHAPING REQUIREMENTS

8

LAND SHAPING REQUIREMENTS

by   C. L. Anderson, USDA-SCS (Retired), Columbia, MO; A. D. Halderman, University of Arizona, Tucson, AZ; H. A. Paul, California State University, Chico, CA; and E. Rapp, University of Alberta, Edmonton, AB, Canada

8.1 INTRODUCTION

The irrigation system designer must consider many soil and topographic factors in selecting the method of irrigation. The suitability of each method (surface, sprinkler, and trickle) must be considered for specific site conditions. Soil and topographic factors that the designer considers in selecting the system include:

1 Topographic features such as surface irregularities, steepness of slope, changes in slope direction.

2 Soil features such as water-holding capacity, intake rate, and depth.

3 Geographic features such as field shape, natural drains, buildings, utilities or obstructions.

Related factors that may influence selecting a surface system and land shaping include:

1 Water supply, quality, delivery flow rate, delivery schedule, and location.

2 Labor requirement and availability of labor.

3 Energy requirement, cost and availability.

4 Cost of system installation, operation and maintenance.

5 Available farm equipment.

6 The farmer's preference.

If a surface irrigation method is selected, the field surface must be given detailed consideration. Rarely is it possible to establish a satisfactory surface irrigation system without some land grading to change the surface topography of a field to a planned grade. Soil is moved from high spots and to low spots to enable efficient irrigation and removal of surface water. Without uniform grades low areas may receive excess water and cause drowning of crops. High areas may not receive sufficient water to avoid reductions in crop yields. In level borders, salts may accumulate in the high areas causing further crop damage and yield reduction.

8.2 SOIL SURVEY AND ALLOWABLE CUTS

Land shaping is normally done only on soils suitable for surface irrigation. Land shaping should never be done without first knowing the soil pro-

(Joins sheet 5)   (Joins sheet 6)

FIG. 8.1 Portion of a typical soils map in a river valley at a scale of 1:20,000. Soils were determined by field investigation and the soil delineation lines drawn on an aerial photograph.

file conditions and the maximum cuts that can be made without permanently affecting agricultural production. A general soil survey map is available for many areas and if not available, the services of a soil scientist should be obtained. In the United States soils information may be obtained from the local Soil Conservation Service office or a state university. The standard soil survey generally delineates areas larger than 2 ha (5 acres). The soil profiles are described to a depth of 150 cm (60 in.). In alluvial soils, sand and gravel pockets may occur within a soil mapping unit of some other soil type. It is often advisable to remove these sand and gravel materials and replace them with soil similar to the rest of the field. An example of a standard soil map in a river valley is shown in Fig. 8.1. A legend provided with the map describes the soils that are designated by the symbols on the map. For example, near the center of the figure is an area with the symbol Sm. The legend shows this to be a Sharkey silty clay loam. The dots indicate that there are pockets of sand that could affect the development of this area for irrigation. A detailed investigation should be made to determine the extent of such material before the land grading begins so that adequate fill material can be made available.

FIG. 8.2 Cross-slope benches on a silt soil with a slope before grading of 4 percent. Benches are 15 m (50 ft) wide with an irrigation grade of 6 cm (0.2 ft) per 30 m (100 ft). Elevation difference between benches is approximately 0.6 m (2 ft).

Sand and gravel pockets that occur near the beginning of the surface irrigation run may greatly limit the irrigation efficiency because of high intake rates.

Soils with deep, well drained subsoils generally have few limitations on depth of cut. Cuts of 3 to 4 m (10 to 13 ft) have been made in soils with deep subsoils without permanently reducing potential crop production. Some shallow soils may be suitable for irrigation and still not have sufficient depth to permit the necessary grading for surface irrigation. Soils that are shallow over bedrock, gravels, caliche, heavy claypan or fragipans may have severe limitations on the permissible depth of cut that can be made. Cross-slope benching sometimes is used on sloping land to reduce the depth of cut needed to obtain suitable field grade. Fig. 8.2 is an example of cross-slope benching. Water delivery is by open ditch with concrete drop structures.

8.3 TOPOGRAPHIC SURVEY

It is extremely important that the entire farm be studied before any grading work is attempted even though land grading is usually done on a field by field basis. The most economical design for an individual field may be undesirable when the entire farm system is considered. The location and elevation of the water delivery and drainage systems, the crops to be grown, the method of surface irrigation to be used and the regular farming operations should be considered by the designer in planning the overall farm development before beginning work on an individual field. A general topographic map of the entire farm showing the location of all physical features can provide much of this information. The designer must select the most desirable location for the various elements of the farm irrigation, water reuse and drainage systems, location of field boundaries, field roads and row direction for each field. With the information several alternative designs can be developed for consideration and selection by the farmer or farm manager. The plan selected will permit land grading to be carried out on individual fields over a period of years if necessary.

The general topographic map is used for planning purposes. The map used for the land grading design is generally made using the grid staking procedure described in Section 8.6. The topographic map may be made by any of the conventional surveying methods, such as plane table, transit, field cross-section, grid system or aerial photography with stereo plotting. The survey method used will depend on the size of the area to be planned, availability and capability of personnel and the time constraints for getting the job done.

The grid system is sometimes used on flat, relatively uniform fields in making the topographic planning map. In this case, the grid pattern is of greater spacing than that used for land grading work, generally 60 to 90 m (200 to 300 ft). If the entire farm is to be graded during the same season it may be desirable to use the grid system on a regular spacing for both the planning and the land grading.

The plane table and transit surveys generally are based on random shots. The elevation points should be selected to reflect changes in land slope. The horizontal distance between shots usually should not exceed 90 m (300 ft). Fig. 8.3 is an example of a random shot transit survey plotted to

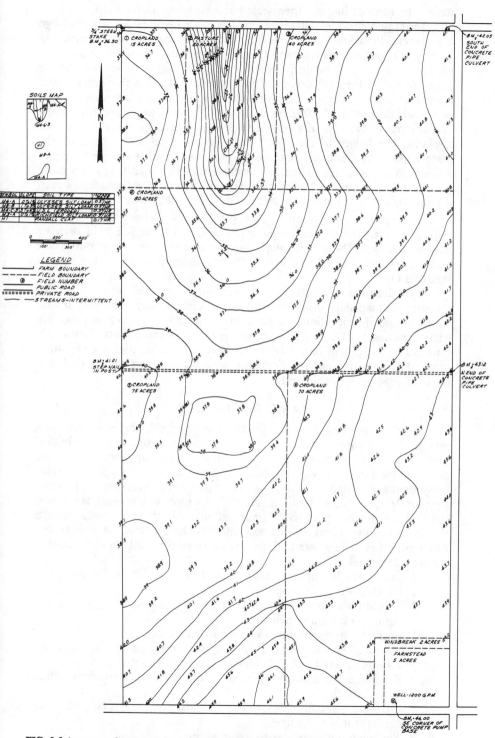

FIG. 8.3 A topographic map made by a random shot survey. Using this information the planner can develop alternative field arrangements as a guide in designing the land shaping.

scale. The contour line are then scaled and sketched on the map.

Aerial photography with stereo plotting is generally used when a large area (several thousand hectares) is to be planned as a unit or where work is to take place on several large adjacent farms at the same time. The photography can be done during the season of low vegetative growth and the maps plotted as they are needed. This procedure requires that vertical and horizontal control points be established before the photos are taken. These points which must be visible on the photos, usually are made with white plastic strips in the form of a cross. Low level photographs are taken from an airplane. The height of the plane above ground level will depend on the camera used and the detail required for the topographic map. For example, a 15-cm (6-in.) focal length lens in the camera at 915 to 1220 m will provide the resolution needed to plot 0.6 m (2 ft) contour lines. The topographic map is made by stereoscopic plotter using stereo pairs of photos.

In general, the topographic map should show contour lines by 0.3-m (1-ft) invervals. On very flat, irregular topography 0.15-m (0.5-ft) increments may be desirable. When the slope exceeds one percent, the planner may desire a contour interval of 0.6 m (2 ft) or greater in order to have a more readable map.

It is important, regardless of the survey method used, that the topographic map show other physical features of the farm. The map should show location and elevation of bench marks, location and source of irrigation water, existing field boundaries, drainage patterns and outlets, farmstead, farm roads, location of both buried and above ground utilities and any other physical features that may affect the planning of the system and the design of the land grading for each field.

8.4 SYSTEM LAYOUT

Prior to designing the land grading for any field, the farm irrigation system, the drainage system, main field boundaries, road and other physical features must be planned. From soil information, topography, and the desires of the farmer, alternative field arrangements are developed. Field arrangement and design are a very important part of the land shaping process. Frequently, inadequate attention and study are given to planning. It is essential that the planner plan with "imagination". All of the potential layouts should be visualized and the one best suited to the site, method of water application, crops to be grown and the farmer's own personal likes and dislikes should be selected.

In making the layout, the planner should consider these factors:

1 Location of water supply.
2 Amount and arrangement of supply ditches and/or pipelines, and automation requirements.
3 Location, amount and size of drainage ditches.
4 Location, size and method of operation of the return-flow system.
5 Ease of access to all fields for all farm equipment.
6 Field location and arrangement to minimize soil variations within each field.
7 Combining small fields into larger fields, especially if large farming equipment is to be used.
8 Reducing noncrop area to a minimum.
9 Elimination of point-rows where possible.

LEVELING LAYOUT ON SLOPING LAND FOR
COMBINED DRAINAGE AND IRRIGATION

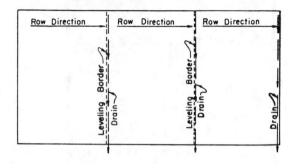

PLAN FOR 80 ACRE FIELD

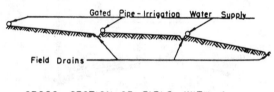

CROSS SECTION OF FIELD WITH A
CONTINUOUS SLOPE IN ONE DIRECTION
WITH DRAINS AT REQUIRED INTERVALS

FIG. 8.4 An example of a 32-ha (80-acre) field laid out in three planes for irrigation and drainage. In some soils this field could be divided into two planes but the earthmoving would be greater.

10 Ease of management of irrigation water and the cultural operations.

11 Current and future crops to be grown.

Certain design criteria must be determined before the field can be subdivided and the shaping design prepared. For example, both irrigation and drainage affect the allowable length of fields. If the field length is too long, water penetrates too deep at the upper end of the furrow or border by the time the water reaches the lower end of the run. Excessively long fields may also create erosion problems during irrigation or rainfall runoff. Short fields require more delivery and drainage facilities for a given area. Locally available irrigation and drainage guides can usually be obtained in the United States from Soil Conservation Service, Extension Service, or a university representative. From these guides and soils information, optimum lengths of fields are selected. Additional information on surface hydraulics is given in Chapter 12 and detailed design criteria are given in Chapter 13. Fig. 8.4 - 8.6 are examples of some of the alternatives that the land shaping designer should consider. Fig. 8.4 shows a field divided into three planes, all sloping in the same direction to keep earth moving to a minimum. Irrigation water is supplied to the upper end of each plane by ditch, buried pipeline, or surface pipe. Drainage is provided at the lower end of each plane. Fig. 8.5 is an example of grading a field away from the center. Water delivery is made

LEVELING LAYOUT ON FLAT LAND FOR
COMBINED DRAINAGE AND IRRIGATION

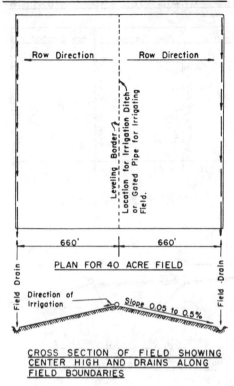

FIG. 8.5 An example of a field laid out in two planes so that the same water delivery system, ditch or pipeline, will be used to irrigate both planes.

to the center and drainage is from each side. The land shaping costs may be greater in this design but water delivery and drainage may be simplified and the farming operations may be more easily performed. The degree and uniformity of slope and soils will also affect the field planning and layout. The field should be divided so that each type of soil can be easily and properly irrigated.

In low rainfall areas where drainage is not needed for rainfall runoff, fields are sometimes shaped to zero grade or nearly so. Fig. 8.6 is an example of a farm layout using a central water delivery system with level fields on each side. Water delivery with this type of layout can be automated rather easily, thus reducing the labor and management requirements.

8.5 OPTIMIZING CUTS AND FILLS

Optimizing cuts and fills has traditionally been equated with minimizing cuts and fills in land grading. It has been defined as minimizing the cost of land grading. The designer must be concerned with minimizing land grading costs, but a design with minimum grading costs may not be the optimum. The layout of the system, the degree of irrigation efficiency attainable, labor requirements, cost of operation and maintenance and farm-

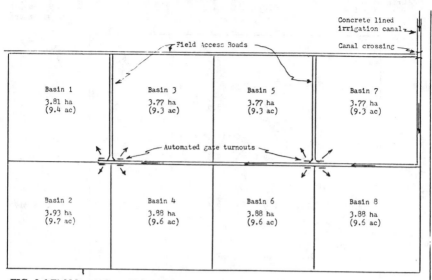

FIG. 8.6 Field layout for a 32-ha (80-acre) field shaped in level basins for flood irrigation. This type of layout can be easily automated.

ability are often of greater significance than minimizing the cost of land grading.

Land grading can be grouped into two general classifications: (a) a plane or series of planes; and (b) warped surfaces.

Land is graded to a plane or a series of planes when the lay of the land, the depth of the soil, and the economic factors are suitable and when required for the irrigation method selected. The current trend is to shape fields as planes for efficient irrigation. Where field shapes cannot be accomplished with planes, warped surfaces may be used. For example, in arid areas where crops are dependent mainly on moisture supplied by irrigation water and either flood or furrow irrigation is to be used, the plane surface is preferred. In humid areas where most of the moisture for the crop comes from rainfall, irrigation is by the furrow method and depth of cut are limited by the soil, the warped surface is often used.

8.5.1 Plane or Series of Planes

The least squares method is a statistical procedure for obtaining the best fit line to a set of points. The procedure eliminates the need for trial and error methods of determining the average elevation. It was first adapted to fitting a plane to the different field elevations of an uneven rectangular land surface by Given (1940). Later the method was extended to include land areas of any shape (Chugg, 1947). The "least squares and average profile method" (Marr, 1957) is the process of determining by the method of least squares the slope of the average profile in the two coordinate directions rather than fitting a plane to the field elevations. The procedure is easy to follow, is practical, and probably the most widely used method of land grading calculations.

8.5.2 Plane of Best Fit (Calculator Method)

This method is a regression analysis in two directions that can be per-

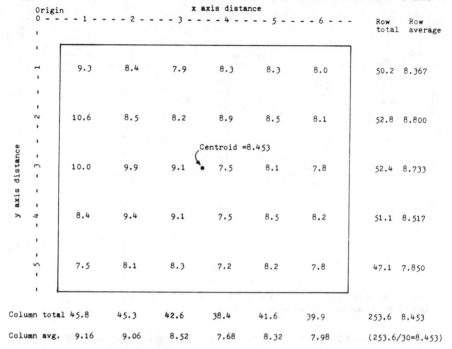

FIG. 8.7 Grid map showing land elevations at grid corners, location and elevation of the centroid, and the average profile figures.

formed with a programmable hand calculator if a regression analysis program is available. Elevations could be used directly eliminating the need for setting stakes described in the example. The example illustrates the use of this method and shows how the calculations are made.

Fig. 8.7 is a base map of a rectangular area which is to be graded as a unit. Stakes are set and the elevations are given at 30-m (100-ft) intervals starting 15 m (50 ft) in from the property lines forming a grid with the elevation known at each grid corner.

Step 1: The elevations are added by horizontal rows and by vertical columns and the average elevation of each row and column determined. The sum of the elevations and the average for each row is recorded to the right and for each column at the bottom of the grid map as shown in Fig. 8.7.

Step 2: Determining the location and elevation of the centroid. On a square or rectangular field the location is at the intersection of diagonal lines drawn from each corner to the opposite corner. In Fig. 8.7 the centroid is on row 3 midway between columns 3 and 4. The elevation of the centroid is the average of all of the grid point elevations. In Fig. 8.7 this elevation is 8.453.

Step 3: Determinine the slope of the lines which most nearly fit the average profiles in the two coordinate directions. According to the least squares method, the slope of the line which best fits the points on each of these two profiles is:

$$G = \frac{\Sigma(S_i H_i) - \dfrac{(\Sigma S_i)(\Sigma H_i)}{n}}{S_i^2 - \dfrac{(\Sigma S_i)^2}{n}} \qquad \dots \dots \dots \dots \dots \dots \dots \dots \dots \dots \dots \dots \quad [8.1]$$

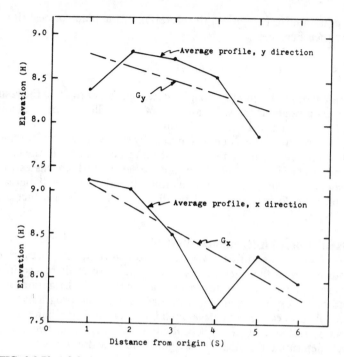

FIG. 8.8 Plot of the average profile and line of best fit in the x and y directions.

where G = the grade or slope of the line which best fits the grid elevations in the direction under consideration; S_i = the distance of grid point i from the origin; H_i = the elevation of grid point i; and n = the number of grid points in the line. If the slope is positive, the grade line slopes upward, and if negative it slopes downward. Fig. 8.8 shows the plotting of the average profiles and the lines of best fit as computed.

Step 4: Using the slope of the lines of best fit in the x and y coordinate directions and the elevation of the centroid, the plane which best fits all of the grid point elevations may be delineated.
The equation of the plane is:

$$H = a + (G_x)(S_x) + (G_y)(S_y) \quad\dotfill\quad [8.2]$$

in which H = elevation of the point under consideration; a = elevation of the point in the plane at the origin; G_x = slope of the best fit line in the x coordinate direction; S_x = distance from the origin to the x coordinate of the point being determined; G_y = slope of the best fit line in the y coordinate direction; and S_y = distance from the origin to the y coordinate of the point being determined.

With the elevation of the centroid known, the elevation at the origin can be calculated with equation 8.2. In the previous example, H at the centroid is 8.453, G_x = - 0.256, S_x = 3.5, G_y = - 0.131, and S_y = 3, then the elevation of the plane at the origin is:

$$a = 8.453 - (-0.256)(3.5) - (-0.131)(3) = 9.742$$

The equation for computing the elevation of the plane of best fit at any grid point then becomes

$$H = 9.742 + (-0.256 \, S_x) + (-0.131 \, S_y).$$

If these calculations are made with adequate significant figures, the resulting plane will require about the same amount of cut as fill and the total amount of earth to be moved will be a minimum.

In some instances the slopes calculated by the least squares average profile method may be either too flat or too steep to suit the crop or the preferred method of irrigation. Another plane may be selected which passes through the centroid but at the desired slope in either direction. This will increase the amount of earth to be moved and in all probability will also increase the length of haul.

8.5.3 Plane of Best Fit (Computer Method)

Optimizing cuts and fills is very adaptable to computers because of the repetitive nature of the calculations and the data are easily placed in an array. Federal agencies and state universities involved in irrigation development and many consulting engineering firms have such programs.

A simple program can be developed based on the optimizing procedure in Section 8.5.2 Plane of Best Fit (Calculator Method) with two changes.

First, determine the slope for irrigated fields as follows:
1 Calculate the slope for each row of stakes using the least squares method.
2 Obtain a weighted average based on the number of stakes in each respective row.

Second, locate the centroid (SCS, 1961).

Two iteration procedures can be used which may minimize cuts and fills a slight amount from the above least squares procedure. These are:
1 After calculating the slope by the least squares method, increase and decrease the slope until the minimum cuts are obtained.

2 A second option is to set maximum and minimum allowable slopes and to make iteration between them until the slope with the minimum cuts is obtained.

Computer time and costs will increase with the iteration procedures. The above described programs minimize cuts and fills but do not minimize costs. Haul distance and type of equipment can affect the land shaping costs.

8.5.4 Computation of Cut and Fill Adjustments

In land shaping it is essential that the volume of material excavated be adequate to make the fills. If the cuts to be made equals the fills without either borrowing or wasting material, the earthwork is in balance. Experience has shown that the volume of cuts computed from the design must exceed the fills by a certain percent. The volume of cut divided by the volume of fill is called the cut-fill ratio (R). This cut-fill ratio generally will range from 1.2 to 1.5 depending on the soil, the field conditions and equipment used. To obtain the desired cut-fill ratio plane of best fit as described in Sections 8.5.2 and 8.5.3 should be lowered from 1 to 2 cm (0.03 to 0.07 ft). From his experience in an area, the designer will generally select the desired value for R. For example, if 30 percent more cut than fill is needed,

$$R = \frac{\text{Cut}}{\text{Fill}} = 1.3 \dots\dots\dots\dots\dots\dots\dots\dots\dots\dots\dots\dots \quad [8.3]$$

The amount to lower the plane to achieve an R of 1.3 often is done by a trial and error procedure. Equation [8.4] can be used to estimate the amount to lower the grade line in lieu of the laborious trial and error procedure.

$$\Delta_g = (\frac{1 + R}{2RN_t})(R\Sigma F - \Sigma C) \dots\dots\dots\dots\dots\dots\dots\dots\dots\dots \quad [8.4]$$

in which Δ_g is the increment the grade line is lowered; R is the cut-fill ratio; N_t is the number of both cut and fill coordinates; ΣC is the summation of cuts before lowering the plane; and ΣF is the summation of fills before lowering the plane.

Equation [8.4] is an approximation and will vary with different soil surface reliefs. Since R is simply an estimate based on experience, the adequacy of the cut-fill ratio should be checked as the land shaping progresses. Using the above formula for the plane of best fit applied to the data in Fig. 8.8, the following computation is made to determine Δ_g.

$$\Delta_g = (\frac{1 + 1.3}{2 \times 1.3 \times 30})[(1.3 \times 7.940) - 7.950]$$

$$\Delta_g = 0.07$$

Fig. 8.9 shows how the plane is adjusted by Δ_g to obtain the desired cut-fill ratio. The top figure at each grid point is the original ground elevation. The second is the elevation of the plane of best fit. The third figure shown with a line drawn through items the cut or fill required to obtain the plane of best fit. The last figure is the cut or fill after the plane has been adjusted by Δ_g to provide the cut-fill ratio (R) desired.

Sometimes it is necessary to use material from the field to construct farm roads or elevated ditches or material excavated for a drain may have to be spread on the field. The adjustment in the plane to allow for this material can be estimated by the following equation:

$$gb = K \frac{V}{A} \dots\dots\dots\dots\dots\dots\dots\dots\dots\dots\dots\dots\dots\dots \quad [8.5]$$

where V is the volume of borrow or waste, A is the area of the field and K is a dimensional constant equal to 1.0 for gb in meters (K = 27 for V in yd³ and A in ft²).

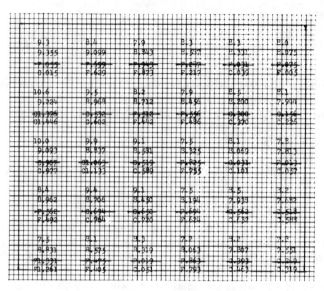

FIG. 8.9 A grid map showing how the plane of best fit is adjusted to obtain the desired cut-fill ratio (R).

8.5.5 Warped Surface

Warped surfaces have two advantages over plane surfaces. First, warped surfaces take maximum advantage of natural topography. As a result, depths of cuts and fills, lengths of haul and construction costs are minimized. Second, less soil is disturbed in construction, and there is less likelihood of cutting into unproductive subsoils.

There are three basic methods used in designing warped surfaces by manual procedures. All are trial and error methods: (a) profile method; (b) plane inspection method; and (c) contour adjustment method.*

The profile method: Using this method the designer works with profiles of the grid lines rather than with elevations. The profiles are normally plotted in one direction with the individual profiles located on the paper so that the datum line for each profile is in the correct position with adjacent profiles. Profiles may be plotted across the slope or down the slope. Fig. 8.10 shows a plotting of the profiles across the slope and Fig. 8.11 shows profiles plotted down the slope. Either method is satisfactory.

Trial gradelines are plotted on each profile based on the design criteria. The balance between cuts and fill is approximated by eye by comparing the areas between the plotted profile and the trial grade line. Usually several trials are necessary before a satisfactory set of gradelines are attained. When the volume of cut and fill is computed further adjustment of the grade lines may be necessary to provide the proper cut-fill ratio for the field.

The plan inspection method. In this method the survey data are plotted on a grid map and, by trial and error, proposed elevations are selected which meet the down-field and cross-slope limitations. The designer must

---

*May be obtained by writing the Superintendent of Documents, U.S. Government Printing Office, Washington, DC 20402; SCS National Engineering Handbook, Section 15 IRRIGATION, Chapter 12, Land Leveling.

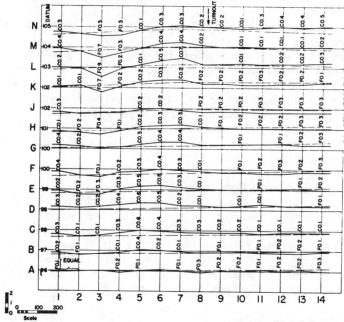

FIG. 8.10 Plotting of across slope profiles for the profile method of warped surface design for land shaping.

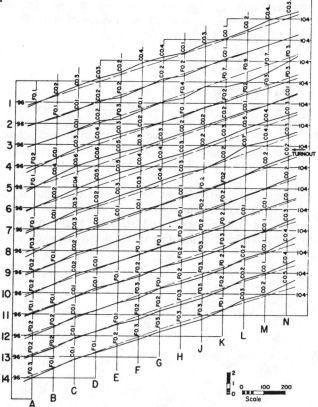

FIG. 8.11 Plotting of down slope profiles for the profile method of warped surface design for land shaping.

simultaneously consider down-field slope, cross slope, earthwork balance and haul distance.

The contour adjustment method: This is a method of adjustment, by trial and error, of the contour lines on a plan map. It is applicable where the cross slope can be made uniform, where grading is done over a period of time, when the stakes cannot be maintained in the field, shaping the land between terraces, and in removing extreme ridges or swales from land to be irrigated by the contour-ditch method. This method requires an accurate contour map of the area with adequate reference points and bench marks for both vertical and horizontal control.

8.5.6 Warped Surface by Computer

Harris, et al. (1966) presented a computer program for a warped surface. In designing warped surfaces, the computer works with two station segments. Row grade and cross slopes are adjusted repetitively until the entire field meet the design requirements. Warped surface design limits can be varied to suit the needs of each individual field. Design limits are maximum and minimum row grades and cross slopes, maximum change in row grade and cross slope at a station and the cut-fill ratio (R).

8.5.7 Earthwork Calculations

The volume of excavation, or cut, is generally used as a basis for contracting and for estimating equipment requirements and job schedule. An accurate method of computing the volume of excavation or fill in land shaping is by use of the prismoidal formula

$$V = \frac{L}{6} (A_1 + A_m + A_2) \dots\dots\dots\dots\dots\dots\dots\dots\dots\dots\dots \quad [8.6]$$

where V is the volume, L is the distance between end planes A_1 and A_2 are the end plane areas and A_m is the middle section plane area in consistent metric or English units. This formula is not commonly used because of the laborious procedure.

Where excavation involves the entire grid, the volume of earthwork could be obtained by multiplying the area of the grid by the average cut at the four corners of the grid. Since cut and fill both often occur within a grid a procedure known as the four-point method is used. This method is based on the equations

$$V_c = \frac{L^2}{K} \frac{\Sigma C^2}{(\Sigma C + \Sigma F)} \dots\dots\dots\dots\dots\dots\dots\dots\dots\dots\dots\dots \quad [8.7]$$

$$V_f = \frac{L^2}{K} \frac{\Sigma F^2}{(\Sigma C + \Sigma F)} \dots\dots\dots\dots\dots\dots\dots\dots\dots\dots\dots\dots \quad [8.8]$$

where V_c is the volume of cut in m³, V_f is the volume of fill in m³, ΣC is the sum of cuts on the four corners on a grid square, ΣF is the sum of fills on the four corners of a grid square, L is the length of the grid square, and K is a dimensional constant equal to 4 in the metric system with C and F in meters and 108 with C and F in feet and volume in cubic yards. Tables have been developed to provide a rapid method of determining the excavation.

Another method, called the summation method, is used to estimate volumes of excavation. This method assumes that a given cut or fill at a grid point represents an area midway to the next grid point. The grid point is considered as being in the center of the grid rather than at the corner. This is the least accurate of the methods presented. In the metric system, the cut in meters multiplied by the area of the grid in square meters gives the excavation in cubic meters. (In the English system, each foot of cut is 370 yd^3 for a 100 ft grid. If the tenths of cut or fill are added, the total is multiplied by 37.) This provides a rapid method of estimating excavation.

8.6 STAKING AND GRADE CONTROL

Land that is to be graded for surface irrigation is usually staked on a grid pattern. The relative elevation of the original ground surface at each stake is the basis for determining the field design, computing the volume of cut and fill, and controlling the grading operation. For ease in computing volume and area the grids should be made square. The stakes may be set at any spacing desired by the designer. The most commonly used spacing is 30 m (100 ft). Larger spacing makes it more difficult for the operator of the grading equipment to maintain the grade from one stake to the next. Closer spacing greatly increases the work of staking, surveying and computing the grading design. A 15-m grid requires four times as many stakes as required for a 30-m grid. Occasionally it may be necessary to insert extra stakes in the grid pattern to record significant topographic changes in the land surface or to show irregularly shaped boundaries.

Before shaping work begins the field should be cleared of vegetation. Heavy crop residues should be removed by mowing and raking or by burning. Crop ridges should be eliminated by disking or smoothing. Sod fields should be mowed and raked but not plowed. Any operation that leaves the surface in a loosened condition should be avoided.

8.6.1 Staking the Field

Wood lath (10 × 45 mm × 1.2 m) are excellent stakes for setting the grid points. They cost less than 25- by 50-mm stakes, are easily set from a standing position and cut or fill markings near the top of the lath are easily seen by the equipment operator. The lath should be sharpened and driven into the ground far enough so that they will remain standing until the grading work is completed.

The following procedure outlines a method of establishing key rows of stakes across the field to be graded. After the key rows are established, the remaining grid stakes can be set by sighting across the stakes in the key rows.

Step 1: If two adjacent sides of the field are straight and form a 90-deg angle row A and line 0 may be placed on the field boundary (See Fig. 8.12). Starting at grid point A-0 using a tape, stakes are set for row A accurately at 30 m (100 ft) along one side of the field. The staked line should be aligned with a surveyor's transit set at station A-0, or by setting several stakes at random intervals along the boundary in a straight line and setting the stakes by visual sighting. Accuracy is essential for satisfactory guide lines.

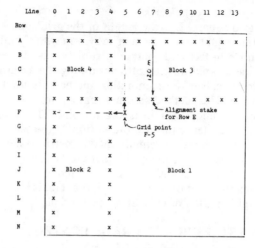

FIG. 8.12 Placement of key rows of stakes for grid staking a field

Step 2: Line 0 along the adjacent boundary is established in a similar manner perpendicular to row A. If the adjacent boundary is not perpendicular a 90-deg angle can be established with a transit or by using tapes to establish a right angle.

Step 3: Establish a stake or range pole approximately mid-way along E and 120 m (400 ft) in from row A. Starting at station E-0, sight on the stake or range pole for alinement, and using the tape for proper spacing, set the stakes in row E parallel to row A.

Step 4: Starting at station A-4, sighting on the stake at E-4 for alignment, set the stakes in line 4 with a tape. All the remaining stakes can be set without the use of a tape.

Step 5: Starting in the area indicated as Block 1 fill in the grid stakes by visually sighting on the stakes in the four guide lines. For example, the stake at grid point F-5 is located by sighting across the stakes at A-5, E-5 and F-0, F-4. Block 2 is filled in next, then Block 3 and Block 4 is completed last.

The example shown in Fig. 8.12 is for a 16-ha (40 acre) field. For larger fields it may be desirable to establish the guide rows near the center of the field to eliminate some of the errors that occur in setting long lines by the visual sighting.

8.6.2 Grade Control

The grid stakes serve as reference points to control the shaping operation. Ground surface elevations are obtained at each grid point using an engineer's level and rod graduated in meters or in feet. On small fields, 16 ha (40 acre) or less, where the view is unobstructed and the variation in elevation does not exceed the length of the rod, the rod readings may be used to indicate the relative elevation of the grid points. The most common practice is to convert the rod reading at the grid points to elevations related to a known or an assumed elevation of a bench mark. Since the bench mark is to provide a permanent reference point for the construction as well as for the survey, it must be a point that will be out of the way of construction equipment and likely to be undisturbed by people or animals.

The procedure for grade control during construction varies with localities. A mark or line may be made across the grid stake at 30 cm (1 ft)

FIG. 8.13 A Hargus marker showing a cut of 0.5 ft (University of Alberta).

above the point where the level rod was placed. The equipment operator can then reference the cuts and fills to a point 30 cm below this mark. Another procedure is to set grade stakes with the top of the stake being the elevation for the finished grade. Both of these are time consuming procedures. A more common procedure is to use the ground level at the stake as the reference point. If the surface is loose it is a good practice to tramp the ground along side the stake where the level rod will be placed. The amount of cut or fill is written with a lumber crayon near the top of the stake so that is is easily read from the tractor. Cuts usually are marked with a red crayon and the fills with blue. Some contractors also like to paint the top of the stakes red in cut areas and blue in fill areas. Red and blue plastic flags are also used for this purpose. Fig. 8.13 shows a Hargus marker manufactured in Canada from tough wet strength kraft paper and formed into a jacket which slips over the stake.

Regardless of the method used to indicate the cut and fill, the operator must carry the design grade from one stake to the next. Frequent checks should be made to see that the design grade is obtained. Some large contractors use an engineer's level to check the work in progress and to assist the operator in obtaining the design grade. Many operators use a small hand level to check the grade between the stakes.

8.7 FIELD OPERATIONS

The land grading work must be planned and scheduled to assure satisfactory field conditions. Some of the factors that must be considered are:
1 Time

The field operations must fit into the cropping sequence. The grading should be scheduled so that the work can be carried out between the harvest of one crop and the planting of the next crop. The time available between crops will help determine the type of earth moving equipment to be used.

2 Weather

While weather conditions cannot always be predicted, there is generally a time in each region when conditions are most favorable for earth moving work. If possible, the grading work should be planned for the most favorable season to avoid long shut down periods. Particularly, field work should not be scheduled for periods of high rainfall, freezing weather, or other weather conditions that may cause excessive delays in earth-moving.

3 Field condition

Some field conditions during land grading can have serious adverse effects on the quality of the land shaping work and crop production after the work is done. Grading should not be done when the fields are excessively wet, except when water leveling of rice fields (See Section 8.7.3). Operating scrapers in wet soils is difficult and can cause serious damage to the soil structure resulting in poor soil conditions for plants and irrigation. Grading of very dry soils should also be avoided. Cuts are difficult to make, fills become loose and powdery, resulting in poor quality work and higher cost of operating the earth-moving equipment.

8.7.1 Types of Equipment

There is a great variety of equipment that can be used for land shaping. Each type of machine has its own capabilities and limitations. The examples of equipment described in this section were selected to illustrate some of the types that are being used.

Most land shaping is done with an earth-moving scraper sometimes referred to as a carryall. Occasionally a motor grader may be used on small fields, for narrow benches or where only minor grading work is needed. It is

FIG. 8.14 A cable controlled scraper pulled by a crawler type tractor (SCS, USDA).

FIG. 8.15 Two scrapers being pulled by a farm tractor. Cut is being made by the front scraper. When it is loaded the second scraper will

also used to shape the ridge on the down-hill side of benches and the slope from one bench to another.

Scrapers are available in a wide range of sizes. For land shaping for irrigation, scrapers ranging in capacity from about 1.5 m³ (2 yd³) to 19 m³ (25 yd³) are used where large quantities of earth are to be moved over an appreciable distance. Prior to the 1960's most scrapers were pulled by crawler tractors, (Fig. 8.14). The crawler tractor had low soil compaction characteristics, provided good traction on a variety of soils and handled scrapers with capacities of 6 m³ (8 yd³) to 15 m³ (20 yd³). It was well-suited for short hauls but its slow speed, about 8 km/h (5 mi/h) limited its suitability. Most of the land shaping is now being done with scrapers pulled with

FIG. 8.16 Scraper with power take-off driven elevator loaders pulled by farm tractors (SCS, USDA).

FIG. 8.17 Twin-motor rubber-tired scraper has high speed, large load capacity (University of Alberta).

rubber-tired wheel tractors. Their speed and maneuverability makes them well-suited for this type of work. Small scraper units, 1.5 to 6 m³, are designed to be pulled with farm tractors, (Fig. 8.15). Fig. 8.16 shows a power take-off operated elevating scrapers pulled by farm tractors. The larger units, 6 to 19 m³, usually are equipped with their own power unit (Fig. 8.17). Most of these large units are owned and operated by contractors. These large units often require a pusher in making heavy cuts. The pusher may be a crawler or a rubber-tired tractor. Usually one pusher is used for two or more scraper

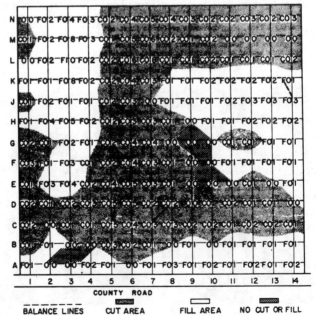

FIG. 8.18 A typical cut-fill map used by the equipment operator. Cuts and fills are in tenths of a foot. The dashed line separates the field in two areas of balanced cut and fill.

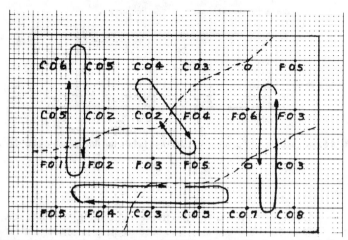

FIG. 8.19 Sketch showing how the haul lanes might be visualized by the operator in determining the movement of earth from a cut area to a fill area.

units depending on the length of haul. Bottomless drag scrapers used for light shaping or finish work are described in Section 8.8.

8.7.2 Operating Procedure

The quality of the land shaping operation depends on the equipment used, condition of the field, and the skill of the operator. There are probably as many different ways of approaching the actual earth moving process as there are equipment operators. The earth-moving work should be carried out so that the grid stakes are not disturbed until the job is ready for the final smoothing operation. The farmer or contractor is normally given a map showing the work to be done. The information shown on the map depends on local custom and contractor preferences. In some cases it may be a map showing just the amount of cut or fill. A typical cut and fill map is shown in Fig. 8.18. Numbers indicate the cut or fill in tenth of a foot. For example, CO5 indicates that the cut is to be 0.5 ft. If the grid point was marked F15, the fill would be 1.5 ft. Some engineers make a detailed study of the contractor's map and prepare earthwork balance areas suggesting the direction and location that earth is to be moved. Most contractors or operators study the map and make their own decision on matching cut and fill areas. Fig. 8.19 is an example of how the earth movement patterns might be visualized by the operator. On large, more complicated, jobs the haul patterns may be sketched out on the contractor's map. In some instances the contractor may want a complete map showing original ground elevations, planned finished grade elevations and the cut and fill figures. Other procedures include such things as indicating the cut with red figures and the fill with blue, or by cross-hatching the cut areas in red and the fill areas in blue.

Cutting and filling is normally done parallel to the grid stake rows. The procedure varies with operator preference and equipment. Some operators cut and fill a strip the width of the scraper adjacent to the stake rows. The same procedure is followed for the opposite side of the lane along the next row of stakes. When these are brought approximately to grade, work is carried on in the intermediate area. The grade between stake rows is carried visually across the lane. This process is continued until the entire field has

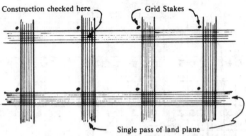

FIG. 8.20 Sketch showing the planing of stake rows before the final construction check is made.

been shaped. Another procedure is to make a cut along the stake row as deep as can be made on the first pass. The second strip is made just far enough over so that the tractor can straddle the strip that is left. The strip that was left is then removed as a third strip. This procesure is repeated across the cut area. Regardless of the procedure followed, the cuts and fills should be made in uniform layers. This reduces uneven setting of the field, facilitates carrying the proper grade from one stake row to the other, permits faster travel, and is easier on equipment. Normally the small area around each stake is left until the shaping is completed. If deep cuts are made leaving a large high mound at the stake, it may be desirable to remove the stake, cut away the mound and replace the stake. Cut areas should be disked as soon as possible after the cuts are made to keep the surface from drying and becoming hard. This will make the preparations for the final smoothing much easier. In order to get a better check on construction, some contractors use a land plane, making one pass down each stake row in both directions. Fig. 8.20 is a sketch illustrating this procedure. The grade is then checked at the intersection of the two passes of the plane.

Normally grades are checked with a permissible tolerance of ± 3 cm (0.1 ft). On level or flat grades it may be necessary to check to a ± 1.5 cm (0.05 ft). In no case should reverse grades in the direction of irrigation be permitted. When the field construction has been approved, the stakes should be removed, the mounds and depressions at the grid stakes graded to the level of the adjacent surface and the field disked or chiseled to prepare it for the final smoothing operation.

Due to the travel of construction equipment and the smooth loose surface that is left after the final smoothing operation, dust can often be a serious problem. Dust affects nearby residents, creates hazardous conditions on nearby roads or highways and, in severe conditions, can damage adjacent farm land. This can be very severe under dry soil conditions and high winds. Land shaping on dry soils should be avoided where possible. In some instances this may require the installatin of temporary irrigation facilities, such as a portable sprinkler system, to provide better working condition and reduce dust problems. If soil moisture is adequate, working the field with small chisels or a spring-tooth harrow may provide sufficient surface roughening to control dust and soil blowing. The finished field should be irrigated and planted as soon as possible. If the planting of the next crop is to be delayed for any length of time, it may be advisable to plant a cover crop to provide adequate protection from wind erosion.

FIG. 8.21 Water leveling of rice field. Note the "wave" of water and soil produced by the tractor and blade (SCS, USDA).

8.7.3 Water Leveling

Contour levees have been used for many years to control flood irrigation on rice fields. The normal procedure is to establish contour earthen ridges, called levees, at 6-cm (0.2 ft) vertical intervals throughout the field. The contour interval may vary depending on the slope of the field and the available irrigation stream. The flood depth will vary from 7.5 cm (3 in.) to 15 cm (6 in.) depending on the slope above the levee and the surface irregularities. Since 7.5- to 10-cm water depth is normally adequate to control weed growth, the volume of water required to fill the levee can be reduced by 30 to 50 percent by removing the slope and the surface irregularities between levees. One of the methods of doing this has been through the use of water leveling.

Rice is normally grown on soil having a restrictive layer with a permeability rate of about 0.5 mm (0.02 in.) per hour. If this restrictive layer is thick enough to support the tractor and other equipment under a water saturated condition, water leveling method can be used. The equipment used for water leveling is generally a bottomless scraper or a blade mounted on the back of a farm tractor (Fig. 8.21). The general procedure is as follows:

1 Smooth the field with a land plane to remove as many of the surface irregularities as possible.
2 Locate and construct the permanent levees. Levees should be straightened by filling across the low areas.
3 Flood as much area as can be leveled in one day to a depth that will cover the high point.
4 Carry out the leveling work using the tractor mounted blade or the bottomless scraper. Tractor movement is carried out to produce water waves that will assist the leveling operation. More than one tractor is often used to speed up the work and to create more wave action.
5 Soil movement can be determined by water depth measurements.
6 After the suspended soil material has settled and the water has become clear, the remaining water should be drained off.

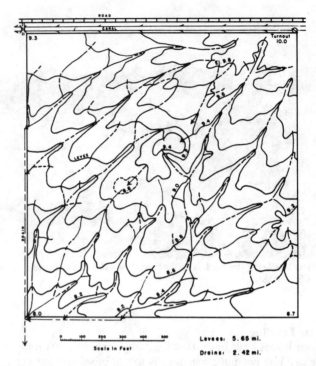

FIG. 8.22 Typical layout of contour levees on a 40-acre field without land smoothing.

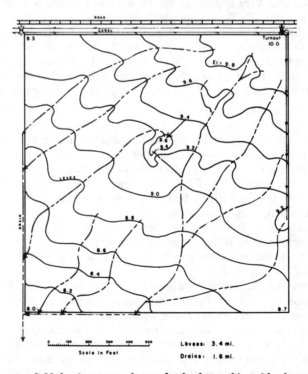

FIG. 8.23 The same field showing contour levees after land smoothing with a large land plane.

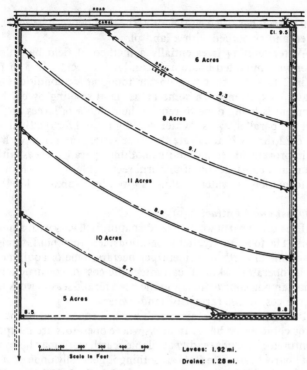

Levees: 1.92 mi.
Drains: 1.28 mi.

FIG. 8.24 The same field after land shaping to straighten and parallel levees followed by water leveling.

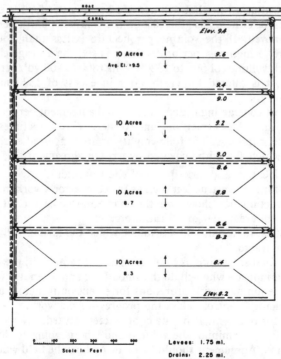

Levees: 1.75 mi.
Drains: 2.25 mi.

FIG. 8.25 The same field after construction of uniform basins. Final smoothing may be laser controlled.

7 When the soil has dried sufficiently, the area between levees should be worked and smoothed with a land plane.

Water leveling is essentially a method of field benching. One tractor with mounted blade unit can level a hectare in 5 to 10 h (an acre in 2 to 4 h) depending on the topographic conditions and the operator's experience. In some areas, land shaping by earthmoving equipment is being done to reduce the number of levees and to construct the parallel levees. Water leveling is used for the final leveling. Figs. 8.22 thru 8.25 illustrate the levee system on a field with no surface improvement, some land smoothing, parallel levees with water leveling and construction of uniform rectangular basins. More detailed information on water leveling is given by Faulkner (1964)

8.7.4 Cost and Contracting

The cost of earth work in land shaping will vary with a number of factors. The topography, field conditions, length of haul, total volume and volume of earth movement per hectare, type of equipment used and the operator's skill all can affect the cost of construction. Competition among contractors, and the need for off-season work often affects the cost to the farmer or land owner.

In most areas the major earth work is contracted for on a cubic yard or cubic meter basis. In areas where operators are inexperiences or contractors are not readily available, earth work may be contracted on an hourly basis. The final smoothing is generally done on a hectare or an acre basis. Some contractors make an estimate of the volume of work that can be done per hour and set the cost per cubic meter or cubic yard at a rate that will provide hourly rates for the equipment being used. Others base the cost on the volume of earth per unit area to be moved. If the volume per hectare is less than 380 m^3 (200 yd^3/acre) a high cost per unit volume is set, the average cost is used for earth volumes of 565 to 1320 m^3/ha (300 to 700 yd^3/acre) and the least cost is quoted for earth work in excess of 1320 m^3/ha (700 yd^3/acre).

Most contracting is done on a verbal agreement between the contractor and the farmer or land owner. This sometimes leads to misunderstandings. The following items are suggested as some of the things which should be covered by written agreement.

1 Field location and description of site conditions.
2 Basis for the construction such as the contractor's work map, special construction not shown on the map, allowance for settlement in heavy fill areas, cut to fill ratio, etc.
3 Facilities, material, and assistance which the farmer will provide to the contractor.
4 If the final smoothing is to be done as a separate job, the agreement should indicate who will do the smoothing and when it is to be done.
5 Completion date with provisions for extension if factors beyond the contractor's control delays the progress of the work.
6 Basis for payment. If by the cubic meter or yard, the method to be used for determining the volume and who is going to do the computation. Agreement should be made on how the field will be checked for completion.

7 Liability insurance.
8 Responsibility in locating and protecting buried utilities.
9 Final clean-up of rubbish, excess materials and equipment, broken
   equipment, etc.
10 Pay-schedule.

8.7.5 Safety

All equipment must be equipped with the required safety devices. All equipment should be operated in a reasonable and safe manner. Safety shields, such as those over power-takeoff shafts, must be ijn place and securely fastened. Special precautions should be taken when operating equipment around electric power lines, buried utilities and communication lines. Buried utilities and communication lines have been damaged and lives lost during land grading operations. The general location of the buried utility should be indicated on the contractor's map. Most utility companies will, on request, stake the exact location of their utility line and indicate the depth of the cover over the line. While some governmental agencies serve written notice to the farmer that he is responsible for notifying the utility company of the work to be done on his farm, all parties involved in the land shaping process must be safety conscious.

8.8 FINAL SMOOTHING

Normally it is impractical and too expensive to finish land surfaces to exact grade with heavy earth-moving machines. The heavy scraper work should provide a field surface so that two or three passes over the field with the finishing equipment such as a bottomless scraper or land plane, leveler, or float will produce the desired uniform surface.

Bottomless scrapers are of two types, the land plane and drag scraper, both of which are designed to remove small irregularities. As the name implies, the bowl or bucket has little or no bottom and earth moving is accomplished by scraping dirt from the high areas and dragging it to the low areas. The machine works best in a loose soil which has adequate moisture so that a smooth, relatively firm surface remains after the smoothing is completed. Dry, powdery soil conditions should be avoided. When such conditions exist the field should be irrigated. As soon as the field has sufficiently dried the surface should be worked with a disk or shallow chisel farm implement before the final smoothing is done.

The land plane, a bottomless scraper with the long frame, is often designed to work either manually or automatically. It is manufactured in lengths up to 27.5 m (90 ft) and with blade widths up to 4.5 m (15 ft). The most popular size, especially for large fields, has a length of 18.3 m (60 ft) and a blade width of 3.7 to 4.3 m (12 to 14 ft) as shown in Fig. 8.26.

The blade on the land plane is mounted midway on the frame and is adjustable vertically so that the depth of cut and the amount of dirt carried by the bucket can be regulated. Some planes are equipped with hydraulic controls so that the tractor operator can control the blade level from the tractor. When in use the blade is set at a level that will maintain about one-third to one-half of a load in the bucket. If the blade is set too low the soil in the bucket will become excessive, spilling around the ends and over the top of the bucket. If it is set too high there will be insufficient soil to adequately fill the

FIG. 8.26 A 18.3 m (60 ft) land plane with a 3.7 m (12 ft) width blade being used to smooth a field after the heavy earth moving work has been done. The tractor must be large enough to permit the blade to cut off the high spots and drag the soil to the low areas (Univ. of Alberta).

low areas. When the adjustment is properly made, the plane will automatically remove the high spots and fill in the depressions nearly one-half the plane's length. Cuts of 6 to 9 cm (0.2 to 0.3 ft) or less can be made within the length of the machine. Obviously, the longest plane will be best. Since the longer machines require a wider turning area, they will leave a wider strip around the boundaries of the field. The smaller the field the greater will be the percent of the field that cannot be properly smoothed. On fields of 8 ha (20 acres) or less and on narrow benches, the smaller and more maneuverable planes may be more desirable.

Since the cut and fill operation by the heavy earth-moving equipment is normally carried out parallel to the rows of the grid stakes, it is generally desirable to carry out the first planing operations diagonally to the grid pattern. It is customary to take care of the final smoothing by running the plane over the field in three directions. Fig. 8.27 shows how this operation is normally carried out. The first two operations are in a diagonal direction,

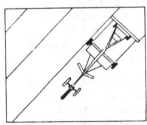
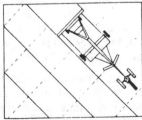
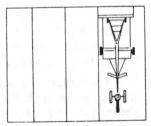

First Pass — Diagonally across area in one direction.

Second Pass—Diagonally across area in opposite direction to first pass.

Third Pass—In direction of greatest slope of the field.

FIG. 8.27 Sketches showing the method of operation for the land plane in the final smoothing.

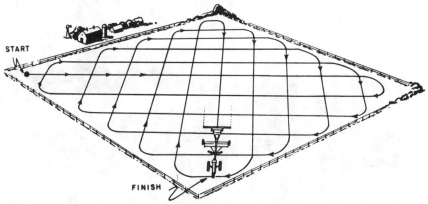

START

FINISH

FIG. 8.28 Sketch of a method of performing two diagonal planings with a minimum of time.

perpendicular to each other, and the third operation in the direction of irrigation. On fields that are shaped to a grade this is usually in the down slope direction. This operation is termed land planing and, when contracted separately from the cut and fill part of the land shaping, is normally contracted on a hectare or an acre basis. In carrying out the smoothing operation considerable time is lost in making the turns at the field boundaries. Fig. 8.28 shows how the first two diagonal planing operations can be carried out to cover the field with 2 passes of the plane in the minimum amount of time. This can reduce the time required by about 25 percent over planing the entire field in one diagonal direction and then repeating the operation at right angles to the first operation.

FIG. 8.29 A battery-operated laser transmitter is mounted on the tripod. The receiving unit mounted on the drag scraper controls the elevation of the scraper blade through the hydraulic control system (SCS, USDA).

FIG. 8.30 A small leveller which was made in a small machine shop using steel pipe for the framework (SCS, USDA).

The drag scraper, a two-wheel bottomless scraper was designed to remove surface irregularities too large to be planed and too small to be taken care of economically with a carryall scraper. This machine is manufactured in a number of widths to serve various purposes ranging up to 5.5 m (18 ft). It is well-suited for handling large volumes of earth over a short haul. Through the use of hydraulic controls, the tractor operator can cut, drag, and unload as desired. When the drag scraper is used with operator control, it is generally necessary to follow up with a land plane to get the desired results.

In recent years the drag scraper has become an excellent final smoothing machine through the use of a laser to control the blade level. Fig. 8.29 is an example of this type of equipment. The laser transmitter is mounted on a tripod and projects a beam from a revolving head. The transmitter can be adjusted so that the laser beam is projected on a level or a graded plane as desired. A receiver mounted on the drag scraper automatically operates the hydraulic system on the scraper holding the scraper blade at the desired elevation of the final field elevation. Using the laser-controlled drag scraper the elevation difference of the field surface on a 15-ha (40-acre) field has been 25 to 30 mm.

Many excellent levelers and floats have been made in farm shops and by commercial companies. There are many variations as to the detail of design. Fig. 8.30 is an example of one of these machines. The earth-moving principles are the same as those of the land plane, and the effectiveness is dependent on the length of the frame. Levelers and floats that can be pulled by the average farm tractor are more important in restoring the smoothness of a field after tillage than in removing the irregularities left by heavy earth-moving equipment. Even wooden floats or drags, improve land preparation. The use of automatic-type levelers and floats each year does much to improve field surfaces for irrigation.

8.9 ANNUAL MAINTENANCE

Land shaping often requires a large investment of capital. Annual maintenance is an important part of the farmer's operation to obtain the

FIG. 8.31 A small farmer-owned land plane approximately 9 m (30 ft) in length. The rubber-tired wheels near the center of the plane are lowered to transport the plane from one location to another (SCS, USDA).

proper return on this investment. Special treatment and annual maintenance is needed to bring land back to full production and keep the surface uniform. Tillage, such as plowing, disking or chiseling, after smoothing mixes the disturbed soil, loosens areas that have been compacted and allows soil aeration. In cold climates leaving the soil surface in a roughened condition over winter will allow the frost action to break down clods and improve infiltration. Plowing should be done with a two-way plow. This plow has two sets of moldboards, one set turns the soil to the left and the other to the right. While one set is in the ground, the other is lifted clear. Some farmers prefer to use sweeps or chisels that cut through the soil, raise it slightly and drop it back in the original position. After major tillage, the field may need refinishing with a float or similar equipment to correct minor irregularities.

Fill areas settle and cut areas "fluff up" leaving the surface uneven following the first irrigations and farming operations. For this reason it is best to plant an annual crop the first year after land shaping. When the crop has been removed, low spots can be refilled and high spots removed for a permanent crop such as grass or alfalfa. If these irregularities are slight land planing may be adequate. Many farmers use a small plane such as shown in Fig. 8.31 to maintain the smooth surface of their fields. The laser-controlled drag scraper is an excellent tool for doing this maintenance work.

If land shaping has exposed infertile subsoils, it is a good plan to plant a green manure crop the first year to add organic matter. Heavy applications of barnyard manure, if available, can be used in place of a cover crop or with the cover crop. This should be followed by the application of a proper commercial fertilizer to all cut areas to improve the fertility. Trace minerals, such as iron or zinc are sometimes deficient in cut areas.

8.10 References

1 Chugg, G. E. 1947. Calculations for land gradation. AGRICULTURAL ENGINEERING 28(10):462-463.

2 Faulkner, M. D. 1964. Leveling rice land in water. Louisiana Sta. Univ., Rice Exp. Sta., Crowley, LA.

3 Givan, C. V. 1940. Land grading calculations. AGRICULTURAL ENGINEERING 21(1):11-12.

4 Harris, W. S., J. C. Wait and R. H. Benedict. 1977. Warped-surface method of land grading. Presented at joint meeting of Southwest and Southwest sections of ASAE in Dallas, TX. Feb. 1965.

5 Marr, J. C. 1957. Grading land for surface irrigation. Div. of Agr. Sci., Univ. Calif. Cir. 438, 55 p.

6 Soil Conservation Service. 1961. National Engineering Handbook, Land leveling. Sect. 15, Chp. 12. U.S. Dep. Agr., Washington, DC.

8.11 Additional Source Materials

7 Acton, F. S. 1959. Analysis of straightline data. John Wiley and Sons, Inc. New York. 267 p.

8 Bamesberger, J. G. 1961. Land leveling for irrigation, USDA Leaflet No. 371.

9 Benedict, R., J. Waite, W. Wilkes, and W. Harris. 1964. Land grading and leveling programs for digital computers. Ark. Agr. Exp. Sta. Bul. 691.

10 Butler, E. D. 1961. Land leveling program in the Arkansas Delta. AGRICULTURAL ENGINEERING 42(3):128-131.

11 Erie, L. J. and A. R. Dedrick. 1979. Level-basin irrigation: A method for conserving water and labor. USDA Farmers' Bul. No. 2261, 24 .

12 Heiple, K. (Not dated). Earthmoving—an art and a science. R. G. Le Tourneau, Inc. Peoria, IL. 59.

13 Houston, C. E. 1966. Trends and costs of land grading for irrigation in California. TRANSACTIONS of the ASAE 9(4):565-570.

14 Hung, J. Y. T. 1972. Locating the cut plane for land leveling. ASAE Paper No. 72-228, ASAE, St. Joseph, MI 49085.

15 Paul, H. A., J. E. Quintero and R. E. Harwood. 1969. Calculos simplificados para lo nivelacion de tierras. Instituto Colombian Agropecuario, Boletin Technico No. 1.

16 Phillips, R. L. 1958. Land leveling for irrigation and drainage. AGRICULTURAL ENGINEERING 39(8):463-470.

17 Rapp, E. 1963. Engineering aspects of land leveling. Can. Dep. Agr. Publ. 1145.

18 Rapp, E. 1964. Equipment for irrigated land development. Can. Agr. Eng. 6:22-25.

19 Raju, C. S. 1960. Land grading for irrigation. TRANSACTIONS of the ASAE 3(1):38-41.

20 Saveson, I. L. 1958. Land forming in alluvial areas. Irrig. Eng. and Maint. 8(1):17-38.

21 Shih, Sun-Fu and G. J. Kriz. 1970. Land forming designs by computer methods. ASAE Paper No. 70-230, ASAE, St. Joseph, MI 49085.

22 Shih, Sun-Fu and G. J. Kriz. 1971a. Computer program for land forming design of rectangular fields. N. Carolina Agr. Exp. Sta. Tech. Bul. No. 205.

23 Shih, Sun-Fu and G. J. Kriz. 1971b. Symmetrical residuals method for land forming design. TRANSACTIONS of the ASAE 14(6):1195-1198.

24 Shih, Sun-Fu and G. J. Kriz. 1971c. Computerized land forming design for an irregular shaped field. ASAE Paper No. 71-709, ASAE, St. Joseph, MI 49085.

25 Smerdon, E. T., K. R. Tefertiller, R. E. Kilmer and R. V. Billingsley. 1966. Electronic computers for least-cost land forming calculations. TRANSACTIONS of the ASAE 9(2):190-193.

26 Sowell, R. S., S. F. Shih and G. J. Kriz. 1971. Land forming design by linear programming. ASAE Paper No. 71-708, ASAE, St. Joseph, MI 49085.

27 Turner, J. H. and C. L. Anderson. 1971. Planning for an irrigation system. Am. Assoc. for Vocational Instructional Materials, Athens, GA.

28 Wood, I. D. 1951. Land preparation for irrigation and drainage. AGRICULTURAL ENGINEERING 32(11):597-599.

chapter 9 ▬▬▬▬▬▬▬▬

FARM WATER DELIVERY SYSTEMS

9

9

FARM WATER DELIVERY SYSTEMS

by   J. A. Replogle, USDA-SEA-AR, U.S. Water Con-
      servation Laboratory, Phoenix, AZ; J. L. Merriam
      (retired), California Polytechnic State University,
      San Luis Obispo, CA; L. R. Swarner, U.S. Bureau
      of Reclamation, Denver, CO; J. T. Phelan
      (retired), USDA Soil Conservation Service,
      Washington, DC

9.1 CHAPTER SCOPE AND LIMITATIONS

In Chapter 3, consideration was given to the source of the farm water supply. When the source exists on the farm either from a well, which represents about 42 percent of the water supplies for irrigated farms (according to the Bureau of the Census, 1973), or is directly diverted from a stream, representing 16 percent, the farm operator can control the timing and quantity of the delivery, limited only by the size of the source. The remaining 42 percent of the water is mostly surface waters delivered to the farms through a supply system operated by a water-user association or irrigation district. This loosely organized farmer neighborhood cooperative or farmer-government organization is an entity whose purpose is to facilitate water delivery, application, and removal. These organizations give the farmer access to a more distant, usually more reliable, and larger water supplies by sharing the construction and operating costs with many other users. But for the more desirable water supply there are usually more restrictions and external control of amounts delivered and of flow rates.

This chapter summarizes the various forms of farm water delivery policies and their impacts on farm labor, crops, and on-farm investment needed to achieve effective water management. Delivery policies and methods are classified in terms of distinguishing physical parameters. Physical features of the systems and hardware items, like automatic gates and flow metering methods, that can significantly contribute to their improved operation are described. The delivery system is purposely viewed as a subsystem of the farm production scheme, as part of the total water handling costs of the farm unit, since essentially all costs are ultimately paid by the farmer.

Requirements needed to improve operations of the entire process, rather than a description of that which exists, are stressed. Thus, undesirable methods or procedures are given short shrift, whereas newer concepts that are proving to be superior are emphasized.

9.1.1 Schedules

The delivery policy used by the water supply organization is called a scheduling system or schedule. This scheduling system may have been established by law or selected based on economic and engineering factors that were important, at least at the time of organization. Schedules consist of the delivery frequency to the farm, the delivery flow rate, and the delivery duration. The relations between these three factors control the capital cost and operating expenses of the delivery system. The same factors bear heavily on the effective and economical use of water, labor, energy, and capital investment on the farm. It may even limit the crops that can be considered.

In general, the farm water delivery policy must be compatible with the distribution system on the farm. In some cases this has required on-farm designs to be adapted to an existing delivery system with the net result that the total production system is severely limited. Often, improving the design and operation of on-farm systems necessitates modifying the water delivery systems.

Districts or associations select the schedule, after considering the legal, physical, and economic factors; and the resulting apportionment of the costs of the water supply, distribution, and application systems become a part of the farm production complex. Thus, increased costs in the distribution system that result in more effective water use and reduce on-farm labor, capital, or energy costs, may be easily justified.

9.1.2 Schedule Types

Schedule types may be broadly classified as demand, rotation, or continuous flow. Combinations of two or more of these methods may be used in any system depending on the location of the farm with respect to the distribution system, the seasonal water requirements, or the available water supply. An alternate classification is to broadly group the schedule types into either rigid (predetermined) schedules or flexible (modifiable) schedules. If flow duration is variable, further combinations can be defined.

Rigid schedules may be:

(a) fixed amount-fixed frequency,
(b) fixed amount-variable frequency, and
(c) varied amount-fixed frequency.

Flexible schedules may be:

(a) demand,
(b) frequency demand (24-h duration),
(c) limited-rate demand.

9.1.3 Primary Features of Rigid-Schedule Irrigation Systems

Fixed Amount-Fixed Frequency Schedule. The fixed amount-fixed frequency schedule includes the rotation and the continuous flow methods. For example, it may consist of full flow in the lateral for 24 h every 3 wk to a particular farm or area, or it may consist of a small flow that is delivered continuously.

This system is primarily for the convenience of control by the water agency or the water users' associations. Once schedules are established, they are not changed. The system capacity may be designed around a minimum peak flow rate that will then be used almost continuously. Capital investment and operating costs are minimal. The on-farm problems, however, may be severe. The rigid frequency and amount schedule often supplies excess water in the spring and late summer, contributing to water waste and drainage problems. Extra labor is usually needed, because the stream is generally too small, requiring continual shifting of flows. The frequency seldom matches the soil or the crop needs. Effective use of the continuous flow method may

FIG. 9.1 Reservoirs in citrus in Coachella, CA, 1971.

require investment in expensive sprinkler or trickle irrigation systems or on-farm reservoirs as the only practical way to effectively use the water, Fig. 9.1.

This method should not be used for conventional surface irrigation, except under extreme conditions when other delivery methods are impractical.

Fixed Amount-Variable Frequency Schedule. The fixed amount-variable frequency schedule is a type of rotation method that again results in a minimum peak-capacity system, but much management is needed to change the frequency to accommodate the seasonal requirements of most crops and soils in the serviced areas. For example, the full flow in the lateral may be delivered for 24 h every 3 wk in the spring and for 24 h every week during the peak summer months. On the farm, the imposed frequency of delivery may or may not match variable soil and crop requirements, resulting in water wastage on some fields and drought damage on others. Since the flow volume is fixed, off-season requirements of less water mean longer time periods between return deliveries, so between deliveries canals may be empty. This schedule may function quite well for perennial crops needing the same depth of water applied at each irrigation. The change in frequency attempts to match the evapotranspiration needs. Also, the periods when the canals are empty may be beneficially used for maintenance operations.

Varied Amount-Fixed Frequency Schedule. The varied amount-fixed frequency schedule is a third rotation type that is also designed to provide a minimum peak flow delivered almost continuously. To deliver off-peak seasonal requirements with the fixed frequency method, district operation management reduces the delivery flow rates. Thus, the canals are seldom empty, but are operated below capacity for most of the season.

On the farm this method may or may not suit a particular soil-crop combination. The rotation frequency is chosen to meet the summer peak evapotranspiration rate and will result in a higher number of seasonal applications than the preceding method. This frequency may be greater than necessary on fine-textured soils, but may result in drought damage to crops on coarse-textured areas. Because the amount is varied, this schedule can be fairly efficient for preennial crops.

Other Features of Rigid Schedules As previously mentioned, fixed amount-fixed frequency delivery schedule is generally the least desirable from the user's viewpoint. The other two scheduling systems can, theoretically, be optimized for a specific crop on specific soil and field conditions. Such conditions are unlikely to exist for much of the service area. On a project or district basis rigid schedules often result in low project irrigation efficiencies, create drainage problems, leach soil nutrients, and waste labor on the farm. They frequently become expensive methods when supply, distribution, application, and limitations on crops are considered in the whole production system under current prices, but, as mentioned previously, may have served well under some previous economic situation.

All of these systems are frequently modified by using reservoirs either on farm or at the district level. These modifications may enable moving the systems into the flexible-group classification.

9.1.4 Primary Features of the Flexible-Schedule Irrigation Systems

Demand Schedule. The demand schedule in which the irrigator may have water as desired—flexible in frequency, rate, and duration—is often too expensive, especially if large rates or complete time flexibility are offered. The capacity of such a system would be too large for reasonable capitalization and operating costs, since it must be large enough to meet the combined probable demand of all the users at any one time.

Frequency Demand Schedule. The frequency demand schedule is one of the modified demand methods that is a compromise between the convenience of the farmer and that of the water agency. The frequency of the deliveries can be varied almost as needed, usually within 1 day of the requested time. Flow rate is as requested with enough water to satisfy almost all needs. The duration of the fixed delivery is usually in 24-h increments. The fixed duration may require an on-farm reservoir to properly manage the water for attaining reasonable on-farm water application and labor efficiencies. (See Fig. 9.1—Reservoirs in citrus fields in Coachella Valley, CA). This schedule requires larger lateral capacities near the far end of the system than needed for a rigid schedule irrigation system, because several users on a lateral may want water simultaneously. System reservoirs near the far end of laterals and submains allow smaller increases in system capacities. At the upper end of the system (near the supply) the design capacity would be about the same for all scheduling methods, since downstream use tends to average out for large numbers of users whether the schedule is rigid or flexible.

Under this schedule the maximum farm delivery rate is as large as usually required by the irrigators to effectively use the water and the labor. This increases the cost of the distribution system, largely due to the increased capacity towards the ends of the laterals. However, increasing a pipe diameter from 300- to 400-mm more than doubles capacity, without a comparable increase in cost.

Limited Rate Demand Schedule. Limited rate demand scheduling is another modified demand system. It allows flexible rate and duration deliveries as well as frequency but with a limit on the maximum flow rate. In some irrigation districts, deliveries are usually made within 1 day after being requested. In other districts, requests are made 3 or 4 days in advance of the desired day, and delivery is usually made within 1 day of the requested time. The maximum rate is set fairly high to economically use labor on the farm, but it is limited by the economics of the total farm and distribution systems. With this schedule the rate may be varied by the irrigator during irrigation, and the duration is controlled by him within the requested 24-h period.

Obviously, this schedule requires operational spillage, or storage, or automation of the supply since flows are varied by the irrigator. Regulating reservoirs established at several locations along the mains or at heads of laterals are satisfactory. They can be filled at night when on-farm demand drops off. This permits mains and submains to operate near peak capacity most of the time. When submains and laterals are automated by use of level-top float controlled canals, or closed or semi-closed pipe lines, operations become very simple. (See Section 9.3). Main canals need to be automated so that gates at all checks can be simultaneously set to pass any flow demanded any place in the system. Peak flow demands are met from the strategically placed reservoirs so main canal flows are relatively stable. Nonautomated sloping main canals that have reservoirs in which what was formerly operational spillage can be accumulated can also function satisfactorily. To do so they must have adequate flow through them to satisfy the day-to-day peak demand in the upper reach, and an adequate automated service area below the reservoir to utilize the operational variations.

On-line system reservoirs may be located slightly above or below the canal and the variations of flows pumped into or out of them. The lifts and quantities are often small. The grade may be reduced on a steep canal to provide the elevation differences needed for the reservoir.

Distributions systems without adequate reservoir capacity and incomplete automation may require the operating agency to impose some limitation as to the magnitude of the farmer-controlled changes on this schedule.

Further Comparisons of the Flexible Schedules. The limited rate demand schedule is highly practical for the water user. It enables the user to irrigate each crop, usually when needed, and to use a stream size that is economical and efficient for the particular situation. Thus, differences in soils and crop requirements can normally be accommodated. Limitations on crop types that can be successfully grown are eliminated. A ditch rider, who is an agency employee, may need to be on call 24 h a day to reduce operational spillage as the farmer changes his flow rate or finished irrigating. This need does not occur when the system has an automated canal or closed or semiclosed pipelines for delivery and suitable reservoirs.

Nearly all large projects in the northwestern United States operate on a type of limited-rate demand schedule. In some cases it is possible to deliver water on a straight demand schedule during part of the season, and change to one of the modified demand systems or even to a rigid schedule during peak crop growth periods. These various projects adopt programs that differ in detail but strive to approximate demand scheduling as closely as practical. The demand type schedules are practical when the supply system has a variable source, adequate capacity, automated canals, and/or closed or semiclosed pipelines. The turnout is kept locked except for the scheduled days. The volume of water is measured by a totalizing meter, but the flow rate and duration are controlled by the irrigator (Merriam, 1973, 1977). With a fully adequate capacity, it is not necessary to schedule the delivery day, thus greatly reducing operation and maintenance costs.

9.2 FLOW CONTROL AND MEASUREMENT TO FARMS

Good irrigation water management requires knowledge of the quantity of water available. Water at rest, stored water, is measured in units of volume—cubic meters (m^3). Water in motion is measured in units of flow rate, that is, volume per unit time—cubic meters per second (m^3/s) or liters per second (L/s).

There are many methods for measuring water deliveries to farms that are commonly used by water agencies. In most cases some form of measurement is used, the quality or accuracy of the measurement notwithstanding. Many of the same measuring methods and devices are also applicable within the farm units.

Generally, the flow meters used for farm deliveries can be classified into two major groups—those for measuring the total quantity of flow (totalizing meters) and those for measuring flow rate. With appropriate auxillary equipment, meters from either group can be converted to the other group.

9.2.1 Totalizing Meters

The most common totalizing meters used are the propeller meters which measure flow in pipelines, and the Dethridge meter which measures flow in canals.

Propeller Meters. Propeller meters are often used to measure discharges from wells and for pipeline deliveries to farms. In such service, they are usually fitted to show both totalized flow quantity and flow rate. The totalized flow is read on a digital accumulator, and the flow rate is read from a needle on a dial. The readout combination resembles an odometer-speedometer on an automobile.

These meters use a multibladed propeller with from two to six blades made of metal, plastic, or rubber, which are geared to the totalizer device to give any desired volumetric and rate units. They are designed and calibrated for operation in closed conduits, usually round pipes, that are flowing full. The propeller diameter usually ranges from 50 to 80 percent of the pipe diameter. The effective low-to-high flow range for an individual meter is about 1:10. Ordinarily, the meters will operate within flow velocities ranging from 0.15 to 5.0 m/s. Meters are commercially available for pipe sizes ranging from 0.05 to 1.83 m in dia. The principle involved is a simple counting of the revolutions of the propeller as the water passes. Changes in the pattern of flow approching the meter or changes in the bearing friction or the resistance of the propeller surfaces affect the accuracy of the meter registration. Routine accuracy of a well-maintained propeller meter is usually considered to be ± 5 percent, unless calibrated in place.

In canals, a suitable length of pipe is usually provided for a portable propeller mechanism that is temporarily installed by an operator. Often a submerged culvert in the canal system serves this purpose. Because of debris problems, this mechanism is continuously attended for the few minutes during the sample measurement and then removed. In this application, the meter is used to measure flow rate which is then assumed to be constant throughout the delivery to produce totalized flow. However, it may be left in place for the entire period of a delivery and actually totalize the flow as is needed for the limited-rate-demand schedule.

Dethridge Meters. Dethridge meters were designed by J. S. Dethridge of Australia in 1910 and have since been used extensively on that continent (Bos, 1976; Replogle, 1970). It is basically an undershot waterwheel turned by the canal water passing through its emplacement—a short concrete section specially shaped to provide only minimum practical clearance of the lower half of the wheel. The Dethridge meter registers totalized volume of delivered water. It is simple and sturdy in construction and operates with a small head loss. It has not been very popular in the United States, probably because of its bulkiness.

9.2.2 Rate Meters

Rate meters most commonly used in canal system deliveries to farms are orifices, weirs, and flumes. Orifice, venturi, sonic, electro-magnetic, and vortex meters are available for pipeline deliveries.

Orifices. Orifices are well defined openings in a plate or bulkhead, the tops of which are well below the upstream water level. If the orifice discharges into the air, it is a free-discharging orifice. If it discharges under water, it is a submerged orifice. Free discharging orifices require considerable head and are usually limited to special locations where head loss is not critical. Submerged orifices require less head and are more commonly used. Orifice shapes vary widely. Circular, sharp-edged orifices are easily machined. Rectangular, variable-area orifices can be formed by partly opening a rectangular gate or a radial Tainter gate. Segmental (part circle) orifices are used in pump discharge pipes because they will pass both air and sand if properly oriented.

In practice, circular, sharp-edged orifices are fully contracted so that the bed and sides of the approach channel and the upstream water surface are remote enough to avoid influencing the discharging jet. Usually, orifices formed by partly opening a gate are suppressed from one or more sides and may require special rating procedures to be sufficiently accurate.

A general disadvantage of orifices is debris clogging that prevents accurate measurements. It is frequently difficult to tell if partial clogging has occurred, especially in sediment-laden water. To prevent canal overtopping, the top of the orifice wall should only extend to the maximum expected upstream water level so that it can act as an emergency weir in case of orifice plugging.

The head-discharge relation for circular, sharp-edged, submerged orifices is

$$Q = C_d A \sqrt{2g(h_1 - h_2)} \dots \dots \dots \dots \dots \dots \dots \dots \dots \dots \quad [9.1]$$

where Q = discharge rate, m³/s, or 10³ L/s; C_d = discharge coefficient, ranging between 0.57 and 0.61; A = area of the orifice opening, m²; g = gravitational constant, 9.8 m/s²; and (h_1 - h_2) = head differential across the orifice, m.

For freely discharging orifices, h_2 = zero and h_1 is measured from the center of the orifice. Orifices should be installed and maintained so that the velocity of approach is negligible. Well-maintained circular orifices can produce accuracies within about ± 1 percent if properly machined and installed.

If the discharge is to be regulated, it is frequently necessary to suppress both the sides and bottom like the rectangular orifice formed by the opening below a sluice gate or a radial gate. The discharge relations for these are less well defined than for the sharp-edged, unsuppressed orifices, but equations for discharge and tables of discharge coefficients are given in several sources (e.g., Bos, 1976). The accuracy of these installations can approach ± 5 percent.

A variation is the Neyrpic module gate which delivers water at a constant flow rate, even though the head may vary a little (Bos, 1976). These modules can also act as a gate.

A related device commonly used for both regulating and measuring flow to farms is the metergate. Basically, it is a submerged orifice or calibrated valve gate operated with a screw lift. The valve gate is at the upstream end of a pipe section that discharges water through the canal bank. Usually, the metergate is placed at right angles to the canal flow from which it diverts water. High velocity flows in the main canal can cause eddies which affect the stability of the discharge. In such cases, an approach section may be needed that is about 5 pipe dia long. Locations of the stilling well intakes and installation precautions are described by Bos (1976). Accuracy of the metergate is about ± 3 to 6 percent.

The constant-head-orifice (CHO) is a combination regulating and measuring structure used on turnouts to farms. It is usually two sluice-type gates in tandem, installed about 1 m apart. The CHO was developed by the United States Bureau of Reclamation and is so named because it is operated by setting and maintaining a constant head differential (h_1 - h_2), across the orifice formed by the upstream gate. The differential head is usually 0.06 m (0.20 ft). To set a required flow, the gate orifice opening is determined and set on the upstream gate. The downstream gate is then adjusted to cause the constant head of 0.06 m. The discharge will then be at the required value to an accuracy of about ± 7 percent. Because of the small head differences, extreme care should be taken in reading heads.

Sharp-Crested Weirs. Sharp-crested weirs are overflow structures made from thin plates whose thickness at the overall edge is less than 2 mm. The weir opening is usually rectangular, triangular, or trapezoidal, although other shapes have been used. These weirs are accurate, i.e., will repeat their calibration to within ± 1 or 2 percent, especially the rectangular and triangular types. The trapezoidal versions are considered to be reliable to ± 5 percent. The most popular trapezoidal style is the Cipoletti weir with 0.25:1 sideslopes (horizontal:vertical).

To obtain high accuracy, the nappe must be ventilated on the downstream side of the weir plate so that the pressure there will be atmospheric. If the weir is not ventilated, the resulting vacuum will cause over-discharging. The downstream water level should be kept at least 5 cm below the crest of the weir. This means that the required head loss always exceed by 5 cm the upstream head over the weir crest. This required high head loss is one of the major limitations to using sharp-crested weirs for irrigation flows, especially in relatively flat irrigated areas.

The discharge through rectangular sharp-crested weirs can be approximated with a power function of the form

$$Q = KLH^{3/2} \qquad\qquad\qquad\qquad\qquad\qquad\qquad [9.2]$$

where K incorporates the gravitational constant, the discharge coefficient, and other effects; L is the weir crest length; and H is the total energy head on the weir (the flow depth plus the approach velocity energy). Methods for determining K are given by King (1963) and Brater (1976). Other expressions dealing more directly with other shapes of opening and the effects of weir suppression are given by Bos (1976).

Critical-Flow Flumes. Critical-flow flumes are rate meters particularly suited for measuring irrigation water in open channels. They consist of a specified construction built into the canal floor and/or sidewalls that raises the upstream flow level above that which would have existed without the constriction. This rise must create enough fall to cause critical flow in the contracted section, or throat, of the flume.

Of particular interest because of the wide applications are the Parshall flumes (Parshall, 1953), the H-flume (Bos, 1976; Kulin and Compton, 1975), the cutthroat flume (Skogerboe et al., 1972), and special trapezoidal and triangular-shaped flumes (Robinson, 1968). These styles, however, permit curvilinear flow to occur in the control section, making them hydraulically complex with their calibrations depending on laboratory ratings. Additionally, all but the trapezoidal and triangluar flumes require depth measurement in a converging flow section, which is not favored by modern hydraulics practice. Because of the complex-hydraulic behavior and method of head sensing, these flumes must be carefully constructed throughout their length, and can become relatively expensive devices.

Another general type, long-throated flumes, including broad-crested weirs with rounded upstream edge, have been used for many years in Europe (Ackers and Harrison, 1963; Bos, 1976). These flumes are called long-throated because the contraction length is about twice the maximum flow depth, causing the flow to be nearly parallel as it passes through the flume throat. The cross-sectional shape can be almost any regular geometric pattern—rectangular, triangular, trapezoidal, circular, parabolic, or complex. The flow depth is sensed in an upstream channel section, again, in an area of parallel flow. The calibrations were based on calculated stage-discharge relations, which used laboratory-determined discharge coefficients and would usually meet the desired ± 5 percent requirement demanded of the flumes. More recent calibration procedures, involving mathematical modeling, use boundary layer development caused by frictional effects and velocity distribution, to improve the reliability of these calibrations to within ± 2 percent (Replogle, 1975; 1977). Simplified field installation techniques have greatly decreased the skill requirements of field construction yet permit accurate determination of individual discharge rating, restoring the original intended accutacy in spite of most field construction anomalies (Replogle, 1977).

An important aspect of flumes is the high submergence limits that can be tolerated as compared with sharp-crested weirs. (Submergence limit is defined here as that point of submergence caused by downstream water level at which the real discharge deviates by 1 percent from that indicated by the observed flow stage). This translates into the minimum required head loss for the measuring device in the canal system. Since sharp-crested weirs can withstand no submergence, the minimum required head loss is equal to the flow depth (actually energy head) over the weir plus about 5 cm, to assure free flow. Flumes, on the other hand, have submergence limits that range from about 50 percent to as high as 95 percent. Thus, the absolute head loss can be relatively low.

The submergence limit of critical-flow devices, like flumes and weirs, is a function of the streamline curvature at the control section and the losses in the downstream expansion. Broad-crested weirs and long-throated flumes, which have straight and parallel streamlines at the control section, can approach the high submergence limit of 95 percent, if part of the kinetic energy is recovered by a proper exit expansion (Bos, 1976). The submergence limits for truncated flumes, i.e., flumes that have no expansion section, are about 70 to 85 percent depending on whether the throat area approximates a rectangular or a triangular shape, respectively, provided that it is long throated.

These submergence limits, even for truncated versions, are higher than those routinely listed for H-flumes, cutthroat flumes, and Parshall flumes, which are 25, 65, and 60 percent, respectively.

The older flumes, however, are quite usable, and many are in continual service and most certainly need not be replaced if they were originally properly installed and subsequently maintained. The preceding discussion attempts to point out that accurate, yet, much less expensive designs are now available for new installations, with the added capability of being tailored to most existing canal systems so that the canals can, in effect, be retrofitted with a low head-loss measuring device that can operate to as high as 95 percent submergence with a single depth reading.

When installing any of the critical-flow flumes, care must be taken to assure adequate free flow discharge at both high and low ranges. Checking at high flow only is risky. For example, a rectangular flume placed in a trapezoidal channel, where the channel controls the flow level and not some type of check gate, may be free flowing at high stages but may be submerged at low flows (Replogle, 1968). Usually, these shape combinations need to be checked for submergence at low flows. On the other hand, a trapezoidal or triangular-shaped flume placed into a rectangular channel will require checking at high flows and will tend to become less submerged at lower flows. Rectangular-shaped flumes, including the Parshall and the cutthroat, are sometimes installed with a vertical offset to avoid submergence at low flows (Replogle, 1968; Humphreys and Bondurant, 1977), however at the expense of higher head loss at maximum flow.

A style of trapezoidal, broad-crested weir particularly suited for slip-formed canals can inexpensively and accuratley measure irrigation flows. The weirs are really long-throated flumes since they have an upstream approach ramp. They are usually truncated at the downstream end of the weir sill although an exit ramp could be built to improve the submergence limit

FIG. 9.2 Type FBIM flume is near the submergence limit due to the added friction of downstream vegetation, but is functioning well.

from about 85 to 95 percent. Like other long-throated flumes, the throat length (sill length in the direction of flow) is about twice the expected maximum flow head over the sill, and the converging section ramp is constructed with a 3:1 slope. The exit ramp, if used, would be on a 6:1 slope.

Construction tolerances are liberal, with the major emphasis placed on

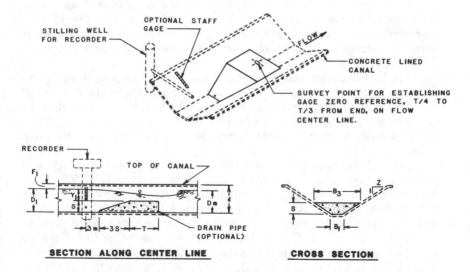

FIG. 9.3 Profile and cross-section of a long-throated flume, broad-crested weir style.

TABLE 9.1. BROAD-CRESTED WEIR DESIGN AND SELECTION

Canal Data				Flow Data for Canal Without Weir in Place			
1. Constructed depth—d	=	m		4. Maximum discharge rate - Q_m=			L/s
2. Bottom width — B_1	=	m		5. Flow depth for Q_m — d_m=			m
3. Sideslopes — Z	=	m		6. Freeboard — f_b=			m

Weir Selection Using Tables				Trial 1	Trial 2	Trial 3
7. Trial flume identification:						
8. Bottom width, B_1 (matches line 2)		B_1 =				m
9. Sill width, B_3		B_3 =				m
10. Sideslopes, Z (matches line 3)		Z =				-
11. Critical section (sill) length, T		T =				m
12. Trial sill height, S		S =				m

Submergence — Freeflow Check, Weir in Place			Trial 1	Trial 2	Trial 3
13. Sill-referenced flow depth for Q_m, (tables)	Y_1 =				m
14. Upstream water depth (line 12 plus line 13)	D_1 =				m
15. Actual increase in water depth: (line 14 minus line 5)	ΔY_1 =				m
16. Required increase in water depth: (15 percent x line 13)	ΔY_2 =				m
17. Submergence check: (line 15 minus line 16)	ΔY_3 =				m

(If negative a higher sill must be used, and repeat lines 7-17 with a new trial).
(If positive, this flume is okay, but a lower sill might also be used).

18. Actual freeboard: (line 1 minus line 14)	f_1 =					m
19. Required freeboard: (20 percent x line 13)	f_2 =					m
20. Freeboard check: (line 18 minus line 19)	f_3 =					m

(If negative, try a lower sill and repeat lines 7-20).
(If positive, this flume is okay, but a higher sill might also be used).

21. Flume selection: If both checks, (lines 17 and 20), are okay, use that trial.
(If one or the other are not met, repeat lines 7-20 for a new trial.)
(If none of the standard flumes works, see text.)

22. Flume selected _____ Location _____

Comments: _____ How were Q_m and d_m determined? _____

making the sill top level, and on careful zeroing of the reference gage. Analysis shows that the error in discharge is almost proportional to the error in sill width, but may be about twice the error that may exist in locating the gage zero (Replogle, 1977; Clemmens and Replogle, 1979).

Sediment movement along the canal floor is interrupted somewhat by these broad-crested weirs. However, the effect is usually minimal. A few meters upstream, the flow is similar to pre-flume installation flows. At the ramp the velocity increases, which helps to carry bedload over the sill. Furthermore, analysis shows that the channel flow area must be obstructed by about 10 percent in the approach section to cause about 1 percent error in final readout.

The high accuracy and low cost of these styles of long-throated devices make them highly desirable for measuring irrigation flows in slip-formed canals. The basic shape conforms to the canal which is used as part of the flume structure. Fig. 9.2 shows a broad-crested weir style of flume satisfactorily operating at near the submergence limit, with a head loss in the canal of less than 4 cm.

Fig. 9.3 shows details for installing the broad-crested weir style and defines the symbols used in Table 9.1 and the headings of Table 9.2. Table 9.1 is a flume selection design sheet that can be used to determine a desired sill size (flume or broad-crested weir size) that will operate in a given canal and flow situation. Precomputed discharge tables for nine selected sizes are provided in Table 9.2. Gages, direct reading in discharge units, can be calculated for marking in a machine shop, by multiplying the depth values in Table 9.2 by the appropriate side-slope (for z = 1, multiply by 1.414; for Z = 1.25, multiply by 1.600; etc.) Field installation techniques are described

TABLE 9.2. FLUME CALIBRATIONS—WATER DEPTH AT GAGE LOCATION REFERENCED TO SILL CREST (cm)

	Canal shape A $B_1=30.5$ $Z=1{:}1$				Canal shape B $B_1=61.0$ $Z=1{:}1$				Canal shape C $B_1=61.0$ $Z=1.25{:}1$		
Flow rate L/s	FA1M $B_3=76.2$ T=61.0 S=22.9	FA2M $B_3=91.4$ T=45.7 S=30.5	FA3M $B_3=106.6$ T=30.5 S=38.1	Flow rate L/s	FB1M $B_3=121.9$ T=91.4 S=30.5	FB2M $B_3=137.2$ T=76.2 S=38.1	FB3M $B_3=152.4$ T=61.0 S=45.7	Flow rate L/s	FC1M $B_3=156.2$ T=106.6 S=38.1	FC2M $B_3=175.3$ T=91.4 S=45.7	FC3M $B_3=194.7$ T=76.2 S=53.3
10	3.89	3.50	3.17	25	5.32	4.94	4.61	50	7.10	6.61	6.17
15	5.06	4.53	4.11	30	5.97	5.53	5.17	60	7.96	7.41	6.93
20	6.06	5.42	4.93	35	6.58	6.11	5.70	70	8.76	8.16	7.64
25	6.93	6.24	5.68	40	7.15	6.65	6.20	80	9.52	8.88	8.31
30	7.75	6.99	6.38	45	7.70	7.16	6.68	90	10.23	9.56	8.95
35	8.52	7.70	7.03	50	8.22	7.65	7.15	100	10.92	10.21	9.57
40	9.25	8.36	7.64	60	9.21	8.59	8.02	150	14.0	13.1	12.34
45	9.94	8.99	8.23	70	10.14	9.46	8.85	200	16.7	15.7	14.8
50	10.59	9.59	8.79	80	11.01	10.28	9.63	250	19.0	17.9	16.9
60	11.81	10.73	9.85	90	11.83	11.06	10.37	300	21.2	20.0	18.9
70	12.94	11.78	10.83	100	12.62	11.81	11.08	350	23.2	22.0	20.8
80	14.0	12.77	11.76	125	14.5	13.6	12.75	400	25.1	23.8	22.5
90	15.0	13.7	12.65	150	16.1	15.2	14.3	450	26.9	25.5	24.2
100	15.9	14.6	13.5	175	17.7	16.7	15.7	500	28.5	27.1	25.8
125	18.1	16.7	15.4	200	19.2	18.1	17.0	550	30.1	28.6	27.3
150	20.1	18.6	17.2	250	21.8	20.6	19.5	600	31.7	30.1	28.7
175	21.9	20.3	18.9	300	24.3	23.0	21.8	700	34.5	32.9	30.1
200	23.6	21.9	20.4	350	26.5	25.2	23.9	800	37.2	35.5	31.4
225	25.2	23.4	21.9	400	28.6	27.2	25.9	900	39.7	38.0	33.9
250	26.7	24.9	23.3	500	30.6	29.2	27.7	1000	42.1	40.3	36.3
275	28.1	26.2	24.6	600	32.5	31.0	29.5	1200	46.5	44.6	38.6
300	29.4	27.5	25.8	700	36.0	34.4	32.9	1400	50.5	48.5	42.8
325	30.7	28.8	27.0	800	39.2	37.5	35.9	1600	54.2	52.2	46.6
350	32.0	30.0	28.2	900	42.1	40.4	38.7	1800	57.7	55.6	50.2
375	33.2	31.2	29.3	1000	44.9	43.1	41.4	2000	60.9	58.8	53.6
400	34.3	32.3	30.4	1100	47.5	45.7	43.9	2200	64.1	61.9	56.7
450	36.5	34.4	32.5	1200	50.0	48.1	46.3	2400	67.0	64.8	59.7
500	38.6	36.4	34.4		52.4	50.5	48.6	2600	69.8	67.5	65.4

Dashed lines indicate limit of most accurate range of flow.

in Replogle (1977) and in Clemmens and Replogle (1980). In summary, the most important dimension on these weirs is the gage zero setting. In practice, it is best to determine if the user normally measures a particular flow rate. If this is the case, then mount the wall gage, using surveying equipment, at the elevation for that flow determined from Table 9.2. If the sidewall then happens to be slightly out of slope alignment, the most usually used flow will be correct, and the errors will be relegated to the lesser used regions of the gage (Clemmens and Replogle, 1979).

Calibrations for these flumes are usually not true power functions of the form

$$Q = CH^n \qquad [9.3]$$

which would plot as a straight line on log-log paper, and therefore are usually presented in table form (like Table 9.2) or as polynomial approximations to the table. Another useful approximation for the flume functions can sometimes be formed by adding a constant, k, to the head reading. The resulting equation is then of the form

$$Q = C(H + k)^n \qquad [9.4]$$

The true calibrations for long-throated flumes in table form can be calculated to within $\pm$ 2 percent by the mathematical modeling procedure presented by Replogle (1975), and the power functions and polynomial functions are then designed to reproduce the tables in a more continuous format rather than at discrete points. Direct equations involving the various geometric variables and friction assessments are not practical for presentation here. Even in the forms used by the mathematical modeling, numerous iterations are involved, making computer techniques the practical method for solution.

Other Meters. Venturi meters produce less pipeline headloss than orifice meters but are more expensive to install. Sonic meters, which are based on detecting changes in the sound transmission-path length due to the flowing fluid, and electro-magnetic meters, which are also based on fluid motion acting on transmitted signals, all require expensive monitoring equipment and can usually be afforded only on main lines in a delivery system and not at each farm turnout. The same is generally true of vortex shedding meters and thermal meters; therefore, these devices are only mentioned in passing. Even the venturi meter is usually considered a main-line device since the straight pipe requirements make it impractical for routine farm turnout use.

9.3 DELIVERY SYSTEMS SUITED TO MECHANIZED AUTOMATED IRRIGATED SYSTEMS—THE APPROACH TO IDEAL SYSTEMS

With the increased appreciation of the value of higher on-farm irrigation efficiency to save water, labor, and energy, there is an increased tendency for mechanizing and automating supply, distribution, and application systems. To obtain these savings, the whole system must be able to deliver water in a way that is flexible in frequency, rate, and duration. For best operation, the flow must be controlled at the point of application by the automatic controls, or mechanized so that flow is easily controlled by the irrigator when he is making on-the-spot decisions. For purposes herein, systems that are self-sequencing once set, or operate with feedback controls without the presence of the irrigator, will be called "automatic." Highly mechanized systems that require monitoring and action by a irrigator-operator will be called "semiautomatic," since certain portions of the irrigation sequence may proceed without attention from the irrigator.

The need to control the rate and duration at the point of application is essential to obtain high irrigation efficiency. For surface methods, large stream sizes reduce the application time and labor requirements. For all irrigation methods, it is essential that the water be turned off when the proper depth has been applied, since additional water is wasted.

The automated application systems have in-field controls responsive to flow volumes, head, duration, soil moisture, etc., to apply the desired irrigation; therefore, the delivery system must also automatically make the desired flow conditions available.

Such automated application systems required a fair amount of supervision and maintenance. They require a large capital investment but have low nonsupervisory labor requirements. Their flow capacity requirements are moderately large. The systems may receive nearly continuous use since odd-hour and nighttime operations are feasible.

The semiautomated systems require the irrigator to be in the field to operate the controls. He adjusts the flow rates, head, durations, etc., as he feels best, or, in a fashion that can be preplanned as is done for the automated systems. However, in addition to doing the irrigating, he can make on-the-spot decisions, repairs, etc. The in-field application equipment for semiautomation is the simple, standard type of controls requiring little maintenance or supervision. Such a system requires larger flow capacity since nighttime operation should be avoided, and the stream size should be as large as the irrigator can handle with good manual equipment and design.

The semiautomated systems, for example, 2- to 5-ha basins supplied through manually operated lift gates, which is practical on low intake-rate soils, require moderate capital investment and have low supervisory and moderatley low nonsupervisory labor requirements (Dedrick et al., 1978). Because of their large capacity and low per hectare labor requirement, they justify a well-trained and paid irrigator. One man can irrigate 150 to 200 ha.

Highly automated microprocessor controlled systems, like the lift gate method with air cylinder actuators (Dedrick et al., 1978), require more capital investment and seem most easily justified economically when replacing systems where very frequent irrigations were necessary, using large quantities of labor, and small flow rates. Mechanization without full automation is more readily justified with more normal frequencies and larger flow rates. A low-cost but automatic system is widely used in New Zealand that consists mainly of drop-gage structures tripped by manually preset timers (Stoker, 1978). Under favorable topographical and soil conditions, as much as 100 ha per manhour can be operated.

The on-farm water delivery system for both the automated and semiautomated systems usually consists of closed or semiclosed pipelines (Merriam, 1973) or a level top canal with automatic controls (Merriam, 1977). They should be supplied from a flexible source like a reservoir to operate effectively.

Closed pipeline systems are easy to operate but line pressure changes as the water level in the reservoir fluctuates and as flow rates are changed. The latter causes head variations due to increased or decreased friction loss. These variations require changes in the field settings for high efficiencies. Since the maximum pressure will be the static head, outlet controls must be capable of regulating flows at the higher pressures as well as the lower pressure at full pipeline capacity. When there are outlets to several laterals which can be used simultaneously, the pipe sizes may be reduced in the downstream direction.

Semiclosed pipeline systems have the pressure broken into small (6.0 m or less) steps by downstream water level float-controlled valves. Other water level control systems exist but are more sophisticated and expensive so they are seldom justified for on-farm distribution systems. The semiclosed systems may be used to automatically provide variable flow, based on downstream rate requirements with small (0.1 to 0.3 m) head variations at turnouts. The float valves eliminate water hammer. The head loss at each float stand is about 1-1/2 velocity heads for full flow.

Canal systems for on-farm deliveries for automatic or semiautomatic applications can be level-top ditches. Such ditches permit a no-flow condition to exist as do the pipelines. The no-flow or variable flow condition is controlled by a float-actuated gate maintaining a constant downstream water level, or by more sophisticated devices better adapted to larger canals. These systems are actuated by variations in the water level downstream created by the variable flow needs in the field. The bottom of the ditch may be parallel to the level top or have some slope. In either case, the water surface level decreases at a turnout, creating a negative wave at critical velocity back to the gate which then supplies the desired flow rate. The initial withdrawal pending the gate operation is taken from canal storage as the water level drops. The canal's cross section usually corresponds to a design velocity of about 0.3 m/s. The level reaches can be a kilometer long for typical on-farm conditions, and longer on large canals. They must be broken into steps on steeper gradients. Mechanical gates, like the Neyrpic AVIS gates, which were developed for this situation, can handle canal drops of about 1.0 m. The AVIO gates for reservoir outlets can control heads up to about 3 m. For greater heads, it may be used with a float valve system like the Harris float valve described by Merriam (1973).

Canals controlled by the Neyrpic gates have small variations in flow depth and thus require only a small freeboard. When reaches are less than about 1 km long, the constructed ditch cross section is about the same as for a sloping canal. Because of the reduced freeboard over normal operations, canal costs are only a little greater (except for lined canals, which can be steeper and handle higher velocities and thus be relatively small). The additional cost for gates makes possible the automatic operation of the delivery systems. Flow measurements need to be made by totalizing meters since flow rates are expected to vary.

The emerging field of microprocessors shows promise of economical applications to canal flow controls and can be expected to improve canal operational capabilities. Labor, water, and energy savings are possible when additional capital is invested to provide automation or semiautomation of the application system and to modify the delivery system to provide for the needed flexibility in frequency, rate, and duration.

9.4 THE USE OF RESERVOIRS IN FARM WATER DELIVERY SYSTEMS

Under many circumstances, small reservoirs are essential elements in a complete farm water delivery system. They may be built and owned by the using farm or may be designed and built by the irrigation district for canal regulation. They may be a practical means for upgrading the scheduling system that can be offered by a district. For example, to upgrade a fixed-amount fixed-frequency schedule system, which would characteristically have laterals that were too small to directly upgrade to one of the demand schedule systems, strategically placed, off-canal reservoirs can frequently be constructed at less cost than enlarging and expanding existing canals and right-of-way. This would be especially true if the existing canal were already lined, and relining costs were to be considered for a larger canal section.

Basically, reservoirs are management facilities whereby a variable or steady flow can be regulated into a different variable or steady flow that is more convenient or efficient in terms of water, labor, energy, and crop production. They may be used in connection with pump or gravity systems and with all irrigation methods, and are often combined with other facilities like pipe lines, wells, pump-back systems, and canal delivery systems.

The most common uses of farm irrigation reservoirs are to provide:
  (a) storage of runoff from rainfall for irrigation during dry periods,
  (b) long term or temporary storage of water that may be available from surface or subsurface sources during periods of low demand or when cultural or harvest operations may prevent irrigation,
  (c) "overnight" storage of the flow from a well or system delivery for use during the day,
  (d) a "re-regulating" capacity needed to adjust flows of an undesirable

size or to match the flow requirements of other elements of the irrigation system,

(e) storage for irrigation tailwater, and

(f) the control needed to maintain a desired surface or subsurface water elevation in an adjacent area.

Usually farm reservoirs are designed to serve several of the above purposes as required by the site conditions, the unique needs of the farm system, and the management goals of the farmer. In some instances, extraneous purposes such as that of providing a source of water for frost control, fire protection, domestic use, spraying operations, recreation, etc., may be included.

In many states, laws and regulations may control, limit, or even require the installation of farm reservoirs. Water rights may be affected or public safety endangered by their construction. All applicable state statutes and regulations must be met, and often perimts are required before construction is authorized.

9.4.1 Use and Management

Classification of reservoirs may be done by major function: (a) long-term storage, (b) temporary storage, (c) "overnight" storage, (d) regulation, (e) tailwater reuse, and (f) water table control.

Long-term storage reservoirs are usually associated with project operation and year-to-year carry over. They have large capacity and basically are needed to provide flexibility in frequency. Their occurrence on-farm is rare though they have been economically used in conjunction with wells where water can be pumped in the spring season, stored, and then used to meet peak summer demands. This permits fewer wells and longer, more economic use of them.

Temporary storage facilities are commonly used in the more humid areas to collect runoff from rainfall for use during occasional dry periods or for frost or temperature control. They are often used to provide a source for sprinkler or gravity systems in areas where "rain-fed" cropping is possible but where irrigation can improve crop quality and yield.

"Overnight" reservoirs are common. They accumulate pump or canal deliveries for 24 h, more or less, to permit applications that are variable in rate and duration. They may have a capacity as small as that needed to hold 12 to 16 h of flow from a well or small surface source with the irrigation applied for only part of the 24-h period. However, this small size lacks reserve for unusual conditions and a larger capacity for up to 48-h storage is frequently more desirable, particularly in conjunction with a semiautomated surface system. They are particularly adapted to facilitating the continuous pumping of a well or the 24-h delivery flow from a water agency (see Fig. 9.1, Reservoirs in citrus fields in Coachella Valley, CA).

The economical size is a function of the duration of irrigation, the area to be covered in a set, the capacity of the distribution system, the economics of labor, and the water supply rate and duration of flow.

Regulating reservoirs are usually small. Their function is to permit relatively small variations in flow and/or duration. They may be needed to accept flow from a water agency or a well to permit small variations in flow to sprinklers as the length of lateral changes or time to move reduces flow needs for short periods. One may be needed adjacent to a well supplying a sprinkler system so that variations in pumping rate due to a falling water table can be equalized by the booster system to provide a steady rate and a constant pressure.

One may conveniently be used to accumulate enough water to provide the larger initial stream desired for furrows. When the reservoir is empty, the cutback stream size would be satisfied by the basic source such as a well, and the reservoir could be used to accumulate the tailwater to be ready for the next set. Extra capacity over the minimum volume is good design.

Tailwater reuse systems require a sump or reservoir at the lower side of a field. The capacity needed may vary from complete storage of the runoff from a field or several fields to a small sump for a recycling system. The later pumping back to an irrigation reservoir for re-regulation is frequently the most desirable.

A reservoir constructed to help maintain a water table is sometimes used in subirrigation systems or with ponded crops such as rice. These have no specific storage requirement, but evaporation may cause significant water loss if the surface area is large.

Classification of reservoirs may also be done by location: (a) gravity reservoirs serve the irrigated area by gravity flow while (b) low level reservoirs require pumped delivery. Preferably, an irrigation reservoir is located at a high point on the farm so that all flow can be taken out by gravity. When water supply agencies deliver water at an elevation below the desired water surface of a reservoir, a small pump to fill the reservoir may be desirable. The cost of pumping may be compensated by the value of the irrigation of additional areas or by the labor and water saved through more efficient application.

Where topography permits on large farms, it may be practical to have two reservoirs at the same elevation that are interconnected by a pipeline or a level top ditch designed to permit flow in either direction. This permits combining deliveries to the middle to obtain large flows for efficient labor utilization from smaller, less expensive conveyance.

A similar condition utilizing the concept of flow to the center from two ends exists when the reservoir is at the high side and a well is located at an alternate side.

A gravity reservoir, in conjunction with large capacity closed or semi-enclosed pipelines, or level top ditches provide the flexible source for deliveries that are flexible in rate and duration and provide large stream sizes that greatly reduce labor costs.

Low level reservoirs may permit the utilization of low swampy areas, may be part of a tailwater reuse system, or may accept gravity delivery from low head supply canals or groundwater or surface flows. In shallow water table areas, they may eliminate the need for lining to control seepage losses. The use of a pump to deliver the water from the reservoir restricts the flexibility of irrigation stream size but permits a wider choice of delivery location.

When a low or swampy area otherwise unusable is converted to a reservoir, better land use may result since the reservoir can be managed to enhance wildlife. Somewhat larger surface areas may be practical when the water table is at the surface or at a shallow depth since the reservoir will have little seepage loss. If the water table is above the bottom of the reservoir but below the adjacent water surface, such side losses as occur may be controlled by placing an interceptor drain ditch or pipe around the periphery of the reservoir and pumping the drain water back into the reservoir. This alternate may be cheaper than lining which may also be subject to flotation when the reservoir is being emptied.

Depending upon the head, pumps used for deliveries from farm reservoirs are usually low lift, high volume propeller pumps or centrifugal pumps. They may feed directly into pipe conveyance systems which in turn may feed sprinkler lines, gated pipes, field ditches, or other irrigation distribution devices. When variable flows are desired in the field, internal combustion, variable speed engines can provide a flexibility of rate better than electric motors. To avoid having to frequently reset engine speed to closely match field demand, it is practical to have an overflow bypass back into the reservoir which is designed to maintain the needed head on the pipeline for delivery. In operation, the pump is set to supply the desired field flow with a small excess overflowed through the bypass standpipe to maintain a constant head. In semiautomated systems with closed or semi-enclosed pipelines or level top canals, a bypass stand beside the pump and reservoir is often used. Without a bypass, the irrigator must take the full unregulated flow pumped and this almost always results in less efficient utilization of water and labor. With a bypass a fine readjustment of the flow rate in the field to match the precise need will cause the small rejected flow to go directly back into the reservoir rather than being wasted in the field or passing on to a tailwater reservoir. The extra pumping cost for the small head on the rejected flow is justified by the flexibility of rate obtained in the field to conserve water there and is less than the cost of pumping it back from the tail water reuse reservoir.

Another variation of a low level reservoir utilizes portable pumps which are moved to the field being irrigated. In extensive level areas, the reservoir can consist of an extensive level-top canal system to provide storage, distribution, and sometimes drainage depending on the drainage water quality. When the ground water table is about the same elevation as the canal, there is little or no seepage loss. Where elevation differences preclude a single level canal system, several steps in the system can be created by the use of mechanical gates, such as the Neypric AVIS automatic constant downstream level canal gates referred to earlier (Merriam, 1977).

It is practical to use such depressed canals for project transmission canals. Pumping costs may be compensated for by the very small or no seepage loss from an unlined canal; and more importantly, by the elimination of the need for on-farm reservoirs to provide flexibility of rate and duration. Pumps, portable or permanent, from the canal may serve small elevated on-farm canals or pipelines. The pumps should have variable flow and bypass capabilities. The elevated canals may be lined.

An indirect way of obtaining on-farm flexibility in rate and duration is through the use of reservoirs in the project supply system. The management and the project canal system must be able to permit variable flows and duration at farm turnouts which are under the control of the irrigator. The farm turnouts can be automated by the use of float valves to pipeline systems or Neyrpic AVIO gates to canals. The irrigator can then use the project system as his reservoir.

With this variable out-take, the canal flows vary. To overcome these variations, the canal system may be automated, or a sloping canal section can terminate in a reservoir for reregulation. The next sections below the reservoir must be treated the same way, and the final section should serve the farms through the system of closed or semiclosed pipelines or level canals. In this way, the project canals and reservoirs can provide the needed on-farm flexibility in rate and duration to permit a limited flow demand schedule.

Flexibility of frequency should still be under the control of the canal operation to match canal capacities. Increased canal capacities may be obtained by canal automation because closer control will reduce freeboard requirements. This permits the limited flow demand type of schedule.

9.4.2 Investigations and Design

The storage basin of a farm reservoir may be formed by constructing a dam to impound water in a valley upstream, usually to store intermittent runoff, by excavating a basin for an "overnight" reservoir, or both.

Special consideration must be given to design if failure of any structural element could cause danger to life or serious property damage. When a sizable structure is involved, state or local codes often include specific requirements for the protection of the public and the environment.

Prior to the design of a farm reservoir, surveys showing topographic conditions, geology, foundation conditions, availability of construction materials, and the associated elements of the irrigated system are needed. A clear definition of the purpose and operational plan for the reservoir must be available. The type of basin, dam, or appurtenant structures must be individually selected for each site on the basis of the above factors.

The capacity of the reservoir must be such that when used in combination with the other elements of the farm irrigation system, it will provide the needed volumes of water at an adequate variable flow rate and in a timely manner. The overnight reservoir typically stores 24 to 36 h of inflow. Additional capacity for on-stream reservoirs must be provided to compensate for the volume of sediment that is expected to be retained in the reservoir during its anticipated life or during the period between cleaning and repair. Off-stream reservoirs are often more practical. Allowances must also be made for evaporation and seepage losses if the storage period is appreciably long or if they are expected to be significantly large.

When the reservoir is an element of a tailwater recovery system, its required capacity will be determined by an analysis of the expected inflow hydrograph and the flow rate and time of operation of the outlet pump.

Provision must be made to bypass the maximum expected flood flow or irrigation flow or both. Overtopping of an earth fill is not permissible. An earth or structural spillway should convey these flows at a safe velocity to a point downstream where the structure will not be endangered.

If the reservoir is formed by the construction of a dam and foundation investigations so indicate, a cutoff will be required under the dam. This cutoff should be made of relatively impervious material keyed into the foundation and extending into the fill. It should be of a width that will permit the effective use of compaction equipment and have side slopes not steeper than 1:1.

The earth embankment should be designed to have slopes that will permit quick emptying of the reservoir without sliding or sloughing. The downstream slope must be stable with allowance made for any toe drainage conditions that are anticipated. The top width should be adequate to permit passage of the equipment and personnel needed for irrigation and farm operations and for maintenance of the works. If the surface area of the pond is large and exposed to strong wind, riprap or slope protection may be necessary to protect the fill from wave action.

Excavated storage basins should be designed with side slopes no steeper than those required to maintain slope stability in the type of material encountered or as needed for the installation and maintenance of a lining. Where surface runoff enters through a natural or excavated channel, entry to the reservoir will need to be protected against erosion by a suitable structure. Erosion protection should also be provided when flow enters the reservoir through a pipe.

The cut-and-fill requirements are usually balanced. It may be practical to over-excavate the bottom to obtain cheaper fill material. The dead storage is seldom detrimental and may be beneficial.

Controlled outlets or pumping facilities are necessary to feed the stored water into the irrigation system. The capacity of the works should be such that the peak period irrigation demand can be met. Structural facilities should be designed for convenience and reliability of operation and be made of materials that are durable and have a life expectancy compatible with the other elements in the irrigation system.

9.4.3 Construction Requirements

Foundation areas of earth fills or structures need to be cleared of trees, logs, stumps, roots, brush, sod, and rubbish. Foundation surfaces should be sloped to not steeper than 1:1. The earth surface should be scarified before placement of fill material and should be sufficiently moist so that the first layer of fill can be compacted and bonded to the foundation.

All excavation and fill should be accurately built to line and grade. Fill material should be selected and be free of roots, sod, frozen soil, large stones, or other detrimental material and be compacted to the density established in the design.

Concrete used in appurtenant structures should be of a quality consistent with the size and requirements of the job. Lining materials should be installed in the manner recommended for the type used. Prefabricated elements of inlets, outlets, measuring devices, etc., should be installed in conformance with industry standards or the specifications of the manufacturer. All work should be finished in a workmanlike manner.

Areas not exposed to prolonged flooding should be seeded or planted to adapted grasses or ground cover to control erosion and reduce maintenance requirements.

Earth fills or entire ponded areas may need fencing to prevent damage from animals or vandalism. Burrowing rodents, crayfish, trees, etc., that could create channels through the embankment should be eliminated or kept under control. In some instances, farm reservoirs can be considered "attractive nuisances" and fencing or restricted access may be required to protect the landowner from damage claims. Facilities needed to protect workmen from injury during operation and maintenance work should be provided.

Construction operations should be done in such a manner that erosion and air and water pollution will be minimized and held within legal limits.

9.4.4 Evaporation and Seepage Losses

The depth of the water lost through evaporation during short periods in shallow ponds can be satisfactorily estimated by assuming it to be equal to the evapotranspiration (E_t) minus precipitation (P). Since the volume of water lost is dependent upon the surface area, evaporation losses can be minimized by providing a maximum water depth and minimum surface area consistent with an economical balance between construction, maintenance costs, and water loss.

Evaporation losses from storage reservoirs that are full the year round may be as high as 2500 mm per year under adverse climatic conditions and little compensating rainfall. At the other extreme, the small regulating reservoir convering a small area may lose 4 to 6 mm per day of operation and up to 10 mm under very adverse climatic conditions. This may represent a very small percentage of the water controlled; e.g., 0.2 ha surface $\times$ 5 mm evaporated per day per 500 m³ of water controlled per day is about 0.2 percent loss. Even if the reservoir is kept full, though used only one-third of the time, it is still a negligible loss.

An overnight storage reservoir with several times the surface area and a moderately severe climatic condition might lose about 2 percent of its controlled water.

Draining an earth-covered membrane-lined reservoir for a few days between uses will save very little water over retaining it partially full. This is because the water in the earth covering above the membrane will evaporate. Uncovered linings such as soil cement may be emptied but may develop cracks due to temperature changes. Most compacted soil and bentonite linings and concrete lining should be kept covered with water to reduce maintenance problems at the expense of the evaporation loss.

Seepage losses are much more difficult to precisely estimate, but in most instances an adequate appraisal can be made when based on the permeability of the foundation materials. Local experience with existing reservoirs built in similar soils in the area are perhaps the best guide as to the need for pond sealing. Often it is practical to delay the installation of seepage control methods until a period of evaluation has determined a need. Seepage losses can seldom be completely eliminated and control methods vary greatly in their effectiveness and cost.

Compaction of the surface of the basin is perhaps the least expensive method for controlling seepage. It is effective if the exposed materials have a wide range of particle sizes and sufficient silt and clay to effect a seal. Exposure to cyclical freezing and thawing or wetting and drying can reduce the effectiveness of the compacted layer.

When the characteristic of the exposed materials make compaction inappropriate, a blanket of suitable material can be spread over the seep areas. Clays and silty clays are excellent and sandy loams are usually satisfactory. Thickness of the blanket should be at least 30 cm if the water depth does not exceed 3 m and an additional 15 cm should be added for every meter of depth over 3 m. The blanket materials should be uniformly spread and compacted under optimum moisture conditions. If the blanket is exposed to alternate wetting and drying or to freezing and thawing, it should be protected with a covering of 30 cm of gravel.

Areas of a reservoir that have well-graded materials but insufficient clay and which are expected to be permanently wet can be sealed by adding bentonite. Bentonite works best in waters with low concentrations of calcium salts, which tend to cause flocculation. It is spread over the exposed soil material, mixed, and compacted at optimum moisture conditions. The amount of bentonite needed per unit area must be determined by a laboratory analysis, but the rate usually varies from 5 to 15 kg/m². Since bentonite shrinks when it becomes dry and forms cracks, it cannot be recommended for areas subject to intermittent flooding. Sometimes, the bentonite layer is covered with a layer of soil 20 to 30 cm thick, which helps minimize the shrinkage and thus permits at least a degree of intermittent flooding without serious deterioration of the bentonite layer.

The use of waterproof linings is perhaps the most effective way to control seepage losses. Concrete paving creates a good lining if contraction and expansion joints are properly placed and made watertight, but is expensive. Soil cement is often practical for sandy soils deficient in clay (Portland Cement Association, 1973). Thin polyethylene, vinyl, and butyl rubber membranes will provide an almost watertight seal if properly installed and protected from deterioration and mechanical damage (ASAE, EP340.1, 1979). Installation recommendations of the material manufacturers should be followed. Membranes are usually protected with a covering of at least 15 cm of earth. Installation costs vary widely and each site must be individually evaluated to determine feasibility. Side slopes as flat as 5:1 may be necessary for construction equipment.

Where the economic justification for the reservoir is principally in labor saving because of the larger stream of water made available, or the seepage loss returns to a shallow ground water basin for repumping, appreciable seepage economically may be permitted. Increased irrigation efficiency made possible by the reservoir may easily save more water than the seepage loss. A 200 mm per day seepage from 0.5 ha reservoir area is less than 7 percent of 150 mm applied on a 10 ha field. However, if the value of the water is high in the terms of its productivity or direct cost, very little seepage can economically be tolerated. In general, the economic value of the water conserved by lining the reservoir, or of preventing a high water table problem, whether for economic, aesthetic, or environmental reasons, must exceed the cost of the lining and its maintenance.

9.4.5 Flow Head and Rate Outlet Controls

With changing water levels in the reservoir, outflow rates may also change significantly, particularly for free surface flow into canals or partly filled, or open (free surface flow) pipeline systems. When a constant outflow rate is desired for these distribution systems, the reservoir outlet control can include a constant level float gate or valve, similar to that previously described for canal control (Section 9.3). Better design utilizing the semiautomated systems takes advantage of the variable flows a reservoir can provide to permit more effective use of water and labor in the field. The same controls will do both jobs—for steady or variable flows.

For either open pipeline systems desiring steady flows or closed and semiclosed pipelines permitting variable or steady flows, a float-controlled valve should be installed (Merriam, 1973). It will control the water level within a range of ± 15 cm. It should be installed in a stand adjacent to the gate stand. Both stands may be combined if reservoir overflow protection is provided other than in the gate stand, and back pressure on the outlet control gate is adequately considered.

For outlets to canals at a constant head, the float-controlled constant downstream level Neyrpic AVIO gate can be used (Merriam, 1977). An upstream gate must be installed for complete cutoff or emergency use.

To set a constant flow rate from the established constant head, orifice type controls are the best. They can often be used as a rate-measuring device as well by using a calibrated gate or Neyrpic modules. Other type submerged flow meters such as venturi, propeller, pitot, etc., are practical but more expensive. Variable length weirs may be used. Totalizing meters are often more desirable. Constant flow rates, however, do not usually fully realize all the advantages of reservoirs.

When the constant downstream level gate is used in conjunction with level-top semiautomatic systems, the flow controls are regulated at the point of application in the field with the float gate providing the desired flow from the reservoir.

Since the smaller size AVIO gates are limited to heads of 1 to 2 m, higher reservoir heads must be reduced. This may be done by installing a float-controlled valve before the gate.

9.4.6 Example of On-Farm Reservoir Installation

Fig. 9.1 pictures "overnight" irrigation reservoirs in citrus groves, Coachella Valley, CA. Such reservoirs are often the heart of a good water delivery system. They can serve 30 to 60 ha, depending on the topography and the distribution system. They can each function to supply variable stream sizes and duration; can accumulate small streams—even for several days— to permit large flows which greatly reduce labor costs; can, in conjunction with semiautomated on-farm distribution systems, reduce labor and make efficient irrigation practical; can make pumping from low-yield wells economical; can facilitate irrigation sets of other than 24-h duration; etc.

Because of varying site conditions, irrigation requirements, and individual preferences, seldom are two reservoirs constructed in exactly the same manner although their function may be identical.

Inlets to reservoirs are sometimes pipes running over the bank. These may extend past the bank (and so require support) to let the inflow fall freely on the pond surface without erosion. They may be laid on the embankment slope and extend to the bottom. Under partially full reservoir conditions, the latter will reduce pumping heads because of siphon action in those cases where pumping is necessary, but the hazard of possible back siphoning and air entrapment must be considered.

The inlet may be combined with the outlet with proper valving, check valve, or overflow arrangements to prevent backflow if the supply is from a source such as a well.

Unless a siphon is used, the reservoir outlet pipe will be under the embankment and must be adequately protected from crushing, piping, corrosion, leaking, etc. This can be done by laying the pipe in a trench and backfilling with 10 cm concrete all around. Depending on load conditions, reinforcing may be needed. For mortared-joint concrete pipes, longitudinal reinforcing should be used.

The outlet controls often consist of an open gate stand adjacent to the reservoir. The top of the stand is about 25 cm below the top of the embankment so that the stand will overflow before the embankment is overtopped. Without this arrangement, an alternate overflow protection must be provided.

The outlet control gate to serve the field is often on the inlet end of the field pipeline leaving the stand open to the reservoir. In such installations an air vent-vacuum relief structure must be placed on the field pipeline about five pipe diameters downstream from the gate. (See Chapter 11.)

When the embankment is higher than about 4 m, in-line gates and necessary vents may be more practical. Such a gate may well be placed within the reservoir and have a long stem up the slope of the enbankment.

The inlet in the reservoir to the outlet pipe can be made with a short vertical section of large diameter pipe to assure slow entrance velocities. Its crest may be at or near the bottom of the reservoir unless the latter has been over-excavated below minimum live storage depth.

Especially with closed distribution systems, provision must be made to prevent trash or debris from entering the system. Often trash problems can be minimized by preventing its entrance into the reservoir by means of suitably designed screens or trash removal devices. Protection for the distribution system inlet can sometimes (as in the Coachella Valley area) be as simple as a properly sized section of hardware cloth rolled into a cone and placed over the reservoir outlet. The mesh size of the screen is as coarse as practical to screen out the debris, commonly 1 to 2 cm. The conical shape provides the needed rigidity and allows some of the debris to fall off. However, occasional cleaning is usually necessary. Should the debris problem be more severe, especially designed trash racks and outlets may be necessary to prevent undue interruption to water delivery.

While the above discussion has been concerned primarily with "overnight" reservoirs, similar variations in design and construction are common in reservoirs having other purposes. Because of the many alternate materials, devices, and types of equipment available, the designer is offered great flexibility in meeting the functional needs of the reservoir and in providing the operating characteristics desired by the irrigator in a practical and economic manner.

9.5 MOSQUITO CONTROL IN FARM WATER DELIVERY SYSTEMS

9.5.1 Basic Principles

Irrigation systems can frequently increase existing mosquito problems or create a new problem unless special control measures are exercised. Mosquito populations can be increased by forming free water surface conditions needed for mosquitos to complete their life cycle of egg, larval, and pupal stages. Favorable conditions include reservoir construction, and conditions that encourage floating vegetation and debris: e.g., standing water for a duration of several days on fields or in tailwater ditches, and seepage areas along canals or through leaky dams. Mosquito control planning should be coordinated with other basic irrigated farm needs such as soil and water conservation, flood control, wildlife management, recreation, and occasionally, hydroelectric power. Reduction or elimination of manmade mosquito sources can be achieved by proper water management and maintenance of facilities by applying knowledge of both irrigation principles and mosquito biology.

9.5.2 Storage Reservoirs

The normal summer fluctuation zone in reservoirs should be completely cleared of vegetative growth, at least annually, before spring reservoir refilling. Clearing should be to the elevation of the normal high water line, except for isolated trees and sparse vegetation that are along abrupt shore lines exposed to wave action. Vegetation can be controlled mechanically or by use of approved chemical measures. Borrow areas or other depressions located in the summer fluctuation zone or outside the reservoir basin should be self-draining or diked and stocked with mosquito-eating fish.

Continuous water-level lowering of the reservoir, during the mosquito breeding season, is effectively employed on large supply reservoirs, but is usually not practical in small reservoirs used as described herein.

9.5.3 Conveyance and Distribution Systems

Lining, or other suitable seepage control measures, should be provided for all sections of canals and laterals located in permeable soil where excessive leakage would cause waterlogged areas or ponds.

Efficient irrigation practices throughout the service area should be encouraged by providing, as nearly as practical, demand scheduling, thus reducing irrigation tailwater ponding and waterlogging.

Turnouts and other hydraulic structures should be designed to prevent retention of ponded water for excessive periods when they are not in use. Depending on local conditions and climatic factors, a pond duration of not more than 3 days will prevent adult mosquito emergence.

Further elaboration on details of mosquito control in irrigation projects is presented in Engineering Recommendation ASAE R267.4 (soon to be replaced by an Engineering Practice under the title of "Principles and Practices for Prevention of Mosquito Sources Associated with Irrigation").

References

1 Ackers, P. and A. J. M. Harrison. 1963. Critical-depth flumes for flow measurements in open channels. Hyd. Res. Paper No. 5, Dept. Sci. and Indust. Res., Hyd. Res. Sta., Wallingford, Berkshire, England. 49 pages.

2 American Society of Agricultural Engineers. 1979. Installation of flexible membrane linings. In AGRICULTURAL ENGINEERS YEARBOOK, ASAE Engineering Practice, EP340.1. p. 479-484.

3 Bos, M. G. (Ed). 1976. Discharge measurement structures. Publ. No. 20. Internatl. Inst. for Land Reclamation and Improvements (ILRI), Wageningen, The Netherlands, 464 p.

4 Brater, Ernest F. 1976. Handbook of hydraulics. 6th Ed. McGraw-Hill, NY. 608 p.

5 Bureau of the Census. 1973. 1969 Census of agriculture. U.S. Dept. of Commerce.

6 Bureau of Reclamation, U.S. Dept. of the Interior, 1961. Design of small dams.

7 Clemmens, A. J. and J. A. Replogle. 1980. Constructing simple measuring flumes for irrigation canals. U.S. Dept. of Agr. Farmers Bul. No. 2268.

8 Dedrick, Allen R., John A. Replogle, and Leonard J. Erie. 1978. On-farm level-basin irrigation—Save water and energy. Civil Eng. 48:60-65.

9 Humphreys, A. S. and J. A. Bondurant. 1977. Cast-in-place, 2-foot concrete trapezoidal flow-measuring flumes. U.S. Dept. Agr. Tech. Bull. 1566, p. 43.

10 King, H. W. 1963. Handbook of hydraulics. 5th ed. McGraw-Hill, NY. 10 chapters.

11 Kulin, G. and P. R. Compton. 1975. A guide to methods and standards for the measurement of water flow. Natl. Bur. of Standards, Spec. Publ. 421, U.S. Dept. of Commerce, Washington, DC. 90 p.

12 Merriam, J. L. 1973. Float valve provides variable flow rate at low pressures. Proc., Am. Soc. Civ. Eng. Irrig. and Drain. Spec. Conf., Fort Collins, CO, p. 385-402.

13 Merriam, J. L. 1977. Level-top canals for semi-automation of on-farm irrigation and supply systems. Proc., Am. Soc. Civ. Eng. Irrig. and Drain. Spec. Conf., Reno, NV. p. 217-224.

14 Parshall, R. L. 1953. Parshall flumes of large size. Bul. 426-A (reprint of Bull. 386). Colo. Agr. Expr. Sta., Colorado State Univ., Fort Collins, CO. 40 p.

15 Portland Cement Association. 1973. Soil primer. Portland Cement Assoc., Old Orchard Road, Skokie, IL 60076.

16 Replogle, J. A. 1968. Discussion of rectangular cutthroat flow measuring flumes by G. V. Skogerboe and M. Leon Hyatt. Proc., Am. Soc. Civ. Eng., J. Irrig. and Drain. Div. 94(IR3):359-362.

17 Replogle, J. A. 1970. Flow meters for water resources management. Water Resour. Bull. 6(3):345-374.

18 Replogle, J. A. 1975. Critical-flow flumes with complex cross section. Proc., Am. Soc. Civ. Eng. Irrig. and Drain. Spec. Conf., Logan, UT, p. 366-388.

19 Replogle, John A. 1977. Compensating for construction errors in critical-flow flumes and broad-crested weirs. In Flow measurement in open channels and closed conduits, Natl. Bureau Stds. Spec. Publ. 484, Vol. I, p. 201-218.

20 Robinson, A. R. 1968. Trapezoidal flumes for measuring flow in irrigation channels. U.S. Dept. of Agr., Agr. Res. Service, ARS 41-141. p. 15.

21 Skogerboe, G. V., R. S. Bennett, and W. R. Walker. 1972. Generalized discharge relations for cutthroat flumes. Proc., Am. Soc. Civ. Eng., J. Irrig. and Drain. Div. 98(IR4):569-583.

22 Soil Conservation Service, U.S. Dept. of Agriculture,.1971. Ponds for water supply and recreation, Agricultural Handbook No. 387.

23 Soil Conservation Service, U.S. Dept. of Agriculture,.1976. Earth dams and reservoirs, Technical Release No. 60.

24 Soil Conservation Service, U. S. Dept. of Agriculture. 1978. National handbook of conservation practices.

25 Stoker, R. 1978. Simple, efficient, automatic irrigation. Irrigation Age 12(6):76-77.

26 Water Measurement Manual. 2nd Ed., revised 1974 (Metric supplement to water measurement manual. 1971. 224 p.) Bureau of Reclamation, U.S. Dept. of Interior.

27 Zimmerman, J. D. 1966. Irrigation, Chapters 16 and 17. Krieger Publ. Co., Huntington, NY (original edition, John Wiley & Sons, Inc., NY) 516 p.

chapter 10

FARM PUMPS

10

10

FARM PUMPS

by   R. A. Longenbaugh, Colorado State University,
     Fort Collins, Colorado, H. R. Duke, USDA-
     SEA/AR, Fort Collins, Colorado

10.1 INTRODUCTION

Pumps are mechanical devices which impart energy to a fluid. They are commonly used on farms with irrigation or drainage to lift water from one elevation to a higher level or to add pressure to the water.

In this chapter, we emphasize how to select the proper pump for a specific job. The theory of how a pump imparts energy to the liquid is rigorously treated by Karassik et al. (1976) or can be found in the books by Davis (1969) or Addison (1966). These references also discuss the theories used to design new pumps. The number of engineers involved in the design and development of new pump models is quite small, but a large number of engineers and irrigation equipment personnel are required to select pumping equipment for a variety of operating conditions.

This chapter briefly describes the types of pumps most commonly used for irrigation and drainage applications. We discuss in detail the performance characteristics of pumps, the determination of the desired discharge, the development of the system head curve and finally pump selection. Information is also provided to determine operating and initial investment costs. Field testing of existing systems is briefly discussed and some remarks are offered on pump maintenance and operation.

Chapters 14, 15 and 16 contain information on the selection and operation of specific irrigation equipment that normally use pumps. Because of these detailed descriptions, this chapter will not describe specific piping arrangements on the discharge side of the pump. The construction of drainage wells is briefly referenced in Chapter 7 and a rigorous treatment of well drilling can be found in the book Ground Water and Wells by Johnson, Inc. (1975) or is included in the books by Todd (1963), DeWeist (1966) or McWhorter and Sunada (1977). An understanding of groundwater hydraulics is important for proper selection of a pump to determine well drawdown and evaluate the possibility of well interference.

10.2 TYPES OF PUMPS

Pumps are generally categorized into one of two basic types based on the method by which energy is imparted to the fluid. In positive displacement pumps, the fluid is physically displaced by mechanical devices such as

347

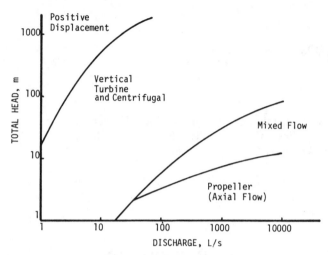

FIG. 10.1 Subclassification of pump types as a function of total
operating head and discharge.

pistons, cams, plungers, screws, diaphrams, vanes or other similar devices.
The discharge of positive displacement pumps is independent of delivery
pressure. Pumps of this type are not commonly used for pumping irrigation
or drainage waters. Fertilizer injection pumps are, however, quite often
positive displacement pumps and thus provide accurate control of the
amount of liquid fertilizer injected into the irrigation water.

The other type of pump is one where the energy is transferred by kinetic
principles such as centrifugal force, viscous forces or momentum. This type
includes popular irrigation and drainage pumps known as centrifugal, vertical turbine, submersible turbine, propeller, mixed flow and jet pumps.
Within these subcategories the size, shape, speed, number of stages and
mode of operation can vary significantly. In general, there are sufficient
models or combinations of pumps to permit the proper selection of a pump
or group of pumps to meet any irrigation or drainage need.

Most of the pumps of the second type can operate over a range of conditions where the discharge rate will vary as the operating head fluctuates. Fig.
10.1 illustrates some generalized operating conditions. Note that the positive
displacement pumps produce the highest operating heads but are usually
limited to small discharge rates. The most common irrigation and drainage
pumps are the centrifugal or vertical turbine pumps and they operate over a
wide range of conditions. For high discharge rates at low heads the propeller
or axial flow pumps are commonly used. Mixed flow pumps use impellers
which are shaped so as to impart the energy by both axial and centrifugal
forces. Many of the new high volume, high efficiency turbine pumps are actually mixed flow pumps.

Since both irrigation and drainage waters often contain some sediment,
it is necessary to use a pump which can cope with this problem. Sediment in
the water will cause wear of any pump and if the sediment can be excluded or
only clear water pumped it is most desirable. Both propeller and centrifugal
pumps can handle a reasonable amount of sediment but will require periodic
replacement of impellers and volute cases. Turbine pumps are more susceptible to sediment. Unfortunately, a fairly large number of the existing wells

produce sand which is then pumped. The pumping of any sediment will reduce the life of the pump and increase maintenance costs. Positive displacement pumps must be used only for sediment-free liquids.

10.2.1 Brief Pump Description

This presentation assumes the readers have some knowledge of pumps. For the beginner it is suggested that reading from Hicks (1957) or Colt Industries (1979) will provice the necessary background information. Both of these books or literature available from each pump manufacturer will provide cutaway drawings and descriptions of the various pump components. A brief description follows of the principle components for centrifugal and turbine pumps.

Centrifugal pumps. A centrifugal pump has one or more impellers which are fastened to a rotating shaft which turns inside a volute case. Water enters the eye of the impeller and is forced outward along the vanes of the impeller by centrifugal energy caused by the rotating impeller. The centrifugal force is converted to a velocity head and thus a pressure as the liquid flows out the discharge port in the volute case.

There are several different types of commonly used centrifugal pumps including: end suction, double suction, multiple stage, and split case. All of these descriptors reference how water enters the eye of the impeller or how many impellers are present. If there are two or more impellers where the discharge from one is allowed to flow into the next, it is called multi-stage. An end suction pump is where water enters the eye of the impeller from only one side, while double suction means water enters the impeller from both sides.

Generally, the impellers are fixed solid onto the driveshaft which is then coupled to the drive unit, electric motor or engine. For electric driven systems, the driveshaft may actually be a single armature shaft for the motor and drive shaft for the impeller.

Because of their compactness and specific construction, centrifugal pumps are often stocked as completely assembled units that can be bought and taken directly to the site requiring only a hookup to the suction and discharge pipes. It is generally not necessary to select and assemble a large number of components for this type of pump.

Centrifugal pumps are often physically located above the water source. Water is permitted to enter the pump through a suction pipe and priming of the pump to remove air bubbles in the pipe on the suction side is usually necessary. The physical limitations on how high the pump can be located above the water source is discussed in Section 10.6.6

Turbine pumps. Turbine pumps are often used to extract water from wells or raise it from rivers, lakes or sumps where it is not practical to use centrifugal pumps. Both the standard vertical turbine and the submersible turbine impart the energy to the liquid by rotating an impeller inside the bowl. When several of these bowl assemblies are directly coupled together so that the flow from the lower impeller moves directly to the next impeller above, then it is a multi-stage pump. One impeller rotating inside a bowl is referred to as a single stage and when these separate stages are coupled together it is multi-stage. For example, a four-stage pump would have four separate bowl assemblies coupled together.

The impellers are directly fixed on a shaft, referred to as a line shaft.

For the vertical turbine pump, this shaft extends vertically upward and is coupled to a motor or gearhead which provides the driving power. The line shaft extends upward through the eye of the impellers. The discharge from the bowl also exits vertically upward and is carried within the column pipe. The line shaft is within the column pipe and is supported by bearings at two to three meter intervals. For oil lubricated pumps there will also be an oil tube surrounding the lineshaft but inside the column pipe.

The column pipe, line shaft and oil tube, if present, extend from the bowls upward to the pump head. The distance from the bowls to the pump head is referred to as the pump setting. The pump head serves as an elbow to redirect the flow horizontally and it supports the weight of the column pipe, bowls, impellers and line shaft. It generally rests on a foundation and also serves as a mounting bracket for the motor or gearhead.

For the submersible turbine pump, a submersible electric motor is coupled to the bowl assemblies below the lowest stage with a mounting bracket. Water is still allowed to enter the first stage from below as for the vertical turbine. The discharge exits the top stage and flows to the land surface in an unobstructed pipe. There are no rotating parts except for the electric motor armature and the short line shaft coupling the impellers.

The submersible turbine has been developed since the early 1960's and is now being used more. Some of the advantages of the submersible are that for smaller sized pumps and deeper settings they are cheaper. They are popular where the entire pumping plant needs to be below ground such as in parks and golf courses. They are also preferred in areas experiencing vandalism. In many instances submersibles can be used in a crooked well where it is impossible to use a lineshaft vertical turbine which has severe problems if the lineshaft isn't perfectly straight.

Some of the disadvantages of submersible turbines are: they are more susceptible to service interruptions due to electric line voltage fluctuations; the motor must have adequate water flow around it to provide cooling; the length of motor below the pump may prevent pumping from the bottom of wells with little saturated thickness; wells that pump sand may cause bearing problems in the pump and motor; submersible electric motors of 50 kW capacity or larger are just now becoming available but are extremely expensive; and if there are any problems with the motor the entire pump and motor must be pulled.

10.2.2 Pump Lubrication

Almost all pumps have moving parts which require some type of lubrication to prevent wear. In some instances the bearings are lubricated and sealed at the time of manufacture. In other instances oil or grease must be added periodically or continuously and even the water itself may be used as the lubricant.

Vertical turbine pumps. The vertical turbine pump may be either oil or water lubricated. For irrigation it is not critical if the water contains an occasional drop of oil, however, for domestic pumps most are water lubricated. An oil lubricated pump should be used if some sand or sediment is being pumped. A light weight turbine oil should be allowed to continuously drip in the annulus between the lineshaft and oil tube. The rate of oil application varies between manufacturers but should be about six drops per minute. Some of this oil will pass out the weep hole located in the casting for the upper stage. The bearings in the bottom of the pump are generally lubricated and sealed when manufactured.

For water lubricated pumps with depths of setting in excess of 13 m it is advisable to include a water prelubrication system to be used just prior to start up. Water lubricated pumps generally use rubber bearings that are lubricated as water flows up the column. For deep pump settings, in excess of 15 m, the time for water to rise in the column pipe may be sufficient to allow the top rubber bearings to overheat and seize on the lineshaft if prelubrication is not practiced.

There are also bearings in a gearhead or motor that need lubrication. Models having either an oil reservoir or grease fitting for use of heavier grease are available. Some of the larger size motors and gearheads have provisions for water cooling where a small part of the pumped water is piped through the motor or gearhead castings to act as a coolant.

Centrifugal pumps. Most centrifugal pumps have a packing gland where the driveshaft exits the volute case. Some of the newer pumps use a mechanical silicate seal to prevent leakage at this point and to allow an occasional drop of water to act as a lubricant. Older models use beeswax impregnated hemp which provides some support to the shaft, acting as a bearing but not permitting large flows. A very small amount of water is permitted to leak through this packing gland also to act as a lubricant. Occasionally packing glands must be tightened or replaced if leakage through them is excessive. Care should be taken to not over-tighten packing glands thus cutting off the lubricating water as this could result in pump failure or motor burnout.

Jet pumps. The jet pump has no moving parts in the well or beneath the water surface. Lubrication of the mechanical parts above ground should be according to manufacturers' recommendations.

Propeller pumps. Because of the shallow setting of most propeller pumps, they usually use a heavy grease that is applied with a grease gun. They often use heavy brass bearings to support the propeller head and rotating shafts. Care must be taken to assure proper lubrication according to the manufacturer's recommendations.

10.2.3 Impeller Vanes

There are three different types of impellers that might be used in centrifugal or turbine pumps. They are open, semi-open and enclosed impellers. These terms refer to whether the impeller vanes are enclosed by shrouds. An open impeller has no shrouds and is most often used for fluids which contain organic matter such as sewage sludge, paper pulp or other liquids containing clogging material. The semi-open impeller has one shroud on the side opposite the intake to the impeller. The semi-open impeller must be adjusted so the operating clearance between the open side of the impeller vanes and the bowl or volute face is within a very close tolerance of about 0.1 mm. The closed impeller has shrouds on both sides of the impeller vanes and clearance adjustment is not so critical. Hicks (1957, pages 19-26) contains a number of excellent pictures showing the different types of impellers and the wear rings.

10.3 SELECTING PUMP DISCHARGE

The major role of pumping in agricultural crop production is to control, as effeciently as possible, the soil-water environment. In humid or lowland areas, this may mean pumping to obtain adequate soil aeration or drainage. In areas of unreliable precipitation, pumps are used primarily to provide adequate soil water. Chapters 6 and 7 detail evaluation of requirements for irrigation and drainage, respectively, and may define minimum pump discharge for a particular application.

Physical, economic and legal constraints result in a maximum desirable pump discharge, as well as a minimum, particularly when applied to irrigation. The maximum discharge of a well is limited by both the aquifer properties and construction and development of the well itself. Thus, it is imperative that a well test be conducted prior to final pump selection. The pumping rate from surface supplies may be limited by ditch or pipeline conveyance capacity. Legal rights to water use limit pumping rates in many areas.

The consequences of improper selection of pump discharge are many. Too large a pumping rate may result in pump surging or pumping sand, both of which will damage the installation. Excessive design discharge also results in larger capital cost for the pump and driver, higher demand charges for electrically powered units, and excessive erosion and runoff, particularly under sprinkler irrigation systems. Selection of the optimum pump discharge requires consideration of both crop water needs and soil physical characteristics.

For drainage conditions, it is generally desirable to pump at some optimum rate to lower the water table a preselected amount while minimizing expenditures for wells, pumps, power, etc. Too low a pumping rate, however, results in insufficient soil drainage and possible salt accumulations.

10.3.1 Crop Water Requirements

Crop water use depends on many factors, including climatic variables, type of crop, and stage of crop growth. Methods of estimating crop water use are discussed in Chapter 6. The seasonal water use is not to be confused with the amount of water that must be supplied by irrigation. In most areas, we expect some precipitation. The amount of this precipitation that is useful for crop production depends on when it falls, how much is lost to surface runoff, and how much can be stored in the root zone of the soil. Thus, the total pumping requirement is less than the total crop requirement by an amount equal to the effective precipitation.

It should also be noted that crop water use is not a constant throughout the growing season, but increases with increasing temperature, solar radiation and wind. It also increases with decreasing humidity, and with an increase in crop cover. As discussed subsequently, it is not usually necessary to pump irrigation water at the peak rate of crop water use.

10.3.2 Soil Water Storage

One of the major functions of soil in the agronomic environment is that of a water storage medium. The details of soil water transport, plant water uptake, and the concept of available water are discussed in detail in Chapter 4. Suffice to say that any soil has a characteristic amount of available water storage capacity that will retain water until used by the plant. This available storage may range from 0.05 cm/cm for very sandy soils to 0.2 cm/cm for clay-loam soils. So long as the crop root zone has developed to significant depth, the total soil water storage capacity is typically from three to twenty times the maximum daily crop water use. Thus, the soil itself serves as a buffer against short-term extremes in crop water use rates, allowing design of the pumping plant at somewhat less than peak crop water use rates.

Heermann et al. (1974) analyzed sixty years of climatic data from eastern Colorado to determine required irrigation application rates for a range of soil properties. Results are presented in terms of net application rate per unit area as a function of available soil water at several probability levels. With this information, the net application rates necessary to avoid plant water stress can be readily determined.

It is important to note, however, that the required irrigation pumping rate must be greater than the net application rate. No irrigation system is capable of distributing water with perfect efficiency or perfect uniformity. Thus, the design pumping rate must be increased so that adequate water is provided in all areas of the field at the level of application efficiency the distribution system is capable of producing. Application efficiencies may range from 90 percent or more for drip irrigation systems to considerably less than 50 percent for gravity irrigation on coarse-textured soils. Thus, the required pumping rate may be two or more times the required net application rate.

10.3.3 Intake Rates

The rate at which soils will infiltrate water is an important consideration for any irrigation system. For gravity irrigation systems, it is probably the most important factor, as intake rate controls advance of water across the soil surface. The influence of intake rates on surface irrigation systems is covered in detail in Chapter 12 of this publication.

For sprinkler irrigation systems, the role of intake rates is somewhat different. Sprinkler systems are capable of high application efficiency and uniformity only if the system, rather than the soil, controls infiltration. That is, the application rate must be less than the soil intake rate throughout the irrigation.

For a given style of sprinkler head, whether conventional or "low pressure," the area wetted by a given head is relatively independent of the discharge of that head. Thus, the application rate is directly proportional to sprinkler discharge. Note that the same is true for moving sprinklers; although total depth of application is dependent on speed of movement, peak application rate is dependent only on discharge rate and radius of throw of the heads.

Thus, for sprinkler systems, too low a pumping rate will provide insufficient water for optimum crop growth, while too large a pumping rate will result in surface runoff. Excessive surface runoff necessitates increased pumping to adequately irrigate the "dry spots" because of the reduced uniformity it produces.

10.3.4 Operating Criteria

The pumping rate is also affected by operational decisions such as the hours of irrigation per day and days per week. Most irrigators anticipate some shut-down time to allow for maintenance of the system. For many of the gravity irrigation systems labor is required to change the water and sometimes pump schedules are modified to coincide with labor availability. When the hours of operation are reduced from a continuous rate then the

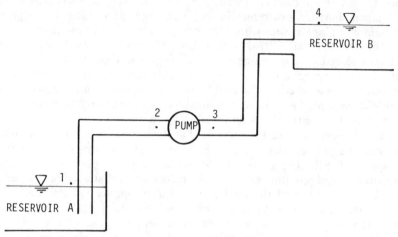

FIG. 10.2 Schematic of generalized pumping conditions.

discharge rate must be increased to provide adequate water during the critical crop period.

Other operating decisions that could impact the design discharge are the cropping pattern and area to be irrigated. Preplant irrigation to fill the root zone could reduce the design discharge by as much as 10 percent for certain crop and soil conditions.

10.4 TOTAL PUMPING HEAD

The energy that the pump imparts to the liquid can be referred to as increased head. It can be expressed as an increase in pressure or energy per unit volume. The most common unit now used by pump or irrigation equipment personnel is a unit of length; ie, meters or feet. The Bernoulli equation written in terms of head rather than pressure for the system illustrated in Fig. 10.2 is:

$$\frac{V_1^2}{2g} + \frac{P_1}{\rho g} + z_1 + H_p = \frac{V_4^2}{2g} + \frac{P_4}{\rho g} + z_4 + H_f \quad \dots \dots \dots \dots \dots \dots \dots [10.1]$$

Where: V_1 and V_4 are velocities, respectively at points 1 and 4, in m/s (ft/s); g is the constant for acceleration due to gravity, 9.81 m/s² (32.2 ft/s²); P_1 and P_4 are presures in the liquid at points 1 and 4, respectively, in Pa (lb/ft²); ρ is the liquid density in kg/m³ (in English units, $\rho g = \gamma$, the specific weight of liquid in lb/ft³); z_1 and z_4 are elevations of points 1 and 4, respectively, in reference to a preselected datum, m (ft); H_p is the energy imparted to the liquid in moving it from point 1 to point 4, m (ft); and H_f is the friction loss in the piping system from point 1 to point 4, m (ft).

A careful selection of the points between which equation [10.2] is written will minimize the number of terms to be computed. For example, when computing the pump energy supplied between points 1 and 4, the $P_1/\rho g$ and $P_4/\rho g$ terms become zero. The velocities at points 1 and 4 in Fig. 10.2 may also be assumed to be zero and equation [10.2] can be rewritten as: $H_p = z_4 - z_1 + H_f$.

The difference in water surface elevations between points 4 and 1 can be easily calculated. It remains to calculate only the friction losses in the connecting pipe. One should also consider the entrance and exit losses as water leaves Reservoir A and is discharged into Reservoir B. For pipe sizes commonly used for irrigation, the error induced by disregarding entrance and exit losses and differences in velocity heads should be less than one meter.

Other points could also be selected between which to compute the energy contributed by the pump. For example, points 2 and 3 might be selected on Fig. 10.2. For these points, equation [10.1] can be rewritten as

$$H_p = \frac{V_3^2}{2g} - \frac{V_2^2}{2g} + \frac{P_3}{\rho g} - \frac{P_2}{\rho g} \quad \dotfill \quad [10.1a]$$

where it is assumed that the elevation of point 2 is the same as point 3 and the friction loss in the pipe between 2 and 3 is negligible. If the pipe size on the inlet side of the pump is equal to the size on the outlet side, then $V_3^2/2g = V_2^2/2g$ and equation [10.1a] further is reduced to one where the energy imparted to the pump is equal to the difference in pressure head across the pump.

The computed or observed values of energy imparted by the pump to the liquid should be the same for points 1 to 4 or 2 to 3. Other points could be selected but care must be taken to evaluate all the terms included in equation [10.1] for any pair of points.

In summary, the head which a pump imparts to a fluid includes the effect of fluid velocity, pressure differences, elevation and friction loss. One parameter which is not illustrated in Fig. 10.2 is the operating head. As discussed in Chapters 11, 14, 15 and 16, it is necessary to have water pressure for sprinklers, gated pipe or drip irrigation systems to work. This pressure can be expressed as an equivalent height of water column. In computing the total head which a pump must provide, it is necessary to add in the required operating head where it is applicable.

10.4.1 System Head Curve

The system head curve graphically illustrates that more head is required to increase the flow or discharge through the system. The total head is independent of the pump with the exception of friction loss that occurs in the column pipe for a vertical turbine pump. Fig. 10.3 is a sketch of a typical system head curve illustrating the parameters which contribute to the total head and how they vary as the discharge increases. For some systems not all of the illustrated parameters would be applicable.

Static lift. The static lift is the vertical distance between the center line of the pump and the elevation of the water source when the pump is not operating. It is computed by subtracting the elevation of the water surface of the source of supply from the center line elevation of the pump. If the water elevation of the source is below the pump elevation, the static lift is positive. It is possible, however, for the pump to be located at an elevation below the water surface elevation, and for that condition the static lift is negative. For this latter condition some books refer to the static lift as a static suction head.

When pumping from a well the static lift is the vertical distance from the center of the discharge pipe to the static water level in the well when the pump is not operating. A good reference point when determining the system head curve for wells is the elevation of the center of the discharge pipe im-

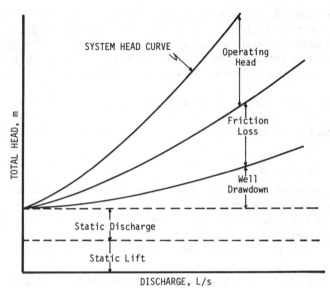

FIG. 10.3 A system head curve illustrating the parameters which contribute to the curve.

mediately adjacent to the pump head. Both the static lift and static discharge head should be referenced to this same point to minimize computational problems. This is also the point where a pressure gage is often located, thus further simplifying the problems in computing the total pumping head.

Static lift is independent of the system discharge and is constant for all Q values. It is possible that the static lift may vary with time; in that case the system head curve is also time dependent. This is discussed more in a later section.

Static discharge. The static discharge head is a measure of the elevation difference between the center line of the pump or discharge pipe and the eventual point of use. When pumps discharge directly into canals a short distance from the pump at the same elevation, the static discharge head is zero. If, however, a pump supplies water to some distant point at another elevation then it is necessary to compute the static discharge head. To obtain this value, subtract the elevation of the pump from the elevation of the final point of delivery.

Similar to the static lift, the static discharge head is independent of the discharge rate but might be time dependent. For example, the static discharge head would change with time if the farmer moved the point of delivery up or down hill or the riser length on the center pivot was changed.

Well drawdown. As a well is pumped the water level declines which is commonly referred to as the well drawdown. The amount that it declines is a function of the pumping rate, Q; the aquifer properties of transmissivity, T; and storage coefficient, S; the well radius, R; and the time the pump is operated, t. The well drawdown, s, can be computed using the nonsteady state equation developed by Jacob which is reported in the publications by McWhorter et al. (1977), Todd (1963), and De Weist et al. (1966). The series equation takes the form:

$$s = \frac{Q}{4\pi T} \left[-0.5772 - \ln u + u - \frac{u^2}{2 \cdot 2!} + \frac{u^3}{3 \cdot 3!} \right] \quad \ldots\ldots\ldots\ldots\ldots\ldots \quad [10.2]$$

where s is the well drawdown in m, u = $(R^2 S)/(4Tt)$, R is the well radius in m, S is the storage coefficient which is dimensionless, T is the transmissivity in m^2/d, t is the time in days since pumping started, and Q is the discharge rate in m^3/d.

Equation [10.2] can be easily solved on an electronic calculator. Generally it is necessary to consider only the first three terms in the series to obtain the required accuracy. This equation can also be used to calculate the water table decline which occurs at some radius from the pumped well. Such calculations are made separately and then superimposed to calculate anticipated water table decline when two or more wells are pumping in close proximity causing well interference.

In reference to equation [10.2], it should be noted that for a particular flow rate and aquifer conditions, the drawdown will continue to increase with time. In actual practice the pumping levels in wells nearly stabilize after a few days of continuous pumping. It is suggested that when one calculates the drawdown a "t" of at least one day be used.

The relationship between well drawdown and discharge is not a linear function. Only for small flow rates is a linear approximation satisfactory. For large irrigation wells pumping at rates in excess of 30 L/s the drawdown will increase exponentially with discharge.

Another excellent way to determine the well drawdown versus discharge relationship is to test pump a well at various rates and observe the drawdown. This is known as a well test. Care must be taken to pump at a constant discharge rate for sufficient time so that the drawdown stabilizes. The best procedure for a well test is to begin pumping at the highest flow rate until the drawdown stabilizes and then successively reduce the pumping rate. This will result in less time required to reach equilibrium conditions for the lower discharges.

Friction loss. When water flows through a pipe system there is a loss of head due to friction. The amount of friction head loss can be determined from the Darcy-Weisbach or Hazen-Williams formulas which are described in Chapters 11, 14, and 15 and in most hydraulic books like those by Karassik et al. (1976) or Davis (1969). Another method for evaluating friction loss is to use friction loss tables, nomograms or curves provided by pipe manufacturers. Friction loss tables for steel pipe can be found in Chapter 11 and in the Colt Industries (1979) and Ingersoll-Rand (1977) hydraulic handbooks. Friction loss tables for plastic, gated aluminum, concrete, or transite pipe are also in Section 11.4 or can be obtained directly from the manufacturer.

As illustrated in equation [10.1], the pump will be expected to add energy to the water to overcome the friction loss. As the discharge increases the velocity also increases linearly, but the friction loss is a function of the velocity squared. For irrigation systems having very long pipelines or undersized pipe, the friction loss can be very significant. Because of current energy costs it is often economically feasible, due to friction losses, to select a larger pipe size if the water velocities are in excess of 1.5 m/s (5 ft/s).

Friction losses must be considered on both the intake and discharge side of the pump. It is necessary to compute the friction loss on the suction side of

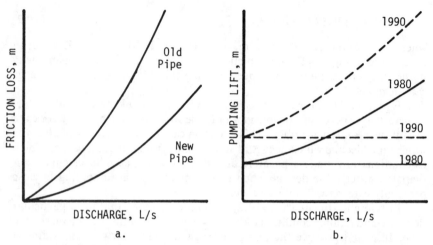

FIG. 10.4 The two illustrations show how the system head curve could change with time due to increased friction in (a) and a lower water table in (b).

centrifugal pumps, especially, to assure that there is sufficient net positive suction head available to prevent cavitation. This will be disussed more later in this chapter.

The friction loss in the pump head and column pipe for a vertical turbine pump must be computed. This can best be done by utilizing manufacturer's tables or curves which report friction losses in m/100 m of the column pipe as a function of the pump discharge rate. Friction loss in the column should be kept below 5 m/100 m. If it exceeds this amount consideration should be given to selecting the next larger column pipe size.

Operating heads. All irrigation systems using pumps require some operating pressure or head with the possible exception of the case where water is discharged directly into an open ditch or partially filled pipeline. The operating pressure and its relation to discharge are discussed in Chapters 14, 15, and 16. Operating pressures for existing systems may be continuously increasing functions as illustrated in Fig. 10.3, or they may be step type functions. An example of an irrigation system with a step type function would be a center pivot with an end gun which periodically starts, and stops.

Proper selection of pumps for a specific system requires knowledge of how the operating head changes. We recommend that the operating head versus discharge relationship be evaluated for the entire range of operating conditions. When the operating head and/or discharge vary significantly, then special attention must be given to select a pump that can satisfy all conditions. Most pumps will not operate efficiently over wide ranges in operating heads. Assuming there is a particular condition that is most prevalent, then the pump must be selected to operate efficiently for that set of conditions and have the capability to operate for all other conditions.

10.4.2 Variations in System Head

The system head curve is definitely time dependent due to variations in well drawdown, friction, operating conditions, and static water level changes. Fig. 10.4 illustrates two conditions which would cause different system head curves.

In Fig. 10.4a, the two curves show how friction losses increase in old steel or iron pipe. Fortunately, aluminum, polyvinylchloride plastic (PVC) and transite pipes, which are commonly used since the late 1970's, may have little change in friction factors with time. However, the friction factor for these materials could change due to deposition of salts. Some steel or cast iron pipes have been nearly closed off by development of oxidized ferrous nodules. In other instances, the pipes have rusted through and begin to leak. Column pipe for vertical turbine pumps is usually steel and is quite susceptible to deposition or failure due to rust.

When the static water level in a well or for a surface supply declines as illustrated in Fig. 10.4b, then the system head curve changes. The condition illustrated is for a well where the water table declines as would be expected in a groundwater mining area. Note that the 1980 and estimated 1990 static lifts are parallel, but the addition of the drawdowns results in curves that are not parallel. This is due to a greater 1990 drawdown caused by the reduction in saturated thickness because of declining water levels.

Proper selection of a pump requires consideration of how pumping conditions will change over the anticipated life of the pump. For example, if additional wells are being drilled and new land irrigated, then one might expect the rate of water level decline to accelerate. On the other hand, if pumping is being limited by water administrators or artificial recharge initiated by a groundwater management district, then static water levels might stabilize or even recover.

The importance of knowing if there will be well interference between the depression cones of two or more wells or the impact of impermeable or recharge boundaries on pumping wells cannot be overemphasized. Failure to consider these factors could result in the installation of an oversized pump which will surge during significant periods of the irrigation season.

10.5 PUMP CHARACTERISTIC CURVES

Pumps, because of their mechanical nature, have certain well defined operating properties. These properties vary between types of pumps, manufacturers and models. Most pump manufacturers publish information which describes how each pump will perform. This information is often referred to as a pump's characteristic curves. Data for these curves are developed by testing a number of pumps of a specific model from which a set of curves or tables is prepared which represent the specific operating conditions. Some manufacturers' curves represent an average performance for the initial test group, however other manufacturers may prepare their curves for the pump having the poorest performance. In some instances the manufacturer may even add a factor of safety to assure that all the pumps will provide a greater head than that published for a specific discharge.

Manufacturers, as part of their quality control, will periodically test a few pumps of specific models for which there are published curves to determine if the curves adequately represent expected performance. Because of minor variations in manufacturing a new pump may not perform exactly as the performance curve indicates. Manufacturers can develop performance curves for a specific pump from a pump test. The cost of an individual pump test is billed to the customer and could range from only a few hundred dollars

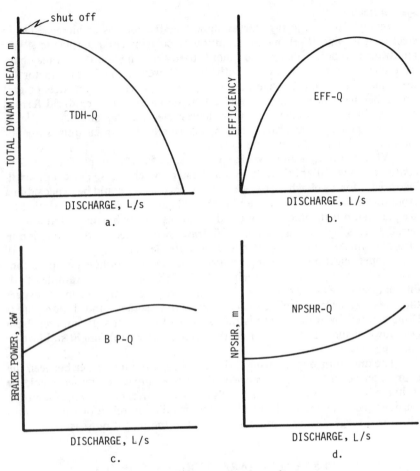

FIG. 10.5 Typical characteristic curves for a pump.

to as much as $1,000 for typical farm irrigation pumps. Most large pumps used in pump stations typical of those designed by the Water and Power Resources Service are individually tested to develop or confirm the characteristic curves.

The performance of pumps will change with time. Since they are mechanical devices they will wear and the rate of wear is dependent upon the environment in which they are operated. For example, when pumping muddy water or water from a well containing sand, both the centrifugal and turbine pumps will be subjected to above normal wear. Replacement of the impeller, wear rings, or even the entire bowl assembly may be required every year if wear is excessive. The best way to evaluate an installed pump's performance is to do a field test as described later in this chapter. The field test should provide information needed for decisions on pump repair.

10.5.1 Pump Characteristic Curves

There are four different characteristic curves that are most commonly provided by a manufacturer. Some manufacturers use tables but the most common approach is to use graphs. The four types of curves are illustrated in Fig. 10.5.

Total dynamic head versus discharge. This curve is often referred to as the TDH-Q curve and relates the head produced by the pump as a function of the discharge. Generally these curves will dip downward to the right although there are some pumps which will have multiple humps. The most common curves for irrigation pumps have shapes similar to those in Figs. 10.5a and 10.6.

The total dynamic head, TDH, which a pump must impart to the liquid can be computed from equation [10.1]. This is also equal to the head from a system head curve for a specific pump discharge, Q. The specific operating point will be illustrated in Section 10.6.

The TDH-Q curve for a single stage pump is shown in Fig. 10.5a. A later section of this chapter describes multiple-stage pumps. The curve can be used to evaluate how the discharge will vary due to fluctuations in the total dynamic head. As the head increases the discharge decreases and vice versa.

If a pump is operated against a closed valve, the head generated is referred to as the shut-off head. Shut-off head is illustrated in Fig. 10.5a. Note that the efficiency of the pump at this point is zero because the pump still requires energy to drive it. For turbine or centrifugal pumps it is necessary to know the shut-off head. The pipe on the discharge side must be capable of withstanding the shut-off head in case a valve is accidently closed on the discharge side.

Efficiency versus discharge. The curve relating efficiency to discharge is labeled Eff-Q curve in Fig. 10.5b. The Eff-Q relationship is also drawn as a series of envelope curves upon the TDH-Q curve in Fig. 10.6. There is generally only one peak efficiency which is related to a specific discharge. If the pump can be operated at this discharge then for a given amount of energy input to the pump, the output work will be maximized.

Efficiencies vary between types of pumps, manufacturers and models. Generally, the larger pumps have higher efficiencies. The efficiency also is related to the types of materials used in construction, the finish on the castings or machining, and the type and number of bearings used. For example, enameled impellers, which are smoother than bronze or steel, will result in a higher efficiency.

Efficiency is defined as the output work divided by the input work. The output work the pump does on the liquid is commonly referred to as the water power, WP (water horse-power in English units).

$$WP = \frac{Q \cdot TDH}{C} \quad\dots\dots\dots\dots\dots\dots\dots\dots\dots\dots\dots\dots\dots\dots\dots\dots \text{[10.3]}$$

where Q = the discharge flow rate, L/s (gpm); TDH = total dynamic head, m (ft); C = coefficient to convert to energy units, 102.0 (3960); and WP = output work, kW (hp).

The input work is generally expressed as the size of power plant which would be required to drive the pump. This is commonly referred to in English units as the brake horsepower, BHP. Brake power in SI units is expressed as kilowatts.

The equation for defining the pump efficiency, E_{pump}, can be developed by dividing equation [10.3] by the input power, BP:

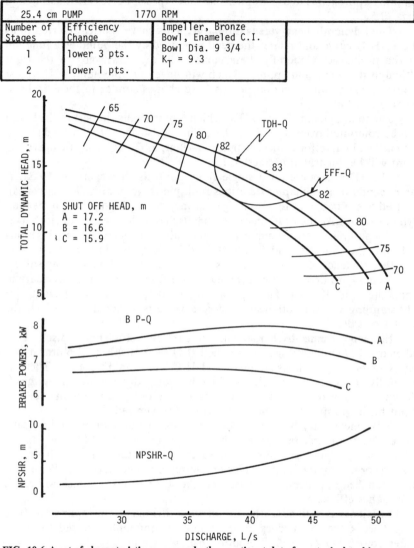

FIG. 10.6 A set of characteristic curves and other pertinent data for a typical turbine pump.

$$E_{pump} = \frac{Output\ Work}{Input\ Work} = \frac{WP}{BP} = \frac{Q \cdot TDH/C}{BP} \dots\dots\dots\dots\dots \quad [10.4]$$

where the terms are defined as before.

Knowing the pumping rate, total pumping head and pump efficiency, it is possible to use Equation [10.4] to compute the input power required to drive the pump. As reported in manufacturer's curves, the pump efficiency for a turbine pump would not include any line-shaft bearing losses.

For a turbine pump the manufacturer's reported efficiency is for a specific number of stages. If, for a specific application, the number of stages differs, then it is necessary to adjust the reported efficiencies upward or downward depending on the number of stages. Fig. 10.6 indicates that efficiency values as graphed must be lowered three percentage points for only a single stage pump, lowered one percentage point for a two-stage pump and would remain unchanged for more than three stages.

Input power versus discharge. As mentioned previously, the input power is referred to as the brake power required to drive the pump. It is expressed as a kilowatt demand (power) and the curve is commonly called BP-Q curve. It should be noted that even at zero discarge when the pump is operating against the shut-off head, an input of energy is needed.

The shape of the BP-Q curve can take several different forms. The most common form for irrigation pumps is similar to the curves of Figs. 10.5c and 10.6. In other instances the BP-Q curve will have the highest kilowatt demand at the lowest discharge rate and the required input power will continue to decline as Q increases. The shape of the BP-Q curve is a function of the TDH-Q and Eff-Q curves. The vertical scale for most BP-Q curves is so small that it cannot be read with an accuracy of more than one or two kW. Manufacturers suggest engineers compute the BP required to drive a pump from equation [10.4] rather than use the curves.

Net positive suction head versus discharge. The fourth characteristic curve is the net positive suction head required, NPSHR, versus discharge relation, NPSHR-Q curve. The NPSHR is the amount of energy required to move the water into the eye of the impeller and is a function of the pump design. This characteristic also varies for different types of pumps, manufacturers and models. Its value is determined by the manufacturer from laboratory tests. The NPSHR is a function of the pump speed, impeller shape, liquid properties, and discharge rate. If sufficient energy is not present in the liquid on the intake side of the pump to move the fluid into the eye of the impeller, then the liquid will vaporize and pump cavitation will occur. Cavitation should be avoided as it physically destroys the pump.

A thorough discussion of how to use NPSHR-Q values is given in a succeeding section. It is only necessary to consider NPSHR for the first stage of a multi-stage pump. However, for other applications of pumps in series NPSHR must be checked for each pump.

10.5.2 Pumps in Series

Two or more pumps may be connected and operated in series. Pumps in series are connected so that the discharge from the first pump or stage is piped into the inlet side of the second pump. If more than two pumps are in series, they are connected so all the flow successively passes from one pump to the next with each pump adding more energy to the fluid. This type of operation would be common where the same discharge rate is needed but a larger head is required than can be produced with a single pump.

A multi-stage turbine or submersible turbine pump consists of single stages connected in series. The same discharge passes through all stages and each one adds additional head to the fluid. Another example of pumps operating in series is where a centrifugal or turbine pump takes water from a stream or well and pumps it to another centrifugal pump which then boosts the pressure for use in a sprinkler system.

For two pumps operating in series the combined head is equal to the sum of the individual heads for a specific discharge as shown in Fig. 10.7. By selecting several discharges and the respective heads, the combined head,

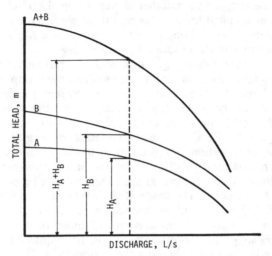

FIG. 10.7 Characteristic curve of TDH-Q for two pumps
operating in series.

which is the sum of H_A plus H_B is obtained to develop the new TDH_{A+B}-Q
curve. For three or more pumps in series the combined TDH-Q curve is
derived by summing the individual heads for specific discharges. If each
pump is identical, like in a multi-stage turbine or submersible turbine pump,
it is only necessary to multiply the head for a particular discharge by the
number of stages to get the combined value.

For pumps in series the flow from one pump is identical to the flow in
the others, i.e. $Q_A = Q_B = Q_C$. The combined input energy for a series of
pumps is the sum of the input energies for each pump for the specific Q. A
combined BP-Q curve can be developed in a similar fashion to the combined
TDH-Q curve.

The equation to compute the combined efficiency values, is:

$$E_{series} = \frac{Q(TDH_A + TDH_B)}{102(BP_A + BP_B)} \quad\dots\dots\dots\dots\dots\dots\dots\dots\dots\dots\dots\quad [10.5]$$

where the Q is in L/s; TDH is in m and BP is in kW.

The methodology requires the selection of a specific discharge; then
determine values for TDH_A, TDH_B, BP_A and BP_B; calculate the new efficiency and then plot this new efficiency versus the selected discharge. Additional
discharge values should be selected and the process repeated. If all the
pumps in series are identical then the Eff-Q curve will be the same as for a
single stage.

As mentioned previously, the NPSHR-Q characteristic must be satisfied
for each stage but a combined curve has no meaning.

10.5.3 Pumps in Parallel

Two or more pumps may be operated in parallel. A typical example
would be where two or more pumps draw water from a single pump sump
and the individual flows are discharged into a single pipeline. Another example would be where several small wells are individually pumped with small
turbine pumps and the flows are discharged into a common pipeline. Pumps

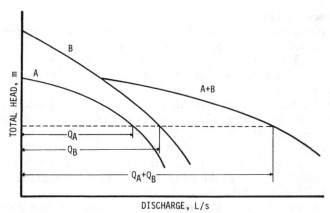

DISCHARGE, L/s

FIG. 10.8 Characteristic curve of TDH-Q for two pumps operating in parallel.

often are operated in parallel where the system requires wide variations in discharges for approximately the same heads. Watering of parks or golf courses often requires pumps both in parallel and in series.

To properly select pumps to be operated in parallel, it is necessary to develop their combined operating characteristic curves. The procedure for developing a combined TDH-Q curve is illustrated in Fig. 10.8. First select a particular operating head and then determine Q_A and Q_B. The combined flow is the sum of the two and it should be plotted against the originally selected head. New values of head are selected and the process repeated. Only if the two pumps are identical will Q_A equal Q_B

To determine the BP-Q relationships it is necessary to add the input energy required by each pump for the same specific head, add BP_A to BP_B, and then plot against the combined flow, Q_{A+B}. Again, it is necessary to repeat the procedure for different heads.

The new efficiency calculated, for two pumps in parallel is:

$$E_{parallel} = \frac{(Q_A + Q_B)TDH}{102\,(BP_A + BP_B)} \quad \dotfill \quad [10.6]$$

where the Q's are in L/s, TDH is in m and the BP's are in kW.

The methodology requires selection of a specific operating head; determining respective values for Q_A, Q_B, BP_A and BP_B; calculate the new efficiency and plot against the combined discharge Q_A plus Q_B. This new Eff-Q curve will likely have more than one peak efficiency value. For more than two pumps in parallel it is necessary to add in the additional discharge in the numerator and BP values in the denominator.

If the pumps are not taking water from a common source, which results in different pumping heads generated by each pump, then the analysis becomes more difficult. In practice there are many situations where two or more pumps are operating in parallel and they are not properly selected. The consequence is that the pumps oppose each other resulting in fluctuating discharges, one pump provides all the water while the other operates inefficiently or may pump water from one pump back down another.

10.5.4 Speed Variations

The performance of a centrifugal or turbine pump is affected by the rotational speed of the impeller. The characteristic curves described in subsection 10.5.3 are determined for a specific impeller speed. The speed is usually noted as on Fig. 10.6 where the pump speed was 1770 revolutions per minute, rpm.

Most electrically driven irrigation pumps are directly coupled to the motor shaft and turn at the same speed as the motor. Common operating speeds for electric motors are 1150, 1750 or 3500 rpm. Most pumps manufacturers have standardized the speeds for which they publish characteristic curves. Many of the smaller pumps operate at 3500 rpm while the larger pumps operate at the lower speeds. The most common speed for electrically driven well pumps is 1750 rpm.

Belt driven electric or internal combustion engine driven pumps can operate at a range of speeds. Speeds for belt driven electric pumps can be varied by changing belt pulleys or by changing to a different speed electric motor coupled with different pulley ratios. Internal combustion engine powered pumps are connected through a driveshaft and gearhead to transmit the power to the pump lineshaft. Speeds for engine driven pumps can be varied by changing the throttle setting, changing the gearhead ratio or a combination of these methods.

Variation in pump speed can be used advantageously to operate a single pump to match two different system head curves or to provide two different design discharges. Reduction in pump speed is also one of the best ways to correct surging pumps. Selection of the proper pump to operate efficiently at two different speeds increases the complexity of pump selection several fold.

Affinity Law. Mathematical expressions known as Affinity Law I are described in the Colt Industries Handbook (1979). It is a series of three equations which relate how the discharge, head and required brake power will vary with a change in pump speed. The equations are:

$$\frac{Q_1}{Q_2} = \frac{RPM_1}{RPM_2} \quad \dots\dots\dots\dots\dots\dots\dots\dots\dots\dots\dots\dots\dots\dots\dots\dots \quad [10.7]$$

$$\frac{H_1}{H_2} = (\frac{RPM_1}{RPM_2})^2 \quad \dots\dots\dots\dots\dots\dots\dots\dots\dots\dots\dots\dots\dots\dots\dots \quad [10.8]$$

$$\frac{BP_1}{BP_2} = (\frac{RPM_1}{RPM_2})^3 \quad \dots\dots\dots\dots\dots\dots\dots\dots\dots\dots\dots\dots\dots\dots\dots \quad [10.9]$$

where Q is expressed as L/s, H is head generated by the pump in m, BP is the input energy required to drive the pump expressed as kW, RPM is the pump speed in revolutions per minute, and the subscripts 1 and 2 relate to respective values of the parameters for the two speeds.

The Affinity Law indicates that the discharge varies linearly with change in speed, the head varies as the square of the ratio of the two speeds, and the brake power has a cubic relationship. A slight increase in pump speed will produce more water at a higher head but will require considerably more power to drive the pump.

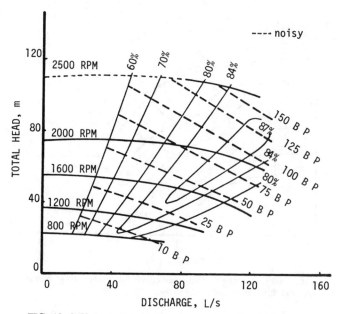

FIG. 10. 9 Characteristic performance curves of a single stage pump for several operating speeds.

The Affinity Law is valid for a wide range in pump speeds because all the physical components have the same dimensions. This law says nothing about how the pump efficiency will change with speed. Generally speaking, pumps that are efficient at one speed will probably be efficient at a slightly different speed. For increases in pump speed the NPSHR increases but can not be determined from these equations.

Characteristic curves. Manufacturers can test their pumps at various operating speeds and determine how the performance characteristics change as the speed changes. They sometimes print complex characteristic curves as a function of pump speed as shown in Fig. 10.9. Note that the drop in the TDH-Q curve from 2000 to 1600 to 1200 rpm is not linear. Also note how the efficiency envelope curves have a shape that is similar to common shapes for system head curves. It is because of this similarity in shape that the efficiency of a pump does not often change with variations in pump speed. In general, the use of equations [10.7], [10.8], and [10.9] to compute new operating points for specific speed changes will result in very similar curves to those obtained by the manufacturers tests.

The input energy, BP, on Fig. 10.9 is plotted as a family of lines which indicate the size of motor or engine required to drive the pump, disregarding all line shaft losses. The particular plot is very useful for interpolating operating conditions and energy requirements for speeds other than those plotted.

10.5.5 Changes in Impeller Diameter

It is possible to trim the diameter of the impeller in both centrifugal and turbine pumps, thus changing the pump's performance. Trimming impellers is a common practice to better match a pump for a specific job. In Fig. 10.6 there are three sets of TDH-Q and BP-Q curves each labeled A, B and C,

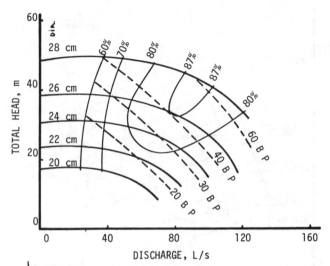

FIG. 10.10 Characteristic performance curves for a single stage pump with several different impeller diameters.

respectively. The A curve is for a full diameter impeller, the B curve is for a small trim in impeller diameter and the C curve is for an even greater trim.

Affinity Law II. Similar to the impact of change in speed on performance there are three mathematical equations known as Affinity Law II which relate the impact of impeller diameter to changes in pump performance. Colt Industries Handbook (1979) describes these relationships in detail and gives examples of how they can be used. The equations are:

$$\frac{Q_1}{Q_2} = \frac{D_1}{D_2} \quad\dotfill \quad [10.10]$$

$$\frac{H_1}{H_2} = (\frac{D_1}{D_2})^2 \quad\dotfill \quad [10.11]$$

$$\frac{BP_1}{BP_2} = (\frac{D_1}{D_2})^3 \quad\dotfill \quad [10.12]$$

where Q, H and BP are as defined before and D is the diameter of the impellers. Note that the variation in discharge is linearly related to change in diameter, the change in pump head is a function of the square of the ratio of impeller diameters and the input energy required varies with the cube of the ratio of diameters.

Pump impeller diameters are only reduced in size by trimming. It is not feasible to increase their diameter by welding on additional metal. The applicability of the Affinity Law II is limited to a maximum trim of about 10 percent of the original diameter. Computed performance for trims in excess of 10 percent could be significantly in error. This limitation on the law must be imposed because trimming physically changes the relationship between physical dimensions of the impeller and bowl or volute case.

Characteristic curves. Manufacturers often test their pumps with various diameter impellers and report the results as in Figs. 10.6 or 10.10. The nonlinearity between successive reductions in diameter due to equal amounts of trim is less obvious in Fig. 10.10 as compared to the impact of speed changes shown in Fig. 10.9. The shape of the efficiency envelopes is again similar to the shape of many system head curves which explains why an efficient pump that has its impeller trimmed will probably still be efficient.

Interpolation of trim amounts from characteristic curves similar to Figs. 10.6 or 10.10 is a common practice. It would be possible to also use equations [10.10], [10.11], and [10.12] to compute specific operating points. The recommended procedure, as described in Section 10.6, is to superimpose the system head curve over the characteristic curves and then select the impeller diameter which will produce the desired discharge.

10.5.6 Hydraulic Thrust

As energy is imparted to the water there are forces which act upon the impeller and other components of the pump. These forces can be sizeable and can stretch or deform drive or line shafts. Bearings must be designed into the pump, motor or gearhead to support these forces. It is beyond the scope of this text to describe how to compute the total thrust loads in a pump. Most pump manufacturers catalogs contain an engineering section with the necessary graphs, tables and sample calculations to illustrate how to compute thrust loads. Lineshaft stretch will be briefly discussed in subsection 10.6.5.

10.6 PUMP SELECTION

Previous sections have described system head and pump characteristic curves. Proper selection of a pump requires knowledge and use of these curves. The normal procedure for selecting a pump is to first determine the system head curve and desired discharge. Then pump manufacturer's catalogs with characteristic curves or tables are used to select pump models for consideration that will operate efficiently at or near the design discharge.

Engineers, pump installers and irrigation equipment salesmen have historically determined the design discharge and the system's total dynamic head for that discharge. These values were the ones used to select the pump. Very few irrigation or drainage systems will have fixed operating conditions. By determining the system head curve or curves for a range of discharges above and below the design discharge, sufficient information will be available to evaluate pump performance for all expected operating conditions.

10.6.1 Matching Pump Characteristic and System Head Curves

It is possible to superimpose a pump's characteristic curves of TDH-Q and Eff-Q upon a system head curve. This is illustrated in Fig. 10.11. The point where the TDH-Q and system head curves cross would be the actual operating point. The pump would produce a discharge of 60 L/s with a dynamic head of 17 m. Note that the operating point is just to the right hand side of the peak efficiency which is very desirable.

In subsection 10.4.1 considerable discussion was presented on how system head curves might change with time, operating conditions or uncontrolled physical processes. It is necessary to estimate the two extreme and average system head curves as illustrated in Fig. 10.12a. Pumps must be

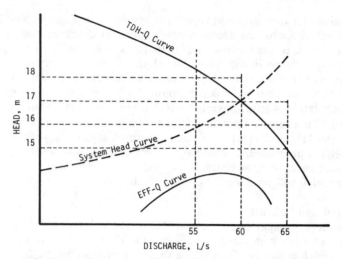

FIG. 10.11 Superposition of the characteristic curves of TDH-Q and
Eff-Q upon a system head curve.

selected to operate efficiently under the predominant operating conditions,
but they must also be capable of satisfying the extreme circumstances. As illustrated in Fig. 10.12a, the pump would have to be able to produce
significantly different dynamic heads for a desired discharge of 60 L/s. In
subsection 10.5.1 we described how the discharge decreases as the required
head increases for both centrifugal and turbine pumps. Referring to Fig.
10.11, the discharge will drop from 60 L/s to 55 L/s when the head is increased 1 m. If the system head decreases from 17 m to 15 m the discharge will increase to 65 L/s.

The variation in discharge rate that occurs due to a change in the system
head curve is illustrated in Fig. 10.12b. The curves sloping upward to the
right are system head curves for May and September. The May curve is lower
due to the higher static water levels prior to the pumping season which then
decline throughout the season to a level represented by the September curve.
The two curves labelled A and B which slope downward to the right are

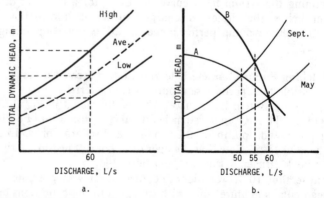

FIG. 10.12 Illustrations of system head curve variations and the impact of these variations on
pump discharge.

TDH-Q curves for two different pumps. For pump A the discharge in May would be 60 L/s but this would drop to 50 L/s by September because of the increased head. The discharge for A throughout the pumping season would range between 50 and 60 L/s.

The discharge for pump B would also be 60 L/s in May but would drop to only 55 L/s in September. Note that the pump B total dynamic head in September would be much higher than for pump A which results in a smaller discharge decline for pump B.

The superposition of the TDH-Q and Eff-Q curves for several pumps being considered upon the system head curves provides a visual means of evaluating how each of the pumps will operate. Care must be taken to assure that the all superimposed graphs have the same vertical and horizontal scales. The pump with the steepest TDH-Q curve will have the least fluctuation in discharge and the pump with the flatest curve will have most variation in discharge. For example, with sprinkler systems the uniformity of water application is directly related to maintaining a constant operating pressure or flow rate at the upstream end of the system. If water is provided from a well where the pumping levels vary over time then it would be desirable to select the pump with the steepest curve to maintain high water application uniformity.

By superimposing both the TDH-Q and Eff-Q curves on the system head curve, it is possible to see how a pump's efficiency will change as the operating points change. For pumps with steeper TDH-Q curves the change in efficiency will often be less but this is not always true. The shape of the Eff-Q curve is not the same for each pump and this will have a large influence on which pump will operate most efficiently over the range in operating conditions.

10.6.2 Selecting the Number of Stages

In subsection 10.5.2 a procedure to obtain the characteristic curves for two or more pumps operating in series was described. By superimposition of these combined curves on the system head curve, as illustrated in Fig. 10.11 and Fig. 10.12, it is possible to evaluate how either a multistage turbine pump or multicentrifugal pump arrangement would operate. When the required pumping heads are greater than what a single stage can develop, then it is necessary to use multistage turbines or two or more centrifugal pumps in series. The question is: how many stages or series pumps are needed.

The selection procedure is similar to that for a single stage pump. It requires that the system head curve and design discharge be known. The manufacturers' catalogs are checked to determine likely pumps that will produce the design discharge efficiently. The selection process is illustrated in Fig. 10.13. The system head curve is the dashed line sloping upward to the right. Single stage TDH-Q curves for pumps A and B are shown in the lower part of the figure. The design Q is 60 L/s with a required system dynamic head of 120 m.

To determine the number of stages of the A and B pumps required, it is necessary to divide the 120-m head requirement of the system by the head produced by each pump for the design Q of 60 L/s. For pump A the head produced by a single stage is 20 m and thus it would require six stages. For pump B the single stage head produced at 60 L/s is 40 m and thus only three stages would be required. Once the number of stages is known the multistage

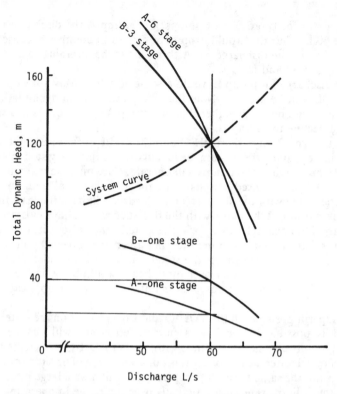

FIG. 10.13 Illustration of how multistage turbine TDH-Q curves are prepared and used to select the best pump.

TDH-Q curve can be developed as discussed in subsection 10.5.2.

Curves of TDH-Q for the six-stage A and three-stage B pumps are plotted on Fig. 10.13. Note that for a single stage pump curve B is steeper than A, but for the multistage curves the six-stage A TDH-Q curve is steeper. The slope of the multistage TDH-Q curve is a function of both the number of stages and the slope of the single stage curves.

For the pumps used in the example illustrated in Fig. 10.13, the division of the 120-m system head by 20-m and 40-m values resulted in exactly six and three stages. Consider pump C which is illustrated in Fig. 10.14. The TDH-Q curve for a single stage of pump C produces 50 m of head at the 60 L/s discharge rate. Division of 120 m by 50 m indicates it would require 2.4 stages. The number of stages must by a whole number. Two stages of pump C would produce a discharge of 57 L/s at 112 m of head. Three stages of pump C would produce 63 L/s at 130 m. Neither the two or three stage model C pump would match the desired conditions.

A solution to the lack of proper match would be to select a three-stage model C pump but trim the impellers. Fig. 10.6 illustrates how the TDH-Q curves change with an impeller trim. The manufacturers curves or the Affinity Law II, equation [10.10], could be used to compute how much trim would be required. Use of the manufacturers curves for specific trim is recommended even though it may be necessary to interpolate to obtain the trimmed diameter.

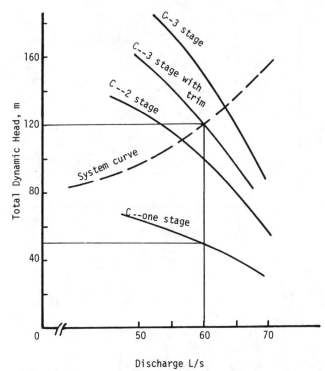

Discharge L/s

FIG. 10.14 Illustration of the effect of trimming the impeller to obtain the desired operating point.

The pump industry has two different approaches for solving the lack of match represented by pump C in Fig. 10.14. One approach is to select the three-stage C full diameter impeller pump and then trim all three impellers to a new diameter which coincides with the desired operating conditions. The other approach is to use manufacturers curves and select, say, two stages with full diameter impellers and trim only one to obtain the match. Some manufacturers recommend one approach over the other. There does not seem to be a concensus as to which is best. Both methods require accurate machining. If only one impeller is trimmed it is recommended that in assembly the stage containing this impeller be the top one on a vertical turbine pump. If the trimmed impeller stage is the lower or first stage there is a higher risk of cavitation problems.

10.6.3 Efficiency Considerations

The importance of efficiency on pump selection must be further emphasized. We stated earlier that only pumps having high efficiencies for the design discharge should be considered as potential candidates. The recommendation that the selected pump should have an operating point slightly on the right-hand side of the peak efficiency is often made. The reason for this recommendation is that as pumps wear and the TDH-Q curve drops, the operating point will move to the left and into the region of higher efficiency. If a new pump is selected that operates on the left-hand side of the maximum efficiency, then as the pump wears the TDH-Q curve lowers and the

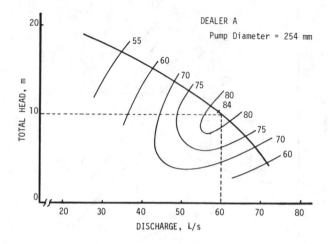

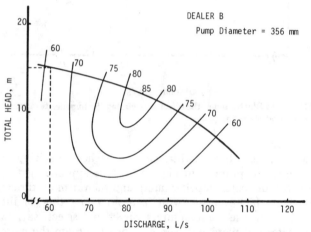

FIG. 10.15 Characteristic curves for two different size pumps showing the operating point for a 60 L/s discharge.

operating point moves to the left into the region of much lower efficiency.

The importance of efficiency on power consumption can be computed with equation [10.4] or equation [10.14] and [10.15] in the next section. A simple example using Fig. 10.15 may also be informative. Assume a farmer contacts two different pump dealers to obtain a pump which will produce 60 L/s (1000 gpm) with a TDH of 30 m (100 ft). Dealer A in Fig. 10.15 selects a pump having a diameter of 254 mm (10 in.) which will produce only 10 m of head at the desired discharge of 60 L/s. Thus, it will require three stages of this pump to match the system requirement. The cost of the three-stage pump is $3000. Note the operating efficiency of this pump is just on the right-hand side of the peak and has a value of 83 percent.

Pump dealer B selects a larger diameter pump, 356 mm (14 in.), which will produce the required discharge of 60 L/s at a 15-m head. To meet the re-quired head of 30 m then it is necessary to have only a two stage pump. The pump cost is $2400 which is lower because of the one less stage. The per stage

TABLE 10.1. COMPARISON OF PERFORMANCE AND
TOTAL PUMPING COSTS FOR THE TWO PUMPS
DESCRIBED IN FIG. 10.15

Item	Pump A	Pump B
Initial investment cost	$3000	$2400
Number of stages	three-stage	two-stage
Size of motor needed	30 kW (40 hp)	37.5 kW (50 hp)
Energy consumed	45,000 kWh	58,900 kWh
Annual power bill*	$2700	$3634

*Assumes 394,560 m³ (320 acre-ft) were pumped and the
cost of power was $0.06/kWh. The annual hours of operation would be 1720.

cost is higher because of the larger diameter. The efficiency of this pump is only 63 percent and it is far left of the peak efficiency.

Since both pumps will produce exactly the same amount of water and meet the system head required, the farmer has historically selected the cheaper pump and been satisfied. Some pump dealers have also emphasized that the one less stage is also better. Little effort prior to 1980 has been made by pump dealers to calculate the impact of the much lower efficiency of pump B on total pumping costs. Table 10.1 briefly compares specific information for the two pumps. Note that the savings in power costs alone is $834 per year which would offset the higher purchase price. In addition, pump B required that a 37.5 kW motor be used instead of the 30 kW motor used on pump A. The difference in motor sizes is all due to the lower efficiency of pump B.

10.6.4 Centrifugal Pump Selection

The selection criteria described previously are applicable for centrifugal pumps. Trimming of the impellers to match the pump to the specific use is quite common. The decisions that must be made for pump selections are: which model of pump to use, how much to trim the impeller, and what speed to turn the impeller. Manufacturer's characteristic curves will contain information needed to make these decisions. It will also be necessary to compute the size of motor or engine needed to drive the pump.

Centrifugal pumps will have a suction and discharge fitting or flange. The pump is connected with standard pipe fittings to both suction and discharge pipes. Care must be taken to minimize friction loss on the suction side. Large friction losses can cause cavitation. Most centrifugal pumps used for irrigation require that a foot valve be installed on the intake end of the suction pipe so that the pump will remain primed at all times. There are some self-priming centrifugal pumps.

10.6.5 Selection of Vertical Turbine Pump Components

The selection of a specific pump model, determination of the number of stages, and need for impeller trim have been discussed previously. Other components that must be selected are the lineshaft, column pipe, pump head and drive units.

Lineshaft selection. It is necessary to select the size of lineshaft which will transmit the required power to the impellers. Lineshafts that are too small will fail due to shear. Manufacturer's curves and tables are used to select the shaft diameter. The variables that must be considered are the rotating speed, RPM; the shear strength of the shaft; and the brake power to be transmitted to the shaft.

The lineshaft is also subjected to vertical forces which stretch the shaft. The vertical forces are called thrust load and are the sum of the weight of the impellers, the weight of the lineshaft, and the hyudraulic thrust on the impellers. Manufacturers provide a thrust constant for each impeller and the downthrust is computed by multiplying the constant by the total dynamic head generated by the pump.

Lineshaft elongation or stretch is computed by using tables or graphs. The critical problem with lineshaft stretch is that it may exceed the clearance an impeller has in the bowl. When this occurs it is either necessary to select a larger shaft with less stretch or pay to have a greater clearance machined in the bowls. Satisfactory operation of a turbine pump requires selection of the proper lineshaft.

A decision on whether a turbine pump is to be oil or water lubricated is needed. See subsection 10.2.2 for a discussion of lubrication. The bearings and lineshaft assemblies are different. For oil lubricated pumps an oil tube is included as part of the column and lineshaft assemblies. The column and lineshaft assemblies may not be interchangeable for different manufacturers.

Column pipe. The column pipe transmits the water from the bowls to the pump discharge head. The major decision for selection is the size and length of the column. The column pipe diameter should be large enough to permit passage of the design discharge with reasonable friction losses. Friction losses in excess of 5 m/100 m should be avoided by chosing a larger column pipe. Friction loss tables express the loss as head loss per 100 m of column pipe as a function of pumping rates, L/s. These tables will be different for water and oil lubricated pumps and for different sizes of lineshafts.

The length of column pipe must be long enough to keep the bowls submerged. This requires that the depth to the water surface in a well or sump be known under pumping conditions. The pumping depth is the sum of the depth to the static water level plus any drawdown as described in subsection 10.4.1.

Suction pipe. It is desirable to have some suction pipe on a turbine pump to cause the water to enter the bottom stage in a vertical direction. The length of suction pipe can vary from less than one to as much as five meters. Long suction pipes are not recommended. It is often necessary to set the pump near the bottom of the well or sump to permit pumping during low water conditions. This physical constraint may limit the length of suction pipe to 20 or 30 cm.

Strainers or screens. When pumping from a well which has a perforated casing it is not common to install a strainer or screen. Pumps operating from open sumps, rivers or lakes should have strainers or screens on the intake side to prevent entrance of objects that might damage the impellers. Strainers or screens may become plugged and when this happens pump cavitations can occur.

Pump discharge head. The discharge head serves three purposes: it supports the total weight of the column, lineshaft and bowl assemblies; it acts as an elbow to redirect the flow from the vertical to a horizontal direction; and it provides a mounting bracket for the motor or gearhead. It is necessary to select a discharge head which is compatible with the size of the column and the physical dimensions of the motor or gearhead. The size of the discharge flange or threaded coupling is also a consideration. It is common to have the same size discharge and column pipes.

The friction loss through the pump head should be less than one meter of head or else a larger pump head should be selected. Manufacturers tables list the friction loss as a function of the pumping rate. Tables also exist listing the physical dimensions. Several different configurations of pump heads exist but the most common are either the type "A" or "H" heads.

Drive units. Selection of either an electric motor or gearhead requires consideration of the physical dimensions and then compatibility with the lineshaft and discharge head. Special short lengths of lineshaft, known as head shafts, may have to be machined to assure that the threaded sections are in the correct location to permit tightening of the lineshaft and impeller adjustment.

Gearheads have different gear ratios which can be installed so as to obtain a change in rotational speed of the driveshaft and lineshaft. Selection of a specific gear ratio depends on the desired speed of the lineshaft and the proposed rotational speed of the driveshaft.

10.6.6 Checks on NPSHA to Prevent Cavitation

Energy is required to move water through the eye of the impeller. This was defined in subsection 10.5.1 as the net positive suction head required, NPSHR. To assure that the required energy is available, an analysis must be made to determine the net positive suction head available, NPSHA. The available head is a function of the system in which the pump operates. It can be calculated for all installations. If the NPSHA does not exceed the NPSHR then the pump will cavitate.

The source of energy to move the liquid into the eye of the pump is either atmospheric pressure or an induced pressure on the intake side of the pump. The equation for computing the NPSHA is as follows:

NPSHA = (barometric pressure, m) – (static suction lift, m)
– (friction loss in the suction pipe, m) – (vapor pressure of the
liquid at the operating temperature, m) . . . .   [10.13]

Values for the atmospheric pressure at various altitudes can be found in the Colt Industries Handbook (1979), Table 24, page 86. The barometric pressure must be divided by the specific gravity of the fluid to obtain an equivalent liquid column height. The vapor pressure and specific gravity of water as a function of temperature can be found in Table 23, page 84, of the same reference.

The barometric pressure head in m, P_b, also can be estimated for a standard atmosphere as: $P_b = 10.33-0.00108$ E where E is the elevation in m. Vapor pressure in m, e, can be estimated from the following tabulation:

°C =	0	5	10	15	20	25	30	35	40	45	50
e =	0.06	0.09	0.13	0.17	0.24	0.32	0.43	0.58	0.76	0.99	1.28 (m)

The friction loss in the suction pipe is a function of the discharge rate through the line and the pipe length, size and fittings. The friction loss must be accurately computed using friction tables or the Darcy-Weisbach or Hazen-Williams formulas.

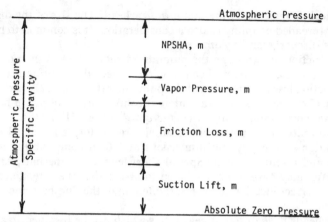

FIG. 10.16 Schematic for NPSHA versus atmospheric pressure, suction lift, friction and vapor pressure.

Example. Compute the NPSHA for a pump that is operating at an elevation of 1524 m (5000 ft) msl where the water temperature is 48.9 °C (120 °F). Assume the discharge to be 63 L/s (1000 gpm) and the suction line is a schedule 40 steel pipe 20.3 cm (8 in.) in diameter and 200 m (656 ft) long. The centrifugal pump will be located 2 m (6.56 ft) above the water surface.

From Table 24 of the Colt Industries Handbook, the atmospheric pressure is 8.60 m (28.2 ft). From Table 23 of the Colt Industries Handbook the vapor pressure for the 48.9 °C water is 1.19 m (3.8 ft) and the specific gravity is 0.989. The friction loss is found, using Table 1, page 47 of the Colt Industries Handbook, to be 1.56 m/100 m length of pipe or for 200 m of pipe this becomes 3.12 m. Substituting the values into equation [10.13], the computed value is:

$$NPSHA = \frac{\text{barometric pressure}}{\text{specific gravity}} - \text{friction} - \text{suction lift} - \text{vapor pressure}$$

$$= \frac{8.60 \text{ m}}{0.989} - 3.12 \text{ m} - 2 \text{ m} - 1.19 = 2.39 \text{ m}$$

A graphical representation of this problem is given in Fig. 10.16.

Knowing the NPSHA for discharge, 63 L/s it is necessary to check if it is greater than the NPSHR value for the same discharge. If not, then the pump will cavitate. To prevent cavitation, the location of the pump can be lowered in reference to the water surface or the suction pipe could be changed to reduce the friction loss. A pump left to cavitate will self-destruct.

Especially for centrifugal pumps where the pump is physically located above the water source it is essential that a check be made to see that the NPSHA is greater than the NPSHR for all operating conditions. Changes in the system head curve with time or a change in water temperature may cause a well-designed pump to cavitate and malfunction.

10.7 POWER UNITS

Electric motors and internal combustion engines are commonly used to convert either electric energy or fossil fuels into mechanical energy. The most common fossil fuels used for pumping are natural gas, diesel, propane, butane and gasoline. Most large irrigation and/or drainage pumps are located at a specific site permitting the use of stationary engines or motors. The portable pumps are generally driven either by an engine mounted on the pumping unit or by the power take-off from a farm tractor.

Rising energy costs have been responsible for doubling or tripling pump operating costs between 1975 and 1980. This trend is predicted to continue with energy costs projected to at least double again between 1980 and 1990. To hold the pumping costs as low as possible it is imperative that the entire pumping plant be efficient. This requires careful selection of the pump and power plant for the specific job.

Previous sections of this chapter have described how to compute the size of power unit required to drive the pump. Oversizing usually results in higher investment costs, higher power costs, lower efficiencies in some instances, and is generally a luxury which most farmers cannot afford. Undersizing, however, usually prevents proper operation of the pump, could also cause lower efficiencies, and usually reduces the life of either the motor or engine because of overheating and excessive wear of bearings and other components.

For centrifugal and short-set turbine pumps it is generally not necessary to compute the energy required to overcome bearing losses in the pump. For turbine pumps having settings of 20 m or more the bearing losses can become significant and must be calculated using manufacturers curves or nomographs. The lineshaft bearing losses must be added to the brake power which is computed with equation [10.4].

10.7.1 Electric Motor Driven Pumps

Prior to 1945, most electrically driven irrigation or drainage pumps were belt driven. Now the most common mode of operation is the direct coupling of the motor and pump drive or line shaft. For many centrifugal pumps the impeller is mounted solidly onto the shaft which supports the armature of the motor. Hollowshaft electric motors are now commonly used with vertical turbine pumps. The advantage of the direct coupled motor and pump over belt drives is that the safety hazards associated with belt drives are eliminated and there is no loss of energy due to belt slippage. Belt driven pumps had one advantage in that pump speeds could be easily changed by selecting other pulleys.

Motor selection. The size of electric motor must be large enough to deliver the required energy to the pump. In years past it was a common practice to overload electric motors 10 to 15 percent above their nameplate rating. In most instances this resulted in satisfactory performance when the motors were the older models with cast iron frames, copper windings and rated service factors of 1.15 or greater. During the late 1970's, the electric motor industry began using lighter motor frames, aluminum windings, different insulation material, and changed the service factor to 1.0 or 1.05. This lower service factor number implies that the motor will no longer stand the 10 to 15 percent overload. In general, motors available on the market in 1980 should not be operated continuously above their rated load.

When motors are used at altitudes over 1100 m mean sea level (msl), most motor suppliers suggest the rated load capacity be reduced to prevent overheating. Motors operating at altitudes above 1100 m may not have suffi-

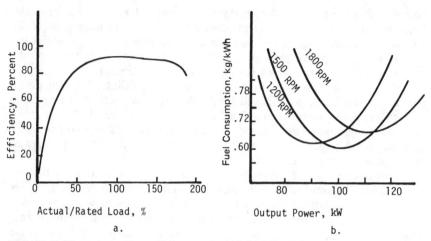

FIG. 10.17 Illustrations of how efficiency varies with load for electric motors and how output power increases with increased engine speed.

cient air passing through them to prevent overheating. Where the motor is operated at high temperatures such as in agricultural fields during July or August, it may also be necessary to derate their nameplate energy output. Manufacturer's literature should be consulted for specific recommendations on how to compute the lower rated power when the motor is to be used at high altitudes or where air temperatures exceed 37 °C.

Most motors will operate under load in excess of their rated output, but this results in much hotter operating temperatures. When a motor operates for significant periods of time at temperatures in excess of its design operating temperature, the life of the motor is reduced significantly. A common rule of thumb is that for each 10 °C above the motor's rated operating temperature, the motor life will be reduced by 50 percent. Overheated motors must be avoided to obtain a long service life.

Motor efficiency. Most motor manufacturers provide data on the operating efficiency of their product as a function of the load factor. A typical curve is shown in Fig. 10.17a which shows that the efficiency is zero at no load but rises steeply so that at 50 percent of rated load the efficiency is over 80 percent. When the load increases to 75 percent of the rated load, the efficiency is near its maximum and will stay at that level until it begins to drop slowly when the motor is loaded to 125 to 150 percent of its rated load. Over a fairly wide range of load, the efficiency is nearly constant.

Smaller electric motors have lower operating efficiencies. For 7.5 kW or smaller motors the maximum efficiency is usually below 88 percent. Larger motors have higher efficiencies but seldom more than 92 percent. For 75 kW sized motors or larger the range in peak efficiencies is from 90 to 92 percent.

Electric motors do not have internal components like engines which would wear resulting in reduced operating efficiencies. Bearings do wear but when this occurs excessive heat is generated and they quickly sieze causing motor failure. It is safe to assume that if a motor runs at all it will be operating at or very near its peak efficiency if the load is between 75 and 125 percent of its rated load.

Overall efficiency. Overall pumping plant efficiency for electrically powered plants is a product of the pump efficiency and motor efficiency:

$$E_{overall} = E_{pump} \times E_{motor} = \frac{Q \cdot TDH}{C \cdot kW} \qquad \dotfill \qquad [10.14]$$

where the terms are the same as described in equation [10.3] and equation [10.4], except that kW is the input energy to the motor measured in kilowatts. This efficiency is also commonly referred to as wire-to-water efficiency. The maximum possible theoretical value for this overall efficiency ranges from 72 to 77 percent. For example, a pump having a maximum efficiency of 84 percent coupled with a 90 percent efficient motor could have an overall efficiency of 76 percent.

As discussed in section 10.6 and again in 10.8, it is now very important to select efficient pumps. Farmers are now being advised that an overall efficiency of at least 65 percent is achievable. If field tests of new pumping plants do not meet this level, then the final payment to pump installers should be withheld until corrections or replacements are made which do provide that level of efficiency.

Motor operating problems. Problems develop when motors are subjected to unbalanced voltages, high voltages or low voltages. All of these problems will result in overheating of the motor, reduced motor life and, in severe cases, the motor will not even operate. Generally, correction of these problems is the responsibility of the electric company. An exception is where wires that are too small are installed to transmit the electricity from the power company's main lines to the motor resulting in excessive voltage drop over the line and low voltage at the motor.

Submersible turbine pumps with the motor located in the well beneath the pump are particularly susceptible to unbalanced voltages or voltage surges. This could result in motor burn-out and expensive repairs.

In ranking the reasons for motor failure the major ones include: overloading, lightning strikes, destruction of insulation by mice, and lack of proper lubrication.

Motor options. There are a number of different types of motors available including: non-weather protected; weather protected; dust proof; explosion proof; low, medium and high thrust motors; ones with reverse ratchets and other less important options. Generally, irrigation and drainage applications will use weather protected motors of medium to high thrust. Most motors used on turbine pumps will also utilize a reverse ratchet to prevent the lineshaft from turning backward when the motor is stopped.

Non-weather protected motors are used but this requires construction of a pumphouse to prevent rain or other moisture from directly entering the motor causing a short circuit. When pump houses are used they must provide sufficient ventilation so the motor can be air cooled. The use of sun shields to prevent sunlight from directly striking the motor and control panels is recommended and this also prevents rainfall or water applied from overhead sprinklers from striking the motor.

Some large motors also have a water cooled option where a very small part of the pumped water is circulated through the motor casting thus dissipating motor heat. Many motors have some type of heat sensing device which stops the motor when it becomes overheated.

10.7.2 Internal Combustion Engine Driven Pumps

Internal combustion engines and even steam engines were used to power pumps as early as the 1920's and 1930's. The Irrigation Survey published in the Irrigation Journal (1979) documents the different types of power used to pump water. In some of the states over 90 percent of the pumps are internal combustion engine driven. In other states, such as California, Oregon and Washington where electricity has been readily available, the pumps are almost all electrically driven.

Natural gas, where it is available, has been a cheap source of power for pumping. Large numbers of pumps in Texas, Oklahoma, Kansas, Nebraska, Colorado, New Mexico, and Arizona have relied on this source of power for the past three decades. Federal regulation of natural gas prices has helped make this a cheap fuel, but threats of deregulation in the 1980's may cause its price to quadruple making it an expensive fuel source.

Diesel, butane, propane, and to a lesser extent gasoline have been used as fuel sources where neither electric or natural gas lines exist. Price structure has also made diesel competitive with electricity in some areas. These fuel sources are used to power most of the portable pumps.

There are direct coupled centrifugal pumps which implies that the pump impeller is mounted on the engine shaft or a shaft which is coupled to the engine by the clutching mechanism. Most turbine pumps powered by engines require the use of a gearhead and driveshaft to transmit the power from the engine to the pump lineshaft.

Engine selection. There are many different sizes, makes and models of engines. Each of these engines has a set of operating characteristics which describe how it will perform over a wide range of conditions. Each engine has its most efficient zone of operation. There are a number of ways to plot engine performance data, but Fig. 10.17b illustrates how fuel consumption and output power are related to engine speed. These are the typical horseshoe curves that illustrate how output power increases as engine speed increases. Note in this case there is a particular engine speed where the engine is most efficient. For the curves in Fig. 10.17 the engine would operate most efficiently at 1500 RPM producing about 100 kW with a diesel fuel consumption of about 0.60 kg/kWh (0.36 lb/BP-h).

The speed of the engine is generally controlled by the throttle setting and the load. Proper selection of the right pump and gearhead is difficult but extremely important to maintain high efficiencies and desired operating conditions. One of the biggest reasons for inefficiencies is misapplication of equipment.

A detailed description of the process for selecting an internal combustion engine and gearhead for a turbine pump is not possible here, but the steps to take are: (a) select the pump that will operate efficiently while producing the desired discharge against the system's total dynamic head; (b) compute the required brake power to drive the pump, equation [10.4], and include the bearing losses; (c) consult engine manufacturers catalogs and select an engine that will produce the desired driveshaft power while operating efficiently; (d) knowing the desired pump speed in rpm and the speed of the selected engine at the most efficient point of operation, then select a gearhead with gear ratios to couple the two units together with each turning at the proper speed.

There are light-weight and heavy duty engines. There are also low, medium and high compression engines. Their performance, cost, efficiency and expected lifetime vary significantly. Lightweight industrial or automotive engines are usually the cheapest, have a short lifetime, are low compression and thus have low efficiencies. The heavy duty engines may initially cost three to five times the cost of the lightweight engines, but they have high compression ratios which permit more efficient burning of the fuel. The rotating speed of the heavy duty engine is generally much slower and the lifetime much longer. Overhauls, as needed, are common with the heavy duty engine, but most lightweight engines are scrapped rather than overhauled because all components are worn to the point that the repair is more costly than buying another new lightweight engine.

Manufacturers' curves for engines generally are for operating conditions at sea level and where the temperature is below 30 °C. It is necessary to derate the engine's performance for both altitude and temperature where field applications are different. The derating of engines is much more important than the previously mentioned need to derate electric motors. To overcome air density differences due to higher elevations and temperatures, turbochargers are used. Manufacturers should be consulted for specific information on how their engines will perform.

Care must be taken using engine performance data to assure that the continuous operating curves are used. Irrigation pumps generally operate for extended periods and thus the commonly published intermittant curves must be derated. Continuous rated power output may be 15 to 20 percent below the rated intermittent expected output.

Engine efficiency. Engine efficiency is defined as the ratio of the output work to the input energy. In Fig. 10.17b, the ordinate is an expression of the amount of output work which could be expected as a function of fuel consumption. Thus, the low point on the horseshoe curve is the point of best efficiency for each speed. This figure illustrates that even under new conditions that there could be a wide range in efficiencies. The efficiency is also significantly affected by timing and carburation adjustments. As the engine components wear the fuel consumption increases, the driveshaft power decreases, and efficiencies drop. Continued maintenance is essential with engines to obtain high efficiencies. This includes changing of spark plugs, points, and air filters plus regular changes of oil and oil filters at recommended intervals.

Engine efficiencies can be determined in the field by installing torque meters in the driveshaft and then measuring the transmitted power to the pump while measuring fuel consumption. Such measurements are needed to determine when equipment changes or overhauls on engines should be undertaken. The use of the torque meter permits separation of the pump and engine efficiencies.

To evaluate the overall engine powered pumping plant performance it is possible to compute an overall efficiency. It is the product of efficiencies of the individual components.

For turbine pumps it becomes:

$$E_{overall} = E_{pump} \times E_{gearhead} \times E_{driveshaft} \times E_{engine}$$

and for direct coupled centrifugal pumps it is:

$$E_{overall} = E_{pump} \times E_{engine}$$

The efficiency may also be evaluated by dividing the energy imparted to the water by the pump by the input energy (IE) to the engine:

$$E_{overall} = C \frac{Q \cdot TDH}{IE} \qquad \dots \dots \dots \dots \dots \dots \dots \dots \dots \dots \quad [10.15]$$

TABLE 10.2. REPRESENTATIVE ENERGY CONTENT OF
FUELS AND ELECTRICITY

Fuel		Energy content*	
Diesel	39,020 kJ/L	36,980 Btu/L	140,000 Btu/gal
Butane or			
propane	26,340 kJ/L	24,960 Btu/L	94,500 Btu/gal
Gasoline	34,560 kJ/L	32,760 Btu/L	124,000 Btu/gal
Natural gas	35,400-37,260	33,550-35,310	950,000-1,000,000
	kJ/m^3	Btu/m^3	Btu/1,000 ft^3
Electricity	3,600 J/kWh	3,410 Btu/kWh	—

*British thermal unit (Btu) = 1,055 joule (J) = 1.055 kJ
kWh = 3,600,000 J = 3,600 kJ = 3,412 Btu
gallon = 3.785 liters (L)

where Q = the discharge flow rate, L/s (gpm); TDH = total dynamic head, m (ft); C = coefficient to convert Q·TDH to kJ/h, 35.30 kJ/h (33.46 Btu/h); and IE = energy in the fuel consumed by the engine, kJ/h (Btu/h).

The engine fuel consumption for diesel, butane, propane and gasoline is usually measured in liters per hour (gallons per hour). Natural gas consumption is reported in cubic meters per hour (1000's of cubic feet per hour). The energy content of the fuels varies from dealer to dealer and with time. Representative values are given in Table 10.2.

For electrical powered pumps, the coefficient C in equation [10.15] is 0.00981 for Q in L/s, TDH in m, and IE in kWh/h (0.0000189 for Q in gal/min and TDH in ft).

10.7.3 Other Power Sources

In some countries, both animals and humans have been used to power a variety of different types of irrigation and drainage pumps. Such practices were common in the 1970's in India, Pakistan and Egypt. The importance of these types of pumps is well recognized, but it is outside the scope of this chapter to describe their use.

Wind power. Wind has been widely used as a power source to provide both domestic and livestock water. It was also used by the Dutch to lift water for both irrigation and drainage. Research is now underway in the United States to evaluate the potential for using wind powered systems to either mechanically lift the water or to generate electricity which is then used by conventional electric pumping plants.

A research study conducted by the Science and Education Administration of the U.S. Department of Agriculture has demonstrated near Amarillo, Texas, that wind will provide a significant portion of the energy required in that area to irrigate crops. A back-up power source, however, is needed. A clutching mechanism was developed which permits the transfer of wind energy from the windmill to a gear train to power the vertical turbine pump when wind speeds are adequate. The clutching mechanism and special gearhead will also drive the turbine pump from either an electric motor or an internal combustion engine when wind energy is lacking. During some periods of operation part of the energy is derived from both the wind and the standby power supply.

Another study funded by the U.S. Department of Energy that was underway in 1979 involved mapping high wind zones and relating these areas to crop irrigation needs. This should further define the feasible regions where wind may economically provide energy for irrigation. The 1979 cost of equipment to utilize wind for powering large irrigation pumps in the Central Great Plains was prohibitive.

Solar energy. The sun's energy has been harnessed in at least three sites in the United States to power irrigation pumps. One of the systems installed at the University of Nebraska farm near Lincoln utilizes photovoltaic cells to generate electricity which is either stored in batteries or used directly to drive an electric motor. Two other systems, a small one near Willars, New Mexico, and a much larger unit on a farm near Collidge, Arizona, use parabolic reflectors to concentrate the sun's energy on a pipe containing a liquid which then collects and stores the energy for later use. The stored energy is utilized to drive a turbine which generates electricity or else it is used to drive a freon engine.

On-going research at the three solar irrigation pumping sites is continuing in 1980 and efforts are underway to improve system reliability. Standby power is still needed but it has been possible to utilize significant quantities of solar energy. As of 1979, the costs associated with utilization of solar energy are prohibitive for individual farmers.

Hopefully both wind and solar power will be developed to provide some of the energy needed by pumps. Consideration is also being given to burning on-farm refuse or coal to meet the energy requirements.

10.8 ECONOMIC CONSIDERATIONS

The decision on whether to install an irrigation or drainage system should include an economic analysis. Pumps, wells and irrigation equipment are expensive. The returns from crop sales or benefits obtained from drainage must be large enough to repay investment and operating costs.

The cost of pumping includes both fixed and variable costs. Fixed costs are those which occur independent of the yearly hours of operation and include such items as initial investment costs, taxes, insurance and annual payments to repay connect charges for electric or natural gas service. The variable costs include those items which the farmer buys each year to keep the system operating. Variable or operating costs are dominated by energy costs, but also include the cost of lubricants, labor, purchase of repair parts or filters, overhaul costs, and regular maintenance costs.

Prior to 1975 the annual amortized investment costs were larger than the annual variable costs. The cost of pumping equipment has increased significantly since 1975 but the cost of power has tripled or quadrupled. Variable or operating costs now exceed the amortized investment costs. The decision on which model of pump to buy should include a complete economic analysis of both fixed and operating costs. Some of the more expensive pumps may have significantly lower operating costs due to higher operating efficiencies. Power costs are a function of the individual rate structures, the amount of energy used, and demand charges.

10.8.1 Investment Costs

The cost to initially purchase the equipment is the investment cost. Items to be considered in determining the dollar investment include the cost of the well, pump, gearhead, motor, engine, driveshaft, and fuel tank. The cost to be used should be the actual installed cost. Each of the above mentioned items had an expected life which may be different.

Although the total investment cost can be obtained by adding the costs of the individual components, it is more desirable to obtain the annualized investment cost. The annualized investment cost can be added to the annual operating cost to obtain an annual pumping cost. Annualized investment costs can be computed by several methods. One approach is to take the cost of a particular item and divide it by its expected lift to obtain an average annual cost. This method disregards interest on the investment.

A second method to obtain annualized investment costs is to compute an amortized cost considering the expected life of the item and the interest rate. Capital recovery factors, CRF, can be computed from the equation:

$$CRF = \frac{I(I+1)^n}{[(I+1)^n-1]} \quad \dots\dots\dots\dots\dots\dots\dots\dots\dots\dots\dots\dots\dots \quad [10.16]$$

where: I is the current interest rate, expressed as a fraction, for which money can be borrowed and n is the anticipated life of the equipment. Values for the CRF as a function of interest rates and years of life can be tabluated. The Sprinkler Irrigation Handbook (1977), page 459, has such a table.

To use the CRF to obtain an annualized cost it is necessary to know the initial cost, the expected life of the item, and the appropriate interest rate. The annualized cost is obtained by multiplying the CRF by the initial cost of the item. To obtain the total annualized cost of a pumping plant it is necessary to compute the annualized cost of each component and then sum these numbers to obtain the desired annualized investment cost.

To obtain expected lifes for various pump components it is necessary to obtain manufacturers recommendations. Expected lifes for some items such as engines, gearheads and pumps may be expressed as an expected total hours of operation. For example, some of the lightweight engines may have expected lifes of only 3,000 hours while some of the heavy duty engines would be expected to operate for 15,000 to 20,000 hours with minor and/or major overhauls. If manufacturers recommendations are not available for expected lifes, then they must be estimated. Table 10.3 contains some estimated life periods for various pump components (also see Table 3.1).

Pump and well lifes can be significantly impacted by the quality of the water being pumped. Poor quality water can cause corrosion or increase oxidation of ferrous components. Deposition of salts can plug well screens and pump impellers. The life of unused pumps that are submerged in water may

TABLE 10.3. ESTIMATED USEFUL LIFE (YEARS)
OF VARIOUS PUMP COMPONENTS

| | Annual hours of use | | | |
	500	1000	2000	3000
Well	25	25	25	25
Pump	15	15	15	10
Gearhead	15	15	15	10
Drive shaft	15	15	7	5
Engine	15	15	10	7
Gas line	25	25	25	25
Engine foundation	25	25	25	25
Electric motors	25	25	25	25
Electric controls and wiring	25	25	25	25

even be shortened because of corrosion, deposition, and encrustation.

A third method used to obtain an annualized cost would utilize a graduated depreciation schedule. This form is often used to permit maximum credit of investment costs on federal or state income tax return. The most common method is the use of the CRF approach (also see section 3.7).

10.8.2 Other Fixed Costs

Taxes, insurance and energy hookup costs must be paid annually and are independent of the pumps operation. The taxes are generally assessed as an increase in property tax when the irrigation or drainage system is installed. Increased property taxes are often estimated as one percent of the initial investment costs. Pump taxes may also be assessed based on the volume of water pumped but this type of tax should be treated as a variable cost.

Insurance against fire, vandalism, and acts of nature (lightning) should be purchased. Liability insurance is also recommended. This cost is often incorporated into the insurance policy for the farm. The cost is often considered negligible or entered as a flat annual fee ranging from $100 to $250.

Both electric and natural gas companies have established repayment policies to recover the cost of connecting new services. In some instances, the cost of extending service lines is billed as a one-time cost at the time of hookup. For this type of charge the annualized cost could be computed using the CRF method where the expected life would be 25 to 30 years.

In other instances the power companies compute the total cost of the new service and then divide this by either a five or ten year recovery period. Customers are required to sign a contract promising to pay this annual hookup cost in addition to the regular power bills.

10.8.3 Variable Costs

The annual operating or variable costs are dominated by the cost of power. There are two methods commonly used for estimating annual power costs. One method requires estimating the energy demand which would be expressed as kilowatt-hour per hour for electricity, cubic meters of natural gas consumption per hour, or liters per hour comsumption for diesel, gasoline or propane powered engines. This demand is then multiplied by the annual hours of operation to obtain the total annual energy consumption. For natural gas and electric systems it is then necessary to apply the rate structures for the respective utilities to obtain the annual power cost. For diesel, gasoline and propane powered plants the cost is computed by multiplying the cost per liter for the fuel times the total liters of fuel consumption.

The level of energy demand can, as mentioned above, be estimated by solving either equation [10.14] or equation [10.15] for the kW or IP terms. It would be necessary to know the discharge rate of the pumping plant, Q, in L/s; the total dynamic head, TDH in m; and estimate an overall efficiency level for the pumping system, $Eff_{overall}$. Maximum, acceptable and observed values of overall efficiency are tabulated in Table 10.4.

The second method utilizes tables that tabulate the cost to pump a given volume of water, 1000 m^3 (ac-ft), against a specific pump head, TDH. The tables are a function of overall efficiency and the average cost of power, $/kWh or $/1000 m^3 of natural gas or $/liter of the other fuels. Total annual power costs can be calculated by estimating the annual volume to be pumped

TABLE 10.4. TYPICAL VALUES OF OVERALL EFFICIENCY
FOR REPRESENTATIVE PUMPING PLANTS,
EXPRESSED AS PERCENT*

Power source	Maximum theoretical	Recommended as acceptable	Average values from field tests†
Electric	72-77	65	45-55
Diesel	20-25	18	13-15
Natural Gas	18-24	15-18	9-13
Butane, propane	18-24	15-18	9-13
Gasoline	18-23	14-16	9-12

*Ranges are given because of the variation in efficiencies of both
pumps and power units. Especially the difference in efficiency for
high and low compression engines used for natural gas, propane
and gasoline must be considered. The higher value of efficiency can
be used for higher compression engines.
†Typical average observed values reported by pump efficiency test
teams.

and the total pumping head, TDH. These two numbers are then multiplied
together and this product is multiplied by the appropriate number from the
tables.

Table 10.5 tabulates values for electric powered plants and Table 10.6
tabulates values for natural gas powered plants where the energy content of
the gas is 33,540 Btu/m^3 (950 Btu/ft^3). Table 10.6 can be used for natural
gas having a higher or lower energy content by utilizing the equation at the
bottom of the table.

This second method assumes an average cost of energy, \$/kWh or
\$/1000 m^3 of natural gas. Most electric or natural gas company rates now are
either declining block or have a combination of cost of service, demand and
energy costs. To utilize the table method it is necessary to use an average cost
of the power knowing that this average cost is a function of the rate structure,
total energy consumed, size of pumping plant and period of use.

Other variable costs such as lubrication and labor are often expressed as

TABLE 10.5. COST OF PUMPING WATER WITH ELECTRICITY AS A FUNCTION OF
OVERALL PUMPING PLANT EFFICIENCY AND ELECTRICAL ENERGY COST

Overall efficiency, percent	Consumption kWh/ac-ft per ft head	Cost of electrical energy, ¢ / kWh									
		3¢	4¢	5¢	6¢	7¢	8¢	9¢	10¢	11¢	12¢
		- - - - - - - - - - - cents per acre-foot per foot of head* - - - - - - - - - - -									
20	5.12	15.36	20.47	25.59	30.71	35.83	40.95	46.07	51.18	56.03	61.42
25	4.09	12.28	16.38	20.47	24.57	28.66	32.76	36.85	40.95	45.04	49.14
30	3.41	10.24	13.65	17.06	20.47	23.89	27.30	30.71	34.12	37.54	40.95
35	2.92	8.77	11.70	14.62	17.55	20.47	23.40	26.32	29.25	32.17	35.10
40	2.56	7.68	10.24	12.80	15.36	17.92	20.47	23.03	25.59	28.15	30.71
45	2.27	6.82	9.10	11.37	13.65	15.92	18.20	20.47	22.75	25.02	27.30
50	2.05	6.14	8.19	10.24	12.28	14.33	16.38	18.43	20.47	22.52	24.57
55	1.86	5.58	7.45	9.31	11.17	13.03	14.89	16.75	18.61	20.47	22.34
60	1.71	5.12	6.82	8.53	10.24	11.94	13.65	15.36	17.06	18.77	20.47
65	1.57	4.72	6.30	7.87	9.45	11.02	12.60	14.17	15.75	17.32	18.90
70	1.46	4.39	5.85	7.31	8.77	10.24	11.70	13.16	14.62	16.09	17.55
75	1.36	4.09	5.46	6.82	8.19	9.55	10.92	12.28	13.65	15.01	16.38

*To convert the cost to cents per 1000 m^3 per meter of head, multiply the tabulated values by
2.661 (1 acre-ft = 43,560 ft^3 = 1233.5 m^3, 1 ft = 0.3048 m).

TABLE 10.6. COST OF PUMPING WATER WITH NATURAL GAS AS A FUNCTION OF OVERALL PUMPING PLANT EFFICIENCY AND NATURAL GAS COSTS*

Overall efficiency percent	Consumption ft³/acre-ft per ft head	Cost of natural gas, $/1,000 ft³							
		$2.00	$2.25	$2.50	$2.75	$3.00	$3.25	$3.50	$4.00
		dollars per acre-foot per foot of head†							
20	19.40	0.0388	0.0437	0.0485	0.0534	0.0582	0.0631	0.0679	0.0776
18	21.56	0.0431	0.0485	0.0539	0.0593	0.0647	0.0701	0.0755	0.0862
16	24.25	0.0485	0.0546	0.0606	0.0667	0.0728	0.0788	0.0849	0.0970
15	25.87	0.0517	0.0582	0.0647	0.0711	0.0776	0.0841	0.0905	0.1035
14	27.71	0.0554	0.0623	0.0693	0.0762	0.0831	0.0901	0.0970	0.1108
13	29.85	0.0597	0.0672	0.0746	0.0821	0.0896	0.0970	0.1045	0.1194
12	32.33	0.0647	0.0727	0.0808	0.0889	0.0970	0.1051	0.1132	0.1293
11	35.27	0.0705	0.0794	0.0882	0.0970	0.1058	0.1146	0.1234	0.1411
10	38.80	0.0776	0.0873	0.0970	0.1067	0.1164	0.1261	0.1358	0.1552
9	43.11	0.0862	0.0970	0.1078	0.1186	0.1293	0.1401	0.1509	0.1724
8	48.50	0.0970	0.1091	0.1213	0.1334	0.1455	0.1576	0.1698	0.1940
7	55.43	0.1109	0.1247	0.1386	0.1524	0.1663	0.1801	0.1940	0.2217
6	64.67	0.1293	0.1455	0.1617	0.1778	0.1940	0.2102	0.2263	0.2587
5	77.60	0.1552	0.1746	0.1940	0.2134	0.2328	0.2522	0.2716	0.3104
4	97.00	0.1940	0.2183	0.2425	0.2668	0.2910	0.3153	0.3395	0.3880

*Assumes 31,780 Btu/m³ (900,000 Btu/1000 ft³) energy content of natural gas. For other energy contents, multiply the other energy content in Btu/m³ and divide by 31,780.
†To convert the cost to $/1000 m³ per meter of head, multiply the tabulated values by 2.661.

a cost per hour of operation. The labor included is only the labor required to service the pump and its power plant. For engine operated plants it is necessary to check the oil level and other operating conditions daily as well as perform periodic oil changes and other maintenance at recommended intervals. To obtain specific dollar values for these two costs it is necessary to determine from manufacturer's recommendations how often maintenance should be performed and then multiply this by the per unit cost and the number of maintenance periods in a year.

Other costs for unexpected repairs or parts must be estimated from previous experience. Periodic minor or major overhauls can be predicted as a function of hours of operation. The overhaul cost for engines is generally pro-rated over all the years between the overhaul periods.

10.8.4 Total Pumping Costs

The total pumping costs can be expresed as either an annual cost based on current values or it can be expressed as total life cycle cost. The annual cost would be the sum of the annualized investment cost plus the other annual fixed costs plus all the variable costs for a year. The life cycle costs would be the sum of the initial investment costs plus the sum for each year of the other fixed and operating costs. In the life cycle cost analyses it is necessary to include an interest charge on the invested money.

The point which is being made is that decisions on which pump to buy should be based on either an annual cost, including both fixed and operating costs, or else use the life cycle costs. Because of the dominating cost of power it is no longer an acceptable practice to chose the pump solely on initial purchase costs (also see section 3.7).

10.9 FIELD TESTING

Field testing of both new and old pumping plants is recommended to evaluate performance. New plants should be tested to confirm that the pump is operating as expected. If pumps are not operating as expected then it would be possible to make corrections or adjustments while the equipment is still under warranty. The pump installer should be held responsible to make sure the pump performs properly.

Tests on older pumping plants are needed to evaluate how either the pumping system or pump performance has changed. Periodic tests every few years are recommended. Pumps having low efficiencies or ones that are not producing the expected discharge or head should be considered for repair or replacement. A cost analysis must be made to determine the economic feasibility of repair versus continued operation with the current pump.

Data to be collected during a field test should include the discharge, Q; the pumping lift, static lift, discharge head and input power to the motor or engine. For engine driven plants it is desirable to use a torque meter or other similar device which measures the power transmitted in the driveshaft, thus permitting a separate analysis of engine and pump performance.

Equipment used in field tests must be well maintained and calibrated regularly. The lack of periodic calibration can result in misleading data that could be used to make costly decisions. Pressure gages should be cross-referenced or calibrated at weekly intervals and flow measuring devices should be checked as often as possible but at least once a year.

10.10 MAINTENANCE

Pump maintenance includes providing the proper lubrication to the pump and drive units. Manufacturer's recommendations should be followed. Other items for consideration include the periodic adjustment of impeller clearance in turbine pumps, the tightening of packing glands for centrifugal pumps, the periodic maintenance and adjustment of engines to obtain good efficiency, and the need to check the pump's operation to see if conditions are changing.

In the colder regions where freezing occurs, it is necessary to drain the pumps or provide protection if the pumps must operate during freezing temperatures. The motor, pump head, gearhead and engine should be periodically cleansed of dirt and grease so that these units can dissipate the heat they generate. Rodent protective screens must be installed on all electric motors and these screens kept clear to permit air to circulate through the motor.

A good maintenance program should be implemented. Although there is a cost associated with such a program it will maintain high pump efficiencies, help to reduce power costs, improve dependability of the equipment, reduce operating costs and provide extended pump life.

References

1 Addison, H. 1966. Centrifugal and other rotodynamic pumps. Chapman and Hall, London, 565 p.

2 Colt Industries. 1979. Hydraulic handbook. Colt Industries—Fairbanks Morse Pump Division, Kansas City, KS, 248 p.

3 Davis, C. V. 1969. Handbook of applied hydraulics. McGraw-Hill Inc.

4 De Wiest, R. J. and S. N. Davis. 1966. Hydrogeology. John Wiley and Sons, Inc., 463 p.

5 Hicks, T. G. 1957. Pump selection and application. McGraw-Hill, 422 p.

6 Heermann, D. F., H. H. Shull, and R. H. Mickelson. 1974. Center pivot design capacities in eastern Colorado. Proc. Am. Soc. Civ. Engr., J. Irrig. and Drain. Div., 100(IR2):127-141.

7 Ingersoll-Rand. 1977. Cameron hydraulic data. Ingersoll-Rand Company, Woodsliff Lake, NJ.

8 Irrigation Journal. 1979. Irrigation survey. Irrig. J. 29(6):58 A-H.

9 Johnson Division UOP. 1975. Groundwater and wells. Johnson Division UOP Inc., Saint Paul, MN, 440 p.

10 Karassik, I. J., W. C. Krutzsch, W. H. Fraser, and J. P. Messina. 1976. Pump Handbook. McGraw-Hill.

11 McWhorter, D. B. and D. K. Sunada. 1977. Groundwater hydrology and hydraulics. Water Resour. Publ., 290 p.

12 Sprinkler Irrigation Association. 1977. Sprinkler irrigation. Irrig. Assoc., Silver Springs, MD, 615 p.

13 Todd, D. K. 1963. Groundwater hydrology. John Wiley and Sons, Inc., 336 p.

chapter 11

FARM WATER
DISTRIBUTION SYSTEMS

11

Chapter 11

FERMENTED
DISTRIBUTION SYSTEMS

11

FARM WATER DISTRIBUTION SYSTEMS

by   E. G. Kruse, SEA-AR, USDA, Fort Collins, CO;
A. S. Humpherys, SEA-AR, USDA, Kimberly,
ID; and E. J. Pope, SCS, USDA, Lincoln, NE

11.1 INTRODUCTION

This chapter describes systems used to convey irrigation water from the farm supply—a canal, reservoir or well—to the fields. Such systems generally have a capacity less than about 0.3 m³/s (10 ft³/s). Lined or unlined conveyance ditches or pipelines should be well designed and maintained to minimize water loss by seepage and to allow efficient irrigation. Generally, the pipelines or ditches that offer greater seepage control, ease of water diversion, and reduced maintenance have a higher initial cost.

When buried pipe is used instead of open ditches to convey irrigation water, the conduit can follow the most direct route from the water supply to outlet points, rather than following field contours. Weed problems and loss of productive land are eliminated because crops can be planted up to or over the buried pipeline and the pipe does not interfere with planting, cultivating, and harvesting. Seepage and evaporation losses are eliminated when water is transmitted in a well constructed pipeline. Pipelines present fewer safety hazards than open ditch systems. Pipelines may not be desirable, however, if the irrigation supply contains large amounts of sediment and the flow conditions in the line are such that the sediment would tend to settle out and reduce the carrying capacity of the line.

Portable pipe systems laid on the soil surface have many of the advantages of buried pipe systems. If adequate labor is available, they can be removed from fields while cultural operations are in progress.

Seepage losses in unlined project canals and farm ditches often range from one-fourth to one-third of the total water diverted. In extremely sandy or gravelly ditches, half the irrigation supply can be lost in seepage on the farm. Reducing seepage by using improved conveyance facilities may increase water available for crop needs, allow irrigation of additional land, and prevent water-logging or salination of soils below the ditches.

Unlined ditches, because of their potential for high seepage and because of maintenance expense, should only be considered for temporary use, or in cases where money is not available for lined ditches or pipelines.

Design considerations for open ditches and low pressure farm pipelines are discussed in this chapter. In areas with a history of failure or rapid deterioration of pipelines or lined ditches, the guidelines in this chapter should be evaluated and modified on the basis of local experience. The design of farm pipelines with high pressure ratings is discussed in Chapter 14, Dynamics of Sprinkler Systems.

11.2 UNLINED DITCHES AND STRUCTURES

Unlined ditches are used worldwide, primarily because of low cost and ease of construction. They may be either permanent sodded ditches or temporary earth ditches.

11.2.1 Soil Limitations

Unlined ditches can generally be used in any soil that is suited to crop production. Special precautions may be needed in erodible soils to prevent structure and ditch washouts. In addition, seepage losses may be excessive in sandy and gravelly soils.

Seepage losses are influenced by many factors, but depend primarily upon soil permeability. Losses can be very high in non-cohesive, coarse-textured soils and low in fine textured clay soils. The permeability of soils containing clay fractions will be less when in equilibrium with water having a high sodium to calcium-plus-magnesium ratio than with water having a low sodium to calcium-magnesium ratio. Water can be lost through shrinkage cracks in certain clay soils.

Light and medium textured soils with a low clay content usually erode easily, are subject to piping around structures, and may be unstable in newly constructed ditches. Ditches should be designed using maximum permissible velocity criteria according to soil texture to minimize erosion. Sodded ditches are quite stable.

11.2.2 Ditch Design and Capacity

Ditches normally are designed for a capacity equal to the crop water requirements during peak demand plus irrigation and operational losses. Water losses vary with the irrigation distribution system, method of irrigation, soil, crops grown, farm management practices, and other factors, which are discussed in other chapters. With proper design and management, water application efficiencies of at least 60 percent should be obtainable with conventional furrow and border irrigation systems. Level basin and systems with tailwater recovery are capable of 80 to 90 percent efficiencies. For reasons of economy, ditches should not be oversized. However, they should be large enough to provide adequate water for crop requirements under existing and anticipated irrigation methods. The amount of surface water intercepted from precipitation and irrigation waste must also be considered in determining ditch capacity.

Flow Principles. Detailed principles of open channel flow are discussed in hydraulics texts, and only those applicable to farm ditches are mentioned here. Flow is uniform if the velocity and depth do not change throughout the length of channel considered. Flow is nonuniform or varied if the depth changes with distance along the channel. Varied flow can be gradual or rapid depending on the distance over which the change occurs. A hydraulic jump is an example of rapidly varying flow. Flow is spatially varied if the discharge varies along the channel. This type flow occurs in a ditch with spiles or siphon tubes that divert water along the length of the ditch. Flow is steady if the velocity and depth are constant with respect to time. Flow in irrigation ditches is predominantly steady and uniform. Uniform flow can be tranquil, critical, or rapid, depending upon whether the slope is mild, critical or steep. Unlined irrigation ditches are always designed with low velocities so that the flow is tranquil, with Froude number less than 1.0.

$$\text{Froude number} = V/\sqrt{(gy)}$$

where
   V = the mean flow velocity,
   y = flow depth and
   g = gravitational acceleration, all in a compatible system of units.
   Slope.   Slope is the gradient of the ditch, and, for uniform flow, is also the energy loss per unit of length (the energy gradient). Therefore, for uniform flow in irrigation ditches, the slopes of the energy grade line, water surface, and channel bottom are equal. The slope influences flow velocity and must be low enough so the velocity does not cause scouring. Keeping velocities higher than a certain minimum is also desirable to discourage weed growth and siltation. The minimum slope for an unlined farm ditch should be 0.0004 (0.04 m/100 m or 0.04 ft/100 ft). At this slope, water control is excellent, but there may be siltation if the water carries sediment. Most farm ditches are designed with a slope of 0.001. Siphon tubes or spiles can be used, but checks may be needed at 150 to 200 m intervals. Steeper slopes can be used with very small ditches without serious erosion. For slopes greater than 0.002, use of siphon tubes is difficult, and checks are needed at frequent intervals (see Section 11.2.5 for a discussion of check and drop structures). Where unlined ditches must follow steeper field slopes, drops must be used to control the gradient.

Maximum permissible or nonerodible velocities vary with soil texture. For most farm ditches they range from 0.5 m/s for fine, non-colloidal sand to 0.75 m/s for firm loams and fine gravel and may exceed 1 m/s in colloidal and gravelly soils. Permissible velocities for specific soil materials are summarized in Table 11.1 (Chow, 1959).

Ditch Cross Section.   If water is delivered at a fixed rate throughout the season, the size of ditch needed will be different than if water is delivered periodically in larger streams. The size of stream and thus the size of ditch depends upon the water available, soil, topography, method of irrigation, crop, irrigation system, and farm management practices. Irrigation with borders or basins and with high intake rate soils requires larger streams than furrow irrigation and low intake rate soils. Care should be taken so that the ditches are not designed too small, because even with good maintenance, a ditch may become partly clogged with sediment, grass, weeds, or trash.

Fig. 11.1 shows a typical design ditch cross section. The ditch side slopes depend primarily upon the stability of the soil. Small ditches in stable soils with bottom widths of 30 to 50 cm (12 to 20 in.) commonly are designed with side slopes of 1:1 or 1 1/4:1 (horizontal:vertical). Larger ditches are designed with side slopes of 1 1/2:1 to 2:1, depending upon the soil. Side slopes for berms and in fill should not be steeper than 1 1/2:1 for any ditch. A general guide as to suitable slopes than can be used in various soils is given in Chow (1959, Chapter 7B).

TABLE 11.1. MAXIMUM PERMISSIBLE CANAL VELOCITIES* (FROM CHOW 1959)

Original material excavated	Clear water, or water transporting non-colloidal silts and sands	
	m/s	ft/s
Fine sand, non-colloidal	0.46	1.50
Sandy loam, non-colloidal	0.53	1.75
Silt loam and alluvial silts, non-colloidal	0.61	2.00
Ordinary firm loam volcanic ash, and fine gravel	0.76	2.50
Stiff clay and alluvial silts, very colloidal	1.14	3.75
Graded, loam to cobbles, non-colloidal	1.14	3.75
Graded, silt to cobbles, colloidal	1.22	4.00
Coarse gravel, non-colloidal	1.22	4.00
Cobbles and shingles	1.52	5.00
Shales and Hardpans	1.83	6.00

*For straight, aged channels of small slope and flow depths less than 1 m.

Flow Resistance. When water is turned into a ditch, uniform flow becomes established after a short transitory zone. Uniform flow parameters at a section are evaluated by combining the continuity equation

$$Q = AV^* \dots\dots\dots\dots\dots\dots\dots\dots\dots\dots\dots\dots\dots[11.1]$$

with a flow resistance equation. The Manning formula is the most widely used for farm ditch design:

$$V = CR^{2/3}S^{1/2} \dots\dots\dots\dots\dots\dots\dots\dots\dots\dots\dots[11.2]$$

---

*Symbols in this chapter represent quantities as defined in the List of Frequently Used Symbols, pg. 817, unless otherwise indicated. Consistent units of measurement should be assumed by the reader. Thus, for Q, discharge, in m^3/s, A, area, has units of m^2 and V, velocity, of m/s. In English notation, assume Q in ft^3/s, A in ft^2 and V in ft/s unless stated otherwise.

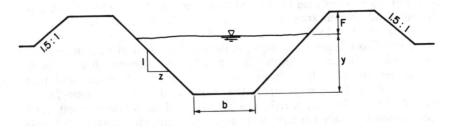

Flow Area $= (b+zy)y$

Wetted perimeter $= b + 2y\sqrt{1 + z^2}$

Hydraulic radius $= \frac{A}{P} = \dfrac{(b+zy)y}{b + 2y\sqrt{1+z^2}}$

F = Freeboard

FIG. 11-1 Ditch cross section and nomenclature.

TABLE 11.2. VALUES OF MANNING n FOR UNLINED DITCHES (FROM CHOW, 1959)

Type of Channel and Description	Minimum	Normal	Maximum
a. Earth, straight and uniform			
1. Clean, recently completed	0.016	0.018	0.020
2. Clean, after weathering	0.018	0.022	0.025
3. Gravel, uniform section	0.022	0.025	0.030
4. With short grass, few weeds	0.022	0.027	0.033
b. Earth, winding and sluggish			
1. No vegetation	0.023	0.025	0.030
2. Grass, some weeds	0.025	0.030	0.033
3. Dense weeds or aquatic plants in deep channels	0.030	0.035	0.040
4. Earth bottom and rubble sides	0.028	0.030	0.035
5. Stony bottom and weedy banks	0.025	0.035	0.040
6. Cobble bottom and clean sides	0.030	0.040	0.050

where

for SI units

$C = 1/n$

and for inch-pound units

$C = 1.486/n$.

R = the hydraulic radius in meters or feet and
S = the energy gradient.

The Manning n, the resistance coefficient, has identical values for both systems of measurement. For irrigation ditches with a given boundary condition, the Manning n is assumed constant. In general, those conditions which increase turbulence and retard the flow, especially surface roughness and vegetation, increase the roughness coefficient. Vegetative growth is one of the disadvantages of an unlined ditch. Root masses or plant stems and leaves in the channel can greatly increase the value of n and cause it to change during the season. Values of n for different unlined ditch conditions are shown in Table 11.2 (Chow, 1959).

When the slope, normal depth and cross section for a given ditch are known, the discharge can be calculated directly by combining equations [11.1] and [11.2].

$$Q = CAR^{2/3} S^{1/2} \dots\dots\dots\dots\dots\dots\dots\dots\dots\dots\dots\dots\dots\dots [11.3]$$

However, use of the Manning formula in designing farm ditches often requires one or more trial solutions. For new ditches, the cross section must be assumed. Then velocity and carrying capacity are computed and compared with the design discharge. New values of flow depth are tried, until the computed discharge matches the design discharge.

The following example illustrates solution of a typical problem where ditch dimensions must be determined:

Example 1

Given: $Q = 150 \text{ L/s} = 0.15 \text{ m}^3/\text{s}$
$S = 0.001$
1:1 side slopes

Find: Bottom width and normal flow depth for a ditch that will be well maintained and weed free

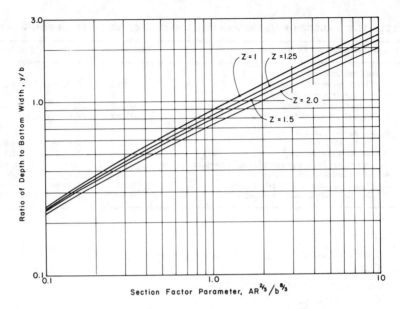

FIG. 11.2 Section factors for solution of Manning Formula.

Assume: Manning n = 0.025 for a straight, clean but weathered earth
        ditch
Solution: 1 Use standard bottom widths compatible with construction
            equipment. For this example use b = 0.3 m.
          2 Assume a normal depth and calculate A, R and V.
          3 If calculated Q = AV is not equal to the design Q, assume
            different value for normal depth and repeat calculations.
          4 Final values will be:

$$b = 0.3 \text{ m}$$
$$y = 0.45 \text{ m}$$
$$A = 0.34 \text{ m}^2$$
$$R = 0.21 \text{ m}$$
$$V = 0.45 \text{ m/s}$$

To avoid trial and error solutions, alternative methods using charts,
tables or nomographs have been developed. These are explained in the
following paragraphs.

Section Factor. The section factor $AR^{2/3}$ as defined by Chow (1959)
is a very useful tool in uniform flow computation and ditch design. Rear-
ranging equation [11.3] gives:

$$AR^{2/3} = \frac{Q}{C\sqrt{S}} \quad \dots\dots\dots\dots\dots\dots\dots\dots\dots\dots\dots\dots\dots\dots\dots\dots\dots [11.4]$$

Thus, when the discharge, slope and roughness are given, the section
factor can be calculated and the normal depth can be determined. To
simplify computation, dimensionless curves of $AR^{2/3}/b^{8/3}$ versus y/b are
shown in Fig. 11.2.

TABLE 11.3. CONVEYANCE FACTORS (K_d) AND AREAS (A) FOR SOLUTION OF THE MANNING FORMULA FOR TRAPEZOIDAL CHANNELS
n = 0.025   side slope 1:1

y_1	b = 0.2 m		b = 0.3 m		b = 0.4 m		b = 0.6 m		b = 0.8 m	
	A	K_d	A	K_d	A	K_d	A	K_d	A	K_d
m	m^2	m	m^2	m	m^2	m	m^2	m	m^2	m
0.20	0.0800	0.710	0.1000	0.949	0.1200	1.196	0.1600	1.703	0.2000	2.223
0.22	0.0924	0.861	0.1144	1.138	0.1364	1.425	0.1804	2.015	0.2244	2.621
0.24	0.1056	1.029	0.1296	1.347	0.1536	1.675	0.2016	2.354	0.2496	3.049
0.26	0.1196	1.215	0.1456	1.575	0.1716	1.948	0.2236	2.718	0.2756	3.508
0.28	0.1344	1.418	0.1624	1.824	0.1904	2.243	0.2464	3.107	0.3024	3.998
0.30	0.1500	1.642	0.1800	2.093	0.2100	2.559	0.2700	3.524	0.3300	4.517
0.32	0.1664	1.884	0.1984	2.384	0.2304	2.901	0.2944	3.969	0.3584	5.068
0.34	0.1836	2.147	0.2176	2.697	0.2516	3.265	0.3196	4.440	0.3876	5.650
0.36	0.2016	2.430	0.2376	3.033	0.2736	3.654	0.3456	4.939	0.4176	6.266
0.38	0.2204	2.736	0.2584	3.392	0.2964	4.068	0.3724	5.469	0.4484	6.911
0.40	0.2400	3.064	0.2800	3.774	0.3200	4.508	0.4000	6.024	0.4800	7.592
0.42	0.2604	3.413	0.3024	4.181	0.3444	4.973	0.4284	6.613	0.5124	8.303
0.44	0.2816	3.788	0.3256	4.163	0.3696	5.466	0.4576	7.226	0.5456	9.046
0.46	0.3036	4.185	0.3496	5.071	0.3956	5.985	0.4876	7.874	0.5796	9.825
0.48	0.3264	4.607	0.3744	5.554	0.4224	6.530	0.5184	8.552	0.6144	10.639
0.50	0.3500	5.054	0.4000	6.064	0.4500	7.106	0.5500	9.260	0.6500	11.484
0.52	0.3744	5.525	0.4264	6.602	0.4784	7.712	0.5824	10.001	0.6864	12.366
0.54	0.3996	6.024	0.4536	7.166	0.5076	8.343	0.6156	10.773	0.7236	13.282
0.56	0.4256	6.549	0.4816	7.759	0.5376	9.006	0.6496	11.578	0.7616	14.239
0.58	0.4524	7.101	0.5104	8.381	0.5684	9.699	0.6844	12.418	0.8004	15.227
0.60	0.4800	7.682	0.5400	9.032	0.6000	10.423	0.7200	13.291	0.8400	16.252
0.62	0.5084	8.283	0.5704	9.713	0.6324	11.176	0.7564	14.196	0.8804	17.316
0.64	0.5376	8.928	0.6016	10.424	0.6656	11.962	0.7936	15.139	0.9216	18.417
0.66	0.5676	9.592	0.6336	11.166	0.6996	12.783	0.8316	16.116	0.9636	19.557
0.68	0.5984	10.290	0.6664	11.938	0.7344	13.636	0.8704	17.126	1.0064	20.732
0.70	0.6300	11.015	0.7000	12.743	0.7700	14.516	0.9100	18.175	1.0500	21.949

Conversions: Length, m x 3.281 = Length, ft
Area, m^2 x 10.765 = Area, ft^2

Example 2

An alternate solution for Example 1 is:

Given: $Q = 150 \text{ L/s} = 0.15 \text{ m}^3/\text{s}$
$\quad\quad\; S = 0.001$
$\quad\quad\; z = 1.0 \text{ (1:1 side slope)}$

Find: Bottom width and normal flow for a ditch that will be well maintained and weed free.

Assume:
$$n = \frac{1}{C} = 0.025$$

Solution: 1 Use standard bottom width compatible with construction equipment. Use b = 0.3 m.

2 Compute section factor from equation [11.4]

$$AR^{2/3} = \frac{Q}{C\sqrt{S}} = 0.1186$$

3 Calculate

$$AR^{2/3}/b^{8/3} = 2.94$$

TABLE 11.4. CONVEYANCE FACTORS (K_d) AND AREAS (A) FOR SOLUTION OF THE
MANNING FORMULA FOR TRAPEZOIDAL CHANNELS
n = 0.025   side slope 1.25:1

y_1	b = 0.2 m		b = 0.3 m		b = 0.4 m		b = 0.6 m		b = 0.8 m	
	A	K_d	A	K_d	A	K_d	A	K_d	A	K_d
m	m^2	m	m^2	m	m^2	m	m^2	m	m^2	m
0.20	0.0900	0.812	0.1100	1.052	0.1300	1.300	0.1700	1.808	0.2100	2.327
0.22	0.1045	0.992	0.1265	1.271	0.1485	1.559	0.1925	2.150	0.2365	2.756
0.24	0.1200	1.193	0.1440	1.514	0.1680	1.844	0.2160	2.524	0.2640	3.219
0.26	0.1365	1.417	0.1625	1.782	0.1885	2.156	0.2405	2.928	0.2925	3.719
0.28	0.1540	1.664	0.1820	2.074	0.2100	2.496	0.2660	3.364	0.3220	4.254
0.30	0.1725	1.936	0.2025	2.394	0.2325	2.864	0.2925	3.832	0.3525	4.826
0.32	0.1920	2.233	0.2240	2.740	0.2560	3.261	0.3200	4.333	0.3840	5.435
0.34	0.2125	2.556	0.2465	3.114	0.2805	3.688	0.3485	4.868	0.4165	6.081
0.36	0.2340	2.906	0.2700	3.518	0.3060	4.145	0.3780	5.438	0.4500	6.766
0.38	0.2565	3.284	0.2945	3.950	0.3325	4.634	0.4085	6.042	0.4845	7.490
0.40	0.2800	3.690	0.3200	4.413	0.3600	5.155	0.4400	6.683	0.5200	8.253
0.42	0.3045	4.126	0.3465	4.907	0.3885	5.709	0.4725	7.359	0.5565	9.056
0.44	0.3300	4.591	0.3740	5.433	0.4180	6.297	0.5060	8.073	0.5940	9.900
0.46	0.3565	5.088	0.4025	5.992	0.4485	6.919	0.5405	8.825	0.6325	10 784
0.48	0.3840	5.616	0.4320	6.584	0.4800	7.576	0.5760	9.614	0.6720	11.711
0.50	0.4125	6.177	0.4625	7.210	0.5125	8.268	0.6125	10.444	0.7125	12.680
0.52	0.4420	6.771	0.4940	7.872	0.5460	8.998	0.6500	11.313	0.7540	13.693
0.54	0.4725	7.399	0.5265	8.568	0.5805	9.764	0.6885	12.222	0.7965	14.749
0.56	0.5040	8.062	0.5600	9.301	0.6160	10.569	0.7280	13.172	0.8400	15.849
0.58	0.5365	8.761	0.5945	10.072	0.6525	11.412	0.7685	14.165	0.8845	16.994
0.60	0.5700	9.495	0.6300	10.879	0.6900	12.294	0.8100	15.199	0.9300	18.185
0.62	0.6045	10.266	0.6665	11.726	0.7285	13.217	0.8525	16.277	0.9765	19.422
0.64	0.6400	11.076	0.7040	12.611	0.7680	14.180	0.8960	17.399	1.0240	20.706
0.66	0.6765	11.923	0.7425	13.537	0.8085	15.184	0.9405	18.565	1.0725	22.037
0.68	0.7140	12.809	0.7820	14.503	0.8500	16.231	0.9860	19.776	1.1220	23.416
0.70	0.7525	13.736	0.8225	15.510	0.8925	17.320	1.0325	21.032	1;1725	24.844

Conversions: see Table 11.3.

4 From Fig. 11.2 find y/b = 1.48, from which
   y = 0.3(1.48) = 0.444 m.
The final ditch depth will be the flow depth, y, plus freeboard. The velocity
can be determined from either equation [11.1] or [11.2] using the relationships shown in Fig. 11.1 and compared with the permissible velocity given in
Table 11.1.

   Conveyance Factor. Equation [11.3] can also be written $Q = K_d S^{1/2}$
where $K_d = CAR^{2/3}$, the conveyance factor (Chow, 1959, Section 6-1).
Values of conveyance factor are listed in Tables 11.3 to 11.5 for several
common ditch cross sections. Similar tables can be generated for other side
slopes and bottom widths and for inch-pound rather than metric units.

   If the Manning n for the channel in question is different than 0.025,
the K_d value can easily be corrected by the ratio of actual n to 0.025 (see
Example 3). Channel design using conveyance factor tables is especially
convenient when the designer wants to consider several alternate bottom
widths before settling on the final cross section.

   Table 11.3 can be used to solve the problem of Example 1 as follows:
   Example 3
   Given: Q = 150 L/s = 0.150 m^3/s
          S = 0.001
          1:1 side slopes

TABLE 11.5. CONVEYANCE FACTORS (K_d) AND AREAS (A) FOR SOLUTION OF THE MANNING FORMULA FOR TRAPEZOIDAL CHANNELS
n = 0.025   side slope 1.5:1

y_1	b = 0.2 m		b = 0.3 m		b = 0.4 m		b = 0.6 m		b = 0.8 m	
	A	K_d	A	K_d	A	K_d	A	K_d	A	K_d
m	m^2	m	m^2	m	m^2	m	m^2	m	m^2	m
0.20	0.1000	0.910	0.1200	1.152	0.1400	1.399	0.1800	1.906	0.2200	2.425
0.22	0.1166	1.118	0.1386	1.399	0.1606	1.687	0.2046	2.278	0.2486	2.882
0.24	0.1344	1.352	0.1584	1.675	0.1824	2.006	0.2304	2.685	0.2784	3.379
0.26	0.1534	1.614	0.1794	1.980	0.2054	2.356	0.2574	3.128	0.3094	3.917
0.28	0.1736	1.904	0.2016	2.316	0.2296	2.739	0.2856	3.607	0.3416	4.496
0.30	0.1950	2.223	0.2250	2.684	0.2550	3.156	0.3150	4.125	0.3750	5.118
0.32	0.2176	2.573	0.2496	3.084	0.2816	3.607	0.3456	4.681	0.4096	5.782
0.34	0.2414	2.955	0.2754	3.518	0.3094	4.094	0.3774	5.277	0.4454	6.490
0.36	0.2664	3.320	0.3024	3.987	0.3384	4.618	0.4104	5.914	0.4824	7.242
0.38	0.2926	3.818	0.3306	4.492	0.3686	5.180	0.4446	6.592	0.5206	8.040
0.40	0.3200	4.302	0.3600	5.033	0.4000	5.780	0.4800	7.312	0.5600	8.884
0.42	0.3486	4.822	0.3906	5.612	0.4326	6.420	0.5166	8.076	0.6006	9.774
0.44	0.3784	5.378	0.4224	6.230	0.4664	7.100	0.5544	8.884	0.6424	10.713
0.46	0.4094	5.973	0.4554	6.888	0.5014	7.822	0.5934	9.737	0.6854	11.700
0.48	0.4416	6.606	0.4896	7.586	0.5376	8.586	0.6336	10.636	0.7296	12.736
0.50	0.4750	7.280	0.5250	8.326	0.5750	9.394	0.6750	11.581	0.7750	13.823
0.52	0.5096	7.994	0.5616	9.109	0.6136	10.246	0.7176	12.575	0.8216	14.961
0.54	0.5454	8.750	0.5994	9.935	0.6534	11.143	0.7614	13.616	0.8694	16.150
0.56	0.5824	9.549	0.6384	10.806	0.6944	12.087	0.8064	14.708	0.9184	17.393
0.58	0.6206	10.392	0.6786	11.722	0.7366	13.077	0.8526	15.849	0.9686	18.689
0.60	0.6600	11.279	0.7200	12.684	0.7800	14.115	0.9000	17.042	1.0200	20.039
0.62	0.7006	12.212	0.7626	13.694	0.8246	15.202	0.9486	18.286	1.0726	21.444
0.64	0.7424	13.191	0.8064	14.751	0.8704	16.338	0.9984	19.583	1.1264	22.905
0.66	0.7854	14.218	0.8514	15.857	0.9174	17.525	1.0494	20.934	1.1814	24.423
0.68	0.8296	15.293	0.8976	17.013	0.9656	18.763	1.1016	22.339	1.2376	25.998
0.70	0.8750	16.416	0.9450	18.220	1.0150	20.054	1.1550	23.799	1.2950	27.632

Conversions: see Table 11.3.

Find: Bottom width and normal flow depth options for a ditch which will be well maintained and weed free
Assume: Manning n = 0.025
Solution:

$$K_d = Q/S^{1/2} = \frac{0.150}{\sqrt{0.001}} = 4.743$$

Interpolating from values in Table 11.3, find options:

b = 0.3 m,	y = 0.45 m,	A = 0.34 m^2,	$V = \dfrac{Q}{A} = 0.44$ m/s	
b = 0.4 m,	y = 0.41 m,	A = 0.33 m^2,	V = 0.45 m/s	
b = 0.6 m,	y = 0.35 m,	A = 0.33 m^2,	V = 0.45 m/s.	

If an n other than 0.025 is assumed, say 0.020, the section factor ($AR^{2/3}$) must also be reduced by the ratio (0.020/0.025), so that the conveyance factor (K_d) will remain unchanged. Values of the new section can easily be

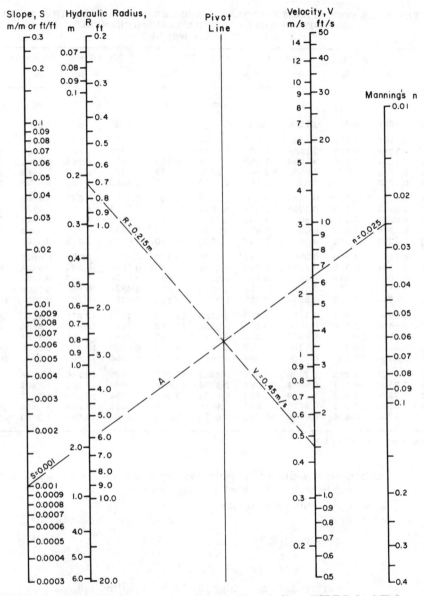

FIG. 11.3 Nomograph for solution of the Manning Formula [from USDT-Federal Highway Administration, 1973].

found from Table 11.3. Entering this table with $K_d' = K_d(0.020/0.025) = 3.795$ gives:

$$b = 0.3 \text{ m}, \qquad y = 0.40 \text{ m}, \qquad A = 0.28 \text{ m}^2, \qquad V = 0.54 \text{ m/s}$$

$$b = 0.4 \text{ m}, \qquad y = 0.37 \text{ m}, \qquad A = 0.28 \text{ m}^2, \qquad V = 0.54 \text{ m/s}$$

$$b = 0.6 \text{ m}, \qquad y = 0.31 \text{ m}, \qquad A = 0.28 \text{ m}^2, \qquad V = 0.54 \text{ m/s.}$$

**TABLE 11.6. CAPACITIES OF EARTH DITCHES WITH 1:1 SIDE SLOPES
AND MANNING n = 0.025 (FROM UNIVERSITY OF IDAHO, 1958)**

Water depth	Grade, m/m							
	0.0005		0.0010		0.0020		0.0040	
	V	Q	V	Q	V	Q	V	Q
m	m/s	m^3/s	m/s	m^3/s	m/s	m^3/s	m/s	m^3/s
Bottom Width 0.15 m								
0.1	0.13	0.003	0.19	0.005	0.27	0.007	0.38	0.010
0.2	0.19	0.013	0.27	0.019	0.38	0.027	0.54	0.038
0.3	0.24	0.032	0.33	0.045	0.47	0.064	0.67	0.090
Bottom Width 0.30 m								
0.1	0.15	0.006	0.21	0.008	0.30	0.012	0.42	0.017
0.2	0.21	0.021	0.30	0.030	0.42	0.042	0.60	0.060
0.3	0.26	0.047	0.37	0.066	0.52	0.094	0.74	0.132
0.4	0.30	0.084	0.43	0.119	0.60	0.169	0.85	0.239
Bottom Width 0.45 m								
0.1	0.16	0.009	0.22	0.012	0.32	0.018	0.45	0.025
0.2	0.23	0.030	0.32	0.042	0.45	0.059	0.64	0.084
0.3	0.28	0.062	0.39	0.088	0.56	0.125	0.79	0.177
0.4	0.32	0.109	0.45	0.154	0.64	0.218	0.91	0.309

Conversions: Length, m x 3.281 = Length, ft
Velocity, m/s x 3.281 = Vel, ft/s
Discharge, m^3/s x 10^3 = Discharge, L/s
Discharge, m^3/s x 35.3 = Discharge, ft^3/s

Another method of solving the Manning formula is by use of the nomograph shown in Fig. 11.3. This nomograph is most useful for computing the flow capacity of a ditch when the ditch dimensions are known:

Example 4
Given: S = 0.001
1:1 side slopes, z = 1
b = 0.30 m
ditch depth = 0.55 m

Assume: Manning n = 0.025
Solution:

Allowing 0.10 m freeboard, maximum flow depth, y = 0.45 m

$$R = \frac{A}{P} = \frac{(b+zy)y}{b + 2y\sqrt{1+z^2}} = \frac{0.338}{1.57} = 0.215 \text{ m. (See Fig. 11.1)}$$

On the nomograph, draw line A connecting S = 0.001 with n = 0.025. Then draw line B through R = 0.215 m and intersecting line A at the pivot line. Read V = 0.45 m/s. Ditch capacity, Q = VA = 0.45(0.338) = 0.15 m³/s.
Estimates of the capacities of earth ditches can be obtained from Table 11.6. Computed capacities of unlined ditches are seldom exact because of factors that can change capacity with time. Therefore, estimates of ditch size and capacity from Table 11.6 will often be accurate enough for farm design.

11.2.3 Elevation and Freeboard

Elevation Above Field Surface. The elevation of a ditch relative to the ground surface depends upon whether the ditch is used for conveyance or distribution, the method of irrigation, the topography, and the soil. Ditches used for conveyance only can carry the water either below or above ground elevation. They should be at an elevation low enough to allow free flow through a measuring structure at the diversion point. Ditches used for distribution should be constructed so that the water surface can be checked up to an elevation at least 10 to 15 cm above the field surface. Furrows or corrugations, which are cut below the ground surface, require only small streams of water and therefore lower elevations in the supply ditch. Water depths on borders and basins next to the ditch are often as great as 15 cm. In this case, the water surface in the ditch must be maintained at higher elevation.

In order to maintain a uniform ditch grade where the field topography is uneven, or where the field slope is very flat, the ditch may carry water partly in fill at an elevation above the field through part of its length. Such ditches may be difficult to hold in loose sandy soil, particularly when they are new. In these soils, the ditch should be constructed to maintain the water surface below ground elevation except when checked to a higher elevation for distribution during irrigation.

Freeboard. Freeboard varies somewhat with the type of ditch, its size, and use. Permanent ditches should have a freeboard of 30 cm to provide for checking the water and to allow for settling. Minimum freeboard for ditches carrying 30 to 150 L/s (1 to 5 ft³/s) should be 10 to 15 cm (4 to 6 in.) and for ditches carrying 150 to 300 L/s (5 to 10 ft³/s), 15 to 20 cm (6 to 8 in.). Smaller ditches require a minimum of 10 cm (4 in.). Where ditches are subject to excessive clogging (from weed growth, sediment, or trash), to livestock trampling, to rodent damage, or to other factors that could cause overtopping, the freeboard should be increased accordingly.

11.2.4 Ditch Construction

Most ditches can be constructed to follow the general slope of the field surface. Contour ditches for flood irrigation are laid out approximately parallel to keep the lengths of run between them as nearly the same as possible. However, if the ditch grade is different from that of the field, constructing a pad to the desired grade may be necessary. If an earth fill is used, it should be constructed in layers 15 to 20 cm thick and well compacted to prevent settling. Side slopes should be no steeper than 1 1/2:1. Earth fills are always subject to seepage losses and washouts caused by rodents. A pipeline should be used where water is to be carried over a depression requiring earth fill more than about 1 m deep.

Permanent ditch banks or berms and those in fill should be at least 50 to 60 cm wide at the top. The width of temporary ditch berms should equal or exceed the flow depth and not be less than 30 cm.

Most ditches are excavated in either the natural soil or in raised pads or embankments with mechanical ditchers or excavators, which can be adjusted for different bottom widths and depths. Earth for fill should not be taken from borrow areas alongside the ditch. Depressions thus created collect seepage, and weed growth is hard to control.

11.2.5 Control Structures

Functional Requirements. One of the most important factors in the efficiency and ease of operation of irrigation systems is the character of the control structures used. Because surface irrigation characteristically has a high labor requirement, good control structures are needed to reduce labor and to simplify irrigation. Use of automatic and mechanical structures is increasing.

Temporary structures are installed for one season's use or are moved during each irrigation. Permanent structures usually permit better water control with less labor than temporary structures. Cost can often be minimized by combining two or more functions in a single structure. For example, checks, drops, turnouts, divisors, and measuring structures can be used in various combinations. Wooden structures are being increasingly replaced by cast in place or precast concrete and modular type metal structures, which are efficient and can be installed by the farmer. Names of the various structures referred to in this chapter are those in common use in the USA. The same structure may be referred to by different names in other countries. Flow measuring structures that are also suitable for use in farm ditches are described in Chapter 9.

Sediment deposited in irrigation ditches and structures necessitates frequent ditch cleaning and often results in inaccurate flow measurement. Trash in irrigation water is a source of weed infestation on the farm and clogs irrigation structures. When trash is a problem, special structures are often needed to remove it from the irrigation water supply. Structures in unlined ditches interfere with ditch cleaning and, if possible, should be designed to allow ditch-cleaning equipment to pass through.

Inlet or Receiving Structures. Canal, lateral, and reservoir turnout structures that deliver water to farm ditches are discussed in Chapter 9. These are permanent structures considered part of the canal delivery or supply system. They normally consist of an adjustable flow control and a measuring device.

A stilling basin or energy dissipator is required where water is received into a ditch directly from a pump or pipe outlet. This can be a rubble or riprap lined basin to provide a stilling pool deep and long enough to dissipate turbulence before water enters the unlined ditch section. Minimum required water depths in the basin for different drop heights from cantilevered pipe outlets are 30 cm for drops to 30 cm high, 45 cm for 30 to 60 cm drops and 60 cm for 60 to 90 cm drops. A short section of concrete-lined ditch with the invert placed below grade by the above amounts can also be used. A structure similar to the combination concrete pipe outlet and division box shown in Fig. 11.4a is very effective. The floor baffle-type pipe outlet dissipator (Vanoni and Bostrom, 1943) shown in Fig. 11.4b works well for high velocity pipe discharges into a ditch. Specially designed energy dissipating structures may be needed where water discharges at very high velocities into a ditch.

Drop Structures. Grade control structures are required to prevent erosive velocities from occurring on steep slopes. The water is lowered over drops and conveyed downslope in a stairstep manner. The energy of the falling water must be dissipated to prevent erosion of the downstream channel or undercutting of the structure. Although many different types of drop structures are used, they basically consist of either vertical or inclined drops and a stilling pool or other energy dissipating means. They are often

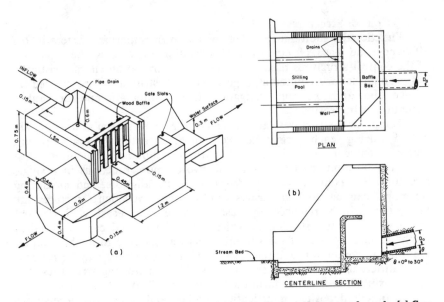

FIG. 11.4 Structures for dissipating energy where pipes discharge into open channels. [a] Combination pipe outlet-division box [from USDA, SCS, 1978]. [b] floor baffle-pipe outlet dissipator [from Vanoni and Bostrom, 1943].

combined with checks as illustrated by the typical structures shown in Fig. 11.5.

Drop heights in conveyance ditches should be limited to a maximum of 1 m (3 ft); recommended height is less than 0.6 m. Drops in field distribution ditches should be limited to 30 cm (1 ft); recommended height is from 15 cm to 23 cm. Where distance between drops is 100 m or more, crest elevation should not be lower than the bottom of a stable ditch 100 m upstream. Where distances are less than 100 m, the crest elevation should not be lower than the sill of the next upstream structure nor less than 10 cm above the apron of the next upstream structure. Design details for drop structures in unlined ditches can be obtained from agencies such as the Soil Conservation Service. Other design information is available from Booher (1974); Gilden and Woodward (1952); Herpich and Manges (1959); Jensen et al. (1954); Kraatz and Mahajan (1975); Portland Cement Association (1960); Robinson et al. (1963); and Skogerboe et al. (1971).

Humpherys and Robinson (1971) concluded from a study of small drop structures that (a) prefabricated structures were low in cost and easy to install but frequently too small for adequate erosion control, (b) wide stilling basins with low end sills caused the least turbulence and erosion, (c) when adequately designed, head wall structures with a rectangular notch and gravel-lined stilling basin were the most economical and effective. Where ditch slopes require drop structures to be closely spaced, use of concrete ditch lining or buried pipelines is usually an economical alternative.

Another commonly used drop for small ditches, particularly where a combined road crossing is needed, is shown in Fig. 11.6. A concrete pipe may be used in place of the corrugated metal pipe shown. This drop is rather easily plugged by trash, and riprap is needed at both the inlet and outlet.

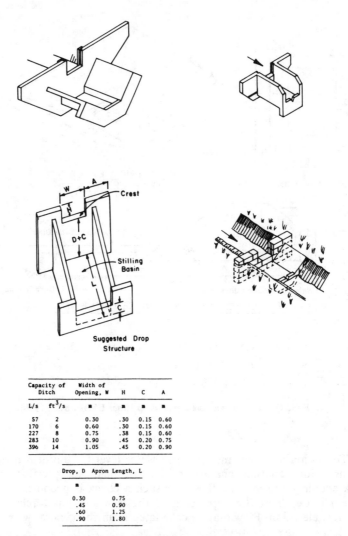

Capacity of Ditch		Width of Opening, W	H	C	A
L/s	ft³/s	m	m	m	m
57	2	0.30	.30	0.15	0.60
170	6	0.60	.30	0.15	0.60
227	8	0.75	.38	0.15	0.60
283	10	0.90	.45	0.20	0.75
396	14	1.05	.45	0.20	0.90

Drop, D	Apron Length, L
m	m
0.30	0.75
.45	0.90
.60	1.25
.90	1.80

FIG. 11.5 Common types of drop structures used in farm irrigation ditches [from USDA, SCS, 1978; Colorado State College, 1944; and Portland Cement Association, 1960].

Check Structures. A check is any structure used to maintain or increase the water level in an open channel above the normal flow depth. The drop structures in Fig. 11.5 have flash board slots and can be used as checks. Other designs are given by the USDA Soil Conservation Service†; Booher (1974); Gilden and Woodward (1952); Herpich and Manges (1959); Jensen et al. (1954); Kraatz and Mahajan (1975); Portland Cement Association (1960); Robinson et al. (1963); and Skogerboe et al. (1971). Commercial prefabricated and modular metal, concrete, and masonry structures are commonly used.

---

†Detailed irrigation structure design information applicable to a given state is available from the USDA Soil Conservation Service State Engineer. Design information for readers outside the USA available from USDA, Soil Conservation Service, Washington, DC.

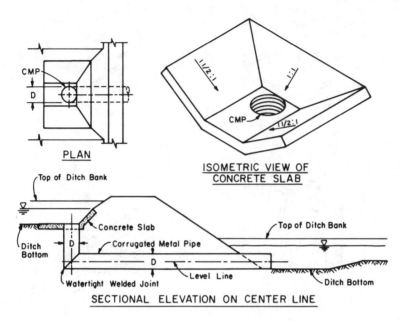

PLAN

ISOMETRIC VIEW OF
CONCRETE SLAB

SECTIONAL ELEVATION ON CENTER LINE

NOTE: Capacities range from 65 L/s (2.3 ft³/s) for 254 mm (10 in) pipe and 0.3 m (1 ft) drop to 229 L/s (8.1 ft³/s) for 381 mm (15 in) pipe and 0.9 m (3 ft) drop.

FIG. 11.6 Corrugated metal pipe drop [from USDA, SCS, 1978].

To maintain a constant upstream water level for uniform diversion of discharge into another channel or onto a field, an overflow weir-type check should be used. The flow over such a check is proportional to $y^{3/2}$, where y is depth of flow upstream of the weir measured relative to the weir crest elevation. Flow may be calculated from the general weir equation

$$Q = CLy^n \dots\dots\dots\dots\dots\dots\dots\dots\dots\dots\dots\dots\dots\dots\dots [11.5]$$

where the value of n for most overflow checks is approximately 1.5, and c varies from 1.55 to 1.82 for SI units (meters, seconds) and 2.8 to 3.3 for inch-pound units. When the crest length L is large, variations in discharge result in relatively small changes in the upstream water level. Equation [11.5] is satisfactory for design purposes but not for exact water measurement through the opening.

When the water level or flow rate is to be controlled downstream, a sluice gate or orifice-type check is desirable, because Q is proportional to $\sqrt{y}$ and thus the discharge is more constant. The discharge through an orifice may be determined from the general equation

$$Q = CA(2gy)^{1/2} \dots\dots\dots\dots\dots\dots\dots\dots\dots\dots\dots\dots\dots [11.6]$$

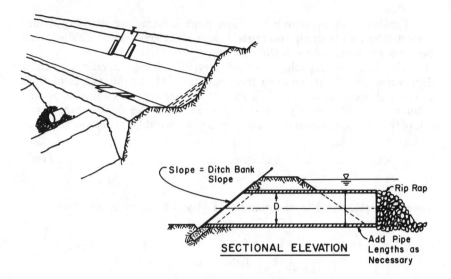

SECTIONAL ELEVATION

FIG. 11-7 Precast concrete farm turnout [from USBR, 1951].

where

C = coefficient of discharge,
A = area of orifice opening,
g = acceleration of gravity, and
y = the head causing flow.

For free flow, y is the upstream head measured from the center of the orifice opening; for submerged flow, y is the difference between upstream and downstream water levels. The coefficient of discharge C ranges from 0.6 to approximately 0.8 for any consistent set of units, depending on the position of the orifice in relation to the sides and bottom of the structure and the roundness of the orifice edge (Brater and King, 1976).

Gates for automatically controlling the water level in canals, which are discussed in Chapter 9, can also be used in some farm ditches (Thomas, 1960). Semi-automatic timer-controlled checks have been used to a limited extent in farm ditches but are still in the experimental stage. They are normally used with timer-controlled outlets for releasing flow onto a field or into another ditch (Calder and Weston, 1966; Evans, 1977; Haise and Hall, 1971; Hart and Borrelli, 1970; Humpherys, 1969; and Robinson, 1972). Automation of farm water control structures can reduce the irrigation labor requirement and improve the efficiency of surface irrigation.

Outlets and Discharge Controls. Discharge control devices are used to control the release of water from a farm ditch into basins, borders, furrows, or another irrigation ditch. A field turnout may be a fixed opening in the side of a ditch or one equipped with check boards, gates or other devices to adjust the opening area. If only a portion of the total flow is to be delivered through a given turnout, a more constant discharge is obtained by using an orifice-type device rather than an overflow or weir-type structure. One of the most frequently used turnouts is a concrete or metal pipe with a slide gate on the inlet, as shown in Fig. 11.7. For unlined ditches the headwall and slidegate are usually vertical. The capacity of pipe turnouts can be determined from equation [11.6], where C = 0.80, the value for a short tube.

Erosion is often a problem when large streams are discharged into noncohesive soils from pipe turnouts. Erie et al. (1969) describes a dissipator box that effectively controls this erosion.

The free flow capacity of broad crested weir-type openings of fixed dimensions may be determined from equation [11.5] using values of C as given by Brater and King (1976). Flow from turnouts into a field is usually submerged. The submergence is d/y where d is the downstream depth above the weir crest elevation. Submerged discharged Q_S is determined from

$$Q_s = CQ \quad\dotfill [11.7]$$

where Q is calculated from equation [11.5] and C varies as follows:

d/y	0.0	0.2	0.4	0.6	0.8
C	1.00	0.98	0.90	0.78	0.59.

Leakage from field turnouts is often a problem. Hanson (1966) reported that a large proportion of the water lost in irrigation ditches was from gate leakage. Leakage is particularly large from check board type structures when the water level remains above the bottom of the opening. Leakage can be minimized by using rubber seals as reported by Robinson (1970).

Siphon tubes are widely used to distribute water from a ditch onto a field. They are usually made from aluminum or plastic and are available commercially in different diameters and lengths. They can be obtained in sizes to control streams as small as 0.06 L/s (1 gal/min) or as large as 56 L/s (900 gal/min). The larger sizes are used to flood border strips or check basins. Checks must be closely spaced in head ditches with considerable slope to maintain a nearly constant operating head on each tube for uniform discharge. The discharge depends on the operating head, the tube diameter and length, the number and degree of bends in the tubes and the roughness. Siphon tube discharge is determined by the general orifice equation [11.6] in which the coefficient of discharge C can be evaluated from

$$C = C_O \sqrt{\frac{D^{4/3}}{Kn^2 C_O^2 \, L + D^{4/3}}} \quad\dotfill [11.8]$$

where
C_O = discharge coefficient for the tube entrance determined from laboratory tests for specific tubes (approximately 0.83)
D   = inside diameter of tube (millimeters or inches)
n   = roughness coefficient, 0.008 for aluminum tubes up to 7.6 cm (3 in.) diameter and 0.012 for aluminum tubes over 10 cm (4 in.) diameter

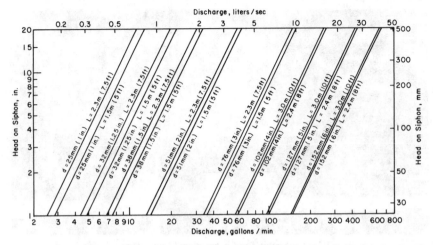

FIG. 11.8 Discharge of aluminum siphon tubes at various heads [L = length of tube, d = nominal diameter]. [From USDA, SCS, 1962]

K   = 1.25 x 10⁶ when D and L are in millimeters and meters, respectively; K = 5,087 when D and L are in inches and feet
L   = length of tube (meters or feet).

The discharge for aluminum siphon tubes operating at different heads is shown in Fig. 11.8 (USDA, SCS, 1962). Discharge capacities for plastic tubes have been determined by Tovey and Myers (1959).

Automatic discharge control devices for field turnouts are being developed to improve irrigation, but are not widely used at present. Calder and Weston (1966); Thomas (1960); and Robinson (1972) describe some that are commercially available. Remotely controlled pneumatic valves for farm turnouts were developed by Haise et al. (1965); Haise and Kruse (1969); Payne et al. (1974); Erie and Dedrick (1978); and Dedrick and Erie (1978). Humpherys (1978) developed timer-controlled gates for border turnouts where the turnout must first be automatically opened to begin irrigation and then closed to terminate irrigation while the water level in the head ditch remains constant. A mechanized traveling dam can be used to irrigate fields having contour ditches. The dam travels slowly along the ditch and diverts water onto the field on the low side of the ditch as it moves. It is used mostly on meadows or pastures where the ditch banks are sodded or the soils are stable.

Division Boxes. It is often necessary to divide water from a farm lateral into two or more ditches for distribution to different parts of the farm or to other farms. This may be accomplished with a divisor at the ditch junction. For accurate flow division it is best to measure the flow in each channel. Some divisors are designed to give a fixed proportional flow division, whereas others have a movable splitter to change the flow proportions. Divisors that give accurate proportions often divide the flow at a control section where supercritical flow exists, such as at the nappe of a free overfall (Fig. 11.9). Flow can be divided accurately without creating supercritical flow if: (a) the approach channel is long and straight for at least 5 to 10 m upstream so that the water flow approaches the divisor in parallel paths without cross currents, (b) there is no backwater effect

FIG. 11.9 Simple fixed proportional flow divisor.

that would favor one side or the other, and (c) the flow section of the structure is of uniform roughness. Detailed divisor designs are described by Kraatz and Mahajan (1975).

Debris and Weed Seed Control Structures. Trash and weed seed in the water cause problems in irrigation systems. Besides spreading weeds, trash and debris plug siphon tubes, pipe gates, and sprinkler nozzles. Trash racks are needed at entrances to pipe and underground crossings. Various types of screens and trash racks have been developed for screening irrigation water (Bergstrom, 1961; Couthard et al., 1956; and Pugh and Evans, 1964). One of the simplest devices is a screen placed below a weir or drop where at least 20 cm of free fall is available, Fig. 11.10. Both

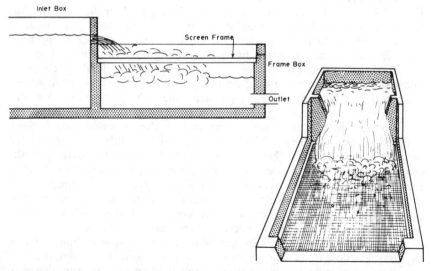

FIG. 11.10 Structure with fine mesh screen for trapping debris in irrigation stream [adapted from Bergstrom, 1961].

water-powered and electrically driven self cleaning screens are available commercially (Fig. 11.12).

11.3 LINED DITCHES

Lining ditches is an effective way to prevent ditch erosion, control rodent damage, and reduce seepage at reasonable costs. Lining also reduces maintenance, controls weed growth and ensures more dependable water deliveries. The seepage reduction helps protect neighboring land from water-logging and salt accumulations. Linings must be properly designed and installed to avoid damage from vegetative growth, fluctuating water tables, livestock traffic, or freezing and thawing.

11.3.1 Lining Materials

Selection of the lining material should be governed by availability of the material and installation equipment, ditch size, climate, foundation conditions, and type of irrigation stream, whether continuous or intermittent. Concrete is probably the most popular lining material, but asphaltic materials, bricks, membranes, metals, chemical sealants, and impermeable earth materials are also used. All of these materials make good linings if they are properly selected for the site conditions and are installed correctly.

Lined ditches can be installed in any soil, but many soils have limitations that must be overcome before a durable low-maintenance lining can be installed. Rigid linings, such as concrete, should be installed on drained soils, particularly in the colder climates. Where soil drainage is poor, the concrete must be thicker or steel reinforcement added. Thickness required for buried flexible liners also depends upon the expected subgrade materials (Section 11.3.5).

11.3.2 Capacity

Ditch capacity considerations are given in Section 11.2.2. After the necessary capacity has been determined, ditch size can be designed using the Manning formula, (equation [11.2]) with a coefficient of roughness as follows:

Material	Manning "n" for capacity
Slip formed concrete	0.015
Brick	0.017
Shotcrete	0.016
Concrete panels	0.016
Sheet metal, flex membrane	0.013
Compacted earth	Same as unlined ditches; see Section 11.2.2

11.3.3 Elevation and Slope

Lined ditches should be designed so the water surface at field turnout points is high enough to provide the required flow onto the field surface. If ditch checks or other control structures are used to provide the necessary head, the backwater effect must be considered in computing freeboard requirements. The required elevation of the water surface above the field surface will vary with the type of turnout structure used and the amount of water to be delivered. A minimum head of 12 cm should be provided.

TABLE 11.7. SUGGESTED MINIMUM CONCRETE MIX REQUIREMENTS
(FROM ASAE, 1978e)

Class of climate (Freeze-thaw cycles/year)	Type of concrete	Compressive strength @ 28 days		Cement content*	
		MPa	lb/in.^{2}	kg/m^3	bags/yd^3
Mild (20 or less)	Non Air-entrained	20.7	3000	307	5.50
Moderate (20 to 80)	Air-entrained	20.7	3000	321	5.75
Severe (over 80)	Air-entrained	24.1	3500	362	6.50

*For concrete containing type V, sulphate-resistance Portland cement, increase the cement content 20 percent. Up to 15 percent of the cement, by weight, may be replaced with pozzolans.

The required freeboard varies with the size of ditch, the velocity of the water, the horizontal and vertical alignment, the amount of storm or waste water that may be intercepted, and the change in the water surface elevation that may occur when control structures are operating. The minimum freeboard for any lined ditch should be 7.5 cm.

For concrete lining, design velocities in excess of 1.7 times the critical velocity should be restricted to straight reaches that discharge into a section or structure designed to reduce the velocity to less than critical velocity. This will avoid unstable surge flows. The maximum velocity in these straight reaches should be 4.6 m/s. Maximum allowable velocities in ditches with sprayed-on membrane linings should be limited to 1 m/s. Buried membrane installations are only effective where the flow velocity is limited to that allowable for the covering earth material (Section 11.2.2).

11.3.4 Concrete Lining and Construction

Portland cement concrete linings can withstand high stream velocities and therefore are particularly suitable for controlling erosion as well as preventing seepage. They are more resistant to mechanical damage than most other linings. However, their use is limited to sites having non-expanding soils and good internal drainage. Concrete linings must be specially designed and protected when installed on poorly drained soils, in areas subject to severe frost heaving, or on soils having a high sulphate-salt concentration.

Concrete used in ditch linings should be plastic enough for thorough consolidation but stiff enough to stay in place on the side slopes. Where slipforming is used to construct a trapezoidal section, the side slopes should be no steeper than one vertical to one horizontal for heights greater than 0.9 m.

Concrete compressive strength is a good indicator of durability. The more severe the climate, the greater the compressive strength should be. Table 11.7 lists suggested minimum mix requirements. On sites where the sulphate concentration in the soil solution is more than 0.1 percent, concrete linings of special sulphate-resistant cement (type II, IIA or V) should be used.

TABLE 11.8. SUGGESTED MINIMUM THICKNESS PORTLAND CEMENT
CONCRETE SLIPFORM DITCH (FROM USDA, SCS, 1971)

Design velocity		Climatic area*		
		Mild	Moderate	Severe
m/s	ft/s	Minimum thickness—cm (in.)†		
less than 1.8	< 6	4.0 (1.6)	5.0 (2.0)	6.5 (2.6)
1.8 to 2.7	6 to 9	5.0 (2.0)	5.0 (2.0)	6.5 (2.6)
2.7 to 3.7	9 to 12	6.5 (2.6)	6.5 (2.6)	6.5 (2.6)
3.7 to 4.6	12 to 15	7.5 (3.0)	7.5 (3.0)	7.5 (3.0)

*Climatic areas—
    Mild—Average January temperature above +4.5 °C.
    Moderate—Average January temperature between −4.0 and +4.5 °C.
    Severe—Average January temperature below −4.0 °C.
†Assumes high quality concrete placed on firm subgrade and protected from external hydrostatic pressure, uplift by expansive clays, and frost heave.

Only type V cement should be used when soil sulphate concentrations exceed 0.3 percent. The cement content should be increased 20 percent for concrete containing type V, sulphate-resistance Portland cement.

Aggregates should conform to ASTM (1977a), although locally available aggregate materials may be used if they are well graded, clean, and durable. The maximum aggregate size should not exceed one-third of the lining thickness. The water used in the concrete must be clean and free of acids, alkalis, oils, sulphates, and other harmful materials. A good rule is that water should be suitable for drinking. See ASAE (1978c) or USDA-SCS (1971) for additional information on concrete design and preparation.

The thickness of ditch linings must be established from engineering considerations on each job. Location, ditch size, velocity, subgrade conditions, operation, and climate should be evaluated in establishing the thickness required (Table 11.8).

To control cracking caused by shrinkage and temperature change, contraction joints, at least 6 mm wide, should be cut transversely in the concrete to a depth of about one-third the thickness of the lining. These joints should be spaced uniformly, at most 9 m apart. Construction joints should be butt type, formed square with the lining surface and at right angle to the ditch. Construction joints abutting structures should be faced with a suitable expansion joint material.

The concrete should be prevented from drying for at least 5 days after it is placed. Exposed surfaces should be kept continuously moist by sprinkling, flooding, or fog spraying, or by covering with moistened canvas, cloth mats, straw, sand, or other approved material. Concrete may be coated with curing compound instead of the continuous application of moisture. The compound should be sprayed on the moist concrete surfaces as soon as free water has disappeared, but should not be applied to any surface until patching, repairs, and finishing of that surface are completed. The curing compound should be thoroughly mixed immediately before applying, and should provide a continuous membrane that does not check, crack, or peel.

An irrigator who wishes to excavate and line his own ditches can use concrete panel-formed lining. This lining requires only semiskilled labor and a minimum of equipment, and, for small jobs, it may cost less than slipforming. After excavating the ditch to grade, guide forms are set about 3 m (10 ft) apart. Concrete is then poured in alternate 3 m sections. When the concrete is set, the forms are removed and the skipped panels are poured. The bottom in each section is poured first, and then fresh concrete is screeded up the slope.

11.3.5 Other Rigid Lining Materials

Brick ditch linings are no longer used in the United States but are still constructed in some parts of the world where labor is plentiful. The bricks may be laid in single or double layers, often in a herringbone pattern. Cement mortar is used for bedding, bonding and sometimes for plastering the exposed surface. Brick lining does not require expansion or contraction joints, can be easily repaired, and, where labor is inexpensive, is more economical than concrete.

Pneumatically applied mortar or shotcrete is sometimes used for lining small ditches, repairing old linings, or connecting structures to linings. Special machines are required, and the mortar mix must be carefully controlled. The equipment ordinarily does not pass particles larger than 5 mm; thus shotcrete requires a greater proportion of cement than does concrete. Because the lining is thinner than concrete lining, shotcrete lining is not recommended for cold climates. Pneumatically applied mortar may be applied with or without reinforcement depending on the physical requirements of the individual project. The lining is usually thin, with little structural strength, and should not be applied on a base that is subject to structural cracking. The mortar can be applied readily on various surfaces regardless of shape or inclination from the horizontal.

Prefabricated metal liners can also be used to control seepage and erosion in small irrigation ditches. They come in convenient lengths for easy handling. The sections are battened together, and a special sealant is used to make the joints watertight. The top edge of the lining should be firmly anchored in the ditch bank. These liners have an advantage in isolated areas where bringing in other kinds of lining material is difficult. Construction and maintenance costs are low, and such liners should provide many years of trouble-free service. In areas where the water has a high concentration of salt or other chemicals injurious to the metal, metal liners should not be installed without a coating specifically formulated to protect them from these chemicals. If freezing temperatures are common, the foundation must have adequate drainage to prevent water from accumulating under the lining.

Asphalt linings for seepage control can be installed as asphaltic concrete or in sheets, planks, or membranes. Asphaltic concrete consists of sand and gravel bound together with asphaltic cement. It is similar to Portland cement in many respects, but it is not as durable and is less resistant to high velocity flow and mechanical damage. The subgrade must be sterilized to prevent vegetation from growing through the lining.

Cold mix asphaltic concrete linings tend to develop shrinkage cracks that must be sealed. The surface of cold mix material deteriorates faster with time due to exposure and erosion than does hot mix. Hot mix asphaltic concrete has given fairly satisfactory results, but special equipment is needed to blend and place it. It can be placed by a slip form or heated screed moved slowly along the ditch by winch or tractor. Asphalt concrete linings should be compacted to less than 5 percent voids to be essentially impermeable.

Prefabricated asphalt planks are not often used for the smaller farm ditches. The planks are 0.6 to 1.2 m wide by 2.4 to 3.7 m long and 6 to 12 mm thick. They should be laid in warm weather so that they soften enough to conform to the shape of the ditch. Generally they are installed transversely. If they are lapped, the lap is downstream so that water puts no stress on the joints. If the sheets are butted, a cap strip is used to cover the joints. Plank ends should be buried in the berm for anchorage. Unless the lining is well anchored, it sags on the slope and wrinkles.

11.3.6 Other Linings

Buried membrane, compacted earth, or clay linings are seldom used in farm-sized ditches. If these linings are considered economically feasible, design information should be obtained from appropriate government agencies or consulting engineering firms.

11.3.7 Structures

Most of the drop, check and turnout structures described in Section 11.2.5 are also applicable to lined farm ditches. If the lining material is erosion resistant, stilling ponds can be made smaller or eliminated. Drop structures are not needed for velocity control in lined ditches unless the estimated velocities exceed the limiting values given in Section 11.3.3.

Straight ditch sections lined with concrete, metal, or asphalt can be used to transmit water at supercritical velocity. This velocity must be reduced before bends, distribution reaches, or turnouts into erosive channels. Energy-dissipating structures for such channels must be individually designed. The general configuration of a suitable structure and laboratory tests for discharges of 0.05 to 0.60 m³/s (1.6 to 20 ft³/s) have been published by the U.S. Bureau of Reclamation (1963).

11.4 LOW PRESSURE PIPE SYSTEMS

11.4.1 Applications

Low pressure pipe systems for irrigation water distribution have been used extensively since the 1950's. The availability of relatively low-cost, light-weight rigid plastic pipe has made buried pipe systems especially popular. The use of buried pipe allows the irrigator to eliminate pad construction for ditches, to take the most direct route from water supply to outlet points, and to avoid weed problems and loss of productive land because crops can be planted up to or over the pipelines. Seepage and evaporation losses are eliminated when water is transmitted in a well constructed pipeline. Portable pipe systems laid on the soil surface have many of the advantages of buried pipe systems. If adequate labor is available, they can be removed from fields while cultural operations are in progress.

Pipelines may not be desirable for irrigation water transmission if the irrigation supply contains large amounts of sediment and the flow conditions in the line allow the sediment to settle out and reduce the carrying capacity of the line. Pipelines are safer than open ditch systems—small children cannot fall into the water stream and tractors cannot be driven into the channel with the chance of overturning. Some pipelines constructed with large diameter risers do require safety precautions, however—either adequate height above ground surface or a screen or cover to keep children from entering.

Three general types of onfarm irrigation pipelines are used. The first is the completely portable surface system where water enters the line at the supply, a well, reservoir, or ditch turnout, and the water is applied to the field from the open end of the pipeline or from gated outlets distributed along the line. The second system is a combination of buried and surface pipe where buried permanent line is used to transmit water from the source to the field to be irrigated. Then water is taken to a gated surface pipe through one or more risers. The third system, generally used for border dike or basin irrigation, eliminates the need for surface pipe. Water is released onto the portion of the field to be irrigated from risers on the buried line.

11.4.2 Location

The irrigator can exercise much flexibility in locating surface pipelines between the source of water supply and the point of water application. The lines should be located to provide a minimum of interference with traffic on farm roads or with field cultural operations. A pipe network should be designed so that the shortest possible length of pipe is used to deliver water to all fields to be irrigated. More care is needed in locating buried systems. If possible, the line should be located where it can be easily buried with uniform trench depth and the minimum necessary cover. It is best to avoid routes that will be crossed by heavy surface traffic. Also, since air relief must be provided at high points in a buried pipeline, the number of these high points should be minimized. Trenching across fields to be irrigated should be avoided if possible. Irrigation water may enter poorly compacted backfill in the trench and cause piping along the line, simultaneously causing incomplete irrigation of the field being watered.

11.4.3 Pipeline Capacity

An irrigation pipeline must be sized carefully to obtain the best operating economy when both initial and operating costs are considered. Too large a pipeline requires excessive initial investment; too small a line may require excess energy use when water has to be pumped to overcome pipeline friction losses. The size must be adequate to deliver enough water to meet crop needs.

The Darcy-Weisbach Formula (Brater and King, 1976) expresses head loss of turbulent flow in pipelines on a rational basis:

$$H_\ell = f \frac{L}{D} \frac{V^2}{2g}, \dots\dots\dots[11.9]$$

where
  H_ℓ = the loss of head in equivalent height of water in a length of pipe L,
  D = the inside pipe diameter,
  V = the mean velocity,
  g = the gravitational acceleration, and
  f = a resistance coefficient.
Equation [11.9] is dimensionally consistent and can be used with the same

f values for either inch-pound or SI units. Values of f have been related to boundary roughness dimensions for certain types of pipe surfaces and determined empirically for others and are tabulated in most hydraulic handbooks.

Two other formulas (Brater and King, 1976) are used extensively for determining friction losses in irrigation pipelines. Resistance coefficients for these formulas are readily available, having been determined empirically for tubing commonly used in irrigation. The first of these formulas, the Hazen-Williams, can be written:

$$V = CR^{0.63}S^{0.54} \quad \dots\dots\dots\dots\dots\dots\dots\dots\dots\dots\dots [11.10]$$

where
    R = the hydraulic radius and
    S = the friction slope in meters per meter or feet per foot.
For SI units (R in mm),
    C = $0.0109C_1$,
where
    C_1 = the Hazen-Williams resistance coefficient.
When R is in feet,
    C = $1.318C_1$.
Values of C_1 range from 144 to 146 for aluminum tubing. C_1 values for other materials are given in standard hydraulic handbooks.

The Scobey formula for riveted steel pipe has also been used to compute head loss in aluminum pipe:

$$S = 10^{-3}CV^{1.9}D^{-1.1} \dots\dots\dots\dots\dots\dots\dots\dots\dots\dots\dots [11.11]$$

For SI units (V in m/s, D in mm),
    C = $5162K_S$
where
    K_S = the Scobey resistance coefficient.
For V in ft/s and D in ft,
    C = K_S.
The exponents in the Scobey equation may have different values for other pipe materials (Brater and King, 1976). Recommended values for C_1 and K_s are given in the following sections, as different pipe materials are discussed.

Energy losses through fittings and valves also have to be considered in the design of an irrigation pipeline. These so-called "local" losses are frequently estimated by applying a coefficient to the velocity head at the fitting. The sum of all local losses is then added to the estimate of pipe wall friction to give total loss in the pipeline. Local loss coefficients to be used in the equation:

$$H_f = \frac{KV^2}{2g} \quad \dots\dots\dots\dots\dots\dots\dots\dots\dots\dots\dots [11.12]$$

TABLE 11.9. RESISTANCE COEFFICIENT K FOR USE IN FORMULA $H_f = K \dfrac{V^2}{2g}$ FOR FITTINGS AND VALVES (FROM USDA, SCS, 1968)

Fitting or valve	Nominal diameter						
	75 mm (3 in.)	100 mm (4 in.)	125 mm (5 in.)	150 mm (6 in.)	175 mm (7 in.)	200 mm (8 in.)	250 mm (10 in.)
			Standard pipe				
Elbows:							
Regular flanged 90 deg	0.34	0.31	0.30	0.28	0.27	0.26	0.25
Long radius flanged 90 deg	0.25	0.22	0.20	0.18	0.17	0.15	0.14
Regular screwed 90 deg	0.80	0.70					
Tees:							
Flanged line flow	0.16	0.14	0.13	0.12	0.11	0.10	0.09
Flanged branch flow	0.73	0.68	0.65	0.60	0.58	0.56	0.52
Screwed line flow	0.90	0.90					
Screwed branch flow	1.20	1.10					
Valves:							
Globe flanged	7.0	6.3	6.0	5.8	5.7	5.6	5.5
Gate flanged	0.21	0.16	0.13	0.11	0.09	0.075	0.06
Swing check flanged	2.0	2.0	2.0	2.0	2.0	2.0	2.0
Foot	0.80	0.80	0.80	0.80	0.80	0.80	0.80
Strainers-basket type	1.25	1.05	0.95	0.85	0.80	0.75	0.67
			Other				
Inlets or entrances:							
Inward projecting	0.78	All diameters					
Sharp cornered	0.50	All diameters					
Slightly rounded	0.23	All diameters					
Bell-mouth	0.04	All diameters					
Sudden enlargements	$K = \left(1 - \dfrac{d_1^2}{d_2^2}\right)^2$ where d_1 = diameter of smaller pipe						
Sudden contractions	$K = 0.7\left(1 - \dfrac{d_1^2}{d_2^2}\right)^2$ where d_1 = diameter of smaller pipe						

are listed in Table 11.9. The coefficients in Table 11.9 have been determined for standard pipe diameters. Coefficients for special irrigation pipe dimensions and materials are not available. However, since local losses occur as a result of flow separation and changes in velocity, and are not greatly affected by the roughness of the fitting, values in Table 11.9 should be valid for irrigation pipe fittings with similar geometry.

Where the available head at the intake is limited, for example, from a canal or a reservoir, pipeline capacity must be great enough so that adequate water can be delivered to meet crop requirements. It generally will cost less to use large pipe with the available head than to use smaller pipe and booster pumps for such systems. When an entire irrigation system is to be designed, including pump, power unit, and piping, irrigation pipe sizes should be chosen so as to minimize total annual costs. Both fixed and operating items must be considered, including the initial cost of the piping, the number of years' service expected from the pipe, the initial cost and expected years of operation of different sized pump units, the energy cost of pumping, and the rate of return desired on the funds invested. Keller (1975) has presented a procedure for constructing design charts from which economical pipe sizes can be selected, based on the above considerations for non-looping distribution systems having a single pump station. More complicated systems can be designed using procedures given by authors that Keller cites.

11.4.4 Surface Systems

Portable surface pipelines for low pressure irrigation water distribution have been constructed of aluminum, plastics, and flexible rubberized fabrics; aluminum is by far the most common. All can be used solely for transmission of water or can be provided with gated outlets for water distribution to fields. Low pressure aluminum pipe is generally available in 127, 152, 203 and 254-mm (5, 6, 8 and 10-in.) nominal diameters. The pipe is rolled from 1.295-mm (0.051-in.) aluminum sheet and closed with a welded seam. Minimum standards for pressure rating, deflection, and denting resistance of aluminum irrigation tubing are published by ASAE (1978b). A rolled or cast female coupling is provided on each pipe length to contain flexible, pressure sealing gaskets. Mechanical clamps are sometimes provided for positive pipe coupling. Aluminum gated pipe sections of large diameters are generally manufactured in 6-m (20 ft) lengths, whereas those of smaller diameters are 9 m long. Two men are needed to pick up or lay these lines if the pipe is to be moved from one field to another. Commercial machines are available for picking up aluminum pipe sections, loading them on the trailer and then unloading the line in a new location.

Rigid, extruded polyvinyl chloride (PVC) surface pipelines are also available. The PVC compound contains an ultraviolet inhibitor that allows extended exposure to sunlight without deterioration. Diameters and couplings are compatible with aluminum gated pipe.

Rubberized fabric tubing is flexible and folds flat when not under pressure. Sections generally are available in 15-m (50-ft) lengths and in diameters from 100 to 300 mm (4 to 12 in.). Under intermediate internal pressures, the pipe cross section will not be round; therefore, in determining flow rates through this type of tubing, the relation of cross-sectional area to pressure needs to be considered as well as the resistance to flow and the hydraulic head available (Humpherys and Lauritzen, 1962).

Conduits of thin-walled, flexible vinyl tubing for water transmission and distribution to furrows became available in the mid-1970s. This tubing is sold in lengths of several hundred meters, packaged in compact rolls. Wall thickness is 0.25 mm (0.010 in.) or less and diameters from 250 mm to 500 mm (10 to 20 in.) are available. A tubing installation implement, which mounts on the three-point hitch of a farm tractor, can plow a shallow groove to provide a bed for the tubing and unreel and lay the tubing simultaneously. Plastic fittings are available for couplings, tees and connections to the water supply. Gates are inserted in the field, at the exact locations necessary to best serve furrows or corrugations, using a simple insertion tool.

Flow capacity of the flexible vinyl tubing is about one-fourth less than aluminum tubing of the same diameter, because at the low allowable operating pressures the tubing is not fully round. When gated sections are laid on sloping land, earth can be shoveled on the tubing as necessary to maintain enough pressure in upper reaches for adequate gate flow.

Initial cost of flexible vinyl tubing is a fraction of that for comparable aluminum or rigid plastic pipe. However, the material is not durable and usually has to be replaced each irrigation season.

TABLE 11.10. FRICTION LOSS IN ALUMINUM PIPE AS ESTIMATED BY
HAZEN-WILLIAM FORMULA.* JOINT LOSSES ARE INCLUDED.
ASSUMES 9-m (30-ft) PIPE LENGTHS (ADAPTED FROM USDA, SCS, 1979)

		Nominal Outside Diameter								
		102 mm (4 in.)		127 mm (5 in.)		152 mm (6 in.)		203 mm (8 in.)		254 mm (10 in.)	
L/s	ft³/s	G†	NG	G	NG	G	NG	G	NG	N	NG
		Head loss (m/100 m or ft/100 ft)									
10	0.35	1.89	1.80	0.65	0.62	0.27	0.26	0.07	0.07		
15	0.53	4.07	3.86	1.40	1.31	0.59	0.55	0.15	0.14		
20	0.71	6.94	6.57	2.41	2.26	1.01	0.94	0.26	0.24	0.09	0.09
30	1.06	14.86	14.02	5.16	4.82	2.18	2.01	0.57	0.52	0.20	0.18
40	1.41			8.87	8.26	3.74	3.45	0.98	0.89	0.35	0.31
50	1.77					5.71	5.25	1.50	1.35	0.54	0.48
75	2.65					12.24	11.23	3.24	2.92	1.16	1.03
100	3.53							5.57	5.00	2.01	1.77
125	4.41							8.54	7.65	3.08	2.71
150	5.30							12.02	10.74	4.35	3.82

*Local loss coefficient
K_c = 0.3 for gated pipe, 0.2 without gates
C_1 = 145 for 102-152 mm pipe
C_1 = 144 for 203-254 mm pipe
For 6-m (20-ft) lengths, increase loss by 10 percent for gated pipe and by 7 percent without gates.
For 12-m (40-ft) lengths, decrease loss by 5 percent for gated pipe and by 4 percent without gates.

†G = Gated, NG = not Gated

Inlets to portable pipelines, particularly aluminum lines that have low
resistance to crushing, must be designed with care. If subatmospheric
pressure occurs in the line, pressure differential can flatten long lengths
of pipe. Subatmospheric pressures might occur if a deep well pump that
supplies a pipeline without check valves is suddenly stopped and the
open valves on the pipeline are inadequate for vacuum relief. Rapid
closing of an upstream valve in a line carrying high velocity flow may also
lower pressures below the valve. Likewise, pressure surges in the line can
cause joints to separate or pipe sections to rupture. Valves controlling
flow into surface pipelines should be opened carefully so that the lines fill
slowly before they are brought up to full pressure. If water is to be released
from the pipeline through gates, one set of gates should be open at the time
the line is filled. Likewise, shutting off the flow to the pipe before all gates
are closed will help prevent a vacuum from developing in the line. If
the pipeline is supplied directly from a pump, a stand pipe near the pump
outlet with diameter at least half the pipe diameter serves to even out
pressure surges and to relieve vacuum as well. A check valve near the pump
discharge prevents flow of water from the line back through the pump if the
pump stops. Low pressure portable pipelines are generally not installed over
terrain with wide variations in elevations. However, if extreme high points
exist in the line, some method of air release at these points may be needed
to avoid restriction in pipe capacity.

The Hazen-Williams Formula (equation [11.10]) is used for computing
head loss in aluminum irrigation pipe. The Soil Conservation Service
(USDA-SCS, 1976) recommends a value of C_1 = 144 for gated pipe in
9.1-m (30-ft) lengths. Additional, local head losses are estimated for each
pipe joint, the amount of loss depending on whether the pipe is fitted with

TABLE 11.11. MULTIPLE OUTLET
FACTOR FOR GATED PIPE
(USDA, SCS, 1976)

Number of outlets	Multiple outlet factor
5	0.44
10	0.39
15	0.37
20	0.36
30	0.35
50	0.34
100	0.34
more than 100	0.33

gates or not. Values of friction loss for aluminum pipe with and without outlet gates are given in Table 11.10 for pipe sizes commonly available in the U.S.A. The head loss values in these tables include a component to account for local losses at the pipe joint. If equation [11.11] is used for aluminum pipe, $K_S = 0.40$. $C_1 = 150$ for PVC transmission pipe and 130 for PVC pipe with gates.

Water is commonly released from surface pipelines through small gates installed in the pipeline at intervals equal to the furrow or corrugation spacing. Head loss in a section of pipe discharging water through open gates can be estimated by first determining the head loss as if the pipe were transmitting the full flow from one end of the discharging section to the other, and then applying a correction factor that depends on the number of gates open (USDA, SCS, 1976). Multiple outlet factors are presented in Table 11.11. As an example, assume a 600-m length of 254-mm aluminum gated pipe (6-m lengths) discharging 50 L/s through 100 gates spaced 0.75 m apart. The maximum head loss in the line will occur, of course, when the 100 gates farthest from the inlet are discharging. From Table 11.10, the head loss in 254-mm gated aluminum pipe carrying 50 L/s is 0.54 m/100 m. For 6-m pipe lengths, the SCS recommends increasing the estimated friction loss by 10 percent. Thus, head loss per unit length due to wall friction is:

$$S_w = 1.10(00.0054) = 0.00594$$

From Table 11.11 the multiple outlet factor for 100 gates = 0.34. The total head loss in the 600-m line is the sum of that in the lengths with closed and open gates or: $H_1 = 525(0.00594) + 75(0.34)(0.00594) = 3.27$ m.

Pipe Gate Discharge. Pipe gates are supplied in many sizes and styles by a number of manufacturers. Gate discharge is governed by the orifice flow equation, equation [11.6], when there is no flow in the pipeline past the gate. When there is flow, gate discharge decreases as velocity increases. Since the discharge coefficient varies significantly for different gate configurations, discharge information should be obtained directly from the manufacturer of the gates used.

When gated pipe has to be laid on too great a slope, head is greatest on the downslope gates. These gates must be nearly closed to maintain constant furrow streams along the pipeline. With the small gate openings, velocities are large and may cause erosion damage where the streams impinge on the furrow surface. Flexible tubes or "socks" are often attached

TABLE 11.12. PRESSURE RATINGS* FOR NON-THREADED THERMOPLASTIC PIPE (ASAE, 1978c)

SDR†	Polyvinyl chloride compounds								Polyethylene compounds								Acrylonitrile-butadiene-styrene					
	1120 1220 2120		2116		2112		2110		3406		3306		2306		2305		1316		2112		1210	
	lb/in.²	kPa‡	lb/in.²	kPa	lb/in.²	kPa	lb/in.²	kPa	lb/in.²	kPa	lb/in.²	kPa	lb/in.²	kPa	lb/in.²	kPa	lb/in.²	kPa	lb/in.²	kPa	lb/in.²	kPa
5.3																						
7.0									200	1379	200	1379	200	1379	160	1103						
9.0									160	1103	160	1103	160	1103	125	862						
11.5									125	862	125	862	125	862	100	689						
13.5	315	2172	250	1724	200	1379	160	1103	100	689	100	689	100	689	80	552	250	1724	200	1379	160	1103
15.0																						
17.0	250	1724	200	1379	160	1103	125	862	80	552	80	552	80	552	62	427	200	1379	160	1103	125	862
21.0	200	1379	160	1103	125	862	100	689									160	1103	125	862	100	689
26.0	160	1103	125	862	100	689	80	552									125	862	100	689	80	552
32.5	125	862	100	689	80	552	63	434									100	689	80	552	63	434
41.0	100	689	80	552	63	434	50	345									80	552	63	434	50	345
51.0	80	552	64	441	50	345	40	276									64	441	50	345	40	276
81.0	50	345	40	276	30	207	25	172									40	276	30	207	25	172
Lo Hd§	40	276																				

*Static or working pressure plus surge at any point for water at 23 °C (73.4 °F). If surge is not known, allowable working pressures are lower (ASAE, 1978c).

†SDR = Standard Dimension Ratio as defined in section 11.4.5.

‡kPa = kilopascals, kN/m².

§Maximum pressure rating varies from 273 to 438 kPa (39.6 to 63.5 lb/in.²).

to each gate to minimize erosion. High velocity streams flowing into the socks are stilled and discharged gently into the furrows. Simple butterfly valves or overflow stands (similar in principle to those shown in Fig. 11.16) may be installed at intervals along the pipeline to maintain more nearly constant heads. Orifice plates can also be placed in the bell ends of the pipe to dissipate excess energy. The size and number of orifices used depends on the head to be dissipated.

11.4.5 Buried Low Pressure Systems

Plastic Pipe. The ready availability and easy handling of light-weight semi-rigid plastic piping has made buried low pressure pipe systems an attractive alternative to lined or unlined ditches for irrigation water transmission. The plastic compounds commonly used in constructing such pipe are polyvinyl chloride (PVC), acrylonitrile-butadiene-styrene (ABS) and polyethylene (PE).

PVC pipe is manufactured in two general size classifications: "iron pipe size" (IPS) and "plastic irrigation pipe size" (PIP). IPS pipe has the same outside diameter as iron or steel pipe of the same nominal size. PIP sizes were developed primarily for irrigation use with size designations developed jointly by manufacturers and the Soil Conservation Service. PIP is smaller in actual diameter than IPS pipe of the same nominal diameter and therefore has a lower flow capacity. Difference in capacity ranges from about 18 percent for 100-mm (4-in.) pipe to 8 percent for 300-mm (12-in.).

Plastic piping materials are commonly available in a range of pressure ratings. The ratio of the allowable tensile stress in the pipe wall to the maximum pressure that the fluid in the pipe can exert continuously with a high degree of certainty that the pipe will not fail is related linearly to the standard dimension ratio (the ratio of pipe diameter to minimum wall thickness). Plastic pipe commonly used for irrigation is manufactured with several Standard Dimension Ratios (SDR). Schedule 40, 80 or 120 PVC pipe, with dimensions corresponding to steel pipe, is also available. For a given pipe material and standard temperature, pipes of all diameters with the same SDR will have the same pressure rating. For PVC and ABS the average outside diameter is used in calculating SDR. For PE the average inside diameter is used. Polyvinyl chloride pipe for low head systems is usually labeled "low head", "SDR = 81" or "SDR = 64." Pressure ratings for pipes of different materials and a range of SDR's are listed in Table 11.12. Operating temperature also affects pressure ratings. For rating service factors at temperatures above 23 °C (73.4 °F) see Standard S376 (ASAE, 1978c).

Head losses in low head plastic pipe are determined using the Hazen-Williams equation (equation [11.10]). A resistance coefficient of $C_1 = 150$ is commonly used. Local losses are calculated using the same K factors as listed in Table 11.9. Head loss values for PIP and IPS plastic pipe sizes commonly available in the U.S.A. are tabulated in Tables 11.13 and 11.14.

Trenches for plastic pipe should have a relatively smooth, firm, continuous bottom and be free of rocks. Where rough rock edges cannot otherwise be avoided, the trench should be overexcavated and filled to the bedding depth with sand or finely graded soils. The trench below the top of the pipe should be just wide enough to provide room for joining the pipe and compacting the initial backfill. Minimum and maximum trench widths for low-head and SDR-81 pipe are listed in Table 11.15.

TABLE 11.13. FRICTION HEAD LOSS IN PVC OR ABS PLASTIC IRRIGATION
PIPELINES.* PIP PIPE; SDR = 21†

Flow rate (Q)		PIP PIPE, NOMINAL DIAMETER				
		101.6 mm (4 in.)	154.2 mm (6 in.)	203.2 mm (8 in.)	254.0 mm (10 in.)	304.8 mm (12 in.)
L/s	ft³/s			Head loss (m/100 m or ft/100 ft)		
6.0	0.21	0.73	0.11	0.03		
8.0	0.29	1.25	0.18	0.05		
10.0	0.35	1.88	0.27	0.07		
15.0	0.53	3.99	0.58	0.14	0.05	
20.0	0.71	6.80	0.99	0.25	0.08	0.03
25.0	0.88	10.27	1.49	0.37	0.13	0.05
30.0	1.06		2.09	0.52	0.18	0.07
40.0	1.41		3.56	0.89	0.30	0.12
50.0	1.77		5.39	1.34	0.45	0.19
60.0	2.12		7.55	1.88	0.64	0.26
80.0	2.82		12.87	3.21	1.08	0.45
100.0	3.53			4.85	1.64	0.67
150.0	5.30			10.29	3.47	1.43
200.0	7.06				5.92	2.43
250.0	8.83				8.94	3.68
300.0	10.59				12.54	5.15
350.0	12.36					6.85

*Table based on Hazen-Williams equation; C_1 = 150.
†To find friction head loss in PVC or ABS pipe having a standard dimension ratio other
than 21, the values in the table should be multiplied by the appropriate conversion factor
shown below:

SDR No.	Conversion Factor
13.5	1.34
17	1.13
21	1.00
26	0.91
32.5	0.84
41	0.785
51	0.75
81	0.694
Lo Head (50 ft)	0.677

The pipe trench should be excavated deeply enough so that from 0.75 to 1.20 m (30 to 48 in.) of cover can be placed over low head irrigation pipes. This cover will normally protect the pipe from traffic crossing, freezing temperatures, or soil cracking, but is not so great that it will cause excessive soil loading on the pipe. Low areas can be crossed with shallow trench depths and then extra fill placed over the pipeline to provide the minimum cover depth. In such cases, the fill should have a top width of no less than 3 m (10 ft) and 6:1 side slopes. The pipe should be uniformly supported over its entire length on firm stable material in the trench.

Low head (50 ft head) PVC pipe is supplied only with solvent welded joints. Pipe suppliers furnish assembly instructions suitable for "do it yourself" installation. Solvent-welded PVC pipe is often flexible enough to be assembled alongside the trench and then lowered into the trench after the joints are well set. Polyvinyl chloride pipe with SDRs of 81.0 and lower can be obtained with solvent weld or gasketed joints. Inexperienced workers can assemble leak-free pipelines more easily with gasketed than with solvent weld joints.

Wherever the flow direction changes, reaction forces on the piping may

TABLE 11.14. FRICTION HEAD LOSS IN PVC OR ABS PLASTIC IRRIGATION
PIPELINES*. IPS PIPE; SDR = 21†

Flow rate (Q)		IPS PIPE, NOMINAL DIAMETER					
		101.6 mm (4 in.)	127.0 mm (5 in.)	154.2 mm (6 in.)	203.2 mm (8 in.)	254.0 mm (10 in.)	304.8 mm (12 in.)
L/s	ft^3/s			Head loss (m/100 m or ft/100 ft)			
6.0	0.21	0.48	0.17	0.07			
8.0	0.29	0.82	0.29	0.12			
10.0	0.35	1.24	0.44	0.19	0.05		
15.0	0.53	2.62	0.94	0.40	0.11	0.04	
20.0	0.71	4.47	1.60	0.68	0.19	0.06	0.03
25.0	0.88	6.75	2.42	1.03	0.28	0.10	0.04
30.0	1.06	9.46	3.39	1.44	0.40	0.14	0.06
40.0	1.41		5.77	2.45	0.68	0.23	0.10
50.0	1.77		8.72	3.71	1.02	0.35	0.15
60.0	2.12		12.23	5.20	1.43	0.49	0.21
80.0	2.82			8.86	2.44	0.84	0.36
100.0	3.53				3.69	1.26	0.55
150.0	5.30				7.82	2.67	1.17
200.0	7.06					4.56	1.99
250.0	8.83					6.89	3.00
300.0	10.59					9.66	4.21
350.0	12.36					12.85	5.60

*Table based on Hazen-Williams equation; C_1 = 150.
†To find friction head loss in PVC or ABS pipe having a standard dimension ratio other than
21, the values in the table should be multiplied by the appropriate conversion factor shown
below:

SDR No.	Conversion Factor
13.5	1.35
17	1.13
21	1.00
26	0.91
32.5	0.84
41	0.785
51	0.75

exist. These forces, if unrestrained, may cause the joints to separate. Concrete thrust blocking is commonly used to provide resistance to pipe movement where side thrust might occur. Concrete pads, poured around pipe
fittings in the trench, allow forces on the pipe to be resisted by a large area of
soil along the trench surface. Standard S376 (ASAE, 1978c) illustrates
fittings requiring thrust blockings and approved configuration of the

TABLE 11.15. MAXIMUM AND
MINIMUM TRENCH WIDTHS FOR
LOW-HEAD AND SDR-81 PIPE
(ASAE, 1978c)

Pipe size		Trench width			
		Minimum		Maximum	
mm	in.	mm	in.	mm	in.
102	4	406	16	762	30
152	6	457	18	762	30
203	8	408	20	762	30
254	10	559	22	762	30
305	12	610	24	762	30
356	14	660	26	762	30
381	15	686	27	762	30

TABLE 11.16. HEAD LOSS IN CONCRETE PIPE WITH CONCENTRIC GASKET JOINTS*

Flow rate (Q)		203 mm (8 in.)	254 mm (10 in.)	305 mm (12 in.)	356 mm (14 in.)	381 mm (15 in.)	406 mm (16 in.)	457 mm (18 in.)	533 mm (21 in.)
L/s	ft³/s				Head loss (m/100 m or ft/100 ft)				
15.0	0.53	0.14							
20.0	0.71	0.25	0.07						
30.0	1.06	0.55	0.17	0.06					
40.0	1.41	0.98	0.30	0.11					
50.0	1.77	1.54	0.46	0.18	0.08				
60.0	2.12	2.21	0.67	0.25	0.11	0.08			
70.0	2.47	3.01	0.91	0.34	0.15	0.10	0.07		
80.0	2.82	3.93	1.19	0.45	0.20	0.14	0.10	0.05	
90.0	3.18	4.97	1.51	0.57	0.25	0.17	0.12	0.07	
100.0	3.53		1.86	0.70	0.31	0.21	0.15	0.08	
150.0	5.30		4.18	1.58	0.69	0.48	0.34	0.18	0.08
200.0	7.06			2.80	1.23	0.86	0.61	0.32	0.14
250.0	8.83			4.38	1.92	1.34	0.95	0.51	0.22
300.0	10.59				2.76	1.92	1.37	0.73	0.32

*Computed from Manning Formula, n = 0.011.

blocks and gives tables for estimating thrust magnitude. Water packing can be used to settle the backfill around low head pipelines. Before backfilling the pipe is filled completely with water and all joints observed for leakage. The pipe should remain filled with water during the entire backfill operation to keep it from partially collapsing from the weight of the saturated fill and from floating. After the pipe is proven to be leak free, 30 to 45 cm (12 to 18 in.) of backfill are placed over the pipe and water is added to the trench until the fill is thoroughly saturated. The wetted fill is then allowed to dry until firm enough to walk on before the final backfill is placed.

Polyethylene pipes are joined using rigid insert fittings and clamps or bell joints with rubber gaskets. Polyethylene only up to 150 mm (6 in.) in diameter is available for buried lines. For details of plastic pipe pressure rating, material characteristics, and installation specifications and procedures, see Standard S376 (ASAE, 1978c).

Protection from vacuum and pressure surges is essential for buried thermoplastic pipelines. See Section 11.4.6 for details.

Concrete Pipe. Non-reinforced concrete was the most commonly used material for irrigation pipelines before thermoplastic compounds came into use. Concrete pipe sections are readily available in diameters from 150 to 600 mm (6 to 24 in.). Most types of concrete pipe are designed to be joined with rubber gaskets; others are joined using a cement mortar. Concrete pipelines should be designed so that the maximum working head does not exceed 5 m (15 ft) above the pipe centerline. Friction losses in concrete pipe may be computed with the Scobey formula, equation [11.11], using a retardance coefficient, K_S, of 0.31 for mortar joints and 0.37 for pipes with rubber gasket joints and a very smooth internal finish. Table 11.16 lists head losses for pipe with rubber gasket joints, computed with the Manning equation, equation [11.2], with a resistance coefficient, n = 0.011.

Trenches for concrete pipe should be excavated deep enough to permit a 0.6 m (2 ft) minimum depth of cover on the pipe. Where trenches are excavated in soils containing rock or in soils subject to appreciable swelling and shrinkage, the trenches should be overexcavated and backfilled with stable materials to provide a firm, uniform base. Trench widths just adequate to allow room for pipe installation provide maximum support for the finished pipeline. Water must be drained from the trench before mortar joints are constructed.

All pipe joints should be clean and free of loose or defective material before joining. Mortared joints must be constructed with extreme care to prevent subsequent leakage. Bonding techniques are given in specification C118 (ASTM, 1977b). Pipe with mortar joints must be partially backfilled immediately, while the mortar is still plastic, to avoid loosening the joints. Backfilling may be completed either immediately after the joints are made or after a delay of at least 30 h. The pipeline should not be filled with water before backfilling is completed.

Pipelines should be tested for leaks by observing the trench after two weeks of continuous water in the pipe. All visible leaks should be repaired. The pipe should be demonstrated to function at design capacity with no objectionable surges, water hammer, or overflow from vents or stands.

Concrete pipelines are ordinarily connected to structures of other material using a Portland cement base mortar and techniques similar to those used for joining mortar joint pipe sections. The connection of the pump stand to the pump must be made so that pump vibrations are not transmitted to the pipeline or its structures.

Nitrogen fertilizers should be used with care in concrete pipelines. When ammonia is added to hard waters, calcium carbonate may precipitate and adhere to the walls of the pipe. Ammonium sulphate concentrations should not exceed 0.1 percent and the line should be flushed immediately after use. Fertilizer manufacturers should be consulted on damage prevention measures.

For design and installation details for concrete pipe, see S261.5 (ASAE, 1978d), Pillsbury (1952), or Portland Cement Association (1952).

Cast-In-Place Concrete. Cast-in-place concrete pipe is no longer used much for farm irrigation pipelines. This type of pipe is limited to operating heads of 5 m (15 ft) or less. It is useful for the same types of application as concrete tile pipelines, especially in areas of heavy traffic where plastic pipelines would not be suitable. Design and construction details may be obtained from offices of the Portland Cement Association or the American Concrete Institute (ACI, 1969).

Asbestos-Cement Pipe. Asbestos-cement (AC) pipe is frequently used for high pressure water lines; however, it is available for lower pressure application as well. This pipe is manufactured from a mixture of Portland cement, Portland blast furnace slag cement, and asbestos fiber. The mixture is placed under pressure on a mandrel to provide a dense pipe with a smooth interior surface. A special grade of AC pipe is made for irrigation application. Nine pressure ratings are available ranging from FT (Fluid Transmission)-30 through FT-90. The numbers represent one-tenth the minimum hydrostatic bursting pressures in pounds per square inch. Sizes from 150 to 610 mm (6 to 24 in.) are available. For the lower pressure ratings, nominal size is equal to the internal diameter. Either the Hazen-Williams formula, equation [11.10] or the Scobey formula, equation [11.11], can be used for computing head loss in AC pipe. Resistance coefficients, $C_1 = 140$ or $K_s = 0.32$ are recommended. Friction losses for FT class asbestos cement pipe are given in Table 11.17. The AC pipe is commonly manufactured in 4-m (13-ft) lengths. Both ends of each pipe length are machined at the factory for precise insertion into a separate rubber gasketed coupler. This assembly must be done within the field trench because the gasketed joints and rigidity of the pipe allow little deflection at each joint.

TABLE 11.17. FRICTION LOSS IN ASBESTOS CEMENT PIPE, FT-30 THROUGH
FT-50* (ADAPTED FROM USDA, SCS, 1968)

Flow rate (Q)		152.4 mm (6 in.)	203.2 mm (8 in.)	254.0 mm (10 in.)	304.8 mm (12 in.)	406.6 mm (16 in.)
L/s	ft³/s			Head loss (m/100 m or ft/100 ft)		
8.0	0.28	0.14	0.03			
12.0	0.42	0.30	0.07			
16.0	0.56	0.51	0.13	0.04		
20.0	0.71	0.77	0.19	0.06	0.03	
25.0	0.88	1.17	0.29	0.10	0.04	
30.0	1.06	1.63	0.40	0.14	0.06	
40.0	1.41	2.78	0.68	0.23	0.09	
50.0	1.77	4.21	1.03	0.35	0.14	0.04
100.0	3.53		3.73	1.26	0.52	0.13
150.0	5.30		7.91	2.66	1.09	0.27
200.0	7.06			4.53	1.86	0.46

*From Hazen-Williams Formula (equation [11.10], $C_1 = 140$)

Factory-made Tees, elbows, tapped couplings and reducer couplings and surface risers, as well as adaptors to steel pipe sections, are available for AC pipelines. Asbestos-cement pipe has considerable resistance to crushing. Pipelines, however, should be installed with a minimum of 2/3 m (2 ft) of well-compacted cover. Newly installed AC pipelines can be tested in the same way as high pressure thermoplastic lines.

Other Pipeline Materials. In recent years, buried irrigation pipelines have been constructed from a number of materials other than concrete or plastic. Steel pipe and vinyl-coated aluminum are examples. The widespread availability of thermoplastic piping, however, has made use of these materials economically infeasible. New piping materials are also being continually developed. Examples of these, which have not yet proved to be competitive with thermoplastics, are epoxy-bonded aggregate, fiberglass, fiberglass encased PVC and foam-cast plastic pipe.

11.4.6 Structures

All buried pipelines require an inlet and one or more outlet structures. In addition, structures for air release, excess pressure relief, or head control may be necessary (Robinson et al., 1963; Seipt, 1974). Frequently two or more functions can be combined in a single structure. Construction methods and materials are similar for any given structure, regardless of pipeline material. Reinforced concrete pipe is often used for inlet and head control risers because of its durability and resistance to impact damage during farming operations. Special openings in pipe sections, beveled ends, etc. can be specially ordered from pipe casting plants near the pipeline site. These modifications are most easily made while the concrete is green.

Inlets. Two types of inlet structures are shown in Fig. 11.11 and 11.12. Both involve vertical risers to connect with the surface supply. Where all downstream outlets can pass small pieces of trash without clogging, the trash bars shown in Fig. 11.11 can be used to keep large trash and animals out of the pipeline. For smaller outlets, especially on gated pipe, and a water supply containing trash, a finer inlet screen is necessary. The screen shown in Fig. 11.10 can also be used for pipeline inlets. Fig. 11.12 shows an inlet structure with a self-cleaning trash excluder. Screens may be obtained with perforations as small as 3 mm (1/8 in.). Self-cleaning screens require

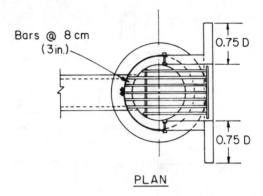

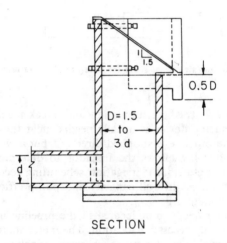

FIG. 11.11 Pipeline inlet with trash grate [from USDA, SCS, 1978].

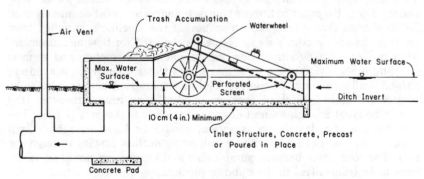

FIG. 11.12 Pipeline inlet with water-powered trash screen. Similar units are available with electric motor drives.

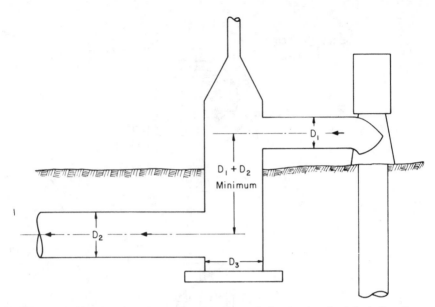

FIG. 11.13 Surge control/air release stand for pump discharging into buried pipeline [from USDA, SCS, 1978].

daily visits to remove trash accumulation and check mechanical functions.

A flow measuring device is usually needed near the pipeline inlet, especially where the pipeline is supplied directly from a well or an irrigation district canal. Metering allows the irrigator to determine the volume of water applied. It is essential if irrigation scheduling recommendations are to be followed. Changes in well capacity and pump efficiency can also be easily detected by metering.

The flow meter needs to be located in the pipeline upstream from the first outlet at an easily accessible location. The meter should be downstream from the air removal vent at a point where the pipeline is flowing full. A straight length of pipe, equivalent to 10 diameters, should precede the meter. The meter should be installed so that it can be easily removed for cleaning and servicing. Propeller or turbine type flow meters are most commonly used for measuring flow in irrigation pipelines. If the pipeline is supplied from an open channel, a critical depth flume may be used for flow measurement. See Chapter 9 for details of flow measurement with meters and flumes.

Pump Stands. When a pipeline receives water directly from a pump, a stand similar to that shown in Fig. 11.13 is necessary. Pump stands can be constructed from reinforced concrete or steel. The inside diameter (D_3) should be great enough to limit downwind velocity to 0.6 m/s (2 ft/s). The vent pipe, if of reduced size as shown, should be large enough to limit velocities to 3 m/s (10 ft/s) with the full pump outflow flowing through the vent. The connection between pump outlet and stand must not allow vibrations to be transmitted to the stand or pipeline.

Vents. Vents similar to that shown in Fig. 11.14 are necessary to allow air to escape and to prevent subatmospheric pressures from developing and causing plastic pipelines to collapse. Vents should be at the downstream end of each lateral, at summits in the line, at points of sudden, extreme increases in pipeline grade, downstream of stands with downward veloci-

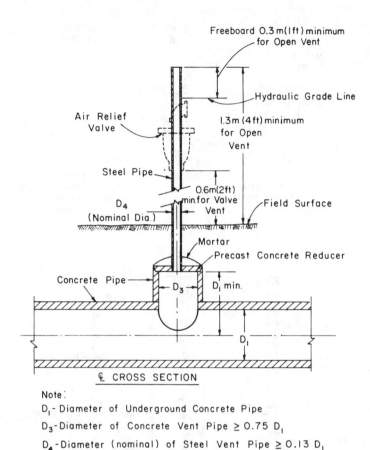

Freeboard 0.3 m (1 ft) minimum
for Open Vent

Hydraulic Grade Line

Air Relief
Valve

1.3 m (4 ft) minimum
for Open
Vent

Steel Pipe

0.6 m (2 ft)
min. for Valve
Vent

D_4
(Nominal Dia.)

Field Surface

Mortar

Precast Concrete Reducer

Concrete Pipe

D_3

D_1 min.

D_1

℄ CROSS SECTION

Note:
D_1 - Diameter of Underground Concrete Pipe
D_3 - Diameter of Concrete Vent Pipe $\geq 0.75\ D_1$
D_4 - Diameter (nominal) of Steel Vent Pipe $\geq 0.13\ D_1$

FIG. 11.14 Air vent for low pressure pipelines. Alternate air relief valve shown with dashed lines [from USDA, SCS, 1978].

ties great enough to entrain air (0.3 m/s or 1 ft/s), and downstream from line gates or valves. Combination pressure-vacuum relief valves can be used in place of open vents. Such valves should have outlet diameters at least one-third the pipeline diameter.

The sudden release of large amounts of accumulated air can damage a pipeline structurally. Momentary pressure rises greater than four times the operating pressure can occur when automatic air release valves suddenly close, abruptly reducing the velocity of the water behind the air body (Ball, 1974). The open air vents, Fig. 11.14, provide a practical means of preventing air accumulations in low-pressure pipelines. When air must be released through self-actuating or manual valves, however, studies indicate that the optimum valve diameter is equal to 0.01 pipe diameters, so that release is not too rapid. Also, the velocity at which an empty pipeline is filled with water should not exceed 0.7 m/s (2 ft/s).

Gate Stands. Gate stands can provide for head regulation, water release, vacuum relief, and air release, all in a single structure. Two styles are shown in Fig. 11.15 and 11.16. These structures may be used with or without the turnout. A gated bypass may be necessary to drain the line at the end of the irrigation season, permit flushing of sediment, and provide for

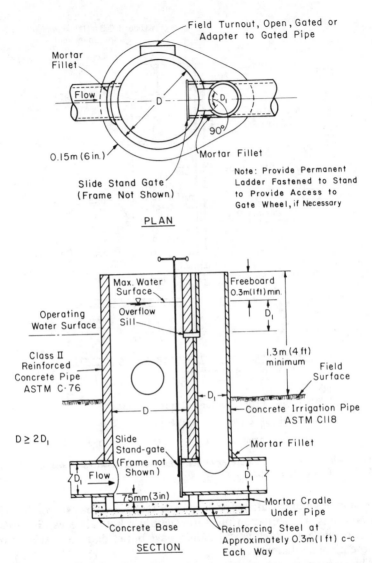

FIG. 11.15 Overflow-type pipeline riser for head control, water release, and/or pressure relief.

coarse flow adjustment. The stand must be large enough to provide access for inspection and repair of the slide gate. The section open to downward flow must be large enough to prevent air entrainment, or a downstream vent must be provided.

Fig. 11.17 illustrates another type of gate stand. This stand allows water to be checked so that outlets upstream of the stand can flow at the desired pressure. The stand also provides pressure relief for the downstream pipeline segment. A stand similar to that of Fig. 11.17 can be used to automatically control head on the downstream pipeline, by replacing the manual valve with a float valve, so that increasing water level in the stand reduces the valve opening and vice versa. Automatic float valves are commercially available in sizes from 100 to 600 mm (4 to 24 in.) with capacities from 45

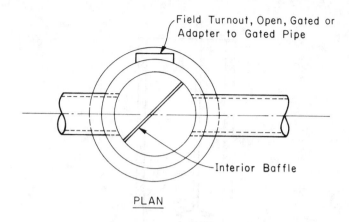

Field Turnout, Open, Gated or Adapter to Gated Pipe

Interior Baffle

PLAN

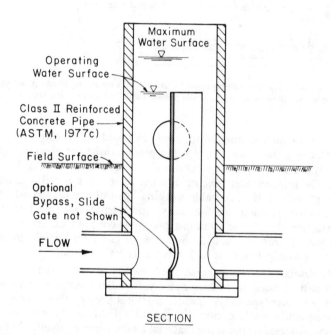

Maximum Water Surface

Operating Water Surface

Class II Reinforced Concrete Pipe (ASTM, 1977c)

Field Surface

Optional Bypass, Slide Gate not Shown

FLOW

SECTION

FIG. 11.16 Standpipe for pressure relief, head control, and/or water release from low pressure pipelines.

to 1,350 L/s (1.6 to 48 ft³/s). A series of automatic float valves in line may require careful design to prevent unstable valve action and pressure surges.

Open air vents and stands should be designed to extend at least 30 cm (1 ft) above the design hydraulic grade line of the pipeline and at least 1.25 m (4 ft) above ground surface. Maximum stand height must not exceed the allowable head on the pipeline material.

Pressure Relief Valves. Open stands can be replaced by pressure relief valves (Fig. 11.18), so that the pipeline need not be open to the atmosphere. Pressure relief valves must be marked with the pressure at which they start to open. Adjustable valves should be capable of being locked or sealed so

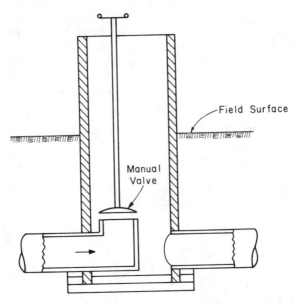

FIG. 11.17 Standpipe for low pressure pipelines. Valve controls operating head on turnouts in pipe upstream of stand. Standpipe height limits pressure on downstream pipe section.

that the proper pressure setting is not easily changed. The valve should be capable of releasing the design flow of the pipeline without elevating pipeline pressure more than 50 percent above the permissible working head of the pipeline. The number and spacing of relief valves on a pipeline is determined by the grade and the allowable working head of the pipe. One valve, at the lowest point, is adequate for many pipelines. If examination of the hydraulic head line shows that allowable pressures could be exceeded, additional relief valves must be added. Some manufacturers supply pressure relief valves that also provide air and vacuum relief functions.

Turnout Risers. The risers described earlier in this section under Gate Stands can be used to release water to the field surface or to gated pipe sections. Another type of riser, which does not project above the ground surface, is frequently used, because it obstructs field machine travel less. This riser is constructed of the same material as the pipeline itself. Water release is controlled by an "alfalfa valve," which can release water directly onto the field surface or, through a hydrant, to gated pipe. No automatic head regulation is provided in this type riser. Discharge from alfalfa valves can be computed from the formula for orifice flow (equation [11.6]). The discharge coefficient can be estimated from $C = 0.79V^{-0.123}$ for V in m/s or $C = 0.9V^{-0.123}$ for V in ft/s (Marr, 1967).

Procedures and equipment have been developed for automatic timed water releases from alfalfa valves, using auxiliary pneumatic valves (sometimes in gated pipe hydrants) and low pressure air supply. Such systems can increase irrigation efficiencies and reduce labor requirements on many fields, if they are carefully designed and if the field is otherwise prepared for efficient irrigation (Fischbach and Gooding, 1970; Haise et al., 1965, 1978; Haise and Kruse, 1966).

Safety. All stands should be high enough, or they should be covered,

FIG. 11.18 Commercially available pipeline valves for [a] air release and vacuum relief and [b] combination air, vacuum and pressure relief. Photo courtesy of Hydro Engineering, Inc.

so that small children cannot enter the pipeline. Covering also prevents wind-blown trash and small animals from entering the pipeline.

11.5 OPERATION AND MAINTENANCE

11.5.1 Unlined Ditches

Ditch maintenance consists mostly of cleaning ditches and controlling weeds and rodents. Ditches need cleaning at least once a year to remove sediment deposits and excess vegetation, and to reshape the ditch. It is usually desirable to allow grass to grow along the ditch banks to stabilize the ditch. However, all vegetable growth must be controlled during the irrigation season to prevent interference with normal ditch operation and to maintain ditch capacity. Weeds can be controlled by periodic burning or with herbicides. If chemicals are used, they should be chosen carefully and competent local advice sought to prevent accidental damage to crops, animals, or aquatic life. Weeds along ditch banks are a major source of weed infestation of crops, so they should not be permitted to go to seed.

Rodents are a major cause of piping and ditch washouts and should be controlled. Rodent activity near structures can sometimes be discouraged by mixing coarse sand and gravel with the backfill.

Ditches should be inspected each year for erosion, bank scouring, weak or low spots in the ditch bank, structure cracking and deterioration, or other conditions that may need repairs. Ditches should be stable; if scouring or erosion is occurring, additional grade control or energy-dissipating structures may be needed. Ditches adjacent to structures can often be stabilized by placing riprap or coarse pit-run gravel in the ditch.

Cracks in concrete and masonry structures should be caulked to prevent water entry and minimize freeze damage. Masonry structures are particularly susceptible to cracking in cold climates. Metal structures should be painted or a protective coating applied when needed. Rubber-type seals normally deteriorate with time and should be replaced as needed. Metal gates should be lubricated and, where possible, left partially open during the winter. This prevents them from sticking or rusting closed and helps to keep them in good operating condition. Ditches and structures should be drained during the non-use season. Water ponded in or near structures contributes to frost damage and structure deterioration.

11.5.2 Lined Ditches

If lined ditches are not properly maintained, the result can be extensive damage to the ditch lining, structures, and field topography, as well as substantial economic loss to growing crops. Material needed to perform maintenance should be stockpiled on the farm. The small investment required for the stockpiled materials will be more than offset by the capability to make timely.repairs.

The ditches should be inspected routinely and maintenance requirements assessed. The timing of the inspections depends somewhat on the type of ditch lining and the soil and climatic conditions at the site. However, as a minimum, a thorough inspection should be made before every irrigation season.

Where rigid linings are used in northern climates the lined ditch should be inspected just before ground freezing to make sure that drainage is adequate.

The lifespan and annual maintenance costs of irrigation ditch lining will vary depending on site conditions and the quality of the initial construction. In the absence of reliable local data, a lifespan of 15 yr and annual maintenance costs of 5 percent of the installation cost are recommended for nonreinforced concrete, flexible membrane, and galvanized steel ditch linings.

11.5.3 Pipelines

Leak Testing and Repair. All buried low pressure irrigation pipelines should be tested for leaks before the trench is filled. The pipeline should be filled with water and slowly brought up to operating pressure with all turnouts closed. Any pipe lengths or joints showing leakage should be replaced and the line retested. The water should remain in plastic pipelines throughout the backfilling process, because the internal pressure helps prevent pipe deformation from soil loading and equipment crossings.

Pipelines should be inspected for leakage at least once a year. Leaks may be spotted from wet soil areas above the line that are otherwise unexplained. Small leaks in concrete pipelines can be repaired by carefully cleaning the pipe exterior surrounding the leak, then applying a patch of cement mortar grout. For larger leaks, one or more pipe sections may have to be replaced. Longevity of concrete pipelines can be increased by capping all openings during cold winter months to prevent air circulation.

Small leaks in plastic pipe, except at the joints, can sometimes be repaired by pressing a gasket-like material tightly against the pipe wall around the leak and clamping it with a saddle. Leaky joints or damaged pipe sections must be replaced with a new pipe splice. The old pipe section is joined into the gap using steel compression fittings (Dresser type coupler) because the pipe cannot be displaced horizontally to join the original belled joints.

Where water is supplied from a canal to portable surface pipe, sediment often accumulates in the pipe. This sediment should be flushed out before the pipe is moved. Otherwise, the pipe will be too heavy to be moved by hand and may be damaged if it is moved mechanically.

Buried plastic pipelines can be expected to have a usable life of about 15 yr if well maintained. The annual cost of maintenance can be estimated as approximately 1 percent of the installation cost.

References

1 American Concrete Institute. 1969. Specifications for Cast-In-Place Nonreinforced Concrete Pipe. Am. Conc. Inst. J., pp. 245-269.
2 American Society of Agricultural Engineers. 1982a. Agr. Eng. Yearb., St. Joseph, MI. S359.1: Trapezoidal Flumes for Irrigation Flow Measurement, pp. 542-544.
3 ASAE, 1982b. S263.2: Minimum Standards for Aluminum Irrigation Tubing, pp. 517-518.
4 ASAE, 1982c. S376.1: Design, Installation and Performance of Underground, Thermoplastic Irrigation Pipelines, pp. 493-503.
5 ASAE. 1982d. S261.5: Design and Installation of Nonreinforced Concrete Irrigation Pipe Systems, pp. 504-507.
6 ASAE. 1982e. S289: Concrete Slip-Form Canal and Ditch Linings, pp. 482-486.
7 American Society for Testing Materials. 1981a. Ann. Book of ASTM Standards C33-81: Specifications for Concrete Aggregates, Pt 14, 15.
8 ASTM. 1981b. C118-80: Specifications for Concrete Pipe for Irrigation or Drainage, Pt 16.
9 ASTM. 1981c. C76-80: Specifications for Reinforced Concrete Culvert, Storm Drain and Sewer Pipe, Pt 16.
10 Ball, J. W. 1974. Problems Encountered with Air Entrapped in Pipelines. Tech. Con. Proc., Sprinkler Irrig. Assoc.
11 Bergstrom, W. 1961. Weed Seed Screens for Irrigation Systems. Pacific Northwest Coop. Agr. Ext. Serv. (WA, OR, ID) Ext. Publ. Bull. 43, 7 pp.
12 Booher, L. J. 1974. Surface Irrigation. FAO Agri. Dev. Paper No. 95, Rome, 160 pp.
13 Brater, E. F., and H. W. King. 1976. Handbook of Hydraulics. McGraw-Hill Book Company, 6th ed., NY.
14 Calder, G. G., and L. H. Weston. 1966. Automatic System of Farm Irrigation. New Zealand J. Agr. 112(2).
15 Chow, Ven Te. 1959. Open Channel Hydraulics. McGraw-Hill Book Company, NY. 680 pp.
16 Colorado State College of Agricultural and Mechanical Arts, Extension Service. 1944. Drops for Farm Ditches. 3 pp. (mimeographed)
17 Coulthard, T. L., J. C. Wilcox, and H. O. Lacey. 1956. Screening Irrigation Water. Agr. Eng. Div. Bull. A.E. 6, Univ. of Brit. Columbia, 14 pp.
18 Dedrick, A. R., and L. J. Erie. 1978. Automation of On-Farm Irrigation Turnouts Utilizing Jack-Gates. TRANSACTIONS of the ASAE 21(1):92-96.
19 Erie, L. J. and A. R. Dedrick. 1978. Automation of an Open-Ditch Irrigation Conveyance System Utilizing Tile Outlets. TRANSACTIONS of the ASAE 21(1):119-123.
20 Erie, L. J., J. A. Replogle, and O. F. French. 1969. Non-Erosive Field Turnout Structures to Distribute Large Streams of Irrigation Water. Pacific Coast Reg. Meeting, ASAE Paper No. PC69-108, Phoenix, AZ.
21 Evans, Robert G. 1977. Improved Semiautomatic Gates for Cut-Back Surface Irrigation Systems. TRANSACTIONS of the ASAE 20(1):105-108, 112.
22 Fischbach, P. E. and R. Gooding II. 1970. An Automated Surface Irrigation Valve. AGRICULTURAL ENGINEERING 52(11):584-585.
23 Gilden, R. O., and G. O. Woodward. 1952. Low Cost Irrigation Structures. Univ. of Wyo. Agri. Ext. Serv. Circ. 122, 16 pp.

24 Haise, H. R., and T. H. Hall. 1971. Semiautomatic Timer-Operated Check Gates for Intermittent Irrigation. Colo. State Univ. Exp. Sta. Prog. Rpt. PR71-3, 1 p.
25 Haise, H. R., and E. G. Kruse. 1966. Pneumatic Valves for Automation of Irrigation Systems. Proc. 6th Congr. Internatl. Comm. on Irrig. and Drain. Specialty Sessions Rept. No. 1:S-1 to S-8.
26 Haise, H. R., and E. G. Kruse. 1969. Automation of Surface Irrigation Systems: The State-of-the-Art. Proc. Am. Soc. Civ. Engr., J. Irrig. and Drain. Div. 95(IR4):503-516.
27 Haise, H. R., E. G. Kruse, and N. A. Dimick. 1965. Pneumatic Valves for Automation of Irrigation Systems. USDA-ARS 41-104. 21 pp.
28 Haise, H. R., E. G. Kruse, M. L. Payne, and H. R. Duke. 1978. Automation of Surface Irrigation: Summary of 15 Years USDA Research and Development at Fort Collins, CO. USDA, SEA, Prod. Res. Rept. 179, 60 p.
29 Hanson, E. G. 1966. The Seepage Problem Defined. ASAE Paper No. 66-728. ASAE, St. Joseph, MI 49085.
30 Hart, W. E., and J. Borrelli. 1970. Mechanized Surface Irrigation Systems for Rolling Lands. Univ. of Calif., Water Resources Center, Contr. No. 133, 93 pp.
31 Herpich, R. L., and H. L. Manges. 1959. Irrigation Water Control Structures. Kans. State Univ. Dep. of Agr. Eng. Agri. Exp. Stat. Contr. No. 82, 12 pp.
32 Humpherys, A. S. 1969. Mechanical Structures for Farm Irrigation. Proc. Am. Soc. Civ. Engr. J. Irrig. and Drain. Div. 95(IR4):463-479.
33 Humpherys, A. S. 1978. Improving Farm Irrigation Systems by Automation. Preprint, 10th Congr. Internatl. Comm. on Irrig. and Drain. Athens, Greece, Rpt. 5, Q. 35, pp. 35.90-35.98.
34 Humpherys, A. S., and C. W. Lauritzen. 1962. Shape Factors for Hydraulic Design of Lay-Flat Irrigation Tubing. TRANSACTIONS of the ASAE 5(2):175-179.
35 Humpherys, A. S., and A. R. Robinson. 1971. Field Evaluation of Drop-Check Structures for Farm Irrigation Systems. USDA-ARS 41-180. 42 pp.
36 Keller, J. 1975. Economic Pipe Size Selection Chart. Am. Soc. Civ. Engr., Irrig. and Drain. Div. Spec. Conf. Proc., Logan, UT. pp. 109-121.
37 Kraatz, D. B. and I. K. Mahajan. 1975. Small Hydraulic Structures. Food and Agri. Org. of the U.N., Irrig. and Drain. Paper No. 26, Parts No. 1 (407 pp) and 2 (293 pp), Rome, Italy.
38 Marr, J. C. 1967. Furrow Irrigation. Univ. of Calif. Agri. Exp. Sta. and Ext. Serv. Man. 37, Davis, CA.
39 Payne, M. L., H. R. Duke and H. R. Haise. 1974. Further Developments in Automation of Level Basin Irrigation. Presented at ASAE Winter Meeting, Chicago, IL.
40 Pillsbury, A. F. 1952. Concrete Pipe for Irrigation. University of California Agr. Exp. Sta. Circ. 418:51 pp.
41 Portland Cement Association. 1952. Irrigation with Concrete Pipe. Skokie, IL, 55 pp.
42 Portland Cement Association. 1960. Concrete Structures for Irrigation Uses. Skokie, IL, Bull. #IS 163 W, 8 pp.
43 Pugh, W. J. and N. A. Evans. 1966. Weed Seed and Trash Screens for Irrigation Water. Colo. State Univ. Agr. Exp. Sta. Bull. S-522S. 18 pp.
44 Robinson, A. R., C. W. Lauritzen, D. C. Muckel, and J. T. Phelan. 1963. Distribution, Control and Measurement of Irrigation Water on the Farm. USDA Misc. Publ. #926,27 pp.
45 Robinson, E. P. 1970. Water Tight Seals for Flow Control Structures. State Rivers and Water Supply Commission of Victoria, 10 pp.
46 Robinson, E. P. 1972. Automatic Control of Farm Channels. 8th Congress, ICID, Varna, Bulgaria, Question 28.2, Report No. 21: pp. 28.2.293-28.2.296.
47 Seipt, W. R. 1974. Waterhammer Considerations for PVC Piping in Irrigation Systems. TRANSACTIONS of the ASAE 17(3):417-423.
48 Skogerboe, G. V., V. T. Somoray, and W. R. Walker. 1971. Check-Drop-Energy Dissipator Structures in Irrigation Systems. Water Management Tech. Rep. 9, Colorado State University, Ft. Collins, CO.
49 Thomas, C. W. 1960. World Practices in Water Measurement at Turnouts. J. Irrig. and Drain. Div., Proc. ASCE, 86(IR2):29-52.
50 Tovey, R., and V. I. Myers. 1959. Evaluation of Some Irrigation Water Control Devices. Univ. of Idaho, Agr. Exp. Sta. Bull. 319.
51 University of Idaho, College of Agriculture, Extension Service. 1958. Irrigation Handbook (mimeographed).
52 U.S. Bureau of Reclamation. 1951. Irrigation Advisors Guide. 216 pp.

53 U.S. Bureau of Reclamation. 1963. Hydraulic Design of Stilling Basins and Bucket Energy Dissipators. Engineering Monograph No. 25 (Revised), Denver, CO.

54 U.S. Department of Agriculture, Soil Conservation Service. 1962. Measurement of Irrigation Water. National Engineering Handbook, Section 15, Chapter 9, U.S. Government Printing Office, Washington, DC. 72 pp.

55 U.S. Department of Agriculture, Soil Conservation Service. 1968. Sprinkler Irrigation. National Engineering Handbook, Section 15, Chapter 11, Washington, DC. 82 pp.

56 U.S. Department of Agriculture, Soil Conservation Service. 1971. Irrigation Ditch and Canal Lining, Non-reinforced Concrete. Engineering Standard 430-A National Engineering Handbook.

57 U.S. Department of Agriculture, Soil Conservation Service. 1978. Engineering Design Standards. (These standards are being continuously revised and updated. Detailed irrigation structure design information applicable for a given state is obtainable from the USDA Soil Conservation Service State Engineer. Design information for readers outside the US available from USDA, Soil Conservation Service, Washington, DC).

58 U.S. Department of Agriculture, Soil Conservation Service. 1979. Furrow Irrigation. National Engineering Handbook, Section 15, Chapter 5 (Draft), Washington, DC. 175 pp.

59 U.S. Department of Transportation, Federal Highway Administration. 1973. Design Series No. 3. 105 pp.

60 Vanoni, V. A. and J. T. Bostrom. 1943. Development of a Baffle Type Energy Dissipator for Pipe Outlets. Soil Conservation Service, Office of Research, Cooperative Laboratory, California Institute of Technology.

chapter 12

HYDRAULICS OF
SURFACE IRRIGATION

12

12

HYDRAULICS OF SURFACE IRRIGATION

by D. L. Bassett, Washington State University; D. D.
Fangmeier, University of Arizona, Tucson,
Arizona; T. Strelkoff, University of California,
Davis, California, private consultant

12.1 INTRODUCTION

Irrigation has been practiced from ancient to modern times in many areas of the world. Farmers in Egypt, China, India, and countries of the Middle East are known to have irrigated agricultural lands at least 4000 years ago. The Greek historian Herodotus of 450 BC visited Babylonia when parts of that country had already been in cultivation for 2000 years and later described its vigorous irrigated agriculture served by a vast network of canals, the largest of which was impassable except in boats (Thorne, 1951). As early as 200 BC the Hohokam farmers diverted water from what is now known as the Salt River through 200 km of hand dug canals in and around the site of modern Phoenix, Arizona (Salt River Project, 1960). Virtually all of this early irrigation is believed to have been accomplished by surface means, there being little indication that other systems were available. This dependence upon surface methods continued until shortly after the close of World War II when technology, energy, and industrial support became available to sustain a rapid rise in sprinkling. This long history of surface irrigation, marked as it is with successes and failures, is valuable. Evidences of sustained irrigation illustrate the contribution that successful irrigation can make to society. Evidences of salted soils and abandoned systems emphasize the need to use surface irrigation wisely or face similar consequences.

About two-thirds of the 25 million hectares (61 million acres) of irrigated land in the U. S. is now served by surface means (Irrigation Journal, 1979) in spite of the popularity of sprinkling. The percentage is even higher in some other parts of the world. Anticipated shortages and high cost of energy required to drive alternative systems suggest that surface irrigation will remain popular, especially if abuses often associated with it can be avoided. The material in this chapter is presented to hasten the development of practices that will reduce such abuses and allow surface irrigation to meet modern agricultural and societal needs more fully.

Improvements in surface irrigation are most likely to result from skillful combination of experience and thorough understanding of the processes involved. While experience has been accumulating for centuries, theoretical analyses of the processes involved are relatively recent. Modern mathematical and computational aids are just now beginning to yield new insight into irrigation system performance. Practical applications of these studies are expected soon. This summary of theory and resulting models should help investigators use these new tools in their quest for improvement in practices.

This review of surface irrigation begins with a description of certain concepts involved. The independent variables with which the investigator must be concerned are then identified. This is followed by an explanation of the equations which describe flow in surface irrigation, in both complete and simplified forms. Techniques for solving the equations in their several forms are mentioned. Mathematical and computer models now available that use the equations to simulate irrigation are reviewed. The chapter concludes with a brief suggestion of the potential contribution that model simulations can make in the overall effort to improve water use efficiency in surface irrigation. From the material presented the reader should gain a general understanding of the current status of hydraulic models of surface irrigation flows. Those who wish to achieve a working competence in the subject will find it necessary to consult the references cited for additional detail and to become actively involved in development or use of such models.

12.2 BASIC CONCEPTS OF SURFACE IRRIGATION

12.2.1 Description of Flow

Surface irrigation is accomplished by one of several application methods, including borders, furrows, checks, and basins. in each case water moves over the land surface in open channel flow. The water may be directed in small earth channels called furrows or corrugations or it may move as shallow overland flow over a carefully smoothed soil surface as in border irrigation. Aside from differences in channel geometry and boundary conditions, the basic flow characteristics are much the same in all of the surface methods. Significantly, the geometry and boundary conditions in border irrigation are mathematically simpler than in the furrow methods. For this reason more work has been done to analyze flow in borders, and more is known about that flow. Consequently, the discussions in this chapter will emphasize border irrigation flow unless otherwise noted. Hydraulically, checks and basins can be considered as special cases of border flow.

In border irrigation, water moves over the soil surface as shallow overland flow. The soil surface is permeable to water. The length of the border strip parallel to the flow is usually great with respect to the width. The surface may have a small or zero slope longitudinally and is normally level in the transverse direction. Surface water depth is usually very small with respect to border width. The land surface may or may not be vegetated. The flow situation is similar in irrigation furrows, except the channel is narrow, and the influence of the channel walls on flow retardance and infiltration may be significant.

Water is typically introduced at one end of the border or furrow (the higher end) from a head ditch or pipe. Within a border the water immediately spreads across the width of the border to permit uniformity of stream size in all units of width. The water then advances down the border or furrow in overland flow behind a distinct wetting front. As flow occurs some water leaves the surface stream as infiltration into the soil. This rate of infiltration usually decreases with time at each point in the field. Therefore, if the inflow stream size is held constant with time, the flow rate and depth at any point downstream will gradually increase with time. On the other hand, the flow rate at any time decreases with greater distance down the channel. The rate of advance of the wetting front down the channel will necessarily decrease with time and distance as more soil becomes available to absorb water and reduce the overflowing surface stream.

Under normal conditions flow continues in the above manner until the advancing front reaches the lower end of the border or furrow. At this time water begins either to leave the field as surface runoff or to form surface storage if runoff is prevented by blocking the end. At some later time, presumably when sufficient water has been introduced to meet the gross water requirement, the inflow is stopped. Surface flow continues, but water depth and velocity decrease, beginning at the upper end. When combined surface flow and infiltration reduce the depth of water at the upper end to zero, a recession or drying front is formed. This recession front proceeds downstream until it reaches the lower end of the channel or a similar front that can form at the lower end of the channel and move upstream. At this time no water remains on the surface, and the irrigation is complete.

At the same time water is advancing over the soil surface in open-channel flow, water which has infiltrated into the soil is moving downward, and in the case of furrows outward, into the soil mass. As irrigation proceeds, the depth of water that has entered the soil at a given point increases, but at a decreasing rate. It is possible that a greater depth of water will enter the soil than can be held in the root zone. Such excess water is considered lost to plants as deep percolation.

It will be noted that sub-surface flow has so far not been a major part of studies of surface-irrigation hydraulics. Emphasis has instead been placed on surface flow, for this determines the intake opportunity times throughout the field and to that extent the characteristics of the irrigation that results. In other words, the small effect of depth of flow on the volume infiltrated per unit wetted area of infiltrating surface has been ignored. This uncouples the equations governing the surface flow and the flow in the porous medium except insofar as the former determines the infiltration time for each element of infiltrating-surface area.

12.2.2 Water Volume Balance

It is useful to recall that a volume balance of water in an irrigation system must exist at all times. All water introduced into the border or furrow must at any time be accounted for in surface flow, infiltration or runoff, assuming that water lost in evaporation from the surface stream during the time of irrigation can be neglected. The requirement of such a balance can be useful to the investigator by providing a check on the accuracy or validity of his work. Much of the complexity in studies of this subject arises from the need to predict the several volumes of water with sufficient accuracy to evaluate or maintain the necessary balance in the computations.

In field trials the volumes of water which enter and leave the field are usually measured by standard water measuring devices. The water volume in surface flow can be measured during an irrigation by simultaneous measurement of flow cross section at selected stations throughout the length of the field. The volume of water that infiltrated into the soil during the irrigation can be determined by noting changes in soil moisture content before and after the irrigation including water percolating below the root zone. Alternately, the physical requirement of volume balance allows determination of infiltrated volume by subtracting surface and runoff volumes from the inflow volume.

Where irrigation is simulated with computer or other models, the investigator must provide for the tabulation of the water volumes mentioned above. This is usually straight forward when some form of finite differencing is used. It is usually desirable to determine and sum all of the volumes for comparison with the inflow volume, rather than to determine all but one volume and then infer its value by difference. The latter approach may obscure errors in the calculation of one or more of the volumes.

12.2.3 Flow Characteristics

In this, as in many hydraulics studies, it is useful to classify the flow regime in order to select equations and assumptions that apply. Classification of flow in surface irrigation is complicated by the existence of different flow regimes at different times and places in a single irrigation. For instance, the active flow in the upper reach of the channel while inflow occurs is markedly different from the shallow, low velocity flow that occurs during recession, and also from the deep, low velocity flow that occurs when water is ponded on the surface. Modelling of surface flow is greatly complicated by attempts to follow these changes in regime with time and distance. Most studies to date have therefore been based upon the single flow regimes felt to exist during the major part of a typical irrigation. In so doing it is recognized that accuracy of description may be sacrificed as flow transition zones are encountered. In view of the difficulty of specifying parameters such as roughness and infiltration, improvement in this respect is probably not yet justified.

The active channel flow felt to dominate the irrigation process is, because of infiltration of water into the soil, gradually varied and unsteady free surface flow. The state of flow is most likely turbulent or transitional (dye traces break up due to eddies generated by surface and vegetative roughness elements) even though Reynolds numbers well below 1000 are common. Since Froude numbers are usually well below unity, the flow is typically sub-critical. Flow regimes other than these are most likely to occur just behind the wetting front during advance, just ahead of the drying front during recession, and near certain boundaries. Investigations to date suggest that any error resulting from use of the regimes above is probably much less than that introduced with some of the input information.

12.2.4 Terminology

Discussions in this chapter use many terms common in surface irrigation. Those whose meaning is taken from established disciplines will be assumed understood. For convenience, however, those terms whose meaning is unique in surface irrigation or whose meaning could be misunderstood will be explained. Terms are grouped with others of common association. Dimensional units are given in parentheses wherever partinent.

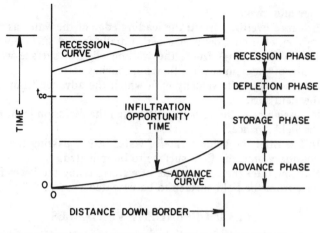

FIG. 12.1 Definition sketch showing certain surface irrigation terms.

Phases of a typical irrigation (See Fig. 12.1).

1 **Advance phase:** that portion of the total irrigation time during which water advances in overland flow from the upper field boundary toward the lower field boundary.

2 **Storage (continuing) phase:** that portion of the total irrigation time between the end of advance and inflow shut-off. If shut-off occurs first, this phase is of zero duration.

3 **Depletion phase:** that portion of the total irrigation time between inflow shut-off and the beginning of recession at the upper field boundary. This is also know as the recession lag time.

4 **Recession phase:** that portion of the total irrigation time between the beginning of recession at the upper field boundary and the disappearance of the last water from the field surface.

Boundaries.

1 **Upper field boundary:** the upstream end of the field to be irrigated, where irrigation water is introduced.

2 **Lower field boundary:** the lower or downstream end of the field to be irrigated, where any surface outflow leaves the field.

3 **Upper computational boundary:** the most upstream point in the field at which surface flow is simulated and at which the upstream boundary conditions are applied. This boundary will coincide with the upper field boundary except during the recession phase when the upper computational boundary moves progressively downstream with the trailing edge of the surface stream.

4 **Lower computational boundary:** the most downstream point in the field at which surface flow is simulated and at which downstream boundary conditions are applied. This boundary will coincide with the lower field boundary except during the advance phase when the lower computational boundary moves progressively downstream. It can also move upstream from the lower field boundary during periods of downstream recession.

Water movement.

1 **Advance (wetting) front:** the leading edge of the water as it advances in open channel flow over a bed initially free of surface water.

2 **Recession (drying) front:** the trailing edge of surface water as the depth of water in surface flow approaches zero.

3 **Advance rate:** the velocity with which the advance front moves over the field surface.

4 **Recession rate:** the velocity with which the recession front moves over the field surface.

5 **Inflow rate:** the volume rate of water inflow passing the upper field boundary onto the field surface to be irrigated.

6 **Outflow rate:** the volume rate of water passing the lower field boundary from the field surface to be irrigated.

12.3 INDEPENDENT VARIABLES

A theoretical study or simulation of overland flow in surface irrigation requires that the major factors affecting such flow be identified and incorporated into the analysis. It is useful to separate constants from variables. When a factor can be assumed to have one value throughout a single irrigation and for all other irrigations, it is considered here to be a constant. If the value of a factor can change within an irrigation or between irrigations, it is treated as a variable. If the value of a variable during an irrigation is determined before the study begins, the variable is considered to be independent. If the value of a variable cannot be determined in advance but results from the irrigation process, the variable is considered dependent.

Values the dependent variables assume during an irrigation are often the solution information desired by the investigator. Which variables are of interest to the investigator may change with irrigations but may include, among others, velocity of advance and recession, time of advance and recession, depth and velocity of surface flow, depth and volume of water infiltrated, volumes of surface runoff and deep percolation, recession lag time, and total irrigation time. It is assumed that the dependent variables of interest and the values they take can be displayed by the investigator at will. No further attention is given them here.

It is necessary that consideration be given to the independent variables involved since the investigator must supply this information before the analysis begins. The values assigned to these variables must be carefully chosen since they affect the solution, often very significantly. The values assigned presumably represent the field conditions of interest to the investigator.

Experience to date indicates that the major independent variables common to all investigations are inflow rate, land surface slope, hydraulic roughness, and soil infiltration characteristics. These are discussed below.

12.3.1 Inflow Rate

It is assumed here that water is introduced to the land to be irrigated at the inlet or upper end of the field, i.e., the upper field boundary. If border irrigation is being simulated, the flow rate is expressed as flow rate per unit width of channel, with the dimensions of $L^3/T \cdot L$, or L^2/T. This implies a line source of water at the inlet end that provides each unit width of border the same flow rate. Where furrow irrigation is involved, the inflow rate is expressed as flow rate per furrow, L^3/T. In most studies to date the inflow rate has been assumed constant with time, although this is not necessary. In fact, one of the important contributions of irrigation modelling is the ability to evaluate the effects of variable flow rates on system performance. If inflow rate is to be varied with time, the investigator must supply the inflow hydrograph, or make provisions for its generation as the simulation proceeds.

Studies to date suggest that irrigation performance is quite sensitive to the inflow rate. Values of the inflow rate should therefore be chosen with care, simulating field conditions with at least the accuracy obtainable with common flow measuring devices.

12.3.2 Land Surface Slope

It is assumed here that the land to be irrigated either slopes downward from the inlet end or is level. Downward slope is considered positive. Zero slope (i.e., level land) is admissible in certain simulation schemes but not in others (as will be shown later), depending generally on whether normal depths are assumed in any of the computations. Negative slopes may be mathematically admissible but are seldom practical and are not considered further here.

Unless otherwise specified, land slope is considered uniform throughout the field length and free of the minor irregularities found in nearly all fields. The slope need not be uniform, however, if the investigator makes provision for specified changes in slope. This ability to evaluate the effects of variable land surface slope is another of the advantages of irrigation modelling.

In border irrigation, cross slope within the border is nominally zero. The land slope specified is the longitudinal or irrigating slope down the length of the border strip.

While land slope clearly affects overland flow in irrigation, there seems little reason to use slope figures that have greater precision than can be obtained with standard surveying techniques. Slope is expressed in dimensionless form, L/L.

12.3.3 Hydraulic Roughness

As in other open channel flow situations, some measure of surface resistance to flow in irrigation channels is required. While the selection of roughness values for such common surfaces as concrete canal linings becomes commonplace, the selection of a corresponding value for small earth irrigation channels is far less certain. Describing roughness in such channels is complicated because the flow is retarded not only by the bed but also by vegetation, usually the crop grown. This vegetation may project vertically into or through the flowing water. The effect of these resistance elements may change with time and distance within an irrigation as clods dissolve, as vegetation bends from the force of the water, and as water depth and velocity change. The drag on the flow will also change between irrigations as cultivation and crop growth occur. The selection of a mathematical roughness expression to use in the simulations is also uncertain, since none of the common expressions were developed expressly for this flow situation.

While theoretical studies have been made to describe dual level (soil and vegetation) drag, most investigators to date have used existing, accepted roughness expressions. Collins and Bassett (1964) obtained satisfactory results using the friction factor f found in the Darcy-Weisbach head-loss equation. Schreiber and Bassett (1967) used an empirical power-law relationship for the Chezy C. Bassett and Fitzsimmons (1976) developed a model of the complete border-irrigation process in which surface roughness can be entered either as the Manning n or the Sayre-Albertson Chi. Strelkoff and Katopodes (1977b) and Katopodes and Strelkoff (1977a) also obtained satisfactory results with the Manning n.

Regardless of the parameter used to describe the resistance to flow, selection of specific values to use in a given situation is not well established. The work of Roth et al. (1974) at the University of Arizona is useful in identifying the range of values of roughness that may be encountered in non-vegetated and in vegetated channels. More work is needed to assist the designer in his selection of roughness expression and of values to use in simulating irrigation flow.

Studies to date suggest, fortunately, that surface retardance is not as important in determining irrigation performance as some other independent variables. Furthermore, it is now evident that satisfactory results have been obtained with such common expressions as the Manning n and modelling work can proceed effectively while improved roughness expressions are being developed. Selection of specific roughness values will probably remain dependent upon experience. A reasonable initial trial value would be one felt to exist early in the irrigation, but after the wetting front has passed.

12.3.4 Soil Infiltration Characteristics

Since the purpose of irrigation is to increase the moisture content of the soil, the channel in surface irrigation must be permeable to water and water must enter the channel bed and walls. The rate of this infiltration clearly influences the surface flow and the entire irrigation performance. The rate is bounded by two conceptual limits—zero for an impermeable bed and some high nearly constant value for a highly porous bed. Normally, however, the infiltration rate decreases with infiltration time (i.e., the time water has been available for infiltration at a point in the channel). Several expressions have been proposed for this rate, and the investigator should use an expression that describes as nearly as possible the field conditions being simulated. In most previous studies and in this chapter the simplest of these expressions is used, namely

$$I = k' \tau^{a'} \quad \dots\dots\dots\dots\dots\dots\dots\dots\dots\dots\dots\dots\dots\dots\dots\dots\dots \text{[12.1]}$$

in which I is infiltration rate, (L/T); τ is infiltration time, (T); and k' and a' are constants for a particular soil and condition; generally a' has a negative sign. When integrated over time, the above equation yields the depth of water that will have infiltrated into the soil up to that time,

$$z = k \tau^a \quad \dots\dots\dots\dots\dots\dots\dots\dots\dots\dots\dots\dots\dots\dots\dots\dots\dots\dots \text{[12.2]}$$

in which z is the depth of water infiltrated, (L), and k and a are again soil constants.

Values of the soil constants for either the rate or depth equation can be obtained from existing information sources such as the USDA Soil Conservation Service or can be measured in the field. Since infiltration rate is known to have a marked effect on irrigation system performance, care in selecting appropriate constants is justified. The investigator should be aware, however, that the infiltration rate may change substantially within a single field, within a growing season and between seasons as crops change (also see Section 4.3).

12.4 FLOW EQUATIONS

It is now known that surface water flow in irrigation can be described by the equations of Saint-Venant. Simulations of irrigation can be obtained by using the equations in their complete form or in certain instances one of three simplified forms, the latter offering economies of solution. The investigator must select the form of the equations to be used. To facilitate this selection, the complete and three simplified forms of the equations are described in the next four sections.

12.4.1 Complete Hydrodynamic Equations

The equations describing the flow of water over a soil surface express two physical principles, conservation of mass and Newton's second law, force equals mass times acceleration. These well know partial differential equations are known as the Saint-Venant equations and are derived in texts and papers on open-channel flow (e.g., Henderson, 1966; Strelkoff, 1969). The mass-conservation equation is

$$\frac{\partial Q}{\partial x} + \frac{\partial A}{\partial t} + I_x = 0 \dots\dots\dots [12.3]$$

and the equation of motion is

$$\frac{1}{g}\frac{\partial V}{\partial t} + \frac{V}{g}\frac{\partial V}{\partial x} + \frac{\partial y}{\partial x} = S_0 - S_f + \frac{I_x V}{2gA} \dots\dots [12.4]$$

In equations [12.3] and [12.4], x is distance, (L); t is time, (T); Q is flow rate, (L^3/T or L^2/T in border irrigation); A is cross-section area of flow, (L^2); I_x is volume rate of infiltration per unit length of channel, (L/T); g is the ratio of weight to mass, (L/T^2); V = Q/A is average velocity in the flow cross-section, (L/T); y is flow depth, (L); S_0 is channel bottom slope; and S_f is channel friction slope.

Equation [12.3] is a volume-conservation equation and implies the assumption that the water density is constant. It states that for a thin slice of space cutting across both surface-water and infiltrated-water profiles (Fig. 12.2), the time rate of increase in profile depths equals the negative of the gradient of discharge in the surface stream. It is evident that

$$I_x = \frac{\partial A_z}{\partial t} = B_z \frac{\partial z}{\partial t} \dots\dots\dots [12.5]$$

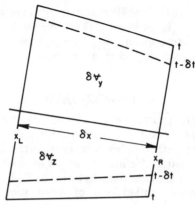

FIG. 12.2 Elementary slice through surface and subsurface profiles.

in which A_z is the volume of water infiltrated per unit length of channel. The second part of equation [12.3] is a definition equation for the depth of infiltration z, which depends upon the volume infiltrated and an arbitrarily designated infiltration width B_z. In a border, B_z is most logically set equal to the border width; then z represents the volume of water infiltrated per unit plan area. In furrow irrigation, B_z can be taken as the furrow spacing, and z would then represent the volume infiltrated, per unit plan area of field.

In equation [12.4], the depth gradient represents the unbalanced hydrostatic pressure force on the surface water contained in the slice of Fig. 12.2, the bottom slope is the component in the direction of flow of the gravitational force on the same element, and S_f is the hydraulic drag (bed and vegetation) each expressed per unit length of channel, and each in ratio to stream weight per unit length of channel. The remaining terms then represent the inertial reactions to this unbalanced resultant force, also expressed per unit length of channel and in ratio to stream weight per unit length. The local acceleration $\partial V / \partial t$ is a direct measure of the unsteadiness of the flow. The convective acceleration $V \partial V / \partial x$, reflects the non-uniformity, and I_x-$V/2gA$ represents a net acceleration stemming from removal of zero-velocity components of the surface stream at the bed by infiltration (Strelkoff, 1969).

The major assumptions made in the derivation of equation [12.4] are that the pressure distribution normal to the channel bed is hydrostatic, that the slope of the bed is sufficiently small that the cosine of the angle of inclination is essentially unity, and that the velocity distribution is essentially uniform in a cross section. The major requirement for the first assumption to be valid is that the streamline curvature be negligible. In the context of surface irrigation, this assumption is certainly reasonable throughout most of the flow system. However, there is some question regarding the reasonableness of this assumption in the tip region of the advancing stream. This is addressed further in Section 12.4.5. The other assumptions, too, are reasonable. In any event, the inaccuracies caused by these assumptions would be far less than those stemming from imprecise specification of infiltration or hydraulic drag on the flow.

The total drag on the flow must be determined empirically (Section 12.3.5), customarily through experiments with uniform flow. Thus, the assumption is made that the unsteady, nonuniform flow in the irrigation stream experiences the same resistance as a steady uniform flow at the same depth and discharge. Errors from this source are also much smaller than those incurred in predicting resistance even in uniform flow. All of the uncertainty is lumped into the Chezy C_h, defined by the following expression for S_f,

$$S_f = \frac{V|V|}{C_h^2 R} \quad \dots\dots\dots\dots\dots\dots\dots\dots\dots\dots\dots\dots\dots\dots\dots\dots\dots \quad [12.6]$$

in which R is the hydraulic radius, (L). The Chezy C_h is most commonly expressed through the Manning formula

$$C_h = \frac{C_u}{n} R^{1/6} \quad \dots\dots\dots\dots\dots\dots\dots\dots\dots\dots\dots\dots\dots\dots\dots\dots \quad [12.7]$$

in which n is the Manning n, and C_u is a units coefficient ($C_u = 1.0$ in the SI system, $C_u = 1.4859...$ in the English System). Contrary to the concept of the Manning formula, in which n is supposed to be a function of absolute roughness geometry alone, in the shallow, often vegetated channels used in surface irrigation, n depends heavily on depth of flow.

Generally speaking, it can be stated for practical purposes that the mathematical models of the surface-irrigation process, consisting of solutions of equation [12.3] and [12.4], are as precise as the input information provided them. In practice, because inflow rate and bottom slope are relatively easily measures, the main source of input error lies with the roughness and infiltration data.

12.4.2 Zero-Inertia Approximation

At the very low velocities normally encountered in surface-irrigation streams, changes in velocity are negligibly small, compared to the force terms in the equation, and all inertial (acceleration) terms can be deleted. It was shown by a formal order-of-magnitude analysis (Katopodes and Strelkoff, 1977b), that for normal Froude numbers below about 0.3, the forces acting on the surface stream are essentially in balance.

The zero-inertia model, discussed in Section 12.5.3, thus comprises a numerical solution of equation [12.3] and equation [12.4] with all three acceleration terms deleted from equation [12.4], as follows,

$$\frac{\partial y}{\partial x} = S_0 - S_f \quad \dots\dots\dots\dots\dots\dots\dots\dots\dots\dots\dots\dots\dots\dots\dots\dots \quad [12.8]$$

The partial derivative notation is still used in equation [12.8], because depth is a function of both distance and time. At the same time it should be recognized that equation [12.8] is actually an ordinary differential equation representing conditions at one particular instant of time, and independent of time rates of change.

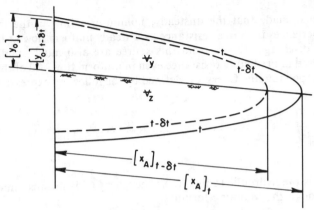

FIG. 12.3 Surface and subsurface profiles at two successive instants.

12.4.3 Normal-Depth Approximation

If the bottom slope is sufficiently steep,* the depth gradient of equation [12.8] is much smaller than either of the right-hand terms. The latter are then in essential balance. Thus

$$S_f = S_0 \qquad \dotfill \qquad [12.9]$$

everywhere, and the flow is at normal depth. Through equation [12.6], depth and discharge are simply related, as follows,

$$Q = AC_h \sqrt{R} \sqrt{S_0} \qquad \dotfill \qquad [12.10]$$

in which the first three terms are known functions of depth.

Any prescribed relation between depth and discharge coupled to a volume-conservation equation such as equation [12.3] yields a kinematic-wave flow model (Lighthill and Whitham, 1955). When the depth-discharge relation is one based on normal depth, the result is the so-called normal-depth model, described further in Section 12.5.4.

12.4.4 Assumed Average Surface Depth

The preceding mathematical equations comprise, in each case, an equation of continuity or volume-balance equation, equation [12.3] and one or another equation of motion, equations [12.4], [12.8], or [12.9]. To constitute the various models, these are applied to elements of the surface stream, like that shown in Fig. 12.2, and solved to predict the flow profiles at successive instants of time. Still further approximation and simplification is achieved by applying equation [12.3] to the entire flow profile, shown schematically in Fig. 12.3. Indeed, equation [12.3] integrated over the length of the advancing stream, becomes

$$Q_0 = \frac{d\Psi_y}{dt} + \frac{d\Psi_z}{dt} \qquad \dotfill \qquad [12.11]$$

*The question is viewed formally in Section 12.5.7.

in which Q_0 is the stream discharge at its upper end and V_y and V_z are the surface volume and infiltrated volume, respectively. A further integration, over time, yields for constant Q_0,

$$Q_0 t = V_y + V_z \quad \dotfill \quad [12.12]$$

The surface volume can be expressed in terms of the flow area A_0 at the upstream end of the stream, the advance x_A of the stream down the field, and a shape factor r_y defined by the following equation

$$V_y = A_0 \; x_A \; r_y \quad \dotfill \quad [12.13]$$

Assuming a constant shape factor is equivalent to assuming a power-law description of the surface profile, with fixed exponent. In the more sophisticated models, the shape of the surface profile comes out as part of the solution of the continuity and motion equations. In the assumed average surface depth approach, the shape factor is set prior to computation to a reasonable, arbitrary constant, while the motion equation is used, in the form of equation [12.10], at the upper end of the stream only, to determine A_0. Conditions at the upstream end are thus assumed at normal.

Equation [12.3] now has the form

$$Q_0 t = A_0 \; r_y \; x_A + V_z \quad \dotfill \quad [12.14]$$

If it is assumed, following Section 12.3.4, that infiltration depth is a known function of intake time alone, then V_z is a known function of the advance function $x_A(t)$. Thus equation [12.14] represents a relation between the two variables x_A and t, which in principle can be solved (See Section 12.5.5).

12.4.5 General Initial and Boundary Conditions

The governing equations in whatever form they are used describe a relationship between the time and distance rates of change of the unknown variables, depth and discharge. Before a solution for the actual values of these quantities at various times and locations can be determined, their initial state must be specified, as well as their values at the two ends (boundaries) of the surface stream. These values are known as the initial and boundary conditions. Thus, the solution develops in response to the equations governing the flow and to the initial and boundary conditions.

If an existing computer program is being used to simulate an irrigation, the user should become familiar with the initial and boundary conditions employed, and operate the program accordingly. Alternately, the user could program other conditions that would more nearly describe the specific situation. There is some flexibility in the selection of conditions, providing of course they describe the flow situation under investigation. The information on initial and boundary conditions which follows is not intended to describe the only acceptable conditions. What is given are certain sets of conditions, which the authors have used successfully. The reader may be guided by them, but is not limited to them.

Upstream boundary conditions: During the irrigation, the inflowing discharge Q_0 is assumed known. This will usually be constant until cut off, at

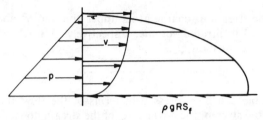

FIG. 12.4 Distribution of pressure (p), velocity (v), and resistance (ρgRS_f) on tip element.

which time it drops to zero. This discharge hydrograph constitutes the boundary condition at the upper end of the irrigation stream.

During recession, the discharge at the trailing edge of the surface stream is zero, and can usually be assumed zero at the upstream computation boundary if that boundary is located sufficiently close to the receding tip. In the course of the calculations the computed depth at the upstream computational boundary is monitored until it drops below a predetermined level, say zero, or a small positive number. Then the computational boundary with its zero-discharge condition is moved a step downstream, and the depth there is computed as time proceeds until it too drops below the prescribed minimum. Thus, the movement of the trailing edge in recession parallels the movement of the upstream computational boundary. Unless the infiltration rate at the time of recession is still large, it is not necessary to define the exact time of passage of the trailing edge past any given station, because the ultimate depth of infiltration there changes but little with changes in recession time.

Downstream boundary conditions: The downstream boundary condition is more difficult to define. During advance, the depth and discharge at the leading edge of flow are both zero, but the velocity is not. The approximate profile shape very close to the leading edge of the stream can be deduced from the following argument. While the details of the flow at the very front of the wave are obscure, it is a reasonable postulate that the streamlines intersect the wave front as shown in Fig. 12.4. There is a velocity gradient in the surface stream, with velocity increasing with distance from the bed, because of bed resistance and infiltration. This would explain the rolling advance of the front evident in the field, as opposed to bulk sliding, which might at first seem to be the consequence of the flow pattern postulated in Fig. 12.4. This distribution makes equation [12.4] applicable all the way to the stream front.

The average fluid velocity in the surface stream changes but slowly with both time and distance. On the other hand, examination of equation [12.6] shows that as the wave front is approached, and the velocity approaches the wave-propagation speed w (Fig. 12.4), the hydraulic radius in the denominator approaches zero and S_f grows without bound. Because velocity changes are gradual, all acceleration terms in equation [12.4] are negligible by comparison, as is the bottom-slope term. The hydrostatic pressure force essentially alone balances the resistance force. Thus, in the limit, as $x \rightarrow x_A$, equation [12.4] reduces to the following variables-separable ordinary differential equation,

$$\frac{dy}{dx} = -S_f \quad \text{or} \quad C_h^2 R\, dy = -w^2\, dx \quad \dots\dots\dots\dots\dots\dots\dots [12.15]$$

in which the left side is a known function of depth. For the plane two-dimensional flow assumed for border irrigation and the Manning formula used for C_h, equation [12.15] is readily solved, as follows

$$\frac{C_u^2}{n^2} y^{\frac{7}{3}} = w^2 \left(x_A - x\right) \dots\dots\dots\dots\dots\dots\dots\dots\dots\dots\dots\dots [12.16]$$

It follows that the surface profile close to the wave front follows a 3/7 power law.

For many forms of the infiltration function, the limiting profile of infiltrated water as $x \rightarrow x_A$ is also a power law. The infiltration time t for a point x behind the wave front, which is propagating at the speed w is

$$\tau = \frac{x_A - x}{w} \dots\dots\dots\dots\dots\dots\dots\dots\dots\dots\dots\dots\dots\dots [12.17]$$

Thus, for an infiltration function of the general form

$$z = k\tau^a + b\tau \dots\dots\dots\dots\dots\dots\dots\dots\dots\dots\dots\dots [12.18]$$

in which k, a, and b are constants, substitution of equation [12.17] produces, as $x \rightarrow x_A$,

$$z = \frac{k}{w^a} \left(x_A - x\right)^a \dots\dots\dots\dots\dots\dots\dots\dots\dots\dots\dots\dots [12.19]$$

since a < 1.

Worthy of note, the flow velocity at any station behind the front must reflect the infiltration downstream from the station as well as the front-propagation velocity. Thus velocity variation as $x \rightarrow x_A$ is given by

$$V = w \left(1 + \frac{zB_z}{A}\right) \dots\dots\dots\dots\dots\dots\dots\dots\dots\dots\dots\dots [12.20]$$

in which zB_z is the volume of water infiltrated per unit length of stream. Equation [12.20] is derived from the consideration that the discharge across the given station, in part, moves the surface profile ahead and, in part, infiltrates into the ground. It is assumed that near the wave front, profile shape changes very slowly. The velocity given by equation [12.20] is of passing interest but was not used in establishing the surface profile in the tip region in any of the studies summarized here. In practice, the precise shape of the profile does not appear to be of major importance in overall flow simulation.

Once the irrigation stream reaches the end of the field, the downstream boundary condition changes in response to the physical circumstances there.

If the downstream end has been blocked to prevent runoff, the discharge is zero, which constitutes the downstream boundary condition. If the surface flow drains freely off the downstream end of the field, as into a drainage ditch, a free-overfall boundary condition is appropriate. In the event that the full Saint-Venant equations are used, a free overfall is characterized by critical depth. Discharge and depth are then related by the equation

$$\frac{A^3}{B} = \frac{Q^2}{g} \quad \dots\dots\dots\dots\dots\dots\dots\dots\dots\dots\dots\dots\dots\dots\dots\dots\dots \quad [12.21]$$

In the zero-inertia approach, critical depth at an overfall is zero (Strelkoff and Kotopodes, 1977a). At this singular point, the mathematical function representing velocity becomes infinite. But as long as inertial effects are indeed negligible throughout the main body of the flow, just a short distance upstream from the brink computed velocities agree well with physical velocities. Near the brink, computed depth is related to discharge by the differential equation

$$A^2 \, C_h^2 \, R \, dy = - \, Q^2 \, dx \quad \dots\dots\dots\dots\dots\dots\dots\dots\dots\dots\dots\dots\dots \quad [12.22]$$

In a border, with roughness expressed by the Manning equation, this has the solution

$$y = \left[\frac{13 \, n^2}{3 \, B^2} \frac{Q^2}{C_u^2} (L - x) \right]^{\frac{3}{13}} \quad \dots\dots\dots\dots\dots\dots\dots\dots\dots\dots\dots \quad [12.23]$$

In models based on normal depth, no special downstream boundary condition is necessary or permissible. There, like everywhere else in the surface stream, depth and discharge are related by the normal-depth condition, namely

$$A^2 \, C_h^2 \, R = \frac{Q^2}{S_0} \quad \dots\dots\dots\dots\dots\dots\dots\dots\dots\dots\dots\dots\dots\dots\dots \quad [12.24]$$

Initial conditions: Initial conditions for the numerical solution are provided at the first time step $t_1 = \delta t$ (Fig. 12.5). At this time, the extent of advance and the depth of flow can be found from equation [12.3] coupled with the assumption that the entire surface profile at this small time follows a 3/7 power law, and that the entire subsurface profile is also a power law, of degree a. In consequence, the volume balance equation takes the form

$$r_Q \, Q_0 \, t_1 = \frac{1}{1 + \frac{3}{7}} A_0 \, x_1 + \frac{1}{1 + a} B_z \, z_0 \, x_1 \quad \dots\dots\dots\dots\dots\dots \quad [12.25$$

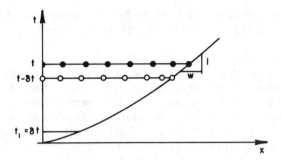

FIG. 12.5 Advance trajectory and solution steps.

in which $r_Q < 1$ is a factor, set arbitarily, designed to account for the fact that the discharge does not rise instanteously to its final value Q_0. A typical value for r_Q is 0.7. With the infiltration function known, equation [12.25] contains two unknowns, y_0 and x_1. A second equation is supplied by the first of equations [12.15]. The assumption of a power-law surface profile allows the derivation therein to be expressed directly in terms of upstream depth y_0 and stream length x_1. Thus

$$\frac{3}{7}\frac{y_0}{x_1} = \frac{Q_0^2}{A_0^2 C_{h_0}^2 R_0} \quad \ldots\ldots\ldots\ldots\ldots\ldots\ldots\ldots\ldots\ldots\ldots\ldots\ldots [12.26]$$

The nonlinear, algebraic equations [12.25] and [12.26] are solved numerically by Newton's method to yield initial values of Y_0 and x_1, as well as the rest of the profile for use as needed in developing the solution at the next time step, $t_1 + \delta t$.

It is not necessary that the initial profile be specified with great exactness. Inasmuch as the solution is governed by the differential equations and boundary conditions, the initial profile need only be correct enough to blend smoothly with the subsequantly computed profiles. It can be expected, however, that the first 10 steps or so will be in error, because of the imprecise initial profile.

12.5 SURFACE IRRIGATION MODELS

A description of surface irrigation hydraulics provides a useful, even necessary, framework for mechanisms that can tell us quickly, reliably and economically where irrigation water will go under given conditions. Such mechanisms are referred to here as models. In the last decade a number of models have been proposed by the authors and others. The early models tended to treat only a portion of the irrigation process, attempting to demonstrate the validity of the basic equations and to identify suitable solution techniques. More recent models, building upon the early experiences, describe the complete irrigation process. These later and more useful models are discussed in the following sections.

The models are first classified, then explained generally to help the reader select a model or portion thereof for a certain use. Model details and computer programs are not included here for space reasons. These can be obtained as needed from the original source references.

12.5.1 Classification

All mathematical models of the surface-irrigation process utilize the volume balance or continuity equation, and differ from one another primarily in the choice of motion equation. Thus the principal classes are (a) the full hydrodynamic model, utilizing the complete Saint-Venant equations. Equations [12.3] and [12.4], (b) the zero-inertia model, which solves equations [12.3] and [12.8], (c) the normal-depth model, equations [12.3] and [12.9], in which the depth at key points in the surface flow is assumed at normal, while the profile shape is established arbitrarily (albeit reasonably) before the start of the calculations. Still further simplification can be made, on the assumption that the advance function is a power law in time.

The governing equations in most cases are too complex to solve analytically. Generally, a numerical solution is sought at a succession of time instants, small increments dt apart, the solution at the end of the time step stemming from that at the beginning. Furthermore, on each successive time line (See Fig. 12.5), the solution—values of depth and discharge—is sought at discrete stations along the flow, shown as heavy dots in Fig. 12.5. The stations at the previous time step, at which values of depth and discharge are known, are shown as open circles. The locations of these two sets of stations are determined by the particular scheme used in solving the equations. This is discussed further in the next sub-sections.

Within each major category, models can be classified in accordance with the solution technique used, whether differential or integral forms of the governing equations are used, whether the equations are locally linearized or remain fully nonlinear, whether first- or second-order accuracy is achieved, and so on. With all theoretically correct (technically, convergent) numerical solution techniques, the results will approach a limit as step sizes in time and distance are reduced. Furthermore, all models in any given major category should converge to the same results. In practice, computational experiments are performed with a number of different step sizes in order to discern the effect of step size. A step size is then chosen commensurate with the desired accuracy.

12.5.2 Full Hydrodynamic Model

The more extensively tested models in this class include those presented by Bassett and Fitzsimmons (1976) and by Katopodes and Strelkoff (1977a). While they differ in significant detail, both models use the complete Saint-Venant equations and the methods of characteristics for solution. Comments on the general structure and performance of models in this class follow.

Retention of all terms in the Saint-Venant equations appears to lead to an accurate but computationally expensive and delicate model. For example, with too large a time step irregular profiles occur, which eventually leads to anomalous results and computer terminated execution. The accuracy of these models under a wide range of slope, roughness, inflow rate and infiltration characteristics makes it possible to regard them as standards against which the more approximate models can be compared.

In an effort to get results as accurate as possible, the approach used to solve the Saint-Venant equations is the so-called method of characteristics.

Characteristics are curves in the x-t plane (See Figs. 12.6a-d) along which the partial differential equations become ordinary differential equations, and along which the physical solution actually propagates (Stoker, 1953; Strelkoff, 1970). In the latter paper Strelkoff discusses the possibility of expressing the Saint-Venant equations in characteristic form, the mathematical and physical implications of using this form, and a formal derivation of the characteristic equations. Suffice it to say, for present purposes, that multiplication of equation [12.3] by $\pm \sqrt{g/AB}$, in which B is top width, and addition to equation [12.4] multiplied by g leads to the following, characteristic, pair of equations

$$\frac{\partial V}{\partial t} + (V \pm c)\frac{\partial V}{\partial x} \pm \frac{g}{c}\left[\frac{\partial y}{\partial t} + (V \pm c)\frac{\partial y}{\partial x}\right] = g(S_0 - S_f) +$$

$$\frac{I_x}{A}(\frac{V}{2} \pm c) \dots\dots\dots\dots\dots\dots\dots\dots\dots\dots\dots\dots\dots\dots\dots\dots [12.27]$$

in which the celerity c is given by

$$c = \sqrt{gA/B} \dots\dots\dots\dots\dots\dots\dots\dots\dots\dots\dots\dots\dots\dots\dots [12.28]$$

These are equivalent to the following set of four ordinary differential equations

$$\frac{d}{dt}(V + \omega) = g(S_0 - S_f) + \frac{I_x}{A}(\frac{V}{2} - c) \dots\dots\dots\dots\dots\dots [12.29]$$

valid on

$$\frac{dx}{dt} = V + c \dots\dots\dots\dots\dots\dots\dots\dots\dots\dots\dots\dots\dots\dots\dots [12.30]$$

and

$$\frac{d}{dt}(V - \omega) = g(S_0 - S_f) + \frac{I_x}{A}(\frac{V}{2} + c) \dots\dots\dots\dots\dots\dots [12.31]$$

valid on

$$\frac{dx}{dt} = V - c \dots\dots\dots\dots\dots\dots\dots\dots\dots\dots\dots\dots\dots\dots\dots [12.32]$$

In equation [12.29] and [12.30], the Escoffier stage variable ω, (Escoffier and Boyd, 1962) given by

$$\omega = \int_o^y \frac{g}{c} dy \dots\dots\dots\dots\dots\dots\dots\dots\dots\dots\dots\dots\dots\dots [12.33]$$

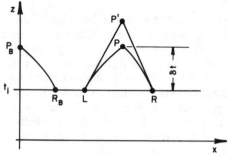

FIG. 12.6a Characteristics.

is just another function of depth, like area, top width, and celerity. In a rectangular channel, such as a border, $\omega = 2c$.

Equation [12.30] and [12.32] are differential equations defining the characteristic curves, and equation [12.29] and [12.31] are the characteristic relations, each valid along its own characteristic curve. The essence of the solution can be discerned from Fig. 12.6a. Since the profile is known at some time t_i, the slopes of the characteristic curves $(V + c)$ and $(V - c)$ emanating from two points L and R, respectively, would be known, as would the values of $(V + \omega)$ and $V - \omega)$ there, respectively. The characteristic curves intersect at some point P, a time increment δt away from t_i. An approximate location of this point P', is obtained by assuming the initial slopes persist all the way to the intersection. The initial rates of change (at t_i) of $(V + \omega_i)$ of $(V - \omega)$ along LP and RP, respectively are known. Thus, approximate values of $(V + \omega)_{P'}$ and $(V - \omega)_{P'}$ can be obtained at P', as, then can $V_{P'}$ and $\omega_{P'}$ individually, and hence, $y_{P'}$.

Greater accuracy can be achieved with a given separation of points L and R, by taking into account the curvature of the characteristic curves LP and RP, and the nonconstant rate of change of $(V + \omega)$ and $V - \omega)$, respectively, along them. Second-order approximation (as opposed to the first-order solution just outlined) is achieved by assuming that the average slope of the characteristic curve between L and P is given by the numerical average of the slopes at the two ends. Similar reasoning holds for the curve between R and P, and the rates of change of $(V + \omega)$ and $(V - \omega)$; average values are approximated by numerical averages. The result is a set of nonlinear algebraic

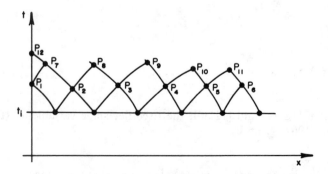

FIG. 12.6b Solution by characteristics.

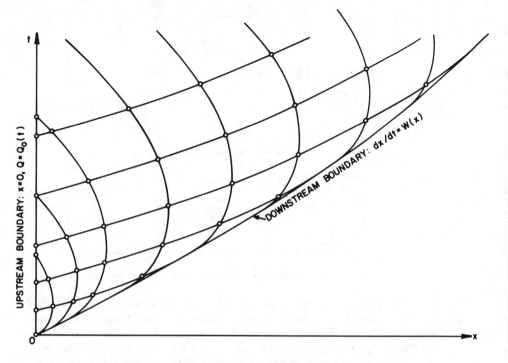

FIG. 12.6c Characteristics net in surface irrigation.

equations which must be solved iteratively to yield the location of point P and the values of all variables there.

At a boundary point, such as P_B, only one characteristic curve leads back to the time line t_i of known conditions. Equation [12.31] and [12.32] are pertinent along this line, and this supplies two of the relations needed to solve for the location on the t axis of P_B, and the depth and velocity there. The third relation, needed to close the system is given by

$$A_{P_B} V_{P_B} = Q_0 \dotfill [12.34]$$

in which the right-hand side is the boundary condition. As in the case of the interior point, the time increment and depth and velocity values can be found by a direct, first-order solution, or, with greater accuracy, by an iterative second-order procedure.

Once a series of points P_1, P_2, . . .P_6 (Fig. 12.6b) have been computed, the values there can be used to calculate points P_7-P_{12}, and so on. In the context of surface-irrigation stream flow, the network so generated has the appearance of Fig. 12.6c, extracted from Sakkas and Strelkoff (1974). In this work the advance phase of a border irrigation was computed by an essentially second-order characteristics scheme utilizing the network of characteristic curves. Practically useful data, consisting of the values of the variables at specified times, were computed by linear interpolation from those at the grid points.

The procedure is fundamentally very accurate, but inconvenient because the solution is found at odd locations and times. A more practical scheme

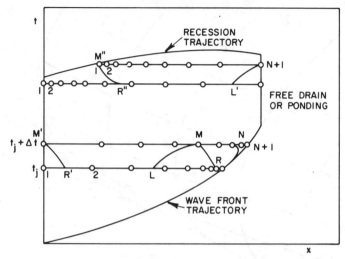

FIG. 12.6d Computational grid—method of specified time lines.

generates data at specified points on specified time lines. Approximate first-order schemes were developed in the pioneering work of Bassett and Fitzsimmons (1976), and Kincaid et al. (1972). A second-order model was presented by Katopodes and Strelkoff (1977a). Fig. 12.6d, extracted from that work, shows the prescribed node locations, irregularly spaced along the wave in order to place the greatest concentrations of points in the regions of greatest change of the variables.

While the equations and solution procedures are the essence of models in this class, numerous supporting provisions are required. These include procedures to select boundary conditions, structure the iterative calculations, tabulate the various volumes of water involved, determine the times and velocities of advance and recession and perform the summary calculations indicating irrigation performance, include the efficiency terms of interest. Usual read, write and control provisions are needed to operate the rather lengthy and complicated programs. A digital computer with considerable storage and software capability is required.

In general, models in this class are capable of sufficient accuracy to serve as a standard of comparison for the simplified modles, and can produce very practical and useful information. Accuracy can be influenced, however, by the programmer through such things as his selection of grid sizes and arrangement of node points. Cost of computation depends on the irrigation conditions imposed and the rate structure of the machine used. In general, however, the computation cost using models in this class will be substantially greater than would be incurred by using the simplified models described below.

12.5.3 Zero-Inertia Model

Work to date suggests that models in this class may become the method of choice for routine computer modelling of surface-irrigation phenomena. The potential for accuracy is good, because Froude numbers in surface irrigation are typically quite low. At the same time, the diffusive character of the governing equations is conducive to stability of computation. Computation cost is generally much less than that incurred with fully dynamic models.

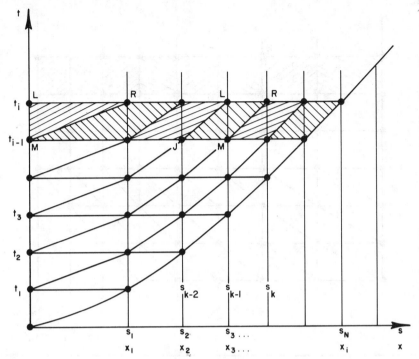

FIG. 12.7a Computational grid during advance.

The first operational zero-inertia model of the complete irrigation process reported by Strelkoff and Katopodes (1977a) remains a useful pattern. While the details are presented in the cited paper and subsequent discussions, sufficient information is presented here to acquaint the reader with the approach used in this benchmark work. Volume- and time-integrated forms of the governing equations are employed in a fully implicit numerical scheme of solution. To cut down on execution expense, the equations are locally linearized, thus obviating the need for iterative solution. To maximize the accuracy of the linearization, the computational network is chosen in such a way as to minimize the change in the linearized variables. This leads to an oblique grid during advance (Fig. 12.7a), a mixed oblique and rectangular grid if cut off occurs before t_L, the time the stream reaches the end of the field, and a rectangular grid for $t > t_L$ (Fig. 12.7b).

Fig. 12.7c illustrates a volume element (over which the governing equations are integrated) at the beginning and end of a time interval δt. The logitudinal extent of the element is defined by two flow cross sections, U upstream, and D downstream, each of which moves in a prescribed fashion over the duration of the time increment. The top of the element always coincides with the free surface of the irrigation stream, the bottom corresponds to the infiltration profile at depth z, previously defined. The movement of U and D are determined by the type of grid desired. For an oblique grid cell, as used during advance, the element configurations appear as in Fig. 12.7d. A rectangular cell corresponds to the successive configurations of Fig. 12.7e.

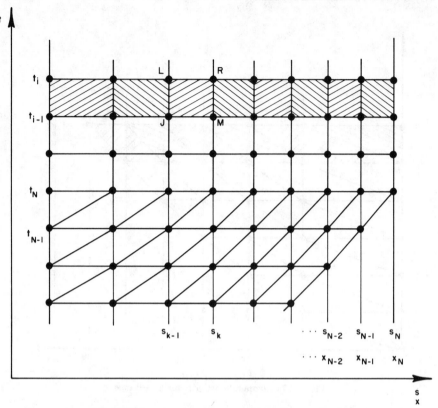

FIG. 12.7b Computational grid—nearly stationary phase.

Each of these are thus special cases of the general configurations shown in Fig. 12.7c. With reference to the latter, the continuity equation [12.3] integrated over the element length and time increment can be written as follows

$$\left(\overline{Q}_{r_U} - \overline{Q}_{r_D} \right) \delta t = [(\widetilde{A} + \widetilde{A}_z) \, \delta x]_t - [(\widetilde{A} + \widetilde{A}_z) \, \delta x]_{t-\delta t} \quad \cdots \cdots \, [12.35]$$

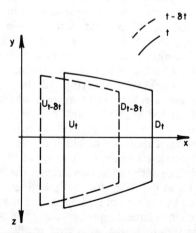

FIG. 12.7c Computational element—general.

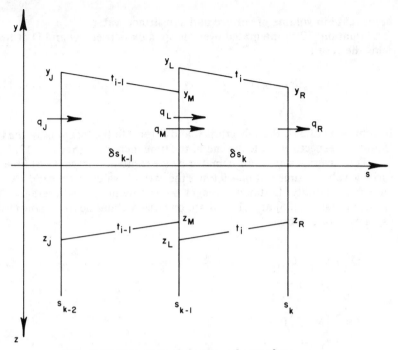

FIG. 12.7d Computational element—advance phase.

The variable Q_r represents the discharge actually crossing the face U or D. If the face were moved at the fluid velocity, Q_r would equal zero. The bar over a variable represents a time average over δt. The tilde over a variable represents a distance average over the span length δ_x pertinent to the given time. The subscripts t-δt and t refer to the beginning and end, respectively, of the time interval. Equation [12.35] simply states that the net volume of water added over the time increment δt to the region between U and D equals

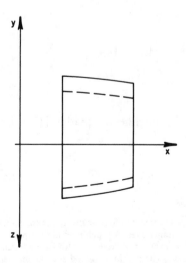

FIG. 12.7e Computational element—nearly stationary phase.

the increase in volume of surface and subsurface water.

Equation [12.8] integrated over the span δx between U and D at time t yields the result

$$y_R - y_L = (S_0 - \tilde{S}_f)\,\delta x \quad\dotfill\text{[12.36]}$$

In equation [12.36], the subscripts L and R refer to the locations of the faces U and D, respectively at t, the end of the time increment (See Fig. 12.7a, b).

Time averages are approximated by a weighted average: $(1-\theta)$ times the variable value at time $t-\delta t$ plus θ times the variable value at time t; $\theta \geq 0.5$ for numerical stability. Distance averages are simply numerical averages.†

Equation [12.35] and [12.36] are linearized about nearby known values as follows for example,

$$Q_R = Q_M + \delta Q_R \quad\dotfill\text{[12.37]}$$

$$Q_L = Q_J + \delta Q_L \quad\dotfill\text{[12.38]}$$

$$y_R = y_M + \delta y_R \quad\dotfill\text{[12.39]}$$

$$y_L = y_J + \delta y_L \quad\dotfill\text{[12.40]}$$

while

$$S_{f_R} = S_{f_M} + \frac{\partial S_f}{\partial Q} M\,\delta Q_R + \frac{\partial S_f}{\partial y} M\,\delta y_R \quad\dotfill\text{[12.41]}$$

$$\frac{\partial S_f}{\partial Q} = \frac{2 S_f}{Q} \quad\dotfill\text{[12.42]}$$

$$\frac{\partial S_f}{\partial y} = S_f \left(-\frac{2B}{A} - \frac{2 dC_h}{C_h dy} - \frac{1 dR}{R dy} \right) \quad\dotfill\text{[12.43]}$$

In a border with Manning roughness, equation [12.43] reduces to

$$\frac{\partial S_f}{\partial y} = -\frac{10}{3}\frac{S_f}{y} \quad\dotfill\text{[21.44]}$$

†Because of the very large gradients in S_f near the upstream computational boundary with large bottom slopes, a better distance average is achieved in the upstream-most cell by setting $S_f = S_{f_R}$. Retaining the numerical average in these cases leads to irregular, sawtooth profiles and erratic computation of recession. The heavily weighted average on the other hand, yields valid results in more cases.

Similar expressions are obtained for S_{f_x}. The resultant set of simultaneous linear algebraic equations is solved efficiently by a double-sweep technique devised by Preissman, as reported by Liggett and Cunge (1975). This model also used the same general operating provisions as the fully dynamic model described above.

12.5.4 Normal-Depth Model

The substitution of equation [12.10] into equation [12.3] leads to a single partial differential equation in one unknown. This too has a characteristic form, which consists of a single pair of ordinary differential equations, one of which defines the single family of characteristic curves, and the other a characteristic relation valid along those surves. This set of circumstances defines the computed motion of the surface stream as a kinematic wave (Lighthill and Whitham, 1955).

The application of kinematic-wave theory to border-irrigation flow was studied by Smith (1972). The essence of the technique is presented below; for further details, the reader is referred to the original paper. In a prismatic channel of given shape in which the conveyance $K^2 A^2 C_n^2 R$ is a function of flow area alone, the partial differential equation resulting from combination of the continuity equation with the normal-depth condition is

$$\frac{\partial A}{\partial t} + \sqrt{S_0} \frac{dK}{dA} \frac{\partial A}{\partial x} = -I_x \quad \text{............................[12.45]}$$

The equivalent pair of ordinary differential equations is:

$$\frac{dA}{dt} = -I_x \quad \text{...[12.46]}$$

valid along

$$\frac{dx}{dt} = c(a) = \sqrt{S_0} \frac{dK}{dA} = \frac{Q}{A}\left(\frac{5}{3} - \frac{2}{3} R \frac{dp}{dA}\right) \quad \text{..................[12.47]}$$

in which the last equality holds with the Manning formula for roughness; p is the wetted bed perimeter of the flow. In a border, with Manning roughness, the celerity c is just 5/3 the flow velocity. The righthand side of equation [12.47] is a function of flow area alone. Equation [12.46], on the other hand, shows that flow area varies along a characteristic curve only as a result of infiltration. Thus, if infiltration were negligible, the pattern of characteristic curves (straight lines in this case) would appear as in Fig. 12.8a. Each of the characteristic lines terminates abruptly at the front of the wave.

The sudden change in depth at this point is a consequence of the normal-depth assumption. Its existence is easily proved by noting that the righthand side of equation [12.47] is an increasing function of area, and also

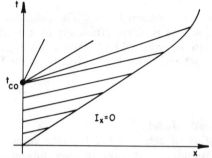

FIG. 12.8a Family of characteristics—zero infiltration.

Q. Thus, if the discharge at x = 0 were to increase gradually, the corresponding characteristic lines would converge as shown in Fig. 12.8b, and if the channel were long enough, they would inevitably intersect. Intersection points could be found from the geometrical and roughness properties of the channel and from the time rate of change of discharge. Since each characteristic is carrying its own value of area, intersection implies a discontinuity in the latter. Furthermore, a sudden change in discharge, as at the start of an irrigation, would lead to immediate formation of the abrupt front described. A simple volume balance in the region of the front leads to the conclusion that front-propagation speed w equals water velocity Q/A just behind the front. Thus, the trajectory of the front appears as in Fig. 12.8a.

The inclusion of infiltration into the problem leads to a gradual reduction in depth along the characteristic, as can be seen from equation [12.45]. This leads to a pattern of characteristic curves like those shown in Fig. 12.8c, drawn from Smith (1972).

Fig. 12.8c also serves to illustrate the numerical solution of the advance phase of a surface irrigation. Equations [12.46] and [12.47] are integrated, and the integrals are approximated numerically to yield a pair of nonlinear algebraic equations. These are solved iteratively to yield values of flow area at interior points in the region of flow in the x-t plane. Over a short length of

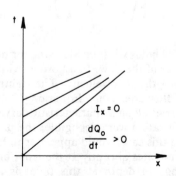

FIG. 12.8b Converging characteristics with increasing inflow.

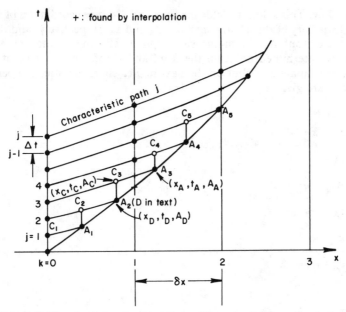

FIG. 12.8c Definition sketch for characteristics—grid construction in surface irrigations.

any given characteristic j, spanning the distance between x_k and x_{k+1} (Fig. 12.8c), the integral of equation [12.47] is approximated by the expression

$$\frac{x_{k+1} - x_k}{t_{k+1} - t_k} = \frac{1}{2}(c_{k+1} + c_k) \qquad [j = \text{const.}] \quad \cdots \cdots [12.48]$$

while the integral of equation [12.45] is given by

$$\frac{A_{k+1} - A_k}{t_{k+1} - t_k} = -\frac{A_{z_{k+1}} - A_{z_k}}{\tau_{k+1} - \tau_k} \qquad [j = \text{const.}] \quad \cdots \cdots [12.49]$$

in which τ is the infiltration time for the point in question. In this pair of equations, t_{k+1} and A_{k+1} are the basic unknowns; x_{k+1} and x_k are known from the given spacing of node points. The correct value of A_{k+1} (and hence t_{k+1}) is approached iteratively in successive approximations‡, using the secant method (Gerald, 1970); equation [12.10] is used in equation [12.47] to relate Q and A.

The trajectory of the abrupt front is obtained by simultaneous solution of three equations relating to the triangular elements of the x-t plane shown

---

‡t_{k+1} is eliminated from equation [12.48] and [12.49], and the remaining unknowns are each expressed in terms of surface depth y_{k+1}. The result is a single, nonlinear algebraic equation in y_{k+1}.

in Fig. 12.8c. This solution yields the time and distance coordinates of a point on the trajectory of the advancing front, t_A and x_A, respectively, and the flow area at the front A_A. With the subscript D referring to the neighboring previously determined point on the front and C referring to a point at the same x-coordinate as D but on the next higher characteristic, the pertinent equations are given by

$$\frac{x_A - x_C}{t_A - t_C} = \frac{1}{2}(c_A + c_C) \quad\dotfill\quad [12.50]$$

$$\frac{A_A - A_C}{t_A - t_C} = -\frac{A_{z_C}}{\tau_C} \quad\dotfill\quad [12.51]$$

$$\frac{x_A - x_D}{t_A - t_D} = \frac{1}{2}(w_A + w_D) \quad\dotfill\quad [12.52]$$

in which $x_C = x_D$, and $w = Q/A$. Before equation [12.50-52] can be solved, a preliminary calculation using equations similar§ to equations [12.48] and [12.49] must be made to yield A_C and t_C on the basis of values at the nearest upstream node point. Example solutions obtained in this manner are presented in the graphs of Section 12.5.7.

Analytic solutions to the kinematic-wave equations can also be obtained if the infiltration function is sufficiently simple. Computations of advance and recession on the assumption of a constant infiltration rate were made by Cunge and Woolhiser (1975).

The final paragraphs of this section deal with the applicability of the normal-depth model. Of the models discussed to this point, the present one is by far the cheapest to execute. Consequently, its range of application becomes a matter of some importance. It is at once evident that the model is useless in the case of dead-level fields, and in diked borders or furrows. In the first case, normal depth is infinite, and so, computations based thereon are meaningless. In the second case, the flow in the field is clearly influenced by the downstream boundary conditions. As the theory, illustrated by Figs. 12.8a-c, shows, conditions on a characteristic are determined solely by the upstream boundary condition and by infiltration along the channel bed. No downstream boundary condition can be imposed on the flow. Thus, neither ponding nor outflow at any depth other than normal can be modelled with the normal-depth approach. As shown in Section 12.5.7, the minimum value of slope at which the normal-depth model is useful, depends upon bed roughness, inflow rate, and infiltration characteristics. These combine into dimensionless parameters that govern its applicability.

It is possible that the range of applicability could be extended by fitting a kinematic-shock profile to the front of the wave, previously computed

§The subscript $k+1$ is replaced everywhere by C.

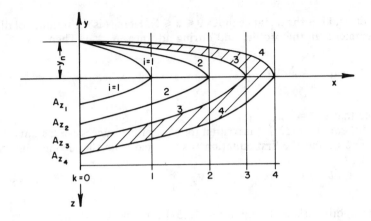

FIG. 12.9a Successive surface and subsurface profiles.

strictly on the basis of a normal depth-discharge relationship as described above. This theory is based on the fact that at the wave front, where the depth drops rapidly to zero, the depth gradient, if not the other terms on the left side in equation [12.4], cannot be considered negligibly small. A reasonable profile, replacing the abrupt front and termed a kinematic-shock profile, can be obtained by retaining all of the terms in equation [12.4], postulating that the profile moves with little change of shape, and at a velocity equal to the velocity of the abrupt front (Henderson, 1966). Such a profile, fitted to the normal-depth profile in such a way as to retain the appropriate volumes above and below the surface would probably lead to a somewhat greater advance rate, as is the case with an impermeable bed (Strelkoff et al., 1977).

12.5.5. Assumed-Average-Surface-Depth Model

In view of the fact that in this model the upstream depth before cut off is normal, and that during runoff, the downstream depth is normal, it must be subject to the same restrictions on applicability as the normal-depth model described in the previous section. In addition, there is some inaccuracy in the assumption that the shape factor r_y is constant, and chosen almost arbitrarily. Still, once all these assumptions have been made, the execution costs of this model are far less than any of the others discussed so far. In fact, in some cases, a solution can be found without recourse to a digital computer at all. This makes the method ideal for quick, rough calculations, provided the bed slope is large enough to allow assumption of normal depth where required.

The essence of the technique is described next. For details beyond the scope of this presentation, the reader is referred to Hart et al. (1968) for computation of advance, and to Strelkoff (1977) for depletion, runoff, and recession. Following the approach of Hall (1956), advance is computed in a series of increments δx_{A_i} occurring over constant intervals of time δt. In the present discussion, and in reference to Fig. 12.9a, i is the time index, so that $t_i = t_{i-1}$

$+$ δt, and k is the distance index $0 \leqslant k \leqslant N$. Here N is the number of distance increments in the profile, and during advance, $N = i$. Then

$$x_{A_i} = \sum_{k=1}^{N} \delta x_k \dots\dots\dots\dots\dots\dots\dots\dots\dots\dots\dots\dots\dots\dots [12.53]$$

Note that $\delta x_k = \delta x_{A_k}$.

If equation [12.14] is written for each of two successive instants of time, t_{i-1} and t_i, and the first equation is subracted from the second, the result is

$$Q_0 \, \delta t = A_0 \, r_y \, \delta x_{A_i} + V_{z_i} - V_{z_{i-1}} \dots\dots\dots\dots\dots\dots [12.54]$$

The righthand side of equation [12.54] for $i = 4$ is shown shaded in Fig. 12.9a. At any time step i, V_z can be approximately computed by the summation

$$V_{z_i} = \sum_{j=1}^{i-1} \frac{A_{z_{i-j}} + A_{z_{i-j-1}}}{2} \delta x_{i-j} + r_{z_T} A_{z_1} \delta x_{A_i} \dots\dots\dots [12.55]$$

Equation [12.55] was derived on the assumption that A_z varies linearly over all the distance steps but the last. There, a tip shape factor r_{z_T} is applied. It was shown in Section 12.4.5 that for an infiltration function that approaches a power law with exponent a at small infiltration times, the theoretical value of the shape factor for small δx_A is given by

$$r_{z_T} = \frac{1}{1+a} \dots\dots\dots\dots\dots\dots\dots\dots\dots\dots\dots\dots\dots [12.56]$$

The surface-stream shape factor r_y, assumed for the entire profile, is given an arbitrary value based on the results of many computer experiments with the more sophisticated models which actually compute the surface profile. A reasonable value for general use is

$$r_y = 0.8 \dots\dots\dots\dots\dots\dots\dots\dots\dots\dots\dots\dots\dots\dots [12.57]$$

which is adequate as long as the surface volume is not too large compared to the infiltrated volume. The subscript notation for A_z in equation [12.55] is a short-hand expression for infiltrated volume per unit length under the assumption that this is a function of infiltraton time τ alone, i.e., $A_z = A_z(\tau)$. Then for any subscript p, the accumulated area is given by

$$A_{z_p} = A_z \, (p\delta t) \dots\dots\dots\dots\dots\dots\dots\dots\dots\dots\dots\dots [12.58]$$

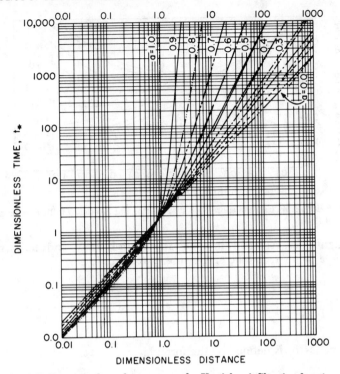

FIG. 12.9b Dimensionless advance curves for Kostiakov infiltration function.

Equation [12.54] solved for δx_{A_i}, yields the latter explicitly, in terms of the increments $\delta x_{A_{i-1}}$, $\delta x_{A_{i-2}}$, ... determined previously, i.e.,

$$x_{A_i} = \frac{Q_0\,\delta t - \sum\limits_{j=1}^{i-1}\frac{1}{2}\left(A_{z_{i-j}} + A_{z_{i-j-1}}\right)\delta x_{i-j}}{A_0\,r_y + A_{z_1}\,r_{z_T}} \quad \ldots\ldots\ldots [12.59]$$

The first increment x_{A_1} is found from

$$x_{A_1} = \frac{Q_0\,\delta t}{A_0\,r_y + A_{z_1}\,r_{z_T}} \quad \ldots\ldots\ldots\ldots\ldots\ldots\ldots\ldots\ldots [12.60]$$

Advance curves for border irrigation computed on the basis of these assumptions with $z = k\tau^a$ are presented in Fig. 12.9b. The scales of the graph are dimensionless, in order to present the results of all possible combinations of slope, discharge, roughness, and infiltration characteristics on a single page. It is evident that at small time and large time, advance follows a power law, $x_A = ft^h$ (f and h are constants), since in these ranges the curves

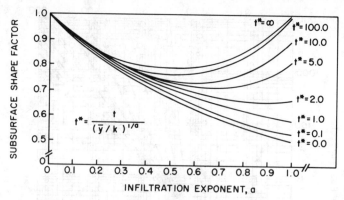

FIG. 12.9c Relationship between subsurface shape factor and Kostiakov infiltration exponent as a function of time.

are straight lines. At small time the exponent h is unity, and the corresponding shape factor r_z for the entire subsurface profile is

$$r_Z = \frac{1}{1+a} \qquad \dots\dots\dots\dots\dots\dots\dots\dots\dots\dots\dots\dots\dots\dots\dots [12.61]$$

This shape factor is defined by the formula

$$V_z(t) = r_z A_z(t)\, x_A(t) \qquad \dots\dots\dots\dots\dots\dots\dots\dots\dots\dots\dots\dots [12.62]$$

At large times, the exponent is

$$h = 1 - a \qquad \dots\dots\dots\dots\dots\dots\dots\dots\dots\dots\dots\dots\dots\dots\dots\dots [12.63]$$

while the corresponding shape factor is

$$r_Z = \frac{a\pi(1-a)}{\sin a\,\pi} \qquad \dots\dots\dots\dots\dots\dots\dots\dots\dots\dots\dots\dots\dots\dots [12.64]$$

The variation of r_z with a at intermediate times is shown in Fig. 12.9c.

Still further simplification in computing the advance function is achieved by assuming a priori a constant value of r_z in equation [12.62] and substituting into equation [12.14] for a direct computation of x_A for any t. The value $r_z = 1/(1+a)$ underestimates, somewhat, the infiltrated volume. This compensates to some degree for the overestimated surface volume stemming from the assumption that upstream depth rises immediately to normal depth. Actually, the upstream depth rises gradually and approaches normal sooner or later, depending upon whether the slope is relatively steep or flat.

As regards the matter of depletion and recession, different analyses are appropriate for freely draining fields or fields diked at the downstream end to prevent runoff. In the case of a freely draining border, it is assumed that at

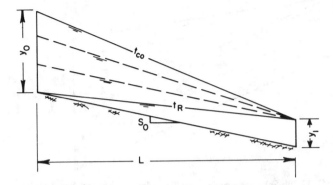

FIG. 12.10a Schematic surface profiles—depletion phase.

·   the time of cut off, the surface profile is approximated by the straight line labeled t_{co} in Fig. 12.10a. The depth upstream, y_0, is at normal depth for the just-cut-off inflowing discharge per unit width q_0. (Hydrodynamic analyses of the type discussed in Section 12.5.2 indicate that at the low Froude numbers typical of border irrigation, there is no significant dynamic negative wave propagating downstream after sudden cut off. The upstream depth reduces gradually, by surface drainage and infiltration.) The unit discharge, q_1, at the downstream end (runoff) is smaller than q_0 by the overall infiltration rate (per unit width) q_{INF}. It is assumed for purposes of computation that during the course of the depletion phase,

1 the depth y_1 at the downstream end is constant and at normal for the runoff q_1 pertaining to the time t_R marking the end of the depletion phase, the beginning of recession:

2 the combined runoff and total infiltration rates continue to add up to the original inflow rate q_0, i.e.,

$$q_1 + q_{INF} = q_0 \quad \dots\dots\dots\dots\dots\dots\dots\dots\dots\dots\dots\dots\dots\dots\dots\dots [12.65)$$

3 the surface profiles at t_R, as well as at t_{co} are straight lines, as in Fig. 12.10a.

These assumptions provide the necessary means for estimating the time t_R at which the upstream depth reduces to zero and recession starts. The duration between the time of cut off t_{co} and t_R is the time it takes to remove the triangular volume of length L and upstream height $y_0 = y_n(q_0)$ by infiltration and drainage at the combined rate, q_0, equal to the given initial unit inflow rate to the border. Thus, with equation [12.65] in force in a border of length L

$$t_R = t_{co} + \frac{y_n(q_0)L}{2q_0} \quad \dots\dots\dots\dots\dots\dots\dots\dots\dots\dots\dots\dots\dots\dots\dots [12.66]$$

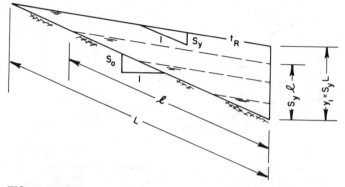

FIG. 12.10b Schematic surface profiles—downstream end free draining.

At the start of the recession phase, it is assumed that the rate of change S_y of depth with distance is uniform over the length of the border, and is

$$S_y = \frac{y_1(t_R)}{L} = \frac{1}{L} y_n(q_1) \quad \dots\dots\dots\dots\dots\dots\dots\dots\dots\dots\dots\dots\dots [12.67]$$

$$q_1(t_R) = q_0 - \overline{I}(t_R)L \quad \dots\dots\dots\dots\dots\dots\dots\dots\dots\dots\dots\dots\dots [12.68]$$

In equations [12.67] and [12.68], the argument t_R indicates that the functions to which it pertains vary with time, and it is the value at time t_R that is required. In equation [12.68] $\overline{I}$ = the average infiltration rate in the border. For purposes of computation, this is assumed to equal the numerical average of the infiltration rates at the upper and lower ends of the border, i.e.,

$$I(t_R) = \frac{I(t_R) + I(t_R - t_L)}{2} \quad \dots\dots\dots\dots\dots\dots\dots\dots\dots\dots\dots\dots\dots [12.69]$$

in which the infiltration rate at the downstream end at time t_R is $I(t_R - t_L)$, because the infiltration time did not start until t_L, the time at which the advancing stream reached the end of the border.

The simple assumption that S_y remains constant as recession proceeds (Fig. 12.10b) provides a realistic gradual decrease in normal depth and outflow rate with time, at the downstream end of the field. Probably the most important physical circumstance required for the validity of the assumptions made concerning the surface stream is a bottom slope sufficiently large, that the assumed water surface never approaches horizontal, i.e.,

$$S_0 - S_y > 0 \quad \dots\dots\dots\dots\dots\dots\dots\dots\dots\dots\dots\dots\dots\dots\dots [12.70]$$

In terms of an average infiltration rate, $\bar{I}$, over the inundated portion of the field, the time rate of change of surface volume is expressed by the following differential equation in l, the length of the inundated portion of the field

$$\frac{d}{dt}\left(\frac{\ell^2 S_y}{2}\right) = -\bar{I}\ell - \frac{C_u}{n}S_0^{1/2}\,(S_y\ell)^{5/3} \quad \dots\dots\dots\dots\dots \quad [12.71]$$

The second term in the righthand side of equation [12.71] is the runoff rate at the downstream end, under the assumption that the depth there, $y_1(t) = s_y l(t)$, now variable with time, is always at normal depth. The Manning formula was used, but any other could be substituted, with a concomitant change in the difficulty of solution of the differential equation.

Simplification and division of the equation by $S_y l$ yields the following nonlinear ordinary differential equation:

$$\frac{d\ell}{dt} = -\frac{\bar{I}}{S_y}\,(1 + C\ell^{2/3}) \dots\dots\dots\dots\dots\dots\dots\dots\dots\dots\dots\dots\dots\dots \quad [12.72]$$

in which

$$C = \frac{C_u S_0^{1/2}\,S_y^{\,5/3}}{n\bar{I}} \dots\dots\dots\dots\dots\dots\dots\dots\dots\dots\dots\dots\dots\dots\dots \quad [12.73]$$

If $\bar{I}$ is considered a general function of time, equation [12.72] can only be solved numerically, stepwise in time. For present purposes, the additional assumption is made that I is constant. For many soils, I does indeed approach constancy with large infiltration times, so in many cases, the assumption will prove reasonable. Then equation [12.72] takes on variables-separable form. Indeed, following integration, it appears as follows:

$$-\frac{S_y}{I}\int_L^\ell \frac{d\ell}{1 + C\ell^{2/3}} = \int_{t_R}^t dt \dots\dots\dots\dots\dots\dots\dots\dots\dots\dots\dots \quad [12.74]$$

Introduction of a new variable

$$\varsigma = C^{3/2}\ell \dots\dots\dots\dots\dots\dots\dots\dots\dots\dots\dots\dots\dots\dots\dots\dots \quad [12.75]$$

allows equation [12.74] to be written in the form

$$t = t_R + \frac{S_y}{\bar{I}\,C^{3/2}}\,[R_{rec}(C^{3/2}L) - R_{rec}(C^{3/2}\ell)] \quad \dots\dots\dots\dots \quad [12.76]$$

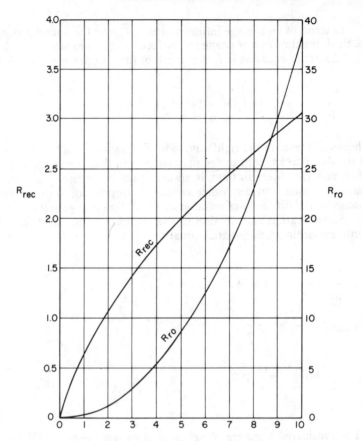

FIG. 12.10c Generalized recession and runoff functions.

in which the recession integral, R_{rec}, is given as a function of its upper limit by

$$R_{rec}(U) = \int_0^U \frac{d\zeta}{1 + \zeta^{2/3}} \quad \dots\dots\dots\dots\dots\dots\dots\dots\dots\dots\dots\dots \quad [12.77]$$

For convenience, the integral in equation [12.77] has been computed (through appropriate expansions) between the limits 0 and 10 and has been graphed in Fig. 12.10c. From this graph and equation [12.76], the entire recession curve for any given set of irrigation parameters can be computed. The computation of runoff is performed in similar fashion. First the volume that runs off between t_L and t_R is computed. This is given by the expression

$$V_{ro}(t_R) = q_0 t_{co} - \frac{S_y L^2}{2} - \frac{z(t_R) + z(t_R - L)}{2} L \dots\dots\dots\dots [12.78]$$

Then, any time after recession starts, the runoff volume is given by the equation

$$V_{ro}(t) = V_{ro}(t_R) + \frac{S_y}{C^3} [R_{ro}(C^{3/2}L) - R_{ro}(C^{3/2}\ell)] \quad \dots\dots\dots [12.79]$$

in which the runoff integral

$$R_{ro}(U) = \int_0^U \frac{\zeta^{5/3}}{1 + \zeta^{2/3}} d\zeta \quad \dots\dots\dots\dots\dots\dots\dots\dots\dots\dots [12.80]$$

has been computed as detailed in Strelkoff (1977), and graphed in Fig. 12.10c.

The total volume of runoff at time t_E when the irrigation is over is

$$V_{ro}(t_E) = V_{ro}(t_R) + \frac{S_y}{C^3} R_{ro}(C^{3/2}L) \quad \dots\dots\dots\dots\dots\dots [12.81]$$

Similar analyses can be made for the ponded case.

12.5.6 Inter-Model Comparisons

In view of the enormous differences in computing costs associated with the various models, some knowledge of the relative accuracy of each would be very desirable. Unfortunately, there has been, to date, insufficient experience with all of the models to form a complete picture of what can be expected in the way of accuracy for any given set of input data. In the present section, one comparison of the various models with each other and with a complete field test (advance and recession) will be presented. A second comparison with field data on advance in a relatively flat border will then be given. A more systematic study of relative accuracy of various models in the computation of advance only will be presented in Section 12.5.7.

The complete field test under discussion was performed at the University of Arizona, under the test number AR-15, as a contribution to Regional Research Project W-65. The particulars, as reported by Bassett (1973) and Roth et al. (1974) follow: $q_0 = 3.28$ L/sm (0.0353 cfs/ft); $S_0 = 0.00101$; $L = 91.4$ m (300 ft); $n = 0.024$ m$^{1/6}$ (ft$^{1/6}$); $k = 56.2$ mm/h^a (2.21 in/h^a); $a = 0.2716$; and $t_{co} = 38$ min; downstream boundary condition = free draining.

Fig. 12.11a compares observed field data with Bassett's (1973) first-order model and Katopodes' and Strelkoff's (1977a) second-order model. Agreement of both models with experiment is quite satisfactory, with the advance, of course, showing better agreement that recession. The difficulty of accurately observing recession is well known. The slight irregularities, ridges, and depressions in a soil surface over which water has been running result in islands and ponds in the vicinity of the trailing edge. Furthermore, the flow near the rear of the stream in recession is extremely shallow, so that exact positioning of the trailing edge is impossible. Many experimentalists simply

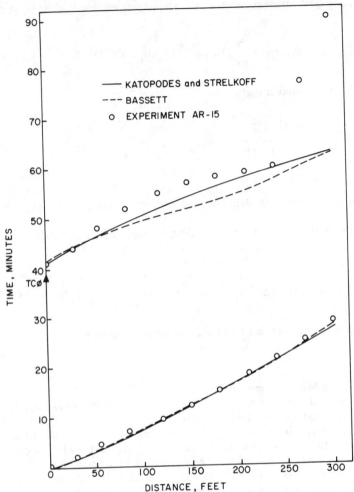

FIG. 12.11a Advance and recession trajectories—hydrohynamic models.

estimate the location of the trailing edge roughly, by eye, as that section at which, say, one-half or three-quarters of the soil surface in a transverse line is uncovered. Errors amounting to 10 percent of field length are easily made. Fortunately, while accurate estimates of recession (both in the field and computationally) are the most difficult to make when infiltration rates are very low, it is precisely in these same circumstances that substantial error in recession trajectories can be tolerated. For with a low infiltration rate, even large errors in the recession curve lead to but small errors in computed infiltration depth.

Fig. 12.11b compares the zero-inertia model with the hydrodynamic model and experiment. Agreement is satisfactory, in view of the aforementioned comments regarding recession. The cost of computation is about 1/10 that of the hydrodynamic models.

Application of the approximate, algebraic model based upon assumed water-surface-profile shape and normal depth at key points is illustrated in

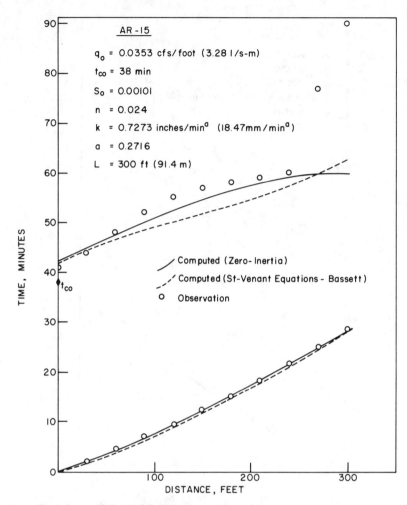

FIG. 12.11b Advance and recession trajectories—zero-inertia models.

Fig. 12.11c. This model, which does not need to be run on a computer because points on the advance and recession curves are determined explicitly from algebraic formulas, shows good agreement in this case.

Because of its dependence on normal-depth conditions, however, the algebraic model gives poor results on a flat slope. Fig. 12.11d illustrates the problem for a field test run by the Agricultural Research Service, U. S. Department of Agriculture, Twin Falls, Idaho, also as a contribution to Regional Research Project W-65 (Bondurant, 1971). In this test $q_0 = 4.226$ L/sm (0.0455 cfs/ft); $S_0 = 0.000051$; $n = 0.26$; $k = 72.6$ mm/h^a (2.86 in/h^a); and $a = 0.67$. The magnitude of the error incurred by the algebraic model suggests a fundamental error in its assumptions. Comparison of the assumed surface profile with that computed by more sophisticated models (say the zero-inertia model, which continues to agree well with experiment) shows, that under the given conditions, the upstream depth increases very slowly, and after an hour's time is only about 50 percent of normal depth.

As regards the accuracy of the algebraic model, it can be said that Fig. 12.11c illustrates the effect of assuming a surface-profile shape, while Fig.

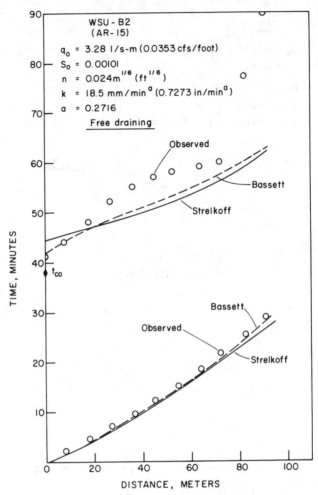

FIG. 12.11c Advance and recession trajectories—assumed average surface depth (moderate slope).

12.11d shows the effect of prematurely assuming upstream depth equal to normal depth. In the latter case, the kinematic-wave model described in Section 12.5.3 would also be heavily in error. This question is addressed in greater detail in the next section.

12.5.7 Dimensional Analysis of Advance

In studies of surface irrigation hydraulics, as in many other engineering analyses, the use of dimensionless variables offers certain simplifications, including a reduction in the number of variables and the number of tests required. The authors are now exploring the use of dimensionless variables, the first work having been released by Katopodes and Strelkoff (1977b). This section considers only the advance phase.

The number of independent parameters affecting advance of an irrigation stream down a border (q_0, S_0, n, k, a [assuming an infiltration formula

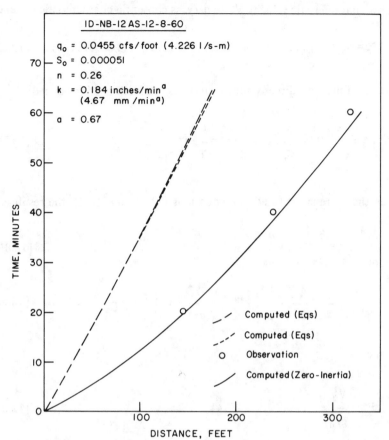

FIG. 12.11d **Advance and recession trajectories—assumed average surface depth (zero slope).**

of the type $z = k\tau^a$])can be reduced to three, and ultimately to just two by expressing the governing equations, boundary conditions, and solutions in dimensionless form.

A surface depth, characteristic of the irrigation, is chosen first. With the discussion restricted to sloping borders, normal depth y_n for the given conditions is an appropriate characteristic depth. Next a characteristic time T is defined. One reasonable choice involves the infiltration rate; then the characteristic time is defined as the time required to infiltrate a depth equal to normal depth, i.e.,

$$T = \left(\frac{y_n}{k}\right)^{1/a} \quad \ldots\ldots\ldots\ldots\ldots\ldots\ldots\ldots\ldots\ldots\ldots\ldots\ldots\ldots\ldots [12.82]$$

Then a characteristic velocity $V_n = q_0/y_n$ is chosen; a characteristic distance follows,

$$X = V_n T \qquad \dots\dots\dots\dots\dots\dots\dots\dots\dots\dots\dots\dots\dots [12.83]$$

Next, the following dimensionless variables (starred) are introduced:

$$y^* = \frac{y}{y_n} \, ; V^* = \frac{V}{V_n}; x^* = \frac{x}{X}; t^* = \frac{t}{T}; \tau^* = \frac{\tau}{T} \qquad \dots\dots\dots\dots\dots [12.84]$$

When these are substituted into equations [12.3] and [12.4], the result is

$$\frac{\partial y^*}{\partial t^*} + V^* \frac{\partial y^*}{\partial x^*} + y^* \frac{\partial V^*}{\partial x^*} + a\tau^{*\,a-1} = 0 \qquad \dots\dots\dots\dots\dots\dots\dots\dots [12.85]$$

$$F_n^2 \frac{\partial V^*}{\partial t^*} - V^* \frac{\partial V^*}{\partial x^*} - \frac{V^*}{2y^*} \tau^{*a-1} + \frac{\partial y^*}{\partial x^*} = P \left(1 - \frac{V^{*2}}{y^{*4/3}} \right) \qquad \dots\dots\dots [12.86]$$

in which

$$F_n^2 = \frac{V_n^2}{g y_n} \qquad \dots\dots\dots\dots\dots\dots\dots\dots\dots\dots\dots\dots\dots\dots [12.87]$$

and

$$P = \frac{S_0}{y_n/X} \qquad \dots\dots\dots\dots\dots\dots\dots\dots\dots\dots\dots\dots\dots\dots [12.88]$$

The upstream boundary condition during advance is

$$V^* y^* = 1 \qquad \dots\dots\dots\dots\dots\dots\dots\dots\dots\dots\dots\dots\dots\dots [12.89]$$

Thus, the solution of the system equation [12.85-89] depends wholly on the values of the three independent parameters a, P, and F_n.

An examination of equation [12.86] suggests that at the low characteristic Froude numbers encountered in border irrigation F_n^2 will be very small and, as a consequence, the effect of all the inertial terms will be negligible, i.e., that the left side of equation [12.86], except for the depth-gradient term, can be discarded. Fig. 12.12a and 12.12b illustrate the effect of Froude number on the advance curve for two different values of P, the

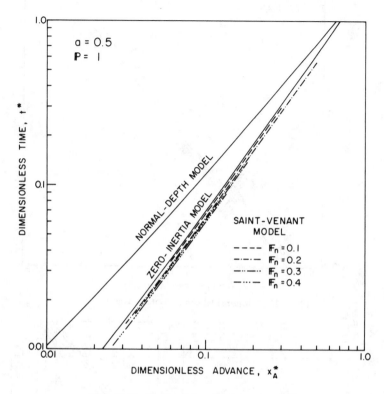

FIG. 12.12a Intermodel comparisons—small P.

parameter that embodies the relative effect of bottom slope. Inasmuch as normal Froude numbers in border irrigation typically run below about 0.2, these graphs can be viewed as justifying, on quantitative grounds, the zero-inertia model.

Note, in passing, the increasing exactness of the normal-depth model with increasing P and x*. This matter is discussed in detail in Katopodes and Strelkoff (1977b), which is also the source of the figures in this section. In particular, Fig. 12.13a-c show at various values of a how advance approaches normal-depth advance, as P and x* become large.

The use of these curves (of which a full set $0.1 \leqslant a \leqslant 0.9$ at increments of 0.1, appear in the aforementioned paper) is shown by means of the following example, for the conditions of test run ID-NB-12AS-12-8-60 discussed in the previous section. In this case

a   = 0.67
P   = 0.069 (equation [12.88])
T   = 563 min(equation [12.82])
X   = 439 m (1440 ft) (equation [12.89])
Y_n = 0.3253 m (1.067 ft)

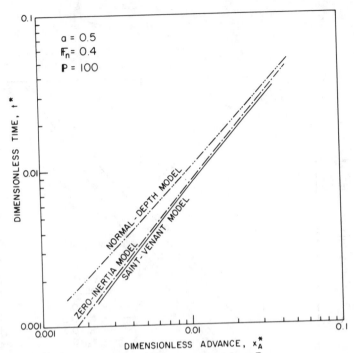

FIG. 12.12b Intermodel comparisons—large P.

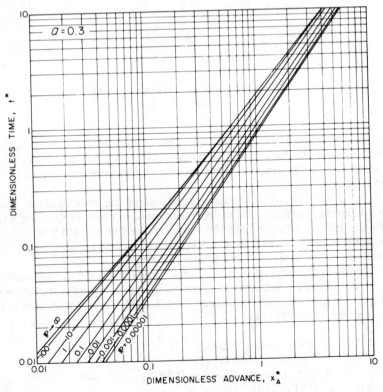

FIG. 12.13a Dimensionless advance—a = 0.3.

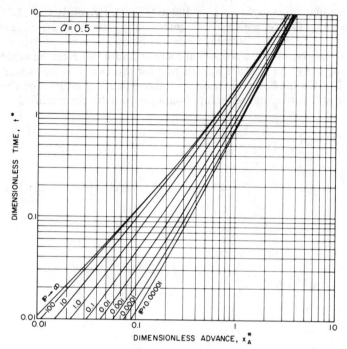

FIG. 12.13b Dimensionless advance—a = 0.5.

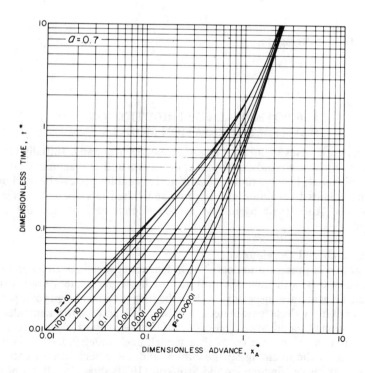

FIG. 12.13c Dimensionless advance—a = 0.7.

Assume that t_L is sought for L $=$ 100 m (328 ft). Evidently $x_A^* = L/X = 0.228$. From Fig. 12.13c, with a $=$ 0.7, interpolating for the given value of P, one finds $t_L^* \cong 0.1$, and $t_L \cong 56$ min. Since this figure corresponds to a $=$ 0.7 rather than the given 0.67, a somewhat more accurate value can be achieved by modifying the given k and a information for this run to values which include a $=$ 0.7. On the basis that the accumulated depth of infiltration should be the same at t_L with either set of input data, we solve for the modified k, k_m, as follows

$$k_m t_L^{a_m} = k\, t_L^{a} \quad\dots\dots\dots\dots\dots\dots\dots\dots\dots\dots\dots\dots \text{[12.90]}$$

$$k_m = 4.67(56)^{0.67 - 0.70} = 4.14 \text{ mm/min}^a = 0.236 \text{ mm/s}^a \quad\dots\dots \text{[12.91]}$$

Then

$$T_m = \left(\frac{0.3253}{0.236 \times 10^{-3}} \right)^{\frac{1}{0.7}} = 3.054 \times 10^4 \text{ s} = 509 \text{ min}$$

$X \ = 397$ m

$x_A^* = 0.252$

$t_L^* = 0.12$

$t_L \ = 61$ min

a more nearly correct figure.

12.6 POTENTIAL CONTRIBUTION OF SURFACE-IRRIGATION MODELS

The potential uses of surface irrigation models are many. Initially they will be used in research and development on design and management practices. As the models become more versatile and available they can be used by trained individuals to evaluate and improve individual farm or field irrigation practices. Also, their usefulness as educational tools should not be overlooked.

12.6.1 Research and Development

Evaluation of present practices: The independent variables Q, S_0, n, infiltration constants, and time of cutoff are the basic parameters needed in any model. They are also necessary to describe irrigation conditions of a particular field. Given these parameters, a model will predict the dependent parameters, including advance, recession, infiltration, runoff, deep percolation, efficiency, and uniformity. With a model, each independent variable can be varied without changing the others and the effects on irrigation studied. For example, Fangmeier and Strelkoff (1978) show the effects of n

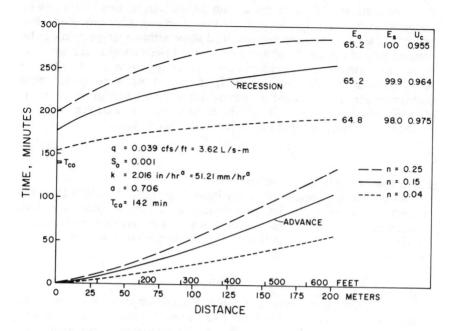

FIG. 12.14 Comparison of computed advance, recession, application efficiency E_a, storage efficiency E_s, and uniformity coefficient U_c, for different values of roughness n.

on advance, recession, efficiency, and uniformity (Fig. 12.14). As expected, advance and recession curves are quite different for each n value, but the effect on efficiency and uniformity is surprisingly small. They also demonstrate the usefulness of models by analyzing the effect of infiltration on efficiency and uniformity, and be presenting curves for obtaining recession-lag time. It would be impractical and perhaps impossible to achieve this with field experiments, that would in any case cost considerably more. Results of such studies would produce many pages of data; however, this can be greatly reduced by using dimensionless parameters. The models can be developed in dimensionless terms and directly provide the dimensionless data. Once obtained, recalculation would not be required and an irrigation designer would enter the data in dimensionless terms to design a system.

Until such results are available, the designer can continue to use current procedures. However, the models can show the limitations of these procedures and so be of great benefit. Current design criteria were developed using certain specified assumptions. The models can determine their range of validity, and predict their accuracy and maximum irrigation performance attainable with a given set of conditions. This alone would be a valuable contribution of the models, and this potential is shown by Fangmeier and Strelkoff (1978) in an analysis of design charts prepared by the U. S. Soil Conservation Service (1974).

Evaluation of experimental concepts: New concepts are continually being suggested to improve irrigation system performance. The potential of these concepts as well as others already in use can in some instances be evaluated with models less expensively than with any other procedure. For example, the result of using variable land slope within a single field can be determined and information gained on how the slope should vary for maximum benefit much less expensively than if field tests were made to get the same information. Variable inflow is used and when controlled with microprocessors may be used to achieve automation and high efficiencies. Models could also be used to study the effect of using various analytical expressions for infiltration and for surface resistance to flow.

12.6.2 Field Irrigation Evaluation

As the models become more available, a potential benefit will be the evaluation of given fields. The performance of current practice on a given field can be shown. The model can then show the best possible irrigation performance for those field conditions and changes needed to achieve that performance.

12.7 SUMMARY

Surface irrigation methods of applying irrigation water are widely used in many parts of the world and will likely remain important as competing demands for energy discourage the use of energy-intensive alternative methods. Competing demands for both energy and water are already encouraging efforts to improve the water use efficiency obtainable with surface irrigation. The ability to analyze and to synthesize the irrigation processes from theoretical considerations without incurring the high cost in money and time required in field trials should provide an important new tool for those seeking to improve the performance of surface irrigation systems.

This chapter contains a summary of recent thought and technical developments regarding the hydraulics of surface irrigation. This summary is prepared to help students, investigators and designers become familiar with some theoretical aspects of the subject and with recent efforts to model the irrigation process.

The chapter begins with a discussion of certain basic concepts in surface irrigation. This is followed by an explanation of the independent variables involved and the information the investigator must provide. The Saint-Venant equations which describe the flow are then presented, together with three successively simpler sets of approximations to the Saint-Venant equations. Recent attempts to model the irrigation process using the four sets of equations are summarized. Suggestions are made regarding the appropriate application of each of the four sets of equations and their respective models. Sample results obtained with each model are presented, demonstrating that current models, though still imperfect, are capable of simulating actual irrigations with accuracy sufficient for most field applications. Finally, suggestions are made regarding some of the many potentially useful applications of operational surface irrigation models.

At the present time the zero-inertia model for border irrigation has had more use than the other three models, appears sufficiently reliable for field applications, is capable of treating a wide range of conditions, and operates at reasonable cost. A similar model for furrow irrigation is now under development.

References

1 Bassett, D. L. 1972. A mathematical model of water advance in border irrigation. TRANSACTIONS of the ASAE 15(5):992-995.

2 Bassett, D. L. 1973. A dynamic model of overland flow in border irrigation. Dissertation, University of Idaho, Moscow. 147 p.

3 Bassett, D. L. and D. W. Fitzsimmons. 1976. Simulating overland flow in border irrigation. TRANSACTIONS of the ASAE 19(4):674-680.

4 Bondurant, J. 1971. Annual Rep. to Reg. Res. Con. W-65 on Hydraulics of Surface Irrigation. U. S. Dep. Agr. Res. Serv., Twin Falls, Idaho.

5 Collins, D. L. and D. L. Bassett. 1964. Terminal profile of water flowing over a porous bed having constant infiltration. TRANSACTIONS of the ASAE 7(4):475-477.

6 Cunge, J. A. and D. A. Woolhiser. 1975. Irrigation systems. Chp. 13, In: Unsteady Flow in Open Channels. Mahmood and Yevjevich (Ed.), Water Resour. Publ., Vol. 2, Fort Collins, Colorado.

7 Escoffier, F. F. and M. B. Boyd. 1962. Stability aspects of flow in open channels. Proc. Am. Soc. Civ. Engr., Hydr. Div. J. 88(HY-6):145-166.

8 Fangmeier, D. D. and T. Strelkoff. 1978. Mathematical models and border irrigation design. TRANSACTIONS of the ASAE 22(1):93-99.

9 Gerald, C. F. 1970. Applied numerical analysis. Addison-Wesley, Reading, Mass., 340 p.

10 Hall, W. A. 1956. Estimating irrigation border flow. Agricultural Engineering 37:263-265.

11 Hart, W. E., D. L. Bassett and T. Strelkoff. 1968. Surface irrigation hydraulics-kinematics. Proc. Am. Soc. Civ. Engr., Irrig. and Drain. Div. J. 94(IR-4):419-440.

12 Henderson, F. M. 1966. Open Channel Flow. Macmillan, New York, 522 p.

13 Irrigation Journal. 1979.29(6):58H.

14 Katopodes, N. D. and T. Strelkoff. 1977a. Hydrodynamics of border irrigation—complete model. Proc. Am. Soc. Civ. Engr., Irrig. and Drain, Div. J., 103(IR-3):309-324.

15 Katopodes, N. D. and T. Strelkoff. 1977b. Dimensionless solutions for border irrigation advance. Proc. Am. Soc. Civ. Engr., Irrig. and Drain. Div. J. 103(IR-4):401-417.

16 Kincaid, D. C., D. F. Heermann and E. G. Kruse. 1972. Hydrodynamics of border irrigation advance. TRANSACTIONS of the ASAE 15(4):674-680.

17 Liggett, J. A. and J. A. Cunge. 1975. Numerical methods of solution of the unsteady flow equations. Chp. 4, In: Unsteady flow in open channels. Mahmood and Yevjevich, (Ed.), Water Resour. Publ., Vol. 1, Fort Collins, Colorado.

18 Lighthill, M. J. and G. B. Whitham. 1955. On kinematic waves. I. Flood movement in log rivers. Proc. Royal Soc. of London, Series A, 229:281-316.

19 Roth, R. L., D. W. Fonken, D. D. Fangmeier and K. T. Atchison. 1974. Data for border irrigation models. TRANSACTIONS of the ASAE 17(1):157-161.

20 Sakkas, J. G. and T. Strelkoff. 1974. Hydrodynamics of surface irrigation—advance phase. Proc. Am. Soc. Civ. Engr., Irrig. and Drain. Div. J. 100(IR-1):31-48.

21 Salt River Project. 1960. Major facts in brief. Phoenix, Arizona. 33 p.

22 Schreiber, D. L. and D. L. Bassett. 1967. Hydraulic description of recession of shallow flow over a porous bed. TRANSACTIONS of the ASAE 10(1):54-61.

23 Smith, R. E. 1972. Border irrigation advance and ephemeral flood waves. Proc. Am. Soc. Civ. Engr., Irrig. and Drain. Div. J. 98(IR-2):289-307.

24 Stoker, J. J. 1953. Numerical solution of flood prediction and river regulation problems. IMM-NYU 200, New York, University.

25 Strelkoff, T. 1969. One dimensional equations of open channel flow. Proc. Am. Soc. Civ. Engr., Hydr. Div. J. 95(HY-3):861-876.

26 Strelkoff, T. 1970. Numerical solution of the Saint-Venant equations. Proc. Am. Soc. Civ. Engr., Hydr. Div. j. 96(HY-1):223-252.

27 Strelkoff, T. 1977. Algebraic computation of flow in border irrigation. Proc. Am. Soc. Civ. Engr., Irrig. and Drain. Div. J. 103(IR-3):357-377.

28 Strelkoff, T. and N. D. Katopodes. 1977a. End depth under zero-inertia conditions. Proc. Am. Soc. Civ. Engr., Hydr. Div. J. 103(HY-7):699-711.

29 Strelkoff, T. and N. D. Katopodes. 1977b. Border irrigation hydraulics with zero inertia. Proc. Am. Soc. Civ. Engr., Irrig. and Drain. Div. J. 103(IR-3):325-342.

30 Strelkoff, T., D. Schamber and N. D. Katopodes. 1977. Comparative analysis routing techniques for the flood wave from a ruptured dam. Proc. Symp. on Dam-Break Flood Routing Models, Bethesda, Maryland, p. 228-291.

31 Thorne, D. W. 1951. The desert shall blosson as a rose. Tenth Annual Faculty Res. Lect., Faculty Assoc., Utah State Agr. Coll., Logan, Utah.

32 U. S. Soil Conservation Service. 1974. Border irrigation. Chp. 4., Sect. 15, Nat. Engineering Handbook.

chapter 13

DESIGN AND OPERATION OF GRAVITY OR SURFACE SYSTEMS

13

13

DESIGN AND OPERATION OF GRAVITY OR SURFACE SYSTEMS

by   W. E. Hart, Wastequip, Inc., Davis, CA; H. G. Collins, SCS, USDA (retired), Portland, OR; G. Woodward, SCS, USDA (deceased), Lincoln, NE; and A. S. Humpherys, SEA-AR, USDA, Kimberly, ID.

13.1 BASIC CONSIDERATIONS

13.1.1 General Description

Surface irrigation uses open channel flow to spread water over a field. The driving force in such systems is gravity and hence the alternate term, gravity flooding. Once distributed over the surface of the field and after it has entered the soil, water is often redistributed by forces other than gravity.

Surface irrigation systems generally require a smaller initial investment than do other types of irrigation systems. However, this is not always the case, especially if extensive land forming is needed for an efficient system. In fact, the need for extensive land forming is one of the main reasons why other types of irrigation systems have been developed.

13.1.2 Types

Prior to the early 1900's, most irrigation systems were of the surface type. Initially, water was allowed merely to spill over the banks of rivers and flood adjacent lands. The resulting distribution of water was usually quite nonuniform. This technique is often called uncontrolled flooding. With slight modifications, it has evolved into the technique known as water spreading. A further refinement, in which man-made ditches carry water along the top edges of fields or strips, is known as contour ditch irrigation or wild flooding.

All other methods of surface irrigation can be classified as controlled flooding. The water is guided down an irrigation slope by channels, which may be as wide as 30 meters or as narrow as several millimeters, or is allowed to flood an essentially level area surrounded by dikes. The major types of such systems are discussed in subsequent sections of this chapter. In general, as the level of sophistication increases, so do initial costs and the potential uniformity of water distribution.

13.1.3 Required Design Variables

Depth of water to be applied. The most important design variable is the depth of water to be applied at each irrigation. This is generally given as an average depth for each field even though the soil-water reservoir may not have been uniformly depleted throughout the field, and the water will not be distributed uniformly over the soil as a result of the irrigation. Most surface irrigation systems in arid and semi-arid areas are designed to raise the soil water content of the root zone to its field capacity even though water may be wasted. This is done in order to utilize water supplies when available, and to reduce the total number of irrigations and hence to also reduce labor. With furrow systems especially, it is sometimes desirable and possible to only partially refill the root zone. In many irrigated areas it is sometimes necessary to apply "excess" water in order to leach out undesirable salts.

Hydraulic variables. As described in the previous chapter, surface irrigation design is a problem in unsteady, nonuniform flow. The main design variables include the field slope and roughness, both of which may vary within a field. Another consideration is the erosiveness of the soil, which will limit maximum inflow rates to a field.

Topographic and related information. The topography of a field limits the types of systems which can be used. Those which have rolling terrain, irrigular shapes and shallow soils may be impractical to irrigate with surface systems. If surface systems are used, they will usually be of the non-sophisticated types with relatively low efficiency and non-uniform water distribution when measured on a field basis. On the other hand, flat terrain, fields of regular shapes and deep soils may be adaptable to a wide range of systems, all of which have the potential for high efficiency and uniform water distribution. Thus, a site under consideration for surface irrigation must be mapped to show field boundaries, land elevations and soil depths and textures (also see Chapter 8). Such information will give land slopes and field shapes directly. It will also assist in determining how much water to apply. In some cases minimum intensity mapping is satisfactory, while in other cases detailed mapping is necessary. The latter is usually necessary on relatively flat terrain that will be land-formed prior to irrigating.

Infiltration. The infiltration characteristic of the soil at each irrigation is a primary input variable. It varies with time and space. It is not at all unusual to have 10-fold variations in infiltration rates throughout a field. Such variations can make the design of an efficient irrigation system extremely difficult, if not impossible.

The design procedures of this chapter follow closely those of the Soil Conservation Service (USDA, 1979) and its method of describing water intake by soils is followed. The basis of the SCS design is to classify soils into intake families. The equation of these families is as follows.

$$F = aT^b + c \quad \dots\dots\dots\dots\dots\dots\dots\dots\dots\dots\dots\dots\dots \quad [13.1]$$

where F is the cumulative intake (mm), T is the time water is in contact with the soil (min), and a, b and c are constants unique to each intake family. Values of the constants are given in Table 13.1, and the families are plotted in Fig. 13.1.

The selection of an appropriate intake family is dependent not only upon the soil, but also upon the irrigation method. Selection can only be made by running field tests. For irrigation systems which have infiltration primarily through a relatively flat, horizontal surface (borders, basins, etc.) ring infiltrometer or basin ponding tests can be run. The measured cumulative infiltration is plotted on Fig. 13.1 and the intake family is that of the curve closest to which the points fall. In these systems, F as defined by equation [13.1], is the depth of infiltrated water.

TABLE 13.1 INTAKE FAMILY AND FURROW ADVANCE COEFFICIENTS

Intake family	a	b	c	f	g
0.05	0.5334	0.618	7.0	7.16	1.088×10^{-4}
0.10	0.6198	0.661	7.0	7.25	1.251×10^{-4}
0.15	0.7110	0.683	7.0	7.34	1.414×10^{-4}
0.20	0.7772	0.699	7.0	7.43	1.578×10^{-4}
0.25	0.8534	0.711	7.0	7.52	1.741×10^{-4}
0.30	0.9246	0.720	7.0	7.61	1.904×10^{-4}
0.35	0.9957	0.729	7.0	7.70	2.067×10^{-4}
0.40	1.064	0.736	7.0	7.79	2.230×10^{-4}
0.45	1.130	0.742	7.0	7.88	2.393×10^{-4}
0.50	1.196	0.748	7.0	7.97	2.556×10^{-4}
0.60	1.321	0.757	7.0	8.15	2.883×10^{-4}
0.70	1.443	0.766	7.0	8.33	3.209×10^{-4}
0.80	1.560	0.773	7.0	8.50	3.535×10^{-4}
0.90	1.674	0.779	7.0	8.68	3.862×10^{-4}
1.00	1.786	0.785	7.0	8.86	4.188×10^{-4}
1.50	2.284	0.799	7.0	9.76	5.819×10^{-4}
2.00	2.753	0.808	7.0	10.65	7.451×10^{-4}

<u>Intake</u> (see equations [13.1] and [13.40]) <u>Advance</u> (see equation [13.35])

$$F = (aT^b + c)\, P/W, \text{ mm}$$

$$T_T = \frac{x}{f}\, e^{(gx/QS^{1/2})}, \text{ min}$$

T = minutes

Q = furrow inflow

$$\frac{P}{W} = \frac{\text{Wetted perimeter}}{\text{Furrow spacing}}$$

S = furrow slope

x = distance

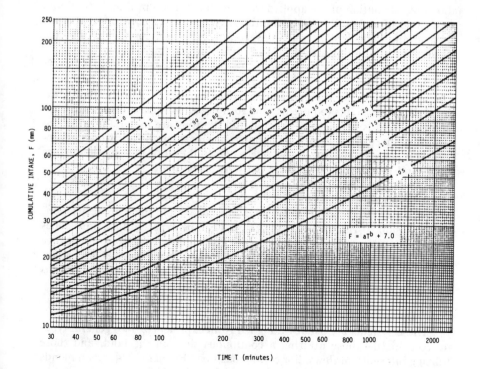

FIG. 13.1 Intake families (USDA, 1979).

Furrow systems require a different approach. An inflow-outflow test is run, and the family is selected using a calculation procedure described in Section 13.6. For furrows, F as defined in equation [13.1] is *not* the depth of infiltration. Rather, the right hand side of that equation must be multiplied by a factor which accounts for the wetted perimeter of the furrow and the furrow spacing. This multiplicative factor is explained in Section 13.6. It is emphasized that a given soil will, in general, be classified into different families for basin type systems and furrow type systems.

13.2 CONTOUR DITCH IRRIGATION

13.2.1 Description

This system, also known as wild flooding, consists of a series of ditches spaced 25 to 100 m apart. These head ditches have a slight slope. The water is removed from the ditches either by making the downslope bank low enough so that the water is not restrained by it when checked, or by making cuts through the downslope bank. In theory the water moves down the slope as a sheet, but in practice it may not. Rather, the water channels and is nonuniformly distributed. In some cases siphons are used for removing water from the ditch, but this is not usual, especially in areas where the land is marginal and the crops grown have a low cash value.

13.2.2 Application

The system is applicable to slopes of 0.5 to 15 percent. It is particularly adaptable to residual soils in foothill areas that have an underlying permeable layer at a rather shallow depth, 0.3 to 0.6 m. This condition allows redistribution of the applied water within the soil profile. It is seldom used on deep sandy soils with high infiltration rates or on clay soils that crack upon drying. The USDA Soil Conservation Service recommends that the system be restricted to soils in the 0.1 to 3.0 intake families.

13.2.3 Advantages

The system has one particular advantage—it is low in first cost. Generally, the systems require no land forming and unlined ditches are common. If a field is on the side of a hill and the soils are underlain by an impermeable layer, then water application efficiencies can be fairly high in properly laid out systems. Successive ditches down the slope pick up the surface runoff and redistribute it to lower portions of the field.

13.2.4 Limitations

Although the range of slopes given was fairly high, there are restrictions due to erosion. If runoff-producing rainfall can be expected then only slopes of 4 percent or less should be considered. Extremely erosive soils will also be carried away by the irrigation practice unless sod-forming crops are grown. High water application efficiencies require frequent short sets which means high labor or an increased cost for automated equipment.

13.2.5 Design

These systems are usually designed from experience. Two general types are recognized—those utilizing a continuous flow on the land, and those utilizing intermittent flows. For continuous flow, Booher (1974) recommends a flow of about 0.7 to 1 L/s per ha (4.5 to 6.5 U.S. gpm/ac). This is equivalent to about 7.4 mm/d (0.29 in./d). For intermittent flow systems, the Soil Conservation Service (USDA, 1967) recommends the use of about 0.028 m³/s (1 cfs), per 30 m (100 ft) of strip width irrigated.

The distance between head ditches, or between a head ditch and the first pickup ditch, would be about 24 to 90 m (80 to 300 ft), varying with the topographic uniformity.

The field to be irrigated usually receives two cultivation treatments other than that of seedbed preparation. Where soils are erosive, seedings are made without cultivation. If knolls and depressions are a problem, and if soil depths are adequate, slight land forming might be advisable. Alternatively, corrugations (Section 13.6) can run directly down the slope.

13.2.6 Headland Facilities

Earthen head ditches are usually used, although concrete could be used if soils are highly erosive or seepage is a problem. Pickup ditches for redistributing water down the slope are usually earthen.

Outlets from earthen head ditches are often mere cuts in the bank, stabilized by sod or rocks, or spiles can be used. Siphons may be used in either earthen or concrete lined ditches. Cast outlets, with grooves for flashboards, can be installed in concrete ditches. Finally, the ditch can be formed so that there will be discharge over the downslope bank when checks (canvas, plastic, or structural) are activated.

Ditch sizing should be in accordance with the recommendations of Sections 11.2 and 11.3. However, when overlfow from the lower bank is planned, this bank must be on a grade to allow approximately equal overflow at each point along it. This grade must usually be determined by trial-and-error, which often requires that ditches be moved. Common grades on checked ditches are 0.0005 to 0.001 m/m.

When multiple cuts or spiles are used, they should be spaced at intervals of approximately 2 to 3 m (6.5 to 10 ft), according to Booher (1974). If desired, the system can be automated and designed according to the techniques outlined in Section 13.9.

13.3 BASIN IRRIGATION

13.3.1 Description

The field to be irrigated by the basin method is divided into level rectangular areas bounded by dikes or ridges. Water is turned in at one or more points until the desired gross volume has been applied to the area. The flow rate must be large enough to cover the entire basin in approximately 60 to 75 percent of the time required for the soil to absorb the desired amount of water. Water is ponded until infiltrated.

13.3.2 Applicability

Most crops can be irrigated with basin irrigation. It is widely used for close-growing crops such as alfalfa and other legumes, grasses, small grains, mint, and rice. It is used for row crops that can withstand some inundation, such as sugar beets, corn, grain sorghum, and cotton, and for other row crops if they are planted on beds so they will be above the water level. It also is well suited to the irrigation of tree crops, grapes, and berries.

This irrigated method is best suited to soils of moderate to low intake rate (50 mm/h or less). It is an excellent way of applying water to soils that have a moderately high to high intake rate, but basin areas may need to be very small.

Basin irrigation is best suited to smooth, gentle, uniform land slopes. Undulating or steep slopes can be prepared for basin irrigation, provided the soils are deep enough to permit needed land leveling.

13.3.3 Advantages

High application efficiency can be obtained easily with little labor. Basin irrigation can be used efficiently by inexperienced workers, and can easily be automated. When basins are leveled with laser-controlled scrapers, basins can be as large as 16 ha (Erie and Dedrick, 1979). Many different kinds of crops can be grown in sequence without major changes in design, layout, or operating procedures. There is no irrigation runoff, there is little deep percolation if no excess is applied, and maximum use can be made of rainfall. Leaching is easy and can be done without changing either the layout or operation method.

13.3.4 Limitations

Accurate initial land leveling is essential and level surfaces must be maintained. Adequate basin ridge height may be difficult to maintain on sandy soils or fine-textured soils that crust or crack when dry. Prolonged ponding and crop scalding can occur if the system is poorly managed. In some areas special provisions must be made for surface drainage.

Drop structures, lined ditches, or pipelines may be required to control water on steep slopes that require benching. Relatively large inflow rates are needed for basins and special structures may be needed to prevent erosion.

13.3.5 Design

Water should be applied at a rate that will advance over the basin in a fraction of the infiltration time to achieve high efficiency. The volume of water applied must equal the average gross irrigation application. The intake opportunity time at all points in the basin must be greater than or equal to the time required for the net irrigation to enter the soil. The longest intake opportunity time at any point on the basin area must be sufficiently short to avoid scalding and excessive deep percolation. The depth of water flow must be contained by the basin ridges.

Design limitations. In theory, maximum depth of flow and maximum deep percolation both occur where water is introduced into a basin, usually considered as a "strip" of unit width for computational purposes. For any given set of site conditions, the depth of flow varies directly and the amount of deep percolation varies inversely with the inflow rate per unit width of basin strip. Thus, if a limit is set on flow depth, deep percolation may be reduced only by shortening the length of the basin strip. If limits are established for both depth of flow and deep percolation, then the design limit for length is determined.

Flow at the head end of basin strips must not exceed some practical depth related to the construction and maintenance of basin ridges.

The average deep percolation (the difference between the net and gross irrigation applications) should be minimized. On some sites excess deep percolation causes acute drainage problems. To avoid this condition, the design efficiency usually should not be less than about 80 percent. This efficiency can be obtained if the time required to cover the basin is not more than 60 percent of the time required for the net application to enter the soil. A design efficiency of less than 70 percent should be considered only for soils having excellent internal drainage. On sites where irrigation water supplies are limited or costly, where subsurface drainage problems are acute, or where crops can be damaged by prolonged surface flooding, design efficiencies in excess of 90 percent are often practical. These efficiencies are easily obtained when laser-controlled scrapers are used (Erie and Dedrick, 1979).

Basin strips usually are designed to be level; however, they may be constructed with a slight grade in the direction of water flow. A slight grade will minimize adverse effects of variations in the finished land surface, such as low areas or reverse grades, which result in a slower rate of advance, reduced efficiency, excessive deep percolation or prolonged flooding that may damage crops. The total fall in the length of the basin strips should not be greater than one-half the net depth of application used as a basis for design. No adjustment is made in the design to compensate for such slight grades.

Drainage facilities may be needed to remove excess water from basins resulting from an accidental overirrigation or heavy rainfall. Large furrows formed when constructing basin ridges facilitate removal of excess rainfall or irrigaton water. They also speed the water coverage rate over the basin and reduce flow depths and deep percolation adjacent to the point or points of water inflow. Surface drainage facilities should be provided for basins on low intake soils, and, in high rainfall areas, on moderate intake soils.

Basin ridges, or levees, should be constructed so that the top width is at least as great as the ridge height. The settled height should be at least equal to the greater of (a) the design gross depth of application, or (b) the design maximum depth of flow plus a freeboard of 25 percent of the maximum depth of flow.

Method of the Soil Conservation Service (USDA, 1974). Design equations are based on equating (a) the volume of water applied to a unit width basin strip during the time period of water advance from the head to the end of the strip, and (b) the volume of intake plus the water in temporary surface storage during the same period.

The designer must know the cumulative intake characteristic of the soil, must select a Manning roughness coefficient (n) appropriate for the crops to be irrigated, and must select the net application depth to be used as a basis for design.

Opportunity Time—The opportunity time required for intake of the selected net application depth can be estimated by solution of the cumulative intake equation in the form

$$T_n = [(F_n - c)/a]^{1/b} \qquad \dots\dots\dots\dots\dots\dots\dots\dots\dots\dots\dots\dots \quad [13.2]$$

where T_n is the time required, also called the net opportunity time (min), and F_n is the desired net application depth (mm).

Advance Time—The time required for the unit inflow rate to advance to the far end of the strip is called the advance time, T_t (min). The required advance time for any desired water application efficiency is determined by multiplying the net opportunity time, T_n, by the efficiency advance ratio, R (Table 13.2).

Water application efficiency is defined as the ratio of average net ap-

**TABLE 13.2 EFFICIENCY AS A
FUNCTION OF THE
EFFICIENCY ADVANCE RATIO
R(R = T_t/T_n)**

Efficiency, E	Efficiency advance ratio R(R = T_t/T_n)
Percent	—
95	0.16
90	0.28
85	0.40
80	0.58
75	0.80
70	1.08
65	1.45
60	1.90
55	2.45
50	3.20

plication (F_n) to gross application (F_g) expressed as percent.

Basin Length and Inflow Rate—The following mass balance equation can be used to estimate length of the basin strip as a function of unit inflow rate (Q_u) and advance time (T_t).

$$L = \frac{6 \times 10^4 \; Q_u \; T_t}{\dfrac{a \, T_t^{\,b}}{1 + b} + 7.0 + 1798 \, n^{3/8} \, Q_u^{\,9/16} \, T_t^{\,3/16}} \quad \ldots\ldots\ldots\ldots\ldots [13.3]$$

where L is the length (m); Q_u is the unit inflow rate (m²/s); T_t is the required advance time for the desired efficiency (min); a, b, and c are constants in the cumulative intake equation; and n is Manning's coefficient.

The design length of the basin strip can be found for any selected inflow rate, efficiency and associated required advance time, by direct solution of equation [13.3]. A similar solution for the unit inflow rate needed for a selected length and efficiency is not possible. A trial and error procedure must be used.

Inflow Time—The inflow time, the time required to apply the gross application onto the basin strip, can be computed from equation [13.4].

$$T_a = \frac{F_n \, L}{600 \, Q_u \, E} \quad \ldots\ldots\ldots\ldots\ldots\ldots\ldots\ldots\ldots\ldots\ldots\ldots\ldots\ldots\ldots [13.4]$$

where T_a is the inflow time (min) for the unit inflow rate Q_u (m²/s), to apply the net application depth F_n (mm) on a basin strip of length L (m), at an efficiency E (percent).

Maximum Depth of Flow—The maximum depth of flow can be estimated from equation [13.5].

$$d = 2250 \, n^{3/8} \, Q_u^{\,9/16} \, T_a^{\,3/16} \quad \ldots\ldots\ldots\ldots\ldots\ldots\ldots\ldots\ldots\ldots\ldots [13.5]$$

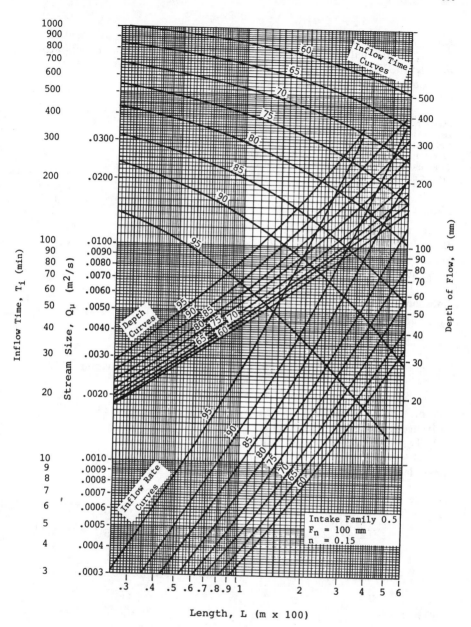

FIG. 13.2 Level basin irrigation sample design (USDA, 1974).

where d is the flow depth at the inlet end of the basin strip (mm). If advance time, T_t is greater than T_a, use T_t in equation [13.5] in place of T_a. The inflow rate for a given maximum depth of flow connot be determined directly. A trial-and-error procedure must be used.

Design Charts—The procedure for design of basins may be simplified by preparation of design charts. A separate chart can be prepared for any combination of Manning coefficient, cumulative intake relationship, and net application depth. Each such chart would describe the relationship between length, time of inflow, inflow rate, and depth of flow for any assumed efficiency. An alternative chart would be to exchange efficiency and net application depth. Fig. 13.1 is a sample design chart.

Computation Example—

Given:

Intake family	0.5
Desired efficiency, E	80 percent
Unit/inflow rate, Q_u	0.005 m^2/s
Maximum depth of flow, d	150 mm
Desired depth of application, F_n	100 mm
Manning roughness coefficient, n	0.15

Find:

Opportunity time required, T_n
Basin length, L
Inflow time required, T_a
Maximum depth of flow, d

Solution:

Opportunity time:

$$T_n = \left(\frac{100 - 7.0}{1.196}\right)^{1/0.748} = 337 \text{ min} \quad \ldots\ldots\ldots\ldots\ldots \text{(equation [13.2]}$$

Ratio, T_t/T_n at 80 percent efficiency = 0.58, from Table 13.2.
Advance time:
$$T_t = (0.58)(337) = 195 \text{ min}$$

Basin length:

$$L = \frac{(6 \times 10^4)(0.005)(195)}{\dfrac{(1.196)(195)^{0.748}}{1 + 0.748} + 7.0 \quad + \quad (1798)(0.15)^{3/8}(0.005)^{9/16}(195)^{3/16}}$$

$$= 359 \text{ m} \ldots\ldots\ldots\ldots\ldots\ldots\ldots\ldots\ldots\ldots\ldots\ldots \text{(equation [13.3])}$$

Inflow time:

$$T_a = \frac{(100)(359)}{(600)(0.005)(80)} = 150 \text{ min} \quad \ldots\ldots\ldots\ldots\ldots \text{(equation [13.4])}$$

Maximum depth of flow:

$$d = 2250(0.15)^{3/8}(0.005)^{9/16}(150)^{3/16} \quad \ldots\ldots\ldots\ldots$$

$$= 144 \text{ mm} \quad \ldots\ldots\ldots\ldots\ldots\ldots\ldots\ldots\ldots\ldots\ldots \text{(equation [13.5])}$$

TABLE 13.3 SUGGESTED BASIN AREAS
FOR DIFFERENT SOIL TYPES AND
RATES OF WATER FLOW (ADAPTED
FROM BOOHER, 1974)

Soil type			
Sand	Sandy loam	Clay loam	Clay
	Unit areas, ha/100 L/s		
0.067	0.20	0.40	0.67

Empirical method of Booher (1974). Basin sizes are suggested for various soil types and inflow rates, as shown in Table 13.3. The areas are approximations and the table is intended to be used as a guide only.

Soils with high infiltration rates, such as sands, require limited basin size even when large flows of water are available. Basins on clay soils can be large or small, depending on the water inflow rate. The objective in selecting the basin size is to be able to flood the entire area in a reasonable length of time so that the desired depth of water can be applied with a high degree of uniformity over the entire basin.

13.3.6 Selection of Headland Facilities

Water may be conveyed to irrigated basins in lined or unlined ditches, or pipelines installed above or below the ground surface. Adequate structures should be provided in the delivery system to permit control and regulation of the water flow. Such structures include checks, checkdrops, valves, or gates. Measuring structures to determine the delivery flow rate are essential for good irrigation management.

Supply ditches. Supply ditches must convey the design inflow rate of each basin, or multiples of the design flow rate where more than one basin is irrigated simultaneously. The water surface in the ditch should be 0.15 to 0.30 m above the ground surface level in the basin, depending on outlet characteristics. Where possible, the ditches should be constructed with a 0.1 percent grade or less to minimize the number of checks and the labor required.

Ditches may be designed and constructed with the water surface below the ground surface where portable pumps, usually of the low head-high volume propeller type, are used to convey water into basins.

Supply ditch outlets. Outlets to release water from ditches into basins may be of several types. Gated rectangular or trapezoidal outlets installed in the side of the ditch are commonly used where the entire ditch flow is discharged into one basin. Gated orifice-type outlets are desirable when more than one basin is being supplied simultaneously from the ditch, to minimize the effect of pressure head differentials on the discharge rate. Outlet gates are not required when the base of the outlet is located at or above the normal water surface elevation in the ditch. Water flows through the outlet into the basin when the water surface in the ditch is raised by regulating control structures in the ditch.

Larger diameter siphon tubes may be used to convey water from the ditch into the basin. Capacity varies with the tube size and the differential head between the water surface in the ditch and the basin. Adjustable gates on the discharge end and a small vacuum pump (removable) and valve on top facilitate removal of air in the tube and initiating flow.

Pipeline outlets. An outlet structure or hydrant is necessary in pipelines to deliver water to the basins. A valve or gate is installed in the vertical riser attached to the pipeline for underground pipelines, or directly to the pipeline on surface installations, to regulate discharge. "Alfalfa" valves are commonly used, and consist of a plate attached to a threaded rod which moves up or down as the handle is turned to regulate flow.

Erosion. The water flow velocity into the basin should not exceed about one meter per second to avoid formation of scour holes or erosion adjacent to the turnout. Turnout structures should be designed with energy dissipation features to limit the discharge velocity.

Outlet location. The number and location of outlets to discharge water into basins varies with the rate of flow required and the width of the basin. One outlet for basin widths up to 60 meters and flow rates up to 0.4 cubic meters per second is common where the outlet incorporates adequate energy dissipation features. Minimizing the number of outlets reduces labor and facilitates use of automatic controls. Spacing the outlets along the basin width, however, may provide a more uniform distribution of water over the basin at the inlet end.

13.4 BORDER IRRIGATION

13.4.1

The field to be border irrigated is divided into graded strips by constructing parallel dikes or border ridges. The ends of the strips are usually not closed. Water is turned in at the upper end and flows as a sheet down the strip. The flow rate must be such that the desired volume of water is applied to the strip in a time equal to, or slightly less than, that needed for the soil to absorb the net amount required. When the desired volume of water has been delivered to the strip, the inflow is turned off. The water not infiltrated is temporarily stored on the ground surface and moves on down the strip to complete the irrigation. Outflow from the strip may be avoided by closing the lower end and ponding the water on the lower reaches of the strip until infiltrated. The discussion in this section follows closely that of the Soil Conservation Service (USDA, 1974).

13.4.2 Applicability

Crop. Border irrigation is suitable for all close-growing, non-cultivated, sown or drilled crops, except rice and other crops grown in ponded water. Legumes, grasses, small gains, and mint are commonly irrigated by this method. It is also used to irrigate orchards and vineyards.

Soils. Border irrigation can be used on most soils. It is, however, best suited to soils with a moderately low to a moderately high intake rate. It is seldom used on coarse sandy soils of extremely high intake rate as excessive deep percolation occurs unless the strip length is very short. Also, it is not suited for use on soils of extremely low intake rate since, to provide adequate intake time without excessive surface runoff, the inflow rate becomes too small to completely cover the border strip.

Slopes. Border irrigation is best suited to slopes of less than 0.5 percent. It can be used on slopes to 2 percent where non-sod crops are grown, and slopes to 4 percent or steeper where sod crops are grown, providing good crop stands are established by supplementary irrigation methods or dependable rainfall. The erosion hazard created by rainfall runoff must be considered in determining the permissible border slopes.

13.4.3 Advantages

Field application efficiency is good to excellent if the border strips are designed and installed properly and good water management practices are followed. Labor requirements are low, and border strip dimensions can be designed for efficient operation of machinery. Within broad limits, border strips can be designed for efficient operation of machinery and for irrigation grades that minimize land leveling costs. In areas where surface drainage is critical, borders provide an excellent means for removing excess surface water rapidly. On some sites, the ends of borders may be closed to reduce or eliminate surface runoff.

13.4.4 Limitations

The topography and soil profile characteristics must not restrict land leveling necessary to eliminate cross slope within feasible border widths, and the achievement of a uniform border strip slope. Small flow rates necessitated by low soil intake characteristics or steeper field slopes require complete elimination of cross slope within the border.

13.4.5 Design

Border design involves balancing the water advance and recession curves to achieve an equal opportunity time for intake at any point along a border strip. On sites suitable for border irrigation, advance and recession curves will be reasonably well balanced if the following two conditions are met: (a) the volume of water delivered to the border strip is adequate to cover it to an average depth equal to the gross application; and (b) the intake opportunity time at the upper end of the border strip is equal to the time necessary for the soil to absorb the net application desired.

Soil intake characteristics. Border design requires knowledge of the cumulative intake characteristics of the soils to be irrigated. The series of cumulative intake equations discussed in Section 13.1.3 are used in this section.

Manning coefficient of roughness. Hydraulic calculations in border design are based on the Manning equation, which includes a coefficient (n) that expresses the flow-retardance effects of different hydraulic boundary

TABLE 13.4. COMMON RETARDANCE
COEFFICIENT USED IN
BORDER DESIGN

Smooth, bare soil surfaces noncultivated, oil-mulch-treated citrus	0.04
Small grain, drill rows parallel to border strip	0.10
Alfalfa, mint, broadcast small grain, and similar crops	0.15
Dense sod crops, small grain with drill rows across the border strip	0.25

conditions. The coefficient varies with crops, stages of crop growth, and degree of roughness of the soil surface.

Table 13.4 contains values of Manning coefficient commonly used in design.

A conservatively high value of n should be used in determining maximum flow depth, and a conservatively low value of n used when determining minimum flow rate.

Design equations.

Inflow Rate—Inflow rate per unit width of border strip (Q_u, m²/s) can be determined for a given net depth of application from equation [13.6].

$$Q_u = \frac{0.00167\, F_n\, L}{(T_n - T_L)\, E} \quad\dotfill\quad [13.6]$$

where F_n is the desired net application depth (mm); L is the length of the border strip (m); T_n is the opportunity time (min) required for the desired application depth; T_L is the lag time (min) that water remains on the head end of the strip after inflow stops; and E is the water application efficiency (percent), the ratio of the desired net application depth to the gross application depth. As time lag is a function of flow rate, a direct solution is not possible, unless the slope exceeds 0.4 percent and the time lag becomes insignificant. A trial and error solution is required when the slope is less than 0.4 percent.

Lag Time—High Gradient Borders—The depth of flow approaches normal depth at the upper end of the border strip in a relatively short advance period on borders with steep slopes. Lag time may be ignored in determination of inflow rate on borders with slopes steeper than 0.4 percent. Lag time for high gradient borders may be computed from equation [13.7].

$$T_L = \frac{Q_u^{0.2}\, n^{1.2}}{120\, S_o^{1.6}} \quad\dotfill\quad [13.7]$$

where T_L is the lag time (min) at the head end of the strip, n is the Manning coefficient, S_o is the border strip slope (m/m), and Q_u is inflow rate per unit width (m²/s).

Lag Time—Low Gradient Borders—Lag time is significant in border strips with slopes of 0.4 percent or less on which slopes the depth of flow may not reach normal depth. Lag time for such low gradient borders may be computed from equation [13.8].

$$T_L - \frac{n^{1.2}\, Q_u^{0.2}}{120 \left[S_o + \left(\dfrac{0.0094\, n\, Q_u^{0.175}}{T_n^{0.88}\, S_o^{0.5}} \right) \right]^{1.6}} \quad\dotfill\quad [13.8]$$

where T_L is the lag time (min), n is the Manning coefficient, Q_u is the inflow rate per unit width (m²/s), S_o is the border slope (m/m), and T_n is the opportunity time (min) required for the desired application depth. Equation [13.8] has been developed from water surface profile computations utilizing incremental values of flow rate, border slope, Manning coefficient, and depth. Lag time may also be estimated from Table 13.5.

Inflow Time—Inflow time (T_a) can be determined by subtracting the lag time (T_L) for a specific inflow rate (Q_u), border slope (S_o) and Manning n from the opportunity time (T_n) required for intake of the desired application depth (F_n), as expressed by equation [13.9].

$$T_a = T_n - T_L \dotfill [13.9]$$

Design water application efficiency. Design water application efficiency, defined as the ratio of the desired net application depth to the gross application depth, must be selected by the designer based on a particular site under a given set of management conditions. Overestimating the efficiency should be avoided. For a given management level, site conditions have a significant effect on the efficiency achievable in border irrigations. Greater efficiency can be expected on gentle slopes than on steep slopes and on soils that have a moderate to moderately high intake rate, than on soils that have either a low or extremely high intake rate. Table 13.6 shows the efficiencies commonly assigned for designing border irrigation.

TABLE 13.5 RECESSION-LAG TIMES, T_L(min), IN LOW GRADIENT BORDERS

Oppor. time, T_n (min)	Border slope, S_o (m/m) 0.0005 Inflow rate, Q_u (m²/s)				0.001 Inflow rate, Q_u (m²/s)				0.002 Inflow rate, Q_u(m²/s)				0.004 Inflow rate, Q_u (m²/s)			
	0.0001	0.001	0.01	0.02	0.0001	0.001	0.01	0.02	0.0001	0.001	0.01	0.02	0.0001	0.001	0.01	0.02
Manning's n = 0.04																
10	1.9	2.2	2.3	2.3	1.1	1.5	1.9	2.0			1.1	1.1				
25	3.1	4.0	4.8	5.1	1.4	2.0	2.8	3.1			1.2	1.4				
50	3.9	5.4	7.1	7.7	1.6	2.3	3.4	3.8			1.3	1.5				
100	4.4	6.5	9.2	10.1	1.6	2.5	3.8	4.3			1.4	1.6				
200	4.8	7.3	10.8	12.1	1.7	2.6	4.1	4.6			1.4	1.6				
500	5.1	7.9	12.1	13.7	1.7	2.7	4.2	4.9			1.4	1.6				
1000	5.2	8.1	12.6	14.4	1.7	2.7	4.3	4.9			1.4	1.7				
2000	5.2	8.2	12.9	14.8	1.7	2.8	4.4	5.0			1.4	1.7				
Manning's n = 0.15																
10	2.5	2.4	2.2	2.1	2.5	2.7	2.7	2.7	1.6	2.1	2.5	2.6		1.1	1.5	1.6
25	6.1	6.3	6.3	6.2	4.4	5.4	6.2	6.4	2.2	3.0	4.1	4.4		1.3	1.9	2.1
50	10.1	11.6	12.5	12.7	5.7	7.7	9.8	10.4	2.4	3.6	5.1	5.7		1.4	2.1	2.3
100	14.5	18.4	21.9	22.7	6.8	9.7	13.4	14.6	2.6	3.9	5.9	6.6		1.4	2.2	2.5
200	18.4	25.3	32.9	35.2	7.5	11.2	16.3	18.1	2.7	4.2	6.4	7.3		1.4	2.3	2.6
500	22.1	32.5	46.3	51.2	8.1	12.4	18.9	21.4	2.8	4.3	6.8	7.7		1.5	2.3	2.6
1000	23.7	36.0	53.6	60.2	8.3	12.9	20.0	22.8	2.8	4.4	6.9	7.9		1.5	2.3	2.7
2000	24.7	38.2	58.4	66.2	8.4	13.2	20.7	23.6	2.8	4.4	7.0	8.0		1.5	2.3	2.7
Manning's n = 0.25																
10	2.4	2.2	1.9	1.8	2.8	2.8	2.7	2.6	2.2	2.7	2.9	3.0	1.2	1.7	2.1	2.3
25	6.5	6.4	6.0	5.8	5.8	6.6	7.0	7.1	3.4	4.4	5.6	6.0	1.5	2.1	3.0	3.4
50	12.3	13.0	13.1	12.9	8.5	10.6	12.5	12.9	4.1	5.8	7.9	8.5	1.6	2.4	3.5	4.0
100	19.9	23.3	25.6	26.0	10.9	14.9	19.1	20.4	4.5	6.7	9.7	10.8	1.6	2.5	3.9	4.4
200	28.1	36.0	43.5	45.5	12.8	18.5	25.6	28.1	4.8	7.4	11.1	12.4	1.7	2.6	4.1	4.6
500	36.9	52.1	70.5	76.5	14.3	21.7	32.2	36.0	5.0	7.8	12.1	13.8	1.7	2.7	4.2	4.8
1000	41.3	61.2	88.0	97.5	15.0	23.1	35.3	39.9	5.1	8.0	12.5	14.3	1.7	2.7	4.3	4.9
2000	44.1	67.3	100.7	113.3	15.3	24.0	37.2	42.4	5.2	8.1	12.8	14.6	1.7	2.7	4.3	4.9

Note: Recession-lag times of less than one minute are omitted

TABLE 13.6 SUGGESTED DESIGN WATER APPLICATION EFFICIENCIES (PERCENT) FOR GRADED BORDER IRRIGATION BY SLOPE AND INTAKE FAMILY

Irrigation slope S_o (m/m)	0.3 F_n^*(mm) 25	50	75	100	0.5 F_n(mm) 25	50	75	100	125	1.0 F_n(mm) 25	50	75	100	125	1.5 F_n(mm) 25	50	75	100	125	2.0 F_n(mm) 25	50	75	100	125	3.0 F_n(mm) 25	50	75	100	4.0 F_n(mm) 25	50	75	100
0.0005	65	65	70	70	65	65	70	70	70	75	75	80	80	80	75	75	80	80	80	75	75	80	80	80	65	70	70	70	65	70	70	70
0.0010	60	60	65	65	65	65	70	70	70	70	70	75	75	75	75	75	80	80	80	75	75	80	80	80	65	70	70	70	65	70	70	70
0.0020	60	60	55	50	65	65	70	70	70	65	65	70	70	70	70	70	75	75	75	70	70	75	75	75	65	70	70	70	65	70	70	70
0.0030	55	55	50		60	60	65	65	65	65	65	70	70	70	65	65	70	70	70	65	65	70	70	70	65	70	70	70	65	70	70	70
0.0040	55	50			60	60	65	60	55	60	60	65	65	65	65	65	70	70	70	65	65	70	70	70	65	70	70	70	60	65	65	65
0.0050	50				60	60	60	55	50	60	60	65	65	65	65	65	70	70	70	65	65	70	70	70	65	70	70	70	60	65	65	65
0.0075					55	55	50			60	60	65	65	65	60	60	65	65	65	65	65	70	70	70	65	70	70	70	60	65	65	65
0.0100					55	55				60	60	65	65	.65	60	60	65	65	65	60	60	65	65	65	60	65	65	65	60	65	65	65
0.0150					55					55	55	60	60	60	60	60	65	65	65	60	60	65	65	65	60	65	65	65	60	65	65	65
0.0200					50					55	55	60	55	50	60	60	65	65	65	60	60	65	65	65	60	65	65	65	60	65	65	65
0.0250										55	55	55	50		60	60	65	65	65	60	60	65	65	65	60	65	65	65		60	60	60
0.0300										55	55	50			55	55	60	60	60	55	55	60	60	60	55	60	60	60		60	60	60
0.0400										50	50				55	55	60	60	55	55	60	60	60			60	60	60				
0.0500															55	55	60	55	50	55	60	60	60									
0.0600															50	50	55	50		55	55	55										

*F_n is the desired net depth of application.

13.4.6 Design Limitations

The design inflow rate, depth of flow, border slope and length should not exceed established limitations.

Maximum flow rates. Flow rates in border irrigation must be nonerosive. The maximum flow rate per unit width should not exceed the flow as given by the following empirical criteria:

For non-sodforming crops, such as alfalfa and small grains:

$$Q_{u\,max} = (1.765 \times 10^{-4})\, S_o^{-0.75} \quad \dots\dots\dots\dots\dots\dots\dots\dots \text{[13.10]}$$

For well-established, dense sod crops

$$Q_{u\,max} = (3.53 \times 10^{-4})\, S_o^{-0.75} \quad \dots\dots\dots\dots\dots\dots\dots\dots \text{[13.11]}$$

where Q_u is the inflow rate (m²/s), and S_o is the border slope (m/m).

The maximum inflow rate for various border slopes and crop conditions are given in Table 13.7.

Maximum depth of flow. The depth of flow at the head end of the border strip must not exceed the border ridge height, less an allowance for freeboard of approximately 25 percent of the ridge height. Flow depths should generally not exceed 150 mm. Greater depth is practical on some soils, but flow depths exceeding 200 or 250 mm should seldom be considered.

Flow depth—high gradient borders. The normal depth of flow at the upper end of border strips with slopes greater than 0.4 percent may be computed from equation [13.12].

$$d_n = 1000\, Q_u^{0.6}\, n^{0.6}\, S_o^{-0.3} \quad \dots\dots\dots\dots\dots\dots\dots\dots\dots\dots \text{[13.12]}$$

TABLE 13.7 MAXIMUM INFLOW RATES, Qu, FOR NON-SOD AND SOD CROP CONDITIONS

Border slope, S_o (m/m)	Crops	
	Non-sod	Sod
	10^{-3} m^2/s	10^{-3} m^2/s
0.0005	52.8	106.0
0.001	31.4	62.8
0.002	18.7	37.3
0.003	13.8	27.5
0.004	11.1	22.2
0.005	9.39	18.8
0.0075	6.93	13.9
0.01	5.58	11.2
0.015	4.12	8.24
0.02	3.32	6.64
0.025	2.81	5.62
0.03	2.45	4.90
0.04	1.97	3.95
0.05	1.67	3.34
0.06	1.46	2.91

TABLE 13.8 NORMAL DEPTH OF FLOW AT UPPER END OF HIGH GRADIENT BORDERS

Border slope, S_o (m/m)	Inflow Rate, Q_u m^2/s	Normal flow depth, d_n Manning's n		
		0.04	0.15	0.25
0.005	0.0005		16.4	22.3
	0.001		24.9	33.8
	0.01	44.8	99.1	134.6
	0.02	67.9	150.2	204.0
0.0075	0.0005		14.5	19.8
	0.001		22.0	29.9
	0.01	39.7	87.7	119.2
	0.02	60.2	133.0	180.7
0.01	0.0005		13.3	18.1
	0.001		20.2	27.5
	0.01	36.4	80.5	109.3
	0.02	55.2	122.0	165.7
0.015	0.0005		11.8	16.0
	0.001		17.9	24.3
	0.01	32.2	71.3	96.8
	0.02	48.9	108.0	146.7
0.02	0.0005			14.7
	0.001		16.4	22.3
	0.01	29.6	65.4	88.8
	0.02	44.8	99.1	134.6
0.025	0.0005			13.8
	0.001		15.4	20.9
	0.01	27.7	61.1	83.1
	0.02	41.9	92.7	125.9
0.03	0.0005			13.0
	0.001		14.5	19.7
	0.01	26.2	57.9	78.6
	0.02	39.7	87.7	119.2
0.04	0.0005			12.0
	0.001		13.3	18.1
	0.01	24.0	53.1	72.1
	0.02	36.4	80.5	109.3
0.05	0.0005			
	0.001		12.5	16.9
	0.01	22.5	49.7	67.5
	0.02	34.1	75.3	102.3
0.06	0.0005			
	0.001		11.8	16.0
	0.01	21.3	47.0	63.0
	0.02	32.2	71.3	96.8

NOTE: Values given in this table are solutions to equation [13.12]. In practice, calculations to ± 1 mm are adequate.

where d_n is normal flow depth (mm); Q_u is the inflow rate per unit width (m^2/s); and S_o is the border slope (m/m). The normal flow depth may also be estimated from Table 13.8.

Flow depth—low gradient borders. The depth of flow (mm) at the upper end of border strips with slopes of 0.4 percent or less may be computed from equation [13.13] (Table 13.9).

$$d = 2454 \ T_L^{3/16} \ Q_u^{9/16} \ n^{3/8} \qquad\qquad\qquad\qquad [13.13]$$

where d is the flow depth (mm), T_L is the lag time (min), Q_u is the inflow rate (m^2/s), and n is the Manning coefficient. The depth of flow in low gradient borders is shown in Table 13.9 for selected values of slope, inflow rate, Manning coefficient, and opportunity time required for the desired net application.

Minimum depth of flow. The flow rate must be large enough to spread over the entire border strip. A smaller flow rate is needed on rough surface strips than is required on adequately graded and smooth strips. The minimum inflow rate per unit width can be computed, using equation [13.14].

$$Q_{u \ min} = (5.95 \times 10^{-6} \ L \ S_0^{0.5})/n \qquad\qquad\qquad [13.14]$$

TABLE 13.9 DEPTH OF FLOW, d(mm), LOW GRADIENT BORDERS

	Border slope, S_0 (m/m)															
	0.0005				0.001				0.002				0.004			
Oppor. time, T_n (min)	Inflow rate, Q_u (m^2/s)				Inflow rate, Q_u (m^2/s)				Inflow rate, Q_u (m^2/s)				Inflow rate, Q_u (m^2/s)			
	0.0001	0.001	0.01	0.02	0.0001	0.001	0.01	0.02	0.0001	0.001	0.01	0.02	0.0001	0.001	0.01	0.02
						Manning's n = 0.04										
10	4.7	17.4	64.4	95.3	4.2	16.3	62.0	92.5	3.6	14.2	55.2	83.1	3.0	11.8	46.7	70.6
25	5.1	19.5	74.0	110.1	4.4	17.2	66.9	100.6	3.7	14.5	57.2	86.3	3.0	11.9	47.4	71.7
50	5.3	20.7	79.5	119.0	4.5	17.7	69.3	104.5	3.7	14.6	58.0	87.7	3.0	12.0	47.6	72.1
100	5.5	21.4	83.4	125.4	4.5	17.9	70.7	106.9	3.7	14.7	58.4	88.5	3.0	12.0	47.8	72.4
200	5.5	21.9	85.9	129.5	4.6	18.1	71.6	108.3	3.7	14.8	58.7	88.9	3.0	12.0	47.8	72.5
500	5.6	22.2	87.8	132.8	4.6	18.2	72.2	109.3	3.7	14.8	58.9	89.2	3.0	12.0	47.9	72.6
1000	5.6	22.3	88.5	134.0	4.6	18.2	72.4	109.7	3.7	14.8	58.9	89.3	3.0	12.0	47.9	72.6
2000	5.6	22.4	88.9	134.7	4.6	18.2	72.5	109.9	3.7	14.8	59.0	89.4	3.0	12.0	47.9	72.6
						Manning's n = 0.15										
10	8.0	29.1	104.3	153.0	8.0	29.8	109.1	161.0	7.4	28.3	107.1	159.4	6.4	25.1	97.3	146.1
25	9.5	35.0	127.5	187.8	8.9	33.9	127.2	189.0	7.8	30.4	117.5	176.2	6.6	25.9	101.7	153.3
50	10.4	39.2	145.1	214.7	9.4	36.3	138.5	206.8	8.0	31.4	122.7	184.7	6.6	26.2	103.6	156.5
100	11.2	42.7	161.1	239.6	9.7	37.9	146.9	220.4	8.1	32.0	126.0	190.1	6.6	26.4	104.6	158.3
200	11.7	45.3	173.9	260.2	9.9	38.9	152.5	229.7	8.2	32.3	127.9	193.4	6.7	26.5	105.2	159.3
500	12.1	47.5	185.4	279.0	10.0	39.7	156.7	236.8	8.2	32.6	129.3	195.7	6.7	26.6	105.6	160.0
1000	12.3	48.4	190.6	287.6	10.1	40.0	158.4	239.7	8.2	32.7	129.8	196.6	6.7	26.6	105.7	160.2
2000	12.4	49.0	193.7	292.8	10.1	40.1	159.4	241.3	8.2	32.7	130.1	197.1	6.7	26.6	105.8	160.4
						Manning's n = 0.25										
10	9.7	34.6	123.3	180.7	9.9	36.4	131.7	193.7	9.5	36.0	133.8	198.3	8.5	32.9	126.3	188.8
25	11.7	42.4	153.0	224.8	11.4	42.6	157.6	233.1	10.3	39.7	151.3	225.9	8.8	34.5	134.8	202.8
50	13.1	48.5	177.1	261.2	12.3	46.7	175.6	261.0	10.7	41.6	161.0	241.5	8.9	35.2	138.6	209.2
100	14.4	54.1	200.9	297.5	12.8	49.7	190.3	284.4	11.0	42.8	167.6	252.4	9.0	35.6	140.9	213.1
200	15.3	58.7	222.0	330.6	13.2	51.8	201.0	301.9	11.0	43.6	171.7	259.2	9.0	35.9	142.3	215.3
500	16.1	62.9	243.0	364.3	13.5	53.3	209.7	316.4	11.1	44.1	174.6	264.2	9.1	36.0	143.2	216.9
1000	16.5	64.8	253.3	381.3	13.6	54.0	213.4	322.6	11.1	44.3	175.8	266.2	9.1	36.1	143.5	217.4
2000	16.7	66.0	259.8	392.2	13.7	54.3	215.5	325.1	11.2	44.4	176.4	267.2	9.1	36.1	143.7	217.8

NOTE: Values given in this table are solutions to equation [13.13]. In practice calculations to ± 1 mm are adequate.

TABLE 13.10. MINIMUM VALUE OF Q_u/L
FOR VARIOUS SLOPES, S_o AND MANNING n's

Border slope, S_o (m/m)	Manning's n		
	0.04	0.15	0.25
	10^{-5} m²/s	10^{-5} m²/s	10^{-5} m²/s
0.0005	0.3324	0.0886	0.0532
0.001	0.4701	0.1254	0.0752
0.002	0.6649	0.1773	0.1064
0.003	0.8143	0.2171	0.1303
0.004	0.9403	0.2507	0.1504
0.005	1.0512	0.2803	0.1682
0.0075	1.2875	0.3433	0.2060
0.01	1.4867	0.3964	0.2379
0.015	1.8208	0.4855	0.2913
0.02	2.1025	0.5607	0.3364
0.025	2.3506	0.6268	0.3761
0.03	2.5750	0.6867	0.4120
0.04	2.9734	0.7929	0.4757
0.05	3.3243	0.8865	0.5319
0.06	3.6420	0.9711	0.5827

Table 13.10 shows the minimum value of Q_u/L for various slopes and Manning n's.

Maximum slope. The maximum allowable slope for a selected net application depth, efficiency, and given intake family can be estimated from equation [13.15] or Table 13.11.

$$S_{o \; max} = \left(\frac{n}{0.0117 \; E} \; \frac{F_n}{T_n} \right)^2 \quad \dots\dots\dots\dots\dots\dots\dots\dots\dots \quad [13.15]$$

Equation [13.15] is based on criteria for minimum depth of flow and does not include slope limitations imposed by erosion hazards due to runoff from rainfall. Although Table 13.11 indicates the theoretical possibility of using border irrigation on very steep slopes, it is better suited to gentle slopes. On slopes over about 4 percent, erosion is an extreme hazard; it is doubtful whether the border method should be considered for slopes in excess of 6 percent.

Maximum length. The theoretical maximum length for open-end borders is limited by the maximum allowable flow rate, as limited by erosion hazard on steep slopes or by the border ridge height on flat slopes. The permissible border length on soils of low intake rate and low slopes, as determined using equation [13.16] may exceed practical limits. The time required to patrol long lengths and the difficulties in determining and making needed inflow rate adjustments usually make these lengths impractical. Border lengths should seldom exceed 400 meters.

$$L_{max} = \frac{Q_u \; E \; (T_n - T_L)}{0.00167 \; F_n} \quad \dots\dots\dots\dots\dots\dots\dots\dots\dots \quad [13.16]$$

DESIGN AND OPERATION OF FARM IRRIGATION SYSTEMS

TABLE 13.11. MAXIMUM SLOPES, $S_{o\ max}$, FOR BORDER IRRIGATION
AS LIMITED BY MINIMUM DEPTH OF FLOW REQUIREMENTS OR
BY A MINIMUM BORDER LENGTH OF 30 METERS

Manning's n values: 0.04, 0.15, 0.25. Column headers under each n are Water application efficiency, E (percent): 50, 55, 60, 65, 70, 75.

Intake family	Net appl. depth, F_n (mm)	n=0.04: 50	55	60	65	70	75	n=0.15: 50	55	60	65	70	75	n=0.25: 50	55	60	65	70	75
0.3	25	0.22	0.18	0.15	0.13			1.12	0.93	0.78	0.66			3.11	2.57	2.16	1.84		
	50	*	*	*	*			0.40	0.33	0.28	0.24			1.11	0.92	0.77	0.66		
	75	*	*	*	*	*		0.25	0.21	0.18	0.15	0.13		0.70	0.58	0.49	0.42	0.36	
	100	*	*	*	*	*		0.19	0.16	0.13	0.11	0.10		0.52	0.43	0.36	0.31	0.27	
0.5	25	0.22	0.18	0.15	0.13			3.08	2.54	2.14	1.82			8.55	7.06	5.94	5.06		
	50	0.09	0.07	0.06	0.05			1.23	1.02	0.85	0.73			3.42	2.83	2.37	2.02		
	75	0.06	0.05	*	*			0.82	0.68	0.57	0.49	0.42		2.29	1.89	1.59	1.35	1.17	
	100	*	*	*	*			0.64	0.53	0.44	0.38	0.33		1.78	1.47	1.23	1.05	0.91	
	125	*	*	*	*			0.53	0.44	0.37	0.32	0.27		1.48	1.22	1.03	0.88	0.76	
1.0	25	0.80	0.66	0.56	0.48	0.41	0.36	6.9†	7.8†	7.84	6.69	5.76	5.02	6.9†	7.8†	8.7†	9.7†	10.7†	11.8†
	50	0.35	0.29	0.24	0.21	0.79	0.56	4.92	4.07	3.42	2.91	2.51	2.18	11.9†	11.3	9.49	8.09	6.97	6.08
	75	0.24	0.20	0.17	0.14	0.12	0.11	3.45	2.85	2.39	2.04	1.76	1.53	9.57	7.91	6.65	5.66	4.88	4.25
	100	0.20	0.16	0.14	0.12	0.10	0.09	2.76	2.28	1.92	1.63	1.41	1.23	7.66	6.33	5.32	4.54	3.91	3.41
	125	0.17	0.14	0.12	0.10	0.09	0.07	2.35	1.94	1.63	1.39	1.20	1.04	6.53	5.40	4.53	3.86	3.33	2.90
1.5	25	1.66	1.37	1.15	0.98	0.85	-0.74	4.2†	4.8†	5.4†	6.0†	6.6†	7.2†	4.2†	4.8†	5.4†	6.0†	6.6†	7.2†
	50	0.75	0.62	0.52	0.44	0.38	0.33	7.1†	8.1†	7.34†	6.26†	5.41	4.7†	7.1†	8.1†	9.1†	10.1†	11.2†	12.3†
	75	0.54	0.44	0.37	0.32	0.27	0.24	7.56	6.25	5.25	4.47	3.86	3.36	8.9†	10.1†	11.4†	12.42	10.71	9.33
	100	0.43	0.36	0.30	0.26	0.22	0.19	6.14	5.07	4.26	3.63	3.13	2.73	10.3†	11.6†	11.83	10.09	8.70	7.58
	125	0.38	0.31	0.26	0.22	0.19	0.17	5.28	4.37	3.67	3.13	2.70	2.35	11.3†	12.13	10.19	8.69	7.49	6.52
2.0	25	2.80	2.31	1.94	1.65	1.43	1.24	3.0†	3.4†	3.8†	4.2†	4.7†	5.1†	3.0†	3.4†	3.8†	4.2†	4.7†	5.1†
	50	1.30	1.07	0.90	7.6	6.6	5.8	5.0†	5.6†	6.3†	7.0†	7.8†	8.11	5.0†	5.6†	6.3†	7.0†	7.8†	8.5†
	75	0.94	0.78	0.65	0.56	0.48	0.42	6.2†	7.0†	7.9†	7.81	6.74	5.87	6.2†	7.0†	7.9†	8.7†	9.6†	10.6
	100	0.77	0.64	0.54	0.46	0.39	0.34	7.0†	8.0†	7.51	6.40	5.52	4.81	7.0†	8.0†	8.9†	10.0†	11.0†	12.0†
	125	0.67	0.56	0.46	0.39	0.34	0.30	7.7†	7.75	6.51	5.55	4.78	4.17	7.7†	8.7†	9.8†	10.9†	12.0†	11.57
3.0	25	1.8†	2.7†	2.3†	2.6†			1.8†	2.1†	2.3†	2.6†			1.8†	2.1†	2.3†	2.6		
	50	2.77	2.29	1.92	1.64			3.0†	3.4†	3.8†	4.2†			3.0†	3.4†	3.8†	4.2		
	75	2.02	1.67	1.41	1.20	1.03		3.7†	4.2†	4.7†	5.2†	5.8†		3.7†	4.2†	4.7†	5.2	5.8	
	100	1.67	1.38	1.16	0.99	0.85		4.2†	4.7†	5.3†	5.9†	6.5†		4.2†	4.7†	5.3†	5.9	6.5	
	125	1.46	1.20	1.01	0.86	0.74		4.6†	5.2†	5.8†	6.5†	7.1†		4.6†	5.2†	5.8†	6.5	7.1	
4.0	25	1.3†	1.5†	1.6†	1.8†			1.3†	1.5†	1.6†	1.8†			1.3†	1.5†	1.6†	1.8†		
	50	2.1†	2.4†	2.7†	2.83			2.1†	2.4†	2.7†	3.0†			2.1†	2.4†	2.7†	3.0†		
	75	2.6†	2.87	2.41	2.06	1.77		2.6†	2.9†	3.3†	3.6†	4.0†		2.6†	2.9†	3.3†	3.6†	4.0†	
	100	2.89	2.39	2.00	1.71	1.47		2.9†	3.3†	3.7†	4.1†	4.5†		2.9†	3.3†	3.7†	4.1†	4.5†	
	125	2.53	2.09	1.76	1.50	1.29		3.2†	3.6†	4.0†	4.5†	5.0†		3.2†	3.6†	4.0†	4.5†	5.0†	

*Not adapted for graded borders.
†Slope limited by minimum border length of 30 m.

where Q_u is the maximum flow rate in m²/s, as determined by equations [13.10] or [13.11].

13.4.7 Design of Borders with No Runoff

Higher irrigation application efficiencies and elimination of surface runoff can be achieved by modification of the border design. This may be accomplished by blocking the end and reducing the inflow rate, or extending the border length and impounding the runoff on the length extension.

Border extensions. The length extension is limited by the lesser of:

1 The length, L_e (m), that can be covered by an impoundment whose maximum depth is equal to the desired net application depth:

$$L_e = F_n/(1000\ S_o) \dotfill [13.17]$$

where F_n (mm) is the desired net application depth and S_o is the border slope (m/m). Removal of all or part of the slope at the lower end of the border by land leveling will increase the length extension as limited by slope.

2 The length that can be adequately irrigated with the volume of runoff from the open-end border strip:

$$L_e = (1 - E/100)\, r_i\, r_n\, L \quad \dotfill \quad [13.18]$$

where L_e is the allowable length extension with end blocks (m), E is the water application efficiency (percent), r_i and r_n are factors that express the effect of intake and roughness on runoff, and L is the normal design length (m).

Empirical values for factors r_i and r_n are given in Table 13.12.

Borders with end blocks (no extensions). On fields where the length of the border is fixed, use of end blocks and elimination of runoff permits reduction of the inflow rate. The reduced inflow rate required can be estimated from equation [13.19].

$$Q_{ue} = \frac{Q_u}{1 + r_i\, r_n\, (1 - E/100)}, \quad \dotfill \quad [13.19]$$

where Q_{ue} is the inflow rate per unit width of border using end blocks (m²/s), Q_u is the inflow rate determined for the border length without end blocks (m²/s), E is the efficiency (percent), and r_i and r_n are empirical factors as given in Table 13.12. Equation [13.19] assumes the reduction in flow rate will not be large enough to result in a significant change in recession-lag time.

TABLE 13.12. INTAKE AND ROUGHNESS
FACTORS FOR ESTIMATING
POTENTIAL RUNOFF

Intake family	Intake factor, r_i (dimensionless)	Manning coefficient (n)	Roughness factor, r_n (dimensionless)
0.3	0.90	0.10	0.80
0.5	0.80	0.15	0.75
1.0	0.70	0.20	0.70
1.5	0.65	0.25	0.65
2.0	0.60		
3.0	0.50		
4.0	0.40		

13.4.8 Sample Calculation
Given:

Intake family	0.5
Net depth of application (F_n)	100 mm
Border slope (S_o)	0.001 m/m
Manning roughness coefficient (n)	0.15
Estimated water application efficiency (E)	70 percent
Allowable flow depth	150 mm
Border length (L)	250 m
Crop	Alfalfa

Find:

Opportunity time required (T_n)
Lag time (T_L)
Design inflow rate (Q_u)
Required inflow time (T_a)
Reduced inflow rate with end blocks (Q_{ue})
Maximum flow rate (Q_u)
Maximum flow depth (d)
Minimum flow rate (Q_u)
Maximum slope (S)
Maximum border length - open end borders
Allowable length extension with end blocks

Solution:

$$T_n = \left(\frac{100 - 7.0}{1.196} \right)^{1/0.748} = 337 \text{ min} \quad \cdots \cdots \cdots \cdots \text{(equation [13.2])}$$

From Table 13.5, the recession lag time, T_L, must be such that $7.5 < T_L < 21.4$. Make a first guess as $T_L = 17$ min. Then,

$$Q_u = \frac{(0.00167)(100)(250)}{(337 - 17)(70)} = 0.00186 \text{ m}^2/\text{s} \quad \cdots \cdots \cdots \text{(equation [13.6])}$$

Revise $T_L = 13.4$ min from interpolation in Table 13.5

$$Q_u = \frac{(0.00167)(100)(250)}{(337 - 13)(70)} = 0.00184 \text{ m}^2/\text{s} \quad \cdots \cdots \cdots \text{(equation [13.6])}$$

Check T_L:

$$T_L = \frac{(0.15)^{1.2} (0.00184)^{0.2}}{120 \left[0.001 + \dfrac{(0.0094)(0.15)(0.00184)^{0.175}}{(337)^{0.88} (0.001)^{0.5}} \right]^{1.6}} \quad . \text{(equation [13.8])}$$

$$= 13.4 \text{ min} \cong 13 \text{ min}$$

Reduced inflow rate with end blocks, without length extension:

$$Q_{ue} = \frac{0.00184}{1 + (0.80)(0.75)(1 - 0.70)} = 0.00156 \text{ m}^2/\text{s} \cdots \text{(equation [13.19])}$$

Border length extension with end blocks:

$L_e = 100/[(1000)(0.001)] = 100$ m, based on slope

$\cdots\cdots\cdots\cdots\cdots\cdots\cdots\cdots\cdots\cdots\cdots\cdots\cdots$ (equation [13.17])

$L_e = (1 - 0.70)(0.80)(0.75)(250) = 45$ m, based on runoff

$\cdots\cdots\cdots\cdots\cdots\cdots\cdots\cdots\cdots\cdots\cdots\cdots\cdots$ (equation [13.18])

Length extension limited by runoff, thus extended length =

250 + 45 = 295 m.

Maximum flow rate:

$Q_{u\ max} = (1.765 \times 10^{-4})(0.001)^{-0.75} = 0.031 > 0.00184$

$\cdots\cdots\cdots\cdots\cdots\cdots\cdots\cdots\cdots\cdots\cdots\cdots$ (equation [13.10])

Maximum flow depth:

$d = (2454)(13)^{3/16} (0.00184)^{9/16} (0.15)^{3/8}$ $\cdots\cdots$ (equation [13.13])

$= 56$ mm < 150 mm

Minimum flow rate:

$Q_{u\ min} = \left[(5.95 \times 10^{-6})(250)(0.001)^{0.5}\right] / 0.15 \cdots$ (equation [13.14])

$= 0.00031 < 0.00184$ m^2/s

Maximum slope:

$S_{o\ max} = \left[\dfrac{(0.15)(100)}{(0.0117)(70)(337)} \right]^2$ $\cdots\cdots\cdots\cdots$ (equation [13.15])

$= 0.0030$ m/m which is > 0.001 m/m

Maximum length—open end border:

$L_{max} = \dfrac{(0.031)(70)(337 - 13)}{(0.00167)(100)} = 4210$ m > 250 m $\cdots$ (equation [13.16])

Preparation of design charts will greatly facilitate use of the design relationships.

13.4.9 Selection of Headland Facilities

Comments applying to basins (Sec. 13.3.6) apply, in general, to border headland facilities.

13.5 CONTOUR LEVEE IRRIGATION*

13.5.1 Description

In contour levee irrigation water is applied to sloping basins. Each basin is bounded by two levees which are on land contours, and two levees (sometimes called checks) which are essentially perpendicular to the contour levees. For non-rice crops, water is introduced into the basin at its highest point until the irrigation requirement has been met, and is then removed through surface drains located along the lower contour levee. For paddy rice, water is usually circulated through the basins throughout most of the season. In contour levee irrigation basins are sloped from contour to contour and emptied primarily by drainage. This is in contrast to basin irrigation (Section 13.3) in which the basins are level and empty by percolation into the soil.

13.5.2 Applicability

Crops. This system is extensively used for paddy rice. It is also used for pasture grasses, hay crops, alfalfa, small grains and row crops which can withstand temporary flooding (e.g., cotton, corn, soybeans, grains and peanuts).

Soils. Soils of medium to fine texture, having waterholding capacities equal to or greater than 100 mm per m, and a total waterholding capacity in the root zone of at least 60 mm are suitable. For all crops other than rice, soil infiltration rates should not exceed Soil Conservation Service family 0.3 (Section 13.1). For rice, only soils whose infiltration rates do not exceed those of SCS intake family 0.1 are suitable.

Slopes. The general land topography upon which contour levee irrigation is applicable must have average slopes of less than 0.5 percent. Slopes within basins are limited on the upper side by soil erosion limitations (usually between 0.05 and 0.3 percent) and on the lower side by the necessity to provide drainage (0.05 to 0.15 percent). Thus, land forming is often needed to alter the slope within basins and to remove minor surface irregularities.

Climate. A large amount of work is necessary in the construction of the levees and checks, and they must be reconstructed after each cultivation. Therefore, except in the case of rice irrigation, unless most early season water is supplied to the crop by rainfall the system is impractical because of the need of frequent reconstruction of levees.

13.5.3 Advantages

The system is applicable to low-intake soils which are difficult to irrigate by other surface methods. It makes maximum use of seasonal rainfall. Water can be uniformly distributed, resulting in high water application efficiencies. The system can be designed to handle high rainfall, with a minimum of soil erosion. Installation costs are low compared to most other methods, especially if little land forming is needed. Large areas can be handled efficiently by a single irrigator.

---

*This Section is based primarily on USDA (1969).

13.5.4 Disadvantages

Crops must be able to withstand up to 12 hours of flooding. The system is applicable to soils of relatively low intake rates only. Some land forming is often needed. Large irrigation streams are necessary, and small net applications are not feasible with the system. Levees, ditches and structures require frequent maintenance.

13.5.5 Design

Irrigator constraints. The design must consider the time schedule of the irrigator, as well as the physical conditions. Thus, if the irrigator will work 12 hours per day (or at least is available for changing water on such a schedule), and he wishes to work only eight days out of the irrigation interval, this will be a constraint on field layout (and inflow rates). Thus, the possible numbers of basins to be irrigated would conveniently be 8, 16, 24, . . . etc. The average basin size would be the field area divided by the number of basins selected.

Land preparation. Although there is no specific need for land leveling of contour-levee irrigated lands, some land forming is usually desirable. Minor irregularities must be removed if the basins are to be uniformly irrigated. Because the normal vertical interval between contour levees is 100 mm, fields which have not been brought to a uniform slope will have basins of varying widths, and these are difficult to irrigate uniformly. The ideal situation is a series of basins of identical size, shape, and slope. Land may be leveled either before or after the levees are installed. In the former case, basins are usually sloping planes, while in the latter they are level.

The levee system can be layed out once the topographic map is completed and the available water stream is known. Limitations on basin size are based on several factors.

Wind considerations. Winds will develop waves in large basins and these waves can cause erosion of levees. A general rule is to limit the maximum length of the basin in a direction parallel to prevailing winds to 120 m.

Drainage considerations. The drainage system of the basins must remove not only the applied irrigaton water which is in excess of that infiltrated or evaporated form the surface, but also storm water which falls in the basin. The maximum drain length should be 200 m. If a basin has only one outlet location, then its maximum length should be 200 m. If the basin drains from more than one outlet, a longer basin can be used. Special climatic conditions, which could cause large runoff rates from the basin, may impose even more stringent requirements.

Placement of contour levees. As the name implies, the levees are on contour. If they are constructed after land leveling, they are put on the existing contour. If they are layed out before leveling, they must be located where the final contours will be.

Vertical interval. The most common vertical interval, V_i, for contour levees is 60 to 120 mm. In flat lands they may be reduced so that basin widths do not become excessive. In steep lands they may be increased somewhat so that basin widths are not too small. A minimum width of 12 m is about optimal. Greater vertical intervals require higher levees to properly contain the water.

Levee dimensions. Levees must be constructed to the following minimum height.

$$H = V_i + F_g + d_f + S \dots\dots\dots\dots\dots\dots\dots\dots\dots\dots\dots \quad [13.20]$$

where H is the levee height, V_i is the vertical interval between levees, F_g is the gross depth of water to be applied, d_f is the freeboard, and S is the allowance for settlement. The freeboard should be no less then 80 mm and the allowance for settling no less than 90 mm.

Side slopes of levees depend to some extent upon the crop to be grown and tillage practices. If the levees are removed and rebuilt frequently, a common practice when cultivating row crops grown in basins, then side slopes should be no steeper than 1-1/2 horizontal to 1 vertical. When levees are permanent (pasture, forage production, etc.) then side slopes should be no steeper than 3 or 4 to 1. This reduces cattle damage and allows levees to be easily crossed by machinery.

Drainage channels. The drainage channel, constructed parallel to and on the upper side of each levee, should be no less than 150 mm deep and have side slopes no steeper than 1-1/2 horizontal to 1 vertical. Storm drainage requirementsɪ may dictate even larger drainage ditches, and crossing by machinery may require different bank slopes. The material excavated in drain construction is used in forming the levees.

Irrigation stream size—non-rice crops. The necessary minimum stream size for all crops except rice is based upon two criteria: (a) The stream must be large enough to meet the entire field's requirement for water—evapotranspiration, intentional leaching required to maintain the salt balance, and unavoidable deep percolation losses; (b) The stream must be large enough (into each basin) to permit coverage of the average size basin in no more than one-fourth the time necessary to infiltrate the required net depth. This latter criterion assures relatively uniform distribution of applied water.

The first criterion is expressed as follows.

$$Q = \frac{A\, F_g}{360\, f\, h}, \dots\dots\dots\dots\dots\dots\dots\dots\dots\dots\dots \quad [13.21]$$

where Q is the flow rate (m^3/s) into the entire field (sometimes called system capacity), A is the field area in ha, F_g is the gross application depth (mm), f is the actual number of days of irrigating, and h is the number of irrigation hours per irrigation day.

The procedure for designing to meet the second criterion is a quasi-rational one. It requires knowledge of the basin size which will be used. In outline form it is as follows:

1 Determine the time, T_n, to infiltrate the desired net application. This is a function of the net application and the soil infiltration characteristics (Section 13.1).

2 Determine the average depth infiltrated during the time required to cover the entire basin. This time is assumed to be $T_n/4$. In making this calculation it is assumed that by the time the basin has been covered, the average depth infiltrated is equal to the value of the infiltration function at time $T_n/8$. Thus,

$$\bar{z}_1 = z(T_n/8) \dots\dots\dots\dots\dots\dots\dots\dots\dots\dots\dots \quad [13.22]$$

where $\bar{z}_1$ is the average depth infiltrated (mm) and $z(t)$ is the value of the cumulative infiltration function at time t. The infiltration function $z(t)$ may be selected from one of the SCS families.

3   Determine the average depth of water in surface storage at the time the basin is just filled. In this calculation it is assumed that the surface storage volume is wedge-shaped, with zero depth at the upper levee and a depth at the lower levee equal to V_i (mm). (The additional water stored in the drainage ditch is ignored.) Thus,

$$\bar{y}_1 = V_i/2 \quad \dots\dots\dots\dots\dots\dots\dots\dots\dots\dots\dots\dots\dots\dots\dots\dots \quad [13.23]$$

where $\bar{y}_1$ is the average depth of water in surface storage (mm).

4   Determine the required flow rate per unit area irrigated to provide the depth of water equal to the sum of $\bar{z}_1$ (mm) and $\bar{y}_1$ (mm) in the time $T_n/4$ (min).

$$q_1 = \frac{40(\bar{z}_1 + \bar{y}_1)}{60\,T_n} \quad \dots\dots\dots\dots\dots\dots\dots\dots\dots\dots\dots\dots\dots\dots\dots \quad [13.24]$$

where q_1 is the required inflow (m³/s per ha).

The above four steps represent a quasi-rational method for two reasons. By requiring the water to spread over the basin in the time $T_n/4$ it is reasonable to assume that the average infiltration depth is somewhere between $z(T_n/4)$ and zero. However, there is no reason to believe this average depth would be $z(T_n/8)$ because both the infiltration and advance functions are nonlinear. In addition, the filling of the basin is through an unsteady wave. Thus, at the time the water reaches the upper levee, the water surface is not necessarily horizontal (although it might approximate that very closely).

5   Determine the area which must be irrigated with each set. The total set time, T_s (min) is computed assuming that the basin is covered in one fourth the net infiltration time (T_n), and each point in the basin receives the desired net application or more.

$$T_s = \frac{5}{4}\,T_n \quad \dots\dots\dots\dots\dots\dots\dots\dots\dots\dots\dots\dots\dots\dots\dots\dots \quad [13.25]$$

This must be less than or equal to 60 h, the irrigation time each day. If this is not the case, the field is not applicable to contour levee irrigation.

6   Determine number of sets per day, N, and total area irrigated per set, A_1.

$$N = \text{Int}\,(60h/T_s) \quad \dots\dots\dots\dots\dots\dots\dots\dots\dots\dots\dots\dots\dots\dots \quad [13.26]$$

where $\text{Int}(x)$ means the integer portion of x.

$$A_1 = A/(Nf) \quad \dots\dots\dots\dots\dots\dots\dots\dots\dots\dots\dots\dots\dots\dots\dots \quad [13.27]$$

7   Determine the system capacity based upon the necessary unit inflow (determined from soil and topographic conditions) and the total area irrigated per set.

$$Q_1 = A_1 q_1 \quad \dots \dots \dots \dots \dots \dots \dots \dots \dots \dots \dots \dots \dots \dots \dots [13.28]$$

8 Complete design by laying out basins according to criteria set forth previously, and the total area irrigated per set. The average basin size must be one, or a simple fraction of, A_1.

Irrigation stream size—rice. Typically, there are three steps in the irrigation of rice. In the *flushing* period, the soil is wet to field capacity after dry planting. The basins may be drained and following that comes *flooding*. In this step all the basins are flooded. They must remain flooded during the growth of the crop, and this is known as *maintaining the flood*. The calculation of appropriate inflow rates for each of these phases follows:

1 **Flushing.** This calculation is carried out according to steps 1, 2 and 3 outlined above, except that the necessary flow rate for one entire basin must be considered. Thus,

$$Q_1 = q_1 \frac{A}{N} \quad \dots \dots \dots \dots \dots \dots \dots \dots \dots \dots \dots \dots \dots \dots \dots [13.29]$$

where Q_1 is the flow rate required during the flushing stage (m³/s), N is the number of basins in the field, and q_1 is as defined previously. The basin size, A_1, is A/N. Clearly, Q, the system capacity, must be greater than or equal to Q_1.

2 **Flooding.** During this period the entire irrigated area is covered with water to some predetermined depth, F_f (often 70 to 100 mm). The duration of flooding may be taken as the time necessary for the crop to deplete one half of the available moisture in the root zone, at the maximum evapotranspiration rate.

$$T_e = \frac{720 \, R_d}{E'_t} , \quad \dots \dots \dots \dots \dots \dots \dots \dots \dots \dots \dots \dots \dots \dots [13.30]$$

where T_e is the duration of flooding (min), R_d is the root zone storage capacity (mm) and E'_t is the peak consumptive use rate (mm/d). The depth of water which must be supplied during this period is equal to the sum of the amount necessary to saturate the root zone, plus the average depth of the wedge-shaped surface storage described above ($\bar{y}_1$), plus the depth F_f, plus the amount lost to deep percolation. The flow rate must be determined for the entire irrigated area.

$$Q_2 = \frac{[R_s + (V_i/2) + F_f + L_p] A}{6 \, T_e} \quad \dots \dots \dots \dots \dots \dots \dots \dots [13.31]$$

where Q_2 is the flow rate into the field during flooding (m³/s), R_s is the depth required to saturate the root zone (mm), and L_p is the estimated deep percolation loss (mm) during the time period T_e.

3 **Maintaining the flood.** The water required in this stage is equal to that lost through deep percolation plus that required for consumptive use.

$$Q_3 = \left(\frac{L_p}{6 T_e} + \frac{E_t}{8640} \right) A \quad \dots \dots \dots \dots \dots \dots \dots \dots \dots \dots [13.32]$$

where Q_3 is the required flow rate (m^3/s) to maintain the flood and E_t is the average consumptive use rate (mm/d).

13.5.6 Sample Problem—Non-Rice Crops
Given:

Intake function	$z = 0.78T^{0.691} + 7.0$ mm
Net depth of application (F_n)	65 mm
Vertical interval (V_i)	60 mm
Estimated water application efficiency (E)	70 percent
Irrigation days (f)	8 days
Irrigation time each day (h)	12 h
Total area irrigated (A)	32 ha
Land slope (average)	0.075 percent

Required:

Total area irrigated at one set (A_i)	ha
System capacity (maximum of Q or Q_1)	m^3/s
Basin dimensions	m $\times$ m

Solution:

The intake function is for a soil with a slower intake than a 0.3 SCS family, so it is satisfactory for basins. Land slope is also within the recommended range.

$$F_g = 100 \, F_n/E = (100)(65)/70 = 93 \text{ mm}$$

$$Q = \frac{(32)(93)}{(360)(8)(12)} = 0.086 \text{ m}^3/\text{s} \dots\dots\dots\dots\dots \text{ (equation } [13.21])$$

$$(1) \quad T_n = [(65 - 7.0)/0.78]^{1/0.691} = 511 \text{ min} \dots \text{ (equation } [13.2])$$

$$(2) \quad \bar{z}_1 = 0.78 \left(\frac{511}{8}\right)^{0.691} + 7.0 = 21 \text{ mm} \dots \text{ (equation } [13.22])$$

$$(3) \quad \bar{y}_1 = 60/2 = 30 \text{ mm} \dots\dots\dots\dots\dots \text{ (equation } [13.23])$$

$$(4) \quad q_1 = \frac{(40)(21 + 30)}{(60)(511)} = 0.0665 \text{ m}^3/\text{s} \dots\dots \text{ (equation } [13.24])$$

$$(5) \quad T_s = \frac{5}{4} \, 511 = 639 \text{ min} \dots\dots\dots\dots \text{ (equation } [13.25])$$

$$(6) \quad N = \text{Int} \frac{(60)(12)}{639} = 1 \dots\dots\dots\dots \text{ (equation } [13.26])$$

$$A_1 = \frac{32}{(1)(8)} = 4 \text{ ha} \dots\dots\dots\dots \text{ (equation } [13.27])$$

$$(7) \quad Q_1 = (4)(0.0665) = 0.266 \text{ m}^3/\text{s} \dots\dots\dots \text{ (equation } [13.28])$$

Thus, the system capacity is governed by Q_1 (not Q) and is 0.266 m³/s. The correct unit stream size is one fourth this (4 ha per basin, 1 basin irrigated at a time).

$$q_1 = 0.266/4 = 0.066 \text{ m}^3/\text{s}$$

With a vertical interval of 60 mm and an average ground slope of 0.075 percent, the width of a basin is

$$W = (60/1000)0.00075 = 80 \text{ m}$$

The length of a basin is

$$L = A_1 /W = 4 \times 10\,000/80 = 500 \text{ m}$$

This exceeds the allowable length for drainage, and half this length could be used if the basins could drain at both ends. The contour levees should be perpendicular to the prevailing wind, if possible.

13.5.7 Sample problem—Rice Crops

Given:

Intake function	$z = 0.78 \ T^{0.691} + 7.0$ mm
Net depth of application (F_n)	45 mm
Vertical interval (V_i)	60 mm
Total area irrigated (A)	32 ha
Number of basins (N)	16
Available water holding capacity of root zone	91 mm
Saturated moisture capacity of root zone	187 mm
Permeability of restricting layer	0.51 mm/h
Peak-period consumptive use rate (E_t')	7.6 mm/day
Average consumptive use rate (E_t)	6.4 mm/day
Estimated deep percolation loss	10 mm/day

Find:

Minimum stream sizes for flushing, flooding and maintaining flood.

Solution:

This soil is just within the 0.1 SCS intake family band, which can be demonstrated by plotting a few points on Fig. 13.1. It is therefore suitable for rice irrigation.

(1) $T_n = [(45 - 7.0)/0.78]^{1/0.691} = 277$ min . . . . (equation [13.12])

$$\bar{z}_1 = 0.78 \left(\frac{277}{8} \right)^{0.691} + 7.0 = 16 \text{ mm} \ \ldots \ (\text{equation} \ [13.22])$$

$$\bar{y}_1 = 60/2 = 30 \text{ mm} \ \ldots\ldots\ldots\ldots\ldots\ (\text{equation} \ [13.23])$$

$$Q_1 = \frac{(40)(16 + 30)(32)}{(60)(277)(16)} = 0.221 \text{ m}^3/\text{s} \ .. \ (\text{equation} \ [13.24]) \\ \text{and} \ [13.29])$$

(2) T_e $= \dfrac{(720)(91)}{7.6} = 8620 \text{ min}$ .... (equation [13.30])

L_p $= \left| \dfrac{(10)(8620)}{(24)(60)} \right. = 60 \text{ mm}$

Q_2 $= \dfrac{[187 + (60/2) + 85 + 60]\ 32}{(6)(8620)} = 0.224 \text{ m}^3/\text{s}$

.... (equation [13.31])

(3) $Q_3 = \left[\dfrac{60}{(6)(8620)} + \dfrac{6.4}{8640} \right] 32 = 0.061 \text{ m}^3/\text{s}$

.... (equation [13.32])

13.6 FURROW AND CORRUGATION IRRIGATION, (USDA, 1979)

13.6.1 Description

Small, evenly spaced, shallow channels are installed down or across the slope of the field to be irrigated. Water is turned in at the high end and conveyed in the small channels to the vicinity of plants growing in, or on beds between, the channels. Water is applied until the desired application and lateral penetration is obtained.

The method is separated into types according to the kinds of crops and size of channel. Furrow irrigation is primarily used with clean tilled crops planted in rows, while corrugation irrigation is associated with noncultivated close-growing crops using small closely-spaced channels aligned down the steepest slope of the field. Corrugations are frequently formed after the crop has been seeded, and in the case of perennial crops, reshaped as needed to maintain the desired channel cross section. Water application principles are the same for both furrow and corrugation irrigation. The primary differences are channel size, shape and spacing, and retardance characteristics. The two terms will be used synonymously in this chapter.

Furrows and corrugations vary in shape and size. Most furrows in row crops are either parabolic in cross section or have flat bottoms and about 2 to 1 side slopes. Typical corrugations have 60-mm bottom widths, 1 to 1 side slopes, and depths of 100 to 150 mm.

13.6.2 Applicability

Most crops can be irrigated by the furrow or corrugation method except those grown in ponded water, such as rice. The furrow method is particularly suitable for irrigating crops subject to injury if water covers the crown or stem of the plants, as the crops may be planted on beds between furrows.

This irrigation method is best suited to medium to moderately fine textured soils of relatively high available water holding capacity and conductivities which allow significant water movement in both the horizontal and vertical directions. The method is suited to fine textured very slowly permeable soils on level sites which permit water impoundment. On sloping sites excessive surface runoff occurs because these soils require very small streams for long periods of time to obtain the desired intake.

The movement of irrigation water applied by furrows or corrugations on coarse textured sands and loamy sands is mainly downward with very little lateral penetration. Efficient furrow or corrugation irrigation on these soils requires very short furrows, small application times, relatively close row spacings and small depths of water application. Furrow grades should be limited so that soil loss from rainfall runoff or irrigation flow is within allowable limits. Furrow grades should generally be 1.0 percent or less, but can be as much as 3.0 percent in arid areas where erosion from rainfall is not a hazard. In humid areas furrow grades should generally not exceed 0.3 percent; however, grades up to 0.5 percent may be permissible if the lengths are sufficiently short. A minimum grade of 0.03 to 0.05 percent in humid and sub-humid areas is necessary to assure adequate surface drainage. Maximum furrow grades for erosive soils can be estimated by the equation:

$$S_{max} = 67/(P_{30})^{1.3} \dots\dots\dots\dots\dots\dots\dots\dots\dots\dots\dots\dots\dots [13.33]$$

where P_{30} is the 30-min rainfall in mm on a 2-year frequency and S_{max} is the maximum allowable furrow grade in percent. Grades on less erosive soils may be increased by approximately one fourth. Cross slopes for furrowed fields with irrigation grades of 0.5 percent or greater should normally be limited to 1.0 percent, and lesser grades to 0.5 percent. Wherever practical, the furrow grade should be uniform.

Corrugations are commonly used on grades of more than 1.0 percent and less than 4.0 percent. They are not recommended in humid areas except for irrigation of perennial crops because of erosion hazards.

13.6.3 Advantages

Moderate to high application efficiency can be obtained if good water management practices are followed and the land is properly prepared. Many different kinds of crops can be grown in sequence without major changes in design, layout, or operating procedures. The initial capital investment is relatively low on lands not requiring extensive land forming as the furrows and corrugations are constructed by common farm implements. Soils which form surface crusts when flooded can readily be irrigated because water moves laterally under the surface. Water does not contact plant stems and scalding is thus avoided. Excellent field surface drainage is obtained when furrow grades are sufficient and adequate outlet facilities are provided. Greater utilization of rainfall may be achieved by irrigation of alternate rows because the remaining available soil water storage capacity is greater than when each furrow is irrigated. Also, the initial intake rate is higher in the non-irrigated furrows.

13.6.4 Limitations

Erosion hazards on steep slopes limit use in climatic areas where precipitation intensities and volumes result in surface runoff, which, when concentrated in furrow channels may cause excessive soil erosion or crop damage from flooding. Surface runoff occurs except where the field is level and water is impounded until intake is completed. Labor requirements may be high as irrigation streams must be carefully regulated to achieve uniform water distribution. Salts from either the soil or water supply may concentrate in the ridges and depress crop yields. Lateral spread of water in coarse-textured soils may not be adequate to entirely wet the soil between furrows. Land leveling is normally required to provide uniform furrow or corrugation grades.

13.6.5 Design

A furrow or corrugation system may be designed only after gathering soils, crops, topography, size and shape of irrigable areas, farm equipment available, farmer operational practices, and farmer personal preferences for the proposed area. The designer must know the intake characteristics and water storage capacities of the various soils, which along with the crop to be grown, will determine the design depth of application and whether furrows or corrugations will be used. The topography will determine the direction and grade of furrows and lengths that will fit individual field boundaries. The farm equipment to be used will determine the spacing and maximum capacity of the furrows. The farmer's operational practices will influence the type of furrow or corrugation system to be designed and the irrigation operating schedules to be followed. Furrow flow rate and time of application are both influenced by the operational method to be used. The designs are based on a normal irrigation and adjusted for variations needed in application time, depth, and flow rates for specific irrigations during the season.

For acceptable uniformity and adequacy of application, the minimum time for water at any point is the time for intake of the net design application. The maximum time is limited by excessive deep percolation. The time water is available for intake at any point, the opportunity time, is the time interval between water advance and recession.

Design assumptions. Development of design relationships requires assumptions for intake vs. time, advance and recession rates, flow retardance, and intake as related to the furrow wetted perimeter. Rate of advance is assumed to be a function of water inflow rate, soil intake characteristics, furrow shape, grade, length, and roughness.

Design limitations. Flow rates into furrows must not exceed the channel capacity as limited by cross-sectional shape and size, slope, and hydraulic roughness. The inflow must advance at a rate which will achieve a reasonably uniform opportunity time throughout the length. Maximum flows are also limited to non-erosive velocities. Erosive soils may erode excessively when the flow velocity exceeds approximately 0.15 m/s while less erosive soils may safely withstand velocities of 0.18 m/s. Velocity and depth of flow for a given cross-section and grade depend on the roughness or retardance of the furrows. Manning roughness coefficients of 0.04 for furrows and 0.10 for corrugations are commonly used in estimating flow velocity.

Recession time, the time for water to disappear at any point after inflow ends is primarily affected by flow rate and by furrow length, shape, and slope for a specific soil. Recession time is relatively short and can be ignored when slopes exceed approximately 0.05 percent. Recession time is a very significant portion of the opportunity time on low gradient (< 0.05 percent) or level furrows. Excess opportunity time results in deep percolation, which should not exceed 20 to 25 percent of the design application depth.

Principles of control. There are three principles of water control which define the type of furrow system. These are (a) gradient with open ends—continuous uniform inflow for the entire irrigation period and recirculation or recovery of surface runoff for reuse; (b) cutback inflow with open ends—reduced inflow rate after water has advanced to the furrow end and continuation of the reduced inflow for the time required to apply the desired application; and (c) level impoundment—impoundment of the the water until intake is achieved, thus eliminating surface runoff. Principle (c) is used for level furrows or where the total fall in the furrow length does not exceed the design depth of application.

Design equations.† Design equations for furrow and corrugation irrigation describe the relationship between length, inflow time, inflow rate, deep percolation, surface runoff, and field application efficiency for selected design values of application depth, soil intake rate, and furrow slope and spacing. Separate design equations and procedures are given for each of the three types. All depths are expressed as equivalent depths over the furrow spacing and unit length to achieve uniformity of expression with other surface irrigation methods where the entire surface is inundated. The water intake per unit length of furrow is directly related to the soil surface in contact with the water, i.e., the wetted perimeter. Intake, however, is in both vertical and horizontal directions in contrast with flooding or sprinkler methods where only vertical intake occurs. The wetted perimeter is increased by an empirical constant to account for horizontal intake caused by soil moisture gradients. This is called the adjusted wetted perimeter. The empirical relationship for adjusted wetted perimeters of typical furrow and corrugation shapes is

$$P = 0.265 \, (Qn/S^{0.5})^{0.425} + 0.227 \quad \dots\dots\dots\dots\dots\dots\dots \quad [13.34]$$

where P is the adjusted wetted perimeter (m), Q is the inflow rate (L/s), S is the slope or hydraulic gradient (m/m), and n is the Manning roughness coefficient. The value of P cannot exceed the furrow spacing W.

The time for water to advance to successive points along the furrow, from regression analysis of trial measurements, is a semi-logarithmic relationship of length, inflow rate, and slope.

$$T_T = \frac{x}{f} e^{\beta} \quad \dots\dots\dots\dots\dots\dots\dots\dots\dots\dots\dots\dots\dots\dots\dots\dots \quad [13.35]$$

where T_T is the advance time (min), x is the distance (m) from upper end of the furrow to point x (the maximum value of x is L, the field length), Q is the inflow rate (L/s), S is the furrow slope (m/m), f and g are advance coefficients varying with furrow intake family, and $\beta = gx/QS^{1/2}$. Intake family and advance coefficients are listed in Table 13.1. The maximum advance distance is reached when the total intake rate along the furrow equals the inflow rate. Because the intake function monotonically decreases, this condition is never reached. However, it is closely approximated and a maximum advance distance can occur.

Gradient furrows opportunity time. The time water is available for infiltration at any point is equal to the inflow time less the time required to ad-

---

†Method of the Soil Conservation Service (USDA) 1979.

vance plus the time water remains after inflow ends.

$$T_0 = T_1 - T_T + T_r \quad \dotfill \quad [13.36]$$

where T_0 is the opportunity time at point x. Inflow time, T_1, is a constant for a specific irrigation. Advance time, T_T, increases at successive points downstream. Recession time, T_r, is assumed zero for gradient open end furrows, whether the inflow rate is constant or cut back. With this assumption and equation [13.35], opportunity time for gradient furrows is

$$T_0 = T_1 - \frac{x}{f} e^\beta \quad \dotfill \quad [13.37]$$

where T_0 and T_1 are in min. (The design inflow, T_1, is the sum of the time to advance to the end, plus the time to fill the root zone.) The average opportunity time, from integration of equation [13.37] between the limits of 0 and x and division by x is

$$T_{(0-x)} = T_1 - \frac{0.0929}{fx \left[\dfrac{0.305\beta}{x} \right]^2} \left[(\beta-1)e^\beta + 1 \right] \quad \dotfill \quad [13.38]$$

where $T_{[0-x]}$ is the average opportunity time (min) over the length x. The average opportunity time for the entire furrow, $T_{[0-L]}$ is determined from equation [13.38] with x = L. The gross water application is

$$F_g = \frac{60\,Q\,T_1}{W\,L} \quad \dotfill \quad [13.39]$$

where F_g is gross application in mm, and W is furrow spacing in meters. Cumulative intake is expressed as an equivalent depth over the furrow spacing and unit length by the equation

$$F_{(0-x)} = (a\,T_{(0-x)}^b + c)\,P/W \quad \dotfill \quad [13.40]$$

where $F_{[0-x]}$ is the equivalent intake depth in mm, $T_{[0-x]}$ is time in minutes, and a, b and c are intake family coefficients as listed in Table 13.1. The opportunity time required for intake of the selected net application depth, F_n, can be estimated by solution of equation [13.2] in the form

$$T_n = \left[(F_n \frac{W}{P} - c)/a \right]^{1/b} \quad \dotfill \quad [13.41]$$

The average intake, $(F_{[0-L]})$, for the entire furrow length is determined by equation [13.40] with time T equal to the average opportunity time $T_{[0-L]}$. Equivalent surface runoff, outflow from the graded furrow, can be estimated as the difference between the gross application, F_g, and the average intake, $F_{[0-L]}$, or:

$$RO = F_g - F_{(0-L)} \quad \dotfill \quad [13.42]$$

where RO is the average surface runoff depth in mm. Deep percolation is the average equivalent depth of water which infiltrates the soil in excess of the design application depth.

$$DP = F_{(0-L)} - F_n \quad \dots\dots\dots\dots\dots\dots\dots\dots\dots\dots\dots\dots [13.43]$$

where DP is deep percolation in mm. When the design application, at the option of the designer, is to be applied at a distance x which is less than the furrow length L, deep percolation is

$$DP = (F_{(0-x)} - F_n) \frac{x}{L} \quad \dots\dots\dots\dots\dots\dots\dots\dots\dots\dots [13.44]$$

where $F_{[0-x]}$ is the average intake (mm) over the length x, as computed from equations [13.38] and [13.40].

The application efficiency is

$$AE = 100 \, F_n/F_g \quad \dots\dots\dots\dots\dots\dots\dots\dots\dots\dots\dots\dots [13.45]$$

where AE is the application efficiency (percent). The equation for efficiency when the design application is at a distance x which is less than the furrow length becomes:

$$AE = 100 \, (F_{(0-x)} - DP)/F_g \quad \dots\dots\dots\dots\dots\dots\dots\dots\dots [13.46]$$

The procedure for design of gradient furrows or corrugations may be simplified by preparation of design charts. Separate design charts such as shown in Figs. 13.3 and 13.4 can be prepared for any combination of roughness coefficient, intake-time relationship, net application depth, and furrow slope. The chart describes the relationship between length and inflow rate with inflow time, runoff, deep percolation, and application efficiency.

Computation example-gradient furrow

Given:

Furrow intake family I_f	0.3
Length, L	275 m
Slope, S	0.004 m/m
Furrow spacing, W	0.75 m
Roughness coefficient, n	0.04
Design application depth, F_n (over full length)	75 mm
Inflow rate, Q	0.6 L/s

Intake and advance coefficient for $I_f = 0.3$, from Table 13.1

a = 0.925	f = 7.61
b = 0.720	g = 1.904×10^{-4}
c = 7.0	

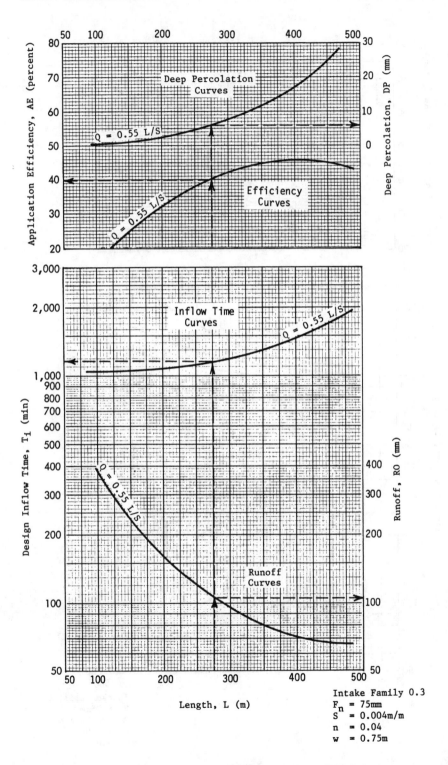

FIG. 13.3 Furrow irrigation design chart (USDA, 1979).

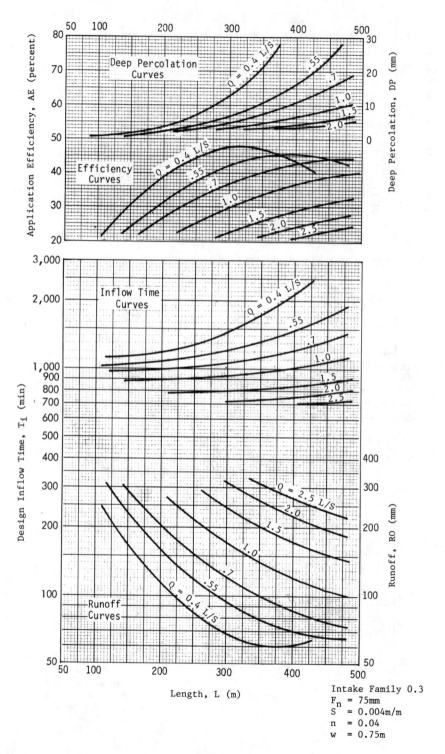

FIG. 13.4 Furrow irrigation design chart (USDA, 1979).

Find:

Design inflow time required, T_1
Surface runoff, RO
Deep percolation, DP
Application efficiency, AE

Solution:
Advance time

$$\beta = \frac{(1.904 \times 10^{-4})275}{0.6\sqrt{0.004}} = 1.38$$

$$T_T = \frac{275}{7.61} e^{1.38} = 143.6 \text{ min} \cdots\cdots\cdots\cdots \text{ (equation } [13.35])$$

Adjusted wetted perimeter

$$P = 0.265 (0.6 \times 0.04/\sqrt{0.004})^{0.425} + 0.227 = 0.40 \text{ m}$$

$$\cdots\cdots\cdots\cdots\cdots\cdots\cdots\cdots\cdots\cdots\cdots\cdots\cdots \text{ (equation } [13.34])$$

Net opportunity time

$$T_n = \left\{ \left[(75 \times 0.75/0.40) - 7.0 \right] / 0.925 \right\}^{1/0.720} = 999 \text{ min}$$

$$\cdots\cdots\cdots\cdots\cdots\cdots\cdots\cdots\cdots\cdots \text{ (equation } [13.41])$$

Design inflow time (sum of T_T and T_n)

$$T_1 = 143.6 + 999 = 1143 \text{ min}$$

Gross application

$$F_g = \frac{60(0.6)(1143)}{(0.75)(275)} = 200 \text{ mm} \quad \cdots\cdots\cdots \text{ (equation } [13.39])$$

Average opportunity time

$$T_{(0\text{-}L)} = 1143 - \frac{0.0929}{7.61(275)\left[\dfrac{(0.305)(1.38)}{275}\right]^2} \left[(1.38 - 1) e^{1.38} + 1\right]$$

$$= 1143 - 47.6 = 1095 \text{ min} \quad \cdots\cdots\cdots\cdots \text{ (equation } [13.38])$$

Average intake

$$F_{(0-L)} = \left[0.925 \, (1095)^{0.720} + 7.0 \right] \frac{0.40}{0.75} = 80 \text{ mm}$$

.... (equation [13.40])

Surface runoff

$RO = 200 - 80 = 120 \text{ mm}$ .... (equation [13.42])

Deep percolation

$DP = (80 - 75) = 5 \text{ mm}$ .... (equation [13.43])

Application efficiency

$$AE = \frac{(100)(75)}{200} = 37.5 \text{ percent}$$ .... (equation [13.45])

Gradient Furrows with Cut-back Inflow
    The volume of surface runoff from irrigation with a constant inflow may
be reduced, and application efficiency significantly improved, by reducing
the inflow rate for a portion of the total application time. This is especially
true for soils having intake rates less than that of the 1.0 Intake Family.
Where provisions are made for re-use of surface runoff, use of the cut-back
inflow method may not be desirable because of complexities in flow regulation and increases in labor requirements.
    The degree of reduction of the inflow rate and the time at which the flow
is reduced is an option of the designer. The following relationships are based
on reducing the initial inflow rate to one-half at the time the initial flow has
advanced to the end of the open-end furrow. Appropriate adjustments are required for a different operating procedure. Advance time (or cutback time) is
computed from equation [13.35], using the initial inflow rate Q. The adjusted wetted perimeter, P_1, under cut-back flow is determined from equation [13.34] using Q/2 as the flow rate. The opportunity time for intake of
the desired net application F_n at length L is calculated from equation [13.41]
after substituting P_1 for P. The total inflow time, T_1, is the sum of T_n and T_T.
The average opportunity time (T_{0avg}) for intake during the advance period is
equal to the absolute value of the second term in equation [13.38] with $x =$
L. The average intake under cut-back conditions is the sum of intake during
the advance period and intake during the remainder of the inflow time during which the inflow rate is reduced to one-half the initial.

$$F_{(0-L)} = \left[a \, (T_1 - T_{0 \text{ avg}})^b + c \right] \frac{P_1}{W} + \left[(a \, T_{0 \text{ avg}}^b + c \right] \frac{(P - P_1)}{W}$$

....[13.47]

The gross application for cut-back conditions is

$$F_g = \frac{60}{WL} \left(QT_T + \frac{Q}{2} T_n \right) \quad \dots\dots\dots\dots\dots\dots\dots\dots\dots\dots\dots\dots [13.48]$$

Calculations of surface runoff, deep percolation, and application efficiency utilize the same equations as for the noncut-back conditions. Design tables or charts for the cut-back inflow method may be prepared for each combination of intake family, net application depth, slope, roughness and furrow spacing. The curves or charts then give the inflow time, cutback time, runoff, deep percolation and efficiency for any combination furrow length and inflow rate. Fig. 13.5 is an example of a cutback furrow irrigation design chart.

Computation example—cutback gradient furrows
Given:
Same as gradient example.
Find:
Same as gradient furrow example plus time of cutback, T_T
Solution:
Cutback is the time of advance at the full flow, T_T, and is equal to that calculated in the previous example.

$T_T = 144$ min

Adjusted wetted perimeter during advance is P as calculated in the previous example.

$P = 0.40$ m

Adjusted wetted perimeter during reduced flow is calculated with the flow equal to Q/2.

$$P_1 = 0.265 \left[\frac{(0.3)(0.04)}{\sqrt{0.004}} \right]^{0.425} + 0.227 = 0.36 \text{ m}$$

$$\dots\dots\dots\dots\dots\dots\dots\dots\dots\dots\dots\dots\dots\dots\dots \text{(equation } [13.34])$$

Net application time is the time water must remain on the surface and is equal to T_n under reduced flow conditions.

$$T_n = \left[\left\{ \left[(75)(0.75)/0.36 \right] - 7.0 \right\} /0.925 \right]^{1/0.720} = 1165 \text{ min}$$

$$\dots\dots\dots\dots\dots\dots\dots\dots\dots\dots\dots\dots\dots\dots \text{(equation } [13.41])$$

Design inflow time (sum of T_T and T_n)

$T_1 = 144 + 1165 = 1309$ min

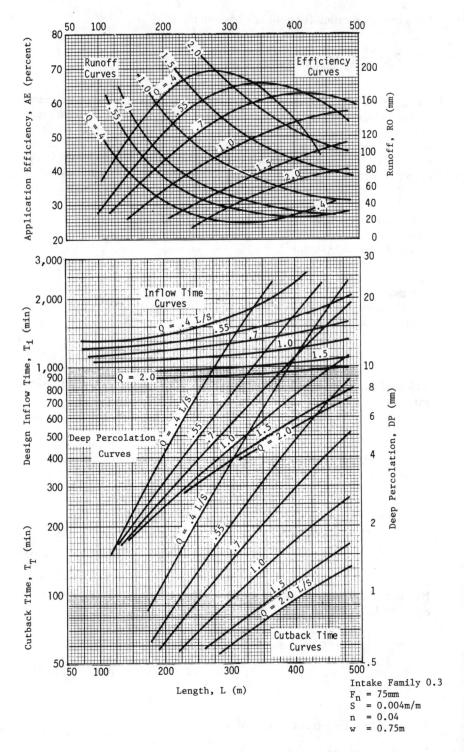

FIG. 13.5 Cutback furrow irrigation design chart (USDA, 1979).

Average opportunity time is the second term of equation [13.38] and was calculated in the previous example as part of $T_{(o-L)}$.

$$T_{0\,avg} = 47.6 \text{ min}$$

Average intake

$$F_{(0-L)} = (0.925\,(1309\text{-}47.6)^{0.720} + 7.0)\frac{0.36}{0.75} +$$

$$(0.925\,(47.6)^{0.720} + 7)\frac{(0.40\text{-}0.36)}{0.75}$$

$$= 79 + 1 = 80 \text{ mm} \quad \dots\dots\dots\dots\dots\dots\dots \text{ (equation   [13.47])}$$

Gross application

$$F_g = \frac{60}{(0.75)(275)}\big[(0.6)\,(144) + (0.6/2)\,(1165)\big] \dots \text{ (equation   [13.48])}$$

$$= 127 \text{ mm}$$

Surface runoff

$$RO = (127 - 80) = 47 \text{ mm} \dots\dots\dots\dots\dots\dots \text{ (equation   [13.42])}$$

Deep percolation

$$DP = (80 - 75) = 5 \text{ mm} \dots\dots\dots\dots\dots\dots\dots \text{ (equation   [13.43])}$$

Application efficiency

$$AE = 100\,(75/127) = 59 \text{ percent} \quad \dots\dots\dots\dots\dots \text{ (equation   [13.45])}$$

Level Impoundment Furrows

Surface runoff is elimated in level furrow systems with diked ends. Water is applied at one end of the furrow at a rate that will provide coverage of the entire length in a relatively short time. The water is then ponded until it infiltrates. The inflow rate should be large enough to advance to the end in not greater than 1.5 times the net opportunity time required for the design application. The rate, however, must not exceed the flow capacity of the furrow nor result in excessive erosion.

The design relationships for level furrows are based on the following conditions or assumptions:

1   The volume of water delivered into the furrow is equal to the average intake over the entire furrow length.

2   The intake opportunity time at the last point covered is equal to the time required for the net application to enter the soil.

3   The longest intake opportunity time at any point along the furrow is such that deep percolation is not excessive.

4   The ends of the furrows are blocked or diked to prevent outflow during the irrigation, and the depth of flow is no greater than can be contained within the furrow.

The inflow depths for level furrows may be approximated by the empirical equation:

$$\text{Inflow depth} = 0.0875 \, Q^{0.342} \quad \dots \dots \quad [13.48]$$

The average hydraulic gradient then becomes:

$$S = \frac{1}{L}(0.0875 \, Q^{0.342}) \quad \dots \dots \quad [13.50]$$

Wetted perimeter, P, is calculated from equation [13.34] and the net opportunity time, T_n, from equation [13.41]. The average opportunity time is the average advance time plus the net opportunity time, or:

$$T_{0 \, avg} = T_n + \frac{0.0929}{fL \left[\dfrac{0.305\beta}{L}\right]^2} \left[(\beta - 1)e^\beta + 1\right] \quad \dots \dots \quad [13.51]$$

Inflow time, T_1, to meet the design assumptions, becomes:

$$T_1 = \frac{PL}{60Q}\left[aT_{0 \, avg}{}^b + c\right] \quad \dots \dots \quad [13.52]$$

The gross application is given by equation [13.37]. Deep percolation, expressed as an average for the furrow length and spacing, is the difference between the gross and net application.

$$DP = F_g - F_n \quad \dots \dots \quad [13.53]$$

Application efficiency is given by equation [13.5].

Charts may also be prepared to facilitate design of level furrows. Fig. 13.6 illustrates a level furrow irrigation design chart.

Computation Example—Level Impoundment Furrows
Given:

Furrow intake family, I_f	0.3
Length, L	275 m
Furrow spacing, W	0.75 m
Roughness coefficient, n	0.04
Inflow rate, Q	1.25 L/s
Design application depth, F_n	75 mm

Find:

Inflow time required, T_1
Deep percolation, DP
Application efficiency, AE

Solution:
Average hydraulic gradient

$$S = \frac{0.0875}{275}(1.25)^{0.342} = 3.43 \times 10^{-4} \text{ m/m} \quad \dots \dots \text{(equation [13.50])}$$

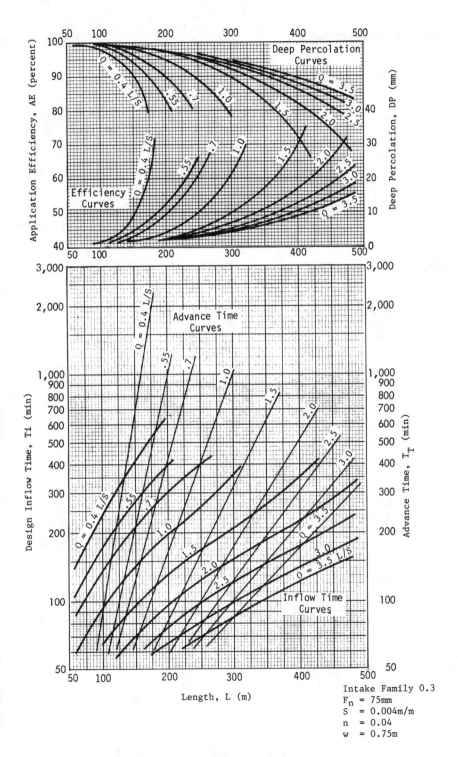

FIG. 13.6 Level furrow irrigation design chart (USDA, 1979).

Adjusted wetted perimeter

$$P = 0.265 \left[\frac{(1.25)(0.04)}{\sqrt{3.43 \times 10^{-4}}} \right]^{0.425} + 0.227 = 0.63 \text{ m}$$

. . . . (equation [13.34])

Net opportunity time

$$T_n = \left\{ \left[\frac{(75)(0.75)}{0.63} - 7.0 \right] / 0.925 \right\}^{1/0.720} = 510 \text{ min}$$

. . . . (equation [13.41])

Average opportunity time

$$\beta = \frac{(1.904 \times 10^{-4})(275)}{1.25 \sqrt{3.43 \times 10^{-4}}} = 2.262$$

$$T_{0\,avg} = 510 + \frac{0.0929}{7.61\,(275) \left[\frac{0.305\,(2.262)}{275} \right]^2} \left[(2.262 - 1)e^{2.262} + 1 \right]$$

$$= 603 \text{ min.}$$ . . . . (equation [13.51])

Inflow time

$$T_1 = \frac{(0.63\,(275)}{(60)\,(1.25)} \left[(0.925)\,(604)^{0.720} + 7.0 \right] = 231 \text{ min}$$

. . . . (equation [13.52])

Gross application

$$F_g = \frac{60\,(1.25)\,(231)}{0.75\,(275)} = 84 \text{ mm}$$ . . . . (equation [13.39])

Average deep percolation

$$DP = (84 - 75) = 9 \text{ mm}$$ . . . . (equation [13.53])

Application efficiency

$$AE = 100\,(75)/(84) = 89 \text{ percent}$$ . . . . (equation [13.45])

Compute advance time, T_T, and ratio (T_T/T_n) to determine if ratio is equal to or less than the 1.5 limit.

$$T_T = \frac{275}{7.61} e^{2.262} = 347 \text{ min} \quad \ldots\ldots\ldots\ldots\ldots\ldots \text{(equation [13.35])}$$

$T_T/T_n = 347/510 = 0.68 < 1.5$, the ratio is acceptable.

13.6.6 Determination of Furrow Intake-Time Relationships

Intake in a furrow or corrugation, unlike other surface irrigation methods where the entire soil surface is in contact with water, occurs through only a portion of the soil surface. This portion is limited to the wetted perimeter which is independent of the furrow spacing. Field measurements are necessary to determine the intake-time relationship for use in furrow design. Field tests have shown that the relationship may be associated with a series of standard intake-time curves or "families," for most soils. These are shown in Fig. 13.1 and their coefficients are listed in Table 13.1. When field tests show dissimilarity with the standard design families, the on-site measured relationship of intake vs. time should be used. Field evaluations of furrow intake-time relationships require measurement of the hydrographs of inflow and outflow from a furrow(s) with a minimum length of 60 to 90 m for high and 150 to 180 m for low intake rate soils. The furrow cross-sections and grade between inflow and outflow measuring points should be reasonably uniform. Soil water conditions should be measured and tests should be run at a level where a normal irrigation application would be needed. Present and past cropping conditions and soil conditions as influenced by cultural operations should be recorded. Monitoring of an entire irrigation set is desirable; however, an alternative is to monitor the first one-fourth to one-half of the total irrigation. Compute the total volumes of inflow and outflow at a minimum of three intermediate times. The average cumulated intake, $F_{[0-L]}$ is determined by the equation:

$$F_{(0-L)} = \frac{1}{LP} (V_{in} - V_{out} - V_s) \quad \ldots\ldots\ldots\ldots\ldots\ldots\ldots\ldots [13.54]$$

where L is the furrow length between inflow and outflow measurement points (m), P is the adjusted wetted perimeter by equation [13.34], V_{in} and V_{out} are water volumes per unit area (L) and V_s is the water volume per unit area in channel storage (L).

The channel volume, V_s, is zero at the end of the irrigation. At intermediate times, V_s may be measured, or estimated by:

$$V_s = \frac{L}{0.305} \left[2.947 (Q_{in} n/S^{1/2})^{0.753} - 0.0217 \right] \quad \ldots\ldots\ldots\ldots [13.55]$$

where Q_{in} is the average inflow rate in L/s. Determine the average opportunity time, $T_{[0-L]}$ that water is available for infiltration. A simple average of inflow and outflow times may suffice where the advance is reasonably linear. Where the advance relationship is curvilinear, determine $T_{[0-L]}$ by averaging several time ordinates at intermediate length points, or by integration of the advance time curve, dividing by the length and subtracting the resultant average time from the average recession time. Fig. 13.7 illustrates a plotting of 4 points representing average cumulated intake and associated average opportunity times, and comparison to a standard intake family. Fig. 13.8 illustrates cumulative inflow, outflow, and intake. Note that cumulative intake

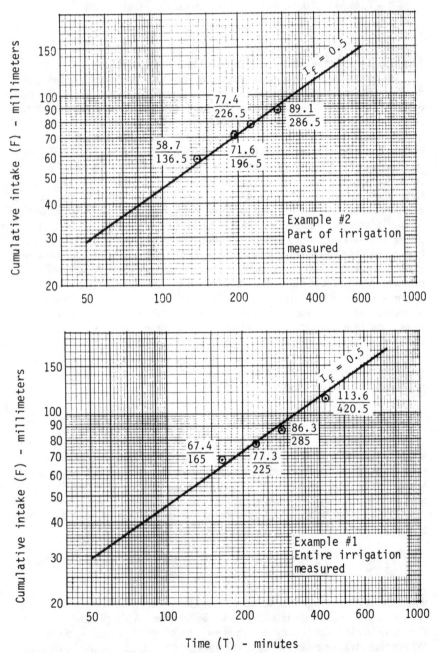

Time (T) - minutes

FIG. 13.7 Examples of cumulative intake vs. time evaluations (USDA, 1979).

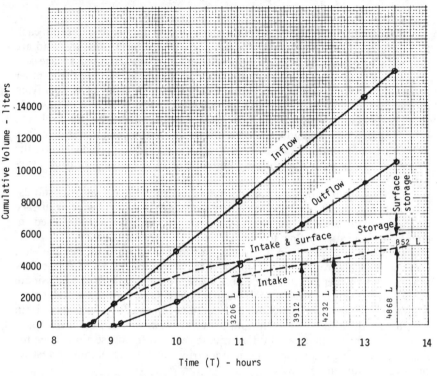

FIG. 13.8 Evaluation example no. 2.

values represent volumes of intake over a width of adjusted wetted perimeters and unit length. Equivalent depth values for the furrow spacing may be obtained by multipying by P and dividing by the furrow spacing W.

13.6.7 Selection of Headland Facilities

Water may be conveyed to fields irrigated by furrows or corrugations in lined or unlined ditches, or pipelines installed above or below the ground surface. Adequate structures must be provided in the delivery system to permit control and regulation of the water flow. Such structures include division checks, checkdrops, inverted siphons, flumes, valves, and gates. Measuring facilities to determine delivery flow rate are essential for proper irrigation management.

Supply ditches. Supply ditches must convey the design inflow rate for the number of furrows irrigated simultaneously. The water surface in the ditch should be 0.15 to 0.30 m above the field surface level. Where possible, the ditches should be constructed with a 0.1 percent grade or less to minimize the number of checks required.

Supply pipelines. Supply pipelines have the same capacity requirements as supply ditches. Pipelines, usually placed underground and either closed or vented to the atmosphere, must be designed with the hydraulic grade line above all points of delivery or have pumping facilities incorporated in the outlets. Non-vented pipelines must have adequate pressure and vacuum release appurtenances, and adequate drainage facilities. The pipe size should be large enough to limit the maximum flow velocity to 1.5 m/s. Pipeline systems may consist of a combination of both underground and surface pipe (Also see Chapter 11).

Supply outlets. Various outlet devices are used to release water into each furrow or corrugation. Outlets of equal size with uniform pressure head are desirable to deliver nearly equal flows to all furrows irrigated at one set. Rates of flow are changed by altering the size of outlets, varying the number of outlets, or changing the operating head. Common types of outlets from ditches are siphon tubes, orifices or spiles, and wier notches. Adjustable gated orifices minimize the effect of pressure head differentials on discharge rate. Ungated orifices and weirs with inverts above the normal water surface in the ditch are used in lined ditches of sufficient depth to permit raising the water surface over the opening by adjustment of regulating control structures.

Outlets from pipelines include hydrants, valves, and vertical stands. Hydrants and stands are used to control discharge into gated pipe or surface ditches which serve a number of furrows. Small valves are used to simultaneously supply flow to several furrows or corrugations.

Gated pipe. Gated pipe, usually portable and rigid or flexible, has uniformly spaced round or rectangular adjustable orifices to discharge flow into individual furrows. Short flexible sleeves may be attached to dissipate energy and minimize erosion at the furrow inlets. Gated pipe, normally placed at the head of a field, may also be located at intermediate locations within a field to reduce furrow length, and to supplement the outflow from upper reaches to achieve the desired inflow rate to the next furrow section. The pipe may, unlike concrete ditches, be temporarily removed to eliminate restrictions on equipment travel.

13.7 WATER SPREADING, (USDA, 1969)

13.7.1 General Description

Water spreading, according to the Soil Conservation Service (USDA, undated),

"is a specialized form of surface irrigation accomplished by diverting flood runoff from natural channels or water-courses and spreading the flow over relatively level areas. The diversion and spreading is controlled by a system of dams, structures, dikes, or ditches, or a combination of these, designed to accomodate a calculated rate and volume of flow"

A major difference between the water spreading system and others discussed in this chapter is that the water spreading system is designed to meet precipitation and runoff conditions of an area and apply runoff to cropped fields while the other systems are designed to deliver water in accordance with plant needs. The systems are commonly designed for 6-hour duration storms of 1.25-year, 2-year or 5-year frequency. They may also be considered as means for controlling runoff to reduce erosion and other damage to the environment (e.g., excessive sediment deposition on range land).

13.7.2 Applicability

Crops. The major crops grown under this system are those found on ranges or pastures. The purpose of the system is to increase the production of forage, hay or seed. Occasionally alfalfa, other legumes, or tame or native grasses are grown under the system. Selection of the crop to be grown will be affected by the dependability of the design precipitation and whether or not ponding must occur.

Soils. Deep, medium to moderately fine textured soils with moderately permeable subsoils and substrata are ideal. Moderately fine to fine textured soils are next best and may be used if land slopes are low enough to allow ponding. Coarse and moderately coarse soils are not suitable because infiltration rates are high and their waterholding capacities are inadequate to maintain crop growth over the infrequent intervals common to this method.

Topography. Land should be smooth and gently sloping. If detention methods are to be used, slopes are limited to 1 or 2 percent so that dikes will not be too widely spaced or too high. If continuous flow systems are used, maximum slopes are limited to 5 percent for uniform topography and 3 percent for undulating topography. Minimum slopes are limited by drainage requirements.

Climate. It is essential that expected runoff events occur at times when the soil can store added water. The water contents of frozen soils or those at field capacity are not increased by a runoff event.

Other considerations. Water must not contain excessive bedload which would deposit in the spreading area. Inaccessible spreading areas must have automatic or semi-automatic systems.

13.7.3 Advantages
Water spreading is an inexpensive means of applying water to an area to supplement rainfall. When properly designed, water spreading systems can result in large returns for relatively small investments.

13.7.4 Disadvantages
Water is applied when runoff occurs. This may or may not coincide with plant water needs, soil water storage availability, or harvest periods. These systems, when developed for inappropriate situations, or when inadequately designed, may damage farm land and property. Concentrated flows may cause serious soil erosion and sediment deposition.

13.7.5 Types of Systems
Designs of these systems can be conveniently divided into flow-type systems and detention-type systems. The flow systems incorporate free drainage from the irrigated area, while detention systems retain the applied water on the irrigated area until it has infiltrated. These two types of systems are further divided into subtypes (Table 13.13).

Spreader-ditch flow systems. These systems distribute a concentrated, short-time flow of water. The ditches are constructed to carry less water when further from the water source, and so the ditch slopes are commonly reduced

TABLE 13.13. TYPES OF WATER SPREADING SYSTEMS

Type	Upper limit of land slope, percent	Maximum flow rate, m^3/s
Flow		
Spreader ditch	5	
Syrup-pan	1.2	0.3
Dike and bleeder	1.2	0.3
Detention		
Manual inlet control	2	
Automatic inlet control	2	

(in equal increments) from about 0.3 or 0.4 percent on the upper end, to 0 percent on the lower end. Water from the upper spreader ditch is collected in the lower spreader ditches and redirected laterally. Thus, although water flows in only one direction in the spreader ditch, it may flow in two directions in the pickup ditches (Fig. 13.9), depending upon land conditions.

Syrup-pan flow systems. In this system there is a single spreader ditch at the upper end of the field, and no pickup ditches. Water spills over the sides of the spreader ditch into the field below. As the water flows down the slope it is intercepted by a contour dike which diverts flow to one end of the field. The dike is broken at that end, and water flows into the next section of the field. It again flows over the field, is picked up by another contour dike and diverted to the opposite end of the field. The system is repeated until the lowest part of the field is reached. Possible ditch and dike layouts and dike construction are given in Fig. 13.10. The maximum distance between dikes is

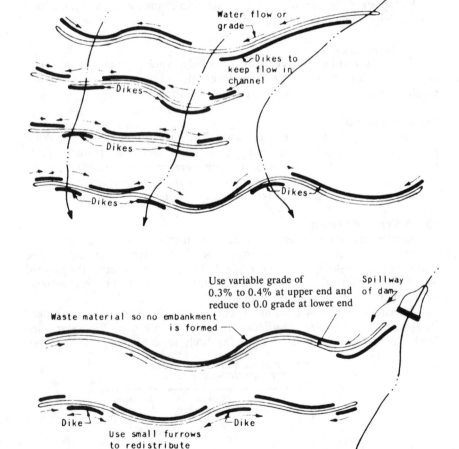

FIG. 13.9 Spreader-ditch flow type systems (after USDA, undated).

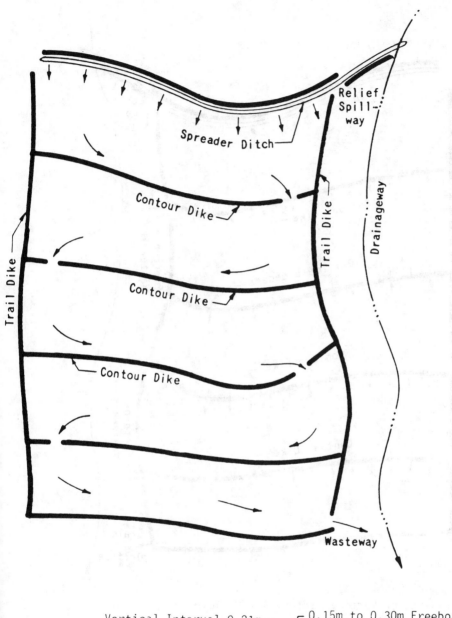

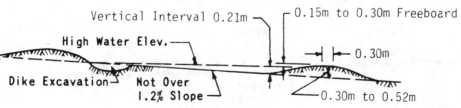

SECTION THROUGH DIKES

FIG. 13.10 Syrup-pan flow type systems.

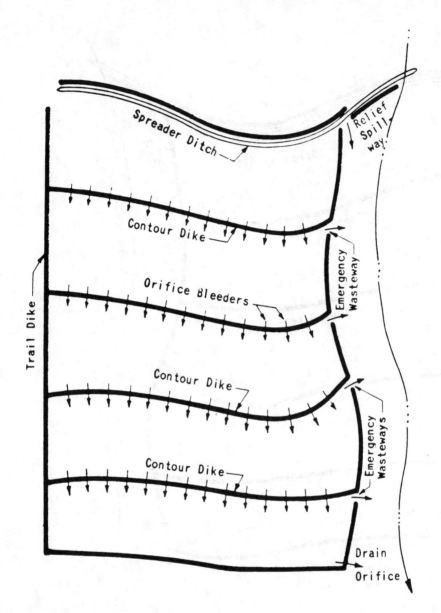

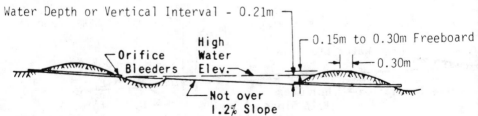

SECTION THRU DIKE AND BLEEDER

FIG. 13.11 Dikes and bleeder flow systems.

the lesser of 90 m, or that length which limits the water level at the upstream side of a dike to 0.2 m. This latter criterion, and the maximum allowable ground slope of 1.2 percent, would limit dike spacings to about 18 m.

Dikes and bleeder flow systems. This is a modification of the syrup-pan system. Water bleeds through dikes to lower portions of the field via tubes (concrete, clay, metal) or weirs placed at intervals along the dikes (Fig. 13.11). An emergency waterway must be provided to allow for stoppage of tubes.

The above three flow systems are designed for a continuous flow of water. Thus, the area covered must be adequate to infiltrate all this water, or some of it will be lost.

Manual inlet control detention systems. These systems are suitable when long duration flows such as from snow melt, are to be used. These detention systems divert flow into each dike individually (Fig. 13.12) until the desired depth has been applied. Under the maximum allowable ground slope of 2 percent, 0.6-m high dikes, spaced at 30-m intervals allow a freeboard of 0.15 m and an application depth of 0.45 m at the lower dike, and a zero application at the upper dike. Checks and turnouts are needed to divert the water from the supply ditch to each dike. A drain is also needed for each dike.

Automatic inlet control detention systems. These systems are basically similar to the manual ones, except that the control structures are designed to allow only the desired depth of water into the diked areas (Fig. 13.13). A vegetated waterway is used for the supply ditch, and grooves or small chan-

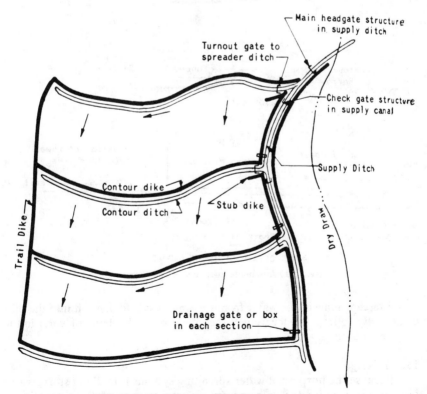

FIG. 13.12 Manual inlet control detention system.

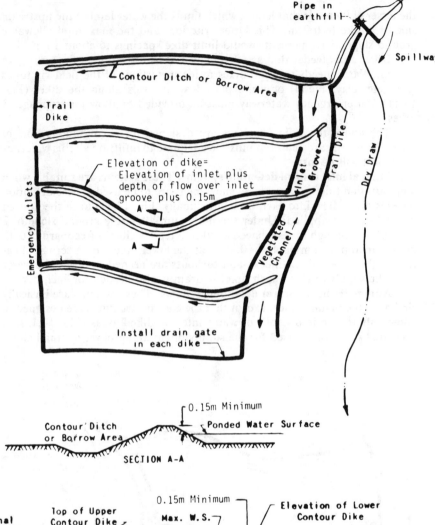

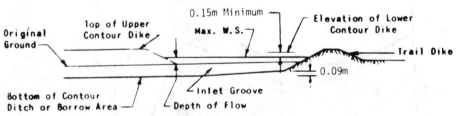

FIG. 13.13 Automatic inlet control detention system.

nels intercept some of the water from the waterway and divert it into the diked area. By appropriate location of the inlet groove, the depth of water in the
channel is limited.

13.7.6 Design

Because the purpose of water spreading systems is to divert storm water
onto agricultural land, it is necessary to know the runoff characteristics of

the area receiving the storm. Standard hydrologic methods can be used to predict volume of storm runoff, V_s, the peak discharge, Q, and the time of concentration of flow, T_c. Consideration must be given to variations in computed runoff rate and volume if there is significant detention storage in canals and reservoirs. All of these calculation are made considering the storm frequencies of Section 13.7.1. The 1.25-year design has an 80 percent chance of occurrence and is called "dependable." The 2-year frequency has a 50 percent chance of occurence and is designated "questionable." The 5-year frequency storm has a 20 percent chance of occurrence and is called "undependable."

Design application depth. In flow systems, the application depth is determined by the duration of flow and the intake characteristics of the soil. The duration of flow is related to the time of concentration, and so a relationship between these three variables can be stated, at least empirically. Table 13.14 can be used to estimate application depths from time of concentration T_c, and infiltration characteristics of the soil, as exemplified by soil type. The table gives the estimated time during which flow would occur over the land, and the depth of water which would be infiltrated under different soil conditions.

In detention systems, the application depth is equal to that which can be held within the plant's root zone (often taken as 1 m). As with flow systems, the total volume of water available determines the area which can be irrigated by a given storm.

Capacity of water supply system. It is not common to use the entire flow from a storm, and the water supply system is designed to convey only that fraction which will be diverted. The required volume to divert, V_d, is

$$V_d = d_a a , \dots\dots\dots\dots\dots\dots\dots\dots\dots\dots\dots\dots\dots\dots\dots\dots\dots\dots\dots\dots [13.55]$$

TABLE 13.14. DESIGN APPLICATION DEPTH FOR
FLOW-TYPE WATER SPREADING SYSTEMS
(AFTER USDA, UNDATED)

Drainage characteristics			Soil		
Time of concentration (T_c) hours	Estimated flow duration, hours	M medium texture*	F Mod. fine texture	H fine texture†	V very fine texture†
			Depth of application, mm		
0.5	8.5	215	120‡	75	40
1.0	9.0	225	130‡	75	40
1.5	9.5	235	130‡	80	40
2.0	10.0	245	140‡	80	45
2.5	10.5	250	140‡	80	45
3.0	11.0	260	145‡	85	45
3.5	11.5	270	150‡	85	45
4.0	12.0	280	150‡	90	50
4.5	12.5	285	160*	90	50
5.0	13.0	290	160*	95	50
5.5	13.5	300	165*	95	50
6.0	14.0	310	170*	99	50

*A free flow system is adequate for these soils.
†Detention type systems should be given first choice.
‡These soils are adapted to either flow or detention type systems.

TABLE 13.15. WATER SPREADING VOLUME RATIOS (AFTER USDA, UNDATED)	
Ratio $\dfrac{V_d}{V_s}$	r_q
0.000	0.000
0.100	0.055
0.200	0.110
0.300	0.165
0.400	0.230
0.500	0.295
0.600	0.370
0.650	0.410
0.700	0.455
0.750	0.500
0.800	0.555
0.850	0.610
0.900	0.685
0.950	0.780
1.000	1.000

TABLE 13.16. WATER SPREADING FLOW RATIOS (AFTER USDA, UNDATED)	
Ratio $\dfrac{Q_d}{Q}$	r_v
0.000	0.000
0.100	0.180
0.200	0.350
0.300	0.505
0.400	0.635
0.500	0.750
0.600	0.840
0.650	0.875
0.700	0.910
0.750	0.935
0.800	0.955
0.850	0.975
0.900	0.985
0.950	0.995
1.000	1.000

where d_a is the design application depth and a is the design spreading area. If the volume of the storm, V_s, is known then the diversion flow can be computed from two other equations. First determine the diversion flow, Q_d, from

$$Q_d = r_q Q, \dots\dots\dots\dots\dots\dots\dots [13.56]$$

where Q is the peak flow, and r_q is an empirical coefficient based upon the ratio V_d/V_s (Table 13.15). On the other hand, if the diversion flow is limited by site considerations, a diversion volume can be found.

$$V_d = r_v V_s, \dots\dots\dots\dots\dots\dots\dots [13.57]$$

where r_v is an empirical coefficient (Table 13.16) based upon the ratio Q_d/Q.

Water disposal. Excess water is returned to the water supply system, and provisions must be made for this. Erosion, other possible damage, and state and local laws must all be considered.

13.7.7 Sample Calculations
Nomenclature:

a	Spreading area, ha
P_n	Design precipitation for n-year storm, mm
R	Runoff from design rainstorm, mm
A	Drainage area, km^2
T_c	Time of concentration, h
d_a	Design application depth, mm
q	Unit peak discharge, m^3/s per 10^3 m^3 of runoff
Q	Peak discharge, m^3/s
Q_d	Diverted flow, m^3/s
R	Design storm runoff, mm
V_s	Storm runoff volume, m^3
V_d	Diverted volume, m^3

Given:

Drainage area	$A = 15.3 \text{ km}^2$
Time of concentration	$T_c = 1.77 \text{ h}$
Maximum available size of water spreading area	$a = 56.7 \text{ ha}$
Design storm runoff	
5-yr (20 percent chance)	$R = 9.65 \text{ mm}$
2-yr (50 percent chance)	$R = 4.32 \text{ mm}$
1.25-yr (80 percent chance)	$R = 1.78 \text{ mm}$
Unit peak discharge	$q = 0.0978 \text{ m}^3/\text{s per } 10^3 \text{ m}^3$
Type of system	Flow
Soil texture on spreading area	F (moderately fine)

Find:

Area, a, for undependable, questionable and dependable supplies (a) using entire flood flow, and (b) for the flood flow of the dependable system.

Solution:

Case (a). Compute, for each frequency, the storm volume, V_s ($= A R$), in 10^3 m^3; the peak discharge Q ($= q A R$) in m^3/s; the design application depth, d_a (Table 13.14, by interpolation); and the maximum spreading area a ($= V_s/d_a$). These results are tabulated below for the entire flood flow.

Class of system	R, mm	V_s, 10^3m^3	Q, m^3/s	d_a, mm	a, ha
Undependable	9.65	150	14.66	132	114
Questionable	4.32	67	6.56	132	51
Dependable	1.78	28	2.70	132	21

The entire flood flow is a dependable supply for 21 ha, a questionable supply for 51 ha and an undependable supply for 114 ha.

Case (b). The peak flow of the dependable system is 2.70 m^3/s. This can be applied to the three cases by computing the ratio Q_d/Q; finding r_v from Table 13.16; determining the volume diverted, V_d ($= r_v V_s$); and computing the area a ($= V_d/d_a$). The results are tabulated below.

Class of system	Q, m^3/s	Q_d, m^3/s	Q_d/Q	r_v	V_s, 10^3m	V_d, 10^3m	d_a, mm	a, ha
Undependable	14.66	2.70	0.18	0.32	150	48	132	36
Questionable	6.50	2.70	0.41	0.65	67	44	132	33
Dependable	2.70	2.70	1.00	1.00	28	28	132	21

13.8 REUSE SYSTEMS

13.8.1 Description

Reuse systems collect irrigation runoff water from a field and make it available for reuse. They consist of collection ditches or diked areas at the lower end of a field, an open channel or pipe drain which directs the collected water to a storage area, and a means for returning the collected water to the same field or delivering it to a different field. The return/delivery system may include a pump and a pipeline, or open channels.

13.8.2 Applicability

Reuse systems are particularly applicable where legal constraints require that water which is delivered to a farm must be used on that farm. This is often the case in groundwater control districts. It is also desirable when surface drainage might otherwise cause inundation and consequent damage to neighboring lands, and where water is pumped from aquifers since re-pumping tail water is less expensive.

Reuse systems are most commonly used with furrow irrigation systems. A reuse system allows large stream sizes to be used throughout the irrigation without excessive loss of water. Providing the economics are favorable, any surface irrigation system might benefit from a tailwater reuse system.

13.8.3 Definitions

Application duration is the elapsed time from the beginning of an irrigation set to the time at which the inlet flow is terminated for that set.

Cutback irrigation results if the initial inflow stream is reduced during an irrigation set.

Cycling-sump systems immediately recycle runoff water, as soon as the small (minimum size) sump accumulates enough water to provide proper pump operation. Pump operations may be as frequent as 15 cycles per hour.

Pumpback systems are reuse systems which deliver runoff by pumping, usually to a point at equal or greater elevation than the collection point.

Sequence systems deliver runoff by gravity to a point at lower elevation than the collection point.

Storage facilities accumulate runoff for later or immediate reuse.

Time of cutback is the elapsed time from the beginning of an irrigation set to the time at which cutback is affected.

13.8.4 Basic Principles of Design

Bondurant (1969) analyzed reuse systems and arrived at general design requirements, many of which are included in the following paragraphs.

Runoff water should be applied to a set different from that on which it occurs (unless cutback is practiced). Recirculating runoff to the same irrigation set that is generating runoff, without substantial reduction of the primary inflow, results only in temporarily storing water on the field. This will not significantly increase the infiltration rate, but will increase the rate of runoff and will probably increase erosion in a furrow. The practice also involves appreciable labor to start new streams from a constantly increasing runoff flow (unless placed in temporary storage), and usually results in different durations of application in various parts of the field.

When computed over the time interval required to irrigate the area contributing to the cycling-sump system, runoff water will have to be returned to the system at the same average rate that it is accumulated if all runoff is to be reused. If temporary storage is provided, stored runoff will eventually have to be recirculated at a rate equal to the average storage accumulation rate to prevent loss by overflow.

Maximum improvement in total water use on the farm will result if stored runoff water is used to achieve a reduced stream size for cutback irrigation; i.e., stored runoff water is pumped from a reservoir to increase the stream size (on another set) during the advance period and pumping is stopped after the set has started to produce runoff. This reduces deep percolation and runoff so that a minimum amount of water must be recirculated.

Reservoir storage can affect maximum savings of labor and water if the runoff water is placed in a reservoir adequate to retain it all for later re-

regulated use. The use of such a reservoir to also contain the initial supply to save labor is usually economical. With such a reservoir, cutback streams are usually not needed, thereby saving labor. Moderate size initial streams can provide reasonably uniform distribution along furrows and more runoff is produced than if a cutback system is used. However, this runoff can be retained on the farm by returning it to the reservoir. Erosion is increased, however, with some soils, and the extra sediment collected in the reservoir must be removed, which increases cost of operation. With erosive soils, stream size must be kept as small as practical to keep erosion to a minimum.

Runoff rate and total volume are necessary inputs to the design of any reuse system. These variables may be estimated with the methods of Section 13.6, and general guidelines for runoff prediction are given by ASAE (1980).

13.8.5 System Design

All systems should be designed in conformance with local regulations (reservoir construction, safety precautions, etc.). Nevertheless, there are certain features of design which are dependent directly on the operation of the reuse system.

There are five main components of a reuse system.

1　A system of drains to intercept and carry the runoff to a desilting basin.

2　A desilting basin to settle out excess suspended matter carried in runoff waters. This may not be needed in all cases, and when it is not needed, the collecting system carries water directly to the storage area.

3　The storage area, which is a sump, dugout, reservoir, pond, etc.

4　A pump with its inlet facilities, power unit and controls (automatic or manual). The inlet facilities should include provision for removing floating debris and weed seeds. Although pumps are not needed in all reuse systems, the functions of the inlet facilities must be met in any case.

5　A conveyance system of pipes or open channels which delivers the water back to the main irrigation system, either to the same field from which runoff originated, or to another field.

The above facilities will be discussed in the following paragraphs. The single most important item is the selection of reservoir size, and that is dependent upon the type of reuse system under consideration. Closely related to reservoir size is pump size, and that item is included with the discussion of reservoir sizes.

Cycling-sump systems. Cycling-sump systems (Davis, 1964) must not cycle more than 15 times per hour to maintain reasonable pumping plant efficiencies and to conform to pump manufacturers' recommendations.

When designing for maximum cycle rate, 15 times per h, and hence minimum reservoir and pump size,

$$S = 60P = 60I_d \quad \dots\dots\dots\dots\dots\dots\dots\dots\dots\dots\dots\dots\dots\dots [13.58]$$

where S is the storage capacity of the sump in liters, P is the pumping rate in L/s and I_d is the design inflow rate to the sump in L/s. If all runoff water is to be reused (no waste)

$$I_d = I_{max} \quad \dots\dots\dots\dots\dots\dots\dots\dots\dots\dots\dots\dots\dots\dots\dots\dots [13.59]$$

where I_{max} is the maximum runoff rate.

TABLE 13.17 MINIMUM SUMP SIZE
FOR TAILWATER RECOVERY
SYSTEMS (AFTER DAVIS, 1964)

Tailwater inflow, L/s	Inside diameter of circular sump			
	Depth in storage in sump between on and off levels, m			
	cm	cm	cm	cm
3.1	64	46	38	33
6.3	89	64	53	46
9.5	109	79	64	56
12.6	127	89	74	64
15.8	142	102	81	71
18.9	155	109	89	79
22.0	168	119	97	84
25.2	178	127	104	89
28.4	191	135	109	97
31.5	201	142	117	102
34.7	211	147	122	107
37.9	218	155	127	109

Sump sizes should conform to those given in Table 13.17 and the following additional restrictions.

1   The inside diameter of the sump should be at least five times the inside diameter of the pump column.

2   The clearance between the sump floor and the strainer must be at least one-half the inside diameter of the pump column.

3   The velocity of inflow to the sump should not exceed 0.3 m/s.

4   The lowest water level should provide submergence over the pump strainer of at least nine times the pump diameter.

The pump should be set off-center in the sump to reduce vortex formation.

Fluctuations in pressure and flow make water deliveries from a cycling-sump system difficult to handle. Thus, such systems must usually deliver the pump flow to regulating reservoirs or major supply reservoirs.

Reservoir systems for continuous pumping. There are two principle procedures used to design the reservoir systems for continuous pumping.

1   Design considering runoff rule only (Davis, 1964)—The size of a reservoir from which accumulated runoff will be pumped continuously is

$$S = V_d \left(1 - \frac{\bar{I}_d}{P} \right) \quad \dots\dots\dots\dots\dots\dots\dots\dots\dots\dots\dots\dots\dots\dots\dots [13.60]$$

where S is the reservoir capacity (not including dead storage), V_d is the design volume of runoff and $\bar{I}_d$ is the average runoff inflow rate corresponding to the design volume of inflow. If all the runoff is to be utilized

$$V_d = V \quad \dots\dots\dots\dots\dots\dots\dots\dots\dots\dots\dots\dots\dots\dots\dots\dots\dots\dots\dots [13.61]$$

where V is the total runoff. Otherwise V_d is limited by some other criteria (e.g., 90 percent of runoff will be recycled, the remainder will be wasted). The average inflow is based on the design volume of runoff and the time over which runoff occurs.

$$\bar{I}_d = \frac{V_d}{60\,T_t} \quad \dots\dots\dots\dots\dots\dots\dots\dots\dots\dots\dots\dots\dots\dots\dots\dots\dots\dots [13.62]$$

where T_r is the duration of runoff in minutes. The time of pumping is

$$T_p = \frac{\bar{I}_d T_t}{P} \quad \dots\dots\dots\dots\dots\dots\dots\dots\dots\dots\dots\dots\dots\dots\dots\dots \quad [13.63]$$

where T_P is the total time (min) which the pump should be run.

2 Design considering reapplication of runoff to the same field (Stringham and Hamad, 1975a, 1975b)—In this case, the first set is irrigated by the supply stream only, and the last set (n + 1) is irrigated by the water stored from runoff. Intermediate sets are irrigated from both sources. The number of furrows irrigated in sets 2 through n is constant, while the number irrigated in set 1 and set (n + 1) is less than in the others.

If the total application time is T, and the duration of each set is t (constant, regardless of source of water) the total number of sets irrigated by the supply is n.

$$n = \frac{T}{t} \dots\dots\dots\dots\dots\dots\dots\dots\dots\dots\dots\dots\dots\dots\dots\dots\dots\dots \quad [13.64]$$

The time T is determined from previous knowledge of the gross water to be applied and the inflow rate of the supply, Q_s. The duration of each set is determined by the designer. If the inflow to each furrow is q, the total number of furrows irrigated by the supply stream, f_s, is

$$f_s = \frac{Q_s}{q} \dots\dots\dots\dots\dots\dots\dots\dots\dots\dots\dots\dots\dots\dots\dots\dots\dots\dots \quad [13.65]$$

and that by the pumped back runoff water, f_p is

$$f_p = \frac{Q_p}{q} \dots\dots\dots\dots\dots\dots\dots\dots\dots\dots\dots\dots\dots\dots\dots\dots\dots\dots \quad [13.66]$$

where Q_p is the pump back flow rate. Clearly,

$$F = n(f_s + f_p) \dots\dots\dots\dots\dots\dots\dots\dots\dots\dots\dots\dots\dots\dots\dots\dots\dots \quad [13.67]$$

where F is the total number of furrows to be irrigated. This is determined from the field size and furrow spacing. It is obvious that there must be a balance between the chosen pump back rate, the application time, and the supply rate.

The volume of water in storage in the reservoir at the end of any given set, i, is V_i (m³).

$$V_i = 3.6 \left\{ \left[iQ_s + (i-1)Q_p \right] R_f - (i-1)Q_p \right\} t, \dots\dots\dots\dots \quad [13.68]$$

where Q_s is the supply flow rate (L/s), Q_p is the pump back rate (L/s), R_f is the total runoff volume expressed as a fraction of the applied volume, and t is the application time for each set (h). The runoff fraction can be determined by one of the methods outlined in Section 13.4. The maximum value of V_i is the necessary reservoir size.

An alternative design is to have a variable number of furrows in each of sets 2 through n to effect a minimum reservoir size (and maximum pump back rate). This design, along with tables and graphs to assist in obtaining a solution, is given by Stringham and Hamad (1975).

Other reservoir systems (Davis, 1964). Design of other reservoir systems involve storage of either the entire runoff flow or part of the runoff flow.

1   Storage of the entire runoff design flow—The most flexible systems are those which store the entire design runoff flow. These allow use of pumps of any convenient size (to deliver cutback or main streams), but they require large reservoirs.

$$S = \bar{I}_d T_t \dots\dots\dots\dots\dots\dots\dots\dots\dots\dots\dots\dots\dots\dots\dots [13.69]$$

No cycling is required during the emptying of the reservoir, and pumping can commence at the convenience of the irrigator.

2   Storage of a portion of runoff design flow—Storage of a portion of runoff design flow allows use of a smaller reservoir than when the entire design flow is stored. Some pump recycling is allowed (in contrast to the systems for continuous pumping).

13.8.6 Equipment

In the following paragraphs, the information on pumps and controls, conveyance systems, desilting provisions, drop structures, high velocity chutes, and protective dikes are taken *verbatim* (except for slight editorial changes) from Fischbach and Bondurant (1970). The information on trash and weed seed screens was provided by Pugh (1975).

Pumps and controls. Single stage reuse type turbines, low lift centrifugal, submerged centrifugal, self-priming centrifugal, or sump pumps are used in reuse systems. The electric driven single stage, reuse type turbine pump makes a convenient pumping plant (Bondurant, 1969) and an overall efficiency of 60 percent or more is easily attained. If the reuse pumping plant is automated with water level controls, absolute fail-safe priming is necessary. Reuse pumps are also powered by internal combustion engines.

Reuse systems are easily adapted to automatic controls. Automatic controls are generally of two types: (a) water level controls in the storage reservoir; and (b) time controls. Water level controls are either air-cell gage switches, float-operated switches, or electrode sensors. The air-cell gage switch uses an air cell located near the bottom of the reservoir connected to a water-level gage switch. The high or low water level contact are adjustable to make the pumping plant start or stop at preset water levels in the reservoir. The float-operated switch turns the reuse pumping plant on and off by the water level float activating a mechanical switch. The electrode sensors use the water as a conductor in the circuit. Time controls (clock operated) are sometimes used on the dugout or larger type reservoirs to turn the reuse pumping plant on and off, or off after an irrigation set has been completed.

Conveyance systems. Most reuse systems will require a return pipe line, either to another field or to the main supply ditch, gated pipe or buried pipe line. The sizes will vary according to the capacity of the reuse pump but probably will be a 100-, 150-, or 200-mm diameter pipe. Pipe lines made from plastic, concrete, asbestos-cement, steel, plastic coated aluminum or fiberglass can be used.

The accessories needed are those which are normally used for pipe lines and pumping plants such as air relief, pressure relief and vacuum relief valves. If the reuse system is connected directly to a gated pipe or pipe line with the main irrigation supply, check valves will be needed on both the reuse and main supply line pump. (Refer to appropriate ASAE standards for design, installation and performance of underground piping systems.)

Desilting provisions. Although surface irrigation is recommended on slopes that do not exceed 1½ percent, many surface irrigation systems are operating today on steeper slopes. Usually, but not always, some erosion takes place on these steeper slopes causing a silt problem. The irrigator may have too long a run for his particular slope requiring a stream size that causes some erosion. Reuse systems operated under these conditions probably will require a desilting basin located ahead of the storage pond, dugout or sump. These desilting basins may need to be cleaned each year or more often with the soil transported back onto the field.

If special desilting provisions must be made, the design criteria summarized by Brown (1950) may be used. The design requires a knowledge of the silt sizes to be removed by settling.

Drop structure, high velocity chutes and protective dikes. The sump or dugout type of storage reservoir needs some means of controlling erosion as the water from the drainage ditch enters the storage reservoir. For some sump-type installations which are constructed of concrete, concrete block, steel casings, etc., the trash screen is attached to the structure and no added drop structure is needed. However, all earth sumps and reservoirs need some structure to prevent serious erosion of the inlet to the reservoir. Cantilevered pipe inlets or most any type of drop structure or high velocity chute works well.

A dike should be constructed around the reservoir to protect it from flood damage due to rainfall or melting snow.

Trash and weed seed screens (Pugh, 1975). The type of screen used will primarily be determined by the design of the reuse systems. Trash in the runoff water must be removed to prevent damage to the return flow pump.

All the water used in a reuse system should be screened before it returns to the field, whether the water source is a reservoir or a cycling-sump system. The screen should be sufficiently small to remove all weed seeds so that fields will not be reinfested with waterborne seeds. Screens are either stationary or moving.

Stationary screens—Stationary screens are the least expensive to construct and maintain. Construction details will vary for each type of screen (Pugh and Evans, 1964).

Horizontal screens set level in both directions are generally best suited to remove both trash and weed seeds. The screen fabric should be 1.6 to 2.4 mesh/mm with 20 m² of area for each m³/s of flow. There must be at least 200 mm of free fall onto the screen to assure self cleaning. The grain of the screen fabric should be set parallel to the direction of the water flow.

Vertical screens placed perpendicular to the direction of flow can be used where there is no fall in the reuse system. The mesh size must be 1.6 to 2.4 mesh/mm to assure removal of the weed seeds. This screen must be manually cleaned and is easily clogged.

Basket outlet screens need at least 80 mm of fall to work properly. The basket can be removed for cleaning, but must be manually cleaned.

Sock-type screens are simple tubes of screen fabric with draw strings on each end. One end is fastened to the outflow pipe and the other is closed to trap the debris. The vertical, basket and sock-type screens are all designed for open ditch type of collection systems. Large screen areas of 65 m² per m³/s of flow are necessary for each of the above screens. These screens also require frequent manual cleaning to be effective. Commercially made filter traps and sand filters will effectively remove small debris and weed seeds, but will only work on the pressure side of the pump system. These filters may be back-flushed manually or automatically if electric power is available.

Paddle wheel screens utilize moving brushes across the screen fabric surface. The screen is fine mesh (1.6 to 2.4 mesh/mm) with a slope of 1 in 10 upward in the downstream direction. Provide a surface area of 20 m² per m³/s of flow.

Backing material and fabric for any of the metallic screens should be the same to prevent corrosion. The plastic or nylon screen fabrics are generally very satisfactory and eliminate any galvanic action.

Moving screens—The most common moving screens are electric or water powered. The electric powered screens are generally used where high flows are anticipated. They use a rotating tubular screen that is self cleaning, using high pressure spray jets to remove the debris.

Rotary cone-shaped drums can be used where the return flow is pumped through the drum. Propellers inside the drum provide the rotation. Screen area necessary per unit of flow depends upon the kind of debris and the effectiveness of the self-cleaning action. Generally, the screen area is similar to that for the stationary screens.

Screens should be kept tight and free from holes. Great care should be taken during the cleaning operation, as fine mesh screens are easily damaged.

All trash and weed seeds should be carefully stored and disposed of to prevent reinfestation of the fields.

13.9 AUTOMATION AND REMOTE CONTROL OF SURFACE IRRIGATION SYSTEMS

13.9.1

Automated irrigation systems reduce labor, energy and water inputs and maintain or increase farm irrigation efficiency. Automation is the use of mechanical gates, structures, controllers, and other devices and systems to automatically divert the desired amount of water onto an agricultural field to satisfy the water requirements of a growing crop.

Border and basin systems are well suited for automation and have received the most attention. Furrow and corrugation systems are much more difficult to automate because water must be uniformly distributed to each furrow or corrugation. A large number of outlets are needed per unit area and each outlet must be relatively inexpensive for an automated system to be cost effective.

Research and development by the USDA, state experiment stations and industry have produced some successful structures, controls, and other devices to automatically control irrigation water on the farm. However, automated surface irrigation is still in its infancy; few commercially produced systems and components are currently available. About two-thirds of the irrigated land in the United States is surface irrigated and most of this and the surface irrigated land in developing countries is suited for automation. To date, automation has been applied to only a small fraction of this large potential acreage.

Irrigation systems and their associated components are classified as either automatic or semiautomatic, depending upon their method of operation. Automatic systems normally operate without operator attention except for periodic inspections and routine maintenance. The irrigator may determine when and how long to irrigate and turn water into the system or start programmed controllers to initiate the automated functions. Fully automatic systems may use soil moisture sensors, such as tensiometers or electrical resistance blocks, to activate electrical controls when soil water is depleted to pre-determined levels. Meteorological data, using climate-based sensors, can also be used to predict when to irrigate and the output from a microprocessor controller can automatically begin irrigation. Once irrigation has been started, water is diverted into the farm distribution system and irrigation is completed without operator intervention. Irrigation duration may be controlled by programmed timers, soil moisture sensors or surface water sensors. Fully automatic systems require a water supply available "on demand" such as from wells or farm reservoirs. Most farm systems, however, do not have the flexibility required for complete automation.

Semiautomatic systems and controls require manual attention at each irrigation and are usually simpler and less costly than automatic systems. Most current semiautomated systems use mechanical or electronic timers to activate control structures at predetermined times. The irrigator usually determines when to begin irrigation and its duration and manually resets or returns the devices to their original positions or moves them from one location to another before the next irrigation. Parts of a given system may be automatic while other parts are semiautomatic or manually operated. The terms automatic or automated are commonly used in practice and in this chapter when referring to automated systems in general and include fully automatic, automatic and semiautomatic systems.

Most automated and some semiautomated system components are remotely controlled by centrally-located controllers. Such systems require communication between the controller and system components located in the field. Communication may be by direct interconnecting electrical wires, by hydraulic or pneumatic conduits, or by radio telemetry. Spurious signals and interference can sometimes be problems when telemetry is used.

13.9.2 Automation Principles and Design Considerations

System components. An automated surface system is similar to a regular system except it must include: (1) a field prepared for controlled irrigation water flow; (2) water supply controls including structures or valves that are automatically controlled; (3) turnout or discharge outlets that deliver a specified flow into each segment of the field being irrigated; and (4) activating mechanisims or devices to open and close gates or valves automaticaly in a selected sequence. A tailwater pickup and water reuse system that automatically recirculates irrigation runoff, or stores it for future use, is also usually needed. The mechanical components and structures that are unique to automated surface systems are discussed later.

Operational sequence. The general operating principles of automated systems are similar for pipeline and open-channel systems even though the structures, valves and other devices are different. Irrigation usually proceeds either downstream or upstream as each field segment is irrigated in sequence. With open distribution channels and an irrigation sequence that proceeds upstream from the lower end of the ditch, the field segment at the lower end of the ditch is irrigated first. Water is automatically checked consecutively at each upstream turnout in the head ditch and diverted onto the field through an automatically controlled gate or a fixed opening. When irrigation proceeds from the upper end of the ditch towards the downstream end, the sequence is reversed. An advantage of irrigation in a downstream sequence is that the ditch can be used to convey water between irrigations before the turnout gates are reset for the next irrigation. Thus, the ditch can be used to convey water to other parts of the field or to carry runoff or excess flood waters. Also, the ditch is naturally drained after each irrigation without dead storage remaining between checks. An advantage of irrigating in an upstream sequence is that in case of a gate failure, only one part of the field is missed because the next gate upstream operates as scheduled. If a malfunction occurs when irrigating in a downstream sequence, water may continue to flow on the same set until the problem is corrected. Another advantage of irrigating in an upstream direction is that flow can be more conveniently diverted into another distribution ditch or channel since the last field segment to be irrigated is at the upper end of the ditch. A disadvantage of irrigating in the upstream direction is that all timers must be operating simultaneously until each segment is irrigated. For long irrigation periods, the total irrigation time may exceed the time capacity of mechanical timers.

Random sequence is possible with automatic systems having gates or valves that can open or close against a head of water.

Irrigation timing. Irrigation duration is usually timed with mechanical (alarm clocks and 24-hour timers), electromechanical or electronic timers. Since electronic timers have become more reliable and can operate for extended time periods using battery power, they are now preferred. Although still commonly used, mechanical timers provide only one timing function, whereas irrigation requires two timing functions—one for the delay until irrigation begins and the other for the irrigation duration. Electronic timers can satisfy these two functions with a single timer and also provide an electrical output to actuate a secondary device for tripping a gate or operating a valve. Most mechanical timers use a direct mechanical linkage to trip or actuate a gate or valve.

Irrigation duration can be controlled with soil moisture sensors such as tensiometers or electrical resistance blocks or with water sensing devices such as floats if certain obstacles are overcome. The sensors can compensate for changes in advance time through a field from one irrigation to the next. One disadvantage of using a sensor to determine irrigation duration is that some means of communication between the sensor and the controller is required. The controller is usually located near the upper end of the field or near the farmstead. Thus, the control system is most feasibly restricted to radio telemetry because direct wire or fluid conduit communication from the lower end of the field is too cumbersome for multiple irrigation sets. Also, satisfactorily terminating furrow irrigation with sensors is difficult because of the interrelationships between stream size, length of run, soil intake rate and sensor location.

Use of sensors to begin irrigation, however, is feasible. With the present state-of-the-art, it is more satisfactory to use sensors to begin an irrigation sequence and then use a timer or volumetric flow measurement to determine irrigation duration. An established sequence must be used once irrigation is started. If irrigation is randomly controlled completely by sensors, management of a stream of water on the farm is very difficult.

Electromechanical and electronic programmed controllers used for sprinkler and drip systems can sometimes be used directly or modified for surface irrigation systems. Some of these controllers do not have the required time duration capacity nor can they be powered from batteries.

Cutback irrigation. One advantage of an automated system is that cutback furrow streams can be used to reduce runoff. Manual cutback is seldom practiced because of the extra labor required and the problem of handling the excess water during the cutback mode. One technique to achieve automatic cutback streams is to pump additional water from a reuse pond only during the initial or wetting phase of the irrigation (Fischbach, 1968). Another method is the split-set technique where the total set or field segment is divided into two parts (Humpherys, 1978a). The first half of the set is irrigated with the entire stream until water runs off the field. The entire stream is then directed onto the other half of the field segment for the same length of time. Water is then reintroduced into the furrows of the first half so that the entire stream of water is distributed across the total set for the remainder of the irrigation. The experimental surge flow concept of Stringham and Keller (1979) is discussed in Section 13.9.5.

Cutback streams from lined ditches can be achieved by constructing the ditch in a series of level bays with spile outlets at equal elevation along the side of the ditch. Water is released sequentially downstream from one bay to the next by timed check gates. As the water advances to the next check, the water level in the upper bay is lowered and flow from the upper bay outlets is reduced (Garton, 1966; Humpherys, 1971; Nicolaescu and Kruse, 1971; Hart and Borrelli, 1972; Evans, 1977).

Flow measurement. Flow measurement or volumetric flow control devices are necessary for an efficient water management system. With volumetric flow control, the volume rather than the time determines the irrigation duration. This method is particularly adaptable to level basin systems having a variable supply flow. With this method, a predetermined volume of water is metered into the basin and the flow measuring or volume control structure is instrumented to terminate irrigation or change irrigation sets when the required volume of water has been applied.

Constraints and limitations. A number of limiting factors need to be considered when designing automated surface irrigation systems. Some of these factors may not be applicable to a specific installation but each can affect the practicality of installing a system.

1 **Flexible water supply.** The degree of automation depends largely on the farm water supply. Semiautomatic controls are usually used when farms receive water on a rotation basis. Water on demand with flexibility in frequency, rate, and duration is needed for fully automated systems. Most existing open channel delivery systems do not have the capability to respond to variable, unscheduled deliveries as automated farm systems accept and reject water. Level top canals equipped with automatic constant water level control gates can help provide the needed flexibility (Merriam, 1974, 1977).

If farm regulating reservoirs or storage ponds are used to accumulate continuous or intermittent canal deliveries, water can be supplied to the farm distribution system with greatest flexibility. Farm runoff can be largely eliminated when a reuse system is used with these reservoirs. Trash and outlet plugging problems are also reduced because water withdrawn from the reservoir is usually cleaner than that supplied by a canal. Farm reservoirs should be located at the upper end of the field or farm. This allows continuous water delivery to the reservoir from a canal while irrigating from the reservoir at a different flow rate. Also, when irrigation is completed, the irrigation system can be shut off without having to make further provision for water being received from the canal. Water delivery from the reservoir to the irrigation distribution system can be efficiently controlled with float valves (Humpherys, 1978b; Merriam, 1973).

2 **Variable soil intake rates.** The operation and management of an automated furrow system is complicated by variable furrows intake rates. One of the primary causes of different intake rates between furrows is unequal tractor wheel traffic. Irrigating every other furrow reduces this problem where alternate furrows receive the same amount of traffic. Variable field slopes also can cause unequal furrow intake rates. Fields that are automated should be planed to a uniform slope to prevent sediment deposition and furrow overtopping. Plant leaves, stems and residues in the furrows and rodent activity affect the furrow intake rate. One of the primary objectives in automating irrigation systems, that of reducing labor, is partially defeated if gates or other distribution outlets must be individually adjusted throughout the season or during irrigations to compensate for variable intake rates. This problem is greatest with easily erodible soils because the stream size must be controlled to minimize furrow erosion. Where a return flow system is used and soils do not erode easily, runoff does not have to be carefully controlled and relatively large streams can be used. Besides reducing the effect of variable intake rates, large streams improve uniformity and increase the probability of all furrows being completely irrigated.

3 **Equipment factors.** One of the constraints to automating surface irrigation systems is the lack of well designed, self-contained, complete commercial systems or system components. Currently, equipment and components needed for an automated system often must be modified or adapted from other uses.

A frequent limitation is the lack of AC electrical power in the field where system components are located. In contrast to sprinkler systems where AC electrical energy is usually provided at a central location, such as for a pump, automated surface systems require very small amounts of energy at a number of locations. Because of this small demand, installing or extending a power line to provide service for automated surface system controls is seldom feasible. Battery or solar-powered electrical components have only very recently become available. Solid state electronic devices, latching relays and latching solenoids that use very little energy and can be powered by batteries are required in most systems. Solar energy will be increasingly used as improved technology lowers component costs.

If used to apply fertilizer, irrigation water may contain high concentrations of soluble salts. System components must be carefully chosen and mounted to avoid corrosion damage. This problem can be minimized by use of plastic, but certain plastics that deteriorate when exposed to direct sunlight and high temperatures should be avoided. Timers and electrical components should be adequately dust and waterproofed. Water-filled or -actuated components also must be protected from freezing. Pneumatic components can be used to avoid frost damage, but maintenance of airtightness is sometimes difficult.

4 Distribution outlets. The cost of automated furrow systems depends upon the number of distribution outlets required. The cost of each outlet must be relatively low for the total system cost to be feasible. For a 400-m length of run, and depending upon the furrow spacing, the number of outlets will vary from about 22 to 32 per ha. By comparison, the number of sprinklers on a mechanical-move center pivot system ranges from about 0.8 to 1.9 per ha.

Tubes, notched outlets, and weir outlets have been used to control flow into each furrow but have not been very satisfactory. Gated pipe currently is the most feasible method of distributing water in an automated furrow system.

5 Erosion. Excessive erosion may occur where the streams from gated pipe strike the soil surface. This problem is aggravated on steeply sloping land that increases pipeline pressure. Fabric tubes can be used over each pipe gate to minimize erosion, but they are a nuisance and add to the total system cost.

Orifice plates or other energy dissipating devices installed in the pipe are sometimes used to control pipe water pressure. Small overflow stands similar to concrete stands used in gravity pipeline systems can also be used to limit the amount of head that can develop in the pipe on steep slopes.

6 Trash and debris. Trash is a common problem where water is supplied from a canal. Clogging of gates and furrow discharge outlets cannot be tolerated in an automated system where an operator is not present to keep outlets clean. Clean water is essential; it is difficult to find a screen that is satisfactory for many canal turnouts, particularly where electric power is not available.

7 Rodents. Automated systems using air supply lines are particularly susceptible to damage by rodents. The most satisfactory solution has been to encase the plastic tubing and other susceptible components in rodent-proof material. Air lines are being installed in concrete lined ditches to reduce this problem. Furrow blockage by gopher mounds may cause nonuniform irrigation with permanent or semipermanent crops such as pasture, alfalfa and orchards.

13.9.3 Pipeline Distribution Systems

Pipeline systems are easier to automate than are open channel systems. Pipelines and associated facilities may be buried or placed on the surface. Many existing systems can be equipped with automated valves and other components to reduce the cost of converting to an automated system. Pipelines can be designed using the criteria presented in Chapter 11. Automated valves, outlets, or both are used to sequence water from one set or field segment to another.

Pneumatic valves. One of the first pneumatically operated valves, developed by Haise et al. (1965), consists of an inflatable O-ring, or doughnut-shaped diaphragm, constructed from a rubber inner tube and supported with a butyl rubber cover. When mounted on an alfalfa valve and in-

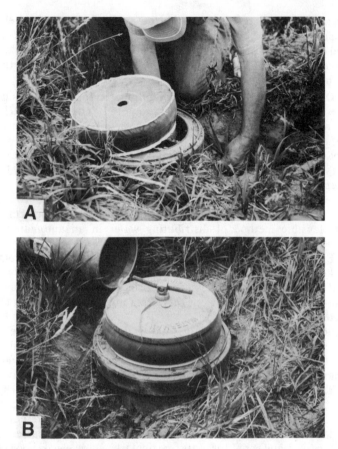

FIG. 13.14 Pneumatic irrigation valve for automating pipelines. Valve being installed on an existing alfalfa valve (A), and inflated valve stopping flow of water (B).

flated with air, the tube forms an annular seal betwen the alfalfa valve seat and lid as shown in Fig. 13.14. In addition to controlling discharge from buried pipelines into borders and basins, it can be used with a hydrant for gated pipe irrigation (Haise et al., 1980), and also can be used with portable controls (Edling et al., 1978). A modification by Fischbach and Goodding (1971), shown in Fig. 13.15, is commercially produced.* The commercial valve has a specially fabricated, pneumatic diaphragm. It is constructed with a male pipe fitting on the bottom inlet and is installed directly on top of a riser from an underground buried pipeline. Pneumatic valves, except the portable models, require an air compressor to provide air for actuation.

Water-operated valves. Self-closing and regulating valves using water for actuation were developed by Haise et al. (1980). These can be used for both on/off and modulating discharge control. Water-filled bladder valves developed by Humpherys and Stacey (1975) are used to control irrigation through gated pipe and buried lateral distribution pipelines. The valves

*Manufactured by the Econogation Valve Co., Humboldt, NE. Trade and company names are shown for the benefit of the reader and do not imply endorsement or preferential treatment of the company or products listed.

FIG. 13.15 Pneumatic valve for automating gated pipe systems.

operate as independent units without an outside energy source. Water from the pipeline is used to close the valve. Valve opening and closing are controlled by battery powered, timer activated, 3-way pilot valves. The valve model shown in Fig. 13.16 is commercially available.†

Buried distribution laterals. A system developed by Varlev (1973, 1978) using buried distribution laterals is used in Bulgaria. This system consists of a buried pipeline with a telescoping riser for each furrow. The risers are extended to the surface hydraulically at the beginning of the irrigation season and pushed down by hand at the end. Water is sequenced from one buried pipeline to another with automatic valves that utilize water pressure in the pipeline for activation (similar to those of Humpherys and Stacey, 1975). A similar system that currently uses flexible risers and calibrated orifices at each furrow outlet was described by Worstell (1976, 1979).

The advantage of a buried system is that field tillage and cultural operations can be performed over the top of the irrigation pipeline. A multi-set

†Manufactured by Hastings Irrigation Pipe Co., Hastings, NE.

FIG. 13.16 Automated valves that use water from the pipeline for operation.

FIG. 13.17 Timer-controlled, semiautomatic drop closed check gate.

concept (Rasmussen et al., 1973) can be used to introduce water into irrigation furrows at one or more midfield locations. This shortens the effective length of run and makes small stream sizes possible. Using small furrow streams reduces erosion and runoff from a field because field runoff occurs primarily from the last subset at the lower end of the field.

Reuse systems. With the appropriate interfacing controls, reuse systems (Section 13.8) can be an important part of an automated pipeline system.

13.9.4 Open Channel Distribution Systems

Simple, timer-controlled gates of various designs have been used in farm ditches for years. Individual farmers have made many of these in their own shops.

Drop open and drop closed gates. The most common gates for open channel systems are the drop closed and drop open types. The drop closed gate is used to divert water directly onto irrigated fields or from one ditch into another. In the open position, it is suspended over a flow opening and, when tripped, falls by its own weight to stop the flow of water (Fig. 13.17). It is

FIG. 13.18 Hydraulically-tripped drop open check gate in a lined ditch.

used in lined and unlined ditches and may be permanently or portably mounted. Semiautomatic drop closed gates and dams tripped by mechanical timers are extensively used in New Zealand with sill- or weir-type side outlets and borders to irrigate pastures (Taylor, 1965; Stoker, 1978).

The drop open gate is hinged so that when tripped, it either falls or swings open to allow water to flow downstream (Fig. 13.18). It is normally used as a companion gate to the drop closed gate. When used in pairs, one gate opens while the other closes to divert water from one turnout or flow opening into another. They may be tripped by different actuating devices including mechanical timers, solenoids, floats, or pneumatic and hydraulic cylinders. Both types of gates were widely used in Hawaii for sugar cane irrigation before the advent of drip irrigation (Reynolds, 1968). Gates of various configurations and designs have been used by different investigators (Calder and Weston, 1966; Kimberlin, 1966; Humpherys, 1969; Hart and Borrelli, 1970; Lorimor, 1973; Evans, 1977; Haise et al., 1981).

Drop closed and drop open gates are sometimes mounted together on the same structure to form a "combination" gate (Humpherys, 1974). These are used as turnouts into a border or basin where it is necessary to first divert water onto the field and then terminate irrigation by closing the opening against a head of water. They are needed where head ditch slopes are relatively flat and there is more than one turnout between each ditch check.

Automated lift-gates and tile outlets. Lift-gates are commonly used in Arizona where large irrigation streams are used. Dedrick and Erie (1978) and Haise et al. (1981) described equipment used to automate lift-gate systems to irrigate level basins. The lift-gates shown in Fig. 13.19 are actuated with air cylinders. Compressed air is diverted through a four-way pneumatically- or electrically-operated pilot valve to the bottom of the cylinder to open the gate or to the top of the cylinder to close the gate. The automated gates are used both as check gates and as diversion structures.

Concrete pipe tile outlets 410 mm in diameter are used with the lift-gates to divert water from a supply ditch into level basins (Erie and Dedrick, 1978;

FIG. 13.19 Automated lift-gate for a level-basin irrigation system.

FIG. 13.20 Pneumatic pillow-disc valve installed on a concrete pipe tile outlet. Note the dissipator box.

Haise et al., 1981). A pillow-disc valve, consisting of a metal ring insert onto which are attached a permanent stop, an air pillow, and a moveable plate or disc, is mounted on the discharge end of the outlet as shown in Fig. 13.20. The plate is forced against the seat by the air inflated pillow to stop the flow of water from the outlet. Concrete energy-dissipator boxes are used at the end of the outlet to control erosion. About 80 kPa air pressure is needed to operate the air pillows, and 345 kPa is used to operate the lift-gates. The air compressor, irrigation controller, and other associated components are located in a control center storage shed where AC electrical power is available.

Traveling dams. Machines that divert water continuously from an irrigation ditch are sometimes used with close growing forage and grain crops where large streams of water are available for surface flooding. These slow-moving commercial machines, powered by small gasoline engines, straddle the ditch and pull canvas, plastic, or rubber dams that cause the water to overflow the ditch banks. The ditches are usually constructed so that the upper bank is higher than the lower bank so that the water always flows over the lower downstream ditch bank.

Other gates and outlets. A number of other different types of gates and outlets have been used to a limited extent. Circular skimming and rectangular weir type outlets with automatic check gates have been used in Australia‡ (Robinson, 1972). Weir-type outlets were described by Sweeten and Garton (1970). Several types of pneumatically- and hydraulically-operated butterfly, modulating, push-off, pillow-disc, self-closing and regulating, and fluidic diverter outlets and check gates were developed by Haise and Kruse (1969) and by Haise et al. (1981). Pillow-disc valves are used as discharge outlets from ditch turnouts, buried pipeline risers and in-

---

‡Metseal automatic irrigation structures developed and licensed by the State Rivers and Water Supply Commission of Victoria, Armadale, Australia.

dividual furrow gates. They consist of a disc that is forced against a seat by an inflated air bladder or pillow positioned directly over the disc. Bowman (1969) tried radio controls with soil moisture sensors and center-of-pressure gates.

13.9.5 Experimental Systems

Most automated systems are still in the experimental stage of development. Several systems have not progressed so far that they have been used but have potential for practical application with further development.

Single-pipe system. Most automated furrow irrigation systems require two parallel pipelines at the upper end of an irrigation run. The conveyance or main pipeline is either buried or placed on the surface. The second pipeline, usually gated pipe, serves as the distribution line. The system could be simplified and the cost reduced if a single pipeline could serve both functions. Several attempts have been made to accomplish this and experimental work is still underway. Reynolds (1968) described the "miniwai" system in Hawaii which uses a membrane installed inside the distribution pipe to cover a group of furrow outlets or openings simultaneously when water flows above the membrane. Water is discharged from the openings when flow inside the pipe is below the membrane. Fischbach used an air cylinder connected to cables or rods to operate sliding pipe gates (Haise and Fischbach, 1970). Stringham and Keller (1979) used a bank of automatically controlled individual pneumatic valves in a single pipe to test the surge flow irrigations concept. Haise et al. (1981) automated individual openings in gated distribution pipe with hydraulic pipe gates and pneumatic pillow-disc valves.

Surge flow. The surge flow concept of automatic cutback irrigation was investigated by Stringham and Keller (1979). Cutback is achieved by using intermittent surge flows. Banks of furrow valves are automatically controlled to be either completely open or closed. When the valves are open half the time and closed half the time, then the full flow running half the time in a given furrow produces about the same average stream size as half the flow running full time. The cycle time can be variable.

Traveling irrigator. A mechanical, continuous-move distribution lateral is being developed for furrow irrigation by Lyle and Bordovsky (1979). This system used a modified rectilinear-move sprinkler lateral. Modifications include replacing sprinklers with drop tubes and orifice-controlled emitters for each furrow and adding the components necessary for propulsion, guidance and control. The system operates on 70 to 170 kPa and moves in a continuous rectilinear pattern through the field. It is used in conjunction with microbasins formed in the furrows. Tillage implements form the microbasins which are used to maximize both irrigation efficiency and rainfall utilization.

Controllers. Solid state electronic technology using microprocessors is developing so rapidly that it is impractical to describe specific units, although some units are available commercially. New concepts include a battery-powered controller capable of integrating flow rates through a nonlinear open channel flow measurement device (Duke et al. 1978). The controller can be programmed to control a large number of turnouts in any sequence and to terminate irrigation after the programmed volume of water has been applied. An experimental soil moisture monitor system that provides a visual indication of soil moisture status and the capability of controlling irrigation automatically was described (Anon. 1975).

13.9.6 Benefits and Costs

Because of the experimental nature and limited use of most automatic surface irrigation systems, measured irrigation efficiency and labor input data are limited. Although there are many potential benefits, the dominant reasons that irrigators automate are convenience and labor reduction. Stoker (1978) reported that with semiautomated border systems and a minimum 0.23 m³/s stream size, 24 ha can be irrigated with one manhour of labor and with a fully automated system using sensors, 81 ha can be irrigated with one manhour. In contrast, only about 0.8 ha/h can be irrigated with nonautomated border systems. With semiautomated level basins in Nevada, labor requirements were reduced 80 percent compared to manual irrigation using graded borders (Kimberlin, 1966), and irrigation efficiency was increased 13 percent by using semiautomation in Hawaii (Reynolds, 1968). Only 1/6 as much labor was required for semiautomated mountain meadow systems as for manual irrigation systems and only half as much water was used (Lorimor, 1973). Fischbach and Somerhalder (1971) reported average irrigation efficiencies of 65 percent for automated gated pipe without a reuse system and 92 percent with a reuse system. The average runoff, 27 percent without reuse, was recovered when the reuse system was used. Humpherys, (1978a) reported a 20 percent decrease in runoff with a corresponding increase in applied water use efficiency for a semiautomated system using gated pipe on plots of sugarbeets and corn as compared with nonautomated plots.

Irrigated sets of either 12 or 24 h are commonly used because of the inconvenience of making irrigation sets more frequent or at time intervals other than 12 or 24 hours. With automation, irrigation sets of different durations can be made as required by soil and plant conditions.

Fertilizer loss can be reduced by automation because of the reduction in runoff and deep percolation and the increased use of reuse systems. Automation can reduce energy costs by making surface irrigation more acceptable than alternative systems that use more energy. Batty et al. (1975) estimated that sprinkler systems use from 4 to 13 times more total energy than do surface systems.

Automation costs are highly variable and depend upon the water supply; degree of automation (automatic or semiautomatic); existing facilities; irrigation method (whether furrow, border, level basin, etc.); and field conditions such as degree of leveling or land surface preparation required, length of run, slope, soil texture, field size and shape, etc. Depending upon local conditions, costs can range from as little as 1/4 to 1/2 the cost of a self-move wheel roll sprinkler system for some of the simple low-cost surface systems to as much as 1-1/2 times the cost of a center pivot sprinkler system for a more elaborate buried lateral multiset system. The cost of a double pipe automated gated pipe system is approximately comparable to that of a self-move side roll sprinkler system.

13.10 References

1 ASAE. 1980. Design and installation of surface irrigation reuse systems. Engineering practice recommended by SW-242, Surface Irrigation Committee of the American Society of Agricultural Engineers.

2 Anonymous. 1975. Go, NoGo. Irrigation Age. 10(1):22-23.

3 Batty, J. C., S. N. Hamad and J. Keller. 1975. Energy inputs to irrigation. Proc. Am. Soc. Civil Engr., J. Irrig. Drain., Div., 101(IR4):293-307.

4 Bowman, C. C. 1969. Semi-automation of irrigation. Internl. Comm. on Irrig. and Drain., 7th Cong. on Irrig. and Drain., Q 24, Rpt. 19, Mexico City, p. 24.271-24.275.

5 Bondurant, J. A. 1969. Design of recirculating irrigation systems. TRANSACTIONS of the ASAE 12(2):195-198, 201.

6 Booher, L. J. 1974. Surface irrigation, FAO (Food and Agriculture Organization of the United Nations) Agr. Develop. Paper No. 95, p. 80-92.

7 Brown, C. B. 1950. Sediment transportation. p. 769-857. In: Hunter Rouse (Ed.). Eng. Hydr., Chpt. XII. Wiley, 1039 pp.

8 Calder, G. G. and L. H. Weston. 1966. Automatic system of farm irrigation. New Zealand J. Agr. 112(2):19, 21, 23.

9 Davis, J. R. 1964. Design of tailwater systems. TRANSACTIONS of the ASAE 7(3):336-338.

10 Dedrick, A. R. and L. J. Erie. 1978. Automation of on-farm irrigation turnouts utilizing jack-gates. TRANSACTIONS of the ASAE 21(1):92-96.

11 Duke, H. R., M. L. Payne and D. C. Kincaid. 1978. Volumetric control for irrigation automation. ASAE Paper No. 78-2545, ASAE, St. Joseph, MI 49085.

12 Edling, R. J., H. R. Duke and M. L. Payne. 1978. Utilization of a portable solid state furrow irrigation valve controller. ASAE Paper No. 78-2546, ASAE, St. Joseph, MI 49085.

13 Erie, L. J. and A. R. Dedrick. 1978. Automation of an open-ditch irrigation conveyance system utilizing tile outlets. TRANSACTIONS of the ASAE 21(1):119-123.

14 Erie, L. J. and A. R. Dedrick. 1979. Level-basin irrigation: a method for conserving water and labor. Farmers Bull. 2261. Supt. of Doc., U.S. Government Printing Office. Washington, DC, 24 p.

15 Evans, R. G. 1977. Improved semiautomatic gates for cut-back surface irrigation systems. TRANSACTIONS of the ASAE 20(1):105-108, 112.

16 Fischbach, P. E. 1968. Design of an automated surface irrigation system. Proc. Am. Soc. Civil Engr., Irrig. and Drain. Spec. Conf., Phoenix, AZ. p. 219-237.

17 Fischbach, P. E. and J. A. Bondurant. 1970. Recirculating irrigation water. ASAE Nat. Irrigation Symp. Paper No. N, Lincoln, NE.

18 Fischbach, P. E. and R. Goodding II. 1971. An automated surface irrigation valve. Agr. Eng. 52(11):584-85.

19 Fischbach, P. E., H. R. Mulliner and J. F. Decker. 1958. Efficient irrigation. University of Nebraska, College of Agriculture, Extension Service, Circular 58-704.

20 Fischbach, P. E. and B. R. Somerholder. 1971. Efficiencies of an automated surface irrigation system with and without a runoff re-use system. TRANSACTIONS of the ASAE 14(4):466-470.

21 Garton, J. E. 1966. Designing an automatic cut-back furrow irrigation system. Okla. Agr. Exp. Sta. Bull. B-651, 20 p.

22 Haise, H. R., E. G. Kruse, and N. A. Dimick. 1965. Pneumatic valves for automation of irrigation systems. U.S. Dep. Agr., ARS 41-104. 21 pp.

23 Haise, H. R. and P. E. Fischbach. 1970. Auto-mechanization of pipe distribution systems. Natl. Irrig. Symp., Univ. Nebr. p. M1-M15.

24 Haise, H. R. and E. G. Kruse. 1969. Automation of surface irrigation systems: The state-of-the-art. Proc. Am. Soc. Civ. Engr., J. Irrig. and Drain. Div., 95(IR4):503-516.

25 Haise, H. R., E. G. Kruse, M. L. Payne and H. R. Duke. 1981. Automation of surface irrigation: Summary of 15 Years USDA Research and Development at Fort Collins, CO. USDA, SEA. Prod. Res. Rept. 179, 60 p.

26 Hart, W. E. and J. Borrelli. 1970. Mechanized surface irrigation systems for rolling lands. Univ. of Calif. Water Resour. Ctr. Contr. 133:93.

27 Hart, W. E. and J. Borrelli. 1972. Distribution channels with multiple outlets. Proc. Am. Soc. Civil Engr., J. of the Irrig. and Drain. Div. 98(IR2):267-274.

28 Humpherys, A. S. 1969. Mechanical structures for farm irrigation. Proc. Am. Soc. Civil Engr., J. of the Irrig. Drain. Div. 95(IR4):463-479.

29 Humpherys, A. S. 1971. Automatic furrow irrigation systems. TRANSACTIONS of the ASAE 14(3):466-470.

30 Humpherys, A. S. 1974. Automatic controls for surface irrigation systems. Proc. IFAC Symp. on Automatic Controls for Agr., Assoc. Comm. on Automatic Control, Natl. Res. Council of Canada, Saskatoon. p. C-6, 1-16.

31 Humpherys. A. S. 1978a. Improving farm irrigation systems by automation. Proc. Intl. Comm. on Irrig. and Drain., 10th Cong., Athens, Greece. Q. 35, R. 5, p. 35.90-35.98.

32 Humpherys, A. S. 1978b. Float valves improve pipeline systems. Irrig. Age 12(7):12, 15.

33 Humpherys, A. S. and R. L. Stacey. 1975. Automatic valves for surface irrigation pipelines. Proc. Am. Soc. Civil Engr., J. of the Irrig. and Drain. Div. 101(IR2):95-109.

34 Kinberlin, L. 1966. Automatic basin systems reduce irrigation water and labor needs. Soil Cons. 32(2):39-40.

35 Lorimor, J. C. 1973. Semiautomatic gates for surface irrigation. Colo. State Univ. Exp. Sta. Progr. Rpt. PR73-77.

36 Lyle, W. M. and J. P. Bordovsky. 1979. Traveling low energy precision irrigator. Proc. Am. Soc. Civil Engr. Irrig. and Drain. Div. Spec. Conf. Albuquerque, New Mexico, p. 121-132.

37 Merriam, J. L. 1973. Float valve provides variable flow rates at low pressures. Proc. Am. Soc. Civil Engr. Irrig. and Drain. Div. Spec. Conf., Fort Collins, CO, p. 385-402.

38 Merriam, J. L. 1974. Discussion: Operation and maintenance of irrigation and drainage systems: Section III-Operation. Proc. Am. Soc. Civil Engr., J. Irrig. Drain. Div. 100(IR3):399-402.

39 Merriam, J. L. 1977. Level top canals for semi-automation of on-farm irrigation and supply systems. Proc. Am. Soc. Civil Engr. Irrig. and Drain. Div. Spec. Conf. Reno, Nev., p. 217-224.

40 Nicolaescu, I. and E. G. Kruse. 1971. Automatic cutback furrow irrigation system design. Proc. Am. Soc. Civil Engr., J. of the Irrig. and Drain. Div. 97(IR3):343-353.

41 Pugh, W. J. 1975. Private communication to William E. Hart.

42 Pugh, W. J. and N. A. Evans. 1964. Weed seed and trash screens for irrigation water. Colo. Agr. Exp. Sta. Bull. 522-S, Colo. State Univ., Fort Collins.

43 Rasmussen, W. W., J. A. Bondurant and R. D. Berg. 1973. Multiset irrigation systems. Internl. Comm. on Irrig. and Drain. Bull., p. 48-52.

44 Reynolds, W. N. 1968. Automatic surface irrigation in Hawaii. Proc. Am. Soc. Civil Engr., Irrig. and Drain. Div. Spec. Conf., Phoenix, AZ, p. 201-218.

45 Robinson, E. P. 1972. Automatic control of farm channels. 8th Cong. Internl. Comm. on Irrig. and Drain., Varna, Bulgaria, Q. 28.2, R.21, p.28.2.293-28.2.296.

46 Stoker, R. 1978. Simple efficient automatic irrigation. Irri. Age. 12(6):76-77.

47 Stringham, G. E. and S. N. Hamad. 1975a. Irrigation runoff recovery in the design of constant furrow discharge irrigation systems. TRANSACTIONS of the ASAE 18(1):79-84.

48 Stringham, G. E. and S. N. Hamad. 1975b. Design of irrigation runoff recovery systems. Proc. Am. Soc. Civil Engr. J. Irrig. and Drain. Div. 101(IR3):209-219.

49 Stringham, G. E. and J. Keller. 1979. Surge flow for automatic irrigation. Proc. Am. Soc. Civil Engr. Irrig. and Drain. Div. Spec. Conf., Albuquerque, New Mexico. p. 132-142.

50 Sweeten, J. M. and J. E. Garton. 1970. The hydraulics of an automated furrow irrigation system with rectangular side weir outlets. TRANSACTIONS of the ASAE 13(6):746-751.

51 Taylor, A. R. 1965. Notable advance in automatic irrigation. New Zealand J. of Agr. 111(4):67-69, 71.

52 USDA. 1967. Planning farm irrigation systems. Chap. 3, Sect. 15 (Irrigation) Soil Conserv. Serv. Natl. Eng. Handb.

53 USDA. 1969. Field engineering manual for conservation practices. Soil Conserv. Serv.

54 USDA. 1974. Border irrigation. Chap. 4, Sect. 15 (Irrigation) Soil Conserv. Serv. Natl. Eng. Handb.

55 USDA. 1979. Furrow Irrigation. Chapt. 5, Sect. 15 (Irrigation) Soil Conserv. Serv. Natl. Eng. Handb.

56 USDA. Undated. Water spreading. Part 14, Eng. Handb. for Work Unit Staffs. Soil Conserv. Serv.

57 Varlev, I. 1973. Surface irrigation underground pipelines with movable risers. TRANSACTIONS of the ASAE 16(4):787-789.

58 Varlev, I. 1978. Reconstruction of open channel systems into automated underground pipe systems for surface irrigation. Proc. Intl. Comm. on Irrig. and Drain., 10th Cong., Athens, Greece. Q. 35, R19, p. 35.261-35.271.

59 Worstell, R. V. 1976. An experimental buried multiset irrigation system. TRANSACTIONS of the ASAE 19(6):1122-1128.

60 Worstell, R. V. 1979. Selecting a buried gravity irrigation system. TRANSACTIONS of the ASAE 22(1):110-114.

chapter 14

FLUID DYNAMICS OF
SPRINKLER SYSTEMS

14

FLUID DYNAMICS OF SPRINKLER SYSTEMS

by   D. F. Heermann, USDA-SEA/AR, Fort Collins,
     CO; R. A. Kohl, South Dakota State University,
     Brookings, SD

The fluid dynamics of sprinkler systems is very complex. The water is moved dynamically from the water source through the pump, into the pipe network of the sprinkler system, through a sprinkler nozzle into the air at a high velocity where it breaks up into droplets and falls to the soil or crop surface and is redistributed. In this chapter, we discuss the flow within the sprinkler-irrigation pipe network and the fluid dynamics of the high velocity jet as it breaks up into droplets and is distributed to the surface of the soil. The significance of drop-size distribution, how it can be controlled, effect on infiltration, interaction with wind and magnitude of evaporation is presented. The importance of the redistribution on the surface and in the soil is discussed. The limitations imposed by infiltration on application rates as they affect design and operation are presented. Design criteria for frost protection, cooling, erosion control, and chemical applications are given. We will emphasize general design considerations with limited discussion on operational criteria. Chapter 15 presents detailed procedures for designing sprinkler irrigation systems. Chapter 18 discusses in more detail the different management strategies for the efficient use of sprinkler systems.

14.1 HYDRAULICS OF PIPE SYSTEMS

Sprinkler systems consist of main lines through which the entire flow is transmitted, laterals into which the flow branches, sprinkler risers, which direct the discharge to the sprinkler heads, and the nozzles, which direct the jet stream into the air. The main line energy losses due to surface resistance can be determined with uniform flow methods, whereas laterals and risers with branching flow require nonuniform flow analysis due to both surface resistance losses and form losses.

14.1.1 Energy Losses in Pipe and Couplers

The Darcy-Weisbach formula, as discussed in Chapter 11, is recommended for calculating energy losses in pipes. The Darcy-Weisbach equation is:

$$h_f = f \frac{L}{D} \frac{V^2}{2g} \quad \dots\dots\dots\dots\dots\dots\dots\dots\dots\dots\dots\dots\dots\dots\dots\dots\dots [14.1]$$

TABLE 14.1. VALUES OF ROUGHNESS e
(EQUIVALENT SAND GRAIN SIZE) FOR VARIOUS
PIPE MATERIALS FOR DETERMINING THE DARCY-
WEISBACH FRICTION FACTOR, f, (from
Albertson et al., 1960)

Material	e
	mm
Plastic	0.003 - 0.03
Commercial steel or wrought iron	0.03 - 0.09
Galvanized iron	0.06 - 0.2
Aluminum	0.1 - 0.3
Concrete	0.3 - 3.0
Riveted steel	0.9 - 9.0
Corrugated metal pipe	30.0 - 60.0

where h_f is the head loss, m; f is the Darcy-Weisbach resistance coefficient; L is the pipe length, m; D is the pipe diameter, m; V is the velocity, m/s; and g is the gravitational acceleration, 9.81 m/s^2.

The friction factor, f, is a function of the relative roughness and Reynolds number and can be found in most fluid mechanics texts. The absolute roughness height is a constant for a given surface and the friction factor is dimensionally consistent. The sprinkler irrigation industry often uses the Scobey or Williams-Hazen formulas for calculating the friction loss in pipes, since these equations use a constant factor for each pipe material. Tables are needed for energy losses with the dimensionally consistent Darcy-Weisbach equation to encourage greater industry use, because when the equation is properly used with the correct roughness height for the material, any variation in pipe diameter does not introduce a slight error, as it does with other commonly used equations.

Table 14.1 presents equivalent sand-grain size for determining the friction factor, f, for various pipe materials.

The friction factor for most commercial materials is represented by the semiempirical equation:

$$\frac{1}{\sqrt{f}} = 1.14 - 2 \log \left(\frac{e}{D} + \frac{9.35}{R_e \sqrt{f}}\right) \quad \dots \dots \dots \dots \dots \dots \dots \dots \dots \dots \quad [14.2]$$

where R_e is the Reynolds number (VD/ν), ν is the kinematic viscosity, m^2/s, and log is the base 10 logarithm. The equation is dimensionless and consistent English or metric units can be used. This equation must be solved using an iteration technique, which can be done with a small programmable calculator. General resistance diagrams for determining the friction factor are readily available in most fluid mechanics textbooks.

The energy loss for pipe with couplers can be determined by increasing the friction factor by 11 percent for aluminum irrigation pipe with couplers spaced 9 m (30 ft). Decreasing the spacing to 6 m (20 ft) would increase the energy loss by 17 percent, while increasing the coupler spacing to 12 m (40 ft) would decrease the energy loss by 8 percent. Lytle et al. (1962) found these losses were the upper limits in their study of various styles of couplers. Either manufacturers data or individual tests sould be obtained for a given coupler to determine energy losses for pipe and couplers.

14.1.2 Energy Losses in Laterals

The energy loss for sprinkler irrigation laterals can be determined by solving the problem as a branching flow system. Chapter 15 presents a simplified procedure for determining the energy loss in laterals with equally spaced and constant-size sprinklers.

The energy loss for center pivot laterals that have either different sizes or variable spaced sprinklers cannot be calculated using the simplified procedure of Chapter 15. Kincaid and Heermann (1970) used numerical techniques to calculate the pressure distribution on a center pivot system with the Darcy-Weisbach equation [14.1]. Their results showed that the head loss in the riser can be approximated by the equation:

$$h_r = \frac{V^2}{2g} \exp (9.2 \, q/Q) \quad\dotfill \quad [14.3]$$

where h_r is the head loss in the riser, m; q is the riser discharge; and Q is the upstream lateral discharge.

The branching loss due to the change in direction and flow into the riser was generally less than 6 kPa and is often neglected in the design of high pressure systems (>400 kPa). The equivalent roughness height of 0.12 mm gave the best agreement with field data for center pivot laterals of steel pipe.

Chu and Moe (1972) derived an analytical approximation for calculating head loss in a center pivot lateral. They assumed an infinite number of tiny sprinklers evenly distributed along the lateral with the discharge proportional to the radius for a uniform irrigation. The total head loss in the lateral is 54 percent of the head loss for the lateral operating as a supply line with the inflow discharge. This supply line head loss can be determined from an energy-loss table or one of the previously discussed equations. The distribution of the pressure loss was approximated by:

$$\frac{h_s - h_L}{h_o - h_L} = 1 - (15/8) \left[\frac{s}{L} - \frac{2}{3} \left(\frac{s}{L}\right)^3 + \frac{1}{5} \left(\frac{s}{L}\right)^5 \right] \quad\dotfill \quad [14.4]$$

where h_o is the pressure at the pivot, kPa; h_s is the pressure at a radius s from the pivot, kPa; h_L is the pressure at the radius L, kPa; s is the radius from pivot, m; and L is the radius of the irrigated area, m.

The approximations were made using Scobey's formula for energy loss with the exponent on discharge equal to 1.9 and 2.0 for the lateral energy loss and pressure distribution, respectively.

14.2 WATER DISTRIBUTION TO THE SOIL

The objective of irrigation is to uniformly apply water over the area for crop use. Sprinkler irrigation systems should be designed to apply water at a rate less than the intake rate to prevent surface runoff. The water application uniformity of a sprinkler system depends less on the dynamics of the individual sprinklers and more on the spacing and operating pressure. The spacing of sprinklers and, on moving systems, the travel speed and path affect irrigation uniformity. The effect of wind and evaporation on uniformity are discussed in a later section.

14.2.1 Single Sprinkler

The design of sprinkler irrigation systems requires, at a minimum, knowledge of the water-distribution pattern from a single sprinkler under no-wind conditions. Single sprinkler distribution patterns are affected by wind and evaporation potential, nozzle pressure, and design of the individual head. Pressure has a very significant effect on the distribution pattern. Low pressure can reduce the jet-stream breakup and cause the spray to be concentrated at a single radius from the sprinkler, resulting in a donut application pattern. High pressures may cause an excessive jet-stream breakup which reduces the radius resulting in a high application rate near the sprinkler. Pins or baffles on the nozzles are used to break up the stream for low pressure, whereas vanes or longer barrel nozzles can straighten the jet and reduce breakup at higher pressures. A more detailed discussion of drop size distribution is presented in Section 14.4.

The distribution of a sprinkler pattern may be presented either as individual measurements on a grid network or as a geometric pattern, which most closely approximates the measured distribution. The geometric pattern can be expressed as a function of discharge and pattern radius.

14.2.2 Moving Sprinklers

The distribution of water from continuously moving sprinklers cannot be represented on grid points as easily as the distribution from stationary sprinklers. The application by moving sprinklers integrates the pattern along its travel path. This integration reduces much of the nonuniformity of application observed with stationary sprinklers. The application by moving sprinklers can be determined by integrating the function for the geometric pattern of a single stationary sprinkler, either analytically or numerically along its travel path. A geometric pattern assumes an idealized condition for the individual sprinkler, but the errors between field measured and integrated idealized patterns are small. Bittinger and Longenbaugh (1962) assumed triangular and elliptical geometric shapes in the derivation of equations for application depths from single sprinklers moving in straight and circular paths. Fig. 14.1 defines variables used in the following equations for determining the application from moving sprinklers. The application depth for a single sprinkler traveling in a straight line with a triangular pattern is:

$$D = \frac{hr}{V} \left\{ (1-m^2)^{\frac{1}{2}} - m^2 \ln \frac{(1-m^2)^{\frac{1}{2}}+1}{m} \right\} \quad \dots\dots\dots\dots\dots \quad [14.5]$$

The natural logrithm is represented by 1n. The velocity, V, would be replaced by ωR when assuming straight line travel on a center pivot system, where ω is the angular velocity of the system.

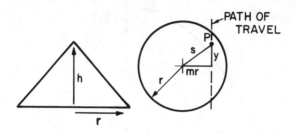

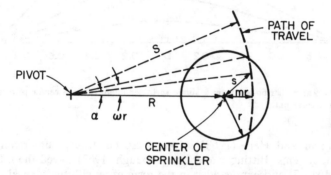

FIG. 14.1 Definition sketch for variables used in modelling sprinkler patterns on a center-pivot irrigation system. Note: R is distance from pivot to sprinkler; V is the travel velocity, and T is time.

Similarly, the application depth for an elliptical sprinkler pattern is:

$$D = \frac{hr\pi\,(1 - m^2)}{2V} \quad\dotfill\quad [14.6]$$

These equations are valid only for linear travel of individual sprinklers, traveling big guns operating at 1/2 or full circle. Heermann and Hein (1968) reported that straight line travel can be assumed for center pivot systems when S > 200 m.

Application depth from a sprinkler with a triangular pattern traveling in a circular path is:

$$D = 2hT - \frac{2h}{r\omega} \int_o^{\omega T} [S^2 + R^2 - 2RS\cos\alpha]^{1/2}\,d\alpha \quad\dotfill\quad [14.7]$$

Similarly, the application depth from a sprinkler with an elliptical pattern traveling in a circular path is:

$$D = \frac{2h}{\omega r} \int_o^{\omega T} [r^2 - S^2 - R^2 + 2\,RS\cos\alpha]^{1/2}\,d\alpha \quad\dotfill\quad [14.8]$$

where S = R + mr.

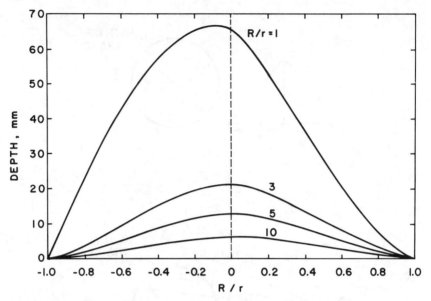

FIG. 14.2 The water distribution from a single sprinkler mounted on a center pivot system traveling in a circular path.

Heermann and Hein (1968) solved these equations numerically for center pivot systems. Bittinger and Longenbaugh (1962) solved the integral of equation [14.7] and expressed it in the form of an elliptic integral of the second kind, which are tabulated in mathematical tables. The advent of modern computer technology makes the numerical integration as convenient as tabulated functions.

A sprinkler moving on a circular path skews the distribution pattern toward the pivot as compared with the same sprinkler moving in a linear path (Fig. 14.2). This skewness causes a larger application depth near the pivot when the individual sprinklers are sized to irrigate the area between ± ½ the spacing of adjacent sprinklers. When $\alpha \leqslant 5$ deg the straight line approximation of the circular path can be used to calculate the distribution of a center pivot system with an error of less than 1 percent and reduce the computer time significantly.

14.2.3 Multiple Sprinklers

Most sprinkler systems use multiple sprinklers for an irrigation set. Even with traveling big guns or with one sprinkler operating at a time, the overlap is important when evaluating the distribution for the entire field. The equations from the previous section can be used to calculate the distribution from each individual sprinkler which are then summed to give the application depths from all sprinklers irrigating a given area. The application depth for a traveling big gun or continuous move linear system with a triangular pattern is:

$$D_y = \sum_{i=1}^{N} \frac{h_i r_i}{V_i} \left[(1 - m_i^2)^{\frac{1}{2}} - m_i^2 \ln \frac{(1 - m_i^2) + 1}{|m_i|} \right] \qquad \dots\dots\dots\dots [14.9]$$

where D_y is the applied depth at a distance Y from the edge of the irrigated field,

$$m_i = min \left(\frac{|Y - X_i|}{r_i}, 1\right)$$

and X_i is the distance from the edge of the field to the travel path for the ith sprinkler.

Similarly the equation for the elliptical pattern is:

$$D_y = \sum_{i=1}^{N} \frac{h_i r_i \pi (1 - m_i^2)}{2V_i} \quad \dots\dots\dots\dots\dots\dots\dots\dots\dots\dots\dots [14.10]$$

The application depth for a center pivot system at a distance S from the pivot for one pass of the system with triangular and elliptical sprinkler patterns, respectively, are:

$$D_s = \sum_{i=1}^{N} 2h_i T_i - \frac{2h_i}{\omega r_i} \int_0^{\omega T} [S^2 + R_i^2 - 2R_i S \cos\alpha]^{1/2} d\alpha \dots\dots [14.11]$$

and

$$D_s = \sum_{i=1}^{N} \frac{2h_i}{\omega r_i} \int_0^{\omega T_i} [r_i^2 - S^2 - R_i^2 + 2R_i S \cos\alpha]^{1/2} d\alpha \dots\dots [14.12]$$

where $(R_i - r_i) < S < (R_i + r_i)$ for the sprinkler to be included in the summation. Again, these equations must be solved numerically and are valid only for no wind conditions and the assumed patterns. Hart and Heermann (1976) showed that the effect of wind on the distribution of center pivot could be approximated by shifting the sprinkler location in the direction of the wind and decreasing the pattern radius.

The application distribution for multiple stationary sprinklers can be determined by summing the application depths at grid points for all sprinklers with patterns overlapping a given point. Hart and Heermann (1976) presented detailed techniques for calculating the system distribution from single sprinkler tests measured on a grid and overlapped with the specified system spacings. The same analysis can be made for assumed geometric patterns of single sprinklers and accumulate the depths for the overlapped patterns of individual sprinklers. Individual sprinkler patterns overlapping at a given point must be summed to determine the distribution of a sprinkler system. An individual sprinkler pattern may need to be rotated or juxtapositioned to compensate for regular shifts in the wind direction. This juxtapositioning of sprinkler patterns may be necessary if regular significant changes in wind direction are expected not only for different sets but even during one irrigation interval.

14.2.4 Measuring Distributions

The distribution of sprinkler systems can be evaluated by measuring an entire system, by measuring the patterns of individual sprinklers, and then by combining, as discussed in the previous section, or by sampling directly. ASAE Recommendations S330 (ASAE Yearbook, 1979) describes procedures for measuring the distribution of a single sprinkler including the format for presenting the data. The test site should be nearly level and the minimum clear distance upwind should be 30 heights of any obstruction for winds 8 km/h or greater. The sprinkler should be positioned at least 60 cm above the collectors, and 90 cm above the surface. Preferably, the sprinkler should be located in the center of the grid of the four adjacent central collectors. A minimum of 80 collectors should receive water during a test. Wind direction and total wind movement at the 4-m height should be recorded for interpretation of the data.

The distribution of a typical sprinkler system can be evaluated with a grid of catch cans or collectors. The grid should be located over a length equal to at least the sprinkler spacing on the lateral and over a width at least equal to lateral spacing.

For linear moving systems, one or two lines of measuring devices perpendicular to the travel path may be more practical and meaningful than a grid system. For center pivot systems, a single linear radial line of equally spaced catch cans may sufficiently characterize its distribution pattern.

Ring and Heermann (1978) recommend using two radial lines 3 degrees apart with 3-m can spacings to evaluate the distribution of a center pivot system. Systems with spray heads may require a closer spacing. The interval between collectors depends on the desired accuracy. Three-meter intervals will provide an adequate accuracy for design or system evaluation. A rough estimate of the distribution may be determined with a 10-m collector spacing.

14.2.5 Uniformity Coefficients

The purpose of the sprinkler system is to uniformly distribute the water to the soil surface of the irrigated area. Sprinkler technology has not been developed to the point where complete uniformity can be obtained. Christiansen (1942) studied distribution patterns of sprinklers and used a numerical expression as an index of the uniformity. The equation is:

$$C_u = 100(1.0 - \frac{\Sigma x}{mn}) \dots\dots\dots\dots\dots\dots\dots\dots\dots\dots\dots\dots\dots [14.13]$$

where x is the absolute deviation from the mean application, m; and n is the number of observations.

Heermann and Hein (1968) modified the uniformity coefficient for evaluating center pivot sprinkler systems where each collector represents a different size segment of a circular irrigated field.

The coefficient of uniformity using this procedure is:

$$C_u = 100 \left[1.0 - \frac{\sum\limits_s S_s \left| D_s - \frac{\sum\limits_s D_s S_s}{\sum\limits_s S_s} \right|}{\sum\limits_s D_s S_s} \right] \dots\dots\dots\dots [14.14]$$

where D_s is the depth applied at a collector and S_s is the distance to equally spaced collectors.

The sprinkler irrigation industry has generally used the Christiansen uniformity coefficient for evaluating and comparing sprinkler irrigation systems. A more complete description of the uniformity can be made when the distribution function is given. The statistical moments can be used to characterize this distribution function quite precisely. The first and second moments, which represent the mean and standard deviation of the distribution, often are sufficent for characterizing sprinkler irrigation distributions. Hart and Heermann (1976) discussed the relationship between the USDA uniformity coefficient, Hawaiian Sugar Planters Association's uniformity coefficient, and a normal distribution for characterizing sprinkler uniformities.

14.2.6 Edge Effects

Most uniformity coefficients for sprinkler irrigation systems generally represent the interior area of an irrigated field. For example with stationary systems, uniformity is generally calculated for either a rectangular or a triangular area in the interior of a field that receives the full amount of overlap. The outer edge of an irrigated field does not always receive the full overlap application from other sprinklers. This includes both the area parallel to the last lateral set on each side of the field and the ends of the field. Hart and Heermann (1976) report about a 10 percent loss of water if the laterals are placed right on the field boundary edge and if they extend to the ends. For a 16-ha-square field, the uniformity coefficients would decrease from 0.964 in the center area to 0.854 for the entire field if the laterals were placed to confine all the water within the field. Heermann and Hein (1968) reported that a center pivot irrigation system field uniformity coefficient of 0.9 was decreased to 0.8 when the tapered outer edge of the distribution is included. Similar reductions in uniformity coefficients for large volume guns also occur. This reduction in uniformity should be considered when evaluating overall field efficiencies. Furrow and border surface systems generally do not have the large tapered edges that effect their uniformities and efficiencies.

14.2.7 Season vs. Single Irrigations

The uniformity coefficients are usually evaluated and reported for a single irrigation using only one set of grid measurements made under a particular set of climatic conditions. Solomon (1978) evaluated the variability to be expected in determining the coefficient of uniformity under nearly identical wind conditions using the same sprinkler nozzle and operating pressure. The standard deviations of coefficients of uniformities from individual, independent tests were 2,4 and 6 when the coefficient of uniformities were 90, 80, and 70 respectively. Since most sprinkler systems are operated to apply several irrigations to a given field during a cropping season, the distribution of cumulative irrigation depths for multiple irrigations is more important than that of a single irrigation. Pair (1968) compared the uniformity from individual tests with the uniformity from accumulated catches from consecutive tests. Table 14.2 (Pair, 1968) shows the improvement in uniformity coefficients by adding successive irrigations. High winds caused a low uniformity for the first irrigation but was compensated by the following irrigations with lower winds, resulting in a higher cumulative uniformity. This does not imply that poorly designed systems should be accepted. It does in-

TABLE 14.2. CHRISTIANSEN'S COEFFICIENT OF UNIFORMITY
FOR EACH IRRIGATION AND FOR ACCUMULATIVE
IRRIGATIONS DURING THE 1965 SEASON FOR A
HANDMOVE SPRINKLER LATERAL HAVING A
12-x 18-m SPRINKLER SPACING (Pair, 1968)

Irrigation number	Plot number										
	1	2	3	4	5	6	7	8	9	10	11
1	49	56	57	48	31	52	52	56	63	59	36
2	85	90	89	87	88	89	87	88	86	87	88
3	73	77	71	75	76	77	77	80	73	81	73
4	75	77	59	71	76	78	72	70	80	73	72
5	81	85	86	84	83	83	86	86	85	82	77
1-2	86	89	88	87	86	86	87	90	89	90	80
1-3	88	89	84	87	86	89	88	90	87	90	86
1-4	87	87	84	85	86	88	88	96	90	90	86
1-5	88	88	85	87	86	88	88	87	89	92	87

dicate that lower uniformities caused by variable wind conditions can be compensated for by consecutive irrigations because of the random nature of wind fluctuations. Normally crop response to limited water applications can be expected to be a function of multiple irrigations. Therefore, multiple irrigation coefficients of uniformities probably are more closely related to water-limited yields than those for individual irrigation nonuniformities.

14.3 REDISTRIBUTION IN THE SOIL

The uniformity of sprinkler irrigation systems normally is evaluated on the basis of distribution at the soil surface. The crop, however, is dependent on available water in the root zone. Water will move within the soil profile due to potential gradients, and plant roots have a horizontal distribution. Hart (1972) analyzed the subsurface redistribution, using finite-difference solutions of the two-dimensional diffusion equation. He concluded that there is substantial redistribution within 1 m and the final soil water distribution is not significantly affected by the nonuniformity on the soil surface within that distance. An estimated 10 percent reduction in total water requirements resulted due to the redistribution of soil water for one test field. Where the nonuniformity exists over larger than 1- to 3-m distances, the subsurface redistribution will have little effect on the final soil-water uniformity.

Seginer (1978) estimated a net loss of $180/ha for a 0.1 decrease in the uniformity coefficient in cotton production due to nonuniform water application. The effective nonuniformity for various crops is a function of the extent of horizontal root development of individual plants. Seginer (1979) analyzed the increase in effective uniformity of water distribution in orchards with various tree and sprinkler spacing and calculated an increase in effective uniformity of 0.1. The "actual" uniformity is appropriate for "point" plants that have a limited horizontal root distribution. He also found that when rows are parallel to the shorter sprinkler spacing of solid set systems, a better effective uniformity can be obtained.

14.4 DROP SIZE DISTRIBUTION

The drop-size distribution of sprinkler spray is of practical importance for two reasons. First, the small droplets are subject to wind drift, distorting

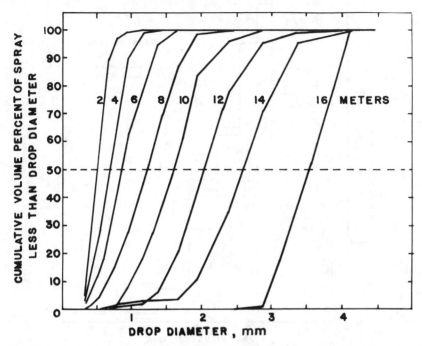

FIG. 14.3 The drop-size distributions collected at 2-m intervals from the main jet of a 3.97 mm (5/32 in.) nozzle operated at 400-kPa (58 lb/in^2) pressure (from Kohl, 1974).

the application pattern. Second, large droplets possess greater kinetic energy which is transferred to the soil surface, causing particle dislodgement and puddling that may result in surface crusting and runoff.

14.4.1 Sprinkler Jet Breakup

A jet of water issuing from a nozzle into the atmosphere eventually breaks into droplets because its initial form is disturbed. The shear of the air against the water surface is not sufficient in itself to disturb the surface of the jet and cause breakup. Turbulent eddies in the water column, no longer having a rigid boundary after emergence from the nozzle, cause the jet surface to deviate and break away from the main stream. Rouse et al. (1952) aptly described the ensuing breakup. "It is only after the surface of the jet has become sufficiently disrupted to produce an appreciable form resistance that the action of the air begins. Such resistance is roughly proportional to the square of the water velocity and to the cross-sectional area of the expanding jet. Thus, as ever more eddies carry water laterally out of the central stream, they are rapidly retarded in their longitudinal course by the surrounding air, but they continue to spread laterally at only slightly diminished speed. Although the outermost fringes of the jet at once form droplets that fall as a spray, the central portion appears simply to disintegrate in midair. In other words, since the equal and opposite reaction to the retardation of the water is the acceleration of the surrounding air, the originally intact but turbulent jet is transformed along its trajectory into an expanding mixture of dispersed water drops and air that travels at an ever-decreasing speed but brings an ever-increasing volume of air into motion."

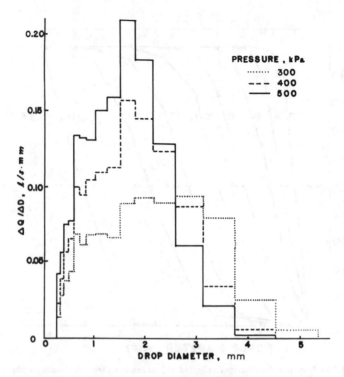

FIG. 14.4 The drop size distributions from a 3.97-mm (5/32-in.) diameter nozzle operated at three pressures (Kohl, 1974).

Water deviating from the jet surface will encounter air at a greater differential velocity than water globules distintegrating near the jet axis where air has been entrained. Merrington and Richardson (1947) have shown that the mean diameter of drops formed from jet breakup is inversely dependent on the jet's relative velocity to the surrounding air. Therefore, water near the periphery of the jet will produce small droplets, whereas the water near the core of the jet with the lowest relative velocity to the air will produce the largest droplets.

At each location along the distintegrating jet, a distribution of droplet sizes is produced. Also, since the speed of the smaller droplets decreases more rapidly than larger droplets, the mean size of drops falling closer to the nozzle will be much smaller than those falling further from the nozzle. This effect is illustrated in Fig. 14.3 where the drop-size distributions at 2-m intervals from the main jet are presented. The volume mean diameter for a drop-size distribution is the drop diameter, above and below which half of the volume of discharge occurs. The volume mean diameters of the distributions in Fig. 14.3 occur at the intersection of the horizontal line through 50 percent on the ordinate and the lines representing the distributions at a given distance.

14.4.2 Drop-Size Distributions

Pressure and nozzle size are the two independent system variables considered in this discussion. Water streamlines prior to the nozzle orifice are significant (the presence of straightening vanes, ring orifice vs. taper bore) but their effects on drop-size distribution were not found in the technical literature.

The effect of pressure on the drop-size distribution from a 4-mm (5/32-in.) nozzle is illustrated in Fig. 14.4. The vertical axis represents the discharge, ΔQ, per millimeter of drop diameter, ΔD, to produce an area under the bar graph proportional to the sprinkler discharge in liters/second. As pressure was increased, the volume of water applied, as larger droplets, decreased while the volume applied as smaller droplets increased to make up the larger total discharge. Since jet velocity is proportional to the water pressure in the supply line, the higher pressures should produce greater relative velocities between the water and the air, resulting in a larger number of smaller droplets. The data plotted in Fig. 14.4 substantiate this hypothesis.

The effect of nozzle size on the drop-size distribution is smaller than that of pressure, as illustrated in Fig. 14.5, where drop-size distributions for three nozzle sizes operated at 400-kPa pressure are presented. The volume of water applied as larger droplet sizes, increased as nozzle diameter increased. Under the same nozzle approach conditions, a small diameter jet will tend to separate and air will be entrained through to its center more rapidly than will a jet with a larger diameter. This process will result in larger differences in relative velocities between water and air in smaller jets. The larger relative velocity and smaller water globules breaking up result in smaller water droplets from the smaller nozzles.

The above data were collected by Kohl (1974) in a building using the

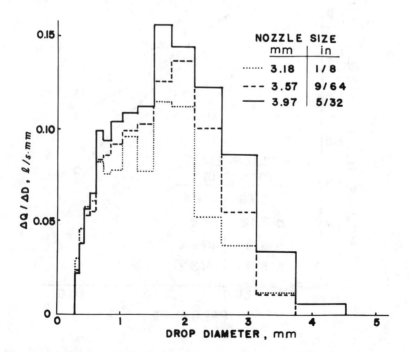

FIG. 14.5 The drop-size distributions from three nozzles operated at 400-kPa (58-lb/in.²) pressure (from Kohl, 1974).

TABLE 14.3. DROP VELOCITIES AT IMPACT
ELEVATION USED TO CALCULATE THE
KINETIC ENERGY OF THE DROP SIZE
DISTRIBUTIONS (from Schladerbush and
Czeratsky, 1957, and Seginer, 1965)

Drop diameter, mm	Impact velocity, m/s	Drop diameter, mm	Impact velocity, m/s
0.33	1.33	1.37	4.33
0.39	1.60	1.66	4.89
0.49	2.03	1.93	5.27
0.57	2.25	2.39	5.93
0.67	2.54	2.88	6.50
0.78	2.89	3.38	6.83
0.93	3.29	4.12	7.12
1.16	3.92	4.93	7.26

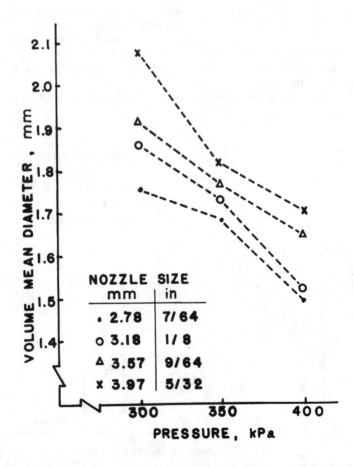

FIG. 14.6 The volume mean diameter of the drop-size distribution as a function of sprinkler pressure for different nozzle sizes (from Kohl, 1974).

flour method, described by Laws and Parsons, 1943. Inoue (1963) found similar trends in drop-size distribution caused by pressure and nozzle size, although his data indicated a greater volume of water falling as mist. He determined drop sizes from stains on filter paper.

The relative importance of water pressure and nozzle diameter on the volume mean diameter of the total spray distributions is illustrated in Fig. 14.6. While the smaller nozzles produce a smaller mean drop size, operating a small nozzle at low pressure can produce larger mean drop sizes than those produced by larger nozzles operating at higher pressure.

14.4.3 Kinetic Energy of Sprinkler Spray

The kinetic energy of the spray as it reaches the land surface was calculated by Kohl (1974) for the measured drop-size distributions, using the velocities at impact elevation listed in Table 14.3. The data of Schladerbusch and Czeratski (1957) were used for velocities of droplets less than 2.5 mm in diameter and the method of Seginer (1965) was used to calculate the velocities of larger droplets.

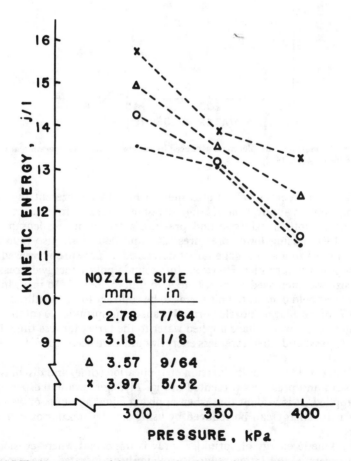

FIG. 14.7 The kinetic energy of the drop-size distributions for an application of 1-mm depth of water to 1 m² (1 dm³; 1 L) as a function of nozzle size and sprinkler pressure (from Kohl, 1974).

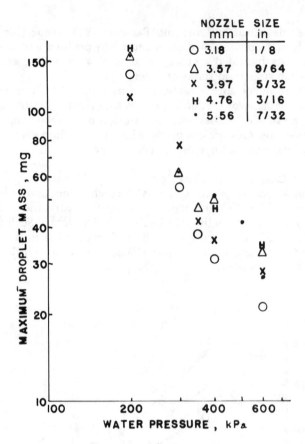

FIG. 14.8 The maximum droplet mass produced by various nozzle sizes operating at different pressures (from Kohl, 1974).

The kinetic energy values presented in Fig. 14.7 represent the energy resulting from an application of 1-mm depth of water to 1-m² surface. Again the effect of both nozzle size and pressure are evident. Schleusener and Kidder (1960), using lower pressures (240 and 280 kPa), also found that energy applied to a strain gage target decreased as pressure increased for a single pass of the sprinkler. However, they found applied energy decreased as nozzle size was increased from 3.97 to 4.76 mm (5/32 to 3/16 in.) while the pressure was held constant. This seems to be contrary to the results shown in Fig. 14.7. Since a larger nozzle normally causes the sprinkler to rotate faster, the larger nozzle would have applied water to the target for less time during the single pass, and, therefore, less energy would have been recorded with the large nozzle.

When considering both duration of application (time) and discharge, all nozzle sizes and pressures presented in Fig. 14.7 were evaluated per unit of water applied. It is evident that when applying equal amounts of water to a field, kinetic energy can be reduced by using smaller nozzles or increasing pressure.

The kinetic energy of sprinkler spray is important where erosion and surface crusting, the latter reducing soil infiltration rates, are problems. Young and Wiersma (1973) found that an 89 percent reduction of rainfall energy, without reduced water application rate, decreased total soil loss from their research plots by 90 to 94 percent depending on soil type. Since the largest droplet sizes have the highest velocities at ground level, and because kinetic energy is a product of droplet mass and its velocity squared, the maximum droplet sizes transfer much more kinetic energy to the soil surface than small droplets. Levine (1952) found a very large increase in aggregate breakdown with increasing drop size, which emphasizes the importance of the largest droplets.

The effect of nozzle size and pressure on the maximum droplet mass is presented in Fig. 14.8. While the maximum droplet mass is similar over a rather large range in nozzle sizes, it is very sensitive to water pressure. Maintaining adequate sprinkler pressure when irrigating soils with crusting problems appears to be an important management factor in minimizing soil crusting.

14.5 SPRAY EVAPORATION

The evaporation of the spray from a sprinkler system can change the coefficient of uniformity and is a direct loss when evaluating the water-application efficiency. In this section, we discuss the magnitude of evaporation that is possible and techniques for measuring evaporation.

14.5.1 Calculated Spray Evaporation

Spray evaporation can be calculated by considering the latent heat loss in reducing the spray temperature to wet bulb temperature, the transient transfer of sensible heat to individual droplets and the transient transfer of water vapor away from the droplets. Water droplets passing through air near terminal valocity rapidly approach wet-bulb temperature. Several researchers (Cline et al., 1969; Pair et al., 1969; Sale, 1965) have verified this fact experimentally for sprinkler spray. Kinzer and Gunn (1951) presented a rigorous development of the equations for the transient transfer of heat to water vapor away from freely falling drops. They verified the equations by means of several meticulous experiments and have presented tables to simplify evaporation loss calculations. These equations are the most accurate available for drops at terminal velocity.

Christiansen (1942) used an approximate expression to calculate water loss from sprinkler spray during flight. He obtained values not exceeding 1 percent loss of the discharge volume. Inoue and Jayasinghe (1962) calculated an evaporation index on the basis of representing the drop-size distribution by an upper-size limit equation. They calculated evaporation losses up to 6 percent. However, half of their drop-size distributions were represented by droplets smaller than 1.2 mm diameter, whereas those of Kohl (1974) and Schleusener (1957) were similar with total volume median diameters nearer to 2 mm. Therefore, Inoue and Jayasinghe's estimates of evaporation loss may be high.

Seginer (1970, 1971, 1973) developed a resistance model to account for evaporation during sprinkling and found that the evaporation amounted to only a few percent of the application rate.

Using the drop-size distribution for a 4-mm (5/32-in.) nozzle operated at 400 kPa (58 lb/in²) pressure from Fig. 14.4, the equation of Kinzer and Gunn (1951), a time of flight of 1.7 s, a relative humidity of 10 percent, and an air temperature of 40 °C resulted in a calculated evaporation loss of 1.5

percent beginning with water at 18 °C. Evaporation could increase to 3.5 percent if the water temperature was 30 °C. However, most irrigation waters are cooler than this and the evaporative potential less, resulting in calculated evaporation losses of less than 1 percent for semi-arid sprinkler irrigation.

Calculated evaporation losses of sprinkler water between the nozzle and crop or soil surface range from less than 1 percent to a maximum of 6 percent, depending on droplet size and atmospheric conditions. Most calculations fall in the lower end of this range, suggesting evaporation losses are about 1 to 2 percent of the amount of discharged water.

14.5.2 Difficulties in Measuring Spray Evaporation

Sprinkler spray evaporation has been measured as the difference between nozzle discharge and the amount of water recovered from a plastic-covered area (Seginer, 1071). Most experiments used a catch device (e.g. rain gage or oil can) placed on a grid to represent the collection area. Since evaporation loss is considered to be the difference between discharge and amount collected in the devices, any loss from the collection device, incomplete catch, or mist carried beyond the collection site would be included in the "measured" loss. Evaporation from the collection device and its effect on reported spray losses have been recognized for a long time. Christiansen (1942) reported relatively high evaporation losses from cylindrical, catch cans. He attempted to remedy the problem by soldering funnels into the top of the cans, but evaporation losses were reduced little. Forst (1963) used a truncated cone or frustum-shaped can in an attempt to reduce the amount of water clinging to the sides of cylindrical containers. Kraus (1966) and Wolfe (1967) attempted to improve the accuracy of cylindrical containers by coating their interior with paraffin to hasten droplet travel down the can walls. Wolfe observed that "the paraffin appeared to make an improvement, but was not entirely satisfactory." Newer, semi-automatic systems (Culver and Sinker, 1966; Hunter, 1966) generally rely on a collection funnel with a sharp top edge. The top edge of some collection funnels is extended upward to form either cylindrical or frustum-shaped walls to minimize splash and evaporation.

Kohl (1972) compared the collection efficiency of devices similar to those referred to above and illustrated in Fig. 14.9 with a modified separatory funnel. While the separatory funnel was not 100 percent efficient due to splash loss, it did illustrate that large evaporation losses occurred from the other collection devices. Table 14.4 presents the catch of these devices as a percent of the catch in the separatory funnels for different application rates. The tests were performed on clear, sunny days, except for several simultaneous runs conducted at night between 0200 and 0500 h at lower temperatures, higher humidities, and without solar radiation to compare with the day runs. The low catch efficiencies (6 to 91 percent) of the tested catch devices was attributed to the long travel time of droplets down hot container walls, the presence or absence of an evaporation suppressing oil, and its magnitude depended on the application rate.

Incoming solar radiation absorbed by the catch can alone could account for the evaporation loss from them. At the time of the tests, horizontal crop surface evapotranspiration was about 1 mm/h. The quart oil can has a shadow area about 1.5 times its open area. While external surface evaporation could use part of this radiant energy, sufficient energy was available to account for the reduced catch in the devices. With a given energy load, the

FIG. 14.9 Types of precipitation collection units compared from left to right: metal quart oil can, Frost can, modified separatory funnel, 76 mm (3-in.) rain gage and inner funnel, and plastic rain gage.

TABLE 14.4. A COMPARISON OF PRECIPITATION
COLLECTION UNITS WITH THE SEPARATORY
FUNNEL UNIT AT VARIOUS APPLICATION RATES
(from Kohl, 1972)

Ave. application, separatory funnel	Oil can	Oil can with oil	Paraffin-coated oil can	Paraffin-coated oil can with oil	Frost can	Frost can with oil	76-mm rain gage	Plastic rain gage	Plastic rain gage with oil
mm/h	-------------	Percent of catch in separatory funnel						-------------	
Day									
0.9	6	69	41	52	10	87	15	36	35
2.2	54	67	63	70	58	90	60	58	59
4.1	62	78			78	88	79	70	70
6.2	76	88			77	90	87	76	74
9.4	81	89			81	91	88	82	81
Night									
2.2	96	97			87	102	96	83	83
4.0	100	101			96	104	98	95	92

lower application rates would result in higher percentages of loss. These data emphasize the difficulties involved in measuring evaporation loss and suggest extreme caution in interpreting published data on "measured" evaporation losses, based on catch-can devices.

George (1955) attempted to minimize the errors due to incomplete catch and mist drift by measuring the electrical conductivity of water caught in funnels. However, clinging droplets and travel time down the funnels still allowed evaporation and salt concentration. He estimated that half of his measured "evaporation loss" occurred in the funnels.

Currently there are no totally acceptable catch devices or methods to accurately measure the evaporation loss of sprinkler spray.

14.5.3 Field Measurement of Spray Evaporation

The most widely publicized "spray evaporation" data set has been that of Frost and Schwalen (1955), which resulted in the development of a nomograph correlating vapor pressure deficit, and wind velocity with nozzle diameter and pressure to estimate evaporation loss.

After the first hundred tests, Frost and Schwalen realized that large losses were occurring from their catch cans. To improve catch-can performance they painted the cans white and put oil into some companion cans to obtain a correction factor to apply to the entire area. This improvement and the use of the overlap application of two sprinklers, as opposed to using a single sprinkler, reduced measured losses from about 30 to about 10 percent with some values down to 3 percent.

Frost and Schwalen also estimated the "average" drop size and computed evaporation loss based upon the total exposed water surface. Even though the "average" drop size and not the volume median diameter was used, which would bias the values toward a much larger exposed surface, they calculated losses "considerably lower than the experimental losses." They believed that the drift of spray carried out of the collecting area by wind accounted for part of this difference.

When analyzing these data, several important points should be emphasized.

1 In view of Table 14.4, considerable loss from the catch cans was probably still occurring, particularly near the outer edges of the pattern where the application rate would be low and near the sprinklers where the smaller drops fall. This could have accounted for a significant discrepancy between calculated and measured losses.

2 Spray reaching the crop (grass) surface outside the collecting area due to wind drift was included as evaporation loss. This may not be a true loss, as will be discussed in the next section.

3 Although the authors pointed out the difference between the computed and measured loss values and the drift beyond their collection site in their original paper (Frost and Schwalen, 1955), their nomograph is still being used.

4 To our knowledge no research has been published in the last 20 years that has significantly improved on Frost and Schwalen's (1955) data.

About 12 years before Frost and Schwalen's research, Christiansen (1942) conducted similar tests with similar cans and obtained comparable values of from 2 to 52 percent loss. He calculated the probable evaporation loss and obtained values less than 2 percent and attributed most of the dif-

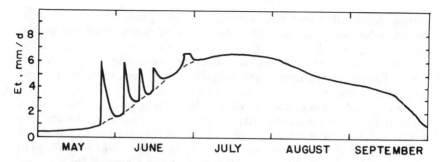

FIG. 14.10 The evapotranspiration curve for potatoes illustrating the increased early season evaporation due to a wet soil surface (from Wright, 1974).

ference between calculated and measured values to evaporation from his cans.

Robinson (1973) placed 250-ml beakers in an alfalfa field to estimate salt concentration increase in sprinkler-applied water due to evaporation. The salt concentration increase within the field indicated a 4.4 percent evaporation loss, part of which would have occurred on the interior walls and liquid surface in the beakers.

Seginer (1967), who compared above canopy sprinkling of corn with small perforated pipe sprinkling under the canopy, obtained a rough estimate of less than 5 percent loss from overhead sprinkling.

14.5.4 Evaporation from Soil and Wetted Vegetation

Overhead sprinkler irrigation wets the crop canopy and soil surface and results in evaporation of water from these surfaces. Whether or not this evaporation should be considered a loss depends on the background evapotranspiration.

A wet soil will lose water or dry rapidly at first as soil water can move to the surface through pores due to a water potential gradient. Once the surface dries, heat energy must be transmitted through the air-dry surface soil to the moist soil below where evaporation can take place. Then the water vapor must diffuse through the dry surface soil to escape into the atmosphere. The greater the depth of dry surface soil, the slower is the loss of water from the soil beneath.

Bare, medium-textured agricultural soils lose water rapidly (within a few days) from the surface 10 cm, slowly (over 2 months) from the next 10 cm, and extremely slowly from below the 20-cm depth (Veihmeyer, 1927).

If soil moisture must be present within a few centimeters of the surface to obtain seed germination and seedling growth, then some base level of evaporation must be considered as a normal part of the seasonal evapotranspiration curve. The evapotranspiration data for potatoes (Fig. 14.10) was obtained from Wright* from his lysimeter site near Kimberly, ID and have been smoothed for illustrative purposes. A base of 0.2- to 0.3-mm/d evaporation is considered part of the initial seasonal curve. Sprinkler irrigations and natural rainfall in late May and through June during canopy development and before full cover increased evaporation after soil

---

*(Personal Communication) J. L. Wright, USDA, Science and Education Administration, Snake River Conservation Research Center, Kimberly, ID.

wetting. After full canopy development, soil wetting does not significantly influence total evapotranspiration. The additional wet surface evaporation during canopy development, whether caused by sprinkling, rainfall or surface irrigation, can be considered as a loss above the normal consumptive use curve. Therefore, frequent, light irrigations will tend to result in more evaporation than less frequent, heavier irrigations.

Wetted-foliage evaporation under sprinkler irrigation has been studied with full-cover crops of alfalfa, oats, sudangrass and ryegrass under arid conditions with the findings that evapotranspiration from the wetted foliage was about equal to that from nonwetted, actively growing foliage with adequate soil moisture (Burgy and Poneroy, 1958; Culver and Sinker, 1966; Frost and Schwalen, 1960; Heermann and Shull, 1976; and McMillian and Burgy, 1960). The water evaporating from the wet canopy compensated for the transpirational requirement, which would normally be withdrawn from the root zone; therefore, it is not a loss.

While the evaporating water reduces soil moisture use, salts from sprinkler water remain behind. Under very low application rates, salt buildup could occur and damage the foliage. Specific ion damage to sensitive crops is discussed in Chapter 5. Usually application rates are high enough to wash the leaves free of salts.

14.5.5 Relative Magnitude of Evaporation Losses

With spray evaporation resulting in less than 2 percent loss, additional wetted soil evaporation normally less than 5 to 8 percent, and additional wetted vegetation evaporation considered to be negligible, evaporation losses are small as compared with the effects of distribution patterns and wind distortion (Section 14.6). Rapid cycling of solid set systems or rapid rotation of circular center pivots could increase evaporation losses from wet soil before canopy closure. However, other evaporation losses are considered insignificant as compared with the effects of nonuniform applications.

14.6 WIND EFFECTS ON SPRAY DISTRIBUTION

Wind speed and direction are the two wind characteristics considered in sprinkler design. Wind speed increases with height above a crop or soil surface and can usually be approximated by a straight line when it is plotted against the logarithm of height. Fig. 14.11 presents data collected by Covey et al. (1958) over short grass at O'Neill, NE, which illustrates wind speed profiles.

Sprinkler spray ejected high into the air will be subject to greater wind speeds and greater pattern distortion than spray nearer the surface. The ideal angle for a sprinkler jet under calm conditions is 32 deg above the horizontal. However, under windy conditions, a lower angle must be used to obtain a good range. Most medium-size agricultural sprinklers are about 25 to 26 deg, whereas larger sprinklers are between 23 and 24 deg.

The effect of riser height is similar to jet angle (Strong, 1961). The higher the sprinklers the greater the pattern distortion because of increasing wind velocity with height. Higher sprinklers allow longer travel time during which the wind can act on the spray. However, Wiersma (1955) obtained opposite results, which may have resulted from operating the sprinklers at pressures far below that recommended by sprinkler manufacturers.

Another wind-speed characteristic, illustrated in Fig. 14.11, is the diur-

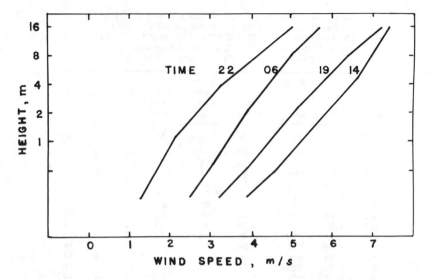

FIG. 14.11 The average variation of wind speed with height and time of day in hours in the air layer near the ground. The observations were taken over short grass at O'Neill, NE, by Covey et al. (1958) (redrawn from Sellers, 1965).

nal reduction in velocity during the night. At most continental locations, night wind speeds decrease sufficiently to effect scheduling successive irrigations during alternate day/night periods. A sprinkler system should be irrigating a given location during the night when the preceding irrigation was during the day and vice versa. The accumulative coefficient of uniformity of two or more irrigations is usually an improvement over one irrigiation during high winds, Table 14.2 (Pair, 1968). Center pivots also should be adjusted to rotate on a multiple of the half-day cycle (i.e. 36, 60, 84, etc, h/revolution) if operated continuously.

Wind direction is the other characteristic that affects sprinkler design. Except in coastal areas or where a local topographic feature exerts a dominating influence, wind direction normally varies, because of the movement of atmospheric pressure centers with their characteristic rotational wind patterns. During the growing season, these pressure centers tend to follow the same routes, resulting in some dominant wind direction quadrants at many locations, especially in open plain areas. Thus, dominant wind directions should be considered in most sprinkler design.

Wind speed and direction must be considered in sprinkler design and not left solely to the sprinkler operator's decision. This means that design is based on expected wind speeds and direction during the irrigation season. Some research reports (Keller et al., 1971; Seginer and Kestrinsky, 1975) have established design criteria on the shift of the center of mass of a sprinkler pattern due to wind velocity and direction. While these data look good, most areas do not have sufficiently predictable winds to incorporate these parameters in design. Therefore, more general approaches must be made. The following guidelines by sprinkler-system type should enable adequate uniformity of water application.

14.6.1 Wind Effects on Measured Distributions for Stationary Systems

Strong (1961) compiled his and other data into design tables, which pro-

TABLE 14.5. PERFORMANCE GUIDES FOR TWO-NOZZLE
SPRINKLERS SHOWING OPTIMUM SQUARE OR
TRIANGULAR SPACING DISTANCES FOR VARIOUS
WIND CONDITIONS UP TO 11 m/s (Christiansen's B-C
Profiles, Reworked from Strong, 1961)

Nozzle sizes	Average pressure	Discharge	Effective diameter	Wind speed range, m/s			
				0-5	2-7	5-9	7-11
mm	kPa	L/s	m	m	m	m	m
3.18 × 2.38	240	0.20	20	12	11	10	9
3.57 × 2.38	260	0.26	21	13	12	10	9
3.97 × 3.18	280	0.38	24	14	13	12	10
4.37 × 3.18	290	0.43	25	15	13	12	10
4.76 × 3.18	310	0.49	27	16	15	13	11
5.16 × 3.18	325	0.57	28	17	15	13	12
5.56 × 3.18	340	0.63	29	17	16	14	12
6.35 × 4.78	410	1.09	34	20	18	16	14
7.14 × 4.78	450	1.35	36	22	20	17	15
7.94 × 4.78	455	1.58	37	22	20	18	16
8.73 × 4.78	480	1.89	38	23	21	18	16
9.53 × 4.78	520	2.18	39	23	21	19	16
9.53 × 6.35	520	2.73	44	27	24	21	19
11.11 × 6.35	550	3.46	50	31	27	24	21
12.70 × 6.35	550	4.06	53	32	29	26	23
14.26 × 6.35	620	5.24	57	34	31	27	24
15.88 × 6.35	620	6.12	60	36	32	29	25

vide guidelines for sprinklers spacing under different wind-speed ranges.
Tables 14.5 and 14.6 provide a range of nozzle sizes and their corresponding
recommended optimum square or triangular spacing distances for various
wind speed ranges up to 11 m/s. Except for the smallest nozzle sizes, where
the average pressure was slightly below recommended current manufacturer
minimums, the pressures are at or slightly above recommended minimum
pressures. The last four columns of Table 14.5 indicate the recommended

TABLE 14.6. PERFORMANCE GUIDE FOR SINGLE NOZZLE SPRINKLERS SHOWING
OPTIMUM RECTANGULAR SPACING DISTANCE FOR WIND CONDITIONS
UP TO 11 m/s (Christiansen's D-Type Profile, Reworked from Strong, 1961)

Nozzle size	Average pressure	Discharge	Effective diameter D	Wind speed range, m/s							
				0-5		2-7		5-9		7-11	
				0.4D	0.75D	0.4D	0.75D	0.4D	0.75D	0.4D	0.75D
mm	kPa	L/s	m	m	m	m	m	m	m	m	m
1.59	210	0.025	17	7	13	6	11	5	10	5	9
1.98	210	0.04	18	7	13	6	12	5	10	5	9
2.38	240	0.07	19	8	14	7	13	6	12	5	10
2.78	280	0.12	20	8	15	7	13	6	12	5	10
3.18	300	0.16	22	9	17	8	15	7	13	6	12
3.57	310	0.20	23	9	17	8	16	7	14	6	12
3.97	320	0.26	25	10	18	9	16	8	15	7	13
4.37	340	0.31	26	10	20	9	18	8	16	7	14
4.76	360	0.38	28	11	21	10	19	9	16	8	14
5.16	370	0.45	29	12	21	10	19	9	17	8	15
5.56	410	0.56	35	14	26	13	23	11	21	10	18
6.35	450	0.77	37	15	27	13	25	12	22	10	19
7.14	480	1.02	38	15	28	14	26	12	23	11	20
7.94	480	1.26	40	16	30	14	27	13	23	11	21
8.73	520	1.55	41	16	31	15	27	13	24	12	21
9.53	550	1.88	42	17	31	15	28	13	25	12	22

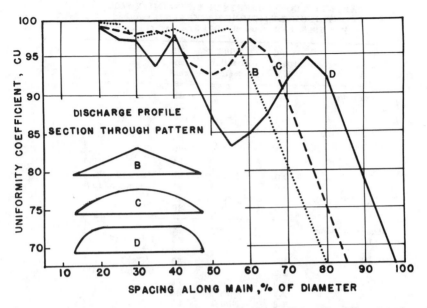

FIG. 14.12 Geometrical sprinkler patterns and uniformity coefficients for different spacings of lines with sprinklers spaced closely along the lines (Redrawn from Christiansen, 1942).

square or triangular spacing, based on an optimum spacing of 0.6 times the effective diameter under low wind-speed conditions. To maintain similar CU values, spacings are reduced with increasing wind speeds. The optimum rectangular spacing is 0.4 by 0.75 times the effective diameter under light wind conditions. The last six columns of Table 14.6 list the reduction in these spacings as wind speed increases.

Strong (1961) listed an effective diameter for the sprinkler pattern which is less than the outer edge of the spray listed as the wetted diameter by sprinkler manufacturers. Strong found this effective diameter a good base from which to design sprinkler spacing under windy conditions. In addition, it allowed the prediction of coefficients of uniformity, CU's, under these windy conditions.

Christiansen (1942) published calculated CU's for six basic sprinkler profiles. Fig. 14.12 is a reproduction of three of these profiles and their corresponding CU's. These curves were developed for spacing increments of 5 percent of the wetted diameter along the lateral and various spacings along the main line. Strong (1961) extended the usefulness of this graph to include various spacings along the lateral, as well as along the main. To do this, he converted the spacing distances to percentages of the effective diameter read off the CU's from Christiansen's chart for the corresponding sprinkler profiles, multiplied the CU's and divided by 100. Strong indicated that he had conducted many tests with actual data and found the above method was reliable and capable of producing results within 3 percent of calculated values from superimposition of profiles, unless extended spacings were used. He assumed that double nozzle sprinklers would fall somewhere between Christiansen's B and C profiles and single nozzle sprinklers would produce the D profile.

TABLE 14.7. COEFFICIENTS OF UNIFORMITY FOR
CONTINUOUSLY MOVING LATERALS (Reworked
from Pair, 1968, 1975, Shearer, 1971, and Hart et al., 1976)

	Wind speed, m/s	Wind direction* degrees	CU, percent	Source
Center pivot	3.2	—	81	Pair, 1968
	2.2	—	86	”
	4.5	—	80	Pair, 1975
	2.3	—	71	”
	1.6	—	86	”
	3.7	—	80	”
	1.3	—	81	”
	0.0	None	89	Hart and
	9.0	270	91	Heermann, 1976
	9.0	90	86	”
	9.0	0	89	”
Straight-moving lateral	2.7	—	89	Pair, 1968
	1.4	—	89	”
	1.3	—	89	”
	2.3	0	82	Shearer, 1971
	2.9	75	92	”
	2.2	75	88	”
	1.8	90	93	”
	1.6	90	92	”

*Zero degrees is at a right angle to sprinkler lateral and in the same
direction as lateral travel, 90 deg is parallel with the lateral or towards the pivot. The symbol — indicates wind direction data were
not available.

14.6.2 Wind Effects on Continuously Moving Laterals

Moving laterals, either straight moving laterals or center pivots, have advantages over set systems in windy conditions. First, the lateral is continuously moving and represents an infinite number of positions along the main line. Second, the sprinkler spacing along the lateral is relatively close, usually 12 m or less for systems in current production. A disadvantage is that sprinklers are usually higher above the soil surface than set systems and, therefore, are subject to higher wind velocities. Nevertheless, moving laterals should be able to achieve high CU's even in moderately windy conditions.

Pair (1968, 1975) reported CU's for both a mechanical drive straight-moving lateral and several different center pivots under low to moderate winds. Shearer (1971) reported similar data, including wind direction for a water-drive straight-moving lateral. These data are summarized in Table 14.7. Hart and Heermann (1976) simulated the effect of 9-m/s wind on a center pivot system by overlapping test distribution patterns obtained under the wind condition (Table 14.7). The main effect of the wind was changing the irrigated area. Wind increased the irrigated area by 19 percent and decreased it by 17 percent when it was parallel away from and toward the pivot, respectively. In general, all of the moving systems maintained good CU's due to their inherent design and operation advantages. To further improve the accumulative CU, the systems should be operated to return to a given site during the opposite time of the day from the last irrigation to take advantage of diurnal wind fluctuations.

14.6.3 Wind Effects on Traveling Gun Sprinklers

Shull and Dylla (1976a, 1976b) studied the effects of wind on traveling gun application uniformity. They found that wind elongated the pattern downwind from the sprinkler, shortened the pattern upwind, and narrowed the pattern at right angles to the wind. The wetted width and wetted distance upwind from the sprinkler decreased at about the same rate as the wind velocity increased.

The wetted distance downwind from the sprinkler increased as wind velocity increased but the increase was proportionately less than the decrease in width and distance upwind. As a result increased wind velocity decreased the wetted area covered by the sprinkler.

Increased pressure in the system increased both the upwind and downwind distances more than the width, with the greatest increase being in the downwind direction. This would be expected as higher pressures would cause greater breakup and, therefore, smaller droplets which will be carried further downwind than large ones in high winds.

The narrowing effect of the wind on the sprinkler pattern made wind direction an important design parameter. Shull and Dylla† obtained data from a 27.5-mm diameter ring nozzle and 22.9-mm and 30.5-mm diameter taper bore nozzles. From these measurements they made the following recommendations:

1 Do not irrigate when wind velocities exceed 4.5 m/s. Irrigate at night if possible, since in most windy areas wind velocity decreases at night.

2 Do not space travel lanes more than 40 percent of the manufacturer published no-wind wetted diameter. A lane spacing of 40 percent of the no-wind wetted diameter gave CU values of 85 percent or more 78 percent of the time, with maximum wind velocity of 4.5 m/s.

3 Orient travel-lane directions normal to the prevailing wind direction. Small wind-direction angles relative to the travel direction reduce the pattern width and thus reduce the application uniformity for a given travel-lane spacing.

4 Do not irrigate when the wind direction is parallel, or nearly parallel to the travel direction. Under their tests a lane spacing of 52 percent of the wetted diameter gave CU values of 85 percent or more 67 percent of the time when the wind direction relative to travel direction was 64 degrees or greater, and the maximum wind velocity was 7.6 m/s. Reducing the spacing to 47 percent of the no-wind wetted diameter resulted in CU values of 85 percent or more 73 percent of the time with the same wind direction restrictions.

These recommendations illustrate the importance of wind direction with respect to travel-lane direction. Every possible effort should be made to design travel-lane direction normal to the prevailing winds at the irrigation site.

Since the wetted area under big gun sprinklers decreases as wind velocities increase (Shull and Dylla, 1976a, 1976b), the application rate also increases and could exceed the infiltration rate of the soil. Extra care is needed in designing a traveling big gun system on soils of marginal infiltration rates.

---

†(Personal Communication) Hollis Shull and A. S. Dylla, USDA, Science and Education Administration, North Central Soil Convervation Research Center, Morris, MN.

14.7 INFILTRATION LIMITATIONS

Stationary sprinkler irrigation systems are designed to meet water use requirements and to prevent surface ponding and runoff. However, moving systems designed to meet water-use requirements often requires application rates that exceed soil intake rates. This section considers intake functions, application rates, and their interaction, and procedures for designing moving systems that have application rates exceeding intake rates.

14.7.1 Basic Intake Rate and Intake Function

A detailed discussion of infiltration was presented in Chapter 4. Intake rates for ponded soil surfaces are not constant but generally decrease with time and approach a basic intake rate asymptotically. The recommended maximum application rates for stationary sprinkler irrigation systems is the basic intake rate.

Moving sprinkler irrigation systems apply water over shorter time periods and the intake rate is generally above the basic intake rate. An intake rate versus time function is needed to design these systems. Dillion et al. (1972) and Kincaid et al. (1969) used the intake rate function of

$$f = K_k t^{-\propto} \dots\dots\dots\dots\dots\dots\dots\dots\dots\dots\dots\dots\dots\dots[14.15]$$

where f is the intake rate, t is the time, and K_k and $\propto$ are constants for a given soil and initial conditions to design center pivot systems. The design technique is discussed in Section 14.7.5.

14.7.2 Measuring Intake Capacities

Field methods for measuring infiltration capacities with a sprinkler and ring infiltrometers are presented in Chapter 4. Dillon et al. (1972) presented an alternative technique for evaluating infiltration rates for center pivot systems. The alternate technique recommends that tests be performed under soil-moisture conditions near the level of the expected irrigations. The soil surface condition should be similar to the conditions expected during the irrigation season. Preferably, the test should be run with a system similar to that planned for irrigation; thus minimizing the difference in intake rates due to drop-size and instantaneous application rate effects. Catch cans are placed near the outer tower at 1.5 m intervals and perpendicular to the sprinkler lateral, which is held stationary. The time for puddling to occur near each of the individual catch cans is observed. The average intake rates are determined by dividing the depth of water in the catch cans by the time of the test. Plotting the times for puddling to occur versus the corresponding application rate gives the limiting intake rate versus time-function for designing moving systems.

14.7.3 Surface Storage and Crop Interception

Design of moving sprinkler systems must consider surface storage and interception by the crop canopy when application rates exceed intake rates. When the surface storage capacity is exceeded, runoff will occur. Dillon et al. (1972) presented a design technique that allowed for the expected surface storage when the applied depth exceeds the intake capacity. Table 14.8 gives the allowable surface storage for various slopes. Interception by crop canopy

TABLE 14.8. ALLOWABLE SURFACE
STORAGE VALUES FOR VARIOUS
SLOPES (Dillon et al., 1972)

Slope, percent	Allowable surface storage, mm
0-1	13
1-3	7.5
3-5	2.5

could slightly increase the allowable surface storage, but is generally not included in the design of systems, which are needed to irrigate bare soils.

Current studies are investigating the use of minimum tillage to increase intake rates and furrow dikes for increasing the surface storage capacity.‡

14.7.4 Application Rates

The application rates for stationary sprinkler systems are generally assumed to be constant over the entire area being irrigated. The average application rate is the total volume applied divided by the area irrigated and the time of application. The actual application rate is not uniform but is higher near the sprinkler, as was discussed in Section 14.2. The rate increases with overlapped sprinklers when adjacent laterals are operated simultaneously. However, the average application rate is acceptable for designing stationary systems.

Moving systems apply water at a rate that varies with time and space. The application rate begins at zero, approaches maximum, and then decreases to zero as the system moves over a point (Fig. 14.13). Linear move

---

‡(Personal Communication) R. Nolan Clark, Agricultural Engineer, USDA-SEA-AR, Southwestern Great Plains Research Center, Bushland, TX.

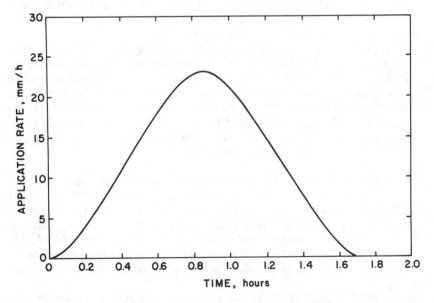

FIG. 14.13 Application rate pattern from moving sprinkler system.

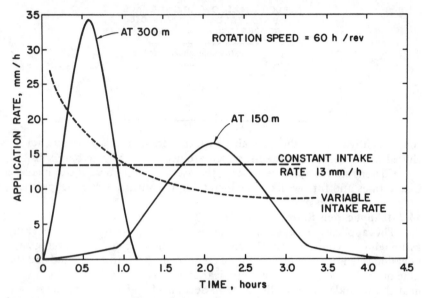

FIG. 14.14 Potential runoff for two distances from pivot of center plot system with constant spacing impact sprinklers.

systems have about the same application rate pattern along the entire length of the lateral. With center pivot systems, the maximum application rate increases directly with distance from the pivot (Fig. 14.14). The area under each application rate-time curve must be equal along the lateral for uniform irrigation. Center pivot systems designed with constant-spaced, constant-sized, impact sprinklers; and spray nozzles have increasingly higher application rates, respectively, for similar capacity systems (Fig. 14.15). Traveling big guns have higher application rates near the travel lane but do not have the overlap from the adjacent sprinklers since two guns are normally not moved simultaneously in an adjacent travel lane.

Application rates for moving systems generally are much higher than those for stationary systems. It is not uncommon to have instantaneous application rates of 60 mm/h with moving systems. Systems with part-circle sprinkler patterns, spraying behind the system, including big guns, have maximum application rates near the leading edge of the application pattern which then decrease to zero. Part-circle sprinklers have the advantage of not applying water ahead of a traveling system but decrease the area being irrigated and increase the application rates.

14.7.5 Relationship Between Application Rate and Intake Rate

Solid or stationary set sprinkler systems are designed so that the average application rate is less than the basic intake rate to prevent surface runoff. The intake rate and its relationship to the application rates at two distances from the pivot on a center pivot system are illustrated in Fig. 14.14. The volume of water represented by the area above the intake rate and below the application rate curves must either be held as surface storage or it will run off.

Kincaid et al. (1969) developed a technique to modify an intake function determined from cylinder infiltrometer measurements to design center pivot

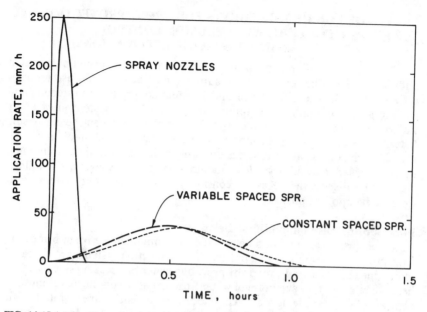

FIG. 14.15 Application rate patterns for typical center pivot systems with constant and variable spaced impact sprinklers and spray nozzles at a distance of 360 m from the pivot.

systems. Their modification assumes that the intake rate at any time depends on the volume infiltrated up to that time. The estimates of runoff with the modified intake function gave a better estimate of measured runoff than that estimated with the unmodified intake function. Field conditions are highly variable and make the validation of runoff estimation procedures extremely difficult.

Center pivot systems with small impact sprinklers or spray-type nozzles have smaller drop sizes, which tend to cause less puddling on the surface than do larger drops. Systems with small drop sizes maintain higher intake rates. The higher intake rates generally do not compensate for the increased application rates due to the decreased time of application because of the smaller sprinkler radius. Instantaneous application rates can approach 250 mm/h with spray nozzle systems, which is much higher than the intake rate of most soils. Potential runoff is reduced by increasing the travel speed and reducing the amount of water applied per irrigation. Field research results indicated considerably less runoff from high pressure impact sprinklers than from low pressure spray nozzle systems (Addink, 1975). Frequent, light irrigations tend to have higher spray and evaporation losses than less frequent irrigations.

Typically design procedures use empirical functions for intake rates. However, with the decreased cost of computing, it is feasible to solve the differential equations for infiltration with variable application rates. The trade-off is in the field data and measurements that are required for the more detailed models versus the less sophisticated empirical infiltration models. Chapter 4 describes several models of soil-water dynamics, which could be used in determining the potential runoff from high application rate sprinkler systems.

14.8 DESIGN REQUIREMENTS FOR FROST PROTECTION, COOLING, WIND EROSION CONTROL, PESTICIDE AND HERBICIDE APPLICATIONS

Sprinkler irrigation systems are often used for other special purposes in addition to providing crop water requirements. Each specific use requires different considerations for the sprinkler system design. Frost protection and cooling generally require low application rates and the system must apply water over large areas. Erosion control is most effective when the entire surface of the soil can be kept continually moist. The application of chemicals through irrigation systems requires uniform irrigations and the system must be safeguarded to avoid possible contamination of water sources. The following sections discuss the necessary considerations for designing systems used for specific applications.

14.8.1 Frost Protection and Bud Delay

Frost protection and bud delay have the common objective of protecting fruit production from freeze damage. Frost protection systems are operated to provide heat to a crop canopy by providing water for freezing since 335 J of heat are released for each gram of water frozen. The ice on the leaves must be kept at 0 °C by applying water continually to the entire area while the air temperature is below 0 °C. The systems must be trouble-free as freeze damage could result if the system does not operate continually and allows the leaf temperatures to drop below 0 °C. Gerber and Martsolf (1966) recommended application rates of 2.3 to 23 mm/h when the winds approach 4 m/s and meam temperatures approach -10 °C. Frost protection is most effective for radiative frosts that occur with low wind conditions. These do not require high application rates.

Sprinkler systems used to delay bud development provide water for cooling critical parts of plants in contrast to supplying heat for freeze protection. The intent is to delay flowering to reduce the probability of frost damage. Bud development is a direct function of the accumulated heating units. By keeping the bud wet during periods when the temperature exceeds 7.2 °C (the wet bud temperature is kept near wet-bulb temperature), the accumulated heat units are less which delays bud development. Mist or fine spray sprinklers operated at pressures up to 165 kPa, umbrella or shrub-type sprinklers operated at pressures of 170 to 200 kPa, and oscillating or impact sprinklers operated at pressures of 250 to 350 kPa have been used for delaying bud development. Griffin et al. (1976) reported that the oscillating or impact sprinklers are probably most suitable for this application. The mist or fine spray sprinklers have generally produced poor results under arid conditions. Systems are typically operated with a 2-minute on and 2-minute off cycle while temperatures are above 7.2 °C. Barfield et al. (1977) did not operate systems they studied during periods of high humidity when the wet bulb was near the dry bulb temperature and the degree of cooling would be minimal. Chesness et al. (1977) reported that operation was normally discontinued between midnight and 0830 when the humidity was high. Application rates range from 3.0 to 3.8 mm/h. Since water must be applied continuously to cool an area, about seven times the amount of water per unit area must be available during a given time period than that normally required for irrigation. To delay bud development, CU's greater than 80 are essential for cooling. The design of systems for bud delay must consider the infiltration rates and drainage needed for the higher application, automatic controls and equipment to operate the sprinklers within the desired parameters (i.e. low humidity and low wind conditions). Barfield et al. (1976), Chesness et al. (1977), Griffin et al. (1976), Larsen et al. (1974), and Stang et al. (1977) reported total application depths from 200 to 1200 mm for delaying bloom 7 to 17 days. Water requirements for evapotranspiration are low during the time bud cooling is needed, and therefore, cooling systems must be used on soils with good internal drainage. High water applications also may leach significant quantities of plant nutrients from the soil. With the right design and operating conditions, significant crop losses due to frost damage can be averted.

14.8.2 Cooling

Sprinkler systems are used to cool the soil for improved germination and seedling protection and have been effective for plant cooling to enhance the quantity and quality of production for green beans, small fruit, apples, cucumbers, melons, tomatoes, lettuce and nursery plant production. Improved yields and quality of hybrid onion seed were enhanced by sprinkling for 5 minutes and allowing 15 to 25 minutes for drying at Kimberly, ID (Wright et al., 1980). Application rates from 1 to 4 mm/h have been used for cooling. Systems can be designed to run either continuously or intermittently requiring slightly higher application rates. Systems are generally required to apply water slightly in excess of the potential evaporation rate but soils with high intake rates and internal drainage are not required as for frost protection systems. The management of these systems is discussed in Chapter 18.

14.8.3 Erosion Control

Sprinkling to control erosion requires keeping the soil surface continually moist. Proper tillage and management of organic matter is generally more effective for controlling erosion than sprinkler irrigation. In areas with high winds and sandy soils, it may be necessary to irrigate daily or even more frequently to keep the surface wet for satisfactory erosion control. Large water capacities are necessary to cover the large areas for erosion protection with sprinkler irrigation systems. Center pivot systems may not be able to apply water frequently enough to keep the surface moist. Solid set systems are more suitable for wind erosion control. Systems should apply water at a rate of 2.5 mm/h for adequate erosion control.

14.8.4 Chemical Application

Sprinkler systems also are used to apply fertilizer, herbicides, and pesticides. Both liquid forms that dissolve and dry powder in water suspensions can be injected through sprinkler irrigation systems. The main consideration in the design of sprinkler systems for chemical application is the method of injection. Chemicals can be injected by: (a) delivering the liquid to the suction side of a centrifugal pump from a supply tank, (b) injecting the chemical into the main lines using a pressure pump, (c) pressurizing the chemical supply tank from the main sprinkler lines and injecting the liquid at the low pressure area of a pitot tube, and (d) pressurizing a supply tank with a pitot tube facing in the upstream direction of flow and then injecting the liquid into the system with a second pitot tube facing downstream. The most important consideration in designing the injection system is preventing contamination of the water supply. An engineering practice titled "Safety Devices for Applying Liquid Chemicals through Irrigation Systems" is currently being developed by the Irrigation Management Committee of ASAE. The proposed engineering practice recommends "fail-safe" interlock between the main irrigation pumping plant and the pump used to inject the chemical into the main line. This would prevent the irrigation pumping plant from stopping and the injection pump to continue to fill the irrigation main line with the chemical solution. It is also important to provide check and vacuum relief valves (anti-siphon devices) for preventing the chemical from draining or siphoning back into the irrigation well or other water supply. The vacuum and check valves must be located between the pump and the point of chemical injection. If water is bled from the main irrigation supply into the chemical supply tank, the connecting line too must be equipped with a check valve to prevent the supply tank from overflowing and contaminating the adjacent area with chemical solution. Many states and local municipalities have codes for proper plumbing of chemical injection systems into irrigation systems.

The CU should equal 80 to 90 for uniform application of the chemicals to the area that is being fertilized or treated with herbicides or pesticides. Nonuniform systems would result in poor placement of the chemicals and, therefore, poor control. In Chapter 18 the operation and management of these chemical injection systems will be discussed.

The sizing of the pump or rate of injection into the sprinkler system should be checked closely so that sufficient quantities of the chemicals can be injected to obtain the desired application rate of the chemical. The rate of injection depends also on whether a continuous injection is made or whether the entire volume of chemical is injected in the beginning or at the end of the irrigation set. Intermittent injection allows the system to be flushed and chemicals to be either flushed from or left on the crop canopy. Center pivot systems require for uniform distribution that the chemical be injected at a continuous rate for an entire irrigation cycle.

References
1 Addink, J. W. 1975. Runoff potential of spray-nozzle and sprinkler center pivots. ASAE Paper No. 75-2056, ASAE, St. Joseph, MI 49085.
2 Agricultural Engineers Yearbook. 1979. ASAE, St. Joseph, MI 49085.
3 Albertson, L., R. Barton, and B. Simons. 1960. Fluid mechanics for engineers. Prentice-Hall, Englewood Cliffs, New Jersey. 561 p.
4 Barfield, B. J., G. M. White, and T. C. Bridges. 1977. Improving water use in sprinkling for bloom delay. TRANSACTIONS of the ASAE 20(4):688-691, 696.
5 Bittinger, M. W. and R. A. Longenbaugh. 1962. Theoretical distribution of water from a moving irrigation sprinkler. TRANSACTIONS of the ASAE 5(1):25-30.
6 Burgy, R. H., and C. R. Pomeroy. 1958. Interception losses in grassy vegetation. Amer. Geophys. Union Trans. 39:1095-1100.
7 Chesness, J. L., C. H. Hendersholt, and G. A. Couvillon. 1977. Evaporative cooling of peach trees in delay bloom. TRANSACTIONS of the ASAE 20(3):446-468.
8 Christiansen, J. E. 1942. Irrigation by sprinkling. Univ. of Calif. Agr. Exp. Sta. Bul. 670. 124 p.
9 Chu, S. T., and D. L. Moe. 1972. Hydraulics of a center pivot system. TRANSACTIONS of the ASAE 15(5):894-896.
10 Cline, J. F., M. A. Wolf, and F. P. Hungate. 1969. Evaporative cooling of heated irrigation water by sprinkler application. Water Resources Res. 5(2):401-406.
11 Covey, W., M. H. Halstead, S. Hillman, J. D. Merryman, R. L. Richman, and A. H. York. 1958. Micrometeorological data collected by Texas A&M, p. 53-96. In: Project Prairie Grass, A field program in diffusion. Vol. 11, Geophys. Res. Papers No. 59. Air Force Cambridge Research Center, Bedford.
12 Culver, R. and R. F. Sinker. 1966. Rapid assessment of sprinkler performance. Proc. Am. Soc. Civ. Engr., J. Irrig. and Drain. Div. 92:1-17.

13 Dillon, R. C., E. A. Hiler, and G. Vittetoe. 1972. Center-pivot sprinkler design based on intake characteristics. TRANSACTIONS of the ASAE 15(5):996-1001.

14 Frost, K. R. 1963. Factors affecting evapotranspiration losses during sprinkling. TRANSACTIONS of the ASAE 6(4):282-283, 287.

15 Frost, K. R. and H. C. Schwalen. 1955. Sprinkler evaporation losses. Agr. Eng. 36(8):526-528.

16 Frost, K. R. and H. C. Schwalen. 1960. Evapotranspiration during sprinkler irrigation. TRANSACTIONS of the ASAE 3(1):18-20, 24.

17 George, T. J. 1955. Evaporation from irrigation sprinkler sprays as determined by an electrical conductivity method. Unpublished thesis, Univ. of Calif., Davis.

18 Gerber, J. F. and J. D. Martsolf. 1966. Protecting citrus from cold damage. Agr. Ext. Serv., Univ. of Florida, Cir. 287, 29 p.

19 Griffin, R. E., J. F. Alfaro, R. E. Anderson, J. L. Ashcroft, R. D. Hill, E. A. Richardson, S. D. Seeley, and D. R. Walker. 1976. Reducing fruit losses caused low spring temperatures. Final Rpt. of Utah Agr. Exp. Sta. to Four Corners Reg. Comm., Oct., Utah State Univ., Logan, UT. 151 p.

20 Hart, W. E. 1972. Subsurface distribution of nonuniformity applied surface waters. TRANSACTIONS of the ASAE 15(4):656-661, 666.

21 Hart, W. E. and D. F. Heermann. Evaluating water distributions of sprinkler irrigation systems. Colo. State Univ. Exp. Sta., Fort Collins. Tech. Bul. 128, 39 p.

22 Heermann, D. F. and P. R. Hein. 1968. Performance characteristics of self-propelled center-pivot sprinkler irrigation system. TRANSACTIONS of the ASAE 11(1):11-15.

23 Heermann, D. F. and H. H. Shull. 1976. Effective precipitation of various application depths. TRANSACTIONS of the ASAE 19(4):708-712.

24 Hunter, E. J. 1966. Visual analysis of sprinkler test results. In: Proc. Annual Tech. Cong., Sprinkler Irrig. Assoc. p. 81-86.

25 Inoue, H. and S. S. Jayasinghe. 1962. On size distribution and evaporation losses from spray droplets, emitted by a sprinkler. Faculty of Agr., Kagawa Univ. Tech. Bul. 13(2):202-212.

26 Keller, J., M. D. Moynahan, and R. L. Ptacek. 1971. Sprinkler profile analysis to predict field performance. ASAE Paper No. 71-756, ASAE, St. Joseph, MI 49085.

27 Kincaid, D. C., D. F. Heermann, and E. G. Kruse. 1969. Application rates and runoff in center-pivot sprinkler irrigation. TRANSACTIONS of the ASAE 12(6):790-794.

28 Kincaid, D. C. and D. F. Heermann. 1970. Pressure distributions on a center-pivot sprinkler irrigation system. TRANSACTIONS of the ASAE 13(5):556-588.

29 Kinzer, G. D. and R. Gunn. 1951. The evaporation, temperature and thermal relaxation-time of freely falling water-drops. J. Meteor. 8(2):71-83.

30 Kohl, R. A. 1972. Sprinkler precipitation gage errors. TRANSACTIONS of the ASAE 15(2):264-265, 271.

31 Kohl, R. A. 1974. Drop size distribution from medium-sized agricultural sprinklers. TRANSACTIONS of the ASAE 17(4):690-693.

32 Kraus, J. H. 1966. Application efficiency of sprinkler irrigation and its effects on microclimate. TRANSACTIONS of the ASAE 9(5):642-645.

33 Larsen, D. C. and W. J. Kocahan. 1974. Using sprinklers to delay bloom for frost protection in apple trees. Paper No. PNW74-44, presented at 1974 Annual Meeting of the Pacific Northwest Region, ASAE, Twin Falls, ID.

34 Laws, J. O. and D. A. Parsons. 1943. The relation of rain-drop size to intensity. Trans. Am. Geophys. Un. 24:452-460.

35 Levine, G. 1952. Effect of irrigation droplet size on infiltration and aggregate breakdown. AGRICULTURAL ENGINEERING 33(9):559-560.

36 Lytle, W. F. and J. E. Winberly. 1962. Head loss in irrigation pipe couplers. Louisiana State Univ. and Agr. and Mech. College, Agr. Exp. Sta., Bul. No. 553, 15 p.

37 McMillian, W. D. and R. H. Burgy. 1960. Interception loss from grass. J. Geophys. Res. 65(8):2389-2394.

38 Merrington, A. C. and E. G. Richardson. 1947. The breakup of liquid jets. Proc. Phy. Soc. (Lonson) 59(2):1-13.

39 Pair, C. H. 1968. Water distribution under sprinkler irrigation. TRANSACTIONS of the ASAE 11(5):648-651.

40 Pair, C. H., J. L. Wright, and M. E. Jensen. 1969. Sprinkler irrigation spray temperatures. TRANSACTIONS of the ASAE 12(3):314-315.

41 Pair, C. H. 1975. Application rates and uniformity of application from mechanical-move sprinkler systems. Proc. 1975 Sprinkler Irrig. Tech. Conf., Atlanta, GA. p. 71-82.

42 Pair, C. H. 1975. Sprinkler irrigation. Sprinkler Irrig. Assoc., 13975 Connecticut Ave., Silver Spring, MD 20906.

43   Ring, L. and D. F. Heermann. 1978. Determining center-pivot sprinkler uniformities. ASAE Paper No. 78-2001, ASAE, St. Joseph, MI 49085.
44   Robinson, F. E. 1973. Increase in conductivity of irrigation water during irrigation. Agron. J. 65(1):130.
45   Rouse, H., J. E. Howe, and D. E. Metzler. 1952. Experimental investigation of fire monitors and nozzles. Am. Soc. Civil Engr. 117:1147-1188.
46   Sale, P. J. M. 1965. Changes in water and soil temperature during overhead irrigation. Weather 20:242-245.
47   Schladerbush, O. H. and W. Czeratzki. 1957. Die Niederschlagsenergie verschiedener Regner und ihr Einflusz auf die Boderverschlammung Landtechnishe Forschung 7(1):25-32.
48   Schleusener, P. E. 1957. Drop size distribution and energy of falling drops from medium-pressure irrigation sprinkler. Unpublished Ph.D. dissertation, Michigan State Univ., East Lansing.
49   Schleusener, P. E. and E. H. Kidder. 1960. Energy of falling drops from medium pressure irrigation sprinkler. AGRICULTURAL ENGINEERING 41(2):100-103.
50   Seginer, I. 1965. Tangential velocity of sprinkler drops. TRANSACTIONS of the ASAE 8(1):90-93.
51   Seginer, I. 1967. Net losses in sprinkler irrigation. Agr. Meteorol. 4:281-291.
52   Seginer, I. 1970. A resistance model of evaporation during sprinkling. Agr. Meteorol. 7:487-497.
53   Seginer, I. 1971. Water losses during sprinkling. TRANSACTIONS of the ASAE 14:656-659, 664.
54   Seginer, I. 1973. A note on sprinkler spray evaporation. Agr. Meteorol. 11:307-311.
55   Seginer, I. 1978. A note on the economic significance of uniform water application. Irrig. Sci. 1:19-25.
56   Seginer, I. 1979. Irrigation uniformity related to horizontal extent of root zone. A computational study. Irrig. Sci. 1:89-96.
57   Seginer, I. and M.Kestrinsky. 1975. Wind, sprinkler patterns, and system design. Am. Soc. Civ. Engr., J. of the Irrig. and Drain Div. 101 (IR-4):251-264.
58   Sellers, W. P. 1965. Physical Climatology. Univ. of Chicago Press. 149 p.
59   Shearer, M. N. 1971. Water distribution from a sprinkler lateral moving continuously in a linear direction. Oregon State Univ. Agr. Exp. Sta. Spec. Rpt. 342.
60   Shull, H. and A. S. Dylla. 1976a. Traveling gun application uniformity in high winds. TRANSACTIONS of the ASAE 12(2):254-258.
61   Shull, H. and A. S. Dylla. 1976b. Wind effects on water application patterns from a large, single nozzle sprinkler. TRANSACTIONS of the ASAE 19(3):501-504.
62   Solomon, K. 1978. Variability of sprinkler coefficient of uniformity test results. ASAE Paper No. 78-2010, ASAE, St. Joseph, MI 49085.
63   Stang, E. J., D. C. Ferree, F. R. Hall, and R. A. Spotts. 1977. Overtree misting for bloom delay in Golden Delicious. Approved for publication as Journal Article No. 70-77 of the Ohio Agricultural Research and Development Center, Wooster, OH.
64   Strong, W. C. 1961. Advanced irrigation design. Agrica and Irrigation, Proc. of an Interna. Irrig. Symp., sponsored by Wright Rain, Ltd., Salisbury, Southern Rhodesia, p. 242-246.
65   Veihmeyer, F. J. 1927. Some factors affecting the irrigation requirements of deciduous orchards. Hilgardia 2(6):125-291.
66   Wiersma, J. L. 1955. Effect of wind variation on water distribution from rotating sprinklers. South Dakota State Col. Agr. Exp. Sta. Tech. Bul. 16.
67   Wolfe, J. W. 1967. Uniformity of distribution of water from small solid-set irrigation sprinklers. Paper presented at the 22nd Annual Meeting, Pacific Northwest Reg., ASAE, Spokane, WA.
68   Wright, J. L., J. L. Stevens, and M. J. Brown. 1980. Controlled cooling of onion umbuls by periodic sprinkling. Submitted to Agronomy Journal.
69   Young, R. A. and J. L. Wiersma. 1973. The role of rainfall impact in soil detachment and transport. Water Resources 9(6):1629-1636.

chapter 15

DESIGN AND OPERATION OF SPRINKLER SYSTEMS

15

DESIGN AND OPERATION OF SPRINKLER SYSTEMS

by   J. W. Addink, Addink Engineering Company, Inc.,
      Lincoln, NE; J. Keller, Utah State University,
      Logan, UT; C. H. Pair (Retired), USDA-SEA/AR,
      Snake River Conservation Research Center,
      Kimberly, Idaho; R. E. Sneed, North Carolina
      State University, Raleigh, NC and J. W. Wolfe,
      Oregon State University, Corvallis, OR

15.1 INTRODUCTION

Sprinkler irrigation is adaptable to many crops, soils, and topographic conditions. Sprinkler systems are classified according to whether the sprinkler heads are operated individually (gun or boom sprinklers), or as a group along a lateral, and according to how they are moved (or cycled) to irrigate the entire field.

Sprinkler laterals can be: (a) periodically moved from one set (irrigation) position to another by hand or mechanically until the entire field is irrigated; (b) set so closely together (solid set) that the field can be irrigated without moving them; (c) continuously moved around a pivot point (center-pivot) to irrigate a large circular area; or (d) continuously moved along a closed or open channel water supply (traveling lateral) to irrigate a large rectangular area.

Gun or boom sprinklers can be periodically moved from one set position to another by hand, or mechanically set closely together so the field can be irrigated without moving them. They also can be mounted on trailers and continuously moved (traveling large gun). Water may be supplied to traveling gun or boom sprinklers from a pipeline through a flexible hose or directly from a ditch.

Periodic move systems are well-suited for irrigation in areas where the crop-soil-weather situation does not require irrigations more often than every 5 to 7 days. For high frequency irrigation, solid-set or continuously moving systems are more adaptable; however, where soil permeability is low, some of the continuously moving systems, such as the center-pivot and traveling large guns, may cause runoff problems. In addition to being adaptable to all irrigation frequencies, solid-set systems also can be designed and operated for frost and freeze protection, blossom delay, and crop cooling.

15.2 SOLID-SET SYSTEMS

As growers attempted to reduce labor and began to use sprinkler irrigation to modify the environment, they turned increasingly to solid-set and permanent irrigation systems. Depending upon the location within the

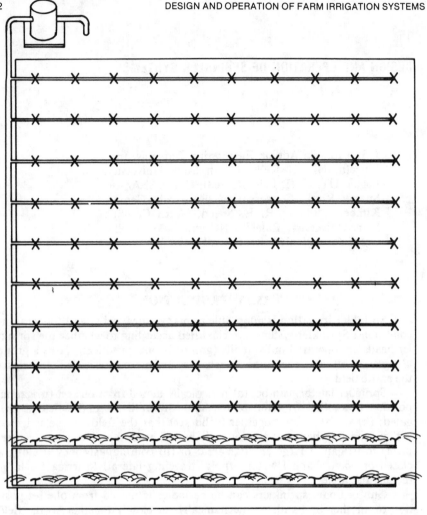

FIG. 15.1. Solid-set system, with o...y enough sprinklers to operate a portion of the system at one time. Sprinklers are moved to irrigate another portion of the field. The "x" marks the spot of each sprinkler position. Quick-coupler riser valves are used at each sprinkler location.

country, different definitions are used to describe these two systems. In this chapter, the following definitions will be used:

Solid-set system: An aboveground portable aluminum pipe system, which is placed in the field or fields at the start of the irrigation season and left in place during the season.

Permanent system: Plastic, asbestos cement, coated aluminum, or coated steel pipe placed underground, with only the sprinklers and a portion of the risers aboveground.

15.2.1 Field Layout

Both the solid-set and the permanent systems have a number of variations. The crops to be grown, mode of operation, available labor and

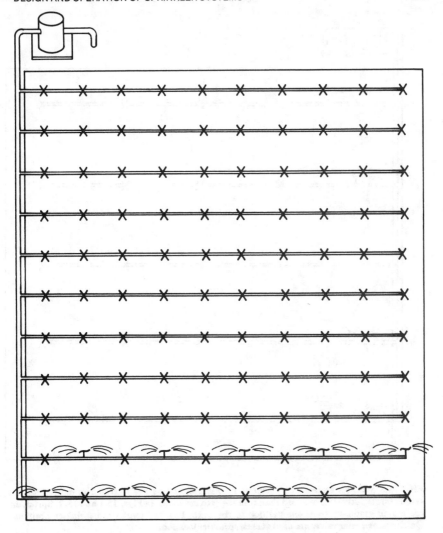

FIG. 15.2 Solid-set system, with every other sprinkler operating on a lateral. This allows the use of a longer lateral using the same capacity sprinkler and pipe size as compared to having every sprinkler operating. The "x" marks the sprinkler positions. Quick-coupler riser valves are used at each sprinkler location.

capital, field shape, size, and topography all influence the selection of the system and the field layout. Solid-set systems normally are laid out with the laterals parallel to the row direction and the mainline perpendicular to the row direction. The mainline may be at one end of the field, or through the field. Most permanent systems also are installed this way, but in orchards and vineyards the laterals can be installed diagonally across the field or at right angles to the row direction. Where possible, mainlines are placed perpendicular to the laterals to facilitate connecting the laterals, but this is not necessary.

With most solid-set and permanent systems, sprinklers are installed on the entire system; however, some growers, in an effort to reduce costs,

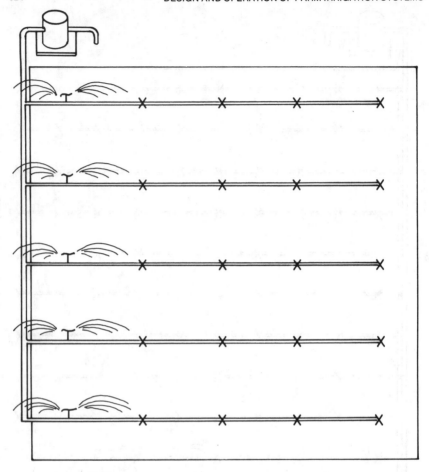

FIG. 15.3 Solid-set system with only one sprinkler operating per lateral. Normally, with this system larger sprinklers are used than with the systems shown in Figs. 15.1 and 15.2. Sprinklers are moved manually from one position to the next. The "x" marks the sprinkler positions. Quick-coupler riser valves are used at each sprinkler location.

purchase only enough sprinklers and risers to irrigate a small area at a time. Sprinklers and risers are moved manually from one location to another across the field. There are several ways a small number of sprinklers may be used:

1   One or more laterals may be operated using every sprinkler on the lateral. Sprinklers then are moved to other laterals, as illustrated in Fig. 15.1.

2   One or more lateral lines may be operated, but with only part of the sprinklers on each lateral operating. For example, every second or third sprinkler may be operated at one time. Then the sprinklers are moved to the next location on the same laterals, as illustrated in Fig. 15.2.

3   One sprinkler on each lateral line may be operated. When only one sprinkler on a line is used, it usually will have a higher capacity than that normally used for a multi-sprinkler lateral, and may be a large gun sprinkler. Sprinklers then are moved to the next sprinkler position as illustrated in Fig. 15.3.

For the three systems described, the sprinklers are moved manually. Quick-coupling riser valves are used to enable moving the sprinklers while the system is operating. The solid-set aluminum pipe system is more adaptable to the movable sprinkler concept than is the permanent-pipe system.

15.2.2 Sprinkler Spacing, Nozzle Size and Pressure

In general, solid-set systems are designed to use low-flow medium pressure sprinklers. However, depending upon the uses for which the system is designed and the type of design used, many sizes of sprinklers are used. Sprinkler spacings will vary from 9 m by 9 m (30 ft by 30 ft) to 73 m by 73 m (240 ft by 240 ft). Sprinkler nozzle sizes can be as small as 1.59 mm (1/16 in.) to as large as 36 mm (1.29 in.), and pressures can vary from 205 kPa (30 lb/in.^{2}) to 585 kPa (85 lb/in.^{2}).

The sprinkler spacing will depend upon the sprinkler and nozzle combination, operating pressure, desired coefficient of uniformity (CU), windspeed, and design use of the system. For certain high-value crops, it may be desirable to design for a high CU. A crop of lesser value may not justify the cost of a high CU design. Since it is not possible to design for all wind conditions, the system should be designed for average conditions. A system designed for frost and freeze protection may not require as high a CU as a system for soil moisture control. A system that is used to supplement rainfall may not need as high a CU as one in which crop production totally depends upon irrigation. As windspeed increases, spacing must be decreased to maintain the same coefficient of uniformity. The following percentages of design diameters should be considered maximum:

Wind conditions	Lateral spacing
No wind	65% of effective wetted diameter
8 km/h (5 mi/h)	60% of effective wetted diameter
8-16 km/h (5-10 mi/h)	50% of effective wetted diameter
> 16 km/h (10 mi/h)	30% of effective wetted diameter

There are three types of sprinkler spacings: square, rectangular, and triangular. Each has its merits, but under variable wind directions a sprinkler spacing approaching a relatively short-sided square generally gives a more uniform distribution. However, to reduce the number of laterals and still have good uniformity, a rectangular spacing is often used. For rectangular spacing, the distance between laterals is increased and sprinklers are moved closer together on the lateral. Some designers use the triangular spacing to try to improve the uniformity of application, but the results vary, especially where spacings as wide as 70 percent of the effective sprinkler diameter are used.

All sprinkler manufacturers should provide performance data for their equipment. The performance data should include the effective diameter under no wind conditions, discharge for various nozzles and pressures, and minimum recommended operating pressure for the various nozzle sizes. Up to a point, as pressure is increased, the effective diameter is increased and more uniform application may result. For a good spray breakup, the pressure should be increased as the nozzle size increases.

15.2.3 Mainline and Lateral Sizes

In sizing main or supply and lateral lines, the designer should not only consider uniformity of application, but also pumping costs and elevation

differences- between the water source and the fields. Many formulas have
been developed to determine friction loss (see Section 14.1.1). One formula
that is generally accepted is the Hazen-Williams:

$$H_f(100) = K \frac{(Q/C)^{1.852}}{D^{4.87}} \quad \dots\dots\dots\dots\dots\dots\dots\dots\dots\dots\dots\dots\dots\dots\dots\dots \quad [15.1]$$

where $H_f(100)$ = a friction loss per 100 m (100 ft) of pipe, m/100 m (ft/100
ft; C = a coefficient of retardation based solely on the character of the pipe
material; Q = the flow of water in the line L/s (ft³/s) (gal/min); D = the
pipe diameter in mm (ft)(in.); K = a constant which is 1.22 x 10¹² for metric
units, 473 for Q = ft³/s and D = ft, and 10.46 for Q = gal/min and D = in.
    Pipe size, type of piping material, coupler design, alignment of pipe,
and distance between couplers will affect the value of C. Normally, as pipe
size increases, C increases. As the number of couplers decreases, C increases.
Piping materials with smoother inside walls will have a higher C value. (For
new or coated steel or aluminum (with couplers) main or supply lines,
C = 120; for cement asbestos, C = 140; and for plastic, C = 150. Also see
Sections 11.4.4 and 14.1.1).
    Normally, flow velocities should not exceed 3 m/s (10 ft/s). However,
allowable pressure loss should not exceed an economically practical value
that is determined by the system designer with the approval of the purchaser.
It becomes a matter of balancing capital cost of the pipe against the pumping
costs caused by friction. Another controlling factor is the use of the system.
Lateral pressures can be controlled partially by take-off valves, and sprinkler
discharge can be controlled by flow-control valves or pressure-regulating
valves.
    It is desirable to have as uniform a discharge as possible, whether the
entire solid-set system, or a segment of the system, is operating. This
requires that pressure loss due to friction in the mainline be minimized, or
that flow control devices be used. Elevation differences within a field must
also be included in any calculations. The Florida Irrigation Society states
that the operating nozzle pressure of any sprinkler in a permanent system
cannot deviate by more than ± 15 percent from the design operating nozzle
pressure.
    For permanent systems in which polyvinyl chloride (PVC) plastic pipe
and asbestos cement (AC) pipe are used for supply lines, velocities should
not exceed 2.25 m/s (7.5 ft/s), and most manufacturers caution against
using velocities in excess of 1.6 m/s (5 ft/s).
    When more than one lateral on a solid-set system is operated at one
time, (especially adjacent to each other) the supply line becomes multi-
outlet. This means more water is flowing through the portion nearer the
pump and friction loss must be computed in each segment.
    In designing aluminum laterals, normally only one—two at the most—
tubing sizes are used. However, for PVC, laterals with several sizes may be
the most economical. When water is being removed at intervals along the
lateral, the friction loss for a given diameter and length of tubing will be
less than if the flow was constant for the entire length. To accurately compute
friction loss in the lateral, start at the last outlet on the line and work back
to the supply line, computing the friction loss between each outlet. This

tedious process has been simplified by a procedure developed by Christiansen (1942). He developed an adjustment factor (F) to correct the friction loss calculated from the general formula that assumes all of the water is carried to the end of the line.

$$H_f(L) = FK \frac{\frac{L}{100}\left(\frac{Q}{C}\right)^m}{D^{2m+n}} \quad \dots\dots\dots\dots\dots\dots\dots\dots\dots\dots\dots\dots\dots [15.2]$$

where $H_f(L)$ = friction loss in lateral of length L with multiple outlets having equal spacings and discharges; K = a constant based on the dimensions used in the formula (equation [15.1]); L = length of pipeline; Q = total flow into the lateral; D = diameter of pipe; m = velocity exponent; n = pipe diameter exponent; and N = number of outlets on the lateral. In the Hazen-Williams formula, m = 1.852 and n = 1.167.

Values of F, assuming that the first sprinkler is one sprinkler riser spacing from the beginning of the lateral, can be computed from the following approximate expression (Christiansen, 1942).

$$F = \frac{1}{m+1} + \frac{1}{2N} + \frac{\sqrt{m-1}}{6N^2} \quad \dots\dots\dots\dots\dots\dots\dots\dots\dots\dots\dots\dots [15.3a]$$

For N > 10, the last term can be omitted.

Jensen and Fratini (1957) modified the above expression for F to account for the first sprinkler being located one-half the sprinkler spacing from the supply line. They assumed that no water flows past the last sprinkler. The modified expression (equation [15.3b]) indicates that the F factor is more than 5 percent larger for N < 20 (Jensen and Fratini, 1957).

$$F = \frac{1}{2N-1} + \frac{2}{(2N-1)N^m}[(N-1)^m + (N-2)^m + \dots + 1^m] \quad \dots\dots\dots [15.3b]$$

Estimates of F values are easy to obtain using equation [15.3a], but these estimates become much more tedious when using equation [15.3b] for large values of N. To simplify their use, F values for m = 1.90 are presented as follows:

Number of outlets (N)	F (equation [15.3a])	F (equation [15.3b])
1	1.0	1.0
2	0.634	0.512
3	0.528	0.434
4	0.480	0.405
5	0.451	0.390
6	0.433	0.381
8	0.410	0.370
10	0.396	0.365
15	0.379	0.363
20	0.370	0.354
40	0.357	0.349
100	0.350	0.347

Detailed tables of F values for other values of m are commonly found in most sprinkler design publications.

Some fundamental relationships between pressure and discharge of sprinklers are presented to illustrate the effect of pressure changes on application uniformity. Sprinkler discharge is a function of the pressure at individual sprinklers

$$q = K\sqrt{P} \qquad\qquad\qquad\qquad\qquad [15.4]$$

where q = sprinkler discharge; K = nozzle discharge coefficient; and P = pressure at sprinkler. Because pressure varies along the lateral due to friction and elevation differences, sprinkler discharge also will vary. However, the ratio of pressure at any point in the lateral to the pressure at any other point will be constant for a given flow.

The pressure ratio on a sprinkler lateral, P/P_0, is the ratio of the pressure at any point on the lateral to the pressure at the end sprinkler. The discharge ratio is equal to the square root of the pressure ratio:

$$\frac{q}{q_0} = \sqrt{\frac{P}{P_0}} \qquad\qquad\qquad\qquad\qquad [15.5]$$

where q = the discharge of any sprinkler whose pressure is P; and q_0 = the discharge of last sprinkler on lateral with pressure of P_0. Thus, with a 20 percent variation in pressure along a lateral, the variation in discharge is about 10 percent. In a constant diameter, multi-outlet lateral, half the pressure loss due to friction will occur in the first 25 percent of the lateral.

To obtain high application efficiencies and also to stay within economical pipe sizes, the variation in lateral pressure must be held to a practical minimum. Variation in pressure should not exceed ± 10 percent of mean lateral design pressure. For practical purposes, allowable pressure loss due to friction can be estimated as 23.4 percent of the required average pressure. For laterals on level fields, this will give an actual pressure drop from the first to last sprinkler of about 20 percent.

Where more than one size of lateral is used, several steps are required to estimate the pressure drop. For a two-size lateral, the first step is to calculate the friction loss for the smallest pipe size (D_2) or the end of the lateral (L_2) using the flow rate entering that section (Q_2).

$$H_f(L_2, D_2) = FH_f \frac{L_2}{100} \quad \dots\dots\dots\dots\dots\dots\dots\dots\dots\dots\dots\dots \quad [15.6]$$

Then the friction loss for L_1 is first calculated by estimating the loss as though the entire lateral had one diameter (D_1) using the flow rate entering the lateral (Q_1)

$$H_f(L_1 + L_2, D_1) = FH_f \frac{(L_1 + L_2)}{100} \quad \dots\dots\dots\dots\dots\dots\dots\dots \quad [15.7]$$

Then the loss for L_2 is computed as though its diameter was D_1 using the flow rate Q_2

$$H_f(L_2, D_1) = FH_f \frac{L_2}{100} \quad \dots\dots\dots\dots\dots\dots\dots\dots\dots\dots\dots \quad [15.8]$$

The value obtained in equation [15.8] is then subtracted from the value obtained for the entire length and D_1 to give the loss in L_1 with diameter D_1 and Q_1

$$H_f(L_1, D_1) = H_f(L_1 + L_2, D_1) - H_f(L_2, D_1) \quad \dots\dots\dots\dots\dots\dots \quad [15.9]$$

The final result for a two-size lateral is obtained by adding the loss obtained for L_2 and D_2 and the loss for L_1 and D_1.

$$\text{Total } H_f = H_f(L_1 + L_2, D_1) - H_f(L_2, D_1) + H_f(L_2, D_2)$$

$$= H_f(L_1, D_1) + H_f(L_2, D_2) \quad \dots\dots\dots\dots\dots\dots\dots\dots \quad [15.10]$$

The total pressure loss due to friction for three or more pipe sizes can be computed in a similar manner.

Example:

Given: Total pipe length $(L_1 + L_2)$ = 283 m (930 ft)

L_1 = 146 m (480 ft), 102-mm (4-in.) O.D. diameter, 99-mm I.D.

L_2 = 137 m (450 ft), 76-mm (3-in.) O.D. diameter, 73.5-mm I.D.

Q_1 = 10.1 L/s (160 gal/min), Q_2 = 5.05 L/s (80 gal/min)

16 sprinklers are used, eight on each section with an average discharge of 0.63 L/s (10 gal/min). Assume the first riser is located one-half the sprinkler head spacing from the beginning of each lateral section.

C = 120 (also see Sections 11.4.4 and 14.1.1)

Results: F, based on equation [15.3b], is 0.37 for N = 8 and 0.36 for N = 16, assuming m = 1.9

For 76-mm (D_2) and Q_2 = 5.05 L/s (80 gal/min)

$$H_f = 1.22 \times 10^{12} \frac{\left(\dfrac{5.05}{120}\right)^{1.852}}{(73.5)^{4.87}} = 2.81 \text{ m/100 m (ft/100 ft) (eq. [15.1])}$$

For 102-mm pipe (D_1) and Q_1 = 10.1 L/s (160 gal/min)

$$H_f = 1.22 \times 10^{12} \frac{\left(\dfrac{10.1}{120}\right)^{1.852}}{(99)^{4.87}} = 2.38 \text{ m/100 m (ft/100 ft) (eq. [15.1])}$$

For 102-mm pipe (D_1) and Q_2 = 5.05 L/s (80 gal/min)

$$H_f = 1.22 \times 10^{12} \frac{\left(\dfrac{5.05}{120}\right)^{1.852}}{(99)^{4.87}} = 0.66 \text{ m/100 m (ft/100 ft) (eq. [15.1])}$$

$$H_f (L_2, D_2) = 0.37 (2.81) \frac{137}{100} = 1.42 \text{ m (4.7 ft) (eq. [15.6])}$$

$$H_f (L_1 + L_2, D_1) = 0.36 (2.38) \frac{283}{100} = 2.42 \text{ m (8.0 ft) (eq. [15.7])}$$

$$H_f (L_2, D_1) = 0.37 (0.66) \frac{137}{100} = 0.33 \text{ m (1.1 ft) (eq. [15.8])}$$

$$H_f (L_1, D_1) = 2.42 - 0.33 = 2.09 \text{ m (6.9 ft) (eq. [15.9])}$$

Total H_f = 2.42 – 0.33 + 1.42 = 3.51 m (11.5 ft) (eq. [15.10])

In terms of pressure, a 1-m pressure head is equivalent to 9.81 kPa (1 ft = 0.433 lb/in.²). Therefore, the friction pressure loss would be 3.51 (9.81) = 34.4 kPa (5.0 lb/in.²).

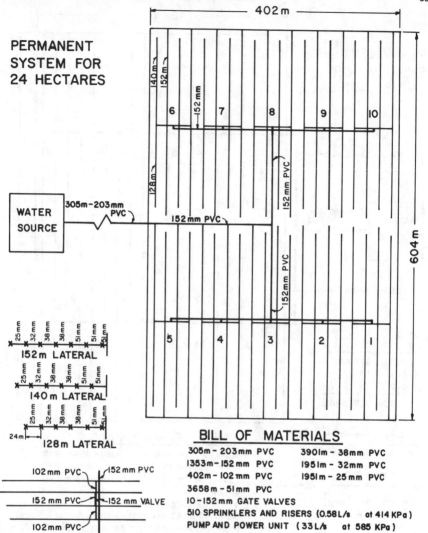

BILL OF MATERIALS

305m - 203mm PVC	390lm - 38mm PVC
1353m - 152mm PVC	195lm - 32mm PVC
402m - 102mm PVC	195lm - 25mm PVC
3658m - 51mm PVC	

10 - 152mm GATE VALVES
510 SPRINKLERS AND RISERS (0.58L/s at 414 KPa)
PUMP AND POWER UNIT (33L/s at 585 KPa)

FIG. 15.4 Permanent system for 25 ha (60 acre) using the lateral design. Two laterals will be operated from each valve. The four sets of laterals marked "1" will be operated simultaneously, then those marked "2" will be operated, etc. through "10".

To simplify the computation of lateral pipe sizes, logarithmic charts, nomographs, tables, slide rules, and pre-engineered lateral tables have been prepared and can be found in many sprinkler design publications (also see Table 11.5, and Sections 11.4.4 and 14.1.1). Estimates of friction losses in laterals have been quite varied. Ree (1959) presented a detailed analysis of the various experiments that had been conducted to determine friction losses and the various friction pressure loss equations.

15.2.4 Operation

The operating mode for a solid-set or permanent sprinkler system

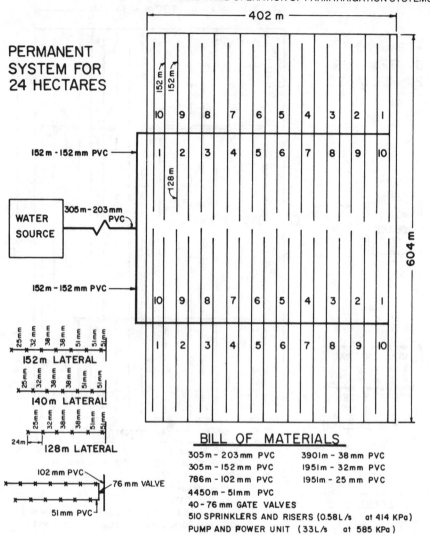

PERMANENT SYSTEM FOR 24 HECTARES

BILL OF MATERIALS

305m – 203mm PVC	3901m – 38mm PVC
305m – 152mm PVC	1951m – 32mm PVC
786m – 102mm PVC	1951m – 25mm PVC
4450m – 51mm PVC	
40 – 76mm GATE VALVES	
510 SPRINKLERS AND RISERS (0.58L/s at 414 KPa)	
PUMP AND POWER UNIT (33L/s at 585 KPa)	

FIG. 15.5 Permanent system for 24 ha (60 acre) using the area design. Eight laterals varying from 128 to 152 m (420 to 500 ft) in length will be operated from each 152-mm (6-in.) valve.

depends upon the design and use of the system, available labor, water supply, and available capital. Either system can be designed on the lateral or area (block) design method. With the lateral design method, individual laterals are controlled by valves and each lateral may be operated as desired. Normally, more than one lateral is operated simultaneously, but the operating laterals usually are widely separated in the field. The lateral design method minimizes the main or supply line pipe size, but it increases the number of valves required and also the time to open and close valves when a manual valve system is used. With the area (or block) design method, a contiguous portion of the field is irrigated at one time. Usually a sub-main is installed to supply water to that portion of the field. Figs. 15.4 and 15.5 show examples of the two types of design.

For frost and freeze protection, the entire system may be operated at one time. Depending upon the crop being protected, the application rate will be 2 to 5 mm (0.08 to 0.18 in.) per h (also see Section 2.8). In the eastern U.S., most orchard systems are designed to apply water over the crop. In the western U.S., both undertree and overtree systems are used; however, with saline water only undertree systems can be used successfully.

Single nozzle, medium pressure sprinklers should be used for frost and freeze protection. For crop cooling and blossom delay, the entire system may be sequenced in alternate on-off modes as one portion of the system may be operated at a time and the operation can be switched to another portion of the system. Sequencing is best accomplished with electric controllers and automatic valves.

If the system is being used strictly for irrigation, only a portion of the system is normally operated at one time. Where several hours are required for irrigation, control may be manual or automatic. For an irrigation system on a shallow-rooted crop grown on a coarse-textured soil, or in a container nursery operation where daily or high frequency irrigation is required, it is best to automatically control the sequencing of the system. Where labor is very limited, automatic control may be desirable regardless of irrigation frequency, however this will increase the initial investment. Conversely, limited capital may require a totally manual system. A limited water supply, such as a well or stream, may mean that only a portion of the system can be operated at once.

15.3 PERIODIC LATERAL MOVE SYSTEMS

All sprinkler irrigation systems in this category have sprinkler laterals which are moved between irrigation settings. They remain stationary while irrigating. The most common system has a single center mainline with one or more laterals which irrigate on both sides of the mainline. If there are multiple laterals, they are spaced equally, so by the time any one lateral reaches the starting position of the lateral ahead of it, the entire field has been irrigated once.

Spacing of the sprinklers on the laterals and the spacing between subsequent sets of each lateral are such that the water distribution patterns from the sprinklers give almost complete overlap. Large systems often require more complex designs with multiple mainlines, although simple systems are possible on rectangular fields up to at least 64 ha (160 acres).

If the mainline is on the ground surface, it usually is made of aluminum. Pipe couplers have rubber gaskets which seal under pressure, but which permit easy coupling and uncoupling. Some couplers are galvanized steel, formed by rolling a piece of steel tubing. Others are cast aluminum. Some are given a pressed fit into the pipe, or welded onto the pipe, while others have a gasket at each end and are clamped or latched onto the pipe.

Valve-tees usually are placed in the mainline at the desired interval for spacing between lateral settings. The valve-tees are controlled by manipulating a valve-opening elbow which makes the connection between the mainline and the laterals. The most common valve spacings are 12.2, 15.2, 18.3, and 24.4 m (40, 50, 60, and 80 ft). Since common pipe lengths are 6.1, 9.1, 12.2, and occasionally 15.2 m (20, 30, 40, and 50 ft), the 12.2-m (40-ft) valve spacing is obtained by using one pipe length per valve. The 15.2 m (50 ft) spacing uses either one pipe length or a 6.1- and a 9.1-m (20- and 30-ft) length. The 18.3-m (60-ft) spacing uses two 9.1-m (30-ft) pipe lengths between valves. The 24.4-m (80-ft) spacing uses two 12.2-m (40-ft) pipe lengths between valves.

There are two types of mainline valve-tees. One type is spring-loaded and requires the spring to be depressed when the valve is in the open position. When in the closed position, the pressure of the water adds to the pressure of the spring to seal the valve. The other type has a valve cap on top which closes against the pressure. In both types, the valve is opened and closed by a hand-operated screw, but for the spring-load type the screw is in the valve-opening elbow.

For many systems, a buried mainline is preferred. The most common materials used for buried mainlines are PVC, AC, and coated and wrapped steel tubing. The pipe should be placed safely below plow depth and should also be below frost depth. If not placed below frost depth, provision for draining the pipe is needed.

With the steel pipe, an advantage is that risers can be welded into place at the factory before the pipe is coated. Unfortunately, it is subject to rust and corrosion. Usually, no coating is placed on the inside. To insure sufficient longevity, it should be 3.4 mm (10-gauge) or heavier.

AC pipe has a long life expectancy (40 yr), if it is adequately protected from excessive water hammer and surge pressures. It is heavy and therefore is manufactured only in relatively short lengths (4.0 m (13.2 ft)).

PVC pipe also has a long life expectancy when protected from excessive hammer and surge pressures. PVC pipe joints can be sealed with cement or rubber O-rings. The O-ring connection has the advantage of permitting a change in pipe length with changing temperature, thus avoiding high tension stresses that could result in separation of a cemented joint. Both PVC and AC pipe require special mainline tees fitted with a riser pipe to reach the valve which is at or slightly above the ground surface. These tees can add significantly to the total cost of the pipelines.

The valves used for buried mainlines are similar to those used for the aluminum surface mainlines. They are attached to the risers by threading or welding.

The design procedure for computing the main and lateral line friction loss is similar for solid-set and periodic lateral-move systems.

15.3.1 Hand-Move Laterals

The first hand-move sprinkler laterals were constructed of light-gauge steel tubing. Essentially, all hand-move sprinkler systems now use aluminum laterals (Fig. 15.6), although some attempt has been made to use plastic. Common pipe diameters range from 51 to 152 mm (2 to 6 in.), and pipe lengths are 6.1, 9.2, and 12.2 m (20, 30, and 40 ft). The most popular aluminum lateral length is 9.1 or 12.2 m (30 or 40 ft). Shorter lengths mean more walking during the move. Longer lengths are more difficult to transport and do not provide proper spacing for the common sprinkler sizes.

Most lateral pipe couplers contain a chevron-type rubber gasket which seals under pressure. They latch automatically when the pipes are pushed together. Many couplers contain an optional adjustment for easier unlatching. On one optional setting, the irrigator must go to the coupler and lift the latch to release it. With a slight adjustment, the coupler can be fixed so it will unlatch automatically when the irrigator lifts or twists the pipe. The automatic unlatch saves some walking, but could increase the hazard of an accidental unlatching. In contrast to the mainline gaskets, the lateral

FIG. 15.6 Photograph of hand-move sprinkler laterals in operation.

gaskets are designed to release their tight contact with the pipe when the pressure on the water is reduced. This permits the water to drain from the pipe when the pressure has been turned off so the pipe can be moved easily to the next setting.

Each coupler is threaded to receive a sprinkler riser pipe, usually 25 mm (1 in.) in diameter. If both the coupler and the riser are aluminum, it is customary to connect them with a zink alloy fitting to avoid thread seizure. The riser should extend at least to the top of the crop canopy, but the uniformity of water distribution is improved if it is extended another 0.5 m (20 in.).

The uniformity of water distribution can usually be improved by using an offset pipe with a 90-deg elbow every second irrigation. The length of the offset should be half of the spacing between lateral settings. Using the offset pipe permits the lateral to be placed midway between the positions used during the previous irrigation. Thus, considering two irrigations added together, a 12.2-m by 18.3-m (40-ft by 60-ft) spacing, for example, is effectively reduced to 12.2 m by 9.2 m (40 ft by 30 ft).

A good procedure for the irrigator to follow when moving the lateral from one setting to the next is to start with the valve opening elbow and the section of pipe connected to it. As soon as these pieces are in place at the new location, the valve is slightly opened so a very small stream of water runs out the end of the first pipe section. As each subsequent section of pipe is put into place, the small stream of water runs through it, flushing out any soil or debris that may have been picked up during the move. Even the last section of pipe with its end plug in place can be connected before the stream of water reaches the end and builds up pressure. Then the irrigator walks back along the lateral, correcting any plugged sprinklers, leaky gaskets, or tilted risers. After returning to the mainline, he opens the valve further until the desired pressure is obtained. A quick check with a pitot gauge on the first sprinkler confirms the valve adjustment. To save time on each lateral move, there is a tendency to completely open the valve and fill the line as quickly as possible. This causes water hammer at the far end of the line, so a surge plug at that end is recommended.

FIG. 15.7 Photograph of a side-roll sprinkler system.

The sprinklers commonly used on hand-move systems may have either one or two nozzles. Typically, individual sprinkler capacities range from about 0.06 to 0.63 L/s (1 to 10 gal/min). Operating pressures range from 240 to 415 kPa (35 to 60 lb/in.²).

Certain crops, such as orchards, require specially designed sprinklers. When the sprinklers are used over the tops of the trees, conventional models can be used. However, when they are used under trees, sprinklers with low water trajectory must be used. Lowering the trajectory reduces the uniformity, unless the spacing is reduced. Hedge-rowed trees present a more difficult problem, especially if it is desirable to irrigate through the skirts of the rows.

15.3.2 Side-Roll

A side-roll or wheel-move system has wheels mounted on the lateral pipes, with the pipe serving as the axle of the wheel (Fig. 15.7). Rigid couplers permit the entire lateral up to 400 m (1/4 mi) long to be rolled forward by applying power at the center or the end while the pipe remains in a nearly straight line. Aluminum pipe having a 100- or 125-mm (4- or 5-in.) diameter is commonly used. To have sufficient strength, the aluminum pipe wall thickness should be at least 1.8 mm (0.072 in.).

The wheel diameter must be large enough so the pipe will pass over the crop without damaging it, and the crop will not prevent the lateral from being rolled to the next position. Among the wheel diameters commonly used are 1.17, 1.47, 1.63, and 1.93 m (46, 58, 64, and 76 in.).

The most popular sprinkler spacing (and pipe length) is 12.2 m (40 ft). The wheels are usually placed in the center of each length of pipe. Thus, a standard 400-m (1/4-mi) lateral contains 32 pipe lengths and 36 wheels because 4 wheels are required for the drive unit. Sometimes an extra wheel is provided for the last pipe section on each end. Normally, the drive unit contains a gasoline engine and a transmission with a reverse gear.

The sprinklers are approximately midway between the wheels on short risers. Often the sprinklers are provided with self-levelers so they will right themselves if the lateral is not stopped where the riser would be exactly upright. A drain valve also is located about midway between the wheels, near the pipe coupler and near the sprinkler. This valve opens automatically when the pressure is off, so the pipe will drain quickly and permit moving the lateral forward to the next set without much time loss. (Attempting to roll the pipeline when it is full of water will damage the equipment.)

The most popular side-roll spacing along the mainline is 18.3 m (60 ft). Two popular operating schemes are used. In one, the lateral is connected to every outlet valve along the mainline, and when the lateral reaches its destination and completes its last set and is empty, it is rolled back to the starting point. In the other, the lateral is connected to every even-numbered outlet valve on the mainline while the lateral is moved across the field, and then connected to the odd-numbered valves while the lateral is moved back to the starting position. For the latter case the lateral interval between irrigations is longer at the two ends of the field than in the center.

As with hand-move laterals, there is a tendency to completely open the hydrant valve and fill the line as quickly as possible, causing a water hammer at the far end of the line. Therefore, a surge plug at the closed end is recommended. The use of offsets, especially for the 12.2-m by 18.3-m (40-ft by 60-ft) spacing, is also recommended.

Today, a lateral with 32 sprinklers is commonly designed with 100-mm (4-in.) diameter pipe, even when the water is introduced into it from one end and the friction loss is 55 to 60 kPa (8 to 9 lb/in.²). However, if the water is admitted to it at the center of the lateral, the friction loss is reduced to about 1/5 as much. A 125-mm (5-in.) pipe would have only about 1/3 the friction loss of the 100-mm (4-in.) pipe when the water is admitted from one end. The best method will depend on the future price and availability of energy.

One type of side-roll system incorporates one sprinkler trailing for each sprinkler on the pipeline. The trailing sprinklers are mounted near the ends of each 9.2- or 12.2-m (30- to 50-ft) section of small aluminum pipes 25 to 40 mm (1 to 1.6 in.) in diameter. The trail-line pipes have water admitted from the side-roll pipe through double-gasket swivel couplers on the lateral. These couplers do not rotate when the wheel line is rolled forward. The trail lines are flexible enough to become essentially parallel to the ground and thus permit the sprinklers to be nearly vertical. When the wheel line reaches the end of the field, the trail lines must be moved to the other side and the position of the sprinklers are shifted so they are upright.

15.3.3 Side-move with Trail Lines

Side-move laterals with trail lines are supported on wheel-mounted A-frames (Fig. 15.8). Thus, the pipe does not serve as the axle of the wheels and can be held higher above the ground. Each A-frame carriage is driven from a drive shaft that extends the length of the pipeline. The drive shaft can be turned from the center of the line or from one end.

The small diameter trail lines can each carry several sprinklers. Usually, short sprinkler risers are used as they are easier to keep upright than tall risers. Outriggers at the last sprinkler on each trail line are used to keep the risers upright; however, these may damage some crops.

FIG. 15.8 Photograph of a side-move sprinkle irrigation system with trail lines in operation.

This system is sometimes called a "movable solid set". It greatly reduces the number of moves necessary to cover the field, thus saving labor. Sprinkler spacings and sizes which give very low application rates at an acceptable uniformity can be used. With the low rates, 24-h set times may be practical for some soils and crops, thus permitting a normal daytime work schedule for the irrigator.

When a trail line system reaches the end of the field, the trail lines must be moved to the start position, a laborious task. They must either be racked up on the A-frames to be returned empty across the field for the next irrigation, or rearranged by hand to the other side of the lateral and placed in position to irrigate on their way back across the field. The wheels on most side-move systems can be turned 90 deg, permitting the lateral to be pulled endwise to another field.

15.3.4 Tow-Move

Tow-move sprinkler laterals have relatively rigid couplers which are fitted with skids or wheels so the line can be moved by pulling it from the end.

The skids consist of a flat metal plate held on the underside of the pipe by one or more clamps. In one type, the skid is placed under the coupler and clamped at both ends. This makes the skid take the major part of the end thrust at the coupler when the pipe is towed. If relatively long sections of pipe are used, a second skid may be needed under the center of each section to reduce abrasion from soil contact. Two or three outriggers along the line are needed to keep the pipe oriented with the skids on the bottom and the sprinklers upright.

For wheel-type units, a pair of wheels mounted on a simple U-frame are clamped to each section of pipe. The wheels are oriented so the entire

FIG. 15.9 Center-pivot sprinkler system.

length of lateral pipe can be pulled endwise. The pipe itself stands only 0.3 to 0.5 m (12 to 20 in.) above the ground. The flexibility of the pipe and the articulation of the couplers permit the lateral to curve slightly while being moved to a new position. In one type, however, the lateral stays straight. The wheels are fixed so they automatically shift to a 45-deg angle from the lateral when pulled in one direction, and then shift back to a 45-deg position the other way when pulled from the other end of the lateral. Thus, by pulling alternately from both ends, the entire length of lateral is shifted the desired distance to the next set position.

The traditional way of moving a tow-move lateral is to snake it across the mainline in an S-shaped curve to a new position on the other side. For the next setting, it is dragged in the other direction across the mainline in an opposite S-shaped curve. With this procedure, each move needs to advance the lateral only half the distance between adjacent sets.

Tow-move systems are the least expensive of the mechanically moved systems. However, they are not used extensively because the moving process is tedious, requires careful operation, and also damages many crops. Tow-move systems have been used successfully in some forage crops and in row crops. The moves are made easier if the mainline is buried. Drain valves on towed pipelines are very desirable, but not as important as on side-roll systems.

15.4 CENTER-PIVOT SYSTEMS

15.4.1 General Considerations

When considering a center-pivot sprinkler system for a field or farm, the suitability of the soil, topography, cropping pattern, equipment costs, maintenance requirements, and operating costs for this type of system should all be thoroughly investigated.

The system must be reliable because in arid areas it will be operated more hours during the cropping season than almost any other piece of farm

FIG. 15.10 Cable suspension, hydraulic piston water-drive unit.

equipment (up to 2,200 or more hours each season). The center-pivot lateral must be repairable in the field because another lateral cannot be quickly substituted for a disabled one. It must be repaired rapidly in case of breakdown on shallow soils, or in fields that have not been well watered. A crop may be lost or severely damaged if the lateral is broken for several days during the crop's maximum water use period. It is important to recommend a center-pivot lateral that has been mechanically proven to operate on soils and topography similar to the area being considered. A local dealer who maintains a supply of parts and provides prompt repair service should be available.

15.4.2 Description of System

A center-pivot system consists of a single sprinkler lateral with one end anchored to a fixed pivot structure and the other end continuously moving around the pivot while applying water (Fig. 15.9). The water is supplied from the source to the lateral through the pivot. The lateral pipe with sprinklers is supported on drive units and suspended by cables as shown in Fig. 15.10, or by trusses as shown in Fig. 15.11, between the drive units. The drive units are mounted on wheels, tracks, or skids that are located 24.4 to 76.2 m (80 to 250 ft) apart along the length of the lateral pipe, which may vary from 61 to 792.5 m (200 to 2600 ft).

Each drive unit has a power device mounted on it that drives the wheels, tracks, or skids on which the unit moves. The rate at which the drive unit and lateral pipe advance around the pivot is determined by the speed of the outermost drive unit. Alignment devices detect any drive units that becomes misaligned. Either the units are speeded up or slowed, as needed. Thus, the advance by the outermost drive unit sets off a chain reaction of advances, beginning with the second drive unit from the outer end and progressing along the lateral to the pivot. Should the alignment system fail

FIG. 15.11 Truss suspension, electric-drive unit.

and any drive unit become too far out of alignment, a safety device stops the whole system automatically before the lateral can be damaged.

There are four methods of powering a center-pivot sprinkler system: hydraulic water drive, which utilizes pistons, rotary sprinklers, or turbines; electric motor drive; hydraulic oil drive, using pistons, rotary motors, or piston-cables; and air-pressure piston drive.

Hydraulic water-driven center-pivot systems are powered by water from the sprinkler lateral pipe with pressures from about 410 to 830 kPa (60 to 120 lb/in.²) at the pivot. Water used to drive the systems is discharged to the field. On the piston-drive systems, each piston-drive unit activates a set of trojan bars. The trojan bars engage wheel lugs to turn the drive unit wheels (Fig. 15.10). The rotary sprinkler and turbine drive systems transmit power to the wheels of each drive unit through a gear box. Other systems use a chain and sprocket mechanism connecting the gear box and the drive wheels.

The electric-drive center-pivot systems (Fig. 15.11) have motors of 0.37, 0.56, 0.75, or 1.12 kW (1/2, 3/4, 1 or 1-1/2 hp) mounted on each drive unit. Most systems operate with 440- or 480-volt, 3-phase, 60-cycle electric power. Electric power is supplied by an engine-driven generator

TABLE 15.1. WATER APPLICATION TIMES FOR DIFFERENT POSITIONS ALONG CENTER-PIVOT
LATERALS, NOZZLE TYPES, AND TURNING SPEEDS

Time for one lateral revolution	Distance along lateral from pivot, meters											
	50.3			100.6			201.2			402.3		
	Sprinkler wetted diameter, meters											
	9.1*	27.4†	24.4‡	9.1*	27.4†	27.4‡	9.1*	27.4†	39.6‡	9.1*	27.4†	53.3‡
hours	Time of water application, minutes											
6	10	31	28	5	16	16	3	8	12	1	4	8
12	21	63	56	10	31	31	5	16	22	3	8	16
24	42	125	111	21	63	63	10	31	45	5	16	30
48	83	250	222	42	125	125	21	63	91	10	31	61
60	104	313	278	52	156	156	26	78	113	13	39	76

*Spray-type sprinkler arrangement
†All medium-sized sprinkler arrangement
‡Small- to large-sized sprinkler arrangement

located at the pivot, or through underground cables which convey electric power to the pivot from commercial sources. A slip-ring connector at the pivot is needed to connect the electric power to wiring on the moving lateral.

In oil-powered systems, the oil-supply and return-flow pipelines extend from the oil pressure pump and oil reservoir to the piston or rotary motors located on each drive unit. The oil pump is powered by an electric motor or internal combustion engine and maintains 4,140 to 13,790 kPa (600 to 2000 lb/in.²) oil pressure in the oil lines.

The cable-drive system has one oil-pressure powered hydraulic cylinder at the pivot point. As the cylinder reciprocates, propelling power is transmitted to each drive unit through a steel cable that extends from the hydraulic piston to the outer drive unit.

15.4.3 Water Application Rates

Water is applied to the soil along a center-pivot lateral at a low rate near the pivot to progressively higher rates toward the outer end. The application rate varies along the lateral because the length of time water is applied per unit length of lateral decreases from the pivot to the outer end. Table 15.1 shows the length of time water is applied at various points along a

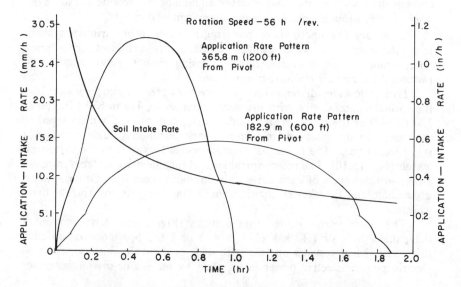

FIG. 15.12 Soil intake and center-pivot application rate patterns.

FIG. 15.13 Small to large sprinkler arrangement on lateral.

center-pivot lateral for different sprinkler arrangements and speeds of revolution. The length of time water is applied at four positions along laterals having three different types of sprinkler arrangements are shown. Fig. 15.12 shows the variation of application rate and the time of application at two locations along the same lateral.

The type of sprinklers, their spacing along the lateral, and the diameter of area covered from an individual sprinkler affect the application rates along a center-pivot lateral (also see Section 14.2). There are three common variations in sprinkler types and arrangements along the lateral, all of which can produce uniform water distribution.

The small to large sprinkler arrangement shown in Fig. 15.13 uses some of the smallest agricultural sprinklers near the pivot, gradually increasing sprinkler size to large sprinklers at the outer end of the lateral, with 35 to 40 sprinklers used on a 400-m (1300-ft) lateral. Recommended pivot operating pressure using this nozzling concept varies from 415 to 690 kPa (60 to 100 lb/in.²).

For the sprinkler arrangement shown in Fig. 15.14, the same medium-sized sprinklers are used, with variations in nozzle size and sprinkler spacing

FIG. 15.14 All medium-sized sprinklers on center-pivot lateral.

FIG. 15.15 Spray-type sprinklers on center-pivot lateral.

along the lateral. The widest spacing of sprinklers is near the pivot and the closest spacing is at the outer end of the lateral. These laterals have 85 to 100 sprinklers. They are normally operated with a pivot pressure of 310 to 520 kPa (45 to 75 lb/in.²).

The third sprinkler arrangement is shown operating in Fig. 15.15. It has fixed sprinklers with spray-type nozzles. Low pivot pressures from 140 to 275 kPa (20 to 40 lb/in.²) are suitable for spray nozzle operations. Fig. 15.16 shows a vertical view of the area wetted by each of the three sprinkler arrangements.

The spray-type center-pivot lateral has the smallest drops, but the highest peak application and the shortest duration of application (Table 15.1). Rates vary from 150 to 300 mm/h (6 to 12 in./h) at the end of a 400-m (1300-ft) lateral. The medium-sized sprinkler-type lateral has the

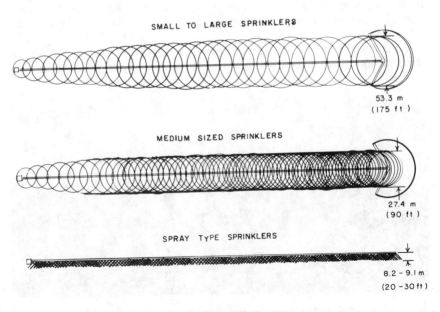

FIG. 15.16 Center-pivot water patterns.

next highest application rates, with a peak varying from 50 to 80 mm/h (2 to 3 in./h). The variable-sized sprinkler-type lateral gives the largest drops, but the lowest peak application rates, from 25 to 40 mm/h (1.0 to 1.5 in./h).

The application rates are determined by the nozzle size, nozzle pressure, sprinkler spacing, length of lateral, and sprinkler types used. Once these items are fixed by the manufacturer, the application rate for that point along the lateral is fixed and will not be changed by varying the speed of lateral rotation. Changing the lateral rotation speed only changes the depth of water applied (also see Section 14.7).

15.4.4 Field Information for Design

A map of the area to be irrigated should show field boundaries, elevations of high and low points, location of any obstructions, water source, and any electricity or natural gas supply locations. In many cases, aerial photos or topographic maps are available for the area.

The pivot position must be in the field so the lateral will not overrun any field boundaries after it is assembled and operating.

The flow rate of the water supply should be measured if there is any doubt about its adequacy. The quality of the water should also be determined, because excessive salts, or specific ions, could cause foliar or soil problems (see Chapter 5).

A schedule of crops to be grown should be obtained. Where crops are rotated, the best estimates of crop types and areas also are needed. A soil survey showing soil types, soil depths, water-holding capacities, and intake rate curves should be obtained for the area to be irrigated (see Chapter 4).

Although center-pivot laterals are manufactured in lengths of 61 to 792 m (200 to 2600 ft), the most common length is 400 m (1300 ft), which irrigates about 53 ha of a 65-ha square area (130 of 160 acres). New corner systems are being manufactured which will irrigate most of the area in the corners of a square area, making a total of 61 ha (150 acres) irrigated in a 65-ha (160-acres) square area (Fig. 15.17).

15.4.5 Sprinkler Design

The center-pivot lateral design should be developed to match: (a) the water intake capacity of the soil; (b) crop water requirements; and (c) the center-pivot lateral characteristics. The design capacity is calculated from the peak water use rate of the crops, the available water capacity of the soil, the time between irrigations, the area irrigated, the water application efficiency, and the lateral operation time to make one irrigation. Crops and climate determine the peak water use rate for an area. Methods of determining crop requirements and peak rate of use are discussed in Chapter 6.

The area irrigated is determined from the length of lateral and the percent of a full-circle operation as follows:

$$A = \frac{\pi L_i^2 P_i}{K} \qquad \dots \dots \dots \dots \dots \dots \dots \dots \dots \dots \dots \dots \dots \dots \dots \dots \dots \dots \dots [15.11]$$

in which A = the area irrigated, ha (acre); L_i = the effective irrigated radius of the center-pivot lateral, m (ft); P_i = the percent of a full circle expressed as a decimal; and K = a constant which is 10,000 for metric units and 43,560 for English units.

FIG. 15.17 Center-pivot corner system irrigating the corner of a field.

Some center-pivot laterals have a folding corner lateral pipe extension system added to the main lateral that folds out as the corner is approached, and folds back toward the main lateral as the near side of the field is approached. One type of corner system is shown in Fig. 15.17. To obtain the area irrigated by one of these center pivots, the four corner areas irrigated must be added to the area calculated from equation [15.11].

The water needed at the pivot to meet the peak water use rate is determined by using the equation:

$$Q = K \frac{E_t A t}{E_i t_1} \quad \dots \dots \dots \dots \dots \dots \dots \dots \dots \dots \dots \dots \dots \dots \dots \dots \dots \dots \quad [15.12]$$

in which Q = the quantity of water L/s (gal/min); E_t = the peak water use rate, mm/d (in./d); A = the area irrigated, ha (acre); t = the time between irrigations, d; E_i = the irrigation water application efficiency expressed as a decimal; t_1 = the lateral operating time for one irrigation, d; and K = a constant, which is 0.116 for metric units and 18.9 for English units.

The water application efficiency varies with the water distribution uniformity along the lateral. It also depends on whether or not the water infiltrates at the point of application to the soil (see Section 14.3). Measured water application efficiencies for one irrigation season varied from 70 to

80 percent (Anderson et al., 1972). The value of Q determined by equation [15.12] should be increased by 5 to 10 percent to allow a safety factor for lateral breakdown during peak crop water use periods.

Assuming the water application rate pattern across a center-pivot lateral is elliptical, as shown in Fig. 15.12, Dillon et al. (1972) developed a formula to estimate the maximum application rate:

$$P = K \frac{Q}{L_g L_1} \dots\dots\dots\dots\dots\dots\dots\dots\dots\dots\dots\dots\dots\dots\dots\dots\dots \quad [15.13]$$

in which P = the maximum application rate of the last few sprinklers, mm/h (in./h); Q = the flow of water into the center-pivot, L/s (gal/min); L_1 = the radius of wetted area at the outer end of the lateral, m (ft); L_g = the effective irrigated radius of the center-pivot lateral, m (ft); and K = a constant equal to 4584 for metric units and 122.5 for English units.

If the water application rate is less than the soil intake capacity, runoff will not occur. If the application rate is higher than the soil intake capacity, a shorter lateral can be used, although it will irrigate a smaller area. On some sloping areas, where the application rate may exceed the soil intake capacity—contour or cross-slope farming—small basins (or pits) between crop rows, or other tillage practices may be used to retain the water at the point of application and prevent runoff.

The average gross depth of water applied by a center-pivot sprinkler system during each revolution of the lateral depends upon the area irrigated, system capacity, and time needed to complete one lateral revolution (also see Section 14.2.2). The gross depth applied is:

$$D = K \frac{Qt_1}{A} \dots\dots\dots\dots\dots\dots\dots\dots\dots\dots\dots\dots\dots\dots\dots\dots\dots\dots \quad [15.14]$$

in which D = the gross depth of water applied, mm (in.); Q = the flow of water into the center pivot, L/s (gal/min); A = the area irrigated, ha (acre); t_1 = the lateral operating time for one irrigation, h; and K = a constant which is 0.36 for metric units, and 0.00221 for English units.

The time to complete one lateral revolution depends on the speed of the drive unit, distance between it and the pivot. The time can be determined by:

$$t_r = \frac{2\pi L_3}{V} \dots\dots\dots\dots\dots\dots\dots\dots\dots\dots\dots\dots\dots\dots\dots\dots\dots\dots\dots \quad [15.15]$$

in which t_r = the time required for the lateral to complete one revolution, h; L_3 = the distance of the end drive unit from the pivot, m (ft); and V = the speed of travel of the end drive unit, m/h (ft/h).

A design procedure proposed by Dillon et al. (1972) is based on detailed soil intake rate determinations:

1   Determine the radius of coverage of system by dividing shortest dimension of the field by two.

2   Determine the peak water use and irrigation efficiency.

3   Determine the water flow required at the pivot.

4   Determine the minimum time required to make one lateral revolution.

5   Determine the net depth of water applied to the root zone per revolution.

6   Determine the minimum design speed of travel of the center-pivot lateral at which potential runoff starts.

7   The maximum depth of water that can be applied to the root zone per revolution is determined. The maximum depth applied should not exceed the root-zone storage capacity.

In accordance with the ASAE Recommendation R264.2, Item 3.2.1, Minimum Requirements for the Design, Installation, and Performance of Sprinkler Equipment, "For crops and soils wholly dependent on irrigation for moisture, the system shall have the capacity to meet the peak moisture demand of each and all crops irrigated within the area for which it is designed. However, if the purchaser deems that an amount of water less than necessary to meet the peak demand is desirable, then the design capacity shall be that stated by the purchaser in writing. Sufficient time must be allowed for moving laterals and for permitting cultural practices on the land. The capacity must allow for reasonable water losses during application periods with the system operating in accord with design conditions."

15.4.6 Operating the System

The responsibility for proper operation of a center-pivot sprinkler system falls on the irrigator. It can be divided into two areas: (a) water management for crops, soils, and weather; and (b) proper system maintenance and repair.

Water management includes timing water applications to meet the water requirements of the crop for the soil-water storage capacity available. The root zone should be filled with moisture before planting unless substantial rain is expected early in the crop season. When irrigating, water penetration depths should be checked at several locations after the operating lateral has just passed to be sure the soil is wet throughout the crop root zone. Center-pivot systems are usually designed to supply water at the peak crop use rate in arid areas when operating continuously. However, most of the time during annual crop seasons the crop water consumption is lower, and intermittent operation of the center-pivot lateral is needed. Systems are sometimes designed for less than the daily peak use rate because the peak use rate often may only occur for several days or a week at a time. Also, rainfall and stored soil moisture can provide some of the water during peak water use periods (Heermann et al., 1974).

When the lateral returns to the starting place, the soil moisture just ahead of the lateral should be checked to determine whether enough has been used to warrant continuous operation. The results of an unexpected system breakdown of a day or more should be considered before delaying an irrigation. Sprinkler heads should be checked periodically for plugging, particularly when the water contains foreign materials.

Most manufacturers provide an operator's maintenance manual which should be obtained and followed. Typical operator's manuals detail steps necessary to prepare the center-pivot system for operation at the beginning

FIG. 15.18 A traveling gun sprinkler unit in operation showing flexible hose and cable.

of each irrigation season, the maintenance necessary during operation, and the preparations that should be made if the system will not be used for long periods.

Each year, before the irrigation season begins, the sprinkler system should be thoroughly inspected for needed repairs and maintenance. All moving parts that require lubrication should be greased with the correct grade, type, and amount of lubricant as specified in the operation manual.

Sprinklers should be examined and repaired or replaced if they are not operating satisfactorily. Any leaks in the lateral pipeline should be repaired. On electrically powered machines, the use of a pressurized spray contact cleaner to clean contact points on electrical controls may prevent system shutdown later in the season. CAUTION: Make sure all electrical power is completely disconnected from the system before cleaning any electrical contact points.

The operator must be careful when operating the system in freezing weather. Ice may accumulate on the center-pivot pipeline and supporting structures. This added weight can rapidly overload and collapse the lateral. As a precaution, low-temperature kill switches are available to shut down operating systems when temperatures drop below freezing.

Center-pivot systems operation is fairly trouble-free on level lands and on uniform sloping lands with slopes up to 10 percent. Undulating topography usually produces more difficulties with more potential for runoff. Bridges may be built across drainage ways, irrigation canals, and other small depressed areas to permit passage of the drive units. On soft or erosive soils, some operators line the tracks with sand where excessively deep tracks are made in the field by the drive wheels.

An excellent job of irrigation is readily obtainable with a properly designed center-pivot sprinkler system. However, the system must be installed on soils having infiltration rates matching or exceeding the lateral's application rate and managed to meet the crop water needs on a systematic and frequent irrigation schedule.

FIG. 15.19 Aerial photograph of a traveling gun sprinkler unit operating along a towpath in a corn field.

15.5 TRAVELING GUN SPRINKLER

Labor scarcity on irrigated farms has accelerated the trend toward mechanized irrigation systems, such as traveling gun sprinklers. Gun and boom-type sprinklers have been changed from intermittent move sprinklers to continuous moving sprinklers mounted on trailers and fed by long flexible hoses. The traveler sprinkler system includes these major components: pump, power unit, main or supply pipe, irrigation hose, traveler unit, and sprinkler.

15.5.1 Description and Field Layout

A traveler system consists of a high capacity sprinkler mounted on a chassis to which a flexible hose, usually 75 to 130 mm (3 to 5 in.) in diameter, and up to 402 m (1320 ft) long is connected (Fig. 15.18). The traveler unit is pulled along selected travel lanes by a cable wrapping on a rotating cable reel. The cable is first unreeled to the opposite end of the lane to be watered and firmly anchored. The cable reel can be powered by a water turbine, water piston, or engine mounted on the traveler. Sometimes an engine is used to power a cable reel set at the end of the travel lane. The cable reel pulls the traveler through the field in a straight line, using the cable as an alignment guide. An aerial view of a traveler in operation is shown in Fig. 15.19.

A reel for handling the irrigation hose between sets and for off-season hose storage is available as a separate unit or as a combination with the traveler. Mainline pipe 100 to 200 mm (4 to 8 in.) in diameter is required to supply water from the source to the flexible hose. Although portable aluminum pipe is the most common type of mainline in use, underground mainline of plastic pipe (PVC) with valves is becoming more popular.

The traveling system is adaptable to many field sizes and shapes with topography from level to rolling and irregular. A traveler can irrigate a rectangular strip as long as 800 m (1/2 mile) unattended. A traveler can also be moved at highway speed to irrigate several fields.

Safety standards for travelers have been established by The Irrigation Association and are being considered for adoption by the American Society of Agricultural Engineers. The "Safety for Self-Propelled Hose-Drag Agricultural Irrigation Systems" includes sections on service, operator controls, power takeoff driven shafts, shields and guards, travel on highways, safety signs, and system load ratings.

15.5.1 System Capacity and Operating Pressure

Traveling sprinklers are designed to discharge from approximately 6 to 63 L/s (100 to 1000 gal/min). The required pumping rate depends on the area to be irrigated, the number of hours per day the system can be operated, the gross peak daily water required, lane spacings desired, and the available water supply. System capacity based on crop water requirements can be calculated from information in Chapter 6. Larger capacity systems allow wider land spacings, but require higher operating pressures.

Normal operating pressure required at the sprinkler varies from 415 to 760 kPa (60 to 110 lb/in.²). Adequate pressure is required to provide droplet breakup suitable for the crop and soil to be irrigated. The effect of operating pressure on drop size is discussed in Chapter 14. Because of the continuous traveling movement, a relatively constant distribution uniformity is maintained over the normal range of pressures. Recommended operating pressure requirements at the gun type sprinkler nozzles are:

Flow range		Pressure recommendation	
L/s	(gal/min)	kPa	lb/in.²
6.3 to 12.6	(100 to 200)	415 to 550	(60 to 80)
12.6 to 18.9	(200 to 300)	485 to 620	(70 to 80)
18.9 to 31.6	(300 to 500)	550 to 690	(80 to 100)
over 31.6	(over 500)	585 to 760	(85 to 110)

15.5.3 Mainline and Irrigation Hose Sizing and Hose Selection and Care

After the system capacity is determined, the mainline and flexible hose size can be determined. Mainline friction loss can be calculated from data in Section 14.11. The best way to select sizes is to calculate fixed and variable costs of several pipe sizes and select the least cost pipe size. However, pipe size is often chosen by rule of thumb.

Because quick coupling aluminum pipe is often used, approximations are:

Flow range		Pipe diameter	
L/s	(gal/min)	mm	(in.)
6.3 to 9.5	(100 to 150)	100	(4)
9.5 to 19	(150 to 300)	130	(5)
19 to 32	(300 to 500)	150	(6)
25 to 44	(400 to 700)	180	(7)
32 to 63	(500 to 1000)	200	(8)

TABLE 15.2. FRICTION LOSS IN FLEXIBLE IRRIGATION HOSE
USED ON TRAVELING SPRINKLER SYSTEM

Flow		Hose size, mm (in.)			
		76 (3)	102 (4)	114 (4.5)	127 (5)
		Friction loss			
L/s	(gal/min)	$\cdots\cdots\cdots\cdots\cdots$ kPa/200 m (lb/in.[2] per 660 ft) $\cdots\cdots\cdots\cdots$			
6.3	(100)	31 (4.5)			
12.6	(200)	113 (16.4)	28 (4.1)		
18.9	(300)	235 (34.1)	59 (8.6)	33 (4.8)	
25.2	(400)		100 (14.5)	57 (8.3)	34 (4.9)
31.5	(500)		150 (21.7)	86 (12.5)	51 (7.4)
37.8	(600)		220 (31.9)	120 (17.4)	72 (10.4)
44.1	(700)			160 (23.2)	95 (13.8)
50.4	(800)			200 (29.0)	120 (17.4)
56.7	(900)			250 (36.2)	150 (21.7)
63.0	(1000)				185 (26.8)

Guidelines for sizing flexible irrigation hose for traveler systems are as follows:

Flow range		Hose diameter	
L/s	(gal/min)	mm	(in.)
3.2 to 9.5	(50 to 150)	64	(2.5)
9.5 to 15.8	(150 to 250)	76	(3.0)
12.6 to 22	(200 to 350)	89	(3.5)
15.8 to 32	(250 to 500)	102	(4.0)
32 to 44	(500 to 700)	114	(4.5)
> 44	(> 700)	127	(5.0)

A more precise way to size the mainline of any type including irrigation hose, is the least-cost method, whereby a size is selected that will provide the least overall cost, considering both fixed and variable costs. Friction losses for flexible irrigation hose are given in Table 15.2.

Irrigation hose should be selected in accordance with standards established by The Irrigation Association and being considered for adoption by the American Society of Agricultural Engineers "Minimum Standards for Irrigation Hose as Used with Self-Propelled Hose-Drag Agricultural Irrigation Systems". Important requirements are adequate pressure rating, tensile strength to withstand pull requirements, maximum elongation of 3 percent, and a tough durable cover which is non-kinking under normal operating pressure. The hose selected should also be repairable in the event of damage.

The end pull force required to drag a hose depends on the hose diameter and length, soil moisture, and soil surface or crop cover conditions. The force will be greatest on wet, tilled, sticky soils, less on firm soils, and still less on wet vegetation or on bare sandy soils.

Hose life can be increased by: Clearing travel lanes of sharp objects that could cut or puncture the hose such as sharp rocks, broken glass and old wire fencing; avoiding running over the hose with equipment or the traveler itself; preventing oil and grease from coming in contact with the hose; expelling water from the hose before winding on a reel; and storing in a cool dry place during the off-season away from electrical equipment and protected from rodents.

TABLE 15.3. MAXIMUM TRAVEL LAND SPACING AS A FUNCTION
OF WETTED DIAMETER AND WINDSPEED

Wetted diameter of sprinkler		Windspeed km/h (mi/h)			
		over 16 (over 10)	8-16 (5-10)	0-8 (0-5)	0 (0)
		Percent of wetted diameter			
		50	60	70	80
m	(ft)	----------------------m (ft)----------------------			
60	(200)	30 (100)	36 (120)	42 (140)	48 (160)
90	(300)	45 (150)	54 (180)	63 (210)	72 (240)
120	(400)	60 (200)	72 (240)	84 (280)	96 (320)
150	(500)	75 (250)	90 (300)	105 (350)	120 (400)
180	(600)	90 (300)	108 (360)	126 (420)	144 (480)

15.5.4 Uniformity and Lane Spacing

One important aspect to consider on any irrigation system is water application uniformity. This is a measure of the evenness of water distribution during an irrigation. High uniformity of water application can be attained by a traveling sprinkler gun because of its continuous movement. Sprinkler irrigation uniformity is often expressed as the coefficient of uniformity (CU). Christiansen's (1942) formula is:

$$CU = 100 \left\{ 1 - \frac{\Sigma(x_i - \bar{x})}{n\bar{x}} \right\} \quad \dots\dots\dots\dots\dots\dots\dots [15.16]$$

where CU is the coefficient of uniformity expressed in percent; x_i = individual water application depths, measured in catch cans in the field; $\bar{x}$ = the mean of all x_i's; and n = the number of readings.

Theoretical water distribution from moving sprinklers has been analyzed by Bittinger and Longenbaugh (1962), Hanson (1971), and Solomon (1971) (also see Section 14.2.5). These analyses indicate that good water distribution uniformity can be achieved with wide land spacing under no wind conditions (also see Section 14.6.2).

Uniformity tests have been run by various researchers (Shearer, 1974; Shull and Dylla, 1976; and Snoozy, 1971). These data show considerable scatter, depending on wind velocity, wind direction, quantity of water output, and pressure. The average windspeed for these tests was approximately 16 km/h (10 mi/h). The average CU's were 70 and 75 percent for line spacings equal to 70 and 60 percent of the wetted diameters, respectively. When operating within the recommended pressure and lane spacing, the average CU's were 77 to 83 percent for lane spacing equal to 70 to 50 percent of the wetted diameters.

Table 15.3, showing maximum travel lane spacing as a percent of wetted diameter for various wind velocities, was developed by Nelson Manufacturing Company (1970). According to Nelson, these spacings should prevent poorly underirrigated areas between travel lanes. Nelson also recommends that direction of travel should be at a right angle to prevailing winds.

A traveling boom system is similar to a traveling gun system except the gun is replaced with a rotating boom having 27-m (89-ft) long arms (pipes) with placement of discharge nozzles spaced along the boom. The nozzle discharge patterns overlap one another. Uniformity tests of a traveling boom and traveling guns (Smith, 1967) indicate that a traveling boom can have wider lane spacings than a traveling gun for the same diameter of throw. This may occur because the rotating boom carries the water as much as 27 m (90 ft) before discharging the water through the end nozzle. The boom height may be adjusted in response to the wind conditions.

15.5.5 Non-uniformity at End of Fields

The continuous movement of a traveler results in a pass of the complete sprinkler coverage pattern throughout the center section of the field, but a complete travel pass is not obtained at the ends of the field. The part of the field not getting a pass of the complete sprinkler pattern varies with the sprinkler coverage diameter, distribution pattern, arc setting, and overthrow beyond the ends of the field. A better job of irrigation can be accomplished by allowing the traveling sprinkler to set 1 h at each end on a 23-h set, and 30 min or longer for an 11-h set on a 400-m (1/4-mi) field.

In addition to end effects, the uniformity is also reduced because of the inadequate overlap along the side of the field. This edge effect can be compensated for by adjusting the sprinkler angle when traveling along the outside lanes.

15.5.6 Uniformity of Travel Speed

The CU's discussed in the previous section were based on a constant travel speed throughout the field. These CU values would decrease if the travel speed of the gun or boom varied from one part of the field to another.

Speed tests have been conducted by several researchers (Addink, 1976; Shull and Dylla, 1975; and Shearer, 1974). Results show some extreme speed variations of up to 60 percent. Most speed variations were much lower. The variations were caused by changes in the winch drum speed and the outside drum or reel diameter as the cable builds up. The drum speed varies because of load differences from one part of the field to another. Even on uniform topography the load will change, because the length of hose being pulled continually changes.

Many travelers have turbine drives. These drives are designed for a certain water flow range and should operate in the design flow range. Some travelers require 32 L/s (500 gal/min) or more of through flow to produce a reasonably constant speed. Several manufacturers make an engine-driven drum.

The winch reel increases in diameter as more cable is wound onto the drum. Variations in reel diameter are large for some travelers. The variation in reel diameter is much greater for a small diameter drum core than for a large diameter drum core. The variation would also increase as the drum width decreases. Some travelers require approximately 50 percent increase in reel diameter from start to finish because the turbine slows down 50 percent from the increased load of the hose.

A speed variation in excess of 10 percent is not recommended for good water distribution. Travelers are available that can be consistently operated within this tolerance.

15.5.7 Operation

Records on labor requirements to move travelers show the time required per move can vary from 0.5 to 4 man-hours (Addink, 1976; Groelz, 1975;

FIG. 15.20 Traveling lateral sprinkler system.

and Snoozy, 1971). Some of the lower figures may not have included travel time to and from the field, pump start-up, or line-filling time. Additional time is used if water remains in the hose. This can be a problem on terraced or steeply rolling land, particularly on short runs in which the hose is not completely stretched out each time. For each situations, an air compressor can be used to purge the water from the hose.

Caution must be used when handling the traveler hose. When extending the hose, it should be as straight as possible to avoid "snaking" when water pressure is applied. Excessive snaking could result in the traveler actually climbing over the hose and causing physical damage to the hose from the seesaw action of the cable. Some sticky soils adhere to the hose and cause pulling forces beyond the safe operation range of the hose. To prevent this, the travel lanes should be planted to grass or a similar foliage to keep the pull force within safe limits.

A properly designed traveling gun system can do an excellent job of irrigation, as with any other type sprinkler, when installed on soils having infiltration capacities matching or exceeding the sprinkler's application rate and when managed to meet the crop water needs on a systematic and frequent schedule.

15.6 TRAVELING LATERAL SYSTEMS

15.6.1 General Description

Because corners are left unirrigated by a typical center-pivot system, several manufacturers have developed sprinkler laterals that move continuously in a rectilinear fashion (Fig. 15.20). The laterals are either fed

FIG. 15.21 Traveling lateral sprinkler system taking water from a ditch.

from a flexible hose, or equipped with a traveling pumping plant to take water from a ditch (Fig. 15.21). Traveling sprinkler laterals can irrigate a rectangular field completely, including the corners.

The flexible hose used for such systems is identical to that used for the traveling guns and boom sprinklers discussed in the previous section. Likewise, a winch and cable system are used to furnish the force to drag the hose, and the problems of maintaining a uniform speed across the field are similar. The winch is powered by a water motor, with the water supplied by the irrigation pump. Because the sprinklers are much smaller than the guns, very little water is wasted at the field edges, and overall uniformities are higher.

The appearance of one traveling system is almost identical to some center pivots. The lateral pipe is mounted on A-frames with size and spacing similar to those used on the center pivots. For one type, each A-frame is driven individually, with the power supplied by the rotation of a long-armed sprinkler located at the A-frame. The sprinkler starts and stops at the command of the sensors, which are continuously checking to keep the pipe in straight alignment. In another type, the A-frames are powered by a single shaft extending the length of the lateral. The shaft, in turn, is powered by an electric motor. The machine operates without sensors to keep each A-frame in alignment.

Most recently, machines have been developed which pump water directly from a ditch. The traveling pumping plant on this machine is also equipped with an electric generator, which provides power for motors installed on each A-frame. These traveling sprinkler laterals are almost identical to electric-driven center-pivot systems' laterals.

15.6.2 Field Layout

Traveling laterals have been tested with lengths up to 400 m (1/4 mi). In some situations, the traveling lateral might be more economical than a center-pivot system for a square 65-ha (160-acre) field. In general, however, the cost of the flexible hose (or traveling pumping plant and ditch), and the extra labor required, have prevented it from replacing the center pivot.

A typical traveling sprinkler lateral would use a single center mainline (or ditch). The lateral would be situated to traverse half the field on one side of the mainline. When it reached the end, its wheels would be turned 90 deg so the lateral can be transported to its beginning position. One estimate sets the labor requirement for these moves at about 6 man-hours per irrigation on a 65-ha (160-acres) field.

Unlike the center pivot, the sprinkler spacing and discharge should be uniform throughout the entire length of the traveling lateral. Therefore, the friction loss can be calculated on the same basis as a hand-move lateral (see Section 15.2.2). The friction loss in a traveling lateral would be only about 63 percent as great as the loss in a center-pivot lateral if the two pipes were the same uniform diameter and discharged the same amount of water.

15.6.3 Sprinkler Spacing, Nozzle Size, and Pressure

For a 65-ha (160-acre) square field, a traveling lateral must have more capacity than a center-pivot system, because it irrigates more of the field. If the soil intake capacity is high enough, the sprinkler size and spacing should be chosen to given the necessary system capacity and to insure adequate uniformity along the lateral. A high soil intake capacity would permit choosing small sprinklers which operate at 275 kPa (40 lb/in.^{2}) or less. Sprinklers with a larger application diameter would reduce the instantaneous application rate.

The traveling lateral has an application rate only slightly more than half the maximum application rate at the far end of a similar center-pivot system. If runoff occurs, the application rate of the traveling lateral can be reduced to any value desired; however, this results in a lower capacity and possibly a lesser amount of land than can be irrigated by the lateral.

15.6.4 Operation

Because labor is involved when the traveling lateral reaches the end of the field or the end of the flexible hose, it is not as adaptable to daily irrigations as is the center pivot. It would normally be operated on about the same irrigation frequency as a hand-move system, thus keeping the manual labor to a minimum. The slower rate of travel also reduces wear on the flexible hose.

The traveling lateral system can obtain a uniformity coefficient between 90 and 95 percent (Pair, 1967; Shearer, 1966, 1971; and Shearer and Wolfe, 1975). Perhaps this is higher than any other sprinkler system in common use. The close spacing of the sprinklers and use of only one sprinkler model and nozzle size contribute to this high performance.

Its limited popularity to date is most likely because of economic considerations, especially the annual cost of the flexible hose. Because of the material cost of the hose and the limited sizes available, they are normally selected with a relatively high friction loss, which in turn tends to place a practical limit on the capacity of the lateral, perhaps not much more than 63 L/s (1000 gal/min). The more recent development of traveling sprinkler laterals have traveling pumping plants and utilize a ditch. This eliminates the restraints of hoses and perhaps will increase the popularity of the traveling sprinkler laterals.

15.7 EFFLUENT IRRIGATION

The passage of Public Law 92-500 requires that "by 1985, there shall not be discharge of pollutants to our streams and rivers". To help meet this requirement, land application of wastewaters is a most cost-effective treatment alternative to conventional wastewater treatment.

One method for land application is sprinkler irrigation. Many wastewaters can be handled with sprinkler irrigation. Normally, wastewaters are divided into municipal, industrial, and agricultural.

Municipal wastewaters can include industrial wastewaters; however, most cities are now requiring rather extensive treatment in the plants prior to discharge into the municipal system.

Industrial wastewaters can require extensive pretreatment, ranging from simple screening to primary and secondary treatment for removing oils, greases, metals, harmful chemicals, pH adjustment, and chlorination.

Agricultural wastewaters include effluents from animal production systems and food processing plants. Most animal wastewaters are nutrient limiting. For land application through sprinkler irrigation, some treatment, such as removal of large fibrous solids, may be required. Agricultural plant wastewaters will generally require more extensive pretreatment, such as removal of solids, greases and oils, and pH adjustments.

Major concerns in designing a sprinkler irrigation system for land application of wastewaters are: (a) the wastewater should be a type that will not destroy or render ineffective the disposal site; and (b) the wastewater should be applied at a rate that does not pollute the ground or surface water, or destroy or render ineffective the disposal site. Oils, greases, and heavy metals can harm the soil and the vegetative cover. The pH of the effluent should be between 5.5 and 7.5. Furthermore, excessive solids can build up a mat on the surface that will destroy the vegetative cover.

Woodland is probably the best disposal site for wastewaters. In woodlands, the soil surface is stable and the surface cover is effective for digesting organic matter. Grassland can also be used; however, it is important to select a grass that is specific for the site. In the mid-Atlantic and southeastern states, a mixture of Reed canary (a water-tolerant grass), tall fescue, and perennial ryegrass has been used. In the coastal areas of the Southeast, for summertime application, coastal Bermuda or one of the other hybrid Bermuda grasses can be used. Corn can also be used, but effluent can only be applied at selected times of the year. Rates of nitrogen that can be applied will vary from approximately 225 kg/ha (200 lb/acre) per year for corn to 785 kg/ha (700 lb/acre) per year for coastal Bermuda. Nitrogen application in excess of plant use can result in leaking of nitrates and pollution of the groundwater.

In designing a sprinkler irrigation system, the effluent, vegetative cover, soil type, and frequency of application should be considered. The well drained, deep sandy, or loamy soils are the better soils for land application. Some soils may require subsurface drainage. Application rate should not exceed the infiltration capacity of the soil, and most recommendations are for a maximum application rate of 6.4 mm/h (0.25 in./h). The total application per week can vary between 25 and 100 mm (1 and 4 in.), with the higher application during the summer months. There should be a rest period between applications, however.

Most land disposal sprinkler irrigation systems use single-nozzle sprinklers. This reduces nozzle clogging problems and also results in a lower application rate. If systems are designed to operate during freezing temperatures, sprinklers that will operate under these conditions should be selected.

Both portable aluminum pipe and buried pipe are used for main and lateral lines. Noncorrosive buried pipe is recommended. (Portable aluminum pipe leaks at the couplers, becomes an obstacle when mowing etc., and is plagued with freezing problems during low temperature operation.)

For systems designed to operate continuously, automation is recommended. However, each component of the system should be checked daily. Automation valves can be operated by air, water, or electricity. The most desirable is either the air valve or water valve unless the water is from a clean source. Electric valves tend to clog if the effluent has more than a minimum solids content. There are several types of pumps: straight centrifugal, self-priming centrifugal, or vertical turbine. The vertical turbine pump has the advantage of not requiring priming. If straight or self-priming centrifugal pumps are used, a flooded suction is recommended. Centrifugal pumps will only handle effluents with 8 percent or less solids. With long fiber solids, such as straw or chicken feathers, a chopper-agitator pump may be needed. Effluents with high solids content may require diaphragm, auger, or piston pumps. Normally, these effluents do not lend themselves to successful sprinkler application. It is desirable to have a duplex pump system, or at least a spare pump and spare pump parts.

Riser pipes should be galvanized and tall enough to clear the foliage, except in woodland where at least a 6-m (20-ft) land should be cleared and seeded to a grass if low risers are used. Riser supports are recommended for overtree sprinklers.

For permanent systems, valves—whether manual or automatic—should be located in a valve box, numbered, and color coded. If the site is in freezing climate, drain valves should be installed to drain the pipe system. The most positive freeze protection system is an air purge system that can be used to clear the pipe of water. Where the system is operated only part of the time, and the wastewater is corrosive or has a high solids content, the system should be flushed with fresh water after each use. Effluents left in the pipes will become septic and create a nuisance. Also, suspended solids will settle and harden at low points in the lines and may cause severe clogging.

In addition to portable aluminum pipe and permanent irrigation systems, center-pivots and traveling gun travelers are used for land application of wastewaters. Both of these systems have fairly high application rates. Effluents with high suspended solids may clog the turbine or piston on water-drive traveling gun sprinklers. Furthermore, the operation of regular sprinklers during windy weather can create severe drifting problems. For this reason, many center-pivot effluent disposal systems are now equipped with spray nozzles directed downward. In center-pivot systems, traction problems can also occur because of the large amounts of water applied.

The design of a sprinkler irrigation system for land application of wastewaters is similar to the design of other types of sprinkler irrigation systems. The designer must follow the rules of good design, keeping in mind that the effluent is not water, but a mixture of water and solids. Many industial and municipal systems are operated continuously; therefore, good maintenance is required. Many wastewaters are abrasive or corrosive and will shorten the life of the system. Special equipment, such as epoxy-coated pump casings and impellers, heavy duty seals, bearings and bushings, and even stainless steel sprinklers, may be required. For automated systems, monitoring equipment, shut-down equipment, slow opening and closing valves, and flow meters are recommended.

A sprinkler irrigation system used for land application of wastewaters should uniformly apply the material without harming the environment. The system must be designed and operated to satisfy the regulatory agencies, owners, operators, and neighbors.

References

1 Addink, J. W. 1976. Traveling big gun irrigation systems. Proc. 1976 Nebr. Irrig. Short Course, pp. 70-82.

2 American Society of Agricultural Engineers. 1978. R264.2, Minimum requirements for the design, installation, and performance of sprinkler irrigation equipment. 1978 AGRICULTURAL ENGINEERS YEARBOOK, pp. 526-528.

3 Anderson, D., and R. J. Brown. 1972. Irrigation adequacy with center pivot sprinklers. American Society Civil Engineers, Irrig. and Drain. Div. Spec. Conf. Proc., pp. 353-358.

4 Bittinger, M. W., and R. A. Longenbaugh. 1962. Theoretical distribution of water from a moving irrigation sprinkler. TRANSACTIONS of the ASAE 5(1):26-30.

5 Christiansen, J. E. 1942. Irrigation by sprinkling. Calif. Agr. Exp. Sta. Bul. 670, 94 p.

6 Dillon, R. C., E. A. Hiler, and G. Vittetoe. 1972. Center-pivot sprinkler design based on intake characteristics. TRANSACTIONS of the ASAE 15(5):996-1001.

7 Groelz, J. 1975. Traveling sprinkler design—labor saving approaches. American Society Civil Engineers, Irrig. and Drain. Div. Spec. Conf. Proc., pp. 123-134.

8 Hanson, R. E. 1971. Water distribution with traveling gun sprinklers. American Society Civil Engineers, Natl. Water Resources Meeting, Phoenix, Preprint, 15 p.

9 Heermann, D. F., H. H. Shull, and R. H. Mickelson. 1974. Center pivot design capacities in eastern Colorado. Proc. American Society Civil Engineers, J. Irrig. and Drain. Div. 100(IR2): 127-141.

10 Jensen, M. C., and A. M. Fratini. 1957. Adjusted "F" factor for sprinkler lateral design. AGRICULTURAL ENGINEERING 38(4):247.

11 Nelson Mfg. Co. 1970. Traveling sprinkler system planning guide. L. R. Nelson Mfg. Co., Inc., TS-270, Peoria, IL.

12 Pair, C. H. 1967. Water distribution under sprinkler irrigation. ASAE Paper 67-709, ASAE, St. Joseph, MI 49085.

13 Ree, W. O. 1959. Head loss in quick-coupled aluminum pipe used for sprinkler irrigation systems. USDA, Agr. Handbook 147, 21 p.

14 Shearer, M. N. 1966. Continuous move laterals apply water uniformly in high winds. Irrig. Eng. and Maint., March, pp. 20.

15 Shearer, M. N. 1971. Water distribution from a sprinkler lateral moving continuously in a linear direction. Ore. State Univ. Coop. Ext. Serv. Spec. Rpt., 10 p.

16 Shearer, M. N. 1974. Big-gun sprinklers: How are they doing? Sprinkler Irrig. Assoc. Newsletter, Vol. IV(2).

17 Shearer, M. N., and J. W. Wolfe. 1975. Performance of continuously moving field plot sprinkler. ASAE Paper 75-2057, ASAE, St. Joseph, MI 49085.

18 Shull, H., and A. J. Dylla. 1976. Traveling gun sprinkler application uniformity in high winds. TRANSACTIONS of the ASAE 19(2):254-258.

19 Smith, N. 1967. Continuous-move sprinkler test. Univ. of Nebr. Agr. Eng. Dep. Student Rpt., 13 p.

20 Snoozy, R. S. 1971. Evaluation of E.Z. Rain irrigation system. South Dakota State Univ. Agr. Eng. Dept. AE709 Spec. Prob., 28 p.

21 Solomon, K. 1971. Traveling sprinkler variables affecting application uniformity. ASAE Paper 71-758, ASAE, St. Joseph, MI 49085.

chapter 16

DESIGN AND OPERATION
OF TRICKLE (DRIP) SYSTEMS

16

DESIGN AND OPERATION OF TRICKLE (DRIP) SYSTEMS

by   T. A. Howell, USDA-SEA-AR, Fresno, CA; D. S.
Stevenson, Agriculture Canada, Summerland,
B.C.; F. K. Aljibury, California State Water
Resources Control Board, Sacramento, CA; H. M.
Gitlin, University of Hawaii, Honolulu; I-Pai Wu,
University of Hawaii, Honolulu; A. W. Warrick,
University of Arizona, Tucson; and P. A. C.
Raats, Institute for Soil Fertility, Haren, The
Netherlands.

16.1 INTRODUCTION

Drip or trickle irrigation is the newest of all commercial methods of water application. It is described as the frequent, slow application of water to soils through mechanical devices called emitters or applicators located at selected points along the water delivery lines. The emitters dissipate the pressure from the distribution system by means of orifices, vortexes, and tortuous or long flow paths, thus allowing a limited volume of water to discharge. Most emitters are placed on the ground, but they can also be buried. The emitted water moves within the soil system largely by unsaturated flow. The wetted soil area for widely spaced emitters (point source) will normally be elliptical in shape as illustrated in Fig. 16.1. Since the area wetted by each emitter is a function of the soil hydraulic properties, one or more emission points per plant may be necessary. In this Chapter, the words "drip" and "trickle" will be used synonymously.

16.1.1 Development of Drip Irrigation

Originally, drip irrigation was developed as subsurface irrigation applying water beneath the soil surface (Davis, 1974). The first such experiment began in Germany in 1869, where clay pipes were used in a combination of irrigation and drainage systems. The first reported work in the U.S. was made by House (Davis, 1974) in Colorado in 1913 who indicated that the concept was too expensive for practical use. Subsequent to 1920, perforated pipe was used in Germany which made this concept feasible. Since then, various experiments have centered around the development of drip systems using perforated pipe made of various materials.

Acknowledgements: The authors gratefully acknowledge the input, both direct and informal, from many colleagues into this chapter. The discussions at many meetings were invaluable in the sharing of experiences and the forming of opinions on this subject. Information developed by the Western Regional Technical Research Committee W-128 and the Southern Regional Technical Research Committee S-143 was very important to completing this task.

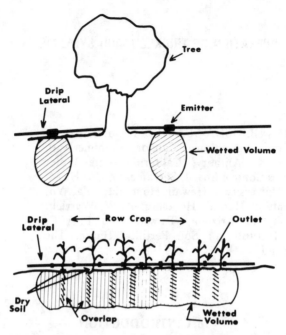

FIG. 16.1 Examples of trickle irrigation soil wetting patterns.

The equipment used in early drip irrigation systems included material used to water greenhouse plants in the United Kingdom in the late 1940's. A few of these early systems are mentioned by Black (1976), Davis (1975) and by Keller and Karmeli (1975). Drip irrigation of outdoor crops dates back to Blass (1964) who used the method in Israel in the early 1960's. The idea of the soil as a storage reservoir was discarded or minimized and replaced with the concept of irrigation keeping up with evapotranspiration on a daily basis.

From Israel the drip irrigation concept spread to Australia, North America, South Africa by the late 1960's, and finally throughout the world. In the few short years since 1970, developments of equipment have tended to rush ahead of studies in understanding the drip concept. In some respects the irrigation hardware overshadowed the fundamental technology of applying water to soils and crops efficiently and adequately.

The development of economical plastic pipe manufacturing after the 1950's made the field use of drip irrigation practical. Publications from Israel, Australia, New Zealand and the United States in the early 1960's described the expansion in the development of drip irrigation. And now, thousands of hectares are drip irrigated in California, Hawaii, Florida, Arizona, Texas and other states. Australia, New Zealand, Israel, South Africa, Canada, and several other countries also have extensive drip irrigated areas.

The low-flow rate emitter proved to be a key to high frequency irrigation in terms of minimum equipment (pipe sizes), minimum power (low pressure), and most importantly, realistic application rates for the short intervals and small areas per emitter. A wide variety of emitters appeared. Some of the early ones were no more than attempts to induce slow leaks in otherwise tight casings or valve-like fittings.

Micro-tubing, which was originally called greenhouse leader-tubing was the first controllable low-flow equipment that Blass worked with when he introduced the idea of irrigating crops on a field scale at frequencies similar to those in greenhouses. The micro-tubing emitters could be adjusted to most types of terrain, compensating through variable lengths and bores for line pressure changes and changes of elevation. For a number of practical reasons, plugging, fragility, and a high labor factor for installation and maintenance, the micro-tubing has largely been replaced by compact emitters that embody the principle of the micro-tube, but not its hydraulic adjustability or facility for locating the drip points away from the pipe. The first successful compact emitter was the spiral tube emitter designed in Israel. This emitter was simply a modification of the micro-tube, or spaghetti tube as it was sometimes called, in which the "tube" was a spiral, cut in plastic and surrounded by a plastic casing.

16.1.2 Advantages of Drip Irrigation

Initially, many claims of advantages for drip irrigation compared to conventional methods were made. Several of the currently recognized advantages are: (a) Application of water at slow rates to limited areas around the trees improved water penetration on problem soils. Field trials have shown that the depth of water penetration could be improved appreciably; (b) Only small areas around the trees are wetted; thus, water savings usually result by reducing the total evaporative surface, reducing runoff (dry soil between rows could store more precipitation) and controlling deep percolation; (c) Further water savings with drip irrigation are obtained when the trees are young because with conventional methods usually a larger area than needed is wetted; (d) Weed growth is reduced because a limited soil surface is wetted; (e) Limited soil wetting also permits uninterrupted orchard cultural operations and minimizes labor scheduling problems; (f) Fertilizers can be injected into the irrigation water; (g) Considerable evidence supported the concept that water availability to plants enhanced plant growth and yield; and (h) Frequent or daily application of water kept the salts in the soil water more dilute and leached to outer limits of the wetting pattern making the use of saline water practical.

Drip irrigation, like other irrigation methods, will not fit every agricultural crop, specific sites, or objectives. It is being used for a variety of crops, climates and soils. Among these crops are almonds, grapes, citrus, stone fruit, avocados, walnuts, pistachios, olives, pecans, apples, pears, figs, vegetable crops, nursery plants, berries, tropical fruit, sugar cane and others.

16.1.3 Disadvantages of Drip Irrigation

Several problems are associated with the drip method. Emitter clogging is considered the most serious problem in drip irrigation. The causes of clogging are attributed to physical, chemical and biological factors. When clogging occurs, the emission uniformity is greatly reduced, and crop damage may occur before the clogging is detected. Improvement in the filtration process and chemical treatment of the water can reduce clogging problems. Rodents, coyotes, dogs, etc., present a problem by damaging the plastic pipe. For crops with high plant densities requiring large amounts of pipe per land unit, drip irrigation may not be economical. In some areas, excess salts accumulate at the soil surface and toward the fringes of the wetted soil. Rain may leach harmful amounts of surface salts into the root zone; drip irrigation should continue during the rain to prevent this problem. Other salinity related problems are discussed in Chapter 5.

16.1.4 Future of Drip Irrigation

The drip method is an acceptable system of irrigating many crops, yet drip irrigation should not be expected to replace other irrigation methods or in some cases to even compete with conventional irrigation methods. The potential for using less water per unit of production may provide the motivation for changing irrigation methods whenever and wherever water costs have very significant effects on profit margins. The rapid expansion of drip irrigation in southern California where water costs are high illustrates this point. Since drip irrigation is not suitable for some crops that are sprinkler or surface irrigated, forages in particular, simple area statistics underestimate the acceptance of drip irrigation over the past decade. If the gross value per unit land area of the various crops that are drip irrigated are compared with those irrigated with conventional systems, the importance of drip irrigation may be seen. Most of the crops irrigated by the drip method yield higher cash returns per unit area compared to some of the crops under conventional irrigation. Developments in the future will probably continue to be concentrated on high value crops, on extending limited water supplies and on the utilization of relatively low quality waters. It is unlikely that expansion of drip irrigation will include solid-stand large-area plants, such as forages and cereals, because the system has little or no advantage compared with conventional sprinkler or surface systems on these types of crops.

16.1.5 Summary

This chapter presents information which is required by engineers to make design decisions and recommendations on maintenance and operation procedures for drip irrigation systems. This chapter is intended to compliment existing works like that of Keller and Karmeli (1975), Goldberg et al. (1976), Davis (1975) and the numerous proceedings from technical meetings such as the 1974 Second International Drip Irrigation Congress (San Diego, CA), the 1975 Drip Irrigation Conference and Trade Show (Fresno, CA) and the 1977 International Agricultural Plastics Congress (San Diego, CA). In addition readers can find the annotated bibliography by Smith and Walker (1975) and the Drip Irrigation Library (Drip Irrigation Information Center and Archives, Bio-Agricultural Library, University of California, Riverside, CA 91502) very useful for reference.

16.2 COMPONENTS AND EQUIPMENT

Up to the point of delivery of water to fields or orchards the components of a drip irrigation system are for all practical purposes the same as for conventional irrigation methods. The flow of water per hectare may be the same or less than that for sprinkler systems depending upon what percentage of the area is not wetted and the efficiency that can be achieved with one system compared to the other. The type of crop, i.e., spacing of plants, has a direct bearing on the capacity of the system per unit area. The flow per hectare is determined by the crop requirement and the application efficiency. The overall capacity of the delivering system to the farm gate is determined by the flow per hectare times the land area in hectares divided by the distribution system efficiency.

16.2.1 System Components

From the source of water to the farm gate, the canals, flumes, or pipes are the main components of any system. Auxiliary components include various combinations of sand separators, screening equipment, intakes, flow regulators or pressure reducers where gravity pressure is obtainable. Booster pumps are used where gravity pressure is inadequate, equalizing reservoirs are used in mountainous terrain. Stilling or settling basins, chlorinators, water meters, valves, gauges, fertilizer injectors and controls are parts of the drip components.

Between the farm delivery and the field drip system some type of fine filtration is required. Most of the filtering devices are simple, but some are elaborate, complete with automatic back-flush devices. The filtering devices must have the capacity for the required flow of water and the ability to remove fine particulate matter down to a size somewhat smaller than the emitter pathways and orifices. Clean water is essential for satisfactory, trouble-free operation of drip systems.

Field systems vary considerably in physical arrangements, but basically consist of a mainline, submains or manifolds, laterals and emitters as shown in Fig. 16.2. Line source applicators are laterals and emitters combined in one unit. Submains provide a means of grouping laterals into zones, the number of which depends upon the area, shape of field, topography, number of settings per day or per interval and limitations to lengths of either submains or laterals. On small areas a single submain may suffice or may be omitted if the mainline can substitute for the submain, but generally cyclical flow will require submains. For the sake of minimizing pipe sizes, both submains and laterals can be arranged to split the flow of water in two directions to minimize friction losses.

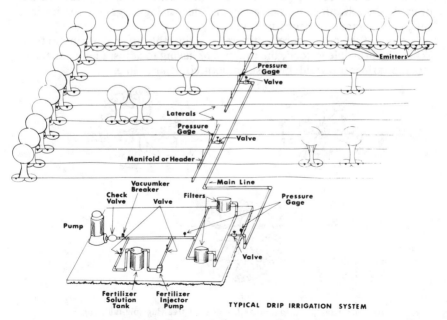

FIG. 16.2 Typical drip irrigation system.

16.2.2 System Layout

Layouts of the drip emitters are generally simple pipe arrangements as seen in Fig. 16.3. The decision as to number and spacing of laterals is governed by the nature of the crop. Each row of plants, whether trees, vegetables, or others will usually require a lateral. In some horticultural crops, double-row planting allows two crop rows to be served by one lateral. The in-row placement of emitters has the distinct advantage of minimum interference with other cultural operations such as cultivation, mowing, spraying and picking. If adequate wetting is not achieved by one lateral line, emitters may be installed on a double line or a loop (sometimes called a pig-tail) around the base of the tree.

16.2.3 Emitters

Decisions about the type of equipment, field layout of equipment, and operation and maintenance of the system then becomes a matter of choice and technical skill. Numerous types of emitters are manufactured. Several types are shown in Fig. 16.4. Five distinct emitter types are:

1 Long path emitters,
2 Short orifice emitters,
3 Vortex emitters,
4 Pressure compensating emitters, and
5 Porous pipe or tube emitters.

A miscellaneous group that includes such things as spitters, foggers, etc. could be added, but these are neither point-, nor line-source applicators, resembling instead small sprinklers. Each of the first four groups includes the point source emitters. The fifth group includes the line source applicators and includes all double walled pipes, soaker hose types, and porous plastic tubes. The fourth group includes relatively new manufactured equipment which offers the advantage of constant flow with changing pressures but discards the advantage of increasing and decreasing irrigation applications simply by alterations of line pressures.

A few of the point source emitters are adjustable up to a maximum flow. While the principle is good, the practice of having to adjust many outlets offers the farmer a formidable task in both effort and time.

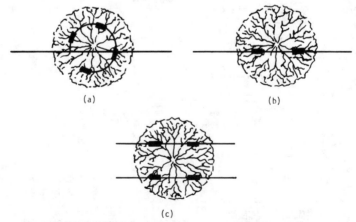

(a)   (b)

(c)

FIG. 16.3 Example drip emitter layouts; (a) tree ring (pig-tail), (b) single lateral with two emitters and (c) dual laterals with two emitters each.

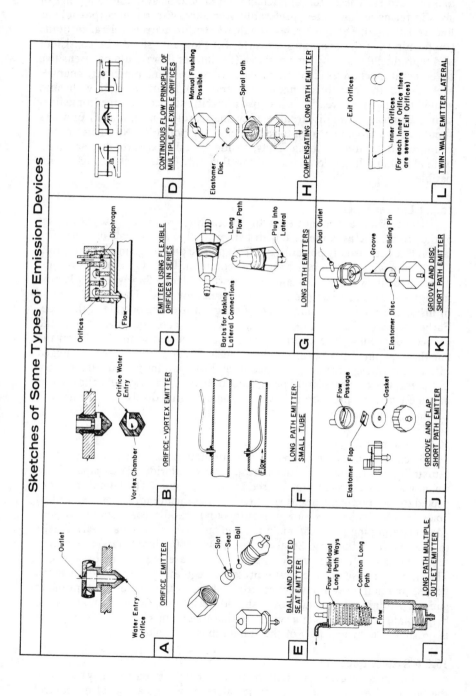

FIG. 16.4 Examples of emitters (after Solomon, 1977).

Emitters should offer a number of features that appeal to the designer and the farmer. These are: (a) Each emitter should be available in a range of sizes in terms of flow rates, preferably over narrow increments (possibly as low as 1.0 L/h); (b) Flow rates should be within narrow tolerance limits (uniformity) for the operating pressure; (c) Flow rate should be consistent and reproducible for long time periods; (d) Emitter flow should be insensitive to temperature changes (changes of water viscosity with temperature changes are beyond the control of manufacturers); (e) Emitters must withstand sunlight and general weathering to give a specific service life-time (some line-source applicators are discardable after one season of use); and (f) Emitters should have a large flow area to reduce clogging potential. No single emitter can meet all the above criteria. Thus, a wide diversity exists for drip emitters. Note also that a large flow area and low flow rate characteristics are incompatible.

16.2.4 Lateral Lines

The emitters are connected to the lateral line or in some cases are a part of the lateral line as is the case with twin-wall pipe and simple orifices. The lateral lines are usually polyethylene (PE) plastic and range in diameter from 8 to 20 mm with 14 mm being the most common. Class C PE pipe is recommended for its weather resistance with carbon black. Two types of PE pipe fittings—barbed and compression—are commonly used. The barbed fittings fit inside the pipe while compression fittings fit over the outside of the pipe. Earlier types of PE developed "stress cracks" where the pipe fit over the barbed connection. The emitters are either connected in the line or on the line through punched or drilled holes. The lateral downstream end is plugged with a pipe fitting or simply crimped over. Provision for periodic flushing of the lateral lines is necessary. Spring loaded flushing valves which allow water passage when the pressure is less than some threshold valve can be used to automatically flush the lines.

16.2.5 Submain Lines

The lateral lines are connected to a submain or in some cases directly to a mainline. Normally these lines are polyvinyl-chloride (PVC) pipe. Occasionally asbestos-cement pipe is used for the mainline. The laterals are connected to the submain using various techniques including: undersized drilled holes in PVC submains (low pressure only), saddle connections, various adapters and pipe tees. The mainline and submain should have valved exits to allow periodic flushing. Small screens (80 to 100 micrometers) are used as safety devices to further protect the laterals in case of pipe line breaks, etc. The submain also may contain pressure regulating valves, flow control valves, manual or automatic control valves, water meter and pressure gauges. Additional manual valves should be located along the mainline to help isolate certain areas of a system so that irrigation with part of the system is possible even when, part of the system may be broken.

16.2.6 System Control

The "head" works consist of the main control station and may consist of the pump, filters, pressure regulating valves, flow regulating valves, control valves, water meters, pressure gauges, automatic controllers or time clocks and chemical injection equipment. Details of these individual components will be discussed later in this chapter.

16.3 HYDRAULICS OF DRIP IRRIGATION SYSTEMS

A drip irrigation system consists of many components as described in Section 16.2. The system may include pumps, filters, meters, valves, automatic timers, mainlines, submain lines, laterals, emitters, and numerous accessory parts. The water moves from the water source by static pressure or through a pump to the mainline, from the mainline to submain line or in small systems directly to the lateral, from the submain line to the lateral and then to the emitters. The emitters distribute the irrigation water to the soil and the plants extract the water from the soil. Fig. 16.1 shows a conceptual view of a drip irrigation system. Hydraulic properties of drip systems are presented by Wu and Gitlin (1973, 1974a, 1974b), Wu and Fangmeier (1974), Howell and Hiler (1974a, 1974b) and Keller and Karmeli (1974, 1975).

16.3.2 Hydraulics of Emitters

Emitters are devices which allow water to flow from the supply to the soil. The hydraulic characteristics of the emitters determine the rate of water flow through the emitter. Many types of emitters have been manufactured to overcome hydraulic limitations. Keller and Karmeli (1975) classified emitter characteristics as follows:

1 Flow regime,
2 Pressure dissipation,
3 Lateral connection,
4 Water distribution,
5 Flow cross-section,
6 Cleaning characteristics,
7 Pressure compensation, and
8 Construction material.

Hydraulically, most emitters can be classified as long-path emitters, orifice emitters, vortex emitters, pressure-compensating emitterts or porous-pipe emitters. The hydraulic characteristics of each emitter are directly related to the mode of fluid motion (flow regime) inside the emitter as characterized by the Reynolds number (R_e) [$R_e = VD/\nu$ where V is velocity, D is diameter and ν is kinematic viscosity]. These flow regimes are usually characterized as (a) laminar $R_e < 2,000$; (b) unstable $2,000 < R_e \leqslant 4,000$; (c) partially turbulent $4,000 < R_e \leqslant 10,000$; and (d) fully turbulent $10,000 < R_e$.

The flow in an orifice emitter is through a small opening. The flow regime is usually fully turbulent. The flow rate is expressed as

$$q = 3.6 \, AC_O \, (2gH)^{\frac{1}{2}} \quad \dots\dots\dots\dots\dots\dots\dots\dots\dots\dots\dots\dots\dots\dots \quad [16.1]$$

where q is the emitter flow rate in L/h, A is the orifice area in mm², C_o is the orifice coefficient (usually about 0.6), H is the pressure head at the orifice in m, and g is the acceleration of gravity, 9.81 m/s². This equation can be utilized to design orifice size for selected operating pressures and emitter flow rates. Since orifice flow is usually fully turbulent, small changes in fluid viscosity caused by fluid temperature changes usually do not affect emitter performance.

The flow in a long-flow path emitter is through a small conduit. For circular conduits the long-flow path emitter flow rate can be expressed as

$$q = 113.8 \, A \, (2gHD/fL)^{\frac{1}{2}} \quad \dots\dots\dots\dots\dots\dots\dots\dots\dots\dots\dots\dots\dots \quad [16.2]$$

using the Darcy-Weisbach equation with q in units of L/h, D is inside diameter in mm, L is pipe length in m and f is the friction factor (dimensionless). The cross-sectional shape of the conduit will affect the hydraulic characteristics of the emitter more in laminar flow regimes than in turbulent regimes. For conduit cross-sections having simple geometric shapes not markedly different from circular, equation [16.2] can be modified by substituting 4r for D where r is the hydraulic radius in mm. For turbulent flow, this use of the hydraulic radius gives fairly accurate results; however, for laminar flow, this method gives results which decrease in accuracy as the shape becomes less circular. Also, for laminar flow the effects of fluid temperature changes can cause significant flow variation through the direct effects on fluid density and viscosity (Parchomchuk, 1976). Changes in emitter flow rates caused by changes in fluid temperature are usually minor if flow is turbulent. Equation [16.2] is useful for estimating emitter dimensions (length and diameter) for selected operating pressure and emitter discharges.

Pressure-compensating emitters attempt to overcome the hydraulic constraints imposed by orifice or long-flow path emitters and to provide a constant emitter flow rate. Usually an elastic material which changes dimensions as a function of pressure is utilized separately or in combination with orifices or small-diameter conduits. These emitters usually allow only small changes in emitter flow rates as pressure is changed. The degree of pressure compensation is durectly related to the emitter elasticity characteristics. Maintaining these elastic characteristics after long exposures has usually been difficult.

Vortex emitters are orifices which have circular cells. The water enters the cell tangentially through an orifice which causes a jet to form vortical flow against the cell wall. The resulting pressure loss is usually greater than a simple orifice of the same dimension. Keller and Karmeli (1975) stated the pressure loss was increased by a factor of 1.73, so that for a given operating pressure and emitter flow rate, the orifice could be larger for a vortex emitter compared to a simple orifice.

Porous-pipe emitters are pipes with many small pores or perforations in the pipe wall that allow water to exit from the pipe. The flow rates from porous-pipes depend on geometry of the material and the applied pressure. Usually prediction of emitter flow rates based on theory is difficult and empirical methods are utilized.

Usually emitter flow rates are best characterized by empirically determining flow rates as a function of operating pressure. This empirical characterization is referred to as an *emitter flow function:*

$$q = K_e H^x \qquad [16.3]$$

where q is the emitter flow rate in L/h, K_e is the proportionality factor that characterized the emitter dimensions, H is the operating pressure head in m and x is the emitter discharge exponent which characterizes the flow regime. The coefficients K_e and x are determined by plotting q versus H on a log-log plot. The slope of the straight line is x, and the intercept at H = 1 is K_e. Fig. 16.5 shows the expected variation in emitter flow rate for variation in pressure and emitter discharge exponent (Keller and Karmeli, 1975). Also for laminar flows and fluid temperatures different from test conditions, the emitter flow variation should be multiplied by the ratio of the fluid kinematic viscosity at the testing temperature to the fluid kinematic viscosity at the operating temperature. Normally, manufacturers can supply the coefficient

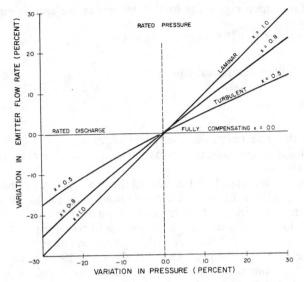

FIG. 16.5 Variation in emitter flow rate resulting from pressure (after Keller and Karmeli, 1975).

of variation (C_v) (ratio of standard deviation to mean) of flow rate for their emitters at design operating conditions. This C_v is an index of manufacturing variability.

16.3.2 Hydraulics of Drip Irrigation Lines

Flow in the drip irrigation lines is hydraulically steady, spatially varied pipe flow with lateral outflows. The total discharge in a drip irrigation line, lateral, submain or main is decreasing with respect to the length of line. The lateral and submain can be considered as having similar hydraulic characteristics and are designed to maintain a small pressure variation along the lateral line. The mainline is designed based on the input pressure, the required pressure and the slope of the energy gradient line which will give a total energy higher than that required at any submain for irrigation.

Friction equations for drip irrigation lines. The drip irrigation lines made of plastic are usually considered as smooth pipes. The Darcy-Weisbach equation is

$$H_f = 6.377 \; fLD^{-5} \; Q^2 \quad \dotfill \quad [16.4]$$

where H_f is friction loss in m, L is pipe length in m, D is diameter in mm, Q is flow rate in L/h and f is the friction factor. Equation [16.4] incorporates an acceleration of gravity of 9.81 m/s². For smooth pipes the friction factor is characterized by the Blasius equation for turbulent flow as

$$f = 0.316 \; R_e^{-0.25} \; (4{,}000 \leqslant R_e \leqslant 100{,}000) \quad \dotfill \quad [16.5]$$

where R_e is Reynolds number. For laminar flow the friction factor is

$$f = 64/R_e \; (R_e \leqslant 2{,}000) \quad \dotfill \quad [16.6]$$

and for the transition region the friction factor may be approximated by

$$f = 3.42 \times 10^{-5} \, R_e^{0.85} \quad (2{,}000 \leqslant R_e \leqslant 4{,}000) \quad \dots\dots\dots\dots\dots \quad [16.7]$$

The Hazen-Williams empirical equation is

$$H_f = 0.628 \, LD^{-4.865} \, [100 \, Q/C]^{1.852} \quad \dots\dots\dots\dots\dots\dots \quad [16.8]$$

where C is the pipe roughness coefficient, equation [16.8] is widely used because of its simplicity. The Hazen-Williams equation, however, has no correction for viscosity.

Hughes and Jeppson (1978) compared the two equations with the results illustrated in Fig. 16.6. Clearly the C value ranges from 130 to 150 depending on R_e in terms of the friction factor. Many pipe manufacturers recommend a maximum velocity of 1.5 m/s in plastic pipe. At this velocity the value of C that compares best to the Blasius equation will depend on pipe diameter with C = 130 for 14- to 15-mm pipe, C = 140 for 18- to 19-mm pipe, and C = 150 for 25- to 27-mm pipe. Underestimating C results in a more conservative friction loss for design purposes.

Energy gradient line for drip irrigation laterals (or submains). The total specific energy at any section of a drip line can be expressed by the energy equation,

$$\bar{H} = z + H + \frac{V^2}{2g} \quad \dots\dots\dots\dots\dots\dots\dots\dots\dots\dots\dots\dots \quad [16.9]$$

where $\bar{H}$ is the total energy, z is the potential head or elevation, H is the pressure head and $V^2/2g$ is the velocity head all expressed in meters. As the flow rate in the line decreases with respect to the length because of emitter discharges from laterals and lateral outflow from submains, the energy gradient line will not be a straight line but a curve of exponential type as express-

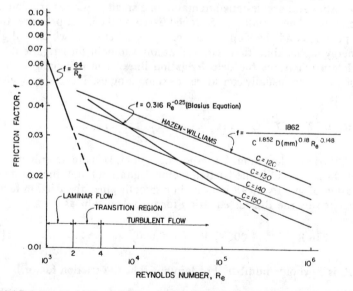

FIG. 16.6 Relationship of Hazen-William "C" to Darcy-Weisbach "f" for 27-mm (1-in.) pipe.

ed by Myers and Bucks (1972) and Wu and Gitlin (1973). The shape of the energy gradient line for level irrigation lines can be expressed by a dimensionless pressure gradient line, since velocity head changes are negligible, as derived by Wu and Gitlin (1974)

$$R_i = 1 - (1 - i)^{m+1} \quad \dots\dots\dots\dots\dots\dots\dots\dots\dots\dots\dots \quad [16.10]$$

in which R_i is $\Delta H_i / \Delta H$ is called pressure drop ratio, m is the exponent of the flow rate in the friction equation, ΔH is the total pressure drop due to friction in meters, ΔH_i is the pressure drop in meters at length ratio i (i = l/L); L is the total length of the line and l is a given length measured from the head end of the line. When the Hazen-Williams formula is used for the pipe flow, the dimensionless energy gradient line can be expressed as

$$R_i = 1 - (1 - i)^{2.852} \quad \dots\dots\dots\dots\dots\dots\dots\dots\dots\dots\dots \quad [16.11]$$

The dimensionless pressure gradient lines for different flow conditions are shown in Fig. 16.7.

The dimensionless pressure gradient line will serve to determine the pressure gradient curve when the total pressure drop is known. The total pressure drop can be determined by using the total discharge as derived by Wu and Gitlin (1974b)

$$\Delta H = a \frac{Q_t^m}{m+1} L \dots\dots\dots\dots\dots\dots\dots\dots\dots\dots\dots\dots\dots\dots\dots \quad [16.12]$$

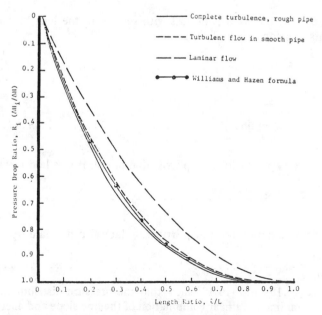

FIG. 16.7 Dimensionless curves showing the friction drop pattern caused by laminar flow, flow in smooth pipe and complete turbulent flow in lateral line (after Wu and Gitlin, 1974).

in which Q_t is the total discharge at the inlet section, a is a constant for a given pipe size, roughness and friction equation and L is total length. Equation [16.12] is strictly applicable to lines with more than 20 outlets.

Pressure variation along a drip irrigation line. If a drip irrigation line is laid on a level ground, the pressure variation along the line will be the pressure gradient curve. If a drip irrigation line is laid on up- or downslopes the pressure variation will be affected by the slopes. When the line is laid upslope it will lose pressure and when the line is laid downslope it may gain pressure. The loss or gain in pressure is linearly proportional to the slope and length of the line and can be shown as follows:

$$\frac{d\overline{H}}{dL} = \frac{dz}{dL} + \frac{dH}{dL} + \frac{d(\frac{V^2}{2g})}{dL} \quad \text{[16.13]}$$

Since the outflow from each emitter is small, the change of velocity head with respect to the length along a drip irrigation line is small and can be neglected.

Therefore, the energy equation can be reduced to

$$\frac{d\overline{H}}{dL} = \frac{dz}{dL} + \frac{dH}{dL} \quad \text{[16.14]}$$

where $d\overline{H}/dL$ is the slope of energy line or energy slope, then

$$\frac{d\overline{H}}{dL} = -S_f \quad \text{[16.15]}$$

The minus sign means friction loss with respect to the length. The dz/dL represents the slope of the line, as in

$$\frac{dz}{dL} = S_O \text{ (downslope)} \quad \text{[16.16]}$$

and

$$\frac{dz}{dL} = -S_O \text{ (upslope)} \quad \text{[16.17]}$$

The pressure variation for a drip irrigation lateral if it is laid downslope is

$$\frac{dH}{dL} = S_O - S_f \quad \text{[16.18]}$$

The pressure variation for a drip irrigation lateral if it is laid upslope is

$$\frac{dH}{dL} = -S_O - S_f \quad \text{[16.19]}$$

Equations [16.18] and [16.19] show that the pressure distribution along a drip irrigation line is a linear combination of the line slope and energy slope. With the knowledge of the dimensionless pressure gradient line, the friction drop at any given length of line can be determined when a total pressure drop (ΔH) is known. If the length of line and slope are known, the pressure head

Pressure loss by friction ───────

Final pressure distribution ──── ────

Pressure loss by slope ── ── ── ──

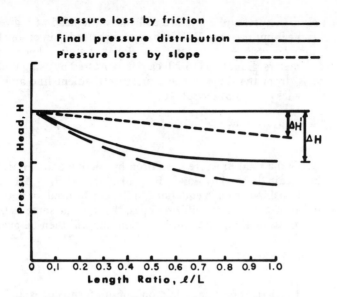

FIG. 16.8 The pressure distribution along a drip irrigation line (downslope) (after Wu and Gitlin, 1974).

gain or drop can be determined. The pressure distribution along a drip irrigation lateral, if an initial pressure is given, can be determined (Figs. 16.8 and 16.9).

The pressure variation along a drip irrigation line as shown in Figs. 16.8 and 16.9 can be expressed mathematically as follows:

$$H_i = H - \Delta H_i \pm \Delta H_i' \dots\dots\dots\dots\dots\dots\dots\dots\dots\dots [16.20]$$

Pressure loss by friction ───────

Final pressure distribution ──── ────

Pressure gain by slope ── ── ── ──

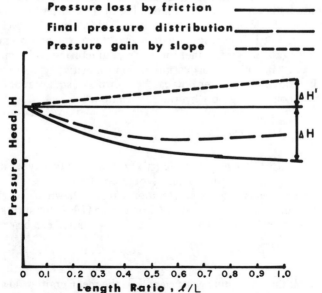

FIG. 16.9 The pressure distribution along a drip irrigation line (upslope) (after Wu and Gitlin, 1974).

in which H_i is the pressure expressed as hydrostatic head at a given length ratio i, H is the input pressure, $\Delta H_i'$ is the pressure head gain or loss by slopes ("+" sign for downslope, "−" sign for upslope) at a given length ratio i all expressed in meters. Equation [16.20] can be expressed by using the pressure drop ratio R_i from the dimensionless pressure gradient line and pressure head gain (or loss) ratio by slopes, R_i',

$$H_i = H - R_i \Delta H \pm R_i' \Delta H' \quad \dots\dots\dots\dots\dots\dots\dots\dots\dots\dots\dots [16.21]$$

in which ΔH is the total pressure head drop by friction and $\Delta H'$ is the total pressure head gain (or loss) by slopes, $R_i = \Delta H_i/\Delta H$ and $R_i' = \Delta H_i'/\Delta H'$. The pressure relationship shown in equation (16.21) can be used for both uniform and non-uniform slopes. For uniform slopes, the R_i' is the same as the length ratio i. The pressure along the drip irrigation line will then be given by

$$H_i = H - R_i \Delta H \pm i \Delta H' \quad \dots\dots\dots\dots\dots\dots\dots\dots\dots\dots\dots [16.22]$$

In case the drip irrigation line is laid on non-uniform slopes and the total length is divided into ten sections and the slope for each section is determined as $S_1, S_2, \dots S_{10}$, the pressure along the drip irrigation line for non-uniform slope can be expressed as,

$$H_i = H - R_i \Delta H \pm 0.1L \sum_{j=1}^{10} S_j \quad \dots\dots\dots\dots\dots\dots\dots\dots\dots [16.23]$$

in which S_j is the slope of the j section along the line using "+" sign for downslope and "−" for upslope.

Emitter flow variation and uniformity coefficient of emitter flow along a drip irrigation line. As shown in equation [16.3] the emitter flow is determined by the hydrostatic pressure at the emitter. This means whenever there is a pressure variation in the drip irrigation line there will be an emitter flow variation along the irrigation line. Emitter manufacturing variation and clogging also result in emitter flow variation. For an orifice type of emitter and uniform line slope, the emitter flow can be shown as a square root function of the pressure considering only hydraulic variation,

$$q_i = K_e (H - R_i \Delta H \pm R_i' \Delta H')^{1/2} \quad \dots\dots\dots\dots\dots\dots\dots\dots [16.24]$$

where q_i is emitter flow at a given length ratio i. For other types of emitters the proper value of x should be used (equation [16.3]).

The degree of emitter flow variation can be shown by a term called Uniformity Coefficient as defined by Christiansen (1942) for sprinkler irrigation. The uniformity coefficient for emitter flow variation can be expressed as

$$C_u = 1 - (\overline{\Delta q}/\overline{q}) \quad \dots\dots\dots\dots\dots\dots\dots\dots\dots\dots\dots [16.25]$$

in which C_u is the uniformity coefficient, $\overline{q}$ is the mean emitter flow and $\overline{\Delta q}$ is the mean absolute deviation from the mean emitter fow. The uniformity coefficient is a quantitative evaluation of the emitter flow variation. There are

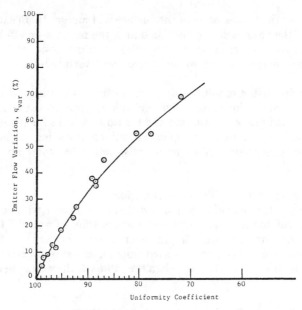

FIG. 16.10 Relationship between emitter flow variation and uniformity coefficient.

other ways of computing the emitter flow variation by comparing the maximum emitter flow with the minimum emitter flow. The one which is commonly used defines,

$$q_{var} = \frac{q_{max} - q_{min}}{q_{max}} \quad \dots \dots \dots \dots \dots \dots \dots \dots \dots \dots \dots \dots \dots \dots \dots \dots [16.26]$$

in which q_{var} is the emitter flow variation, q_{max} is the maximum emitter flow along the line, and q_{min} is the minimum flow along the line. The relationship between the emitter flow variation and the uniformity coefficient is shown in Fig. 16.10. A uniformity coefficient of about 98 percent equals an emitter flow variation of 10 percent, and a uniformity coefficient of about 95 percent equals an emitter flow variation of 20 percent. Keller and Karmeli (1975) proposed another term called emission uniformity to describe the emitter flow variation along a drip line. The emission uniformity is defined as

$$EU = [1.0 - (1.27)(e^{-1/2})C_v] \, (q_{min}/\overline{q}) \quad \dots \dots \dots \dots \dots \dots [16.27]$$

where EU is emission uniformity and e is the numbers of emitters per plant. The emission uniformity increases as the number of emitters per plant increase for constant values of C_v and $q_{min}/\overline{q}$.

16.4 SOIL MOISTURE

The distribution pattern of soil water resulting from trickle sources can be very different from those resulting from the more conventional modes of irrigation. In addition to a generally higher frequency of application, water is added at discrete points on the soil surface rather than over the entire area. If the point is isolated, the soil is wetted in a bulb-like, axially symmetric pattern rather than in a one-dimensional fashion. If the emitters are sufficiently close together the wetter parts of the surface will merge. In an extreme case, if many emitters are close together on a line, the result is an effective line or strip source and the pattern of wetting will be two-dimensional (horizontal in the direction perpendicular to the source and vertical) rather than three dimensional.

In the discussion to follow, the gross features of wetting patterns during infiltration from a point source will be considered first. Next the distributions of pressure head and water content and the flow patterns for steady flows will be discussed. Then more complex situations involving cyclic moisture regimes, heterogeneous and stratified soils, and water uptake by plants will be considered.

16.4.1 Wetting Patterns Durinig Infiltration

An elementary question which is difficult to answer with confidence is "What soil volume is wetted for a given application of water?" One simple answer is found by approximating the wetted volume of soil as a hemisphere and assuming the soil volume is wetted from an initial volumetric water content θ_i to a final water content θ_f. Then the radius of the wetted hemisphere is

$$r = \quad 3qt/[2\pi(\theta_f - \theta_i)] \quad ^{1/3} \quad \dots\dots\dots\dots\dots\dots\dots \quad [16.28]$$

where q is the volumetric rate of application in m³/h, t is application time in h and r is the radius in m. The final water content will correspond to a hydraulic conductivity of the order of 1 mm/d and can be very different for different soils (Rawlins and Raats, 1975, Fig. 16.12B). Such an approximation gives the general picture of what happens under some circumstances (Bar-Yousef and Sheikholslami, 1976). Roth (1974) measured the wetted pattern for three sites of Superstition sand in Southwestern Arizona. In each case, the initial water content was less than 3 percent on a volumetric basis.

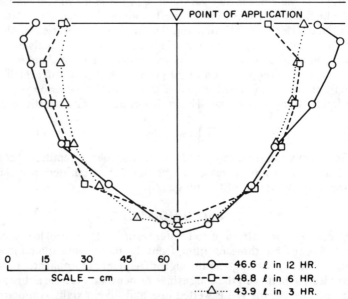

FIG. 16.11 Wetting pattern for equal volume of applied water at three rates (after Roth, 1974).

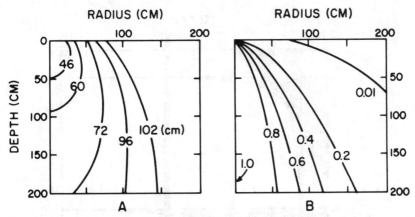

FIG. 16.12 Pressure head (A) and streamline distribution (B) for a single point source under steady conditions (after Raats, 1971).

Water was added at three steady rates of 3.8, 7.6 and 15.2 L/h such that the amount added was about the same at the end of 12, 6 and 3 h, respectively. An excavation was performed exposing a vertical plane passing through the point of application and wetting fronts were measured in each case. The results are shown in Fig. 16.11. Approximately the same volume of soil was wetted in each case. Rather than being in a hemisphere, however, the wetted volume was elongated in the vertical direction. The higher the application rate the larger was the influence of gravity and, as a result, the narrower was the wetted area.

In addition to the wetted volume of soil, the pattern of the wetted surface is also of interest. When the emitter is turned on, the water spreads over the surface and the wetted area gradually expands. However, the wetted area remains finite and tends to stabilize at something on the order of 1-m diameter. The wetted area will depend upon the discharge rate as well as the soil type and the infiltration characteristics. In general, the higher the discharge rate and the lower the infiltrability of the soil, the larger will be the wetted area. In the numerical analysis of Brandt et al. (1971), infiltration from a drip source is modeled by assuming that the water entry zone is saturated. Such a zone of saturation will occur for both line- and point-sources. In their calculations for a line-source on a Gilat loam soil, the width of the saturated zone was approximately 220 and 580 mm, respectively, for two discharge rates of 1.8 and 5.9 L/m per h. The time necessary to reach the maximum wetted area on the surface was on the order of 3 h for the lowest discharge rate to nearly 1 day for the highest rate.

16.4.2 Distributions of Pressure Head and Water Content and Flow Patterns for Steady Flows

After infiltration has occurred for some time, the water content at any position in the vicinity of the source stabilizes. Thus, rather than visualizing the system as we did previously as a wet and dry zone with the wet zone increasing in volume, now consider a spatial distribution of the water content which is not changing with time. The water content increases only at large distances from the source and eventually the increase will everywhere become negligible. The zones of interest are relatively close to the emitter. Such steady flows are easier to deal with from a theoretical standpoint than the

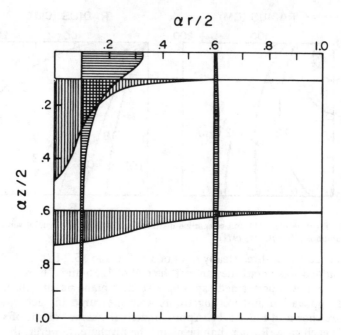

FIG. 16.13 Horizontal velocity proviles at ar/2 of 0.1 and 0.6 and vertical velocity profiles at az/2 = 0.1 and 0.6 for a single point source (after Raats, 1970).

time-dependent case we considered previously. One set of analyses is based on the assumption that the hydraulic conductivity, K, is an exponential function of the pressure head, h, i.e. that K is equal to $K_o \, e^{ah}$. Experimental values of K_o and a are included in several references (Gilley and Allred, 1974; Braester, 1973).

For a single point source, the distribution of the steady-state pressure head is shown in Fig. 16.12A, while the distribution of the streamlines is shown in Fig. 16.12B. The values of the pressure head are based on a hydraulic conductivity of $K = 100 \, e^{0.05h}$ (mm/d), and an input rate q = 0.42 L/h. The lines of constant pressure head (h) would also be lines of constant water content assuming negligible hysteresis effects. The gradients are largest near the source. In Fig. 16.12B, the macroscopic path of the water is along the streamlines. Also, the fraction of the flow occurring between the lines is equal to the difference of the labeled values of the Stokes' stream function. Of course, the cross-sectional area corresponding to a given increment of the radial coordinate increases linearly with the distance from the axis.

For an array of equally spaced line sources at the soil surface, Fig. 16.13 shows the distribution of the vertical and horizontal components of the flux along selected cross sections. The flux profiles show that for small depths the distribution of the flux is far from uniform. The rapid downward movement under furrows in sandy soils was already noted in Chapter 4.

16.4.3 Heterogeneous Soils

Thus far only homogeneous soils were examined, but in many cases stratification is such that it cannot be neglected. In general, the effect of such layering is to retard moisture movement. In the classical experiments by Gardner (Baver et al., 1972), the advance of a wetting front from a two-dimensional furrow was observed. In the case of drip irrigation, the movement of water would be expected to be similar. In the case of a fine soil over a coarser soil, the advance of the wetting front is of course similar to Fig. 16.11 until the transition zone is reached. At that time there will tend to occur a lateral movement along the top of the transition region until the pressure is reduced far enough to permit significant penetration into the coarser material. Without the stratification the lateral spread would be less. An extreme situation with an impermeable layer at some depth centered below each line source and extending over about 50 percent of the area was shown by Maàledj and Malavard (1973, see Figs. 3 and 4). In the case of a coarse material over a finer material, the same sort of phenomenon occurs although for a different reason. In that particular case, the infiltration would also occur as in Fig. 16.11 until the transition zone was encountered but the spreading would occur primarily due to the lower hydraulic conductivity of the finer textured soil. So, again, the result is more horizontal spreading and less vertical penetration within the profile. Due to the fact that the soil water characteristic curves for coarse and fine soils often cross, for low rates and transitions from coarse to fine at some distance from the source the opposite could happen.

Gradual changes of the hydraulic conductivity with depth were studied by Philip (1972) and by Philip and Forrester (1975). For simplicity, they assumed the hydraulic conductivity to be not only an exponential function of the pressure head but also of the depth, i.e., $K = K_o \, e^{a(h+Bz)}$. The distribution of the hydraulic conductivity and the flow pattern around a drip source then becomes a function of $a(1 + B)L$ where L is the characteristic length of the distribution of the drip sources (e.g. their spacing). Compared with a homogeneous soil with the same value of a, it is as if the characteristic length has been multiplied by a factor $(1 + B)$. The distribution of h depends not only on $a(1 + B)L$ but it is also dependent on B itself. This caused a considerable flattening of the contours of equal pressure head.

16.4.4 Cyclical Distribution Patterns

Although some drip sources are operated continuously, in most cases the additions of water are on a cyclic pattern. The resulting moisture regime is somewhat more complicated than those previously discussed. During the wetting part of the cycle, the situation is very similar to the infiltration case discussed earlier. However, when the emitter is shut off, redistribution will occur. Rather than remaining stationary, the movement of the water in response to the hydraulic gradients will continue. While water is being added, the wettest region is adjacent to the emitter. However, when the emitter is closed, this will no longer be the case. In fact, the wettest region within the profile will be somewhere below the emitter at a depth depending upon the time that has passed since the irrigation was cut off. At any particular point within the profile a cyclical pattern will develop with the same frequency as the application. The water content will increase in response to the times when water is added and decrease after the emitter is shut off. The variations of the water content depend upon the frequency of irrigation, the rate of discharge from the emitter and also the location relative to the source. The time at which the increase occurred will also be different within the profile and will tend to lag more at points farther away from the source. Fig. 16.14 shows

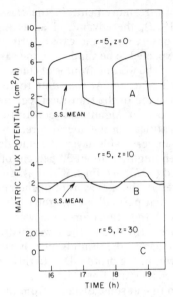

FIG. 16.14 Cyclic pattern of matrix flux potential at three depths near a point source (after Ben-Asher et al., 1978).

calculated results for Sinai sand after Ben-Asher et al. (1978). For the first part of a 2-h cycle, infiltration occurs from a point source at a rate of 0.4 L/h. For point A, (Fig. 16.14), representing a point 5 cm away from the source on the surface, the variation in matric flux potential is 0.6 to 7 cm²/h. the steady-state mean is also shown. This is the value resulting if the intake rate was continuous at 0.2 L/h rather than 0.4 L/h for alternate hours. This corresponds to a pressure head (or tension) variation of −40 to −78 cm. Points B and C, or the deeper points in the profile, show a much smaller variation of time, that is, the fluctuations are damped out around some steady-state average curve. At point B, the pressure head range is about 7 cm of water during a cycle. At point C, the variation is so small that the entire curve is represented by a single line.

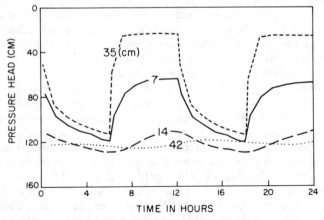

FIG. 16.15 Cyclic pattern of pressure head at four depths below an emitter (after Merrill et al., 1978).

Fig. 16.15 shows cyclic variations of the pressure head at four depths in a heterogeneous column of Pachappa f.s.l. of Merrill et al. (1978). Measurements were made in an undisturbed core 40 cm in diameter and 52 cm deep. The application was equivalent to 10 mm/h steady rate over the entire column, using a cycle of 12 h. Again the variations are most extreme near the surface and are damped with depth. The shallower depths are "wetter" because they are closer to the surface.

16.4.5 Plant Water Extraction

In the design of drip irrigation, the rooting pattern of the particular crop must be taken into account. Differences in emitter spacing as well as irrigation scheduling will occur according to the planting pattern and crop characteristics. For a shallow-rooted crop, for example some vegetables, the emitters should be placed close to the soil surface and one should irrigate frequently for short periods. On the other hand, for a tree crop the rooting volume would be more substantial and perhaps a less frequent irrigation should be in order. In either case, the amount of water present in the profile is going to be less than if the crop were not present.

16.5 CROP WATER REQUIREMENTS

The crop water requirements under drip irrigation may be different from crop water requirements under surface and sprinkler irrigation primarily because the land area wetted is reduced resulting in less evaporation from the soil surface (Aljibury, 1974). Most methods of estimating crop water requirement presently utilized (Doorenbos and Pruitt, 1977; Gangopadhyaya et al., 1966; and Jensen, 1974) provide estimates of evapotranspiration which probably contain a significant soil evaporation component. The evaporation of water from the soil surface is implicitly related to the method of irrigation application and irrigation scheduling. Methods presented by Ritchie (1974) are capable of estimating transpiration and soil evaporation separately. However, direct measurement of plant transpiration under drip irrigation culture has not been conducted for many crops.

Crop water requirements are usually expressed in units of water volume per unit land area (depth) per unit time. Drip irrigation application rates are usually expressed in units of water volume per plant per unit time. Table 16.1 gives a conversion chart to relate emitter flow rate to emitter density (emitters/ha) and crop water use. Table 16.2 gives conversion values for line source applicators. The water volume in L/d per tree for different water use

TABLE 16.1. DAILY WATER QUANTITY IN LITERS PER
EMITTER FOR DIFFERENT ET AND NUMBER OF
EMITTERS PER HECTARE

ET, mm/day	Emitters per ha							
	250	500	750	1000	1500	2000	2500	5000
1.25	50	25	17	12	8	6	5	2
2.50	100	50	33	25	17	12	10	5
3.75	150	75	50	37	25	19	19	7
5.00	200	100	66	50	33	25	20	10
6.25	250	125	83	62	42	31	25	12
7.50	300	150	100	75	50	37	30	15
10.00	400	200	133	100	67	50	40	20

**TABLE 16.2. DAILY WATER QUANTITY IN
LITERS PER METER OF ROW LENGTH FOR
DIFFERENT ET AND ROW SPACING**

Row spacing, m	ET, mm/day			
	5	6.25	7.5	10
0.50	2.5	3.1	3.8	5.0
0.75	3.7	4.7	5.6	7.5
1.00	5.0	6.3	7.5	10.0
1.50	7.5	9.4	11.3	15.0
2.00	10.0	12.5	15.0	20.0

**TABLE 16.3. DAILY WATER QUANTITY
IN LITERS PER TREE FOR DIFFERENT
ET AND TREE SPACING**

Tree spacing, m	ET, mm/day			
	5	6.25	7.5	10
6 × 6	180	225	270	360
9 × 9	405	506	608	810
12 × 12	720	900	1,080	1,440
15 × 15	1,125	1,406	1,688	2,250
18 × 18	1,620	2,025	2,430	3,240

rates and plant spacings is given in Table 16.3. Tables 16.1, 16.2 and 16.3 are included for comparison and conversion and should not imply drip irrigation design or management criteria.

16.5.1 Determination

Crop water requirements can be determined directly from lysimeters (Tanner, 1967) or indirectly from water and/or energy balances. Since the application of water by drip irrigation is usually a three-dimensional pattern, the energy and/or water balance techniques become difficult to apply to drip irrigated fields. Heat pulse velocity movements (Swanson, 1972) appear to reliably estimate daily transpiration of trees; however, calibration is difficult. Crop water requirements are influenced by climate and plant properties.

Methods of measurements. Lysimeters provide the only direct measurement of volumetric water use, but lysimetry is difficult to apply to tree crops due to the large root volume, plant uniformity, and wind interferences. To simulate field growing conditions, a lysimeter must have a soil volume which is large enough to sustain "normal" root and plant growth. The mass of soil in an "appropriately" sized lysimeter might approach 260 metric tons. Therefore, a weighing lysimeter would be expensive and difficult to install. Smaller lysimeters (approximately a meter in diameter) can be utilized to determine water use of young trees under well-irrigated conditions but are limited to a short (several years) study period. Procedures described by Brown et al. (1974) could be used to isolate such undisturbed soil monoliths with intact trees or trees could be transplanted into large soil monoliths. The plant selection is critical since the lysimeter plant must be representative of the surrounding field. Weighing lysimeter accuracy can be significantly reduced by wind movement over a tall plant, such as a tree. The wind forces on a tree can cause limb vibrations which could obscure small mass changes in a sensitive lysimeter. Other lysimeter considerations are discussed by Tanner (1967).

Measurement of water requirements for orchard crops has been performed historically by water balance techniques. A water balance is merely a detailed statement of the law of conservation of mass. The water content of a given soil volume cannot increase without addition from outside, nor can it diminish unless transpired or evaporated to the atmosphere by evapotranspiration process or transported to deeper zones by drainage (disregarding the possibility of lateral inflow-outflow). The integrated form of the water balance equation with the component terms totaled over a certain time period for a specific soil volume is given as

$$P + I - Q - D - E - T \pm \Delta\theta = 0 \quad \dots\dots\dots\dots\dots\dots\dots\dots\dots [16.29]$$

where P is precipitation, I is irrigation, Q is runoff (normally a loss of water, but could be a gain of water in a water-harvesting system), D is drainage (normally a loss of water, but could be a gain from upward flow), E is evaporation, T is transpiration and $\Delta\theta$ is the change in soil water content (positive for decreases; negative for increases) with all terms expressed in depth units. The difficulty encountered in applying the water balance equation to drip irrigation is that the irrigation component is at least two-dimensional and normally three-dimensional; therefore, the measurement of $\Delta\theta$ is rather complex (Ben-Asher, 1979). Additional problems are associated with the drainage component in such a regime. Measurement of soil water content at only one point is virtually useless for estimating evapotranspiration since redistribution of soil moisture may dominate extraction patterns.

The energy balance has been utilized to measure evapotranspiration from crop, forest and range lands. The energy balance is merely a detailed statement of the law of conservation of energy. The integrated form of the energy balance equation is

$$R_N + H - LE + G + M = 0 \dots\dots\dots\dots\dots\dots\dots\dots\dots\dots\dots\dots\dots [16.30]$$

where R_N is net radiation, H is sensible heat from the air, LE is latent heat, G is sensible heat from the soil, and M is a miscellaneous term accounting for photosynthesis, heat storage, etc. Equation [16.30] would apply strictly to a uniform surface with dimensions at least 200 times the height of the measurements and no advection (horizontal transport of sensible heat). Since the water application pattern is multi-dimensional many of the energy terms are thus multi-dimensional. This fact compounds the measurement problems by requireing higher evaluation for the measurements, and therefore, a greater fetch for profile stability.

Transpiration has been measured by various heat pulse techniques (Lassoie et al., 1977). Transpiration rate is related to the sap velocity. Actual amounts of daily water use, however, are a combination of water movement through stems and water stored in internal plant tissues (Swanson, 1972). The sap velocity can be measured by brief applications of heat to the stem and determining the speed of the convective heat transport in the plant stem (Marshall, 1958). If this procedure can be calibrated, the volume of water used by plants could be determined. Several problems exist with this method (Lassoie et al., 1977). To be of use in determining water use under drip irrigation culture, the calibration must be performed on a volumetric basis. An enclosed canopy could be used to perform the calibration periodically on intact trees.

**TABLE 16.4. APPROXIMATE RANGE OF
SEASONAL ET FOR SELECTED CROPS
(after Doorenbos and Pruitt, 1977)**

	mm		mm
Alfalfa	600 - 1,500	Onions	350 - 600
Avocado	650 - 1,000	Orange	600 - 950
Bananas	700 - 1,700	Potatoes	350 - 625
Beans	250 - 500	Rice	500 - 950
Cocoa	800 - 1,200	Sisal	550 - 800
Coffee	800 - 1,200	Sorghum	300 - 650
Cotton	550 - 950	Soybeans	450 - 825
Dates	900 - 1,300	Sugarbeets	450 - 850
Deciduous			
trees	700 - 1,050	Sugarcane	1,000 - 1,500
Flax	450 - 900	Sweet potatoes	400 - 675
Grains			
(small)	300 - 450	Tobacco	300 - 500
Grapefruit	650 - 1,000	Tomatoes	300 - 600
Maize	400 - 750	Vegetables	250 - 500
Oil seeds	300 - 600	Vineyards	450 - 900
		Walnuts	700 - 1,000

Table 16.4 presents a summary by Doorenbos and Pruitt (1977) of the approximate range of seasonal water use by several crops. The actual values will be affected by climate, soil, plant and cultural factors.

Climatic effects. Many microclimatic factors directly or indirectly influence crop water requirements. These climatic factors are not unique to drip irrigated crops; however, drip irrigation culture may interact with several factors. The climatic effects on crop water requirements (Jensen, 1974) have been summarized in several types of potential evaporation equations:

$$E_O = \frac{\Delta}{\Delta + \gamma} [(R_N - G)/L] + \frac{\gamma}{\Delta + \gamma} [2.62 (1.0 + 0.0062 \, u_2)(e_z^o - e_z)]$$

$$\dotfill [16.31]$$

$$E_O = \frac{\Delta}{\Delta + \gamma} [(R_N - G)/L] + \frac{\gamma}{\Delta + \gamma} [\frac{622 \, \rho k^2}{P} \frac{u_z}{(\ln z/z_o)^2} (e_z^o - e_z)]$$

$$\dotfill [16.32]$$

$$E_O = \alpha \frac{\Delta}{\Delta + \gamma} [(R_N - G)/L] \dotfill [16.33]$$

where L is latent heat of vaporization, 2.45 MJ/kg, E_o is potential evaporation rate in mm/d, Δ is the slope of the saturation vapor pressure curve at the air temperature in kPa/°C, and γ is the psychrometer constant in kPa/°C, R_N is net radiation in MJ/m^2 per day, G is soil heat flux in MJ/m^2 per day, u_2 is the wind speed in km/d at a height of 2 m, e_z^o is the saturated vapor pressure in kPa at the air temperature at a height z, e_z is the vapor pressure in kPa at a height z, ρ is density of the air in kg/m^3, k is Von Karman's constant (0.41), P is barometric pressure in kPa, z_o is the roughness length in m and α is a proportionality factor. Equation [16.31] is the Penman (1948) formula; equation [16.32] was presented by Businger (1956) and Van Bavel

(1966); and equation [16.33] is the equilibrium equation presented by Priestley and Taylor (1972) which is more directly applicable to humid climates where α has been found to vary between 1.12 and 1.54 with an average of 1.32.

For daily time periods, the soil heat flux (G) can usually be neglected. The net radiation can be estimated following Ritchie (1974) as:

$$R_N = (1 - \lambda)R_s - [6.39 - 0.01 \, (|T - 7.75|)^{1.8} \, (1.35 \frac{R_s}{R_{so}} - 0.35)]$$

$$\dots\dots\dots\dots\dots\dots\dots\dots\dots\dots\dots\dots\dots\dots\dots\dots\dots\dots\dots\dots [16.34]$$

where λ is the albedo (reflected shortwave radiation), T is mean air temperature in °C, R_s is the incoming shortwave radiation in MJ/m² per day and R_{so} is the solar radiation expected on a clear day in MJ/m² per day. R_{so} depends on latitude and time of year (Budyko, 1963; see also Jensen, 1974). Equation [16.34] is an approximation and its use must be checked with local information where possible. The albedo will depend on soil wetness and plant cover.

Ben-Asher et al. (1978) reported significant reductions in net radiation over a drip irrigated field plot as compared to a sprinkler irrigated field plot. Apparently a larger albedo and increased long-wave exchange (caused by increased surface temperatures) over the dry portions caused the differences noted. Therefore, equation [16.35] might significantly overestimate net radiation over drip irrigated fields. Because a large portion of the soil remains dry in drip irrigation, a shift will occur in the energy balance toward non-evaporative energy dissipation processes (convection and conduction). If the area being drip irrigated is large enough, significant convection heating of the air might result in advected heat transfer which would limit the use of equation [16.33], the Priestley-Taylor equation.

Additional information may be found in Monteith (1976), Jensen (1974), Penman et al. (1967) and Section 6.4 in this book.

Crop effects. Many properties of crops and plant canopies affect water use. The plant species and in many cases plant variety affects phenological development, rooting depth, leaf density and orientation, plant height, and plant morphology.

Radiative transfer depends on many plant and soil parameters. A recent summary by Ross (1976) is a complete account of vegetation effects of radiative transfer in crop stands. Reflection, absorption and transmittance are the main optical properties of plants. However, stand architecture is probably the most important factor in the radiation regime. Architecture includes the distribution (spacing and oration) of plants on the ground, leaf distribution within a plant stand, leaf size and orientation, etc. Cultural practices, such as row direction, plant population, and pruning interact with plant architecture.

Leaf diffusion resistance and canopy resistance depend on leaf morphology, light intensity, water deficits, etc. Under full sunlight and well-watered conditions, stomata are usually fully open and offer limited resistance to transpiration. However, some species (citrus, pineapple, etc.) have been found to have large resistances under those conditions (Ekern, 1965 and Van Bavel et al., 1967).

Plant cover greatly affects water use rates. Ritchie and Burnett (1971) found that a leaf area index (leaf area per unit of ground area LAI) of about

2.7 was required for water use to equal the potential rate. They presented the following relationsip

$$\frac{ET}{E_O} = -0.21 + 0.70 \, LAI^{\frac{1}{2}} \quad \dots\dots\dots\dots\dots\dots\dots\dots\dots\dots\dots\dots\dots [16.35]$$

Equation [16.35] was developed specifically for row crops but is similar to results for other crops. Other research cited 50 percent plant cover as critical for water use to be at the potential rate. Marsh et al. (1974) suggested that 1.7 times the vertical plant projection (plant height times plant diameter) per unit land area per tree for avocados equalled class A pan evaporation.

When plants are exposed to water stress conditions, rooting density and depth are extremely important. Plants with deep roots can utilize a larger soil water reservoir than shallow rooted plants. Black and West (1974) reported that with young apple trees watering only 25 percent of the root zone reduced water use by only 26 percent. Willoughby and Cockroft (1974) found that peach tree roots dynamically adjusted to water states in the soil. Little quantitative evidence is available which relates plant response to the degree or proportion of the root zone wetted. The size of the wetted root zone under drip irrigation depends on the number of emitters per tree, emitter location, application rate and quantity and soil hydraulic properties.

Soil effects. The soil affects crop water use through its radiative properties, influence on plant rooting patterns, and hydraulic properties (see Section 16.4). The albedo of a wet sand following drip irrigation was 0.22 while dry sand at the same site was 0.32 (Ben-Asher et al., 1978). Since only a limited portion of soil surface is wetted in drip irrigation and that portion is usually beneath the crop canopy, the soil albedo will be dominated by values for dry soil which will usually result in a reduction in net short-wave radiation. In the dry soil regions, the net radiation is partitioned mainly into sensible heat since evaporation is small for a dry soil. The sensible heat warms the soil and air. The long-wave radiation is also reduced. The results of Ben-Asher et al. (1978) verified these conclusions for a tomato crop. This reduction in net radiation may result in only small reduction in evapotranspiration since the increased sensible heating of the air might result in increased downwind advected evaporation losses.

In shallow or distinctly layered soils the rooting depth may be limited. In these cases, the crop is quickly susceptible to water stress if the irrigation application rate is inadequate (Phene and Beale, 1976). In these soils the proportion of the root zone wetted should be larger than for uniform soil profiles.

16.5.2 Scheduling

Irrigation scheduling involves two decisions: (a) when to irrigate (timing) and (b) how much to apply (quantity). The reader is referred to Section 18.2 for additional information. These decisions are critical to the management of any irrigation system. The concept of drip irrigation implies a rather high irrigation frequency compared to conventional methods. The key principle of drip irrigation is to maintain a moist segment of the root zone with relatively small applications of water applied continuously or intermittently. The drip irrigation cycle then becomes an infiltration-dominated process as opposed to an extraction-dominated process (Rawlins, 1973) and the input of water should match ET (Stevenson, 1973).

The system must be designed to meet the peak crop use demands. Initially as drip irrigation was being practiced, a flow rate just large enough to meet plant requirements on a continuous basis was deemed desirable. These small flow rates required small orifices or emitters which were susceptible to clogging problems. To minimize clogging, emitter diameters were increased which increased emitter flow rates and changed the irrigation duration from continuous to intermittent. The system applications rate must be maintained less than the soil infiltration rate, however, to avoid excessive ponding or runoff.

Irrigation timing and quantity of application are directly related in most cases. The irrigation application in volume per irrigation divided by the irrigation frequency in days per irrigation should equal the net water requirement rate during that period.

A large quantity of research has been conducted on crop response to irrigation frequency under drip irrigation. Many of those studies can be classified as:

1 crop response to drip irrigation frequency with "adequate' applications;
2 crop response to drip irrigation frequency with "limited" applications; and
3 crop response compared with several irrigation methods.

Several studies reported increased yields as frequency increased for "adequate" applications (Goldberg and Shmueli, 1970; Hiler and Howell, 1973; Freeman et al., 1976; etc.) while other studies reported no increase in yield as frequency increased (Bucks et al., 1973 and 1974a; Revelo et al., 1977; Earl and Jury, 1977; etc.). Soil texture, plant nutrition and salinity (Bernstein and Francois, 1973) certainly influenced these results to some degree. Coarse-textured soils (sands to sandy loams may require frequencies up to one or more irrigations per day while fine-textured soils, daily or bi-daily irrigation has many advantages with regard to nutrient management (Phene and Beale, 1976) and water balance control (Rawlins and Raats, 1975).

When irrigation applications were below optimum levels (limited irrigation quantity), crop yields increased in some studies as irrigation frequency decreased (Earl and Jury, 1977; and Bucks et al., 1973 and 1974b). The crop benefited from the larger, less-frequent irrigations because less water was lost to inefficiencies in application as well as to soil evaporation.

Comparison of application methods in terms of plant response is difficult since the irrigation system and its management are not independent. Under similar management, crop yield is similar to yields under sprinkler or furrow irrigation (Bucks et al., 1973 and 1974b; Ravelo et al., 1977; Bernstein and Francois, 1973; Davis et al., 1977; Freeman et al., 1976; etc.). Various amounts of water savings have been reported but usually water savings depended largely on the inefficiency of the method to which it was compared. With low density crops, drip irrigation will save significant amounts of water, especially when the plants are young and the cover is incomplete.

In addition, the reaction of the crop to the salinity patterns developed by trickle irrigation is important. These items are discussed in Chapter 5.

16.5.3 Instrumentation

Since crop water requirements under drip irrigation contain at least

TABLE 16.5. RANGES OF SEVERAL SOIL
MOISTURE INSTRUMENTS

Instrument type	Usual range, kPa	Comments
Tensiometer	0.0 to 70	Narrow range, simple
Buoyoucos block	20 to > 1,000	Temperature and salinity sensitive
Heat-dissipation	10 to > 600	Expensive, temperature compensated, easily automated
Psychrometer	100 to > 3,000	Salinity dependent, temperature errors, insensitive in lower range

some uncertainty (as discussed previously in this section) and since the system operation contains some uncertainty with regard to emitter flow variations and plugging, the system operator or manager must have means to insure that the crop is receiving a proper amount of water. Instruments that can sense soil or plant parameters have been utilized to check the performance of drip irrigation systems. Other types of instruments have been used to schedule irrigation and/or control system functions. Several types of instruments are discussed by Haise and Hagan (1967) and in Section 18.3.

Various types of soil moisture instruments (tensiometers, Buoyoucos blocks, heat-dissipation blocks, soil psychrometers, etc.) have been used in drip irrigation. Table 16.5 lists the ranges of these devices. Tensiometers, because of their simplicity, availability and the lower range, are well adapted to drip irrigation work. Table 16.6 from Haise and Hagan (1967), lists some suggestions for tensiometer installation. Manufacturers also supply guidelines on installations and use.

Soil moisture devices are used to determine the water movement patterns in the soil and, in some cases, to initiate the irrigation cycle. With proper placement, tensiometers can show locations of over- or under-irrigation. The manager can then adjust the irrigation program to compensate. This system operational check is extremely important since the amount of water in soil storage is usually small and water deficits can develop quickly.

Soil moisture devices are commonly used on automated systems (New and Roberts, 1973). Usually the soil moisture device is used to override a system controller. The controller will be preset to actuate a selected valve for

TABLE 16.6. TYPICAL REQUIREMENTS FOR
INSTALLATION OF TENSIOMETERS AND/OR
RESISTANCE BLOCKS IN AREAS WITH RELATIVELY
UNIFORM CROPS AND/OR SOIL CONDITIONS
(after Haise and Hagan, 1967)

Number of stations	3 to 4
Number of depths	2 to 3 (30, 45, 90 cm)
Depth placement	Top - zone of maximum root activity Bottom - near bottom of active root zone Intermediate - midway between top and bottom positions (a shallow depth may be temporarily needed where seedlings are being established)
Location	Near tree drip line, 60 cm from emitter
Site conditions	Select representative soil and vigorous crop area
Precautions	Avoid trampling area near installation; service tensiometers regularly

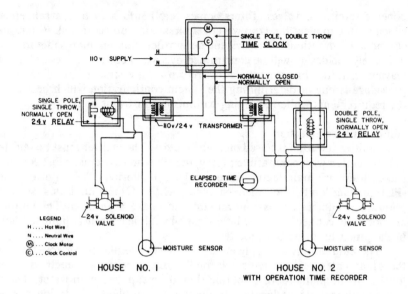

FIG. 16.16 Electrical wiring diagram showing typical components and connections for automatic irrigation control in two greenhouses (after New and Roberts, 1973).

a determined amount of time. If the soil at that station is "too wet", the sensor device opens the valve circuit (no water discharge) and the station is then bypassed. Fig. 16.16 shows a simple control circuit. Other types of control systems have been based on heat-dissipation sensors (Phene et al., 1973) and even automated evaporation pan measurements (Phene and Campbell, 1975).

Evaporation pans are often used to estimate potential evaporation (Haise and Hagan, 1967). Pan exposure and maintenance are important considerations. Several types of evaporation pans have evolved (Bloodgood et al., 1954). A Bureau of Plant Industry (BPI) pan is 1.82 m (6 ft) in diameter, 0.61 m (2 ft) in depth, and installed 0.51 m (20 in.) in the ground with the water surface in the pan at ground level. The Young screen pan (USDA Division of Irrigation) is 0.61 m (2 ft) in diameter, 0.91 m (3 ft) in depth, covered with a screen with 6 mm (0.25 in.) mesh, and installed 0.76 m (30 in.) in the ground with the water surface in the pan at ground level. The U.S. Weather Bureau Class A pan is 1.22 m (4 ft) in diameter, 0.25 m (10 in.) in depth, and mounted approximately 0.15 m (6 in.) aboveground with a water depth of 0.18 to 0.20 m (7 to 8 in.) in the pan. The pans are usually constructed with 22-gage galvanized metal, and the water loss is measured with a hook gage in a stilling well.

The Young screen pan evaporation correlates well with potential evaporation (Richardson and Ritchie, 1973). The Class A pan evaporation is larger than potential evaporation estimates based on climatological data. Campbell and Phene (1976) suggest that installing a screen on a Class A pan will reduce its evaporation to a level closely correlating with potential evaporation.

16.6 DESIGN OF DRIP IRRIGATION SYSTEMS

The design of a drip irrigation system is the integration of physical components (emitters, valves, filters, pipes, etc.) into a system arrangement which is able to meet the crop water requirements subject to soil, water, and equipment limitations. The system design depends on many factors; but ultimately the design will be constrained by several economic factors such as feasibility, initial investment, labor, return on investment, etc. The design procedure is one of determining the system configuration which best meets the requirements while remaining within the constraint boundaries.

16.6.1 Emitter Selection

Emitter selection is based on many factors. The emitter must be selected in conjunction with compatible piping and filtration systems. Most commercial emitters are manufactured from polyvinyl-chloride (PVC), polyethylene (PE), or acrilonitril-butadiene-styrene (ABS). Other materials such as porous ceramic clay, brass, aluminum, etc., have been used. The emitters are normally connected to the lateral pipe by inline coupling, punched holes in the pipe wall or with a pipe tee.

The emitters must supply enough water to the plant root zone to meet the plant-water requirements. Normally, one-third to as much as three-fourths of the plant rooting volume should be supplied with water. The irrigated volume and its location in relation to the plant orientation has not been fully documented for many crops. The larger the percentage of the wetted rooting volume becomes in relation to the volume, the safer the design becomes in case of system breakdown, salinity problems, and when changing from surface irrigation on mature orchards to drip irrigation. However, if the wetted percentage becomes too large, many of the advantages of drip irrigation are lost. The volume of soil wetted depends on the emitter flow rate, irrigation duration, emitter spacing, and soil type (see Section 16.4).

Normally, the emitters are located near the tree drip line or the areas of high-root concentration. Location of emitters close to tree trunks should be avoided. The emitters can be arranged in many possible ways to irrigate the desired percentage of the root zone. Emitters can be on single laterals with evenly spaced emitters, double laterals, laterals with loops, etc. When designing drip systems for beginning orchards, the demands as the orchard matures should be kept in mind. Tree loops can be added to laterals as the trees grow provided the laterals were designed for the extra capacity. Several of the possible emitter arrangements are shown in Fig. 16.3.

The uniformity of water application depends on the emitter hydraulic characteristics, lateral and submain design parameters including energy losses (or gains), filtration, and the manufacturing tolerances in the system components. The effects of manufacturing tolerances on uniformity can be determined if manufacturing statistics are available. The most useful statistical parameter is the coefficient of variation of manufactured emitters at a standard operating pressure. The coefficient of variation is the sample standard deviation divided by the sample mean.

The required emitter flow rate can be computed as follows:

$$q_r = \frac{q_t I_i}{I_t E_i N} \quad\dotfill\quad [16.36]$$

where q_r is the required emitter flow rate in L/h; q_t is the water requirement per plant, L/d, I_i is the irrigation interval, d, I_t is the irrigation time per set, h, E_i is the irrigation application efficiency and N is the number of emitters per plant. The irrigation rate per plant, q_r, is determined by the methods discussed in Section 16.5. The irrigation application efficiency is usually between 0.80 and 0.90 for most drip systems. Keller and Karmeli (1975) suggested that emission uniformity (EU) may be substituted for E_i which is difficult to determine for drip systems. The irrigation time per set is the operation time the lateral is applying water. The irrigation time, I_t, times the number of sets must be less than 24 times the irrigation interval in days. A daily schedule of 22 h allows downtime for system repair and maintenance. The number of sets which can be used daily will depend on the operator's schedule, degree of automation, water supply capacity, maintenance and repair time required.

16.6.2 Lateral Line Design

In lateral line design a first consideration is acceptable uniformity of emitter flow or emitter flow variation. If the manufacturer's variation is not considered, or assumed to be small, the design can be made to achieve a completely uniform emitter flow by using different emitter sizes (Myers and Bucks, 1972) or mibrotube length (Kenworthy, 1972). In general practice the emitter characteristics are usually fixed, and the emitter flow rate is determined by pressure at the emitter in the line. From the study of hydraulics it is certain there will be pressure variations along a drip irrigation lateral due to friction and slope, and therefore, emitter flow variations along a lateral line.

Drip irrigation lateral lines are the hydraulic link between the supply lines (main or submain lines) and the emitters. The emitters can be connected directly to the lateral line (inline or online), mounted on a riser from a buried lateral or attached to the lateral on a tree loop. The lateral line will also have the hydraulic fittings (trees, unions, etc.) to connect to the submain or mainline. In addition, a screen filter at the lateral inlet and some type of flushing valve (manual or automatic) at the distal end of the lateral is desirable.

The lateral pipes range in size from 13 to 19 mm (1/2 to 3/4 in.) nominal size pipe of polyethelene (PE) or polyvinyl-chloride (PVC) material. The material and size of drip irrigation lateral pipe is determined largely by economics. Because drip irrigation requires such a large amount of pipe per unit of land, the pipe cost must be small for the system to be economically feasible. PVC is available in rigid pipe or flexible hose. PE is available in low density (flexible) and high density (rigid) material (Shipston, 1976). Flexible PE pipe is utilized widely for drip irrigation lateral lines because of its lower initial cost compared to PVC pipe. Drip irrigation systems seldom require a pipe pressure rating in excess of 600 kPa. The pipe material selection is based largely on compatibility with pipe and emitter fittings. Generally, PVC requires glued (solvent weld) connections while PE utilizes a mechanical (barbed or compression) connection. Pipe material is also discussed in Section 11.4 and Chapter 15 in this book.

Water temperature and quality, ultra-violet radiation, mechanical stress, and various types of animals can damage plastic pipe and reduce its life expectancy. Additives to the pipe compounds can reduce these damages. Carbon black reduces ultra-violet radiation damage to PE pipe. Higher density PE pipe has greater tensile strength and heat resistance but lower flexibility and stress crack resistance than lower density PE pipe. Stress cracking results from thermal and chemical environments which cause the pipe to lose its flexibility and ultimately fail. Stress cracking can also result from the insertion of emitters into the pipe line when the fitting size is too large for the inside diameter of the pipe.

Plastic pipe used in drip irrigation lateral lines is usually classified as "hydraulically smooth" pipe. The Hazen-Williams roughness coefficient, C, will vary between 130 and 150. The Darcy-Weisbach friction factor, f, can be determined from the relationships for pipes with smooth boundaries. Sections 11.4, 14.1 and 16.3 discuss pipe flow hydraulics.

The design of a drip irrigation lateral involves selecting the pipe size for a given length of run which can deliver the required amount of water to the plants within a desired range of uniformity. Many times the problem reduces to one of determining the maximum lateral length for the given flow conditions and ground slope when the pipe size is limited to a specific size. The lateral must be compatible with the field size and number of sets of submain lines.

Drip irrigation lateral design can be classified into three types of design problems: (I) lateral lengths is unknown but pipe size is constrained; (II) pipe size is unknown but lateral length is constrained, and; (III) neither pipe size nor lateral length are constrained. Usually pipe size is limited to nominal pipe diameters less than 25 mm by economics. Occasionally, tapered laterals (laterals with more than one pipe size) are utilized for long lateral lengths. However, the extra costs associated with tapered laterals for fittings, larger pipe, etc., will usually be only marginally reduced compared to the cost of using smaller diameter pipe for shorter lateral lengths and using more submains. The design information required is ground slope, required emitter flow rate, number of emitters per tree, emitter flow function, tree spacing, and desired uniformity.

In problem type I, the objective is normally to determine the maximum lateral length which can be utilized for the given pipe size while maintaining a specific uniformity value. The maximum length of the lateral is computed for the given conditions using the required pipe size.

The allowable energy loss can be computed using Fig. 16.10 and the following relationships:

$$\frac{P_o}{P_n} = (q_o/q_n)^{1/x} \qquad\qquad\qquad\qquad\qquad\qquad\qquad\qquad\qquad\qquad \text{[16.37]}$$

$$P_o = (P_o/P_n)\,(\gamma)\,[\frac{q_r}{K_e}\,(2 - \frac{q_o}{q_n})]^{1/x} \qquad\qquad\qquad \text{[16.38]}$$

$$P_n = P_o\,(q_n/q_o)^{1/x} \qquad\qquad\qquad\qquad\qquad\qquad\qquad\qquad\qquad \text{[16.39]}$$

$$H_n = P_n/\gamma;\ H_o = P_o/\gamma \qquad\qquad\qquad\qquad\qquad\qquad\qquad\qquad \text{[16.40]}$$

$$\Delta H = H_n - H_o + S_o L \qquad\qquad\qquad\qquad\qquad\qquad\qquad\qquad \text{[16.41]}$$

$$L = 1.70\,\Delta H^{0.35}\,D^{1.71}\,(\frac{Cs}{100\,q_r})^{0.65} \qquad\qquad\qquad\qquad \text{[16.42]}$$

where P is pressure, kPa, q is emitter flow rate, L/h, subscript n is used to denote parameters at the head of the lateral, subscript o is used to denote parameters at the end of the lateral, γ is the specific weight of water, 9.81 kN/m³, H is pressure head, m, ΔH is the pipe friction energy loss, m, D is the pipe diameter, mm, s is the emitter spacing, m, C is Hazen-Williams roughness coefficient, S_o is the ground slope ("+" for down, "−" for up), and L is the lateral length, m. For level laterals equation [16.42] can be solved with an interative solution (successive approximations) or graphic solutions.

Example: What is the length for a 14.7 mm I.D. drip irrigation lateral? A tree loop will be placed at each tree spaced 10 m apart with 4 emitters on each loop. The required emitter flow rate is 4 L/h, emitter flow function $q = 0.63 (P/\gamma)^{0.8}$, ground slope of 1.0 percent. A C_u of 95 percent is desired.

Solution: Consider each tree loop as an emitter with an emitter flow function of $q = 2.52 (P/\gamma)^{0.8}$ and a required emitter flow rate of
$q_r = 16$ L/h.
From Fig. 16.10, $q_n/q_o = 1.2$ for $C_u = 95$ percent;

$$P_o/P_n = P_O/P_n = \left(\frac{q_o}{q_n}\right)^{1/x} = (1/1.2)^{1/0.8} = \underline{0.80}$$

$$P_o = P_O = (0.80)(9.81)\left[\frac{16.0}{2.52}(2.0 - 0.83)\right]^{1.25} = \underline{96\ kPa}$$

$$P_n = P_n = P_O(1.2)^{1/x} = (96)(1.2)^{1.25} = \underline{120\ kPa}$$

$$H_n = P_n/\gamma = H_n = P_n/\gamma = \frac{120}{9.81} = \underline{12.2\ m}$$

$$H_o = P_o/\gamma = H_O = P_O/\gamma = \frac{96}{9.81} = \underline{9.7\ m}$$

$$D = 14.7\ mm$$
$$C = 130$$
$$s = 10\ m$$
$$q = 16\ L/h$$
$$\Delta H = H_n - H_o + S_o L$$
$$\Delta H = 12.2 - 9.7 + (0.01)(L)$$
$$\Delta H = 2.5 + (0.01)L$$

$$L = 1.70[2.5 + (0.01)L]^{0.35}\left[\frac{(130)(10)}{(100)(16)}\right]^{0.65}(14.7)^{1.71}$$

Trial No. 1—Let L = 220 m
 No. 2—   L = 252 m
 No. 3—   L = 259 m
 No. 4— $\underline{\underline{L = 260\ m}}$

An alternative solution method was presented by Wu and Gitlin (1974a) and Wu (1977) in a nomograph form. Figs. 16.17 and 16.18 show the general lateral design nomograph for down and upslope, respectively. The ratio L/H is determined, and the lateral uniformity can be determined if the slope and total lateral flow rate are known. For the above example, using L = 260 m, $\Delta H = 2.5$ m, $S_o = 1.0$ percent (down), H = 11.2 m and Fig. 10.17, the design is well within the desirable C_u range.

In type I design problems, the lateral length which was calculated is used to determine the set length to irrigate the field in the specified time in-

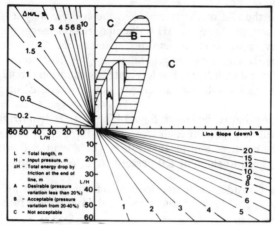

FIG. 16.17 Dimensionless general design chart (downslope) (after Wu, 1977).

tervals. The set size can be determined from the limiting factors of either flow rate or submain length.

The solution to type II problems can be determined by the use of the Wu (1977) general design monograph or by solving equation [16.42] for the pipe diameter, D(mm) as follows:

$$D = 0.73 \, (\Delta H)^{-0.20} \, (L)^{0.58} \, (\frac{100 \, q_r}{C_s})^{0.38} \quad \dots\dots\dots\dots\dots [16.43]$$

The design pipe size must be greater than the computed diameter from equation [16.43] since only discrete pipe sizes are manufactured. In using the design monographs (Figs. 16.17 and 16.18),

$$\Delta H/L = 22(D)^{-4.871} \, (\frac{100 \, q_r}{C_s})^{1.852} \quad \dots\dots\dots\dots\dots\dots [16.44]$$

Type III problems can be solved by using one of the above procedures. The lateral length determined must be suitable to the field dimensions.

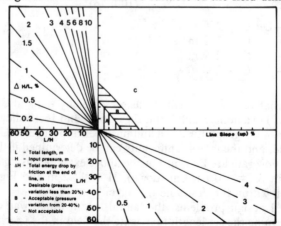

FIG. 16.18 Dimensionless general design chart (upslope) (after Wu, 1977).

Example: An orchard with trees spaced 10 m apart that require four emitters per tree and an emitter flow rate of 4.0 L/h is to be drip irrigated. The emitter flow function is $q = 0.63(P/^g)^{0.8}$. The slope in the lateral direction is 2 percent and the rows are 400 m in length. If only one submain line is used, what would each lateral length and pipe diameter be to maintain the uniformity above 95 percent?

Solution: As in the previous problem, assume each tree loop acts as an emitter with $q_r = 16$ L/h and an emitter flow function of $q = 2.52 (P/\gamma)^{0.8}$. From Fig. 16.10, $q_n/q_o = 1.2$ and from the previous example problem, $P_o = 95$ kPa, $H_o = 9.7$ m, $P_n = 120$ kPa, and $H_n = 12.2$ m. For the downslope side, $\Delta H = 2.5 + (0.02)L$ and for the upslope side, $\Delta H = 2.5 - (0.02)L$. The maximum downslope length for 14.7 mm I.D. pipe is

$$L = (1.70)\,[2.5 + (0.02)L]^{0.35}\,[\frac{(130)(10)}{(100)(16)}]^{0.65}\,(14.7)^{1.71}$$

Trial No. 1—Let L = 200 m
No. 2—   L = 283 m
No. 3—   L = 307 m
No. 4—   L = 313 m
No. 5—   L = 314 m
No. 6—   L = 314 m

Let the downhill lateral length be 310 m and pipe diameter 14.7 mm I.D. The remaining uphill lateral must be 90 m to satisfy the field dimensions. The uphill pipe diameter is

$$D = 0.73\,[2.5 - (0.02)L]^{-0.20}\,[\frac{(100)(16)}{(130)(10)}]^{0.38}\,(L)^{0.58}$$

For L = 90 m, D is 11.6 mm; therefore, a 14.7 mm I.D. pipe is adequate for the uphill lateral. This design can be checked by the Wu (1977) nomographs which show that this design results in an acceptable uniformity coefficient.

16.6.3 Submain Design

The submain line hyraulics are similar to lateral hydraulics as discussed in Section 16.3. Normally, the submain line is designed to allow approximately the same energy loss as compared to the lateral line for level laterals and submain lines. Elevation differences will change this energy loss distribution. Keller and Karmeli (1975) recommended that the lateral energy loss should be 55 percent and the submain energy loss should be 45 percent of the total allowable energy loss.

The submain design depends on the location of flow or pressure regulation. Normally, pressure or flow regulation is located at the inlet of the submain. In this case, energy loss in the submain is directly related to the length of the submain line. The energy loss cannot exceed the allowable limits without lowering uniformity. On particularly steep slopes, each lateral may require individual pressure or flow regulation. In this case the length and diameter of the submain line are determined solely by balancing the energy cost and the pipe cost. Since each lateral in this case is regulated, uniformity is independent of submain energy loss provided that submain losses do not interfere with the flow regulation.

The pipe material normally used for submain and main lines is rigid polyethylene (PE) or polyvinyl-chloride (PVC). These materials prevent emitter clogging from corrosion and scaling in the pipes. Submain and main lines should be provided with either manual or automatic flushing valves. Each lateral connection at the submain should have a secondary filter screen to prevent foreign material from entering the lateral and clogging the emitters. Vacuum control valves should be installed on submain units to prevent soil particles from being sucked back into the emitters when the lines are turned off.

The position of the inlet to the submain line depends on the field slope. Usually laterals are placed on contours if possible with submains running with the prevailing field slope. With sloping submain lines, the inlet is positioned so that the uphill run is shorter than the downhill run. On gently sloping land or level areas the submain inlet should be located near the center of the submain lines.

The submain hydraulic characteristics can be computed by assuming the laterals are analogous to emitters on lateral lines. The hydraulic characteristics of submain and main line pipe are usually taken as hydraulically smooth since PVC and PE pipe are normally used. The Hazen-Williams roughness coefficient is usually between 140 and 150. The energy loss in the submain can be computed with methods similar to those used for the lateral computations. The energy loss at the lateral connection will depend on the type of connection utilized, i.e., tee, elbow, etc. The total submain energy loss should include energy losses through filters, pressure valves, and other minor losses.

Example: Design the submain line for a system with 15 lateral connections spaced 10 m apart, a slope of 5 perc in the submain direction, a lateral flow rate of 500 L/h, lateral inlet pressure is 140 kPa, lateral connections are pipe tees and a screen filter which has a maximum energy loss of 4.0 m will be located at the submain inlet.

Solution: The submain inlet flow rate will be 7500 L/h. The submain length is 140 m. The design must determine the optimum inlet location and pressure or flow regulation position as well as pipe size.

The submain energy loss should approximately equal the lateral friction loss or about 10 to 15 percent of the operating pressure. For this example, assume that the allowable submain friction loss will be 15 percent of 140 kPa or 21 kPa (2.1 m). The downhill submain length by equation [16.42] is

$$L = 1.70[2.1 + (0.05)L]^{0.35} D^{1.71} [\frac{Cs}{100\ q_r}]^{0.65}$$

D(mm)	L(m)
21	52
27	90
35	164

These calculations indicate that a 35-mm (1¼ in.) pipe could be used with the entire submain laid downslope. However, a 27-mm (1-in.) pipe could be used for 90 m of the submain. If the 27 mm size pipe is used, the pipe size for the 30 m of the upslope submain can be estimated by equation [16.43] as equation,

$$D = 0.73(30)^{0.58} [2.1 - (0.05)(30)]^{-0.20} [\frac{(100)(500)}{(149)(10)}]^{0.38} = 22.6\ mm$$

Therefore, the 27-mm (1-in.) pipe can be used for the entire submain with the inlet 30 m from the "high" end. The submain inlet pressure will be 200 kPa (140 kPa lateral, 21 kPa submain friction and 39 kPa filter loss) plus any additional friction across valves or pressure regulating valves. If the submain slope was much larger, individual lateral pressure or flow control would be necessary or the submain length must be reduced.

16.6.4 Mainline Design

Normally, flow or pressure control or adjustment is provided at the submain inlet. Therefore, energy losses in the mainline should not affect system uniformity and mainline pipe size is based on economic comparisons of power costs and pipe costs. The mainline pipe size should be selected to minimize the sum of power costs and capital costs over the lifetime of the pipeline (Wu, 1975).

16.6.5 Water-Supply Manifold

The water-supply manifold consists of the pump, valves, chemical injectors, pressure regulators, filters, water meters, vacuum breakers, automatic controllers, etc. These devices are normally located near the water source. These items are required either individually or collectively for the successful operation of drip irrigation systems. The collection of these devices is referred to as the "control head".

The pump operating range is selected based on the number of operating subunits and their flow rate and the total system head. The system flow rate is the sum of the flow rates to each concurrently operating subunit. The system head is the sum of lateral inlet pressure head, elevation difference from the pump to the highest lateral inlet, and the energy losses between the pump and lateral to include submain friction loss, filter loss, valve loss, and mainline loss. The filter or screen loss will represent energy losses resulting from partially clogged filters and screens. Usually this energy loss will have a fixed range between 30 and 100 kPa depending on the type of filter.

Filters are used to remove solid debris and suspended solids from the water. Filtration is discussed in detail in Section 16.8. Filters require periodic cleaning or back-flushing to successfully operate. The filter selection depends on input water quality and the system flow rate. Automatic back flushing filters and continuously cleaning screens are available. The energy loss across all filters or screens in the system is required to properly select pumps.

The chemical injection system will normally consist of valves, tanks, pumps and/or Venturi injectors. The common modes of chemical injection are by differential pressure and pump injection. The method and rate of injection, chemical concentration, tank capacity, and safety must be considered in the design. Chemical injection is discussed in Section 16.9.

System flow control is developed by strategically located flow and pressure monitoring points and manual and/or automatic valve control. Flow rate or quantity should be monitored at the control head with a water meter. Pressure monitoring stations should be provided at the control head, submain manifold lateral line points, and on both side of filters or screens. Whenever possible, system control should rely on volumetric monitoring since flow rate control by pressure adjustment is usually inaccurate over long time periods. System control by time monitoring is another widely used method of control.

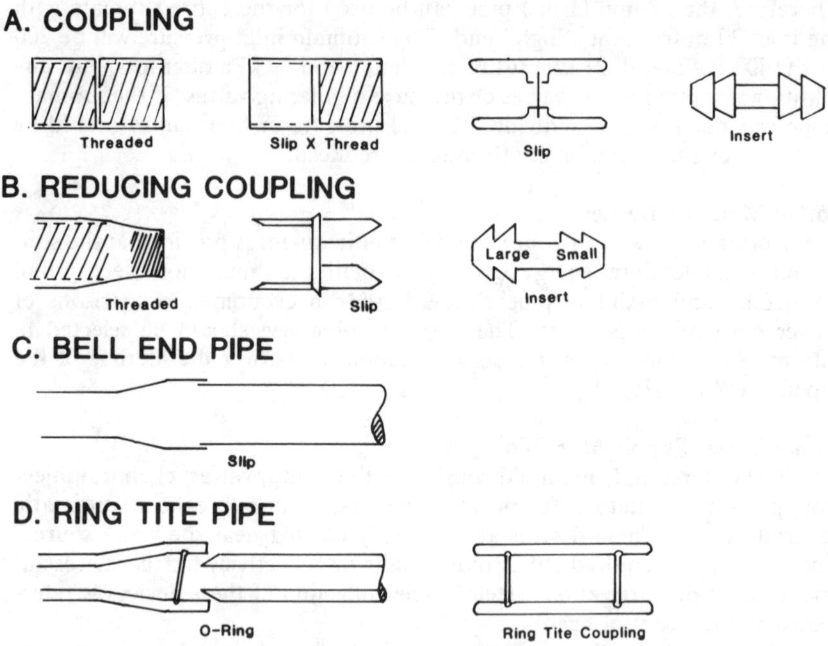

A. COUPLING

Threaded   Slip X Thread   Slip   Insert

B. REDUCING COUPLING

Threaded   Slip   Insert

Large   Small

C. BELL END PIPE

Slip

D. RING TITE PIPE

O-Ring   Ring Tite Coupling

FIG. 16.19 Examples of in-line connections.

For either method of system control (volume or pressure) a series of valves, either automatic or manual, will be required. Automatic valves can be electrically or hydraulically operated. Both types of valves can be controlled from an electronic controller to open and close submains, back-flush filters, flush submains or laterals, to inject chemicals, etc. The control can be activated by time clocks or volumetric accumulators.

16.7 DRIP IRRIGATION INSTALLATION

16.7.1 Commonly Used Parts

Following the design of a drip irrigation system, the necessary parts and connectors for laterals, submains, and main lines must be determined. The parts include joints, connectors, and fittings. Some commonly used parts for a drip irrigation system are listed as follows:

In-Line connection. In-line connection is usually done by using a coupling for polyvinyl chloride (PVC) pipe. If two different size pipes are connected, a reducing coupler is used. When a change of pipe direction (90 deg) is needed, an elbow is used.

Connecting polyethylene (PE) is usually done by insert fittings. There are also reducing insert fittings and elbow (90 deg) insert fittings. Compression PE fittings are available that fit over the outside of the pipe. The compression fittings should have less friction loss and stress cracking problems.

A coupling is designed to have "female" ends which can be either thread or slip. Several typical couplings and insert fittings are shown in Fig. 16.19.

There are two specially designed pipe (PVC) end connections; the bell end connection and Ring Tite*, which are also shown in Fig. 16.19.

*Trade names are mentioned for the convenience of the reader and do not imply endorsement by the U.S. Department of Agriculture.

Branch connection. Two branch connections usually used in a drip irrigation system are a "tee" and a "cross". The tee and cross for PVC pipe are designed having "female" ends which can be thread or slip; they are also designed for connecting different sizes, such as reducing tee or cross. The tee and cross for PE pipe usually have insert ends.

Special fittings. Certain fittings or connectors will be used for connecting valves, gauges, filters, pressure regulators, unions, and end caps or plugs. Parts used for these purposes are as follows:

1 Nipple—A section of pipe with both ends male threaded.

2 Adapter—A section used for changing from one end condition to another, such as from thread to slip, slip to thread, or thread to insert. There are male and female adapters and also insert adapters.

3 Bushing—A fitting used to reduce the size of an end outlet which can be thread by slip, thread by thread, or slip by slip.

The above fittings are usually used in the following installations:

1 Gate valve—a gate valve is designed having both female threaded ends. A nipple or a male adapter is usually connected. PVC gate valves are available, on special order, with slip female ends.

2 Filter—A filter is usually designed having both female threaded ends. A nipple or a male adapter is connected.

3 Pressure gauge—A pressure gauge usually has a 6 mm (¼ in.) male threaded end. A tee connector and a bushing are generally used.

4 Pressure regulator—Pressure regulators are made with different end conditions from different manufacturers. Adapters or nipples are generally used.

5 End plug or cap—End plugs or caps are usually threaded, slip, or insert ends. They are usually designed to match with the end conditions of the pipe. Otherwise an adapter can be connected.

6 Union—a union is designed having either slip or thread female ends. It is installed in-line, connected simply by a pipe, a nipple, or a male adapter.

7 Vacuum breaker—A vacuum breaker is usually designed having a male thread or insert end which is connected to a tee connector and sometimes with a bushing.

8 Drain valve—A drain valve is usually designed having an insert or male thread end. It can be connected directly to a pipe end or connected by an adapter.

Conventional expressions for end fittings. Steel or galvanized pipe is normally threaded. PVC pipe usually cannot be threaded because it is too thin (except schedule 80). The smooth end of a PVC pipe is glued into the smooth socket of some fittings, like a coupling, tee, etc. These smooth sockets on the coupling or tee are designated as "slip" or "s". If the fitting should have threads (it is then a schedule 80 fitting) the threaded end is designated "t". It is conventional to refer to the three ends of a tee in a precise order: the two straight-through ends are described first, and then the side.

16.7.2 Typical Drip Irrigation System Installation Arrangements

A typical drip irrigation system consists of the following items:

1 mainline
2 submain
3 lateral line
4 others such as valves, pressure regulator, filter, pressure guage, etc.

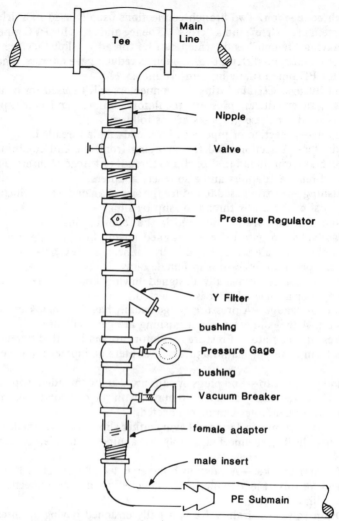

FIG. 16.20 Typical main to submain connection for a drip irrigation system.

When a drip irrigation system is designed, the field installation requires the knowledge of the use of different joints and fittings and the understanding of the types of connections for main, submain, laterals and other special items.

There are, in general, three areas of installation in a drip irrigation system: the main-submain connection and submain manifold, the submain-lateral connection and the lateral end arrangements. They are presented as follows:

Main-submain connection and submain manifold. The main-submain connection is usually a "tee" either thread or slip (for PVC). If the mainline is a PE pipe an insert tee is usually used.

A submain is used to deliver water into laterals and also used as a controller so the field can be irrigated separately under a desired water pressure at any selected time. All the specific items used for control are installed on the submain, which is called a submain manifold. The main-submain connection and a submain manifold are shown in Fig. 16.20.

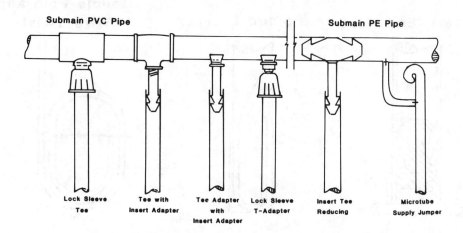

FIG. 16.21 Typical submain to lateral connections.

Submain-lateral connection. A submain can be either PVC pipe or PE tube. Several typical connections for submain to laterals are shown in Fig. 16.21. There are Lock Sleeve tee, tee with insert adapter, tee adapter with insert adapter and Lock Sleeve* tee with adapter insert reducing tee (PE) and microtube supply jumper.

Lateral line end arrangements. The ends of lateral lines can be plugged using an insert plug or simple folding and wrapping. The ends of lateral lines can also be equipped with a drain and flushing valve or flushing valve. Several lateral line end arrangements are shown in Fig. 16.22.

16.7.3 Installation Procedures

The installation of drip systems is quite similar to sprinkler systems. Chapters XI and XVI in Pair et al. (1975) provide adequate instructions. However, extreme precaution must be used to prevent sand, dirt and PVC filings from entering pipelines during installation. All pipelines should be thoroughly flushed before the lateral lines or emitters are installed.

Above-ground laterals should not be rigidly attached or stretched. Usually 3 to 6 percent of extra pipe is included with each lateral to allow "snaking" of the pipe with thermal expansion.

16.8 MAINTENANCE, FILTRATION AND FLUSHING

Plugging of emitters caused by physical, chemical or biological contaminants is universal and is considered to be the largest maintenance problem with the drip irrigation systems. The design of the emitters has been conditioned by recognition of the axiom that the smaller the emitter opening, the more closely the flow rates can be matched with the soil infiltration rates, but results in greater plugging probability. Most emitters have been designed with some degree of compromise between the two divergent features (Solomon, 1977).

A method of evaluating water quality in terms of emitter clogging has not been developed. Many water quality parameters are dynamic and can not be predicted from selected measurements. Bucks et al. (1979) and Bucks and Nakayama (1980) presented a classification system for water analysis to indicate clogging potential and required preventative maintenance that is

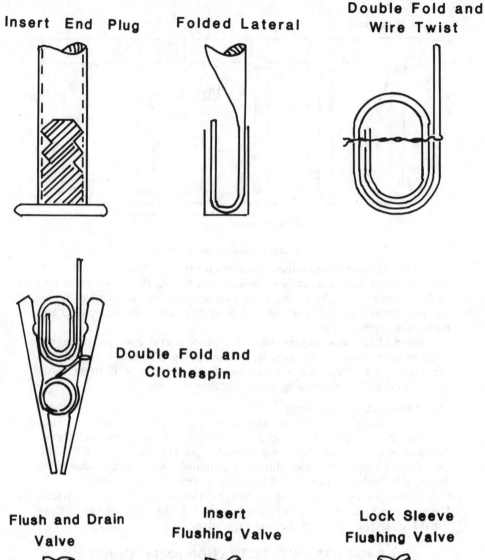

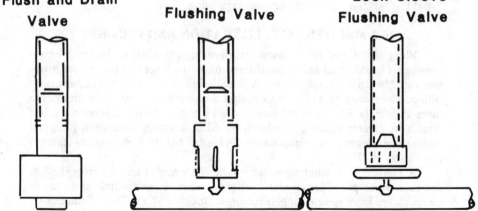

FIG. 16.22 Lateral line end fittings.

TABLE 16.7. TENTATIVE CRITERIA FOR CLASSIFYING
POTENTIAL CLOGGING HAZARD OF IRRIGATION
WATERS TO BE USED IN TRICKLE SYSTEMS
(after Bucks and Nakayama, 1980)

Factor	Clogging hazard		
	Slight	Moderate	Severe
Physical			
Suspended solids (Max. ppm)*	< 50	50-100	> 100
Chemical			
pH	< 7.0	7.0-8.0	> 8.0
Dissolved solids (Max. ppm)*	< 500	500-2000	≥ 2,000
Manganese (Max. ppm)*	< 0.1	0.1-1.5	> 1.5
Iron (Max. ppm)*	< 0.1	0.1-1.5	> 1.5
Hydrogen sulfide (Max. ppm)*	< 0.5	0.5-2.0	> 2.0
Biological			
Bacteria populations (Max. no./mL)†	< 10,000	10,000-50,000	>50,000

*Maximum measured concentration from a representative number
of water samples using standard procedures for analysis.
†Maximum number of bacteria per milliliter can be obtained from
portable field samplers and laboratory analysis.

shown in Table 16.7. The physical, chemical and biological factors are rated
for clogging hazard.

16.8.1 Mechanical Filtration

Mechanical filtration including settling basins, screens, centrifugal sand
separators and cartridge and/or sand filters are used to reduce suspended
particulate matter. These devices are used singularly or in series. Filtration
units may require the addition of booster pumps for proper backwash and
flush operations. Table 16.8 presents filtration media size opening equivalent
to selected particles (Wilson, 1977).

TABLE 16.8. FILTER MEDIA OPENING SIZE EQUIVALENTS
(after Wilson, 1977)

Standard soil particle categories	Millimeters	Inches	Micrometers	Screen mesh*	Sand†
Very coarse sand	1.00 -2.00	0.0393 -0.0786	1000-2000	18-10	
Coarse sand	0.50 -1.00	0.0197 -0.0393	500-1000	35-18	Approx.
Medium sand	0.25 -0.50	0.0098 -0.0197	250- 500	60-35	1/10
Fine sand	0.10 -0.25	0.0039 -0.0098	100- 250	160-60	the
Very fine sand	0.05 -0.10	0.0020 -0.0039	50- 100	270-160	particle
Silt	0.002-0.05	0.00008-0.0020	2- 50	400±270	size
Clay	less than 0.002	Less than 0.00008	Less than 2	—	

*Using market grade wire cloth.
†Assuming round sand; sharp sand approximately 1/12 the particle size.

Settling basins. Settling basins can remove suspended material in water ranging from sand (2000 micrometers) to silt (2 micrometers). The time interval for settling is an important consideration and has a bearing on the size of the basin relative to the quantity of water that must pass through. The finer the sediments the longer the water must rest in the basin. In some cases, two basins side-by-side may be expedient so that withdrawal of water can be alternated. It is impractical to remove particles less than 2 micrometers unless they are coagulated because of the lengthy settling time required. Chemical treatment with alum and/or polyelectrolites may be necessary to flocculate both silt and clay particles. Settling basins must be constructed so that they can be cleaned of sediments at intervals, either mechanically or by water flushing. Once-a-year would be a practical cleaning interval, but depending upon the amount of sediment several times per year may be necessary. In aquatic plant growth, particularly algae, is a problem chemical treatments with chlorine or copper sulfate may be required occasionally.

Sand separators. The operation of a sand separator is based on the principle of centrifugal force or vortex motion, and they are often called cyclonic separators. Dense sediments such as mineral solids are separated from the bulk water, they accumulate and are swept out with a small portion of the water. Several types of these devices are available.

Screens. Screens are probably the most versatile of all components for removing foreign matters from water. They can be constructed in almost any shape, size, and mesh except that there is a lower limit to the opening size. Numerous types of screens are available from giant rotating screens cleaned automatically with jets of water and/or brushes with a small amount of water spillage, to simple screens which are placed in open channels of water and are cleaned by hand at suitable intervals. Simple cylindrical screens set in steel cylinders constructed to safety standards to withstand the pressure are most common. Cleaning of such filters may be automatic or manual. One such system is the double filter in which one filter is operating while the other is being swept clean by rotating jets of water, the wastewater being spilled to the outside as in back flushing. The flow of water is switched from one cylinder to the other automatically by a pressure switch triggered by a pressure build-up caused by the partially clogged screen.

Sand filters. Sand filters encompass a wide range of both filtering and flow capacities. The flow capacity depends largely on the physical dimensions of the filter while the filtering capability depends upon the size of the sand grains, depth of filter bed and flow of water per cubic meter of filtering material.

The filtering material is commonly crushed granite or silica, graded into sizes suitable for each set of specifications for filter requirement and flow capacity of each particular system. The particle sizes range from 200 or 300 to 2000 micrometers or more. Whatever size is used the grains must be large enough that they themselves will not become suspended sediments and pass out of the filter.

Sand filters are cleaned by reversing the flow of water which normally enters at the top and exists at the bottom of the filter in an operation called back flushing in which the water moves upward through the sand, lifting it slightly, and exits at the top carrying accumulated particulates with it. Sand filters are generally inexpensive and easy to operate. Although quite effective in removing most of the suspended material, they will not remove very fine substances or bacteria. Back flushing can be manual or aut, tic. If plug- ging of emitters is still a problem because of the finer materials, further purification might be required on the downstream side of the sand filter.

Cartridge filters. Cartridge fiber filters serve much the same purpose as sand filters, but can be made to filter out finer sediments than the sand. They are made from a variety of materials from paper fibers to fiberglass. Because they may have finer porosity than the sand filters, they must be larger in size or there must be more of them than sand filters for a given rate of water flow and quantity of suspended solids.

Cartridge filters tend to be more useful where the concentration of suspended solids is low because at high concentrations the cartridge must be frequently cleaned and because complete blockage within the filter might oc- cur. Cartridge filters are often the final filtering component before the water enters the system.

16.8.2 Field Inspection

The drip system must be inspected to detect clogged emitters and pipeline leaks or breaks. Filters must be routinely checked (weekly or more often). Water meters can indicate when emitter flow is reducing due to clog- ging. Flow changes as low as 10 percent should be investigated.

16.8.3 Flushing

Flushing individual laterals has been successful in some systems, but flushing tends to be an attempt to cure rather than prevent plugging and if not carried out with careful timing will not always be successful. The lateral itself is a sort of filter since it has much smaller effluent openings than in- fluent openings; and with a gradual reduction in the flow velocity towards the end of the line, particulate precipitation will increase. In addition, there can be mechanical agglomeration of particulates in the lateral and/or growth of micro-organisms, particularly algae, that increase plugging in the lateral. Removing these particles drawn into the lateral will greatly reduce the proba- bility of plugging and lengthen the life of the lateral for irrigation.

Flushing can be done manually, lateral by lateral, by opening the end of the line. Flushing valves are available that can be operated using the irriga- tion water pressure. Any number of laterals can be flushed from one flushing control valve, depending on the system capacity. Flushing velocity should ex- ceed 0.3 m/s (1.0 ft/s) (Shearer, 1977).

Frequency and timing of flushing should be established in each in- dividual case; these practices will depend on water quality, irrigation fre- quency and amount, as well as the limitations in pressure or flow rates of the system. Reasonable flushing frequencies can vary from more than once per irrigation to once a month.

16.8.4 Chemical Treatment

Most surface water supplies for irrigation contain a wide variety of par- ticulate matter ranging from very coarse floating organic materials to very fine mineral sediments. Filtration alone is usually not sufficient to prevent emitter clogging (Nakayama et al., 1978). Such waters will also contain living algae, bacteria, and probably a host of flora and fauna that frequent open ponds, reservoirs, and lakes. All of these materials, some of which may con- tinue to grow in the slow moving water of lateral lines, particularly if nutrients are present, are potential hazards to a system since in the right amounts and conditions they will plug almost any type of emitter. Subsurface waters, namely deep well waters, are normally free of most of these particulates except the troublesome bacteria, predominantly the iron and sulfur bacteria.

Acids. Precipitation of calcium or magnesium carbonate can occur in drip lines and emitters if the water pH is over 7.5. Chemical precipitation can be reduced by adding acids, usually sulfuric or hyperchloric acid, to lower the pH. Phosphoric acid can also be used as water treatment and fertilizer. A reduced pH also aids in bacterial control with chlorination.

Algaecides. Algaecides are commonly used to prevent the growth of algae in water. The most common algaecides are copper sulfate, sulfuric acid and chlorine. Since algae is a photosynthetic plant and must have sunlight, it will not grow in the pipelines and algaecide treatment in the closed system will be of no value unless it disintegrates the algae plant. The treatment may kill the algae that has passed the filtering system, but the algae is still a potential plugging hazard.

Bacteriacides. Bacterial slimes which form within closed pipelines as well as in open water are a hazard in themselves and as a potential coagulant of fine particles. Both algae and bacteria require adequate nutrition and any water that is impoverished of nutrients will exhibit much less of a growth problem than waters carrying loads of nutrient elements.

Chlorine. Chlorine is the most effective and inexpensive treatment for bacterial slimes. The chlorine can be introduced at low concentrations, e.g. 1 ppm, or as slug treatments (Nakayama et al., 1977 and Davis et al., 1977) at intervals as necessary at much higher concentrations , e.g. 10 to 20 ppm for only a few minutes at a time. Most specialists favor slug treatments over continuous treatments. The chlorine is introduced into the system upstream from the filters and can be either sodium hyperchlorite (the ingredient in common bleach) or as chlorine gas. The gas treatment is usually more expensive, more difficult to handle and potentially more hazardous to the operator than sodium or calcium hyperchlorite at similar concentrations. The use of chlorine gas must conform to regulations regarding the safe handling of liquid and gaseous chlorine under high pressure. Calcium hyperchlorite (a solid) can also be used but the calcium tends to produce precipitates that may cause plugging without pH adjustment.

Oxidants. Sodium hyperchlorite acts as an oxidant for precipitation of troublesome iron in the water (Ford and Tucker, 1975; and Ford, 1976). It can be used deliberately to precipitate iron allowing it to settle out before the water is introduced into the system. Chlorine also destroys the iron bacteria in the lines preventing the formation of iron precipitates and slimes within the system. Other oxidants include bromine, iodine, bromine chloride, hydrogen-, calcium-, and sodium-peroxides. Most of these are relatively more expensive than the chlorine compounds.

Flocculants. Flocculants are used largely in conjunction with settling basins. The flocculation of fine materials into large aggregates permits many of them to settle out. With some materials the floccules may be less dense than the individual particles in which cases the floccules are screened or filtered off.

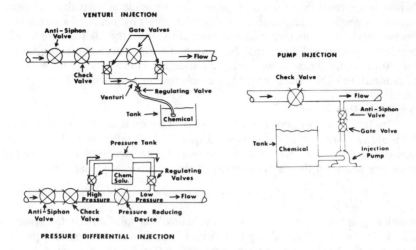

FIG. 16.23 Typical chemical injection methods.

16.9 FERTILIZING THROUGH DRIP SYSTEMS

The application of plant nutrients through the drip irrigation system is convenient and efficient. The application of fertilizers through the irrigation water reduces labor, energy, and equipment cost as compared to the conventional soil surface spreading methods. Furthermore, the required fertilizers can be added in small quantities during the growing season without upsetting or damaging crop production. Less fertilizer can be used because it can be added when or where it is needed. The rates of injection for various crops, soils and climatic conditions need to be established. The use of material that would cause precipitation and clogging of the system should be avoided. In order for the application method to be efficient, the chemicals applied must be distributed uniformly throughout the field. The uniformity of chemical distribution is dependent upon the efficiency of mixing, the uniformity of water application and the flow characteristics of the water and the chemicals within the soil.

16.9.1 Injection methods

Three principal methods used in the injection of fertilizers and chemicals into drip systems are pressure differential, the Venturi (vacuum) and metering pumps as shown in Fig. 16.23. In the pressure differential system, the tank is under pressure (usually mainline pressure). A pressure difference is created by a valve, pressure regulator, etc., between the tank inlet and outlet. The difference in pressure between the connection and the constriction in the flow pipe is sufficient to cause a flow of water through the tank is under pressure, a sealed airtight pressure supply tank constructed to withstand the maximum operating pressure is required. Precise control valves are necessary to maintain a preset injection rate. The Venturi causes a rapid change in velocity. This velocity change creates a reduced pressure (vacuum) which forces fertilizer material into the line. The third method consists of using a rotary, gear or piston pump to inject the fertilizer solution from the supply tank into the pressure line. The pump must develop a pressure greater than that in the pipe line. Since most fertilizers and other chemicals are corrosive, the pump parts must be made of corrosive resistant material. Injection of fertilizers and other chemicals by means of a pump is a precise method of metering these chemicals into the drip system. All chemical injection systems must be equipped with vacuum breaking (anti-siphon) valves upstream from the injection point. This will prevent chemical contamination of the water supply in case of water or power failures.

16.9.2 Injection Rates

The fertilizer injection rate into the system depends on the concentration of the liquid fertilizer and the desired quantity of nutrients to be applied duriing the irrigation. Keller and Karmeli (1975) suggested the following procedure to use in the determination of the fertilizer requirement.

$$q_{fi} = \frac{F_r A}{c \, t_r \, I_t} \quad \dots\dots\dots\dots\dots\dots\dots\dots\dots\dots\dots\dots\dots\dots\dots \quad [16.45]$$

in which q_{fi} is the rate of injection of liquid fertilizer solution into the system, L/h, F_r is the rate of fertilizing (quantity of nutrients to be applied) per irrigation cycle, kg/ha, A is the area irrigated (ha) in the time, I_t, c is the concentration of nutrients in the liquid fertilizer, kg/L, t_r is the ratio between fertilization time and irrigation time, usually taken as 0.8 to allow time to flush the system and I_t is the duration of irrigation, h.

Fertilizer concentration. The concentration of the fertilizer in the irrigation water can range from 4 to 10 ppm. The actual concentration needed depends on the fertilizing material and crop requirement. Therefore, periodic analysis of soils and plant materials should be made, and the rate of fertilization of quantity of nutrients needed per unit area determined.

The fertilizer concentration in the irrigation water which will give F_r is

$$F_c = \frac{100 \, F_r}{t_r \, I_d} \quad \dots\dots\dots\dots\dots\dots\dots\dots\dots\dots\dots\dots\dots\dots \quad [16.46]$$

in which F_c is the fertilizer concentration in the irrigation water, ppm, and I_d is the gross depth of irrigation, mm (in.).

Fertilizer tank capacity. Low cost tanks are practical where an injection pump or a Venturi is used. A large tank provides a good place to store fertilizer for periods of short supply and reduce the labor associated with frequent filling. In such cases an automatic shut-off valve is useful for controlling the timing of injection.

For a pressure differential injection or a Venturi system the fertilizer tank should have enough capacity for a complete irrigation application. This requires a tank capacity C_t of

$$C_t = \frac{F_r A}{c} \quad \dots\dots\dots\dots\dots\dots\dots\dots\dots\dots\dots\dots\dots\dots\dots \quad [16.47]$$

where C_t equals tank capacity in L.

16.9.3 Nitrogen

Nitrogen is the plant nutrient most often applied in drip irrigation systems. Usually nitrogen injection will not significantly increase clogging problems.

Nitrogen sources. Nitrogen is usually applied through the system as anhydrous ammonia, aqua ammonia, ammonium phosphate, urea, ammonium nitrate, calcium nitrate or several other mixtures (Rolston et al., 1979). Careful consideration must be made of the pH in the irrigation water since some nitrogen sources, particularly aqua ammonia and anhydrous ammonia, will increase pH. The increased pH can result in precipitation of insoluble calcium and magnesium carbonates that can clog a drip system. Calcium nitrate is relatively soluble and does not cause a large pH shift. Most ammonium salts are fairly soluble and usually present no major clogging potential. Urea and urea-ammonium nitrate mixtures are highly soluble and usually do not cause large pH shifts.

The nitrogen source readily reacts with the soil depending on several factors. Ammonia volatilization, nitrate leaching and denitrification can occur. Rolston et al. (1979) discuss these interactions in detail and their implications on nitrogen availability to plants.

16.9.4 Phosphorus

Phosphorus can be applied with drip irrigation (Rauschkolb et al., 1976) although clogging may result from precipitation if care is not taken. Phosphoric acid is soluble and with low pH water usually has no clogging problems. Sulfuric acid injection together with phosphoric acid may be sufficient to prevent precipitation of calcium and magnesium especially as the phosphoric acid boundary passes. Inorganic phosphate, orthophosphate and glycerophosphate have been used (Rauschkolb et al., 1976).

16.9.5 Potassium

Potassium can be applied as potassium sulfate, potassium chloride and potassium nitrate. These potassium sources are soluble and have few precipitation problems. Uriu et al. (1977) demonstrated potassium deficiency correction by applying potassium to prune trees by a drip system.

16.9.6 Micronutrients

Iron, copper, zinc and manganese may react with the salts in the irrigation water and result in precipitation. However, the more soluble chelated forms such as iron or zinc EDTA (ethyl-enediamine tetraacetate didhydrate) usually cause little clogging problem.

16.9.7 Plant Requirements

The plant requirements for nutrients under drip irrigation are not widely defined for many crops. The obvious difference between drip application and conventional application is the application efficiency and the ability to apply small quantities on a frequent time scale. Various research reports indicate that fertilizer savings as large as 25 to 50 percent may be possible for some crops (Goode et al., 1977; Black, 1976; Miller et al., 1976; Coston et al., 1978; Phene and Beale, 1976; etc.).

16.10 WEED CONTROL IN DRIP IRRIGATED CROPS

The concentration of the weed problem in the wet zone of the emitter is good and bad. The total weed growth in an orchard or vineyard is less as a result of the weed concentration in the wet zone, however, weed control is more difficult in the somewhat cylindrical pattern around the emitter (Lange et al., 1978). If preemergence herbicides are used during the winter prior to a rainy season, good weed control will be obtained down the tree or vine row for several months. At this time weed seedings begin to germinate in the wet area around the emitter. When these weeds are small they are easy to kill with contact herbicides such as paraquat or dinoseb plus oil (Lange et al., 1976). Both materials are registered for most tree and vine crops.

In most orchards and vineyards using drip irrigation, periodic treatments around the emitter may be required every 3 or 4 wk during the summer or more often if perennials are present. Weeds such as Bermudagrass, Johnsongrass and nutsedge are particularly difficult to control since they grow more luxuriantly in the wet zone than usual. They have also been more difficult to control in the wet zone even where glyphosphate (Roundup®) has been used. Perennial weeds such as perennial bindweed (*Convolvulus arvensis* L.), Johnsongrass (*Sorghum halepense*), Bermudagrass (*Cynodon dactylon*) and nutsedge (*Cyperus rotundus* and *C. esculentus*) are particularly difficult to control selectively in trees and vines.

In row crop land where translocated herbicides can be combined with disking and harrowing prior to planting, a crop can be grown with minimum competition from perennials. Unfortunately eradication of perennial weeds rarely occurs in 1 or 2 yr of spraying and tillage so continuous vigilence is required in order to keep perennial weeds from increasing.

While annual weeds can be controlled by injecting emulsifiable concentrated herbicides in the range of 20 to 100 ppm for 2 h into the drip system toward the end of the irrigation run, perennial weeds have not been affected.

While injection into drip irrigation is simple, accurate and effective on many annual weeds, more data are needed to evaluate the effect of flooding even low amounts of relatively insoluble herbicides into the concentrated rootzone of a drip emitter. This is not the usual method of using even very safe preemergence herbicides like napropamide (Devrinol®), Oxadiazon (Ronstar®), oxyflurfen (Goal®), trifluralin (Treflan®) or EPTC (Eptam®). Therefore, a great deal more work is necessary to evaluate weed control chemical movement in the wet zone and herbicide safety to young trees and vines as well as in row crops where drip irrigation will find a use.

16.11 References

1 Aljibury, F. K. 1974. Water use in drip irrigation. Proc. Second Inter'l. Drip Irrig. Cong., San Diego, CA, p. 341-350.

2 Bar-Yousef, B. and M. R. Sheikholslami. 1976. Distribution of water and ions in soils irrigated and fertilized from a trickle source. Soil Sci. Soc. Am. J. 40:575-582.

3 Baver, L. D., W. H. Gardner, and W. R. Gardner. 1972. Soil Physics. 4th ed. John Wiley and Sons, Inc., New York.

4 Ben-Asher, J., M. Fuchs, and D. Goldberg. 1978. Radiation and energy balance of sprinkler and trickle irrigated fields. Agron. J. 70:415-417.

5 Ben-Asher, J., D. O. Lomen, and A. W. Warrick. 1978. Linear and non-linear models of infiltration from a point source. Soil Sci. Soc. Am. J. 42:3-6.

6 Ben-Asher, J. 1979. Errors in determination of the water content of a trickle irrigated soil volume. Soil Sci. Soc. Am. J. 43:665-668.

7 Bernstein, L. E., and L. E. Francois. 1973. Comparisons of drip, furrow, and sprinkler irrigation. Soil Sci. 115:73-86.

8 Black, J. D. F. 1976. Trickle irrigation—a review. Part two. Hort. Abst. 46:69-73.

9 Black, J. D. F. and D. W. West. 1974. Water uptake by an apple tree with various portions of the root system supplied with water. Proc. Second Inter'l. Drip Irrig. Cong., San Diego, CA, p. 432-436.

10 Blass, S. 1964. Sub-surface irrigation. Nassade 45:1.

11 Bloodgood, D. W., R. E. Patterson, and R. L. Smith, Jr. 1954. Water evaporation studies in Texas. Bul. 787, Tex. Agr. Exp. Sta., 83 p.

12 Braester, C. 1973. Moisture variation at the soil surface and the advance of the wetting front during infiltration of the wetting front during infiltration at constant flux. Water Resour. Res. 9:687-694.

13 Brandt, A., E. Bresler, N. Diner, J. Ben-Asher, J. Heller and D. Goldberg. 1972. Infiltration from a trickle source I. Mathematical models. Soil Sci. Soc. Am. Proc. 35:675-682.

14 Brown, K. W., D. J. Gerard, B. W. Hipp, and J. T. Ritchie. 1974. A procedure for placing large undisturbed monoliths in lysimeters. Soil Sci. Soc. Am. Proc. 38:981-983.

15 Bucks, D. A., L. J. Erie, and O. F. French. 1973. Tickle irrigation on cotton. Prog. Agr. Ariz. 25(4):13-16.

16 Bucks, D. A., L. J. Erie, and O. F. French. 1974a. Quantity and frequency of trickle and furrow irrigation for efficient cabbage production. Agron. J. 66:53-57.

17 Bucks, D. A., L. J. Erie, and O. F. French. 1974b. Trickle irrigation management for cotton and cabbage. Proc. Second Inter'l. Drip Irrig. Cong., San Diego, CA, p. 351-356.

18 Bucks, D. A., F. A. Nakayama, and R. G. Gilbert. 1979. Trickle irrigation water quality and preventive maintenance. Agr. Water Mgmt. 1:149-162.

19 Bucks, D. A. and F. S. Nakayama. 1980. Injection of fertilizer and other chemicals for drip irrigation. Agri. Turf Irrig. Conf. Proc., Irrig. Assoc., Houston, TX. p. 166-180.

20 Budyko, M. I. (Ed.). 1963. Guide to the Atlas of the Heat Balance of the Earth. (Translated from Russian.) U.S. Dept. Com. WB/T-106. 27 p.

21 Businger, J. A. 1956. Some remarks on Penman's equations for the evapotranspiration. Netherlands J. Agr. Sci. 4:77.

22 Campbell, R. B., and C. J. Phene. 1976. Estimating potential evapotranspiration from screened pan evaporation. Agr. Meteorol. 16:343-351.

23 Christiansen, J. E. 1942. Irrigation by sprinkling. Univ. Calif. Agr. Exp. Sta. Bul. 670, 124 p.

24 Coston, D. C., H. G. Ponder, and A. L. Kenworthy. 1978. Fertilizing peach trees through a trickle irrigation system. Comm. in Soil Sci. and Plant Anal. 9:187-191.

25 Davis, K. R., W. J. Pugh, and S. Davis. 1977. Chlorine treatments of drip irrigation systems. 7th Inter'l. Agr. Plastics Cong. Proc., San Diego, CA, p. 113-117.

26 Davis, S. 1974. History of drip irrigation. Agribusiness News 10(7):1.

27 Davis, S. 1975. Drip irrigation. p. 508-520. In: Sprinkler Irrigation, C. H. Pair (Ed.-in-Chief), W. H. Hinz, C. Reid and K. R. Frost (Ed.). Sprinkler Irrigation Assoc.

28 Davis, S., W. J. Pugh, and K. R. Davis. 1977. Drip, subsurface, and sprinkler irrigation compared. 7th Inter'l. Agr. Plastics Cong. Proc., San Diego, CA, p. 73-79.

29 Doorenbos, J., and W. O. Pruitt. 1977. Crop Water Requirements. FAO Irrigation and Drain. Paper No. 24, FAO, Rome, 144 p.

30 Earl, K. D., and W. A. Jury. 1977. Water movement in bare and cropped soil under isolated trickle emitters: II. Analysis of cropped soil experiments. Soil Sci. Soc. Am. J. 41:856-861.

31 Ekern, P. C. 1965. Evapotranspiration of pineapple in Hawaii. Plant Physiol. 40:736-739.

32 Ford, H. W. 1976. Controlling slimes of sulfur bacteria in drip irrigation systems. Hort. Sci. 11:133-135.

33 Ford, H. W. and D. P. H. Tucker. 1975. Blockage of drip irrigation filters and emitters by iron-sulfur-bacterial products. Hort. Sci. 10:62-64.

34 Freeman, B. M., J. Blackwell, and K. V. Garzoli. 1976. Irrigation frequency and total water application with trickle and furrow systems. Agr. Water Mgmt. 1:21-31.

35 Gangopadhyaya, M., G. E. Harbeck, Jr., T. J. Nordenson, M. M. Omar, and V. A. Uryvaev. 1966. Measurement of Evaporation and Evapotranspiration. WMO, Geneva, Switzerland. Tech. Note No. 83 (WMO-No. 201. TP.105), p. 40-61.

36 Gilley, J. R. and E. R. Allred. 1974. Infiltration and root extraction from subsurface irrigation laterals. TRANSACTIONS of the ASAE 17:927-933.

37 Goldberg, S. D., and M. Shmueli. 1970. Drip irrigation—a method used under arid and desert conditions of high water and soil salinity. TRANSACTIONS of the ASAE 13:38-41.

38 Goldberg, D., B. Gornat, and D. Rimon. 1976. Drip Irrigation. Drip Irrig. Sci. Publ., Kfar Shmaryahu, Israel. 296 p.

39 Goode, J. E., K. H. Higgs, and K. J. Hyrycz. 1978. Trickle irrigation of apple trees and the effects of liquid feeding with NO₃ and K⁺ compared with normal manuring. J. Hort. Sci. 53:307-316.

40 Haise, H. R. and R. M. Hagan. 1967. Soil plant, and evaporative measurements as criteria for scheduling irrigation. p. 577-604. In: Irrigation of Agricultural Lands. R. M. Hagan, H. R. Haise, and T. W. Edminster (Ed.). Agron. Monog. No. 11.

41 Hiler, E. A. and T. A. Howell. 1973. Grain sorghum response to trickle and subsurface irrigation. TRANSACTIONS of the ASAE 16(4):799-803.

42 Howell, T. A. and E. A. Hiler. 1947a. Trickle irrigation lateral design. TRANSACTIONS of the ASAE 17(5):902-908.

43 Howell, T. A. and E. A. Hiler. 1974b. Designing trickle irrigation laterals for uniformity. Proc. Am. Soc. Civ. Engr., J. of the Irrig. and Drain. Div., 100(IR4):433-454.
44 Hughes, T. C. and R. W. Jeppson. 1978. Hydraulic friction loss in small diameter plastic pipelines. Water Res. Bull. 14:1159-1166.
45 Jensen, M. E. (Ed.). 1974. Consumptive use of water and irrigation water requirements. A report prepared by the Tech. Com. on Irrig. Water Requirements, Irrig. and Drain. Div., Am. Soc. Civ. Engr., 215 p.
46 Keller, J. and D. Karmeli. 1974. Trickle irrigation design parameters. TRANSACTIONS of the ASAE 17(4):678-684.
47 Keller, J. and D. Karmeli. 1975. Trickle irrigation design. 1st ed. Rain Bird Sprinkler Mfg. Corp., Glendora. 133 p.
48 Kenworthy, A. L. 1972. Trickle irrigation . . . the concept and guidelines for use. Res. Rep. 165, Michigan Agr. Exp. Sta. 19 p.
49 Lange, A., F. Aljibury, H. Kempen, and B. Fischer. 1978. Weed control in drip irrigated orchards and vineyards. Proc. 30th Anniv. Calif. Weed Conf., p. 53-54.
50 Lange, A., H. Kempen, B. Fischer, and J. Schlesselman. 1976. Chemical weed control with drip irrigation. Proc. West. Soc. Weed Sci. 29:53-58.
51 Lassoie, J. P., D. R. M. Scott, and L. J. Fritschen. 1977. Transpiration studies in Douglas-fir using the heat pulse technique. Forest Sci. 21:377-390.
52 Maäledj, M., and L. Malavard. 1973. Resolutions analogiques et numeriques de problemes d'irrigation des sols par canaux equidistants. C. R. Acad. Sci. Paris 276:1433-1436.
53 Marsh, A. W., C. D. Gustafson, S. Davis, R. L. Granson, and R. A. Stroham. 1974. Water use by drip and sprinkler irrigated avocados related to plant cover, evaporation and air temperature. Proc. Second Inter'l. Drip Irrig. Cong., San Diego, CA. p. 346-350.
54 Marshall, D. C. 1968. Measurement of sap flow in conifers by heat transport. Plant Physiol. 33:385-396.
55 Merrill, S. D., P. A. C. Raats, and C. Dirksen. 1978. Flow from a point source at the surface of a heterogeneous soil column. Soil Sci. Soc. Am. J. 42:851-852.
56 Miller, R. J., D. E. Rolston, R. S. Rauschkolb, and D. W. Wolfe. 1976. Drip application of nitrogen is efficient. Calif. Agr. 30:16-18.
57 Monteith, J. L. 1976. Vegetation and the Atmosphere. Vol. 1. Academic Press. 278 p.
58 Myers, L. E., and D. A. Bucks. 1972. Uniform irrigation with low pressure trickle system. Proc. Am. Soc. Civ. Engr., J. Irrig. Drain. Div. 98(IR3):341-346.
59 Nakayama, F. S., D. A. Bucks, and O. F. French. 1977. Reclamation of partially clogged trickle emitters. TRANSACTIONS of the ASAE 20(2):270-280.
60 Nakayama, F. S., R. G. Gilbert, and D. A. Bucks. 1978. Water treatment in trickle irrigation systems. Proc. Am. Soc. Civ. Engr., J. Irrig. Drain. Div., 104(IR1):23-24.
61 New, L., and R. E. Roberts. 1973. Automation drip irrigation for greenhouse tomato production. MP-1082, Tex. Agr. Ext. Serv., 16 p.
62 Pair, C. H. (Ed.-in-chief). 1975. Sprinkler irrigation. Sprinkler Irrig. Assoc., Fourth ed., 444 p.
63 Parchomchuk, P. 1976. Temperature effects on emitter discharge rates. TRANSACTIONS of the ASAE 19(4):690-692.
64 Penman, H. L. 1948. Natural evaporation from open water, bare soil and grass. Proc. Roy. Soc. London (Series A) 193:120-145.
65 Penman, H. L., D. E. Angus, and C. H. M. Van Bavel. 1967. Micro-climate factors affecting evaporation and transpiration. p. 483-505. In: Irrigation of Agricultural Lands, R. M. Hagan, H. R. Haise, and T. W. Edminster (ed.).
66 Phene, C. J., G. J. Hoffman, and R. S. Austin. 1973. Controlling automated irrigation with a soil matric potential sensor. TRANSACTIONS of the ASAE 16(4):733-776.
67 Phene, C. J., and R. B. Campbell. 1975. Automating pan evaporation measurements for irrigation control. Agr. Meteorol. 15:181-191.
68 Phene, C. J., and D. W. Beale. 1976. High-frequency irrigation for water nutrient in humid regions. Soil Sci. Soc. Am. J. 40:430-436.
69 Philip, J. R. 1972. Steady infiltration from buried, surface and perched point and line sources in heterogeneous soils: I. Analysis. Soil Sci. Soc. Am. Proc. 36:268-273.
70 Philip, J. R. and R. I. Forrester. 1975. Steady infiltration from buried surface, and perched point and line sources in heterogeneous soils: II. Flow details and discussion. Soil Sci. Soc. Am. Proc. 39:408-414.
71 Priestley, C. H. B., and R. J. Taylor. 1972. On the assessment of surface heat flux and evaporation using large scale parameters. Monthly Weather Rev. 100:31-92.
72 Raats, P. A. C. 1970. Steady infiltration from line sources and furrows. Soil Sci. Soc. Am. Proc. 34:709-714.

73 Raats, P. A. C. 1971. Steady infiltration from point sources, cavities and basins. Soil Sci. Soc. Am. Proc. 35:689-694.

74 Rauschkolb, R. S., D. E. Rolston, R. J. Miller, A. G. Carlton, and R. G. Burau. 1976. Phosphorus fertilization with drip irrigation. Soil Sci. Soc. Am. J. 40:68-72.

75 Ravelo, C. J. E., E. A. Hiler, and T. A. Howell. 1977. Trickle and sprinkler irrigation of grain sorghum. TRANSACTIONS of the ASAE 20(1):96-99, 104.

76 Rawlins, S. L. 1973. Principles of managing high frequency irrigation. Soil Sci. Soc. Am. Proc. 37:626-629.

77 Rawlins, S. L., and P. A. C. Raats. 1975. Prospects for high frequency irrigation. Science 188:604-610.

78 Richardson, C. W., and J. T. Ritchie. 1973. Soil water balance for small watersheds. TRANSACTIONS of the ASAE 16(1):72-77.

79 Ritchie, J. T. 1974. Evaluating irrigation needs for southeastern U.S.A. p. 262-279. In: Contribution of Irrigation and Drainage to World Food Supply, Biloxi, MS. Am. Soc. Civ. Engr.

80 Ritchie, J. T., and E. Burnett. 1971. Dryland evaporative flux in a subhumid climate: II. Plant Influences. Agron. J., 63:56-62.

81 Rolston, D. E., R. S. Rauschkolb, C. J. Phene, R. J. Miller, K. Uriu, R. M. Carlson, and D. W. Henderson. 1979. Applying nutrients and other chemicals to trickle-irrigated crops. Univ. of Calif., Div. of Agr. Sci., Bul. 1893, 14 p.

82 Ross, J. 1976. Radiative transfer in plant communities. p. 13-55. In: Vegetation and the Atmosphere, Monteith (Ed.), Academic Press.

83 Roth, R. L. 1974. Soil moisture distribution and wetting pattern from a point source. Proc. Second Inter'l. Drip Irrig. Cong., San Diego, CA, p. 246-251.

84 Shearer, M. N. 1977. Minimum screening and automatic flushing. Fourth Ann. Inter'l. Drip Irrig. Assoc. Meet. Proc., Fresno, CA, p. 32-36.

85 Shipston, C. 1976. Familiarization with polyethylene, Drip/Trickle Irrig. 1(1):6-10.

86 Smith, S. W. and W. R. Walker. 1975. Annotated bibliliography on trickle irrigation. Colo. State Univ., Inform. Ser. No. 16., 61 p.

87 Solomon, K. 1977. Evaluation criteria for trickle irrigation emission devices. 7th Inter'l Agr. Plastics Congr., San Diego, CA, p. 97-102.

88 Stevenson, D. S. 1973. Guide to design and operation of trickle irrigation systems. Canada Agr. Publ. Res. Sta., Summerland, B.C. 17 p.

89 Swanson, R. H. 1972. Water transpired by trees is indicated by heat pulse velocity. Agr. Meteorol. 10:277-281.

90 Tanner, C. B. 1967. Measurement of evapotranspiration. p. 534-574. In: Irrigation of Agricultural Lands, R. M. Hagan, H. R. Haise, and T. W. Edminister (Ed.), Agron. Monog. No. 11.

91 Uriu, K., R. M. Carlson and D. W. Henderson. 1977. Application of potassium fertilizer to prunes through a drip irrigation system. 1977 Inter'l. Agr. Plastics Congr. Proc., San Diego, CA, p. 211-214.

92 Van Bavel, C. H. M. 1966. Potential evaporation: The combination concept and its experimental verification. Water Resour. Res., 2:355-467.

93 Van Bavel, C. H. M., J. E. Newman, and R. H. Hilgeman. 1967. Climate and estimated water use by an orange orchard. Agr. Meteorol. 4:27-37.

94 Willoughby, P., and B. Cockroft. 1974. Changes in root patterns of peach trees under trickle irrigation. Proc. Second Inter'l. Drip Irrig. Congr., San Diego, CA, p. 439-442.

95 Wilson, D. L. 1977. Filtration in drip irrigation. 7th Inter'l. Agr. Plastics Cong. Assoc., San Diego, CA, p. 160-165.

96 Wu, I. P. and H. M. Gitlin. 1973. Hydraulics and uniformity for drip irrigation. Proc. Am. Soc. Civ. Engr., J. Irrig. Drain. Div. 99(IR2):157-168.

97 Wu, I. P. and H. M. Gitlin. 1974a. Drip irrigation design based on uniformity. TRANSACTIONS of the ASAE 17(3):429-432.

98 Wu, I. P. and H. M. Gitlin. 1974b. Design of drip irrigation lines. Tech. Bull. No. 96, Hawaii Agr. Exp. Sta., Univ. of Hawaii, 29 p.

99 Wu, I. P., and D. D. Fangmeier. 1974. Hydraulic design of twin-chamber trickle irrigation laterals. Tech. Bull. No. 216, Agr. Exp. Stat., Univ. of Ariz., 12 p.

100 Wu, I. P. 1975. Design of drip irrigation main lines. Proc. Am. Soc. Civ. Engr., J. Irrig. Drain. Div. 101(IR4):265-278.

101 Wu, I. P. 1977. A general drip irrigation design calculator. 7th Inter'l. Agr. Plastics Cong. Proc., San Diego, CA, p. 54-60.

chapter 17

EVALUATING IRRIGATION SYSTEMS AND PRACTICES

17

17

EVALUATING IRRIGATION SYSTEMS AND PRACTICES

by J. L. Merriam (Professor Emeritus), California
Polytechnic State University, San Luis Obispo,
CA; M. N. Shearer, Oregon State University, Cor-
vallis, OR; and C. M. Burt, California Polytechnic
State University, San Luis Obispo, CA.

17.1 PURPOSE

The purpose of evaluating irrigation systems is fourfold: (a) To determine the efficiency of the system as it is being used; (b) to determine how effectively the system can be operated and whether it can be improved; (c) to obtain information that will assist engineers in designing other systems; and (d) to obtain information to enable comparing various methods, systems, and operating procedures as a basis for economic decisions.

Evaluations involve measuring conditions at one or more points in a field selected to be typical or representative. Value judgments or additional measurements must be made to correlate these data to an overall field, farm, or project efficiency. For example, for sprinklers, the pre-nozzle losses, nonuniform pressures, topography influences, etc. usually are not included in estimating the potential application efficiency, and for surface methods, soils typically are assumed to be uniform. The accuracy of most measurements will seldom justify calculating efficiencies to more than $\pm$ 3 percent.

17.2 DEFINITIONS

The on-farm irrigation systems and operations need to be measured to determine the potential efficiency of the systems as designed and the actual efficiency that is obtained with present management. To make and interpret the measurements requires a precise terminology. In order that all comparisons have a common basis, the three performance parameters (potential application efficiency, actual application efficiency, and distribution uniformity) are based on the average depth of water infiltrated or stored in the quarter of the area receiving the least amount of water. The low quarter (LQ) concept was developed by the USDA, Soil Conservation Service, and is recommended as the standard for comparing alternative conditions. For some uniformity studies, Christiansen's uniformity coefficient may be used.

17.2.1 Performance Parameters

Application efficiency (AE) is the ratio of the average depth of the irrigation water infiltrated and stored in the root zone to the average depth of irrigation water applied, expressed as a percent.

721

$$AE = \frac{\text{average depth of water infiltrated and stored in root zone}}{\text{average depth of water applied}} 100$$

$$\cdots\cdots\cdots\cdots\cdots\cdots\cdots\cdots\cdots\cdots\cdots\cdots\cdots\cdots\cdots\cdots\cdots\cdots\cdots \quad [17.1]$$

This term indicates the percentage of the water applied (calculated from nozzle discharge for a sprinkler system) that is stored in the root zone for crop use. It may not be useful in evaluating irrigation system operation in the field because high efficiencies can be obtained by excessive under-irrigation even when water is distributed very non-uniformly.

Actual application efficiency of low-quarter (AELQ) is the ratio of the average low-quarter (LQ) depth of irrigation water infiltrated and stored in the root zone to the average depth of irrigation water applied (calculated from nozzle discharge for sprinkler systems), expressed as a percent. The average LQ depth infiltrated is the average of the lowest one-fourth of the measured values where each value represents an equal unit of area and cannot exceed the soil moisture deficiency (SMD). Values of AELQ indicate both the uniformity of water distribution and adequacy of irrigation. When the LQ value is less than the SMD or the desired management allowed deficiency (MAD), under-irrigation is indicated. The numerical value of the LQ average depth indicates the adequacy of the irrigation.

$$AELQ = \frac{\text{average low-quarter depth of water infiltrated and stored}}{\text{average depth of water applied}} 100$$

$$\cdots\cdots\cdots\cdots\cdots\cdots\cdots\cdots\cdots\cdots\cdots\cdots\cdots\cdots\cdots\cdots\cdots\cdots\cdots \quad [17.2]$$

where the LQ value $\leqslant$ SMD.

Potential application efficiency of low-quarter (PELQ) is the efficiency that is obtainable, expressed as a percent, when the average LQ depth of irrigation water infiltrated and stored just equals MAD.

$$PELQ = \frac{\text{average low-quarter depth of water infiltrated} = MAD}{\text{average depth of water applied}} 100$$

$$\cdots\cdots\cdots\cdots\cdots\cdots\cdots\cdots\cdots\cdots\cdots\cdots\cdots\cdots\cdots\cdots\cdots\cdots\cdots \quad [17.3]$$

PELQ is a measure of how well the system can apply water if management is optimal. The difference between PELQ and AELQ is a measure of management operations. Low PELQ values indicate design problems. This is the only efficiency term that should be used to compare systems or methods.

Distribution uniformity (DU) is the ratio of the average LQ depth of irrigation water infiltrated (or caught) to the average depth of irrigation water infiltrated (or caught), expressed as a percent.

$$DU = \frac{\text{average low-quarter depth of water infiltrated (or caught)}}{\text{average depth of water infiltrated (or caught)}} 100$$

$$\cdots\cdots\cdots\cdots\cdots\cdots\cdots\cdots\cdots\cdots\cdots\cdots\cdots\cdots\cdots\cdots\cdots\cdots\cdots \quad [17.4]$$

The DU indicates the magnitude of distribution problems.

Uniformity coefficient-Christiansen (UC) is the ratio of the average depth of water caught (or infiltrated) minus the average deviation from this depth, divided by the average depth caught (or infiltrated) as would occur with sprinkler patterns overlapped during a complete irrigation, expressed as a percent.

$$UC = \frac{\text{average depth caught - average deviation from average depth caught}}{\text{average depth caught}} 100$$

$$... [17.5a]$$

or

$$UC = (1 - \frac{\text{average deviation from the average depth caught}}{\text{average depth caught}}) ..... [17.5b]$$

Christiansen's coefficient was developed for evaluating the distribution of water from sprinkler systems, but it can be used for all methods if depth infiltrated is used in place of depth caught (also see Section 14.2.5). When evaluating systems requiring large numbers of measured values, the calculating process becomes tedious. It is much simpler for the sprinkler or trickle methods which have nearly a normal distribution pattern, but not for surface methods, to calculate the standard deviation and convert it to approximate UC values for comparison purposes. This can be done with the following conversion:

$$UC_{(\text{approx.})} = 100 - 0.8 \frac{s}{x} .... [17.6]$$

where s is the standard deviation and x is the average catch.

Distribution characteristic (DC) is the percent of the total wetted area under a single non-overlapping sprinkler pattern that received more than the average depth of application. This value is used to characterize sprinklers operated in non-overlapped situations such as orchard sprinklers.

$$DC = \frac{\text{area receiving more than average depth}}{\text{total wetted area}} 100 .... [17.7]$$

Advance ratio (AR) is the ratio of time it takes a stream to advance to the lower end of a field, T_{adv}, to the length of time the water is at the lower end, $T_{o[L]}$ and which is ideally the desired duration of irrigation. It is expressed as a decimal or as a fractional ratio.

$$AR = \frac{T_{adv}}{T_{o(L)}} .... [17.8]$$

Storage ratio (SR) is the ratio of the average depth of water stored in the root zone to the average depth storable. It is used in conjunction with the distribution uniformity and the application efficiency terms. This term is sometimes called storage efficiency. It indicates the adequacy of an irrigation. It is expressed as a percent.

$$SR = \frac{\text{average depth stored}}{\text{average depth storable}} \, 100 \quad \dots\dots\dots\dots\dots\dots\dots\dots \quad [17.9]$$

Soil moisture deficiency (SMD) is the depth of water required to bring a specific depth of soil to field capacity at a particular time.

Management allowed deficiency (MAD) is the percent of the available soil moisture out of the root zone that corresponds to a management allowed crop stress, and is also the resulting calculated SMD at the time to apply an irrigation.

17.2.2 Characteristic Times

Irrigation time (T_i) is the duration of time water should be on the soil surface to replace the SMD.

Application time (T_a) is the duration of time water is flowing into the irrigated area.

Opportunity time (T_o) is the duration of time water is on the soil surface with the opportunity to infiltrate.

Advance time (T_{adv}) is the time it takes water to advance from the near side to the far side of a field.

Lag time (T_L) is the duration of time it takes water to disappear from the upper end of a field after inflow has been turned off $(T_L = T_o - T_a)$.

17.2.3 Characteristic Curves

An **advance curve** is a plot of time after the water had been turned on versus the distance the water front has traversed. (see Fig. 17.1 and 17.4)

A **recession curve** is the plot of the increment of time between when the water is turned on to the time when water disappears from the surface at any distance. (see Fig. 17.1 and 17.4)

An **irrigation curve** is plotted at the time of irrigation (T_i) above the advance curve $(T_i$ plus $T_{adv})$ at any distance to indicate the increment of time needed to infiltrate the water depth which will satisfy the soil moisture deficiency, SMD, at any distance. (see Fig. 17.4)

An **intake rate curve** is a plot of the instantaneous rate that water infiltrates the soil surface versus elapsed time after initial wetting. (see Fig. 17.2)

A **cumulative infiltrated depth curve** is a plot of the accumulated depth of water infiltrated versus elapsed time after initial wetting. (see Fig. 17.2 and 17.11)

An **infiltrated depth distribution curve** is a plot of the depth of water infiltrated at specified locations in a field strip. The depth infiltrated is taken from a 'cumulative infiltrated depth' curve for the time of opportunity at the specified locations for surface irrigation. (see Fig. 17.1 and 17.3) For sprinkler irrigation, it is a plot of depths caught on equal areas arranged in a sequence of descending depths. (see Fig. 17.9)

17.3 FIELD EVALUATION AND INTERPRETATION

The knowledge as to how well an irrigation system is being used (actual application efficiency), or could be used (potential application efficiency), or the confirmation of a design efficiency, can only be obtained by field performance measurements. However, observations of certain performance characteristics may give reasonable approximations and provide guidance for obvious corrective steps. This section presents techniques for visual and measured field evaluations of the principal irrigation methods: furrow, border-strip, and open field-fixed grid sprinklers. The less common methods—basin, pond, trickle, and sprinklers such as center pivot, giant, traveler, perforated pipe, and orchard are briefly commented on. Evaluations of all methods are fully described in *Farm Irrigation System Evaluation: A Guide for Management,* Third edition, (Merriam and Keller, 1978).

The irrigation methods may be broadly classified as: (a) surface methods which are affected by soil and topography (furrow, border-strips, basin, and pond); (b) above the ground methods which are not directly affected by the soil (sprinkle and trickle); and (c) below ground methods (sub-irrigation involving management of the water table and sub-surface irrigation supplying the water through mechanical controls below the surface). Evaluation techniques are not available for this last group.

17.3.1 Furrow Irrigation

Furrow irrigation is adaptable where soils and topography are reasonably uniform. Furrows are sloping channels cut into the soil surface and into which a relatively large initial non-erosive stream of water is turned. As water flows down the furrow, it enters the soil laterally as well as vertically. Water is lost by excess infiltration and by runoff at the lower end. The maximum spacing between furrows is such that the lateral movement will wet across the entire area in less than the irrigation time (T_i) needed to replace the soil moisture deficiency (SMD) at the lower end. Under some conditions wider spacings are used which leave part of the area unirrigated. Also, the type of crop and available machinery often determine the furrow spacing.

The desirable operational practice is to cut back the initial stream size appreciably after it has reached the lower end when enough water is running off to justify the effort. The time of cut-back should be such that the fairly large stream running off at that moment should be about the same size as will be running off at the end of irrigation after having been cut back. With a 1/4 advance ratio, the cut-back should be made at about 2 T_{adv}. This will minimize the volume of runoff. Alternatively a return flow system might be installed which could recover the runoff water whether a cut-back is made or not. Runoff normally should be put into a reservoir from which it can be reused under controlled conditions. Unrecovered runoff is the largest loss in furrow irrigation when a reasonably uniform irrigation is accomplished. A reuse system is almost essential with furrow irrigation to conserve water and labor and should be considered an integral part of an efficient furrow system (also see Sections 13.6 and 13.8).

The shape of the furrow can be a vee and have a relatively small wetted perimeter with a slow intake rate. The smaller stream size flowing at the lower end results in a smaller wetted perimeter. This reduces the intake rate a little so some non-uniformity in intake occurs from the upper to the lower end of the furrow.

The parabolic-shaped furrow, about 0.3-m wide, and 0.08-m deep, has about the same wetted perimeter all along the furrow and resists erosion. Because of its shape, larger streams and wider spacings can be used. It is the most adaptable shape.

The broad furrow, usually 0.4 to 0.8 m wide and 0.08 m deep, has very

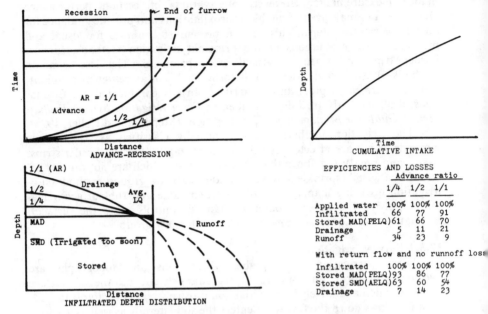

The figure shows Advance-Recession (Time vs Distance) with curves AR = 1/1, 1/2, 1/4, Recession, End of furrow, Advance; a Cumulative Intake plot (Depth vs Time); an Infiltrated Depth Distribution (Depth vs Distance) showing Drainage, Avg. LQ, MAD, SMD (Irrigated too soon), Stored, Runoff; and the following efficiencies and losses table.

EFFICIENCIES AND LOSSES

	Advance ratio		
	1/4	1/2	1/1
Applied water	100%	100%	100%
Infiltrated	66	77	91
Stored MAD(PELQ)	61	66	70
Drainage	5	11	21
Runoff	34	23	9

With return flow and no runoff loss

Infiltrated	100%	100%	100%
Stored MAD(PELQ)	93	86	77
Stored SMD(AELQ)	63	60	54
Drainage	7	14	23

FIG. 17.1 Furrow advance ratio effects and efficiencies.

uniform intake along the furrow providing it is level across to keep the water spread. It is not adaptable to fields with much longitudinal or lateral slope because of the difficulty in keeping the whole furrow bottom covered.

The shape and spacing of furrows has a large effect on T_i and resultant effects on efficiencies. The ability to change furrow shape, spacing, length and flow rate, and the MAD provide more management-controllable functions than are available for other methods.

Observational evaluations permit determining whether or not the furrows are being operated reasonably well. Observations do not require measurements and so cannot provide numerical values or efficiencies, but a numerical check on the SMD to see that it approximates the MAD—is it dry enough to irrigate—should always be done.

Observations should consist of: (a) probing the soil in the furrow near the lower end during and at the end of irrigation to assure that the SMD at the lower end has just been replaced. All water applied after this time is 100 percent wasted unless a return flow system is used; and (b) observing the duration and rates at which water runs off at the lower end provides a measure of runoff loss. Generally, runoff is the largest loss. However, it can be reduced by about one half by utilizing a cut-back of the initial stream, and the loss can be recovered by using a return flow system.

The uniformity of the infiltration is related to the advance ratio, AR (see Fig. 17.1). If the initial stream is large enough to cause the flow to arrive at the lower end in one fourth of T_i (AR = T_{adv}/T_i = 1/4), very high uniformity is assured with distribution uniformity (DU) being between 90 and 95 percent with the higher values occurring on the finer soils. In most cases this requires a large capacity system and much labor and may be uneconomically uniform. It results in a large amount of runoff which may be of the magnitude of 35 percent. This causes a low potential application efficiency

(PELQ). However, if a return flow system is used, PELQ can equal DU and labor is very small. An AR of 1/2 usually results in a DU of about 85 percent and moderate runoff. An AR of 1/1 results in the infiltration of appreciable water too deep at the upper end but very little runoff. ARs of this magnitude will often result in about the highest PELQ when no cutback or return flow system is used.

In summary, high furrow application efficiency is assured when: (a) just enough water is infiltrated at the low end to satisfy the SMD; (b) the initial stream size is cut back to result in little runoff or a return flow system is used to recover runoff; and (c) the AR is low enough to assure reasonable uniformity of depth infiltrated along the furrow.

Measured evaluations permit determining numerical values of DU, AELQ, and PELQ. From these values economic decisions can be made as to where and what changes are desirable. The procedure consists of making field measurements, assembling the information, and then evaluating it.

The test should use existing irrigation practices (SMD, stream size, shape, etc.) so that the AELQ value may be determined. Other furrow shapes and stream sizes should also be tested. The PELQ values are determined by making reasonable calculated modifications of the system or its operation, such as changes in MAD, stream size, furrow shape or spacing, furrow length, use of cut-back streams or return flow system, etc.

Field Procedures. A detailed presentation of techniques and interpretation with illustrations can be found in *Farm Irrigation System Evaluation: A Guide for Management* (Merriam and Keller, 1978). Procedures are outlined below:

1   Determine with a soil auger or other methods the soil uniformity, depth of root zone, and the soil moisture deficiency (SMD) and compare it to the MAD.

2   Measure the inflow rates into the test furrows which, for non-cracking soils, may conveniently be alternate furrows. At least three different stream sizes should be used: too large, too small, and reasonable.

3   Measure the instantaneous intake rate (see Subsection 17.4.2 and 17.4.4) in a furrow with the medium size stream using inflow-outflow measurement. Rates should also be determined for other stream sizes if practical.

4   Measure the advance rate using stakes placed adjacent to the furrows at intervals of 20 to 30 m before the start of the test. The time water arrives at each station, starting with zero time at station zero, is noted for each of the stream sizes on the Water Advance and/or Recession sheet. (Form 1) (This is not the same zero time as for intake.) If it is occasionally impractical to observe this time at the staked station, any time and distance may be noted since the objective of collecting these data is to be able to plot advance curves for several different stream sizes. A minimum of eight points should be obtained. Minor changes in the advance rate are often indicative of variations of soil, slope, or furrow conditions.

5   Determine the depth of penetration and lateral spread of the water during the latter part and at the end of the test by using a probe. It should be rechecked to determine the adequacy of the irrigation by using a soil auger one or two days later. Probings made before the end of the test are necessary to avoid over-irrigation. Water will continue to penetrate deeper after the flow is turned off.

6   Observations of the furrows during and after the test are made

WATER ADVANCE and/or RECESSION DATA

Location _____ Date_____ Soil Texture_____ Moisture Deficiency_____ Crop_____

Comments _____ Observer _____

Adv. or Rec. _____   _____   _____   _____

Identification _____   _____   _____   _____

Stream Size _____   _____   _____   _____

| TIME | | | Station | TIME | | | Station | TIME | | | Station | TIME | | | Station |
watch	diff.	cumu.	feet	watch	diff.	cumu.	feet	watch	diff.	cumu.	feet	watch	diff.	cumu.	feet
			0				0				0				0

FORM 1

relative to erosion and overtopping to estimate maximum usable stream size.

7   The physical condition of the furrow as to shape, wetted width, surface condition, cloddiness, new or reused, slope, moisture, cracking, etc. should be noted.

Assembling the data. Procedures for preparing the data for analysis are outlined below:

1   The advance curves for the several different stream sizes are first plotted on rectangular coordinate paper. This is desirably done while the data are being taken in the field. The shape for uniform field conditions is a smooth upward-curving, sickle-shaped line. It should be plotted through the observed points with very little "smoothing" of the curve shape since minor variations in the smoothness of the curve are indicative of physical changes such as slope, cloddiness, soil, shape, etc. in the furrow. These variations, or lack of them, are part of the diagnostic tools of the evaluation and may indicate where corrective steps need to be taken, or allowances made for the variations.

For subsequent steps, the advance curves need to be extrapolated as shown in Fig. 17.1. For naturally smooth or for arbitrarily drawn smooth curves done after noting the actual deviation, the extrapolation can be done with adequate accuracy by plotting the smooth, uniform condition curve on the same sheet or on three cycle logarithmic paper as the intake rate curve, but with changed units. The plotted portion can be extrapolated using a French curve. The implicit assumption is that conditions for the whole field are uniform and represented by the test area. Short extrapolations can be made on the rectangular coordinate plot. Rate of advance equations can be developed and may be essential for long extrapolations.

2   The intake rate curve plotted from field data (Form 2) for the soil location tested (see subsection 17.4.4), which is presumed typical of the field unless known to be otherwise, is plotted on three cycle logarithmic paper. The difference of flow rates (liters per minute per meter or gallons per minute per 100 feet) between the two measuring stations is plotted at the time (minutes) at which the readings were taken. Since the precision of the measurement is seldom high, the plus or minus range of accuracy must be shown and the line plotted within this range. To try to plot through the

FURROW INTAKE RATE DATA   FORM 2

Location _____ Observer _____ Date _____
Soil: texture _____ classification _____ SMD _____
Furrow: shape _____ wetted width ___ slope ___ condition _____ age _____
Comments

furrow identity _____ type measuring device A _____ B _____ length AB _____

time minutes			station A			station B		intake rate	
1	2	3	4	5		6	7	8	9
watch	diff.	cumu.		flow	rate		flow	rate		

accuracy range

furrow identity _____ type measuring device A _____ B _____ length AB _____

time minutes			station A			Station B		intake rate	
1	2	3	4	5		6	7	8	9
watch	diff.	cumu.		flow	rate		flow	rate		

accuracy range

1. First entry made when stream reaches midway between stations A and B.
    Second entry made when stream is about 3.m past station B. Subsequent
    entries made at increasingly longer intervals of time.
2. Difference in time between successive watch times.
3. Summation of successive time increments. To be plotted vs col. 9.
4,6. Head on Parshall flume, orifice, weir, etc. Show device and units used.
    If a jug is used, show size and time to fill.
5,7. Right column for flow rate. Left column for miscellaneous work entries.
8. Flow rate difference between col. 5 and col. 7 per length AB.
9. Flow rate per unit length. (L/s-m, cfm/ft, gpm/100ft) Plot vs col. 3.
Accuracy range. Show plus or minus limits of accuracy of measurements and
of the resulting flow rates. Show range when plotting.

calculated intake rate points will generally give a broken line which is not valid. Unless some unusual condition occurs, or the duration is very long, the plotted line should be straight. The equation of the intake curve is

$$I = KT^n \dots \dots \dots \dots \dots \dots \dots \dots \dots \dots [17.10]$$

where I is the intake rate in L/min-m, (or gal/min-100 ft) for the length between the stations at the time T in minutes. K is a constant and has the value given by the curve when T equals 1.0 minute, and n is the geometric slope of the line and is negative. It is very desirable to calculate and plot the values for this curve at the time the measurements are taken so that discrepancies can be detected and new readings taken if necessary. The value of K typically changes appreciably with cultural operations, and decreases during the season.

3   The cumulative infiltrated depth curve is developed for the studied furrow shape and spacing, and other spacings as may be desirable to study PELQ. It is done mathematically or graphically by integrating the intake rate equation and plotting it on logarithmic paper with the intake rate curve. For SI units,

$$D_{(S)} = K'T^{n+1} \quad \dots \dots \dots \dots \dots \dots \dots \dots \dots \dots \dots \dots \dots [17.11]$$

where $D_{[s]}$ is depth in mm for the specific furrow spacing S in meters, and K′ = K/(n + 1)S. (For English units, $D_{[s]}$ is depth in inches for a specific furrow spacing S in feet and K′ = K/62.3(n + 1)S.) T is elapsed time in minutes. The cumulative infiltrated depth curve for any specific furrow spacing is easily plotted. The intercept at T = 1.0 min is the value of K′, and the line is drawn upward to the right on the geometric slope of (n + 1):1 (see Fig. 17.2).

4 "Adjusted" intake rate and cumulative infiltrated depth curves are frequently needed. They are necessary for furrows that have been formed in loose soil or soils that crack. They are needed because the infiltration measuring techique described cannot provide rates for the first few minutes of applying water under these conditions. Also, since the technique involves taking the difference between the inflow and outflow measurements, small discrepancies may be magnified.

Assuming the rate and duration of the inflow to the tested furrow can be measured with reasonable accuracy, the total volume of water applied to the test area can also be found to the same degree of accuracy. This can be compared to the average depth obtained using the original cumulative depth infiltrated curve. The latter is then "adjusted" if not reasonably similar, to give a consistent answer. This is done as follows:

An advance curve (extrapolated beyond the end of the furrow) is drawn on rectangular coordinates for the duration T_a (see Fig. 17.2). A recession curve is drawn horizontally (unless known to be appreciably slow) at T_a. The time increment between the advance and recession curves is the opportunity time (T_o) at any station. Convenient, equally spaced stations are selected to give about six or more locations along the extrapolated furrow length. The T_o at each of these locations is noted below the station. Then using the cumulated infiltrated depth curve for the corresponding furrow spacing, the depth infiltrated at each location is determined and an infiltrated depth distribution curve is plotted. The average for each equal section can be calculated or scaled, and their average determined. This should correspond to the actual average depth applied calculated using the inflow rate, the T_a, and the area represented by the spacing and extrapolated furrow length. Alternatively, the area under the infiltrated depth distribution curve for the furrow and extrapolated length can be found and its average determined, or the average line can be drawn (see Fig. 17.2).

If measurements have been carefully made and the soil is uniform for the entire furrow length and does not have a high initial intake rate due to cracks or loose soil, the calculated average depth infiltrated will closely equal the calculated average inflow depth. If they are not reasonably close, adjusted curves should be drawn as described below.

Presuming that the average calculated inflow depth is correct, it is plotted as a point on the cumulative infiltrated depth sheet at a time that is the same time as that at which the average infiltrated depth occured. A line drawn through this point and parallel to the original curve is the adjusted cumulative infiltrated depth curve for that furrow spacing. Its equation may be found and the corresponding intake rate equation calculated and the adjusted intake rate curve drawn. Also a new infiltrated depth distribution curve for the extrapolated lengths should be drawn and used for the evaluation study.

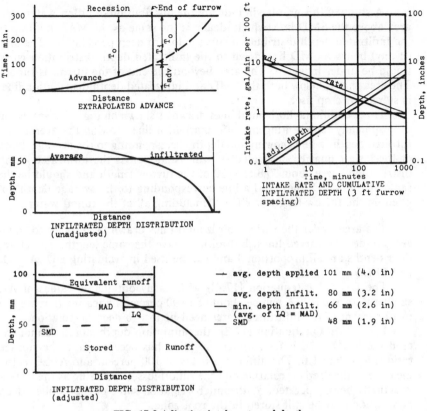

FIG. 17.2 Adjusting intake rate and depth.

Evaluation of information. The basic objective of evaluations is to iden-tify types and magnitudes of losses that are occurring, and then determine how to improve the system and/or its operation. The irrigation conditions can be evaluated by determining certain ratios or efficiencies. The distribu-tion uniformity (DU), which describes the uniformity with which water in-filtrated along the furrow length, is closely related to the advance ratio (AR) (see Fig. 17.1). The actual application efficiency (AELQ), which describes how much of the applied water can be used by the crop at the LQ, permits the determination of the actual losses to deep percolation and to runoff (see Fig. 17.2, SMD). The potential application efficiency (PELQ) indicates what can be done with certain variations in the system or its operation (see Fig. 17.2, MAD). PELQ is the only proper term to use for comparisons with other systems or methods. Low values of AELQ indicate management problems and low values of PELQ indicate system problems. Large values of AR and low ones for DU indicate too long a furrow, too small an initial stream, too small a MAD, or too small a furrow spacing.

Basic to the study of these parameters is the use of the infiltrated depth distribution curve adjusted if necessary. It is most conveniently plotted directly below the extrapolated advance curve to the same horizontal distance scale. On this graph note the actual furrow length as a vertical line at its sta-tion, and the SMD and MAD values horizontally along the length of the furrow. A study of this graph will indicate the distribution of inflow water. The area above the SMD line (which ideally is the same as the MAD line) up to the infiltrated depth distribution curve is lost to deep percolation. The area (if any) below a SMD line down to the infiltrated depth distribution curve would be under-irrigated. The area beyond the end of the furrow is runoff. The area that is below both the SMD and infiltrated depth distribution lines is stored for crop use.

Three additional horizontal lines drawn as shown on Fig. 17.2 are useful in graphically representing the information: a line showing the average infiltrated depth, a line showing the depth corresponding to the average depth of the lowest quarter infiltrated (it will usually cross the infiltrated depth distribution curve about 7/8 to 9/10 of the furrow length and should be the same as the MAD line), and a line corresponding to the average depth applied on the furrow length itself, i.e. including all of the runoff water.

The area under the various horizontal lines drawn correspond to equal areas under the curved lines, but all now have the same length. Therefore, their depths are all proportional and can be used in evaluating DU, AELQ, and PELQ.

The DU value (equation [17.4]) with easily obtainable values of AR smaller than about 1/2, will be about 85 to 90 percent indicating that only 15 to 10 percent of the infiltrated water need be lost to deep percolation.

In the AELQ (equation [17.2]) the minimum depth stored cannot exceed the SMD. If it is less, under-irrigation has occurred and the storage ratio is less than 1.0. The difference between 100 percent and AELQ is the measure of the deep percolation and the runoff losses occurring as the system is actually being operated. With under-irrigation some of the implied deep percolation may be still stored in the root zone.

The PELQ (equation [17.3]) also can be studied utilizing the data from the test. Changes in efficiencies result from changes in operating procedures such as using larger or smaller initial streams which will affect T_{adv} and hence change AR and DU, and to some extent T_i; changes in MAD which will affect T_i and hence modify T_a, DU, and AR; changes in furrow spacing or shape which will affect T_i, T_{adv}, T_a, AR, and DU; and changes in furrow length interact with many of the above. A cut back stream and a return flow system should almost always be studied, especially when high DU values occur which are accompanied by high runoff.

By studying the effects of changing or combining the above management controllable items, the most desirable or economical operation can be determined with a reasonable degree of accuracy. However, it might well not be the highest PELQ depending on various costs of water, labor, and capital. (see *Farm Irrigation System Evaluation: A Guide for Management*, Merriam and Keller, 1978, for illustrations.)

Example.

From the values shown in Fig. 17.2 (adjusted) for an AR of 1/2, SMD = 48 mm, MAD = 66 mm, infiltrated depth = 80 mm, applied depth = 101 mm, the DU, AELQ, PELQ, runoff and deep percolation values without and with a reuse system are as follows:

| Parameter | Values in percent | |
	Without a reuse system	With a reuse system
DU	$(66/82)100 = 82$	$(66/80)100 = 82$
For SMD = 48 mm		
AELQ	$(48/101)100 = 48$	$(48/80)100 = 60$
Runoff	$((101-80)/101)100 = 21$	-0-
Deep percola- tion	$((80-48)/101)100 = 32$	$((80-48)/80)100 = 40$
For MAD = 66 mm		
PELQ	$(66/101)100 = 66$	$(66/80)100 = 82$
Runoff	$((101-80)/101)100 = 21$	-0-
Deep percola- tion	$((80-66)/101)100 = 14$	$((80-66)/80)100 = 18$

17.3.2 Border-Strip Irrigation

Border-strips are adaptable where soils and topography are reasonably uniform. Each unit consists of a sloping strip of convenient width, and needs to be level or nearly level perpendicular to the flow to keep the water spread across it. The sides of the strip must be bounded by low borders (dikes, levees, ridges) of sufficient height to confine the flow. Water is turned in at the upper end at a flow rate that results in sufficient depth to cause it to spread over the entire width. The stream is usually turned off when it has advanced from 0.6 to 0.9 the length of the strip for slow to rapid intake soils. The lower end is irrigated with water draining off the upper end. Water is lost by penetrating too deep, and by runoff of the lower end, although it may sometimes be ponded there if the lower end is diked. The strips should be carefully graded to a uniform slope. Where soils are too shallow or undulating to grade uniformly, the strips may be run approximately down the steepest slope on a variable grade, made fairly narrow, and graded level across. They are then classified as guided rather than graded border-strips and are very successfully used for pasture and some other crops (also see Section 13.4).

Border-strip irrigation is complicated and requires the highest level of management skill of any surface irrigation method to achieve high efficiencies. For high efficiencies, the stream size and the resulting rate of advance must be controlled to match the recession conditions to provide approximately equal infiltration opportunity time at both the upper and lower ends. This must happen while simultaneously infiltrating the desired depth (SMD) of water and also when the water is turned off having it just reaching the proper distance down the strip to limit runoff. The recession curve is the key control since its shape does not change but is unique for the strip so all other things must be adjusted to match it. For many fields the strips are too short causing excessive runoff or poor distribution. On strips that are too short for the MAD and so have lots of runoff, it may be possible to use a cutback stream if the reduced stream size will still cover the entire width, or a reuse system may recover excessive runoff, or they may be converted to basins. Over-irrigation in a properly managed strip will occur only in the central portion. However, it may also occur at the lower end when the stream is too large, or at the upper end when the stream is too small. It is difficult to efficiently apply a shallow or a deep irrigation without correspondingly changing the length of the strip.

Ponding of water at the lower end to prevent runoff may be acceptable on high intake soils but may not be on fine-textured soils where it could cover the plants for a detrimentally long time. Such ponded water may be lost as deep percolation instead of as runoff. A return flow system is often desirable but is not as essential as it is for furrows since runoff is small for proper length and MAD.

Observational evaluations consist of several steps: (a) determining if it is "dry enough to irrigate"; (b) plotting the advance and recession curves to see that they are about parallel which will assure good uniformity, or at a very minimum seeing that the water is at the upper and lower end of the strip for about an equal length of time; (c) noting that the stream was cut off before the water reached the lower end at a time and distance that was not so late as to cause too much runoff, but late enough to provide flow at the lower end long enough to approximate the intake time opportunity at the upper end; (d) at the end of irrigation, and again a couple of days later, checking to see that enough water was infiltrated to satisfy the SMD, "was it wet enough to stop irrigating."

In summary, even though numbers were not obtained that permit studying alternate operations and numerical values of efficiency, good irrigation can be confirmed if: (a) water was on the upper and lower ends of the strip for about the same duration thus assuring reasonable uniformity; (b) water was cut off at about the right time and distance to adequately irrigate the upper and lower ends but avoid excessive runoff; and (c) the SMD was just satisfied assuring the adequacy of irrigation with little excess lost to deep percolation. It is difficult to satisfy all criteria simultaneously.

Measured evaluations permit the determination of numerical values of DU, AELQ, and PELQ, and the losses to runoff and deep percolation so that economic studies can be made of various systems and management practices. The procedure consists of making field measurements under normal conditions of SMD and stream size, with runs of large, medium, and small streams; assembling the data for study, and evaluating it for present conditions (AELQ) and other variations (PELQ).

Field procedures are similar to those used with furrows.

1   Determine with a soil auger or other methods, the soil uniformity, root zone depth, and soil moisture deficiency (SMD) and compare it to the MAD.

2   Place stakes at uniform distances of 20 to 30 m (50 to 100 ft) along a border.

3   In several strips start different stream sizes to provide a range of flows from large to small and including the usual size. (see Section 17.4.3).

4   Note the time water reaches each station for the different stream sizes on the border-strip advance and recession sheet. (Form 3). (It is convenient to plot the advance and recession curve data as they are taken in the field.)

5   Cut off the water before the stream reaches the lower end and note the time and distance. (It is very desirable to cut off one of the medium size streams too soon so that it will not reach the end in order to provide the proper shaped curve for what is usually the extrapolated runoff curve.)

6   Note the time and distance when the water disappears from the surface so that, practically speaking, there is no more opportunity for water to infiltrate. There may be some small puddles or water may be flowing in small

BORDER-STRIP ADVANCE AND RECESSION DATA

Location _____ Observer _____ Date _____

Crop: kind _____ condition _____ retardance _____ root zone depth _____

Soil: texture _____ AMC _____ MAD% _____ MAD _____ SMD _____

Border: spacing _____ condition;_____ Strip: width _____ wetted width _____ slope _____

Comments _____

strip identification			stream size			strip identification			stream size		
advance			recession			advance			recession		
time minutes		station	time minutes		station	time minutes		station	time minutes		station
watch	diff	cumu		watch	diff	cumu		watch	diff	cumu		watch	diff	cumu	
		0	0+00			0	0+00			0	0+00			0	0+00
							0+00								0+00
							0+00								0+00

Note: For border-strip recession data, first time entry should be made when the water is started, second entry at the time water is turned off, and the third entry when the recession starts, and all of these entries should be made at station 0+00.

FORM 3

channels, but a consistent condition for each station should exist at the recorded moment. Water will usually disappear first at the upper end some minutes (or hours on fine textured, low gradient fields) after it is turned off (Lag Time, T_L). It later may disappear simultaneously at locations at both the upper and lower ends on long strips. On a strip that was cut off too soon, it will first disappear at the lower end. (It is sometimes expeditious to simultaneously take recession data on one strip while taking advance data on an adjacent strip using the stream of water just turned off. This is especially true where there is only one stream available.)

7   Cumulative infiltration measurements using at least four cylinder infiltrometers are taken at a representative location (see Section 17.4.4.).

Assembling data. Procedures for preparing data for analysis are outlined below.

1   Plot the advance and recession curve for each test. In addition to plotting them on individual sheets for later evaluation, it is also desirable to plot several curves on one sheet so that the effect of the several stream sizes on the advance curves will be obvious (see Fig. 17.3.). The recession curves should all normally have nearly identical shapes though displaced in starting time. The advance and recession curves should be extrapolated beyond the end of the field until they intersect. The shape of this extrapolated portion should be similar to that obtained from the stream that was cut off too soon.

2   Plot all the individual cumulative infiltrated depth curves from the cylinder infiltrometer data Form 4 on one sheet of 3-cycle log paper similar to Fig. 17-2. They should plot as straight lines essentially all parallel except for possible "dog legs" either steeper or flatter in the early part. They generally will have a large variation in magnitude within the group. A straight line parallel to the general slope of the "non-dog leg" portion is next drawn and

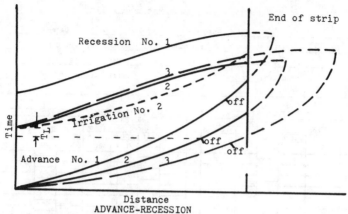

ADVANCE-RECESSION

1. Too small a stream. Cutoff made late in order to adequately
   irrigate the lower end. Upper end over-irrigated, and little
   runoff.
2. Correct stream size. Both ends adequately irrigated. Cutoff
   time and distance correct, and little runoff.
3. Too large a stream. Cutoff made late in order to adequately
   irrigate the upper end. Lower end over-irrigated and much runoff.

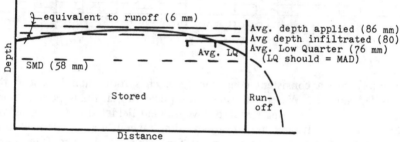

ADJUSTED INFILTRATED DEPTH DISTRIBUTION FOR CORRECT STREAM SIZE

FIG. 17.3 Border-strip evaluation curves.

labeled "typical". This line is not an average of the group, but is representative of the group. If the tested area consisted of soils that crack on drying
and so often have a steep dog leg section, the typical curve should be placed
well above the group sometimes as much as 25 mm at time equals one
minute. The reasons for this are that while the cylinders are hopefully placed
in typical soil conditions, they could not be placed over a "typical crack" and
so are usually placed to avoid obvious cracks, and that the initial infiltration
before the first reading at 20 or 30 s is not measured. The initial very high infiltration rate is therefore not measured even though it exists.

   3   Adjusted cumulative infiltrated depth curve is next drawn. To
develop this curve an extrapolated advance-recession curve representative of
the test is used. The opportunity time (T_o) at six to ten stations equally spaced (or weighted if non-uniform) along the whole extrapolated length is found
and the corresponding depths determined from the typical cumulative infiltrated depth curve. A typical infiltrated depth distribution curve is then
drawn and its average value found by a procedure similar to that illustrated
in Fig. 17.2. The average depth of water applied to the extrapolated length is
calculated. It is usually larger than the typical depth, and often much larger
on soils that crack. An adjusted cumulative infiltrated depth curve is drawn

CYLINDER INFILTROMETER DATA

Location_____ observer_____ date_____

Soil symbol_____ soil texture_____ soil moisture_____

Crop history_____

Remarks_____

CYLINDER ___						CYLINDER ___						CYLINDER ___					
TIME			INFILTRATION			TIME			INFILTRATION			TIME			INFILTRATION		
watch	diff	cumu	depth	diff	cumu	watch	diff	cumu	depth	diff	cumu	watch	diff	cumu	depth	diff	cumu

CYLINDER ___						CYLINDER ___						CYLINDER ___					
TIME			INFILTRATION			TIME			INFILTRATION			TIME			INFILTRATION		
watch	diff	cumu	depth	diff	cumu	watch	diff	cumu	depth	diff	cumu	watch	diff	cumu	depth	diff	cumu

FORM 4

as follows to reconcile the two. At the time the average typical infiltrated depth occurs as found from the infiltrated depth distribution curve, the actual average adjusted depth calculated using the inflow rate, duration of application and the area of the strip did infiltrate. A new curve, the adjusted cumulative infiltrated depth curve, is drawn on the logarithmic paper through that point (time and depth) and parallel to the typical curve.

Evaluation of information. The basic objective of evaluation is to identify the kind and magnitude of the losses and the potential economic value of changing the system and/or its operation. The complicated management of border-strips is studied with two techniques; one, using time vs. distance curves, and the other using depth infiltrated vs. distance curves (see Fig. 17.3).

The time vs. distance relationships are found by plotting the advance and recession curves extrapolated beyond the end of the strip, plus an irrigation curve. This latter curve is drawn parallel to the advance curve and above it by the irrigation time (T_i) the value of which is taken from the adjusted cumulative infiltrated depth curve. This set of curves indicates how long water was and should have been at each point along the strip, whether the time and distance to cutoff were reasonable, whether the border-strip length and MAD were properly interrelated, whether the stream size was correct, whether there were changes in infiltration rate (soil) or slope along the strips as indicated by abnormal shapes of the advance-recession curves, and the time effects of adjusting the management controllable items of stream size, MAD, and length. It also emphasizes the fact that for the border-strip method, the recession curve is the primary limitation as it is unique and has a relatively unchanging shape for each strip. Therefore, all else must be made to conform to it.

The depth infiltrated versus distance (distribution) curve provides the information about the efficiencies and losses. Generally it is most conveniently plotted below the advance-recession curves to the same horizontal distance scale. It is plotted by finding the T_o at various stations along the strip and its extrapolation, then plotting the corresponding infiltrated depths from the adjusted cumulative depth infiltrated curve. A vertical line is drawn at the end of the strip, and for this length, horizontal lines are drawn representing the average depth infiltrated, the SMD, and MAD, the minimum depth (average of the low quarter infiltrated) infiltrated, and a line representing the average depth, including the runoff, applied to just the field length.

The depths represented by these lines are the various items needed for DU, AELQ, and various PELQ's. These terms can be calculated and the various losses and their location found. Based on this information, the economics of various changes in the system can be studied.

Example.

From the values shown in Fig. 17.3, test No. 2 for the correct stream, with SMD = 58 mm, MAD = 76 mm, average infiltrated depth = 80 mm, average applied depth = 86 mm, the DU, AELQ, PELQ, runoff, and deep percolation values are found without and with a reuse system as follows:

	Values in percent	
Parameter	Without a reuse system	With a reuse system
DU	(76/80)100 = 95	(76/80)100 = 95
For SMD = 58 mm		
AELQ	(58/86)100 = 67	(58/80)100 = 72
Deep percolation	((80-58)/86)100 = 26	((80-58)/80)100 = 28
Runoff	((86-80)/86)100 = 7	-0-
For MAD = 76 mm		
PELQ	(76/86)100 = 88	(76/80)100 = 95
Deep percolation	((80-76)/86)100 = 5	((80-76)/80)100 = 5
Runoff	((86-80)/86)100 = 7	-0-

A great deal of interpretive information can be obtained by studying the advance and recession curves and their relative location and shape. A number of illustrations are shown on Fig. 17.4. More management illustrations can be found in *Farm Irrigation Systems Evaluation: A Guide for Management* (Merriam and Keller, 1978).

17.3.3 Basin Irrigation

Basin irrigation in which a level area of uniform soil is surrounded by dikes and into which a known volume of water is turned, has a high potential

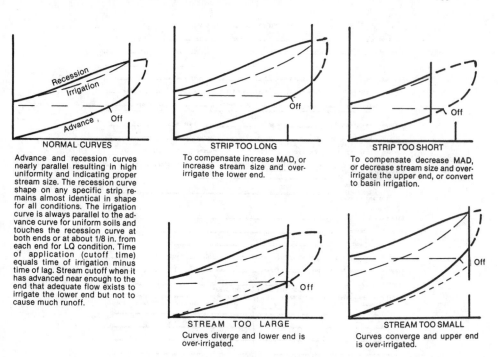

FIG. 17.4 Time and distance curves for border strip advance, recession and irrigation.

application efficiency of about 85 percent limited by the advance ratio and practical land grading uniformity. The dikes can be irregular in direction to conform to soil boundaries. The only significant loss is deep percolation. However, in practice the actual application efficiencies are appreciably lower. This is because all low areas are over-irrigated by their excess stored volume; in order to obtain a small advance ratio, the time of filling is usually too short to obtain accuracy of duration, i.e., two-minutes error during a twenty-minute filling is a 10 percent error; and flow rate measurements are seldom very precise.

Bedded basins may perform like basins. If the channels are small (dead-level furrows) and so provide little storage, the system will operate as a combination of furrows and basins requiring a longer application time than a basin. If so, a higher actual application efficiency may be practical. It may be comparable to that of furrows with no runoff, but management will be quite different, especially the quick filling used for basins cannot be used if the beds are not to be overtopped. DU is directly related to AR when the basin has no low areas.

17.3.4 Pond Irrigation

The physical appearance of a pond-irrigated field is like a basin except that the land surface need not be well graded or level. Its operation is a combination of basin and border-strip methods in that the entire area is flooded, as with basins, but the water is drained off after it has been on the surface for an adequate duration, as with border-strips. The advance and recession curves should be made reasonably parallel, and drainage of the ponded water

SPRINKLER EVALUATION DATA SHEET

Location _____ Observer _____ Date _____

Crop: kind _____ growth stage _____ root zone depth _____

Soil: texture _____ AMC _____ MAD% _____ MAD _____ SMD _____

Design: make _____ model _____ nozzle sizes ____ x ____

    pressure _____ flow rate _____ sprinkler spacing ___ x ___ application rate ___

    riser height _____ lateral pipe diameter _____ lateral slope _____

TEST DATA

Weather: temperature _____ humidity _____ cloudiness _____

    wind velocity   begin _____ during _____ finish _____
    wind direction arrow relative
    to pipe flow direction   begin   during   finish

Sprinkler spacing _____ lateral spacing (move) _____ gauge spacing ____ x ____
Sprinkler number on
    the lateral   ___ ___ ___ ___ ___ ___ ___ ___ ___ ___ end
Sprinkler discharge
    volume caught   ___ ___ ___ ___ ___ ___ ___ ___ ___ ___
    time minutes   ___ ___ ___ ___ ___ ___ ___ ___ ___ ___

Rate caught   ___ ___ ___ ___ ___ ___ ___ ___ ___ ___

Pressure   begin   ___ ___ ___ ___ ___ ___ ___ ___ ___ ___

Pressure   end   ___ ___ ___ ___ ___ ___ ___ ___ ___ ___

Time of test: begin _____ finish _____ ; duration: hr ___ min ___ decimal hour _____

Evaporation gauge depth: begin _____ finish _____ loss _____ ; rate _____

Comments _____

CATCH DATA IN GAUGE CONTAINERS

Show sprinkler numbers and locations   Show depth or volume caught above line
    on lateral   Show calculated rate caught below line

___ ___ ___ ___ ___ | ___ ___ ___ ___ ___
___ ___ ___ ___ ___ | ___ ___ ___ ___ ___
___ ___ ___ ___ ___ | ___ ___ ___ ___ ___
___ ___ ___ ___ ___ | ___ ___ ___ ___ ___
___ ___ ___ ___ ___ | ___ ___ ___ ___ ___
___ ___ ___ ___ ___ | ___ ___ ___ ___ ___
___ ___ ___ ___ ___ | ___ ___ ___ ___ ___

FORM 5

should be started at the proper time.

If filling can be from the high side and draining from the low side, the advance and recession curves can be fairly well controlled by adjusting inflow and outflow rates. If both filling and draining are from the lower side, over-irrigation will occur there, but it may be of little consequence on slow intake rate soils. Efficiencies can be comparable to those found with border-strip irrigation.

17.3.5 Open Field Sprinkler

Sprinkler systems are usually evaluated in the field by determining the uniformity coefficient (UC), distribution uniformity (DU), potential application efficiency (PELQ), and actual application efficiency (AELQ). In addition, measurements are made to determine if the variations between sprinklers along laterals and between laterals are greater than the maximum recommended by ASAE standards. The test evaluations at the tested area do not include line filling and emptying losses, gasket leakage, or non-uniformity of discharge along the lateral and other locations in the field.

Open field (fixed grid) systems include hand move, side roll, end tow, side move with and without trail lines, portable solid set and permanent solid set systems. A sample data sheet evaluating these systems is shown in Form 5. As much data as is practical should be recorded on the form prior to layout of the cans for the can-catch test. This will help determine the number of can-catch tests that should be run. If pressure and discharge data from sprinklers fall within ASAE recommendations, only one can-catch test need be run. However, the two areas between three sprinklers should be used as the test area for single lines and the comparable area between two laterals for solid sets (also see Section 14.2.4).

Can Layout. If an individual lateral is being tested, cans must be placed on both sides of the lateral as shown in Fig. 17.5. The catchments are then mathematically moved to a position representing set B so that the effects of overlap can be calculated. A minimum of 20 cans with a maximum spacing of 3 m must be in the overlapped area to give reasonable uniformity test values.

When sprinklers are operated as a block, cans are spaced equal distances apart between sprinklers as shown in Fig. 17.6. If a triangular sprinkler spacing is used, they are arranged as shown in Fig. 17.7.

Alternate sets which should normally be used for single line sprinklers are shown in Fig. 17.8 This procedure reduces the effective spacing between lateral sets 50 percent when two full irrigations are superimposed. Uniformity can frequently be increased as much as 15 percent through this procedure and larger moves may become practical. To determine the alternate set uniformity, the combined overlapped catch pattern from the design condition is superimposed on itself but offset the half move distance simulating the second round of irrigation. When move distances are not an even multiple of can spacings, catchment cans will not be properly spaced for direct superimposition. Interpolation is then necessary to obtain the overlap effect.

Evaporation from the catch cans varies slightly, being greater for cans near the edge than those nearer the sprinkler which are more thoroughly wet on the outside and contain more water. A test can set outside the catch area with expected typical catch depth placed in it at the start and measured at the end provides an estimate of the loss. All cans should be adjusted relative to this amount if it is important. Evaporation can be retarded by wiping the interior of each can with a cloth moistened with light oil prior to the test (also see Section 14.5).

Selection of cans. Catchments in cans are usually measured volumetrically in milliliters, but can be measured directly as depth in cans with vertical sides. Plastic cups with sloping sides which can be nested for carrying are often more convenient, but are easily blown about. The volume caught in them must be related to depth based on the area at the rim. The tops of the cans must be level and close to the ground.

Duration of tests. Usually tests are run until at least an average depth of 10 mm is caught, but the results will be more accurate if run longer. If tests are to include the overall effect of changes in wind velocity and direction during an irrigation, they should be run for the full period of the irrigation. Long tests with larger can catches improves measurements by reducing the reading error associated with measuring small quantities of water.

Details of each test should be recorded on the data sheet Form 5. Note whether runoff or ponding occurred. Calculations can then be made. Accuracies greater than plus or minus 3 percent should not be expected. It is desirable to convert all test data to one hour duration (rate) for easier comparisons and operation calculations involving various durations.

Distribution uniformity (DU). The distribution uniformity is evaluated by measuring the amount, depth, or rate caught. If water is applied through

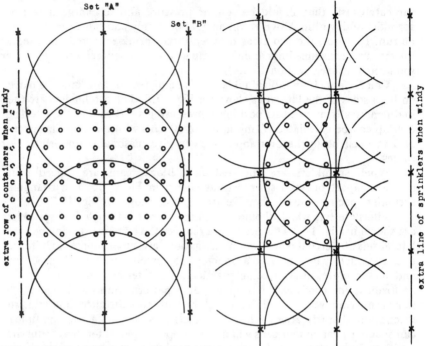

FIG. 17.5 Catch container layout for single lateral sprinkler uniformity test.

FIG. 17.6 Catch container layout for solid set (block) sprinkler layout.

sprinklers at a rate equal to or less than the intake rate of the soil so there is no runoff, can catchments are used to determine the depth of water infiltrated. Can-catch data are mathematically overlapped to simulate infiltration from a complete irrigation at the specific move distance as shown in Fig. 17.5 in which Set "B" is super-imposed on Set "A" (left on right side of test run) and illustrated in Example 1.

Example 1.
Combined Can Catches (* indicates sprinkler location)

	Set A		
*			*
2	6	4	00
78	64	64	74
80	70	70	74
16	18	12	20
62	48	56	50
78	66	68	70
20	26	22	30
32	30	32	34
⑤②	⑤⑥	⑤④	64
48	40	38	48
26	20	18	24
74	⑥⓪	⑤⑥	72
74	64	56	80
8	6	8	6
82	70	64	86
*	Set B		*

NOTE:
1. All readings in mL caught
2. US quart oil cans used.
   (200 mL volume = 25.4 mm depth = 1.00 in depth)
3. Catch time = 1.16 h
4. Sprinkler spacing, 12.2 m × 15.2 m (40 × 50 ft)
5. Circled catches indicate inclusion in LQ.
6. Sprinkler Q = 0.38 L/s (6.0 gal/min)

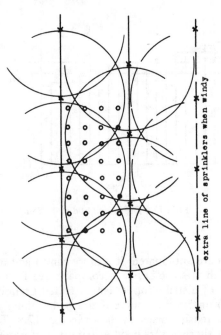

FIG. 17.7 Catch container layout for solid set (triangular) sprinkler layout.

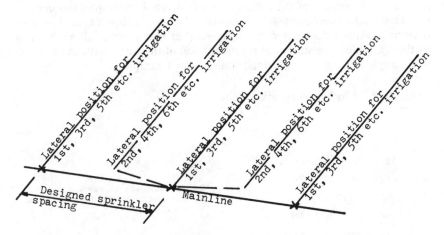

FIG. 17.8 Lateral lines offset half move distance (alternate set irrigation).

$$DU = \frac{\text{average of low quarter catch}}{\text{average catch}} \cdot 100 \ldots \ldots \ldots \ldots \text{(equation} \quad [17.4])$$

$$DU = \frac{(60 + 56 + 56 + 54 + 52)/5}{(\Sigma \text{ of catches})/20} \cdot 100 = \frac{55.6}{68.2} \times 100 = 81 \text{ percent}$$

To simplify calculation for just DU or UC values, volumetric or weight measurement of catches may be used directly in place of depths or rates. However, if several calculations of efficiencies and operations involving

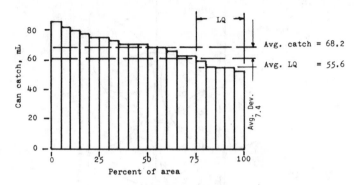

FIG. 17.9 Depth caught distribution.

various durations are anticipated, it is more convenient to convert the initial data to rates which are also easier to compare (see Form 5).

The concept of distribution is demonstrated graphically if a depth infiltrated (caught) distribution curve is drawn by lining up the can catches side by side in decreasing order as shown in Fig. 17.9. This example with a DU of 81 percent shows a very good distribution of water, with only 19 percent of the water infiltrated going too deep when the SMD has been just satisfied. 100 percent of the subsequently infiltrated water goes too deep.

Uniformity coefficient—Christiansen (UC). A common way to show how uniformly water is distributed by sprinklers (but applicable to all methods) is by the UC which is a statistical representation of the catch pattern as shown in Example 2 using the combined values from Example 1.

Example 2. (* indicates sprinkler location)

*			*	
80	70	68	74	
78	66	68	70	
52	56	54	64	
74	60	56	72	
80	70	64	86	
*			*	

NOTE: 1. All readings in mL caught
        2. Sprinkler spacing, (12.2 m × 15.2 m (40 × 50 ft)

The can catches have been arranged sequentially below and on Fig. 17.9 for illustration but this is not essential.

$$UC = \frac{\text{average catch - average deviation from average catch}}{\text{average catch}} 100$$

. . . .(equation   [17.5a])

Can catch	Mean	Deviation from mean
86 mL	68.2 mL	17.8 mL
82	”	13.8
80	”	11.8
78	”	9.8
74	”	5.8
74	”	5.8
72	”	3.8
70	”	1.8
70	”	1.8
70	”	1.8
68	”	0.2
68	”	0.2
66	”	2.2
64	”	4.2
64	”	4.2
60	”	8.2
56	”	12.2
56	”	12.2
54	”	14.2
52	”	16.2
1364 mL		148.0 mL
1364/20 = 68.2 (mean)		148.0/20 = 7.4 (average deviation)

$$UC = \frac{68.2 - 7.4}{68.2} \cdot 100 = 89 \text{ percent}$$

TABLE 17.1. AVERAGE RELATIONSHIP BETWEEN DISTRIBUTION UNIFORMITY AND CHRISTIANSEN'S UNIFORMITY COEFFICIENT

Coefficient	Coefficient values, percent							
Low quarter DU distribution uniformity	39	45	52	59	66	74	83	92
Christiansen's UC uniformity coefficient	60	65	70	75	80	85	90	95

DU values leave about 1/8 of the area at less than the numerator values of LQ average. (see Fig. 17.9)
UC values leave about 1/4 of the area at less than the numerator value of average - average deviation.

Potential application efficiency (PELQ). The PELQ is the only term that is used in design to calculate the gross depth of water to apply at an irrigation. It indicates how effectively a given sprinkler system can apply water under optimum management neglecting prenozzle losses and leaving about 1/8 of the area under-irrigated. (If it is desired to under-irrigate more than the 1/8 of the area resulting from using the average of the low quarter, lighter applications can be made.) The difference between the water applied (as measured at the sprinkler nozzle) and the water caught is an approximation of evaporation and drift losses plus water not accounted for due to part of the area being ungauged (also see Section 14.2). The PELQ indicates how efficiently the tested system can apply water if the system is run the correct length of time to just satisfy the MAD. It is, therefore, a measure of the best that management can do and should be thought of as the potential for the system under the tested conditions to the extent that the tested area is representative of the field.

Example 3.

Combined Can Catch Rates (* indicates sprinkler location)

*			*
8.9	7.6	7.3	8.1
8.4	7.1	7.4	7.6
(5.6)	(6.1)	(5.8)	7.1
8.1	(6.6)	(6.1)	7.9
8.9	7.6	7.1	9.4
*			*

NOTE:
1. All readings in rate mm/h from a test of 1.16 h
2. Sprinkler spacing 12.2 m × 15.2 m (40 × 50 ft)
3. Circled catches indicate inclusion in LQ.
4. Sprinkler Q = 0.41 L/s (6.5 U.S. gal/min)

$$\text{average application rate} = \frac{0.41 \text{ L/s} \times 3600 \text{ s}}{12.2 \text{ m} \times 15.2 \text{ m}} = 8.0 \text{ mm/h}$$

$$= (\frac{96.3 \times 6.5 \text{ gal/min}}{40 \text{ ft} \times 50 \text{ ft}} = 0.31 \text{ in./h}$$

$$\text{PELQ} = \frac{\text{avg. LQ rate or depth of water caught}}{\text{avg. rate or depth applied through sprinklers}} \, 100$$

$$\cdots\cdots\cdots\cdots\cdots\cdots\cdots\cdots\cdots\cdots\cdots\cdots\cdots\cdots\text{(equation} \quad [17.3])$$

$$\text{PELQ} = \frac{(5.6 + 6.1 + 5.8 + 6.6 + 6.1)/5}{8.0} \, 100 \ = \frac{6.0}{8.0} \, 100 = 75 \text{ percent}$$

The water lost to evaporation, wind drift or otherwise unaccounted for is the different between the average rate applied and the average rate caught. This is:

$$\frac{8.0 - 7.4}{8.0} \, 100 = 8 \text{ percent}$$

Additionally, the prenozzle losses from line drainage and leaking gaskets amounting to 3 to 5 percent of the water delivered, need to be considered.

Actual application efficiency (AELQ). The AELQ requires that the soil moisture deficit be known since this is the maximum depth that can be stored. The actual applied depth neglecting prenozzle losses is calculated using the application rate for the duration of the set (T_a).

Example 4.

$$\text{SMD} = 114 \text{ mm}, \, T_a = 23.5 \text{ h}$$

$$\text{AELQ} = \frac{\text{avg. LQ depth of water infiltrated and stored}}{\text{average depth of water applied}} \, 100$$

$$\cdots\cdots\cdots\cdots\cdots\cdots\cdots\cdots\cdots\cdots\cdots\cdots\cdots\cdots\text{(equation} \quad [17.2])$$

$$\text{AELQ} = \frac{114}{8.0 \times 23.5} \, 100 \ = \frac{114 \text{ mm SMD}}{188 \text{ mm applied}} \, 100 = 61 \text{ percent}$$

The difference between AELQ at 61 percent and PELQ at 75 percent (Example 3) identifies a managment problem of the set being too long. If the set were run for

$$T_i = \frac{SMD}{LQ \text{ rate}} = \frac{114 \text{ mm}}{6.0 \text{ mm/h}} = 19.0 \text{ h}$$

or if the SMD were made equal to an MAD of (6.0 mm/h) $\times$ 23.5 = 141 mm (5.6 in.), the AELQ would equal PELQ assuming that wind, humidity, etc. for the day were typically represented by the short duration test.

17.3.6 Center Pivot Sprinklers

The center pivot sprinkler system consists of a long pipe supported on towers fairly high above the ground to pass over tall crops. Each tower has a drive motor so that the entire line rotates about a central pivot point at which water is forced into the pipe and sprinklers. Because of continual movement of the system and the engineered placement and size of the sprinklers, a very high distribution uniformity of about 90 percent is possible. On long lines, 300 m or more, the rate of application at the outer end may be greater than the soil intake rate and runoff may occur. On large fields adapted to this method, there may be appreciable differences in elevation which will affect relative pressures. Because the system rotates, and the fact that the system can be set to pass over different areas during any windy period, the bad effect of wind on the pattern is mitigated. Evaporation losses in the air may be higher than other sprinklers because of the height of the system, the losses are different in the daytime wetted areas relative to the nighttime areas which may be important in some locations. These items need to be considered when estimating the system PELQ.

Because of frequent applications, evaporation losses from the soil may be greater than that of other systems. The SMD is not measurable in the normal manner between irrigations. Over-irrigation and a low actual application efficiency can occur if the system is operated too often. Under-irrigation often is planned so that a soil moisture deficiency gradually accumulates, but because of frequent, even daily irrigation, only slight decreases in yield result with many field crops. Special evaluation techniques normally are required to evaluate the effectiveness of center pivot systems (also see Chapter 14).

17.3.7 Giant Sprinklers

Large diameter nozzles operating at high pressures may discharge water 75 m or more. Because of the long range and high trajectories of these sprinklers, which depend on overlap for uniformity, they are greatly affected by wind and normally have a low distribution uniformity of about 60 percent. However, since they are commonly used to supplement rainfall which applies water uniformly, the low DU may be economical. They are not well adapted to arid-area irrigation. Also because of the large diameter coverage, they create poor application conditions on a wide strip along the periphery of fields where overlap is not feasible (also see Section 14.6).

17.3.8 Traveling Sprinklers

Traveling sprinklers are similar to giant sprinklers, but are often operated to provide part circle coverage. They have the advantage of higher distribution uniformity due to the continuous movement so that DU values of

75 percent or higher are common. They have a rather severe periphery uniformity problem.

17.3.9 Perforated pipe Sprinklers

A pipe with proper spacing and angle of a series of small holes in the top sector can produce a very uniform, nearly rectangular sprinkler pattern. Though it is very difficult to evaluate an extensive pattern, a distribution uniformity of about 90 percent may be probable. This system operates at low to moderate pressures, but it applies water at rates normally in excess of 15 mm per hour making it unsuitable for slow to medium intake rate soils. The periphery problem is the least with this type of sprinkler.

17.3.10 Orchard Sprinklers

Small, usually rotating, low trajectory sprinklers that do not depend on over-lap and which generally cover most of the area between four trees, are easily evaluated. They can have very high efficiencies. Such evaluations cannot be measured by using the PELQ, AELQ and DU concepts because only part of the area is wetted and the pattern tapers to zero depth at the outer edge. They must be evaluated using other terms.

The distribution characteristic (DC) which is the percent of the wetted area which receives more than the average depth infiltrated, measures the uniformity with values of over 60 percent being good. Application efficiency, the average depth stored divided by the average depth applied, measures the efficiency. This value can be very high approaching 85 percent which is higher than can be achieved for most irrigation methods. This is due to the fact that much of the area, 30 to 50 percent, is under-irrigated so very little water is lost to deep percolation, and in-air losses are small in orchards.

17.3.11 Trickle Irrigation

Trickle irrigation systems deliver water slowly, usually at the soil surface, from tubing with many closely spaced very small outlets to create an essentially continuous wetted line, or from more widely spaced emitters to create intermittent wetted spots. In practice, at the peak water use design period they are operated daily and so keep the central portion of the wetted soil bulb well above field capacity on medium and fine textured soils. The distribution uniformity (emission uniformity) of new installations may be close to 90 percent, but it and the potential application efficiency usually decline appreciably with continued use. A more typical value of about 80 percent should be considered. The actual application efficiency will be even lower. Since the SMD cannot be measured, but is estimated from evapotranspiration calculations, more rather than less than adequate water often is applied to assure optimum water. Also, the localized above field capacity condition usually maintained by the frequent application implies that there may be continuous deep percolation. Changes in water temperature and viscosity along laterals affect uniformity of emitter discharge to varying degrees depending on the type of flow, laminar or turbulent, through the laterals and the emitters.

Only the DU can be measured by evaluations so the PELQ and AELQ must be estimated. An evaluation of the system should be made annually because of the rapid changes that occur.

17.4 MEASURING DEVICES AND METHODS USED IN EVALUATIONS

A large variety of flow and pressure or head measuring devices are available. Only a few are adapted for use in the evaluation procedure where fairly small flow rates are typical. Water infiltration into the soil from furrows, sprinkler devices, and ponded conditions need to be measured as does the soil moisture deficiency.

17.4.1 Sprinkler Devices

Pressure. Pressure for most sprinkler tests is measured just outside the sprinkler nozzle so that the flow rate is not reduced. Blocking the flow rate may be of no consequence when there are many sprinklers on one line but when there are fewer than about fifteen, the flow rate change may noticeably affect pressure if the flow is blocked to measure pressure. Also there is a difference in pressure due to friction and shape in the riser and sprinkler so that when flow is stopped, line pressure is not identical to tip pressure and its corresponding jet velocity.

A pressure gauge with a pitot tip is the most practical measuring device. The tip must be placed directly in line with the center of the jet flow. The maximum pressure reading can be found by rotating the pitot tip. The pitot tip should be less than one mm in diameter and have a thin edge to reduce the fanning out of the jet which will wet the operator and make the reading of the gauge difficult.

Flow rate. Flow rate is best obtained by filling a known volume container in a measured time. A loose fitting section of garden hose or other material flexible enough to bend but not collapse can be slipped over the nozzle to deflect the stream into the container. It must not fit tightly around the nozzle or a suction caused by Venturi action may increase the flow rate. Siphoning may also occur. For double nozzle sprinklers, each nozzle may be measured separately and added together. For medium to small nozzles, a can may be held in the jet and the flow deflected back out and caught in the container.

The container can be any wide mouth jar or can of known volume. The volume may be determined from measurements or by weighing the container empty and full of water and computing the volume from the density of water. It is best to use the full volume to the rim of the container when it is overflowing.

The manufacturer's pressure-flow rate tables may be used with caution to estimate flow rates from measured pressures. However the pressures measured, while usually of adequate accuracy, are seldom precise. Nozzles become worn and sprinklers may not be well maintained. Nozzle diameters can be checked with feeler gauges or small drills of known diameter. The pressure-flow rate procedure is acceptable for new nozzles.

Distribution pattern. The pattern of the drops falling from sprinklers is determined by measuring the depths of water caught in small containers. For straight sided containers when an average depth of 25 mm or more is caught, the depths can be measured directly with a scale placed in the center of the container. Straight sided cans need not be the same diameter if depth is measured directly. For volume measurements which will be converted to depths, the top area must be determined. The volume is measured in a graduated cylinder and then converted to depth. Sloping sided containers, usually plastic, can be nested for convenience in carrying. They often need to be fastened down or weighted with a known volume of water to prevent being blown away by the wind.

The containers should be set level if the sprinkler riser is vertical. In open fields with fairly long and high trajectories, the drops are falling nearly vertically. With orchard heads on sloping ground where the riser is perpendicular to the ground, the can tops should be parallel to the ground since with many of the low trajectory jets used in orchards there is appreciable lateral velocity to the drops (also see Section 14.2.4).

Containers should all be the same height and set low to the ground to simulate the ground surface. Vegetation that would interfere with the catch must be removed.

17.4.2 Furrow devices

Intake rates determined from furrow tests are based on a series of inflow minus outflow rates over a specific length of furrow at different times, and the efficiency is based on knowing how much water was applied to the field. For both of these purposes it is essential that the stream flow rates be measured with appreciable accuracy, 5 percent or better if possible.

Flumes. Flumes in general are the most practical device to use except for the flattest grades. The bob-tailed Parshall and the cutthroat flumes require a drop through critical depth and only one depth measurement centrally made in the upper converging section. The drop must be created by some increase in depth upstream which will affect intake rate somewhat. By judicious choice of flume width, ponding may be made reasonable—25 to 40 mm. The flume must be carefully leveled. (Graphs for Parshall flumes are shown in Fig. 17.10.)

For slopes of about 0.002 or less, the ponding may become objectionable because of the increased wetted perimeter of the furrow. Flumes with flaring

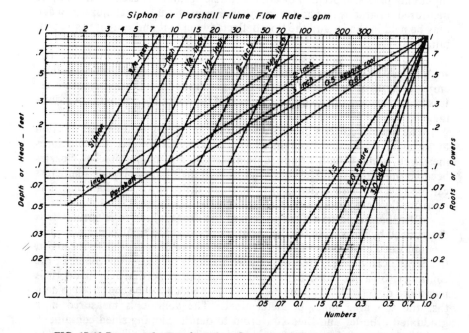

FIG. 17.10 Powers and roots of numbers flow rates of Parshall flumes and siphons.

downstream sections will reduce the ponding and permit submerged flow, require two depth measurements (also see Section 9.2).

The small portable flumes used for evaluations seldom have sight tubes for measuring depths or differences in depths. Staff gauges marked on the side of the flume rarely can be read with adequate precision. The best way to obtain the necessary accuracy is to use a point gauge to measure down from a datum established usually at the top of the flume. The depth should first be measured down to the bottom of the flume and then to the water surface. The difference is then the desired depth. A simple point gauge can be made using any scale with adequately fine gradations and firmly attaching a short piece of wire to it with tape or a rubber band. Precision to 0.5 mm is desirable. A thin stick gauge can be used to measure the depth in the flume directly but it is often hard to read precisely.

The easily-made flat bottomed vertical sided cutthroat flume is satisfactory. Flumes with sloping sides lose accuracy with increasing depth and so are less preferable.

Orifices and tubes. Orifices, short tubes, and irrigation siphons can be used to measure flow by measuring the head and cross-sectional area.

For orifices and short tubes when the head measured from the center of the opening to the upstream water surface for free flows is greater than the diameter, good accuracy is obtainable. However since the flow is a function of the square root of head, appreciable ponding may be needed upstream for fairly small increases in flow. Orifices are adapted for use at the upper end of a furrow when flow is constant and the necessary head can be obtained. A series of orifice plates may be needed to obtain the desired range in flow with reasonable head.

At the outflow station for intake tests the orifices desirably operate submerged with the difference in head being measured. It may even be practical to excavate the bottom of the furrow to obtain adequate submergence if erosion does not make it impractical to keep the depression from filling in. Round orifices do not need full clearance on the bottom so they can be set within a half diameter of the bottom and still be adequately accurate.

Because of the increasing flow rate passing the lower measuring device with time, a series of increasing size plates can be set and the smaller ones pulled out in sequence when head becomes excessive. When the plates are set fairly close together, the increased furrow section length is negligible.

Flow rates in orifices are calculated from the equation

$$Q = CA \, (2gh)^{0.5} \quad \ldots\ldots\ldots\ldots\ldots\ldots\ldots\ldots\ldots\ldots\ldots\ldots\ldots\ldots\ldots [17.12]$$

For sharp edged orifices which exist for holes drilled in the plates C varies from 0.61 to 0.63. A is the orifice area, h is head, and g is the gravity constant all in units consistent with the flow units.

Short tubes (100-mm) with 10-mm radius rounded entrances set into a face plate which can be set across a furrow, function similarly to orifices but have greater capacity for an equal sized opening. Flow for them is given by the equation

$$Q = CAK \, (2gh)^{0.5} \quad \ldots\ldots\ldots\ldots\ldots\ldots\ldots\ldots\ldots\ldots\ldots\ldots\ldots\ldots [17.13]$$

where both C and K are for practical purposes equal to 1.00. The tubes can be circular but are best made rectangular with widths of 30 to 100 mm. The depth should be shallow, e.g. 30 mm, so that they can easily be operated submerged in fairly shallow furrows with very little upstream ponding.

Orifices and short tubes are convenient to use since the plates are easily installed by setting them in a trench if the soil is dry and hard, or they can be pressed into wet soil. For submerged flow they need not be set truly vertical.

Siphons can be used as measuring devices as well as for water distribution. They function similarly to short tubes with a projecting entrance so C and K are less than 1.00, but flows are proportional to the square root of head. Different entrance shapes and tube material will cause minor changes in flow rates so they are usually empirically calibrated. A graph of Head vs Flow is included on Fig. 17.10.

A difference in head is needed to obtain flow rates through irrigation siphons, short tubes, and orifices. For siphons and tubes this can be most conveniently measured by using a 1.5- to 2.0-m length of plastic tubing 2- to 3-mm in diameter. It is helpful to have a short piece of metal tubing on the inlet end of the plastic tubing to hold it in place. Graduations may be marked on the outlet end but a small scale is often as practical. The tubing is filled with water and allowed to flow as a siphon around the equipment being measured to be sure all air is cleaned out. The lower end of the tubing is then held vertically above the downstream water surface for submerged conditions or adjacent to the equipment for unsubmerged outlets. The rise in elevation of the water level in the tubing above the downstream water level for submerged flow, or above the center of flow for the free flowing condition, is the head to be used to calculate the corresponding flow rate.

The difference in head can also be measured using a carpenter's level or a comparable device with a spirit level on it. A stick can then be extended from the upstream to downstream side and set level. The depth to the desired points at each end can then be measured vertically and the difference found. The use of a surveyor's level is practical but seldom convenient.

For the small orifices used in evaluations it is practical to measure the head directly on the orifice plate on one or both sides as needed. It must be measured vertically.

Weirs. Weirs may be used for furrow evaluations but because of decreased accuracy caused by large flow increases with small head increases, and the need to measure head some distance away from the crest, it is more difficult to obtain precision with them.

The 90° V-notch weir is the only practical one used for small flows. The equations for it is

$$Q = C H^{2.5} \dotfill [17.14]$$

where C is 2.5 for flows in cubic feet per second and 1380 for flow in liters per second, and H is in feet and meters respectively.

Special shaped weirs essentially combining the orifice and weir conditions, can be made and individually calibrated. A narrow weir decreasing in width towards the top will provide a flow vs depth relation that is nearly linear. A set of several sizes may be needed. They need to be individually calibrated. They, like the orifice plates, can be set in sequence and as flow and head increase, the smaller upstream plates can be removed.

Volumetric. A volume in a measured time interval is probably the most precise method if the time duration is adequate so that it can be measured precisely. A stop watch measuring the 0.1-second or 0.002-minute is desirable. Containers that fill in less than five seconds tend to cause imprecise values. Such flows usually entrain much air and some overflow must be allowed to compensate. Three or more measurements can easily be made and averaged. It is desirable to have two people cooperating with the watch and container, one of whom should make a countdown to be sure of a common starting time.

The stream flowing into the container must be free flowing. For flow in furrows, a hole usually must be dug to hold the container under the jet from the gated pipe, siphon, or spile.

17.4.3 Border-strip and Basin Devices

Border-strip and basin systems require large streams and correspondingly larger measuring devices.

Siphons and flumes. Where siphons or flumes are used, the earlier discussion under furrows is pertinent except for size.

Weirs. Weirs can be used for flow measurements where the water can be concentrated in one location. For large areas, several weirs may be needed. Rectangular weirs are useful. Cipolletti trapezoidal weirs have no advantage for shallow depths. Weirs must have adequate clearance below the crest on the upstream side to have a standard bottom contraction. Two times the head or more is desirable. To obtain this may require an appreciable ponded depth, or a trench of adequate depth and five times the head in width may be excavated adjacent upstream to the weir. Clearance at the ends of the crest should be about three times the head. The jet from the weir should fall freely so downstream depth of flow will not back up to the crest of the weir though a little drowning-out will not greatly affect accuracy.

The depth over the crest should be not less than about 15 mm. The weirs for this purpose may be quite long and shallow to permit large flows with small heads. They may even have a little horizontal curvature to provide stiffness against overturning.

For shallow wide flows the end contractions are of little importance, but the crest of the weir should be level. If it is not, the length may be broken into sections and a representative head used for each section.

The flow equation used for wide rectangular weirs is

$$Q = K (L - 0.2H) H^{1.5} \quad \dots\dots\dots\dots\dots\dots\dots\dots\dots\dots\dots\dots\dots\dots\dots \quad [17.15]$$

$K = 3.33$ where L and H are in feet and Q is in cubic feet per second. For corresponding SI units K is 1840, where L and H are in meters and Q is in liters per second.

Jet trajectory. Trajectory of a horizontal, or nearly so, jet may often be a simple and practical measuring device where flow from the end of a pipeline is accessible. For full flowing pipes the velocity of flow is equal to four times the distance (L) from the end of the pipe to where the jet has dropped 1.01 feet or 0.308 m which occurs in 0.25 second. The drop and distance would theoretically be measured at the centerline of the jet but the top surface gives reasonable accuracy. The area of the pipe (A) should be calculated in square feet or square meters. The product of the velocity times area gives flow rates (Q) in cubic feet per second or cubic meters per second. Tables are available for pipes flowing partially full.

$$Q = 4 A L \quad \dots\dots\dots\dots\dots\dots\dots\dots\dots\dots\dots\dots\dots\dots\dots\dots\dots\dots \quad [17.16]$$

Vertical jets occur from pipeline valves such as orchard valves. Approximate flow rates can be found for small heights where the pipe rim acts essentially as a weir. When the rise (H) is less than 0.4 of the pipe diameter (d)

$$Q = 8.8 \, dH^{1.5} \quad \dots\dots\dots\dots\dots\dots\dots\dots\dots\dots\dots\dots\dots\dots\dots\dots \quad [17.17]$$

where H and d are in feet and Q is in cubic feet per second.
When the rise H is greater than 1.4 d

$$Q = 5.6 \, d^2 \, H^{0.5} \quad \dots\dots\dots\dots\dots\dots\dots\dots\dots\dots\dots\dots\dots\dots\dots \quad [17.18]$$

For flow in m³/s and d and H in m, then

$$Q = 4.9 \, dH^{1.5} \quad \dots\dots\dots\dots\dots\dots\dots\dots\dots\dots\dots\dots\dots\dots\dots \quad [17.17a]$$

and

$$Q = 3.1 \, d^2 \, H^{0.5}, H > 1.4 \, d \quad \dots\dots\dots\dots\dots\dots\dots\dots\dots\dots\dots \quad [17.18a]$$

For H values between 0.4 and 1.4 d, the flow rate is a little less than that given by either equation. Because of the distortion of the vertical jet by the valves and the pipe therefore not flowing full, both equations probably give higher than true values. The jet is generally rather unstable so values of H are not very accurate.

Other devices. Additional flow measuring devices may sometimes be useful in ditches or pipelines that can be equipped with them. In ditches current meters, surface floats, moving vertical sticks, etc., may be used to determine velocities in several sub-section areas of the cross section and then their product totaled to determine the flow.

In pipelines, a variation of pitot meters are often used. Also visible fluorescent dye at a few parts per million may be injected to observe the velocity. Various other types of meters for rate or total flow volume are available but they are usually permanently installed.

Sometimes measuring devices may be improvised based on some hydraulic characteristic such as constrictions or critical depth; or the stream may be proportionally divided if the total flow is known. The flow measurements may often have to be made in the supply system rather than at the point of application in the field.

17.4.4 Soil Moisture Infiltration and Deficiency Measurements

Efficiency approximations can be made from simple observations of conditions. However in order to obtain numerical values and information for design or comparison, test of the soil infiltration characteristics should be made when conditions representing the usual irrigation conditions exist. This would typically be when the top 10 cm of soil is at about wilting point, which occurs at about a 50 percent MAD. Moist soils or soils that have wet surfaces as from a rain, will have a much slower initial infiltration rate. Soils that are excessively dry may have appreciable open cracks and normally may need to be irrigated before they thoroughly dry out to mitigate cracking. The soil moisture deficiency conditions of the surface and the profile of a tested soil must be recorded as part of the measurements.

There are two different conditions under which irrigation water enters the soil. They require different measurement methods. One condition is called intake which is that from shaped areas such as furrows where the rate depends on the furrow shape and size and involves lateral as well as vertical flow. The second condition is called infiltration which occurs with a flooded surface and the flow is vertical.

Water moves in a soil downward by gravity and capillarity. It also moves laterally and upward by capillarity where shape permits it. Because of the high energy difference between the dry soil near the surface and the free water on it and the corresponding short distance between them, the initial intake rate is very rapid. Gravity plays a relatively small part of the initial rate except in coarse soils. As the length of the capillaries increase and the wetting front reaches moister soil the infiltration rate decreases as the capillarity effect becomes smaller and the gravitational effect plays a proportionally greater part until finally it essentially becomes the only force.

The infiltration rate, or the intake rate for the specific tested length of furrow, generally can be described by the equation

$$I = K T^n + C \quad \dots\dots\dots\dots\dots\dots\dots\dots\dots\dots\dots\dots\dots\dots\dots \quad [17.19]$$

and from the integration the cumulated infiltration depth or the intake depth on unit width of 1.0 foot or 1.0 meter of furrow by

$$D = K' T^{n+1} + CT \quad \dots\dots\dots\dots\dots\dots\dots\dots\dots\dots\dots\dots\dots \quad [17.20]$$

where $K' = K/(n + 1)$.

In this equation K represents an initial intake characteristic of the soil and C represents a more or less stable final rate approximately associated with an unsaturated hydraulic conductivity. T is usually in minutes. For furrows the K values are usually adjusted to a length of 100 feet or to one meter (also see Chapter 4).

For practical purposes in evaluating an irrigation the usual durations permit making C equal to zero and dropping the second term. A plotting of the first term only on log-log paper results in straight lines the intercepts of which at one minute are the values of K and K', and n and n + 1 are the geometric slopes of the lines with n being negative.

Because of various physical conditions in the soil, the plotting of actual infiltration and intake measurements often has a steeper or flatter dog-leg section for the first five to ten minutes, and sometimes more. Steeper initial sections may be caused by plowpans reducing the later flow or by cracks in the soil into which water can enter quickly. As the cracks swell and close the rate decreases rapidly. The magnitude of this effect is greatly affected by the dryness of the soil and resulting size and depth of the cracks.

For ponded conditions using cylinder infiltrometers, graphs with flatter initial slopes steepening after a short time may be caused by air entrapment when initially wetted, particularly in coarse soils, or by the cylinders not having been driven deeply enough—150 mm or more.

Because of soil and sampling differences and infiltration occurring before the first readings can be made, many measured infiltration tests and some intake tests should be "adjusted" to obtain valid numerical values. The "adjusting" process is detailed in the sections on Border-Strips for infiltration and in Furrows for intake. Such adjustment frequently involves changes of more than 100 percent where cracks are prevalent.

Texture often correlates with certain infiltration or intake values of K and n, but structure, the result of natural or cultural processes, often greatly modifies the numerical values from typical textural values.

Furrow intake tests. The intake measurement is done by measuring the inflow and the outflow rate at the two stations. Suitable flow measuring devices are set in several furrows having different typical flow rates. They are set at measured distances apart, 30 m (100 ft) being typical. The setting and flow measurements must be carefully done as described in Section 17.4.2. For slow intake soils the spacing may be longer or two furrows may be combined if each furrow carries about equal flow. If the soil at the upper end of the furrows is typical, the upper device is set at the upper end of the furrow so that the inflow rate can remain constant.

For soils that crack or have a loose layer over a plowpan permitting abnormal lateral flow, adjacent furrows must be run and it may be very difficult to have access to and make measurements at the lower device. Where access becomes too difficult, the entire furrow length can be used. However values can only be obtained after water has advanced the furrow length so the early part of intake is not measured. By utilizing the "adjusting" procedure a curve can be developed but the high initial intake rates will be approximations as the dry dog-leg effect can only be conjectured. This means that the presumed instantaneous intake rate is an average of appreciably different rates. For normal soil conditions, it is more convenient to run alternate furrows to provide a walkway for access to the lower device. Entries called for on the Furrow Intake Test form (Form 2) should all be made. The zero starting time for making measurements is when the stream has reached half way to the second station since this will closely represent an average time on the section. The second rate and time measurement is made simultaneously (or nearly so) at each station when the flow is three to five meters past the second station. Several readings should be taken at short intervals of three to five minutes, and then at increasingly longer intervals for the duration of irrigation (T_i) or the time of advance (T_{adv}) whichever is larger, and at least eight readings should be taken. They must be taken with appreciable precision since the difference between them is presumed to be the intake rate at that moment. The precision of the readings should be determined and recorded so that the "range of accuracy"—the plus or minus of the precision of the readings—can be plotted on the intake rate curve.

Utilizing the "range of accuracy" in plotting the instantaneous intake rate points, is often essential to make the field data useful. Because the difference of inflow minus outflow often may be small (dealing with small numbers), ignoring the accuracy of the two readings may result in misleading numbers. By plotting the points with their plus or minus accuracy to establish a range, a straight line can almost always be drawn. If the points are plotted as the data is taken, poor sets of readings become obvious and can be redone.

Tests should be run on several furrows. The results should not be averaged. Each line should be drawn separately as different conditions may be affecting the results. However they are all expected to be parallel. Soil structure in furrows is often different because of location of tractor wheel travel lanes. In orchards, furrows adjacent to trees may have twice the intake rate of those in central portion. Different stream sizes giving different wetted sections will have some effect, more so in V-shaped furrows than in parabolic or broad ones.

The intake rate plotting and equation are not related to furrow spacing but the plotting of the cumulative intake depth curves and the equations must be related to the specific furrow spacing being studied assuming that water is stored in this width. Since change in furrow spacing is one of the important management controlled variations affecting duration of irrigation, several spacings should be studied for their effect. This is done by finding K_s' values for the various spacings.

Care must be taken with units for the equations. Intake rates for furrows are usually expressed in gallons per minute per 100 feet and liters per minute per meter. There is no width implied for these rates.

When integrating these rates to cumulative depths, area (width) must be considered as well as the conversion constants to obtain the usual depth units of inches and millimeters in the equation

$$D = K_s' T^{n+1} \dots\dots\dots\dots\dots\dots\dots\dots\dots \quad [17.21]$$

The relation of K in gpm/100 feet to K_s' in inches is obtained by dividing by the units conversion constant multiplied by the furrow spacing in feet and the integration constant

$$K_s' = \frac{K \ (\text{gal/min-100 ft})}{62.3 \ (n+1) \ S} \dots\dots\dots\dots\dots\dots\dots\dots \quad [17.22]$$

To convert liters per minute per meter to millimeters depth with a given furrow spacing in meters

$$K_s' = \frac{K(\text{L/min-m})}{(n+1)S} \dots\dots\dots\dots\dots\dots\dots\dots\dots \quad [17.23]$$

Cylinder infiltration test. Where water is applied over the entire soil surface such as for border-strip or basin irrigation, single cylinder infiltrometers provide a comparable vertical flow condition only if their diameter is 250 mm or greater and they are driven at least 150 mm deep. Double ring or buffered cylinders with water level maintained on both sides are sometimes used but seldom provide more accuracy. For soils that do not crack, the infiltration depth from the cylinders may relate fairly closely to the field conditions. More commonly they must be "adjusted" as described in Section 17.4.2 since the small sampled areas seldom represent the actual conditions. Tests must be made when the soil moisture deficiency (SMD) is about the same as when as irrigation should be applied since the infiltration rate varies with moisture content. The rate often also decreases with lateness in the season as the soil structure becomes modified.

At least four cylinders varying in diameter from 250 to 300 mm so that they can be nested for carrying, about 400 mm long, and having a wall thickness of about 1.5 mm, are driven into the ground at least 150 mm. They are usually driven into the soil by a heavy (10 to 15 kg) hammer driving against a 6-mm thick plate about 0.5 m square laid over the top of the cylinder. The operator stands on top of the plate to add weight and drops the hammer to do the driving. The cylinder should be driven straight without wobbling to reduce the size of any cracks around the perimeter. After driving any small crack that does occur should have the soil pressed down to block it.

The test location should be selected to represent as typical a condition as possible and all four, or more, cylinders should be in a small group. Unusual soil conditions such as cracks, cow prints, tractor wheel paths, etc., should be avoided.

The procedure consists of selecting a datum on each cylinder. This may be a point on the top or a mark made inside at a more convenient depth. A scale with a point that will reach the ground from the datum is needed. The point may be a short piece of wire taped to any convenient scale. A point is preferable to a hook which does not permit the measuring of the lowest part of the water depth. High accuracy is not needed for measurements so readings to the nearest 1 mm is adequate.

Before water is poured into a cylinder, protection from possible erosion must be assured. Existing sod may be adequate, or a piece of cloth or paper may be laid in the bottom. The information on the heading of the Cylinder Infiltrometer Data sheet (Form 4) should be filled out and the cylinders identified usually by their diameter. Since several cylinders are to be tested, a stop watch is not desirable nor is it needed for accuracy.

At a convenient full minute water is poured into a cylinder to a depth corresponding to the typical ponded irrigation depth usually about 100 mm, the protective cloth removed, and a depth to water surface measurement made as soon as possible, usually at 15 to 20 s. The depth and watch time are recorded. On very high intake rate soils a second depth reading at 30 s is desirable. Under normal conditions the next reading is taken at 1 min. Subsequent readings should be taken at about 2, 4, 7, 10 min and increasingly longer intervals to provide three or four points for each cycle on the log-log paper. The test duration should be as long as needed to apply an irrigation but may be less as long as a straight line graph is well established. After the first cyclinder is well started, a second may be started.

For some high intake rate soils, it may be necessary to add water. This is done immediately after taking a normal reading. The new refilled depth is recorded as though taken at the same time as the normal one was taken. The change in depth is ignored by leaving it blank on the form.

The cumulative times and depths are desirably plotted for each cylinder as the data is taken. Errors in measurements become obvious and new readings can then be taken. The plotted lines should be straight though often with a flatter or steeper dog-leg section in the first part. The plots from the adjacent cylinders will rarely coincide because of variations in the small sampled areas. They should not be averaged but each should be looked at as identifying some characteristic of the sampled area. It is usual that the "adjusted" curve is above any tested cylinder.

Typical infiltration values for soils are shown in Fig. 17.11. Arbitrarily selected values, but which are representative of many tests, represent various infiltration groups. The USDA Soil Conservation Service intake family curves are shown and also the comparable classification of time-rated curves. The latter are designated as to how many minutes it takes to infiltrate 100 mm (4 in.) which corresponds to a typical depth of irrigation. It therefore has the advantage of having a physically related name. It has a further advantage that variations in the slope $(n + 1)$ of the line do not cause a great effect on the cumulative intake for values in the normal irrigation range of 100 mm plus or minus 35 mm. Broad textural classifications shown must be used with caution since structure has appreciable effect. A great deal of caution must be used when reading the curves for cracking soil.

Soil moisture deficiency. For all evaluations of actual irrigations practices, the soil moisture deficiency (SMD) must be known to obtain the actual

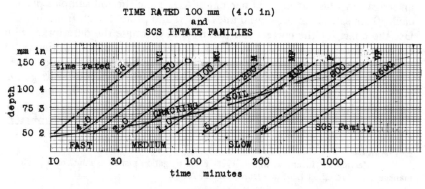

FIG. 17.11

application efficiency, AELQ. The SMD may be estimated using calculated evapotranspiration values, calculated soil moisture sensing devices, etc. It may also be estimated by field observations of the soil profile using devices to make quick measurements of the soil moisture, or by using the feel and appearance of the soil as related to its moisture deficiency. The latter is probably the least complicated and is adequately accurate. The "Soil Moisture Deficiency and Appearance Relationship Chart", (Fig. 17.12) describes conditions. The procedure consists of taking samples of the profile in 30 cm (1.0 ft) increments to the bottom of the root zone and summing the deficiencies.

Skill and confidence in learning to estimate the SMD are developed by sampling the profile in a cropped field where a normal growth pattern of the root system has developed—an orchard with 1.5 to 2 m root zone in uniform soil is ideal. After an irrigation or rain has wet the entire profile, wait until about the top ten centimeters has reached wilting point, about the time for

SOIL MOISTURE AND APPEARANCE RELATIONSHIP CHART
(Approximate relationship of soil moisture deficiency between field capacity and wilting point.)

Moisture deficiency mm/300, in/ft(loamy sand)	Coarse	SOIL TEXTURE CLASSIFICATION			Moisture deficiency mm/300 in/ft
		Sandy (sandy loam)	Medium (loam)	Fine (clay loam)	
mm/300 in/ft					mm/300 in/ft
0 — 0.0	(Field capacity) Leaves wet outline on hand when squeezed	(Field capacity) Leaves wet outline on hand, makes a short ribbon	(Field capacity) Leaves wet outline on hand, will ribbon about 25 mm	(Field capacity) Leaves slight moisture on hand when squeezed, will ribbon out about 50 mm	0 — 0.0
—	Appears moist, makes weak ball	Quite dark color, makes hard ball	Dark color, forms a plastic ball, slicks when rubbed	Dark color, will slick and ribbon easily	—
10 — .4	Appears slightly moist, sticks together slightly				10 — .4
—	Dry, loose, flows through fingers (wilting point)	Fairly dark color, makes a good ball	Quite dark, forms hard ball	Quite dark, will make thick ribbon, may slick when rubbed	—
20 — .8		Slightly dark color, makes a weak ball	Fairly dark, forms a good ball		20 — .8
—		Lightly colored by moisture, will not form a ball		Fairly dark, makes a good ball	—
30 — 1.2		Very slightly colored due to moisture (wilting point)	Slightly dark, forms a weak ball	Will ball, small clods will flatten out rather than crumble	30 — 1.2
—			Lightly colored, small clods crumble fairly easily	Slightly dark, clods crumble	—
40 — 1.6			Slightly colored due to moisture, small clods are hard (wilting point)		40 — 1.6
—				Some darkness due to unavailable moisture, clods are hard, cracked (wilting point)	—
50 — 2.0					50 — 2.0

Field method of approximating soil moisture deficiency (From Merriam, J. L. 1960. TRANSACTIONS of the ASAE 3(1), revised in 1975)

FIG. 17.12 Soil moisture deficiency and appearance relationship chart.

an irrigation when using a 50 percent MAD. Take the soil samples placing each 30 cm sample in sequential piles or row and protect them from the sun. Use the chart for the correct soil texture and correlate the description with the SMD for each sampled layer. It should run the range from wilting point for the top sample to nearly field capacity for the bottom of the root zone.

Sampling to the bottom of the root zone a couple of days after irrigation is the only way to assure that adequate water has been applied to replace the SMD, or whether too much has been applied. Such a check should be made every few irrigations regardless of how the SMD was measured or estimated.

References
1   Merriam, J. L. and J. Keller. 1978. Farm irrigation system evaluation: A guide for management. 3rd ed., Utah State Univ., Logan, 285 p.

chapter 18

IRRIGATION WATER MANAGEMENT

18

18

IRRIGATION WATER MANAGEMENT

by   E. C. Stegman, North Dakota State University,
Fargo, ND; J. T. Musick, USDA-SEA/AR, South-
western Great Plains Research Center, Bushland,
TX; J. I. Stewart, USDA/USAID Kenya
Agriculture Research Institute, Mugaga, Kenya
(formerly University of California, Davis)

18.1 BASIC WATER MANAGEMENT CONCEPTS

18.1.1 Factors Affecting Crop Growth and Yield

For plants to live, they must exchange oxygen (O_2) and carbon dioxide (CO_2) across wet surfaces exposed to the drying effects of the air. As plants exchange gases between the outside air and the interior of their leaves, they inevitably lose water by transpiration. Conditions that favor the exchange of CO_2 and O_2 also favor the loss of water. Thus, water as well as gas exchange, light, and mineral nutrition are essential for growth. Transpiration associates with the plant's capacity to exchange gases and it dissipates, as latent heat, a large part of the absorbed radiant energy not used in the photosynthetic process. This helps the plant avoid damaging effects from high canopy temperatures (Lemon, 1965).

Plant growth develops within a genetic framework in which rate and duration of specific processes are influenced by biochemical compounds such as enzymes and hormones, some of which are adversely influenced by water deficits (Hsiao, 1973). Growth and development processes are also strongly influenced by plant environment. Light and guard cell hydration influence stomatal aperture and diffusion of CO_2 into plant leaves. Plants store only a minor amount of the water they need for transpiration; thus, the storage reservoir furnished by the soil and its periodic recharge are essential in maintaining continuous growth. Temperature affects rate of root water uptake and movement through plants and the rate of biochemical reactions taking place at growing points. Daylength is associated with flowering in many plants. Soil properties affect root extension, availability and uptake of water and mineral nutrients, and the aeration essential for root respiration.

Water either directly or indirectly influences most production factors. Soil water is essential for successful tillage and planting, seed germination, root extension, mineral nutrient absorption by roots, and soil microbiological and chemical processes that result in decomposition of plant residues and mineralization of essential nutrients. Water movement through the root zone of irrigated soils is essential for leaching to maintain a favorable salt balance.

Plant growth and development processes are highly complex and are not fully understood. Even so, man has successfully influenced these processes since the earliest days of crop cultivation. Major improvements have occurred in plant selection and breeding, tillage, fertilization, weed control, disease and insect control, and irrigation. High yields frequently associate with specific management and cultural practices that obtain optimum plant growth in specific environments.

18.1.2 The Soil-Plant Atmosphere System

This section emphasizes interrelationships in the soil-plant atmosphere system that are applicable to development and/or an understanding of water management concepts. More detailed discussions of this subject matter are found in books by: Pierre et al. (1965), Slatyer (1967), Kozlowski (1968a, 1968b, 1972 and 1976), Kramer (1969), Hillel (1971), Meyer et al. (1973), and in reviews by Hsiao (1973) and Begg and Turner (1976).

Atmospheric influences. The evapotranspiration (ET) process is primarily dependent on incoming solar radiation when water is non limiting and plant vegetative cover is complete. The canopy energy balance determines leaf temperature which in turn influences the vapor pressure gradient between the evaporating surfaces within the leaves and the boundary air layer; thus, influencing the transpiration rate (Idso et al., 1966; Gates, 1968). In the absence of significant advective energy, ET proceeds at a rate that is about equal to the evaporation equivalent of the net radiation level (Ritchie, 1971). Pronounced advection from winds in semi-arid to arid climates can, however, occur as warm dry air masses move over irrigated fields (Korven and Pelton, 1967; Rosenberg, 1969). Empirical methods for ET estimation must, therefore, often receive a degree of local calibration to achieve improved accuracy under these conditions (Tanner and Pelton, 1960; Wright and Jensen, 1972; Pruitt and Gupta, 1976).

Periods of high ET demand coupled with depleted soil water conditions may close plant stomata and greatly increase resistances to water vapor loss. Low or high air and soil temperatures can slow or accelerate growth, influence the duration of growing periods, and reduce root water uptake by affecting root permeability (Danielson, 1967). Increased water viscosity at low temperature slows flow rates to root surfaces.

Precipitation contributes very little to crop water supply in desert regions but supplies about one-fourth to one-half of irrigated plant needs in semiarid areas. Many factors such as precipitation intensity and duration, soil intake rate, and timing relative to irrigation events influence precipitation effectiveness.

Growing season rainfall is about 70 to 80 percent effective in semiarid climates (USDA, 1967). In the non-growing season, precipitation storage efficiency depends greatly on the degree of moisture depletion at crop harvest. Musick (1970) found storage efficiency in clay-loam soils of the U.S. Southern High Plains to vary from 40 to 50 percent, when root zones were near the wilting point after harvest, to only 10 percent when near field capacity.

Driving forces and energy gradients associated with water movement. The principle driving force for transpiration is the difference between the vapor pressure at the evaporating surfaces and the bulk air. The resulting transpiration rate is proportional to potential energy gradients that develop in the plant-soil system and it is inversely related to flow resistances in the pathways of water and/or vapor flow. Van den Honert (1948) and Gardner

(1965a) have described this process as analogous to electric current flow in Ohm's law. In steady state form:

$$T_a = \frac{\psi_s - \psi_l}{r_s + r_p} \qquad [18.1]$$

where T_a = actual transpiration rate, cm h^{-1}; ψ_s = soil water potential or energy state with pressure or head units when expressed on a unit volume basis, cm; ψ_l = leaf water potential, cm; r_s = flow resistance in the soil pathway, cm h cm^{-1}; and r_p = flow resistance in the plant pathway, cm h cm^{-1}.

The soil water potential has the components:

$$\psi_s = \psi_m + \psi_g + \psi_o \qquad [18.2]$$

where ψ_m = potential component due to soil matric or capillary forces; this term has a negative sign in unsaturated soils and is positive under saturated conditions (also expressed as ψ_p in Section 4.1.2); ψ_g = potential component due to the gravitational force; this term carries a positive sign when represented as the distance from a reference datum that is located below the point of potential measurement; and ψ_o = potential component due to osmotic effects of solutes in the soil water; this term carries a negative sign indicating potential is below that of pure water.

Likewise, the leaf water potential is expressable as:

$$\psi_l = \psi_t + \psi_o + \psi_m + \psi_g \qquad [18.3]$$

where ψ_t = potential component due to turgor pressure acting outward on cell walls; this term is analogous in soils to a hydrostatic pressure and normally carries a positive sign; ψ_o = potential component due to the presence of solutes in plant cell fluids and is negative in sign; ψ_m = potential component due to matric or capillary forces associated with cell walls and colloidial materials; this term usually is not important in the range of ψ_l that affects transpiration; and ψ_g = potential component due to gravitational force; its magnitude is very small relative to ψ_t and ψ_o and is usually ignored.

Gardner (1964) modeled flow resistance in the soil pathway as:

$$r_s = \frac{1}{BKL} \qquad [18.4]$$

where B = an empirical constant; K = soil hydraulic conductivity, a function of soil water content; and L = root density or total length of roots per unit volume of soil. This density is typically greatest near the soil surface. In a root zone depth divided into fourths, a 40, 30, 20 and 10 percent rule of thumb distribution is often assumed to approximate root density in each respective interval from top to bottom.

The root, xylem, and stomatal systems each offer flow resistances (r_p) in the plant pathway. The influence of r_p relative to r_s is not well understood. Newman (1969) estimted that soil resistance remains smaller than r_p until soil hydraulic conductivity is reduced to about 10^{-7} cm/day. Reicosky and Ritchie (1976) confirmed this level also and they found this conductivity occurs at ψ_m levels of about -1 and -8 bars in sandy and clay soils, respectively (1 bar = 10^5 Pa).

Equation [18.1] and its term expansions illustrate the strong coupling which exists between the plant and soil water flow systems in the transpirational process. Near sunrise, ψ_l begins at an overnight recovery level that is dictated by the prevailing ψ_s distribution in the root zone. In each daytime period, transpiration and resulting cell dehydration lowers ψ_l to produce potential gradients ($\psi_s - \psi_l$) that may overcome the prevailing plant-soil flow resistances to the point where the resulting T_a utilizes the heat energy available for water evaporation. Typically, ψ_l falls to its daily minimum in the mid afternoon period and then proceeds towards recovery as the evaporative demand declines with approaching sunset (Ritchie, 1974). The daily reduction in ψ_l occurs primarily in the ψ_t and ψ_o terms with the most rapid change in ψ_t. This dehydration induced turgor loss can severely wilt stomate guard cells when evaporative demand is extreme and/or soil water potential in the effective root zone is too low. During these periods (particularly in the mid-day hours) stomatal apertures are reduced or nearly closed and resistances to water loss and CO_2 influx are then greatly increased.

Water stress effects on stomatal activity are also species dependent (Hsiao, 1973; Begg and Turner, 1976). Stomata usually close over a narrow range of ψ_l values, but a given magnitude tends not to be unique. Plant age, stress history, or an osmoregulation process (Meyer and Boyer, 1972) may influence stomate closing ψ_l levels in many plant species. Osmoregulation appears to be a mechanism whereby plants can develop a degree of "conditioning" to water stress. Studies (Kassam and Elston, 1974) indicate, that if water stress develops gradually as in field grown plants, a relative increase in cell solute concentration may take place during the stress period. When it occurs, this solute increase subsequently allows development of a more negative ψ_l before stomata close. Thus, after an osmoregulation period plants may tolerate more water stress before stomatal activity affects transpiration or photosynthesis.

Equation [18.1] also shows the plant interaction with soil water takes place via the ψ_s and r_s terms. ψ_s is in turn a function of soil wetness as discussed in Chapter 4. To maintain a given transpiration rate, ψ_l must decrease as soil wetness and ψ_s decrease. The minimum attainable ψ_s depends on the minimum ψ_l capability of plants. The position of the root density parameter (L) in equation [18.4] shows also that a greater soil water depletion to lower ψ_s levels is likely from zones (typically near the ground surface) where root density is highest.

Plant growth responses to water deficits. Water deficits in plants occur as potential gradients develop to move water against flow resistances in the transpiration pathway. These deficits may not limit transpiration until the soil water storage reservoir is considerably depleted, even during relatively high evaporative demand periods (Ritchie, 1973). When water stress induces stomatal movement towards closure, the resulting effects on transpiration and photosynthesis are essentially in phase. But, it is now also known (Begg and Turner, 1976) that cell expansion and division processes require high turgor pressure (ψ_t) levels and that plant growth (the conversion of assimilates to living tissue) ceases at ψ_l levels much above those which cause stomatal closure and a decrease in transpiration rate. During periods (usually parts of each day) when plant growth slows or stops, assimilates are stored for later use. This use can take place in the form of accelerated growth after turgor recovery, such as during the night. During each period of depressed growth, assimilates can be stored for several hours before nonstomatal effects cause photosynthesis to decrease from the potential rate. Thus, repeated brief stresses, such as in mid afternoon maximum ET periods, may only slightly affect yield potential (Hsiao, 1973).

If the day to day ψ_l depressions are too great and/or too long, these stresses normally affect plant parts or organs that are most actively differentiating at the time of the stress period. Growth limiting stress causes reduced leaf area and plant size if it occurs in the early vegetative period, reduced size of reproductive organs or reduced seed set if in the early floral through pollination period and reduced seed size if in the grain filling period. These effects are most pronounced in determinate crops. Indeterminate crops may compensate for earlier stresses by producing more yield later in the season.

Much applied experimentation (see summaries by Salter and Goode, 1967) has shown that determinate crops are most sensitive to water stress in the initial floral through pollination and early grain filling period. Physiologic reasons for this are still not well understood. Hsiao (1973) suggests some reduction in leaf area from stresses in the early vegetative period will often minimally affect yield of a reproductive organ because, (a) reduced leaf area can allow greater canopy light penetration to maintain assimilative capacity, and (b) assimilates accumulated during the day may exceed requirements for growth at night because of other growth limiting factors (such as temperature and nutrition deficiencies).

A lesser stress sensitivity in the latter grain filling period appears due to the ability of plants to translocate earlier accumulated assimilates from the stems and leaves to the grain (Wardlow, 1967; and Gallagher et al., 1975). Physiologic data (Begg and Turner, 1976) support the conclusion that the translocation process is less sensitive to stress than photosynthesis. However, maximum yields are usually associated with a maximized seasonal photosynthesis which requires an optimum leaf area development and the maintenance of a healthy and photosynthetically active plant during the vegetative, reproductive and natural maturation period.

Severe water deficits cause accelerated senescence which is most pronounced in older tissues and leaves. Growth recovery following stress relief is much less likely in older tissue than in young actively growing plant parts.

Soil as a water storage reservoir. Soils retain water against the gravitational force because soil porosities exhibit capillary or matric forces. The root zone water storage capacity depends strongly on the profile textural distribution and the total soil depth proliferated by plant roots. Soil structure affects soil water storage by influencing pore volume and size distribution, resistance to root extension, and profile hydraulic properties. Although these can be massive profile effects, they also occur in layers and lenses that strongly influence the total profile. Under favorable conditions, deep soils provide water for extraction by annual crops in the range of 2 to 3 m, and to greater depths for perennial crops such as alfalfa.

TABLE 18.1. REPRESENTATIVE AVAILABLE WATER HOLDING CAPACITIES FOR VARIOUS SOIL TEXTURE CLASSES*

Texture class	Available water holding capacities
	cm/cm
Coarse sand and gravel	0.02 to 0.06
Sands	0.04 to 0.09
Loamy sands	0.06 to 0.12
Sandy loams	0.11 to 0.15
Fine sandy loams	0.14 to 0.18
Loams and silt loams	0.17 to 0.23
Clay loams and silty clay loams	0.14 to 0.21
Silty clay and clays	0.13 to 0.18

*From Israelson and Hanson (1962) and Sweeney (1970).

Most plant species of agricultural significance can absorb water from the soil pores until the remaining water is held at ψ_m values below -15 bars, the usually defined wilting point (Chapter 4). The upper retention limit is less definable. Water is available to levels of soil wetness where ψ_m approaches zero, however, water also drains readily from the macro pores that wet at this ψ_m. Thus, the upper storage limit (called field capacity; see also Section 4.1.4) is usually best determined by in-situ measurements following a thorough wetting by irrigation or rainfall. During the drainage period growing plants take up part of the gravitational water, thus increasing the available volume range (Miller and Aarstad, 1971, 1973). Jensen (1972) gave procedures (also discussed later) for determining the most representative time to sample a draining soil profile to define an "effective" field capacity.

Approximate upper retention limits are frequently determined by in-situ measurements after drainage has taken place from a well wetted profile for 2 to 4 days. If the upper limit is determined in the laboratory by soil water characteristic analysis, the field capacity is normally selected at water contents in equilibrium with a $\psi_m = -1/10$ bar for sandy soils and with a $\psi_m = -1/3$ bar for fine textured soils.

When field or laboratory measurements are not made, available water storage capabilities can be estimated by determining (as from soil survey data or on site inspection) the textural distribution in the root zone depth and then referring to data as given in Table 18.1. Table 18.1 shows the retention volume for plant available water increases as soil particle sizes decrease from sands to about silt loams. A further particle size decrease to clays increases slightly the void space which stores water in the unavailable range.

Soil as a modifier of root system development and water availability. Both physical and chemical properties of soils limit plant rooting. Physical factors are: (1) naturally occurring dense layers having low porosity or small pore size; (2) soils artificially compacted by agricultural traffic; and (3) high soil strength associated with soil drying. Root proliferation and extension rate into dense soils is also associated with soil water content. Soil resistance to root extension is more directly related to soil strength than to bulk density. Root growth problems are more pronounced under conditions of limited wetting which can result in the subsoil being in the higher soil strength range. Several authors discussed effects of soil physical properties in the ASAE Monograph "Compaction of agricultural soils" (Barnes et al., 1971).

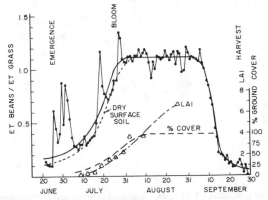

FIG. 18.1. Ratio of daily ET by irrigated field beans to ET by well-watered alta fescue grass [10 to 25 cm tall] used to define potential ET, and leaf area index and percent ground cover by the field beans, Davis, California. Data were determined from field lysimeters in 1968 [from Pruitt, 1971].

Chemical effects on rooting are exhibited in the form of ion toxicities, reduced osmotic potentials due to dissolved salts in the water and by effects on soil physical properties. Ion toxicities and their relationship with pH level were discussed by Pearson (1965). Salinity effects on plants, crop tolerances, and the sodium ion effects on soil properties are discussed in Chapter 5.

Low soil fertility that reduces above-ground growth also reduces root growth. The plant's ability to translocate essential mineral nutrients to root growing points, with the exception of calcium, limits low fertility subsoil effects on root extension. However, calcium-deficient subsoils restrict root extension and limit depth of rooting. Deep tillage, chemical amendments, and leaching have been used on many saline soils to improve the profile for root system development and plant water availability.

Irrigated soils may remain excessively wet for periods of time which can result in poor aeration and restricted root development (Grable, 1966). Restricted oxygen diffusion into poorly drained and excessively wet soils limits root respiration necessary for growth. Root morphology also can be changed and result in large porous roots containing large airfilled pores within the cortex. As the soil dries, roots that have developed under highly hydrated conditions can become poor conductors of water and thus limit water availability and water and nutrient transport into and through the root system.

Leaf canopy development as a determinate of crop water needs. When the soil surface is wet, ET of nonstressed plants proceeds at the potential rate irrespective of leaf canopy development. After the surface layer dries, vapor diffusion rate through the dry layer limits soil evaporation rate and ET is largely influenced by vegetative canopy. Under plant water stress, evaporation from dry soil is low and stomatal aperture largely controls the transpiration rate. Since overlapping of leaves normally occurs as a plant canopy develops, the ET rate does not reach the potential rate until leaf area index (area of one side of all leaves divided by area of ground surface) increases to about 3.0. Because of the nature of the radiation balance of plants with their environment, canopies can reach the potential ET rate at about 75 percent ground cover. Fig. 18.1 illustrates these canopy development effects on ET rate of field beans.

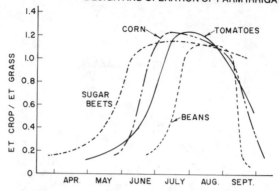

FIG. 18.3. Measured 7-yr mean ET for grass compared with normal ET for other crops [seasonal total for sugarbeets = 850 mm, corn = 710 mm, tomatoes = 650 mm, and beans = 400 mm] as estimated from the 7-yr grass data and the relationships shown in Fig. 18.2, Davis, California [from Pruitt, 1971].

Reduced plant density per unit land area can increase per plant soil water storage. This cultural practice often insures an economic yield in dryland agriculture and it can also moderate plant water stress under very limited irrigation. Crops differ greatly in canopy development with differences in planting dates, spacing, and vegetative development. Figs. 18.2 and 18.3 show canopy effects on ET of lysimeter-grown sugarbeets, corn, tomatoes and field beans at Davis, California compared with fescue grass (Pruitt, 1971).

Root growth characteristics as a determinate of extractable water. In newly planted crops the major root growth occurs during the vegetative development period. Root extension normally slows at the onset of reproductive growth as plants favor translocation of assimilates to the fruit or seed. Length of the vegetative period influences root zone depth. Winter wheat, for example, may extend roots to 2 to 3 m, whereas, spring wheat roots extend to only a 1- or 2-m depth (Kmoch et al., 1957; Hurd, 1968).

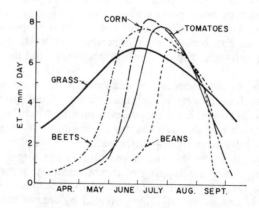

FIG. 18.2. Smooth-curve relationships of the ratio of ET by several crops to ET by 10 to 25 cm tall alta fescue grass, Davis, California. Relationships were determined from field lysimeter data [from Pruitt, 1971].

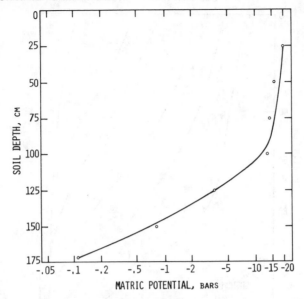

FIG. 18.4 Variation of the soil water matric potential with depth at the lower limit of extractable soil water for grain sorghum [from Ritchie, 1972].

A greater rooting depth increases available water storage capacity, but the storage at lower depths is less extractable because (1) rooting density decreases with depth, and (2) a lower leaf water potential or larger potential gradient is required to extract water (Gardner, 1965b). Fig. 18.4 illustrates this point for a grain sorghum crop that was grown in a Houston black clay soil. At the lower limit of extraction, a matric potential at or below -15 bars is shown to have developed in only the upper 40 to 50 percent of the total root zone depth. In the lower root zone the ψ_m limits are much above this usually assumed wilting point.

The typical root density gradation favors water extraction from the upper root zone in the period immediately following irrigations. As water use proceeds with time, the extraction pattern deepens (Fig. 18.5) and water storage in the lower root zone becomes a factor in moderating plant water stress. However, for maximum yield, irrigations must usually be scheduled before plants begin to extract the deeper water storage. Root zone salinity in the lower depths may further limit water availability.

Root systems are also dynamic in response to water stress. As the upper part of the root system experiences major drying, the most recently developed roots are most active in water uptake and the older roots in dry soil can dessicate and senesce as the plant experiences stress. The older main branching roots tend to become suberized with time and have reduced water uptake ability. Following rewetting, a flush of new growth occurs in the previously dried soil that permits increased extraction ability of the newly stored water.

18.1.3 Relationships Between Crop Yield and Water

Literature on this subject reveals yield relationships to water can range from linear to curvilinear (both concave and convex) response functions. These variations are influenced by the type of water parameter that is chosen, its measurement or estimation accuracy, and the varied influences

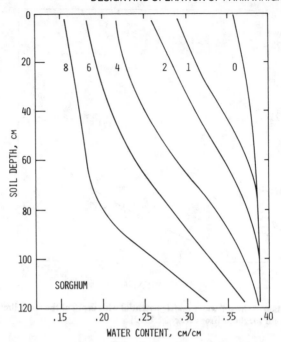

FIG. 18.5 Observed soil water extraction pattern by sorghum roots. Numbers on curves are days after irrigation [from Gardner, 1965b].

associated with site and production conditions. Even different years of the same experiment can result in varied functions. A summary of this vast literature will not be attempted. Instead, material in the succeeding sections will illustrate the more general relationships of crop yields with water when it is expressed as transpiration, evapotranspiration, or field water supply.

Yield vs transpiration or evapotranspiration. When yields are transpiration limited, strong correlations usually occur between cumulative seasonal dry matter and cumulative seasonal transpiration; see reviews by de Wit (1958) and Arkley (1963). These relationships normally originate at or near the origin and rise linearly to some Y_{max} - T_{max} level. The attained T_{max} may equal the potential seasonal transpiration if adequate soil water levels were maintained by rainfall and/or applied irrigation water.

With the close correlation between T and ET, dry matter yield vs cumulative ET usually also plots as a straight line relationship. This relationship tends to pass through or very near the origin for established perennial crops (Scofield, 1945; Stewart and Hagan, 1969) and to the right of the origin for newly planted crops (Hanks et al., 1969; Neghassi et al., 1975). In the latter case, some soil evaporation takes place before newly emerged plants begin accumulating measurable dry matter. Irrigation management and cultural practices in the early season also influence the intercept with the ET axis (Hanks, 1974).

Fig. 18.6 illustrates dry matter vs T and ET relationships for grain sorghum data obtained by Hanks et al. (1969). When reported as in Fig. 18.6, the data primarily reflect the maximum yield potential and the rate of yield increase per unit water use that is attainable for a given set of produc-

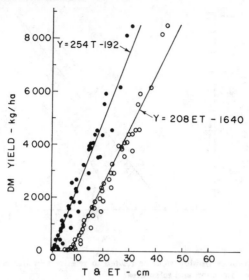

FIG. 18.6 Relationships of grain sorghum dry matter yield to seasonal transpiration [T] and seasonal ET, Akron, Colorado, 1966-67 [from Hanks et al., 1969].

tion inputs (such as a non-limiting nutrient level) in a given climatic setting. Plotting the ratio of Y/Y_{max} with either T/T_{max} or ET/ET_{max} achieves a greater generalization of yield functions. The most consistent relationship occurs between relative dry matter yield (Y/Y_{max}) and relative transpiration (T/T_{max}). This correlation in given years is very near 1:1 (de Wit, 1958) as shown in Fig. 18.7. Shalhavet and Bernstein (1968) demonstrated that this relationship also holds for the effects of soil salinity on transpiration and consequent yield attainment. Thus, if Y_{max}, T_{max} levels are defined or estimable for a given crop-soil-climate-production input setting, the dry matter yield can be estimated for lesser T levels.

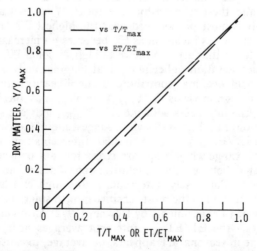

FIG. 18.7 Relative dry matter yield as related to relative seasonal transpiration or relative seasonal evapotranspiration. See Section 18.1.3 for discussion and references.

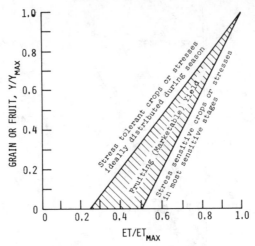

FIG. 18.8 Relative grain or fruiting yield as related to relative seasonal evapotranspiration. See Section 18.1.3 for discussion and references.

When relative dry matter yield (Y/Y_{max}) is plotted vs ET/ET_{max}, this curve again frequently passes thru the origin as a 1:1 fit for perennial crops. When early season evaporation is not a relatively constant ratio of T, the curve shifts from the origin and typically intersects the ET/ET_{max} axis at ratios on the order of 0.05 to 0.10 (Hanks, 1969).

A considerable quantity of transpiration or evapotranspiration must take place during early vegetative growth before the first increment of grain yield is produced. Numerous experiments (Jensen and Musick, 1960; Musick et al., 1963; Jensen and Sletten, 1965a and 1965b; Musick and Sletten, 1966; Hanks et al., 1969; Schneider et al., 1969; Stewart and Hagan, 1969, 1973 and 1975; Musick and Dusek, 1971; Downey, 1972; Hillel and Guron, 1973; Power et al., 1973; Morrey et al., 1975; Stegman and Olson, 1976; Stegman and Bauer, 1977) have been conducted to relate grain or seed yield with the cumulative seasonal ET (a more easily defined parameter than T). Generally, these relationships are linear. When they are curvilinear, the non linearity is most pronounced near the highest ET levels where the prospect of drainage loss inclusion as ET becomes an increasing possibility. At soil wetness levels that are conducive to high seasonal ET, other factors, such as root zone aeration deficiency, plant lodging, and the potential for higher disease incidence, may contribute to a nonlinearity between Y and ET.

When relative grain yields (Y/Y_{max}) are plotted as linear relationships to ET/ET_{max}, the intercepts with the ET/ET_{max} axis are subject to considerable variation. Fig. 18.8 is drawn to suggest these intercepts tend to fall in the range of $ET/ET_{max} = 0.25$ to 0.50. Intercepts tend to occur at the lower end of this range when crops are stress tolerant or when stresses are ideally distributed between the vegetative, reproductive, and grain filling periods of growth. Conversely, intercepts at the high end are likely when stresses are extreme and in the most sensitive growth stages of stress sensitive crops. A review of 14 experiments by Downey (1972), involving a wide range of crops, suggests the ET/ET_{max} intercept averages near 0.40. Thus, a 1 percent decline in seasonal ET appears, on average, to cause about a 1.67 percent decline in yield of crops harvested for their fruiting parts.

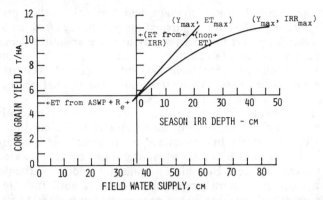

FIG. 18.9 Relations between yield [Y] vs seasonal ET and yield vs seasonal irrigation [IRR] depth, both set within a yield vs field water supply [FWS] functional context [from Stewart and Hagan, 1973].

Yield vs field water supply. The field water supply (FWS) in irrigated fields is derived from the available soil water at planting (ASWP), the effective growing season rainfall (R_e), and the total applied irrigation depth (IRR). Stewart and Hagan (1973) demonstrated that crop yields are typically related to seasonal ET and seasonal IRR as shown in Fig. 18.9. In a given season, the ASWP and R_e components of the seasonal FWS make possible a yield level that is common to both functions. The ET component associated with successive applications of irrigation defines the Y vs ET function above the dryland level, which rises to a Y_{max}-ET_{max} level when the seasonal crop water requirement is fully satisfied. The ET + non ET components of IRR define a Y vs IRR function of convex form. This relationship typifies results of many experiments (Miller et al., 1965; Gerard and Namken, 1966; Stewart and Hagan, 1969 and 1973; Huszer et al., 1970; Yaron, 1971; Stewart et al., 1974; Shipley and Regier, 1975 and 1976; and Grimes, 1977). That is, non-ET losses increase as water is applied to achieve ET_{max} levels due to the inefficiencies of irrigation methods and the inexactness of water scheduling. The amount of water not used in ET, therefore, represents runoff, deep percolation, and/or residual extractable water in the soil when the crop is harvested. It is by this quantity of water that the Y vs IRR function is bent away from the Y vs ET function. The water management implications of these two types of yield functions are discussed further in succeeding sections.

18.2 SCHEDULING IRRIGATIONS TO MEET OBJECTIVES

18.2.1 Maximizing Yield per Unit Area

This objective may be economically justified when water supplies are readily available and irrigation costs are low. All production practices and inputs must be at yield optimizing levels and daily cycles of plant water potentials (see Section 18.1.2) must usually be managed within limits conducive to maximum seasonal net photosynthesis. Precise management criteria for plant water potentials are, however, not well developed. Hsiao (1973) has suggested, that for many plant species, stresses are mild when mid-afternoon ψ_l depressions are not more than about -12 bars; moderate if in the range of -12 to -16 bars, and extreme if below -16 bars.

From an applied water management viewpoint, this production objective is relatively attained when irrigation systems are operated to supply sufficient water for plants to meet the day-to-day evaporative demand and with a frequency that maintains a high soil water potential in the upper root zone. As the requirement for high soil water potential is satisfied, the daily depression in leaf water potential is minimized and net photosynthesis is likely optimized within practical limits.

The requirements of high soil water potentials for maximized net photosynthesis may actually exceed the requirements for maximized transpiration. Ritchie (1974) has estimated that actual net photosynthesis probably falls below potential rates when root zone extractable water is only 20 to 25 percent depleted. But, his experimental evidence (Ritchie, 1973) shows plants can meet most evaporative demands until their root zone extractable water is about 70 to 80 percent depleted or until 40 to 50 percent when expressed on the more traditional root zone available (FC to PWP) water basis. Many applied experiments (summarized by Salter and Goode, 1967) have also shown that for many crops yields will be near their maximum when root zone available water is not depleted by more than 25 to 40 percent between irrigations.

Exceptions to these guidelines can, of course, be cited. A crop like cotton may require programs of stress management in each growth stage which have been locally calibrated to most consistently produce the desired yield and quality levels (Stockton et al., 1967; Erie and Longenecker, 1968; and Grimes and Dickens, 1977). In many soils, about 50 to 60 percent depletion of plant available water should be permitted in the effective root zone prior to irrigations. Timing of the last seasonal irrigation is determined largely by soil textures in the root zone and the given climate. The last seasonal irrigation is often also critical to the degree of maturity attainment in the harvest period. Another example is alfalfa that is grown for seed production (Keller and Carlson, 1967). Irrigations should minimize water stress prior to the blossom period. Thereafter, irrigation scheduling should permit plant available water to approach near exhaustion at about the time of seed harvest.

Irrigation scheduling for avoidance of yield limiting stress must also be tempered by the operational practicalities of specific irrigation system types. With trickle systems it is possible to apply water at a daily frequency and to maintain very high soil water potentials (-10 to -20 cb) in a desired portion of the root zone. High frequency irrigations are not practical for "set type" sprinkler or most gravity systems due to prohibitive labor requirements or unacceptable levels or non-ET losses. The scheduling concepts that follow are necessarily somewhat system specific.

A root zone water content near field capacity at planting insures rapid early season growth and normal root development. Pre-plant irrigations are therefore often desirable in arid and semi-arid climates. Post season or fall irrigations can serve the same purpose, however, water storage efficiencies are usually lower, averaging often below 40 percent (Musick et al., 1971). In subhumid and humid climates off-season precipitation may refill the root zone to near field capacity in most years. Pre-season irrigation scheduling should also consider the probabilities of rainfall and allow appropriate storage capacities to minimize drainage loss.

Irrigation with solid set, center pivot, and automated short-run gravity systems can be applied at relatively high frequency. If the zone of expected

TABLE 18.2. ALLOWABLE ROOT ZONE WATER DEPLETION BETWEEN
IRRIGATIONS FOR NEAR MAXIMUM YIELD AS APPLIED TO SCHEDULING OF
"SET TYPE" SPRINKLER AND NON-AUTOMATED GRAVITY SYSTEMS
(ADAPTED FROM HAGAN AND STEWART, 1972; AND TAYLOR, 1965)

Crop	Available water depletion	Root zone depth normally irrigated in deep soils
	percent	cm
Alfalfa	30-50	120-180
Beans, dry	50-70	60-90
Corn	40-60	75-120
Cotton	50-65	90-120
Deciduous fruit	50-70	120-180
Potatoes	25-50	60-90
Sugar beet	30-60	90-120
Grain sorghum	50-70	90-120
Soybean	50-60	60-90
Wheat	50-70	90-120
Vegetable crops	25-50	60-120

root development is near field capacity at planting, irrigations can normally be delayed to about 25 to 40 percent available water depletion (see Table 18.2 for effective root zone depths in deep soils). Thereafter, irrigations should be scheduled on a high frequency basis to meet expected ET demands. Frequently irrigations are programmed to satisfy the projected ET for each forthcoming weekly interval. For example, if the expected ET in the next 7 day period is 44 mm, this amount, plus an allowance for irrigation inefficiency, would be applied. Two field revolutions would be typical in this interval with a center pivot system. This process is continued week by week to near physiologic maturity of the crop with periodic adjustments as needed for rainfall occurrences and errors in projection of ET demand.

With high frequency irrigation, a particular definition of an allowable depletion level is relatively less important. Rather the high irrigation frequency satisfies the day to day ET demand and a high soil water potential is maintained in the upper root zone (Rawlins and Raats, 1975). Once irrigations are begun the root zone water deficit oscillates about the deficit that existed at the time of initial irrigation.

The tendency in scheduling set type sprinkler (towline, sideroll laterals, etc.) and conventional gravity systems is to allow greater relative root zone water depletions prior to each irrigation. Each water application normally eliminates the prevailing water deficit. This operating procedure minimizes the seasonal labor requirement, but the probabilities of growth limiting stress increase and yields may fall below those obtained with higher frequency irrigations.

A great number of studies have defined allowable soil water depletions as applied to particular crop-soil-climate combinations and the use of either set type sprinkler or conventional gravity systems. Taylor (1965) and Hagan and Stewart (1972) published summaries of these data and an abbreviated summary is given in Table 18.2.

18.2.2 Maximizing Yield per Unit Water Applied

As irrigation water supplies become more limited or as water costs increase in an area, the management objective may shift to optimizing production per unit of applied water. Fig. 18.9 in Section 18.1.3 shows func-

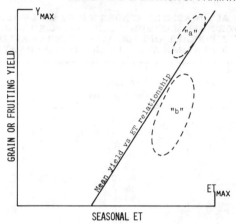

FIG. 18.10 Schematic of irrigation water management effects on water yield relations. Yields plotting in or near region "a" frequently reflect a maximal Y/IRR attainment. Yields plotting in region "b" are indicative of sub-optimal water management regimes.

tional relationships of crop yield with seasonal ET and with seasonal irrigation (IRR) are separated by the seasonal non-ET losses. Typically, these losses increase with increasing seasonal irrigation amount; giving the Y vs IRR function a convex shape that curves away from the Y vs ET function. It can be visualized in Fig. 18.9 that the ratio of Y/IRR will increase as non-ET losses are reduced. In fact, with accurate scheduling methods and efficient irrigation systems, the Y/IRR ratio will likely maximize near the Y_{max}, ET_{max} yield potential. Thus, with an unlimited water supply, the scheduling goal should be to minimize the seasonal non-ET losses.

Scheduling concepts for the limited water supply case range widely from localized qualitative criteria, that are suited to specific crop-soil-climate settings, to more general quantitative methods. All methods attempt to maximize Y/IRR at yields that are only slightly below levels attainable with full irrigation. Often as irrigations are scheduled to maximize Y/IRR with a limited water supply, the attained yield levels will plot in a region (identified by letter "a" in Fig. 18.10) slightly above the mean Y vs seasonal ET function but below the Y_{max}, ET_{max} level. This plotting position is an indication that the scheduling method has successfully reduced the seasonal evaporation loss with minimal reduction in seasonal transpiration (Downey, 1972; Hanks, 1974; Stewart et al., 1975). For very limited water supplies, Y/IRR will also usually be near maximum if the attained yields plot on or above the mean Y vs ET line as defined for a given crop-soil-climate setting.

Fig. 18.10 further illustrates that yield vs seasonal ET data can plot in region "b" which is considerably below the normal or mean relationship. Y/IRR ratios are very low in this region. The associated irrigation regimes are characterized by an abundant water supply in the early vegetative period that is followed by a severe stress in the most sensitive growth stage (12 leaf to blister kernel in corn). Water supply may again be adequate in the latter growth stages. This irrigation regime typically results in a seasonal ET accumulation that is disproportionately high for the level of yield attainment.

Summaries of scheduling concepts are given in the following sections for maximizing the Y/IRR ratio or to obtain yield levels that tend to plot in or near region "a" of Fig. 18.10. Pertinent literature citations can be reviewed for more detail.

TABLE 18.3. GROWING PERIODS IN WHICH AN ADEQUATE WATER SUPPLY
IS MOST CRITICAL FOR MAXIMIZING Y/IRR

Crop	Growth period most sensitive to water stress	Growth interval in which irrigation produces greatest benefits	References
Sorghum	boot-heading	boot-soft dough	Musick and Grimes, 1961; Herpich, 1974; Hay and Pope, 1975
Wheat	boot-flowering	jointing-soft dough	Robins and Domingo, 1962; Schneider et al., 1960; Salter and Goode, 1967
Corn	tassel-pollination	12 leaf-blister kernel	Robins and Domingo, 1953; Howe and Rhoades, 1955; Salter and Goode, 1967
Cotton	1st bloom-peak bloom	1st bloom-bolls well formed	Levin and Shmueli, 1964; Grimes and Dickins, 1977
Dry beans	flowering-early podfill	axillary bud-podfill	Robins and Domingo, 1956; Salter and Goode, 1967
Potatoes	tuberization	tuberization-maturity	Robins and Domingo, 1956; Delis and Tizio, 1964; Salter and Goode, 1967
Soybean	flowering-early podfill	axillary bud-podfill	Brady et al., 1974; Hay and Pope, 1975; Shaw and Laing, 1966
Sugarbeets	no critical stages	WUE is maximized when water depletion is limited to about 50% available water depletion	Hobbs et al., 1963; Larson and Johnson, 1955; Haddock, 1959
Alfalfa	no critical stages	WUE is maximized by irrigating to realize full growth potential from start of spring growth until water supply is depleted	Hobbs et al., 1963; Hanson, 1967; Stanberry et al., 1955

Critical growth stage irrigation. For grain sorghum, that is grown on deep silt loam or clay loam soils in the Southern High Plains of the United States, Musick and Grimes (1961) recommend a preplant irrigation to fully recharge the soil profile. They further recommend a second irrigation at early boot and a third irrigation at mid bloom for the best distribution of a limited water supply.

Similar patterns of water management can be applied to many other crops, i.e., a pre-season irrigation to be followed by water applications in the most stress sensitive growth stages. Table 18.3 provides summaries of these growth stages for a range of crops.

ET deficit irrigation. Woodruff et al. (1972) developed this scheduling concept—which is also called "planned or programmed soil moisture depletion" (Fonken and Fischbach, 1974) and "deficit high frequency irrigation" (Miller, 1976)—for corn irrigation in Missouri; an area with significant summer precipitation. They observed that high corn yields occurred naturally when precipitation amounts and frequency were sufficient to maintain the upper root zone increment (0 to 30 cm) within the tensiometer range, even though soil water from the lower depths was gradually depleted. These water depletion patterns frequently also occur in fields that are irrigated by center pivot sprinklers with average application rates that are less than the average mid season evapotranspiration rates (see Fig. 18.14 in Section 18.3.2).

Experimental testing of deficit irrigation scheduling in the U.S. Central Great Plains (Fischbach and Somerholder, 1974; Fonken et al., 1974) has indicated this concept works well when (a) it is applied to deep soil profiles with available water capacities greater than 0.13 cm/cm of soil depth; and (b) it is used with irrigation systems that can be operated to apply frequent irrigations in controlled amounts. In these situations a reduced system pumping capacity can be used and a decrease in average annual irrigation amount is possible with minimal loss of yield potential.

The scheduling procedures associated with this method include: (a) Filling the maximum expected root zone to field capacity at or near planting; (b) maintaining low soil water deficits in the early season when system pumping capacity can satisfy the ET requirements; (c) irrigating frequently ($<$ 7 day intervals) in the peak ET period to maintain high soil water potentials in the 0 to 30 cm root zone increment. This management criteria keeps the zone of greatest nutrient supply moist and it also minimizes the magnitude and duration of daily ψ_l depression; (d) applying sufficient irrigation amounts in the grain filling period of seed crops so that these irrigations plus the earlier stored water in the lower root zone can about supply the late season ET requirement that is not satisfied by rainfall.

In many areas it may be advisable to run computer simulation tests with water balance models to determine the probable risk of yield loss for various combinations of pumping capacities, soil types, crops, etc.

Stress day index method. Hiler and Clark (1971) introduced this concept as a semi-quantitive basis for irrigation timing. This method initiates irrigations when a calculated stress day index (SDI) approaches defined critical levels in specified growth stages.

The stress day index (SDI) is determined as:

$$SDI = \sum_{i=1}^{n} (SD_i \times CS_i) \quad\quad\quad\quad\quad\quad\quad\quad\quad\quad\quad\quad\quad [18.5]$$

where n represents the number of growth stages considered; SD defines the degree and duration of plant water deficit in growth stage i; and CS defines the crop yield susceptibility in a given growth stage to a given water deficit. The SDI can be calculated on a daily basis, a growth stage basis, or for an entire season. Hiler and Clark (1971) obtained inverse linear relationships of yield vs seasonal SDI for several crops.

Alternative measurements can quantitatively characterize SD; such as the absolute value of leaf water potential or soil water potential, leaf diffusion resistance, percent available soil water depletion and others (Lewis et al., 1974). Data are also available for approximate CS estimates as summarized in Table 18.4, but different data sources (Hiler and Clark, 1971; Lewis et al., 1974; Hiler et al., 1974), may indicate differing magnitudes for the same growth stage of a given crop. These discrepancies appear to be caused by (a) differing definitions of growth stage lengths; (b) differing methods of SD definition from study to study; and (c) defining CS values at different levels of SD or water deficit. This latter factor suggests CS values should be defined as complete functions of SD in each growth stage.

Despite the problem of CS definition, Hiler et al. (1974) have demonstrated that significant improvements in water use efficiency are possible. Applied application is most easily achieved by converting a "standard practice" to the SDI concept. For example, if a water balance procedure is being used to time irrigations at an available water depletion of 50 percent, this depletion criteria can also serve as the SD indicator.

TABLE 18.4. CROP SUSCEPTIBILITIES FOR VARIOUS IRRIGATED CROPS*

Crop	Growth Stage	CS
Corn	Vegetative	0.25
Corn	Silking and tasseling to soft dough	0.50
Corn	After soft dough	0.21
Cotton	Prior to flowering	0.00
Cotton	Early flowering	0.21
Cotton	Peak flowering	0.32
Cotton	Late flowering	0.20
Grain sorghum	Vegetative—6 to 8 leaf stage	0.25
Grain sorghum	Boot	0.36
Grain sorghum	Heading to soft dough	0.45
Grain sorghum	After soft dough	0.25
Peanuts	Vegetative to peak flowering and early pegging (30-50 days after emergence)	0.36
Peanuts	Peak pegging and nut development (60-80 days)	0.24
Peanuts	Late nut development and maturation (90-110 days)	0.12
Rice	Vegetative	0.17
Rice	Reproductive and ripening	0.30
Southern peas	Vegetative—prior to flowering	0.13
Southern peas	Flowering and early pod formation	0.46
Southern peas	Pod development to maturation	0.43
Soybeans	Vegetative	0.12
Soybeans	Early to peak flowering	0.24
Soybeans	Late flowering and early pod development	0.35
Soybeans	Late pod development to maturation	0.13

*From Hiler et al. (1974).

Irrigations should begin when the daily SDI value reaches a predetermined critical SDI_0. In theory, SDI_0 should depend on the availability and/or cost of water. However, in practice SDI_0 is often set equal to the seasonal average SDI which would result from the scheduling or irrigations in each growth stage at SD levels typical of the "standard practice". Table 18.5 illustrates an example SDI_0 calculation, assuming a 50 percent available water depletion criteria as a "standard practice" for grain sorghum. The SDI_0 is computed as 16.5 percent. The SD levels that signal an irrigation need are then obtained by dividing this SDI_0 level by the respective CS values. The computed SD levels are 66, 45, 36 and 66 percent for each respective growth stage. From Table 18.5, one should note that the seasonal SDI for both the "standard practice" and SDI methods of scheduling should be approximately the same. Therefore, yield attainments should also be similar. The SDI method improves Y/IRR in most climates by reducing the seasonal IRR.

Optimal sequencing of ET deficits. Stewart et al. (1975) found an optimal linear yield vs seasonal ET relationship will prevail for a given crop variety whenever ET deficits are optimally sequenced, assuming also that other factors are nonlimiting. They assumed that ET deficits accumulate whenever the actual ET rate falls below the potential rate. These ET deficits cause a reduction in yield below a determinable maximum yield level (Y_{max}) (see Fig. 18.9). Also, a minimum yield reduction (below Y_{max}) takes place when ET deficits are optimally sequenced. Hence, the yield level is the maximum attainable for the attained ET accumulation.

A suboptimal ET deficit sequence results in a seasonal ET accumulation that is excessive for the corresponding yield attainment. These yield vs ET data tend to plot below the optimal yield vs ET relationship as in region "b" of Fig. 18.10.

TABLE 18.5. AN EXAMPLE SDI_0 CALCULATION FOR SUBSEQUENT DETERMINATION OF SD AT IRRIGATION TIME IN EACH GROWTH STAGE OF GRAIN SORGHUM

Growth stage	CS*	SD†	SDI	SDI_0‡	SD at irrigation time §
		Percent AW	Percent AW	Percent AW	Percent AW
6 to 8 leaf	0.25	50	12.5	16.4	66
Boot	0.36	50	18.0	16.4	45
Heading to soft dough	0.45	50	22.5	16.4	36
After soft dough	0.25	50	12.5	16.4	66
			65.5		

*Determined from Table 18.4.
†Assuming SD = percent available water depletion.
‡SD_0 = Avg. SDI = 65.5/4 = 16.4 percent.
§Determined from SD = SDI_0/CS.

Stewart et al. (1975) found optimal and suboptimal ET deficit sequences tend to be identifiable, one from the other. Thus, if irrigations can be scheduled to avoid a suboptimal ET deficit sequence, farmers can be assured of producing the maximum yield for a given water supply.

To avoid suboptimal ET deficit sequences Stewart et al. (1975) have determined the approximate upper limits of ET deficit by growth stage for corn. These data, in Table 18.6, indicate the optimal yield vs ET relationship will persist if: (a) ET deficits are not greater than 50 percent of the potential ET in the vegetative period prior to tassel; (b) ET deficits in the pollination (tassel-blister kernel) period are not greater than 1.4 times the ET deficit in the prior vegetative period. If no "conditioning" (see prior discussion on osmoregulation) ET deficit occurs in the vegetative period,

TABLE 18.6. UPPER LIMITS OF ET DEFICIT INTENSITY (PERCENT) IN THE POLLINATION AND GRAIN FILLING PERIODS OF CORN, BASED ON LEVELS OF ET DEFICIT EXPERIENCED BY THE CROP IN PRIOR GROWTH PERIODS*

Vegetative period→ (Weeks 5-8 incl.)	Pollination period (Weeks 9-12 incl.)	Pollination period→ (Weeks 9-12 incl.)	Grain filling period (Weeks 13-17 incl.)	
If expected ET deficit intensity percent is as follows	The permissible† upper limits in this period are:	If expected ET deficit intensity percent is as follows	The permissible upper limits in this period are:	
			If veg. period ET deficit <38%	If veg. period ET deficit >38%
0	0	0	32	65
10	14	14	60	65
20	28	28	87	87
30	42	42	87	87
40	56	56	87	87
50	65	65	87	87

*From Stewart et al. (1975).
†Permissible means if these upper limits of ET Deficit Intensity (percent) are observed, the ET Deficit sequence will be optimal, i.e., the expected yield will be maximized for the available water supply. ET deficit is expressed in percent of seasonal ET_{max}, see also Fig. 18.9.

none is tolerable in the pollination period; (c)ET deficits do not exceed 32 percent of the potential ET in the grain filling period when no prior deficit occurs in earlier periods. However, if "conditioned" by a pollination period deficit exceeding 28 percent, a deficit up to 87 percent is tolerable in the grain filling period.

Table 18.6 sets forth the upper limits of tolerable ET deficit intensity in any given growth period, based on the levels of ET deficits in previous periods. These data have not been widely tested and may be most applicable to deep soils and warm-arid climates.

Stewart et al. (1977) have more recently suggested a simplified management criteria by noting that the maximum yield for a given seasonal ET deficit level tends to occur when deficits are spread as evenly as possible over the growing season. Thus the main factors to consider in implementing this principle are the length of the irrigation season (dependent on crop variety and soil characteristics), the minimum depth of water that can be uniformly applied with a given irrigation system, and the available seasonal IRR depth. Scheduling is based on the concept of high frequency irrigation, i.e., applying small depths (or the minimum feasible for gravity methods) per irrigation at essentially evenly timed intervals.

18.2.3 Maximizing Net Profit

This water management objective, while the principle motivation of a farming enterprise, has traditionally not been strongly identified with day-to-day irrigation scheduling.

A number of reasons can be given: (a) the variable costs of water application have traditionally been low, (b) prices of production inputs and outputs are uncertain, (c) a number of inputs are stochastic or random rather than controlled or predictable, and (d) suitable production or crop-water response functions are not available for day-to-day management decisions. Consequently, profit potential is frequently gaged in the preplant planning period with a budget type of analysis. In this analysis, estimates are made of the total annual production costs (fixed and variable) for a range of cropping possibilities given the local constraints imposed by soils, climate, water availability, marketing opportunities, and processing or storage facilities. Net income potential is estimated by comparing cost estimates with gross income estimates. The latter are obtained by multiplying expected prices against established or realistic yield potentials for each considered crop. By planting those crops with the highest net income potential, profit may be approximately maximized, or at least an acceptable return may be realized. Irrigations may be scheduled to achieve yields very near the Y_{max} or yield potential level, particularly if the variable costs of irrigations are low and/or irrigation efficiencies are high.

With appropriate input data, other profit maximizing water management concepts are available. Two are briefly discussed.

Marginal value vs marginal cost analysis. Yield response to applied irrigation depth (IRR) is a diminishing return function as shown in Fig. 18.11. Thus, at some point on this curve further water application cannot be justified economically. Fig. 18.11 illustrates a simple graphic analysis to determine the maximum water use efficiency and maximum profit levels of water application. The maximum water use efficiency occurs at the point where a line from the origin becomes tangent with the Y vs IRR function. This IRR amount is the minimum application that should be considered even if a water shortage requires a reduced planted area.

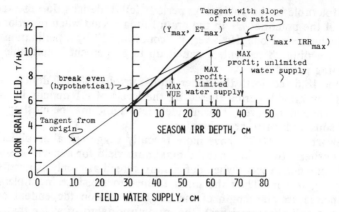

FIG. 18.11 Graphic analysis of a yield vs seasonal irrigation [IRR] function to determine IRR levels that are likely to maximize water use efficiency [WUE] and profit [from Stewart and Hagan, 1973].

The optimum economic level of production for an unlimited water supply is estimated by equating the marginal value of yield improvement with the marginal cost of further water application, as expressed by:

$$\frac{d(Y)}{d(IRR)} = \frac{P_{(IRR)}}{P_{(Y)}} \quad \text{...............................} \quad [18.6]$$

where $P_{(Y)}$ is the crop price, and $P_{(IRR)}$ is the variable cost of water application. In applying the slope of the price ratio $(P_{(IRR)}/P_{(Y)})$ to the given Y vs IRR function, care must be taken to fit the actual graph scale. In Fig. 18.11 the hypothetical prices, selected for illustration, are $3.09 per ha-cm (100 m³) of IRR and $54.36/t (metric tonne) of corn. The maximum profit occurs where the price slope is tangent to the yield function. In this example, the slope is 3.09/54.36, which maximizes profit at a yield of 11.2 t/ha. This yield is near the Y_{max} level due to the low $P_{(IRR)}$.

The maximum profit level for a limited irrigation water supply can also be estimated from a Y vs IRR function as given in Fig. 18.11. "Limited" is here defined as a quantity inadequate for a given farmer to irrigate all of his irrigable land at the level which will maximize net profit per hectare. Another stipulation is that unirrigated cropping is unprofitable.

First the most profitable irrigation level per unit of applied water (IRR) must be found. Second, the number of hectares which can be irrigated at that level with the available (limited) water supply is determined. In Fig. 18.11 the yield required to break even (returns equal costs) is hypothetically taken to be 7 t/ha of corn. A line is then drawn from the intersect of this yield level with the zero irrigation level to where it becomes tangent with the Y vs IRR function. This tangency point determines the most profitable irrigation level. This amount in Fig. 18.11 is 26 ha-cm/ha. Then, if the total supply of irrigation water were 2600 ha-cm, 100 ha could be irrigated. If irrigable land exceeds that, the rest is best left fallow. If, however, unirrigated farming is profitable, the diminishing returns form of the Fig. 18.11 curve dictates a simple answer for the limited water supply case. All irrigable land should be irrigated, with the available water supply distributed equally to each unit of area. It should further be noted that these solutions (for limited water supply) assume the unit water cost to be constant. Often this is not true, so the solutions must be tempered with reality.

Y vs IRR functions typically also assume other production inputs are at some fixed level. Usually in this type of analysis, the nutrient level is assumed to be near optimal or not yield limiting. In actual practice, the nutrient level is frequently based upon soil test data and the consequent fertilizer recommendation for the established yield goal.

When studied experimentally, a degree of substitutability is normally found between fertilizer and water inputs to crop producton (Jensen and Sletten, 1965a and 1965b). Hence, two-variable production functions can be presented as isoquants of increasing yields (Egli, 1971; English and Dvoskin, 1977) and an economic optimum is achieved by maximizing the objective function:

$$Y \cdot P_{(Y)} - (F \cdot P_{(F)} + IRR \cdot P_{(IRR)}) \quad \dots\dots\dots\dots\dots\dots\dots\dots\dots\dots \quad [18.7]$$

where F is the amount of fertilizer and $P_{(F)}$ is its unit price. The other variables are as previously identified. As additional variable inputs are considered, the profit maximizing mix becomes more complex and methodologies such as linear programming are applicable.

A problem with marginal value vs marginal cost analysis as applied to yield functions is that it provides only general guidelines for water management. These guidelines are most applicable to the average or "normal" climatic conditions in a given region and, therefore, may not apply to specific sites or specific years. In addition the guidelines are seasonal in nature, i.e., they indicate only the seasonal IRR depth most likely to maximize net profit. Decision making data are not provided for dealing optimally in a water scheduling sense with the probable economic consequences of specific water shortages during the growing season.

Simulation models. Hillel (1977) has defined computer simulation as "the construction of mathematical models to imitate the essential features and behavior of a real system, the adaptation of such models to solution by computer, and the study of the properties of such models in relation to those of the prototype system". Detailed simulation models to evaluate water management effects on irrigated crop production are invariably complex. In depth treatment can involve interrelationships of management with water supply, cultural practices, alternative costs, soil types, crop responses to soil and atmospheric environment, climate, plant nutrient levels, etc. Thus, modeling usually involves considerable development and operational cost.

In recent years, numerous simulation models have been developed to generally address the goal of profit maximization. Methodologies such as dynamic programming are frequently utilized to illustrate how optimal water scheduling or allocation strategies can be derived under conditions of stochastic variables. Models, dealing with optimization of on farm water management, have been developed by Hall and Butcher (1968), Dudley et al. (1971), Stapleton et al. (1973), Yaron and Strateener (1973), Howell et al. (1975), Mapp et al. (1975), and Trava et al. (1976).

Applications of these complex simulation models in day-to-day water management practices are, however, not yet feasible because of unlikely economic justifications. A second major problem commonly sited by model developers is the need for more detailed production functions. These inputs may be forthcoming from plant growth modeling activities that are designed to evaluate water management strategy effects at any crop growth stage (Curry, 1971; Curry and Chen, 1971; Splinter, 1974; Curry et al., 1975; Childs et al., 1977; and Barfield et al., 1977).

18.2.4 Minimizing Energy Requirements

With a history of low energy costs and plentiful supply, modern irrigation (particularly in the U.S.) has evolved into an energy intensive technology. On many irrigated farms water delivery to the plant root zone requires more energy than all other farm operations combined (Smerdon, 1974; Batty et al., 1975). Only very recently (since about 1972) have energy flows and requirements been studied in irrigated agriculture.

Dvoskin and Heady (1976), estimated that up to 90 percent of the irrigated area in high lift regions of the Northwestern United States could be lost with only a 10 percent energy cut. Williams and Chancellor (1975) similarly estimated cuts in energy for irrigation in arid areas to be much more serious in economic consequence than cuts in energy for tillage and harvesting operations. Energy conservation must obviously be an essential consideration in planning future cropping systems.

Energy conservation techniques. Pumping energy requirements can be calculated (Batty et al., 1975) as:

$$PE = \frac{0.0271 \, A \, D \, H}{E \cdot E_p} \quad \dots\dots\dots\dots\dots\dots\dots\dots\dots\dots\dots\dots\dots\dots\dots\dots \quad [18.8]$$

where: PE = pumping energy, kWh; A = area irrigated, ha; D = net depth of irrigation, mm; H = total dynamic pumping head, m; E = irrigation or water application efficiency, decimal form; and E_p = pumping efficiency, decimal form.

Equation [18.8] shows PE can be lowered by, (a) reducing net depth of irrigation, (b) reducing total pumping head, (c) increasing irrigation efficiency, and (d) increasing pumping efficiency. The combined effects of changes in D, H, E_p and E can be evaluated with the relationship (Gilley and Watts, 1976):

$$PES = \frac{(PE_1 - PE_2)}{PE_1} \cdot 100 = (1 - \frac{D_2}{D_1} \cdot \frac{H_2}{H_1} \cdot \frac{E_{p1}}{E_{p2}} \cdot \frac{E_1}{E_2}) \, 100 \dots\dots\dots\dots \quad [18.9]$$

where PES is the potential energy savings in percent relative to before and after changes in each parameter. The subscripts 1 and 2 indicate the before and after values, respectively. For realistic improvements (Gilley and Watts, 1976; Chen et al., 1976) of $D_2/D_1 = 0.80$, $H_2/H_1 = 0.85$, $E_{p1}/E_{p2} = 0.87$, and $E_1/E_2 = 1.0$ an energy reduction of 41 percent is possible. Without a pressure reduction ($H_2 = H_1$) the energy savings would be 32 percent.

Large reductions in system head (H_2/H_1) may not be possible where systems have large pumping lifts. Improvements in irrigation efficiency for sprinkler systems above present levels will also probably not be very significant.

Runoff re-use systems can improve surface irrigation efficiency by about 30 percent (Fischbach and Somerhalder, 1971; Watts, 1975). Irrigation scheduling also offers considerable potential for energy savings. Some studies (Stegman and Ness, 1974; Heermann, 1975) have indicated water balance methods could save between 15 and 35 percent of the water normally pumped in semi-arid to sub-humid climates.

Scheduling to reduce peak energy demand. In agricultural areas, the peak seasonal demand, on electrical power generating systems for extensive pumping, can require a much greater generating capacity than would otherwise be needed. Economics dictate that this excess generating capacity be used in the off pumping season for home heating, etc. Otherwise, power districts must often pay penalties to their wholesale supplier when winter loads fall below a specified percentage of the summer peak. Experimental programs (Stetson et al., 1975) have shown that an integration of irrigation scheduling with peak demand can offer a way to control some of these increased costs for electricity.

A scheduling program for reducing peak demand must begin with a detailed power supply analysis to determine the hours during the day and the number of days annually when demand should be controlled. Once the power supply analysis indicates a need for demand control, key areas or substations can be selected, a method for controlling demands established, and the inducement to obtain off-peak irrigation determined.

Several methods can be used singly or in combination to control irrigation loads such as: (a) motor controllers that activate by radio signals from the power supplier, (b) time clocks (however, they lock the irrigator into a fixed schedule), (c) an honor system of irrigation with dual type demand meters serving as a check, and (d) a commercial scheduling service or one developed by a power supplier. The scheduling service approach can also fit in with the other methods.

Stored soil moisture is the key to effective off-peak scheduling. Soil profiles should be irrigated to near field capacity in the off-season and maintained near this level in the early season when irrigation requirements are low.

In determining whether an irrigation system can be operated to reduce peak demand use, the following factors should be considered: System capacity, soil type and available water holding capacity, irrigation method, crops, and the associated effects of moisture deficit on yield potential in the peak water use period. For soils that hold 0.13 cm/cm or more of available water in the U.S. Central Great Plains, Addink and Stetson (1975) suggest pumping capacity criteria as given in Table 18.7 for 53-ha center pivot systems. These guidelines are estimated to provide sufficient capacity and operating time to prevent yield loss, because of excessive moisture deficit, in 9 out of 10 yr. For surface irrigation systems in the same area, minimum capacities of 0.9 to 1.4 Ls^{-1} ha^{-1} are recommended on efficient to average systems for successful off-peak scheduling.

18.3 ALTERNATIVE IRRIGATION SCHEDULING TECHNIQUES

Water management objectives typically lead to some form of timing criteria for water application; such as, allowable soil water depletion, soil

TABLE 18.7. CENTER PIVOT PUMPING
CAPACITY REQUIREMENT FOR
OFF-PEAK SCHEDULING*

System capacity		Potential percentage of time off
L/s	gal/min	percent
38	600	10
44	700	20
51	800	30
57	900	40
63	1000	45

*From Addink and Stetson (1975). These
capacities are most applicable to areas with
climates similar to the Central Great Plains
of the U.S. and deep profiles of medium or
finer textured soils.

water potential, ET deficit, stress day index, leaf water potential, etc. With systems that permit a high degree of control over water application, irrigation scheduling can include the concepts of both timing and the water amount to apply. In addition, the decision to irrigate must also include consideration of other factors such as the inherent operational practicalities of the irrigation system type, the institutional and legal constraints associated with the water supply, and the scheduling of farming operations.

Numerous procedures and devices are available for irrigation scheduling. A range of these methods are listed in Table 18.8 under classifications of: (a) soil based measurements, (b) plant based measurements, (c) computed root zone water balance, and (d) evaporation devices. Brief listings of advantages and disadvantages are given for each technique.

Despite their availability, farm manager/operators have not been greatly receptive to any one particular scheduling method. Many factors contribute to this problem and Jensen (1975) has cited the following as probably the most significant: (a) the cost of irrigation water is often low relative to costs of practices that would improve water management; (b) yield reductions caused by delayed irrigations, improper fertilization, and excessive irrigations are not easily recognized or quantified; (c) decision making data are often not available to those making water management decisions on a day-to-day basis; (d) irrigation management decisions are generally made by busy people with limited technical background and training in the management of a complex crop-soil-climate system; (e) traditional scheduling methods have tended to imply the training of every farmer/manager to become a specialist in irrigation water management.

As an alternative to "on farm" scheduling methods, the Irrigation Scheduling Service concept has evolved in recent years. This concept brings the professionally trained and experienced irrigationist to the farmer on a continuing basis. This approach permits a more rapid transferral of new technology to the points of application. It also effectively addresses scheduling as a decision making process for the modern irrigated farm.

The electronic computer with its rapid data processing capabilities makes large scale irrigation scheduling possible. Scheduling services have expanded steadily since Jensen (1969) developed an applied water balance program. Applied use of this program has demonstrated that benefits from

TABLE 18.8. SUMMARY OF SCHEDULING TECHNIQUES

Observed or measured parameters	Required instruments or procedures	Principal advantages	Principal disadvantages	References
Soil based measurements				
Appearance and feel	Hand probe	Simple	Time consuming; approximate; requires interpretative skills	Merriam, 1960; and Hansen, 1962
Electrical resistance	Porous blocks	Provides indirect measure of soil water content	Requires careful installation, calibration, and frequent readings; not sufficiently sensitive in coarse textured soils; short block life; need multiple sites	Schearer, 1963; Fischbach, 1965
Soil matric potential	Tensiometer	Measures fundamental parameter affecting soil water flux	Requires pre-installation preparation; careful installation, frequent readings and service; need multiple sites	Fischbach and Schleusener, 1961; Bauder and Lundstrom, 1977
Plant based measurements				
Appearance	Eye	Simple	Yield potential is usually affected before color and other changes are observed	Haise and and Hagan, 1967
Leaf water potential	Pressure chamber or thermocouple psychrometer	Indicates integrated effect of aerial and soil environment on degree of plant hydration; is correlated with metabolic processes: a fundamental parameter affecting water flux	Subject to large diurnal variation; method is time consuming; requires sampling skill; data not easily interpreted	Gardner, 1965a; Scholander, 1965; Barrs, 1968; Cary and Wright, 1970; Ritchie and Hinckley, 1975; Stegman et al., 1976
Leaf transpiration resistance	Diffusion porometer	Measures relative stomatal opening	Same as for leaf water potential	Ehrler and VanBavel, 1968; Kanemasu et al., 1969; Clark and Hiler, 1971; Teare and Kanemasu, 1972; Kanemasu et al., 1973
Leaf temperature	Non-contacting thermometers	Can be sensed remotely	Application methods not well developed	Gates, 1968; Idso et al., 1977; Jackson et al., 1977
Computed water balance				
Root zone water content	Water balance models	Simulates hydrologic cycle; offers wide flexibility in degrees of application; well suited to projection of water needs	Affected by accuracy of inputs and degree of approximation applied to processes that are modeled; usually requires periodic field checks and updating of the output parameter	Jensen, 1969; Jensen et al., 1970 and 1971; Corey and Franzoy, 1974; Heermann et al. 1976; Lundstrom and Stegman, 1977; Wilcox and Sly, 1974
Evaporation devices				
Free water, porous bulb, or porous plate evaporation	Evaporation pans; atmometers	One site can serve large area	Requires frequent service and data collection; requires careful siting and site maintenance; requires calibrations for each crop	Hagood, 1964; Hargreaves, 1968; Shannon, 1968; Pelton and Korven, 1969; Korven and Wilcox, 1965; Krogman and Hobbs, 1976

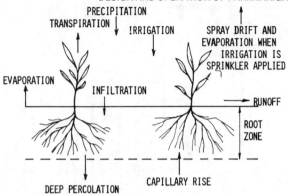

FIG. 18.12 Schematic representation of the root zone water balance components.

improved water management are closely related to optimum management of all production inputs. Consequently, many consulting firms also provide services in plant nutrition, pest management, agronomic guidance, and engineering to improve system operations.

In the future, the prospect of automated scheduling is likely. Integrated circuit and microprocessor electronic innovations will probably result in "black box" units that will artificially simulate plant growth and the root zone water balance. Sensors in the field will provide the needed feedback to these units, and when available at reasonable cost such devices will greatly enhance crop production with modern irrigation systems.

Of the available scheduling methods for the individual manager, it is the authors' viewpoint that water balance models, ranging from relatively simple methods to the more complete, can offer the most effective management techniques. Thus, water balance methods are emphasized in the succeeding sections.

18.3.1 Water Balance Techniques

for the Individual Farm Manager/Operator

Fig. 18.12 illustrates the components of a root zone water balance in irrigated fields. This balance for irrigation scheduling may be computed as:

$$D_i = D_{i-1} + (ET - P_e - IR)_i \quad \dots \dots \dots \dots \dots \dots \dots \dots \dots \dots \dots \dots [18.10]$$

where: D_i = soil water depletion in mm on day i, D_i = 0 when available water content is at field capacity; D_{i-1} = soil water depletion on day i-1; P_e = effective precipitation on day i; IR = net irrigation depth on day i, and ET = estimate of evapotranspiration on day i.

For computational ease, equation [18.10] is usually solved in a tabular format. Column headings are respectively, the date, precipitation amount, net irrigation, ET estimate, and D. On each successive day a new D_i is computed. Fig. 18.13 gives a visual illustration of the water balance scheduling process. At point:

(1) An initial estimate of D is established at planting or near crop emergence. This estimate is usually based on hand probing of the field as discussed in Section 18.3.4.

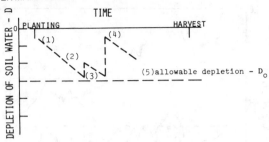

FIG. 18.13 Schematic of the water balance scheduling process. See Section 18.3.1 for discussion.

(2) D is increased daily by addition of the estimated daily ET_i.

(3) D is reduced by the amount of P_e on days when rainfall occurs or by the amount of IR when irrigations are applied. When rainfall or irrigation amounts exceed the current deficit level, D is reduced to zero. The excess rainfall or irrigation is assumed lost to deep percolation.

(4) Irrigation is often terminated to allow some storage capacity for expected rainfall.

(5) An allowable depletion (D_0) is selected by considering the management objective, the operating practicalities for the given irrigation system, and the available water holding capacities of the coarser textured soil types in the given field.

ET estimates for this type of water balance are usually based upon locally calibrated techniques, such as: (a) evaporation pans with associated crop coefficients (Hagood, 1964; Jensen and Middleton, 1970); Hargreaves, 1968; Pelton and Korven, 1969); (b) average seasonal ET distributions (Erie et al., 1965; Black and Brosz, 1972; Sutter and Corey, 1970); (c) simplified methods for estimating ET (USDA, SCS, 1976; Vitosh, 1977; Lundstrom and Stegman, 1977); (d) ET forecasts by cooperative extension service or irrigation district staffs (Jensen, 1975); (e) selection of appropriate fixed daily ET rates (Woodruff et al., 1972).

18.3.2 Water Balance Techniques for Commercial Scheduling Services

The USDA model. The general procedure most widely used by irrigation scheduling services is the USDA computer program as released in 1970 and modified in 1971 by Jensen et al. (1970, 1971). This program was developed cooperatively with farm managers and service groups, thus enabling the incorporation of farm and service manager reactions during the formative stages.

With the USDA model, soil water depletion is calculated as:

$$D = \sum_{i=1}^{n} (ET - P_e - IR + W_d)_i \dots\dots\dots\dots\dots\dots\dots\dots\dots\dots\dots[18.11]$$

where D is the root zone water depletion in mm (after a thorough irrigation D = 0); P_e is the effective daily precipitation (i.e., precipitation minus runoff); IR is the daily net irrigation; W_d is the daily drainage loss from the root zone or upward movement from a saturated zone (for upward movement the sign is negative); and i is the time increment (usually in days) for water balance summation.

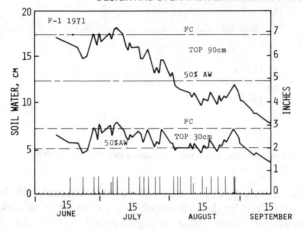

FIG. 18.14 Seasonal distribution of soil water levels that are often typical of center pivot irrigation [from Heermann et al., 1976].

Scheduling services can operate most economically when amounts and occurrence dates for rainfalls and irrigations are observed locally by each farmer. Irrigation amounts should be metered to obtain accurate measurement of gross applications. The gross irrigation depths must be adjusted for irrigation efficiency to obtain the net depth, IR.

ET in equation [18.11] is an estimated term. These estimates can be made with methods as given in Chapter 6. Most services use either the Jensen-Haise or a modified Penman equation for estimating the daily potential or energy limited evapotranspiration, ET_p. Crop coefficients are applied to the daily ET_p for estimating the actual crop ET. For the future, new improvements (Tanner and Jury, 1976; Kanemasu et al., 1976) are being developed so that evaporation and transpiration can be estimated separately. Greater water balance accuracy will then be possible during the period of crop cover development.

Weather data input to any of these methods should come from stations located at representative sites within an irrigated area. Frequent calibration checks will insure that reliable data are obtained from the instruments at these sites.

Drainage from unsaturated soil profiles strongly depends on soil water content above field capacity. In applied applications of equation [18.11], W_d is often estimated to equal the amount by which an irrigation or rainfall exceeds the prevailing D_i. The errors associated with this simplification may be quite acceptable in arid and semi-arid areas, particularly, if irrigations are carefully controlled. Fig. 18.14, for example, illustrates seasonal distributions of soil water levels that are often typical of center pivot irrigation. The two curves show irrigations can be scheduled to maintain relatively high available water levels in the top 30 cm of root zone and at the same time permit a programmed depletion of soil water from the lower root zone. In studies by Heermann et al. (1976), this type of irrigation scheduling resulted in little drainage loss and crop yields were not depressed relative to wetter treatments.

Fig. 18.15, in contrast, illustrates a seasonal distribution of soil water levels typical of conventional gravity irrigation. Soil water content in the entire root zone is raised to levels above field capacity at the time of each irri-

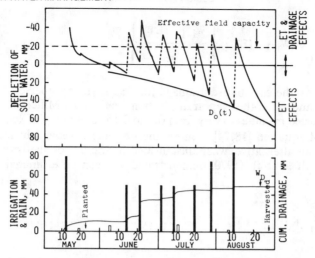

FIG. 18.15 A seasonal distribution of soil water level as typically associated with gravity irrigation systems that are managed to eliminate the prevailing depletion level at each irrigation.
From Jensen [1975].

gation. Then, from the time each irrigation is terminated to the time when
root zone drainage loss becomes negligible, evapotranspiration also occurs
simultaneously. The ET loss in these periods must be considered part of
the plant available water. To account for this additional available water,
attempts can be made to define an "effective" field capacity as indicated
in Fig. 18.15.

Miller and Aarstad (1971), however, found that the time to sample a
soil profile after irrigation for determining the "effective" field capacity is
affected by ET rates, soil properties, and the initial penetration depth by the
applied water. The most practical alternative is to apply daily estimates of
W_d to equation [18.11] after each irrigation until drainage loss is negligible, i.e., $\leqslant 0.1$ mm/day. Ogata and Richards (1957) showed the water
content for a draining soil profile can be expressed empirically by:

$$W = cW_o t^{-m} \quad \dotfill [18.12]$$

where W is the water content (mm); W_o is the water content when $t = 1$ day;
m is a constant for the given soil profile; c is a dimensional constant for
t^m; and t is the time after irrigation has stopped. The drainage rate is then
given by:

$$\frac{dW}{dt} = mW \left(\frac{W}{cW_o}\right)^{\frac{1}{m}} \quad \dotfill [18.13]$$

For daily time increments, Jensen (1972) determined the cumulative drainage is approximated with the equation:

$$W_d = \sum_{i=1}^{\infty} m(W_{i-1} - (ET)_i) \left\{ \frac{W_{i-1} - (ET)_i}{W_o} \right\}^{\frac{1}{m}} \dots \dots \dots [18.14]$$

where W_d = the cumulative drainage (mm); i is the number of days after irrigation, and ET is the evapotranspiration for the day. Miller and Aarstad (1974) reported m values to vary from 0.1 to 0.15 for a range of soil textures.

Use of equation [18.14] requires the field measurement of m and W_o values. As an alternative, when unsaturated hydraulic conductivity data are available, Black et al. (1970) demonstrated W_d can be estimated in sandy soils as:

$$W_d = \sum_{i=1}^{\infty} (K (\theta_a)_i - ET_i) \dots \dots \dots \dots \dots \dots \dots \dots [18.15]$$

where θ_a = W/L is the average soil water content through depth L on day i. L includes the deepest rooting depth of the crop. Equation [18.15] will estimate the drainage loss in finer textured soils also if the hydraulic gradient in the root zone remains close to unity during the drainage period.

A negative drainage loss or upward flow occurs when the water table is relatively close to the soil surface or close to the bottom of the root zone. Theoretical solutions to flow equations for this process have been given by Philip (1957) and Gardner (1958). For applied scheduling, Jensen (1972) suggested using:

$$W_d = \left\{ 1 - \frac{AW - a_4}{100 - a_4} \right\} \left\{ \frac{z_c}{z_w - z_r} \right\}^n \cdot ET \dots \dots \dots \dots \dots [18.16]$$

where z_c is the effective height of the capillary fringe above the water table; z_w is the depth to the water table; z_r is the depth of the roots; AW is the available water to z_r in percent; a_4 is a constant (about 25); and n is a constant for the given soil profile (expected to vary between 1 and 3).

With the USDA scheduling program, the days to next irrigation are computed as:

$$N = \frac{D_o - D}{E \left(\dfrac{dD}{dt} \right)} \dots \dots \dots \dots \dots \dots \dots \dots \dots \dots \dots \dots \dots \dots \dots \dots [18.17]$$

where N = the estimated days; D_o = the current optimum depletion of soil water; D = the estimated depletion to date; and E(dD/dt) = the expected rate of soil water depletion until the next irrigation is needed.

E(dD/dt) is often estimated from a calculation of the mean ET rate for the previous 3 to 5 days or it is estimated from advance 2 to 5 day weather forecasts. When N is more than 5 days into the future, expected potential evapotranspiration rates may be estimated from the long term mean levels of each climatic parameter at given days t. Assuming no rainfall, the advance estimates of daily (E(dD/dt)) are calculated as:

$$ET_i = K_c \cdot E(ET_p)_i \quad \dots\dots\dots\dots\dots\dots\dots\dots\dots\dots\dots\dots\dots [18.18]$$

where K_c is the prevailing value of the crop coefficient (see Chapter 6) on each day i; and $E(ET_p)$ is the expected value of potential evapotranspiration.

In climates where rainfall is significant, E(dD/dt) may include an expected rainfall value so that E(dD/dt) is calculated as the expected $(ET - P_e)_i$. Published functions for estimating expected rainfall are available for many areas in the U.S.

The amount of water to apply at irrigation is usually calculated as:

$$W_I = D_o/E; \quad D_o > D \dots\dots\dots\dots\dots\dots\dots\dots\dots\dots\dots\dots\dots\dots [18.19]$$

or

$$W_I = D/E; \quad D > D_o \dots\dots\dots\dots\dots\dots\dots\dots\dots\dots\dots\dots\dots\dots [18.20]$$

where E is the attainable irrigation efficiency, expressed as a decimal. These calculations are most applicable to conventional gravity methods and set type sprinkler systems. With automated or highly mechanized systems, such as the center pivot type, irrigations may be applied so that the prevailing D is not entirely eliminated at each water application (see Section 18.2.1). When necessary, W_I is also adjusted for the necessary leaching requirement as discussed in Chapter 5.

18.3.3 Communication of Scheduling Service INPUT-OUTPUT

The success of a computerized irrigation scheduling program depends highly upon the input of accurate data and the subsequent proper interpretation and application of its output. Applied experience (Corey and Franzoy, 1974) has indicated that a scheduling service should combine computer efficiency with the judgement of its field technicians.

Irrigation management usually begins when the crop is planted and then it is continued to the last seasonal irrigation. Equation [18.11] is updated at least weekly and fields are commonly inspected at weekly to biweekly intervals. Input data from each client can be transmitted by telephone to data recording systems. Output from the service is normally mailed or given orally when the technician makes his visit. The latter method provides more direct contact with the service and it is often the preferred method.

To eliminate scheduling service dependence on their clients for weekly input data, services can simply compute updates of ET accumulations from the climatic data input to their models. These ET data (usually for each new week) can be provided to each technician in the form of general tables with a printed "visit sheet" for each farm. This sheet is completed at the time the technician visits the farm, i.e., the water balances are updated manually for each field, given the new ET accumulation for the past week and the farmer's report of irrigation and rainfall amounts. Fig. 18.16 gives an example of this type of visit sheet.

FIELD NOTATIONS

Operator—Henness Farms
Location—Casa Grande, Arizona

Date 8/12/74
Farm No. 4208
Weather: Sl. Cloudy, Warm
Air Temp.: 27 (0700)

Field no.	Crop	Last irrig.	Root depth	Hold cap	Depletions Opt	Depletions Pres.	Irrig. date	Amt. to apply
21	Cotton	8-8	120 cm	5.1 cm	8.1 cm	4.6 cm	8-17	Solid
25	Cotton	8-6	137 cm	4.1 cm	7.1 cm	6.1 cm	8-14	Alt. row
32	Cotton	8-4	137 cm	4.3 cm	7.6 cm	7.1 cm	8-13	Alt. row

There was only 2 ft penetration in 21. It will hold ±5 days. You can go from 32 to 25 to 21. You can set the last irrigation on 21 around the 28th of August. The others will follow.

FIG. 18.16 An example "Farm Visit Report" for communication of irrigation scheduling recommendations by a scheduling consultant to a client farm.

For high frequency water application with center pivot systems Heermann et al. (1976) developed an output format that includes calculation of soil water depletion at both the normal start and stop positions, Fig. 18.17. The weekly water balance printout indicates the earliest irrigation starting dates that will avoid deep percolation losses for assumed application depths per revolution and expected rainfall amounts. "No later than" starting dates are also provided so that the stop position is irrigated before soil water depletion exceeds the safe allowable level.

Region: Crook
Farm: Condon Number 2
Corn
Date: Aug. 10

Day	Water used	Irrigation and rains	Irrigation dates	Depletion Where system Starts	Depletion Where system Stops
Aug. 4	0.38 cm	0.00 cm		0.38 cm	0.38 cm
Aug. 5	0.56 cm	0.00 cm		0.91 cm	0.91 cm
Aug. 6	0.66 cm	2.00 cm	Started	0.00 cm	1.60 cm
Aug. 7	0.69 cm	0.00 cm		0.69 cm	0.23 cm
Aug. 8	0.48 cm	0.00 cm		1.14 cm	0.71 cm
Aug. 9	0.64 cm	2.00 cm	Started	0.00 cm	1.35 cm
Aug. 10	0.71 cm	0.00 cm		0.71 cm	0.03 cm

MAXIMUM USEFUL RAIN AND
IRRIGATION AMOUNTS
LARGER AMOUNT WILL BE LOST

Date	Amount
Aug 11	1.47 cm
Aug 12	2.24 cm
Aug 13	2.97 cm
Aug 14	3.68 cm
Aug 15	4.27 cm
Aug 16	4.83 cm
Aug 17	5.36 cm
Aug 18	5.89 cm
Aug 19	6.43 cm

IF THE SYSTEM APPLIES 2.0 cm AND
MAKES A REVOLUTION IN 51 h, THE
RECOMMENDED STARTING TIMES ARE

Amount of rain	Start	No later than
No rain	Aug 13	Aug 19
0.64 cm	Aug 14	Aug 20
1.27 cm	Aug 15	Aug 22
2.54 cm	Aug 17	Aug 24

ASSUME THE SYSTEM WAS STARTED
AUG 13 THE NEXT STARTING
TIMES ARE

Amount of rain	Start	No later than
No rain	Aug 16	Aug 22
0.64 cm	Aug 17	Aug 23
1.27 cm	Aug 18	Aug 24
2.54 cm	Aug 20	Aug 27

FIG. 18.17 An example of detailed scheduling recommendations to a client farm as applied to high frequency center pivot irrigation [From Heermann et al., 1976.

18.3.4 Measurements that Assist Applied Water Balance Scheduling

Soil water content. Water balance computations are normally begun from measurements of initial soil water depletions. Subsequent measurements are also periodically needed to verify the predicted deficits. As discrepencies develop, due either to errors in input data or model simplifications, the necessary corrections are made.

With experience, technicians can achieve adequate accuracy by using the simple feel or appearance method to judge soil water content. This method requires a slotted soil probe of the Oakfield type (Haise and Hagan, 1967). Soil water depletion relative to field capacity is estimated by reference to guides such as given in Table 18.9. Each successive sample from the soil profile at a given site is first identified for textural class. Then the sample is manipulated as suggested by the guide and its water deficit is estimated in mm per unit length of soil column. The total root zone deficit is computed by adding the estimated deficits for each successive profile increment.

The initial soil water depletion is established by sampling at several places in each field. Particular attention should be given to the coarsest textured soils. Thereafter, with irrigation systems that achieve high application uniformity, field sampling can usually be confined to a "key area".

Other methods (see Section 4.1.8) are also available for field application. Electrical resistance blocks have long been available for the indirect measurement of soil water content. This method requires the use of small gypsum blocks and a portable read-out meter. The gypsum blocks are commercially available and they consist of pairs of stainless steel wires or wire grids which are embedded within each block. When placed in the soil profile the moisture content of the porous gypsum is controlled by the soil matric potential surrounding each block. The read-out meter measures either the electrical resistance or conductance between the embedded electrodes. This parameter is correlated to the block wetness and in turn the soil wetness surrounding each block. However, because the block porosity is predominantly fine, the sensitivity of this method to moisture change is not generally as accurate as with tensiometers in the 0 to -80 cb range of ψ_m.

Fischbach (1977) has published a detailed update on the preparation, installation, and scheduling use aspects of electrical resistance blocks. The placement of blocks is very similar to that described below for tensiometers.

The neutron method (see Chapter 4) has also been available for many years. In field use, access tubes must be driven to the desired depth in the early growing season. Moisture determinations are thereafter made whenever desired by lowering a probe into the access tube. Data are obtained on a read-out electronic counter system and converted to percent water by volume by using calibration curves. Care must be exercised to maintain near natural growing conditions around each access tube site. Repeatability and accuracy of the data have made this instrument an excellent method for research applications. The equipment is, however, expensive and data collection is time consuming. A radiation health hazard is a possibility unless the instrument is properly operated and maintained.

Soil matric potential. Soil matric potential (ψ_m) can be sensed on a continuing basis during freeze free periods with tensiometers as described in Chapter 4. The tensiometer has a working range of 0 to -80 cb. Since plants can obtain soil water that is held at potentials much below this magnitude, the tensiometer is most useful in soil textures where about 50 percent or more of the plant available water is within the tensiometer range. Approximate relationships between soil matric potential and available water level are given in Fig. 18.18 for a range of soil textures.

TABLE 18.9. GUIDE FOR JUDGING HOW MUCH OF THE AVAILABLE MOISTURE HAS BEEN REMOVED FROM THE SOIL*

Soil moisture deficiency	Feel or appearance of soil and moisture deficiency in centimeters of water per meter of soil			
	Coarse texture	Moderately coarse texture	Medium texture	Fine and very fine texture
0% (Field capacity)	Upon squeezing, no free water appears on soil but wet outline of ball is left on hand 0.0	Upon squeezing, no free water appears on soil but wet outline of ball is left on hand 0.0	Upon squeezing, no free water appears on soil but wet outline of ball is left on hand 0.0	Upon squeezing, no free water appears on soil but wet outline of ball is left on hand 0.0
0-25%	Tends to stick together slightly, sometimes forms a very weak ball under pressure 0.0 to 1.7	Forms weak ball, breaks easily, will not slick 0.0 to 3.4	Forms a ball, is very pliable, slicks readily if relatively high in clay 0.0 to 4.2	Easily ribbons out between fingers, has slick feeling 0.0 to 5.0
25-50%	Appears to be dry, will not form a ball with pressure 1.7 to 4.2	Tends to ball under pressure but seldom holds together 3.4 to 6.7	Forms a ball somewhat plastic, will slick slightly with pressure 4.2 to 8.3	Forms a ball, ribbons out between thumb and forefinger 5.0 to 10.0
50-75%	Appears to be dry, will not form a ball with pressure† 4.2 to 6.7	Appears to be dry, will not form a ball† 6.7 to 10.0	Somewhat crumbly but holds together from pressure 8.3 to 12.5	Somewhat pliable, will ball under pressure† 10.0 to 15.8
75-100%	Dry, loose single-grained, flows through fingers 6.7 to 8.3	Dry, loose, flows through fingers 10.0 to 12.5	Powdery, dry, sometimes slightly crusted but easily broken down into powdery condition 12.5 to 16.7	Hard, baked, cracked, sometimes has loose crumbs on surface 15.8 to 20.8

*From Israelson and Hanson (1972).
†Ball is formed by squeezing a handful of soil very firmly.

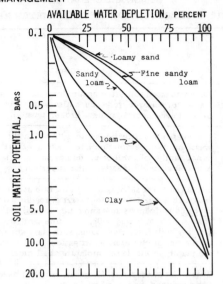

AVAILABLE WATER DEPLETION, PERCENT

FIG. 18.18 Water retention curves for several soils plotted in terms of percent available water removed [From Taylor, 1965].

Tensiometers sense soil matric potentials at point locations. Thus, when used as a scheduling tool, they are usually placed in pairs and at several sites within a given field. At each site a tensiometer is located in the zone of greatest root density (30 to 50 cm) for the timing of irrigations. Some established timing guidelines are given in Table 18.10. A second tensiometer is usually placed at about twice the shallow tensiometer depth (60 to 90 cm). This tensiometer senses when water, during irrigation, has penetrated to this depth. This penetration indicates the need to terminate irrigation. Soil variability usually requires the establishment of about two to four sites per 65 ha to obtain a representative measure of the soil water status.

Tensiometers have long been used effectively by trained technicians and irrigation scientists. Jensen (1975) reported, however, that they are not routinely used by farm managers. Key reasons appear to be: (a) they require considerable time for preparation, installing, recording of observations, and removal from each field; (b) they may interfere with cultivation and may be a nuisance to service; and (c) interpretation of the sensed matric potential can be confusing due to poor soil contact, leaks, and limited tensiometer range.

For specific details on their use, preparation, installation, etc., the following references are available (Fischbach and Schleusener, 1961; Bauder and Lundstrom, 1977).

Leaf water potential. Field observations of plant water potential can be made by measuring the negative hydrostatic xylem sap pressure with a pressure chamber as first described by Scholander et al. (1965). Commercial instruments* are available for research purposes. The use of instruments of this type by farm operators will probably not be widespread for reasons similar to those cited for tensiometers. Commercial scheduling services will

*Soil Moisture Equipment Corp.; Campbell Scientific Inc.; no trade name endorsement is intended.

TABLE 18.10. INTERPRETATION OF TENSIOMETER READINGS*

	Dial reading in centi-bars	Interpretation
Nearly saturated	0 10	Near saturated soil often occurs for a day or two following irrigation. Danger of water-logged soils, a high water table, poor soil aeration, or the tensiometer may have broken tension, if readings persist.
Field capacity	11 20 30	Field capacity. Irrigations discontinued in this range to prevent waste by deep percolation and leaching of nutrients below the root zone. Sandy soils will be at field capacity, in the lower range with clay soil at field capacity in the upper range.
Irrigation range	40 50 60	Usual range for starting irrigations. Soil aeration is assured in this range. In general, irrigations start at readings of 30-40 in sandy textured soils (loamy sands and sandy loams). Irrigations usually start from 40-50 on loamy soils, (very fine sandy loam and silt loams). On clay soils (silty clay loams, silty clays and etc.) irrigations usually start from 50-60. Starting irrigations in this range insures maintaining readily available soil moisture at all times.
Dry	70 80	This is the stress range. However, crop not necessarily damaged or yield reduced. Some soil moisture is readily available to the plant but is getting dangerously low for maximum production. Top range of accuracy of tensiometer, readings above this are possible but the tensiometer will break tension between 80 to 85 centibars.

*From Fischbach and Schleusener (1961).

likely not use these instruments either because of the need to make leaf xylem pressure measurements at rather specific or consistent times (usually the early morning or mid afternoon). Economic considerations would limit both the number of technicians and the time to be spent in each field.

Receptiveness of plant stress criteria for irrigation scheduling, therefore, appears keyed to development of models (Stegman et al., 1976; Childs et al., 1976; Gregory, 1978; Reicosky and Lambert, 1978) for estimating leaf xylem pressures or water potentials. Such models could be added to water balance scheduling programs to achieve a capability for advance estimation of both the soil water depletion and the expected plant water stress levels. Irrigation would then be programmed to prevent or minimize occurrences of leaf water potentials considered critical or damaging to yield potential.

Definitions of critical leaf water potentials as irrigation timing criteria are not well developed. Some studies (Musick, 1976; Stegman and Bauer, 1977) of yield loss due to stress show little yield reduction may occur when mid afternoon leaf xylem pressures average above -11 to -13 bars for sugarbeets and -14 to -16 bars for grain sorghum. Further relationships between the day-to-day diurnal patterns of leaf water potential and ultimate crop yield are likely in the near future as plant growth models are developed and tested.

18.4 RELATED MANAGEMENT OBJECTIVES

18.4.1 Salinity Control

Irrigation waters invariably contain dissolved salts. Thus, salinity control measures are required. Typically, irrigation water is applied to achieve a desired production and/or economic objective. Then, in principal, additional water applications must be scheduled for leaching to provide the only practical way to maintain favorable salt balances in crop root zones. Management and guidelines for salinity control were reviewed by Ayers and Westcott (1976).

Chapter 5 gives procedures for calculating minimum leaching fractions. These fractions are dependent on the salt concentration in the irrigation water and the salt distribution that given crops can tolerate in their root zones without significant yield loss.

If soil water deficits are replaced frequently, plant roots near the soil surface will have a steady supply of water with an electrical conductivity near that of the applied water. Crop yields are, therefore, usually more affected by the irrigation water salinity level than by the soil water salinity in the lower root zone; particularly, if the minimum leaching requirement is maintained (Bernstein et al., 1973 and 1975). Recent research findings (van Schilfgaarde et al., 1974) indicate that leaching fractions can usually be much lower (in the range of 0.05 to 0.20) than previously recommended. These reduced requirements contrast with estimated actual leaching fractions ranging from 0.30 to 0.60 for irrigated areas of the western U.S. (Jensen, 1975). In most areas this leached water percolates to drains or a shallow ground water zone where it ends up in return flow drainage.

Given this wide disparity between required and actual leaching fractions, it is probable that improved water scheduling (via modern scheduling service concepts) can play a significant role in reducing return flows. Other factors that are likely to have an impact in the direction of reduced leaching fractions are: (a) increased competition amongst water users as demand grows and ground water depletion continues in some regions; (b) higher pumping costs, (c) increased use of limited irrigation in water deficit and high water cost areas, and (d) increased environmental concern regarding salts and nutrients in return flow drainage.

Precise attainment of minimum leaching fractions is a complex problem (van Schilfgaarde et al., 1974). For example, a 1 percent error in estimating ET can cause a 20 percent change in leaching if the leaching target is 5 percent. The average leaching fraction for a field is also strongly dependent on the uniformity at which water is applied since the least watered area must also receive the minimum leaching requirement (LR). For a field that is non-uniformly watered, the average field leaching fraction can be estimated as (Jensen, 1975):

$$LF = 1 - \alpha_d (1 - LR) \quad \dots\dots\dots\dots\dots\dots\dots\dots\dots\dots\dots\dots\dots [18.21]$$

where, α_d = the distribution uniformity. α_d is defined as the ratio of the average water depth applied to some fraction (usually about 1/4) of each field area that receives the least amount of water to the average applied to the entire field. If $\alpha_d = 0.89$ (corresponding to a $U_c \cong 95$ percent) and LR = 0.10 then the the LF average for the entire field would be 0.20 to achieve an LR = 0.10 on the least watered area.

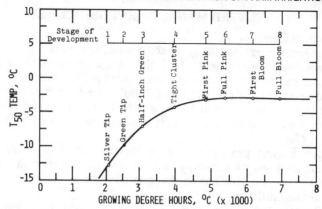

FIG. 18.19 Lethal T_{50} temperatures as related to the phenoclimatology of Red-Delicious apples [from Griffin, 1976].

Willardson and Hanks (1976) in leaching tests with a solid set sprinkler system ($U_c \geq 0.88$) concluded that it is unrealistic to expect to maintain average leaching fractions of less than 0.10 to 0.15 on a field scale. They recommend that it is more practical to irrigate without leaching for several seasons while monitoring the soil salinity status and then to leach periodically.

Other management problems make low leaching fractions difficult to achieve such as rainfall events after irrigation, irrigating to germinate seeds in soils with the lower profile near field capacity, and irrigating shallow rooted crops with gravity systems. These, and other examples that could be mentioned illustrate that salinity control must often involve an appropriate selection of management alternatives, such as: (a) improved irrigation scheduling, (b) modernization of existing irrigation systems, (c) growth of salt tolerant crops, (d) selection of new irrigation methods with improved application control and distribution uniformity, (e) changing cultural practices to minimize salinity effects on seed germination and growth, (f) providing drainage where a saline water table influences the plant root zone, (g) considering deep tillage for increasing water intake and profile conductivity, and (h) using good quality water when available. Details on most of these management techniques are discussed in Section 5.5.

18.4.2 Frost Control

Frosts are of two types: (a) advective cooling resulting from the passage of a cold air mass below 0 °C having winds in excess of about 6 to 8 m/s; and (b) radiative cooling on clear nights when the air mass temperature is above freezing. Irrigation provides more effective protection during radiation induced frosts than during advection frosts which are usually also of much longer duration (Wheaton and Carolus, 1975).

Plants on clear nights radiate energy to a cold sky and, therefore, may be cooled to 0.5 to 2.0 °C below the ambient air temperature (Landers and White, 1967). In addition they withdraw energy from the contact air layer which results in the immediate surrounding air being cooler than the bulk air above the canopy. Leaves, blossoms, and young fruit are the plant parts most sensitive to frost damage and are usually killed by temperatures of -1 to -3 °C (Burke et al., 1976). Lethal temperatures of more hardy parts such as buds of deciduous fruit trees are related to stage of development (see Fig. 18.19 as an example for Red Delicious Apples).

TABLE 18.11. SPRINKLING RATE IN MM PER HOUR NECESSARY
FOR COLD PROTECTION*

Temperature of a dry leaf in °C†	Wind speed in m/s				
	0 to 0.5	0.9 to 1.8	2.2 to 3.6	4.5 to 6.3	8.1 to 9.8
− 2.8	2.5	2.5	2.5	2.5	5.1
− 3.3	2.5	2.5	3.6	5.1	10.0
− 4.4	2.5	4.1	7.6	10.0	20.3
− 5.6	3.0	6.1	12.7	15.2	30.5
− 6.7	4.1	7.6	15.2	20.3	40.6
− 7.8	5.1	10.0	17.8	25.4	50.8
− 9.4	6.6	12.7	22.8	33.0	66.0
−11.7	8.6	17.8	30.5	43.2	86.4

*From Gerber and Harrison (1964).
†The temperature of a dry leaf is the expected minimum leaf temperature on an unprotected leaf. This will range from 0.5 °C below air temperature on nights with light wind to 1.5 to 2 °C on very calm nights.

Irrigation provides frost protection by the release of 80 cal/g of latent heat as water freezes. The high heat conductivity of ice provides for an efficient transfer of this energy to plant surfaces. Thus, as long as an adequate amount of water is applied the plant surface temperature will remain near 0 °C.

Irrigation for frost control is primarily restricted to sprinkler application where water is sprayed uniformly on the plant parts subject to damage. Chances of plant breakage from ice loading are high if duration of sprinkling is long or if application rates and wind speeds are relatively high. If too little water is applied, damage from freezing is worse than if no sprinkling had occurred (Gerber and Harrison, 1964). The cause for this is made more obvious by an examination of the physics involved. When ice forms energy is released, but when evaporation occurs energy is required. About seven times as much energy is used in evaporation as is given off by freezing. Thus, for every gram of water evaporated about 7 g must be frozen to offset the energy loss. If the evaporation loss becomes dominant due to insufficient water application, the plant will rapidly approach the wet bulb temperature. Icicle formation on uniformly ice covered plants provides a simple visual indication that more water is being applied than is immediately freezing. A clear rather than milky-white appearance of the ice provides additional evidence that the plant is not being refrigerated. Table 18.11 lists required application rates as field tested in Florida (Gerber and Martsolf, 1966) under various temperatures and wind speed conditions. The most practical rates range from 2.5 to 7.6 mm/h. Repeat frequency of leaf or foliage wetting must be at least once each minute. Sprinkling must begin by the time the wet bulb temperature reaches 2 °C above the lethal temperature of the plant parts to be protected. Once in operation sprinkling must continue until the wet bulb temperature is back above the lethal temperature by about 2 °C. In practice, systems are usually operated until the plant is free of ice, due to rising air temperature.

Accurate thermometers should be placed in good instrument shelters or radiation screens and used at plant level. Temperature alarm systems are a good idea to warn of impending dangerous temperature levels. Weather forecasts provide a general alert to potential for frost but are usually not sufficiently site specific.

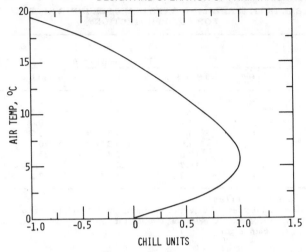

FIG. 18.20 Conversion of ambient air temperatures to chill units for modeling the winter rest period of deciduous fruit trees [from Griffin, 1976].

Frost protection by preseason sprinkling to delay bud development.
It is well known that deciduous trees enter a period of winter rest following their loss of leaves in the fall season. During rest, the trees are incapable of growth. After rest, the buds begin growing in the early spring that eventually leads to blossoming and leafing of the trees. The bud growth rate is temperature dependent. Cool early spring temperatures delay blossoming; conversely above normal temperatures accelerate bud development and the trees blossom early.

Alfaro et al. (1974) have researched the concept of sprinkling fruit trees during the bud growth period to deliberately cool the buds by evaporation; thereby, slowing development and delaying the critical blossoming stage until freezing injury is less likely to occur. To properly control and time the sprinkling operation, empirical relationships are available to model the phenology of various deciduous fruit trees during both their rest and subsequent bud development period.

Richardson and Griffin (1976) have estimated the date of winter rest completion by accumulating chill units, where one chill unit equals 1 h exposure at 6 °C. Fig. 18.20 illustrates the chilling contribution is less than one as temperatures drop below or rise above this optimum value. As an operational procedure, chill unit accumulation should begin during late summer when chill units are still negative. The summation of positive chill units to complete rest vary for different fruit species (Griffin, 1976). For example, the chill unit requirements are about 1234, 880, 860, and 720 for Delicious apples, Bing cherries, Elberta peaches and apricots, respectively.

After rest completion, bud development proceeds whenever the air temperature exceeds about 5 °C. Relationships of phenologic stages to accumulated growing degree hours (GDH), since end of rest, have also been established (Richardson and Griffin, 1976) for major fruit crops as illustrated in Fig. 18.19 for Delicious apples. For these relationships, growing degree hours are accumulated when air temperatures are in the range of 5 to 20 °C. The growing degree hours accumulated each hour are equal to the hourly average temperature minus the base temperature of 5 °C.

With relationships of bud development to GDH accumulation, management consultants can predict stages of bud development from temperature records. Irrigations are then scheduled for freeze protection when temperatures fall to levels that would be lethal for the predicted stage of bud development (see the T_{50} relationship to stage of growth as an example in Fig. 18.19).

A computer program (available from Utah State University; Griffin, 1976) can provide growers with a weekly printout that predicts current stage of bud development, dates to begin sprinkling, blossom dates, etc., and when to discontinue sprinkling so that their trees will blossom on or near a desired date. Research has shown that maximum bloom delay is achieved when sprinkling begins just after rest completion. Sprinkling should occur when air temperatures exceed 7 °C. Systems are normally managed on an on-off cycle which just maintains a continuously wet tree. In early spring, 2.0 to 2.5 mm/h rates are adequate, but as the season progresses and temperatures rise the amount of water must be increased.

18.4.3 Plant and Soil Cooling

As discussed earlier in Section 18.1.2, water potentials in plants under stress can be increased by increasing soil water potentials or soil water availability. However, on severe heat stress days, leaf water potentials may be reduced to growth limiting levels (Hsiao, 1973) even when soil water potentials are near zero. Under these conditions, the internal water status of plants will improve only with an artificial reduction of the environmental heat stress.

Plant and soil cooling is a relatively recent innovation in irrigated crop production. Significant increases in yield and/or quality have been demonstrated for such crops as snap beans, peas, potatoes, cucumbers, squash, tomatoes, strawberries, tree fruits, lettuce and grapes (Wheaton and Kidder, 1966; Howell et al., 1971; Peterson and Weigle, 1970; Bible et al., 1968; Gilbert et al., 1971; Unrath, 1972; Chesness and Braud, 1970). Increased crop yields are attributed to improved conditions for plant growth and the reduction of such stress induced problems as dehydration of fruit, dropped blossoms, and the "burning off" of young seedlings at or near the soil surface.

Degree of cooling achieved. Conventional low application rate and low pressure sprinkler irrigation equipment can achieve ambient air temperature reductions in the range of 3 to 7 °C (Chesness et al., 1977). From data representative of a solid set cooling system (application rates near 0.66 mm/h and Cu > 0.70) Hobbs (1972) determined that the potential air temperature reduction during sprinkling can be estimated from three ambient indices with the regression relationship:

$$\Delta t = 1.87 - 0.107x_1 + 0.114x_2 + 0.180x_3 ; r = 0.84, SD = 1.74 \ldots \ldots \text{[18.22]}$$

where: Δt = estimated air temperature reduction, °C; x_1 = prevailing relative humidity, percent; x_2 = prevailing ambient air temperature, °C; and x_3 = prevailing wind velocity, ms^{-1}.

Evaporative cooling on wetted surfaces typically results in leaf temperature reductions of about 2.0 to 2.5 times the attained air temperature reduction. Reduction to the wet bulb temperature limit is possible (Chesness et al., 1977) when plant surfaces are kept uniformly and continuously wet.

Soil temperatures near the surface are usually reduced to only half the air temperature reduction during sprinkling. However, much greater reductions (> 10 °C) have been achieved at the soil surface to enhance seedling survival (Robinson, 1970).

Crop cooling with sprinkled water also causes near immediate reduction of plant water deficits as reflected by measurements of leaf water potential. Howell et al. (1971) found leaf water potential can increase by 3 to 5 bars within 5 min after cooling is initiated.

Application rates, operation requirements and system needs. Very low application rates are needed for cooling due to the high latent heat of evaporation for water (564 cal/g).

Actual rates in reported studies (Bible et al., 1968; Gilbert et al., 1971) have ranged from about 0.7 to 4 mm/h. Maximum cooling occurs when sprinkling rates range from 1.0 to 1.5 times the prevailing potential evapotranspiration rate (Chesness et al., 1977). Rates near the upper end of this range allow for cooling inefficiency as caused by leaf runoff, non-uniformity of water application, and partial wetting of the leaf canopy. Theoretical models (Barfield and Walker, 1974; Westerman et al., 1976) are also available for predicting application rates and resulting leaf temperatures. Their applied usefulness may be limited, however, for lack of suitable input data.

Sprinkling should be continuous during the heat stress period if rates are near the potential evapotranspiration rate. Cyclic sprinkling of about 10 to 15 min on and an equal time off increases efficiency of water use as rates significantly exceed the potential evapotranspiration rates.

Timing of water application for plant cooling has largely been based on arbitrarily assumed critical temperatures. Sprinkling is usually begun at threshold temperatures near 27 to 29 °C for cool climate crops such as potatoes and at 30 to 32 °C for warm season crops (Chesness et al., 1977).

Crop cooling demands the use of solid set systems which can be operated to maintain continuously wet plants for maximum effectiveness. Below canopy irrigation is only about 1/3 as effective as overtree sprinkling (Lomas and Mandel, 1973).

18.4.4 Wind Erosion Control

Irrigation to control wind erosion is primarily associated with

(1) water effects on the cohesion of soils to form surface crusts and clods,

(2) surface soil protection by plant growth and residues following harvest, and

(3) more intensive cropping that reduces the length of nongrowing periods on cultivated cropland such as summer fallow on dryland. Wind erosion on irrigated land was reviewed by Mech and Woodruff (1967).

Wind erosion primarily occurs when the soil surface is bare or mostly bare and is in a fine, loosened condition because of tillage or winter freezing and thawing. Problems occur mostly during fall through spring, the nongrowing and early growth period of summer crops. Spring erosion problems can be critical during seedling development of row crops on sandy soils because of physical wind damage and/or sand blasting injury. Irrigation of medium- to fine-textured soils results in consolidation of loose surface soils after drying and the development of surface crusts that are largely erosion resistant. The cohesive forces from wetting causes agricultural soils to be wind erosion resistant while the surface is in a wet condition.

However, as the surface dries, sandy soils resume their erodibility. The finer texture range of the coarse textured soils, such as fine sandy loams, can be tilled following irrigation while in a moist condition to produce clods that are erosion resistant until broken down by water or tillage.

Irrigated crops harvested for grain, such as corn, sorghum, and wheat, produce more than adequate residues to control erosion. Generally, surface residues are adequate when the tillage system involves the use of disks, chisels, sweeps and rod weeders but are usually inadequate after moldboard plowing. Leveling or smoothing of irrigated fields by a land plane results in a loose, erodible surface soil condition. In areas subject to wind erosion, it is important to roughen the soil by chiseling or other tillage or by bedding for the next crop immediately after leveling or smoothing. The bedding as a tillage operation can increase surface clods, and in addition, the bed-furrow surface configuration is more erosion resistant than a flat soil surface.

18.4.5 Herbicide and Insecticide Applications

Herbicide application with irrigation systems is becoming an increasingly attractive production practice (Carey, 1973); Ross, 1974; Siefert, 1976). It is seen as a means whereby multiple use may be made of irrigation equipment. Elimination of trips across fields with ground or airplane application systems results in both economic and energy savings.

Pre-plant and pre-emergence herbicides generally (Fischbach and Martin, 1975; Heikes, 1976) perform well when applied with sprinkler irrigation systems (particularly with center pivot and solid set sprinkler). Sprinkler systems with C_u values in the 80 to 90 percent range compare very favorably in application uniformity with other application methods. Best results are achieved when the herbicide application is distributed in the surface 5 to 8 cm of soil with water applications ranging from 6 to 19 mm. Application of pre-emergence herbicides must take place immediately after planting and before new weeds germinate for maximum effectiveness (Fischbach and Martin. 1975).

Success is also keyed to selection of appropriate herbicides for expected weed spectrums in given crops and with application at the right time and in the right concentrations. Current recommendations regarding new herbicides and associated rates and application practices are typically available from the U.S. Cooperative Extension Service in each state, from chemical companies, and in farm magazine articles. Federally (U.S.) registered herbicides are available for sprinkler applications to such crops as corn, sorghum, potatoes, sugarbeets, alfalfa, rice, citrus, nut crops, and ornamentals. Several evaluated herbicides for control of grass type weeds are EPTC (trade names = Eradicane and Eptam), Butylate (Sutan +), Alachlor (Lasso) and Vernolate (Vernam). Atrazine (AAtrex, Atrazine 80W, AAtrex 4L and Atrazine 4L) is widely used to achieve broad spectrum broadleaf control in corn, or it can be mixed with either butylate or Alachlor for combined grass and broadleaf control. Herbicides can successfully control late season grassy weed development in furrow irrigated crops. Another example is the use of multiple applications of EPTC on alfalfa for season long control of barnyard grass, foxtail, and nutsedge.

Principle disadvantages for inclusion of herbicides with water applications are: (a) herbicide application may have to take place when no additional soil water is needed; (b) ground water pollution is a possibility if antipollution devices fail on systems supplied from pumped wells; (c) sprinkler applications during windy conditions can result in serious nonuniformity, a problem that is more severe with fixed and portable systems; (d) sprinkler drift losses may become excessive when wind speeds exceed about 7 m/s; (e) wind effects can also cause poor weed control on the leeward side of bed or hill planted crops; and (f) wettable powder forms of herbicide are difficult to keep in suspension during injection.

Herbicide application with surface irrigation systems can work, however, this type of application has a more limited potential. Light water application is more difficult and profile drainage loss typically exceeds that of sprinkler methods. Also, surface systems frequently have poor distribution uniformity with length of run. Water re-use systems are mandatory to re-cycle end of field runoff back to the field to prevent contamination of surface or ground water.

Furrow methods may not leach sufficient herbicide to the tops of hill or bed planted crops. Lack of sufficient turbulence in the irrigation water to keep herbicides in suspension may be an additional problem.

Insecticides in field crop production are primarily "target" applied after the insect buildup occurs, usually during the crop growing season. Application in irrigation water has limited potential for success and many insecticides pose greater environmental hazards than herbicides. Therefore, little work has been done on developing practices for insecticide application in irrigation water. Drip irrigation systems do, however, have potential for applying systemic insecticides in orchards and vegetable crops.

18.4.6 Plant Disease Control

Irrigation may influence plant diseases by

(1) maintaining higher humidity within the plant canopy,

(2) maintaining wetter soil conditions,

(3) spreading disease organisms by sprinkler droplet splash and reusing surface runoff, and

(4) preventing some stress related diseases and physiological disorders.

Irrigation and wet soil effects from poor drainage as they affect plant diseases were discussed in reviews by Baker and Snyder (1965), Zentnyer (1966), and Menzies (1967).

High humidity and sometimes free water on plant foliage are necessary for secondary infections of many fungus and bacterial foliage diseases to build up to damaging levels. Irrigation, especially sprinkler irrigation, might be expected to increase such diseases. However, observations and farming experiences indicate that irrigation associated infections of this type are slight. Irrigation primarily takes place during warm and mostly clear weather. Spore germination is favored by cool, cloudy, wet weather. Spore germination is favored by cool, cloudy, wet weather and apparently irrigation does not provide the favorable microclimate effect long enough for major secondary infections to develop. However, diseases such as bacterial blight on beans and leaf spot on sugarbeets may increase in severity when irrigation is applied soon after rain or is prolonged in application.

Sprinkler droplets can cause spores to dislodge from infected plants and increase spreading of diseases to other plants. It also tends to wash fungicide residues from foliage and can necessitate more frequent application. Also, wet soil conditions following irrigation necessitates careful timing of spray application by ground equipment. Movement of water can disseminate disease producing agents not only within a field but in surface runoff and drainage waters that are reused on other fields. Water movement may be important in spreading soil-borne rot-type disease organisms but it is normally not the causative factor associated with major spread of diseases.

In dense growing crops, irrigation can increase diseases of vegetable fruit in contact with the soil such as fruit rots of tomatoes, strawberries, and melons; bottom rot of lettuce, and clerotinia rot of beans. The disease-producing environment is the high humidity associated with the low dense crop and the wet soil. Damage can be reduced by using wider rows or trimming plants to increase air movement between rows.

Root rots such as rhizoctonia and verticillum and fusarium wilts are not appreciably affected by irrigation in the normal range of soil water management but may be more severe under excessive irrigation and under some stress conditions. Root pruning of sugarbeets caused by large shrinkage cracks developing in swelling clays can increase rhizoctonia infection by providing entry points into roots.

Some stalk rot infections (such as charcoal rot of sorghum which reduces grain filling, causes premature senescence, and increases lodging) are increased by plant water stress during grain filling. Normal irrigation during grain filling usually provides adequate control of charcoal rot. Other physiologically induced disease problems associated with plant water stress that can be adequately controlled by irrigation are internal drought spot damage to potato tubers, blossom end rot of tomato, and black heart disease of celery. Diseases associated with irrigation are likely to be more widespread on vegetables and more severe in the humid higher rainfall areas.

References

1 Addink, J. W., and L. A. Stetson. 1975. Irrigation scheduling for off-peak power use. Presented at Am. Soc. Civ. Engr. Irrig. and Drain. Spec. Conf., Logan, UT. Aug. 15-17.

2 Alfaro, J. E., R. E. Griffen, J. Keller, G. R. Kanson, J. L. Anderson, G. L. Ashcraft, and E. A. Richardson. 1974. Preventative freeze protection by preseason sprinkling to delay bud development. TRANSACTIONS of the ASAE 17(6):1025-1028.

3 Arkley, R. J. 1963. Relationship between plant growth and transpiration. Hilgardia. 34:599-584.

4 Ayers, R. S., and D. W. Westcot. 1976. Water quality for agriculture. Irrig. and Drain. Paper 29, Food and Agr. Org. of the UN, Rome, Italy.

5 Baker, K. F., and W. C. Snyder (Eds.) 1965. Ecology of soil borne plant pathogens, prelude to biological control. Univ. of Calif. Press, Berkeley.

6 Barnes, K. K., W. M. Carlton, H. M. Taylor, R. I. Throckmorton, and G. E. Vanden Berg (Eds.). 1971. Compaction of agricultural soils. ASAE Monograph, St. Joseph, MI 49085. 471 p.

7 Barfield, B. J., and J. N. Walker. 1974. Development of prediction relationship for water requirements with irrigation cooling. Univ. of Kentucky Water Resources Res. Ins., Res. Rpt. No. 70.

8 Barfield, B. J., W. G. Duncan and C. T. Haan. 1977. Simulating the response of corn to irrigation in humid areas. ASAE Paper No. 77-2005, ASAE, St. Joseph, MI 49085.

9 Barrs, H. D. 1968. In: Water deficits and plant growth, T. T. Kozlowski (Ed.), Academic Press, New York. pp. 235-368.

10 Batty, J. C., S. N. Hamad and J. Keller. 1975. Energy inputs to irrigation. Proc. Am. Soc. Civ. Engr., J. Irrig. and Drain. Div., 101(IR4):293-307.

11 Bauder, J., and D. R. Lundstrom. 1977. Tensiometers their use, installation and maintenance. North Dakota Agr. Exp. Sta. Ext. Cir. AE 100. 6 p.

12 Begg, J. E., and N. C. Turner. 1976. Crop water deficits. Adv. in Agron. 28:161-217. Academic Press, Inc., New York.

13 Bernstein, L., and L. E. Francois. 1973. Leaching requirement studies: sensitivity of alfalfa to salinity of irrigation and drainage waters. Soil Sci. Soc. Am. Proc. 37:931-943.

14 Bernstein, L., L. E. Francois and R. A. Clark. 1975. Minimal leaching with varying root depths of alfalfa. Soil Sci. Soc. Am. Proc. 39:112-115.

15 Bible, B. B., R. L. Cuthbert and R. L. Carolus. 1968. Response of some vegetable crops to atmosphere modifications under field condition. Proc. Am. Soc. Hort. Sci. 90:590-594.

16 Black, S., and D. Brosz. 1972. Irrigation scheduling through moisture accounting. South Dakota State Univ. Ext. Cir. 686.

17 Black. T. A., W. R. Gardner and C. B. Tanner. 1970. Water storage and drainage under a row crop on a sandy soil. Agron. J. 62:48-51.

18 Brady, R. A., L. R. Store, C. D. Nickell and W. L. Powers. 1974. Water conservation through proper timing of soybeans irrigation. Soil and Water Conserv. 29(6):266-268.

19 Burke, M. J., L. V. Gusta, H. A. Quamme, C. J. Weiser and P. H. Li. 1976. Freezing and injury in plants. Ann. Rev. Plant Physiol. 27:507-528.

20 Carey, P. J. 1973. How efficient are sprinkler applied herbicides? Agrichem. Age. pp. 6-8.

21 Cary, J. W., and J. L. Wright. 1970. Response of plant water potential to the irrigated environment of southern Idaho. Agron. J. 63:691-695.

22 Chen, K. L., J. W. Wolfe, R. B. Wensink and M. A. Kizer. 1976. Minimum energy designs for selected irrigation systems. ASAE Paper No. 76-2037, ASAE, St. Joseph, MI 49085.

23 Chesness, J. L., and H. J. Braud. 1970. Sprinkling to reduce heat stressing of strawberry plants. AGRICULTURAL ENGINEERING 51(3):140-141.

24 Chesness, J., L. Harper and T. Howell. 1979. Sprinkling for heat stress reduction. In: B. J. Barfield and J. Gerber, (Eds.), Modification of the Aerial Environment of Plants, ASAE Monograph, ASAE, St. Joseph, MI 49085. 530 p.

25 Childs, S. W., J. R. Gilley and W. E. Splinter. 1977. A simplified model of corn growth under moisture stress. TRANSACTIONS of the ASAE 20(5):858-865.

26 Clark, R. N., and E. A. Hiler. 1971. Plant measurements as indicators of moisture. ASAE Paper No. 71-231. ASAE, St. Joseph, MI 49085.

27 Corey, F. C., and G. E. Franzoy. 1974. Irrigation management in the central states. ASAE Paper No. 74-2562, ASAE, St. Joseph, MI 49085. 9 p.

28 Curry, R. B. 1971. Dynamic simulation of plant growth, II. Incorporation of actual daily weather data and partitioning of net photosyntheses. TRANSACTIONS of the ASAE 14(6):1170-1174.

29 Curry, R. B., C. H. Baker and J. G. Streeter. 1975. Soymod I: A dynamic simulator of soybean growth and development. TRANSACTIONS of the ASAE 18(5):963-968, 974.

30 Danielson, R. E. 1967. Root systems in relation to irrigation. In: Irrigation of Agricultural Land, R. M. Hagan et al., (Eds.). Am. Soc. Agron. Monog. 11, pp. 390-424.

31 Delia, B. R., and R. Tizio. 1964. Studies on water requirements of horticultural crops, I: Influence of drought at different stages of potato on the tubers yield. Agron. J. 56:377.

32 DeWit, C. T. 1958. Transpiration and crop yields. Inst. of Biol. and Chem. Res. on Field Crops and Herbage. Wageningen, The Netherlands. Versl. Landbouwk. Onderz. No. 64.6's-Gravenhage.

33 Downey, L. A. 1972. Water-yield relations for nonforage crops. Proc. Am. Soc. Civ. Eng., J. Irrig. and Drain. Div. 98(IR1):107-115.

34 Dudley, N. J., D. T. Howell and W. F. Musgrave. 1971. Optimal intraseasonal irrigation water allocation. Water Resource Res. 7(4):770-788.

35 Dvoskin, D. and E. O. Heady. 1976. U.S. agricultural production under limited energy supplies, high energy prices, and expanding agricultural exports. Misc. Rpt. Iowa State Univ. Ctr. for Agr. and Rural Develop., Ames, IA.

36 Egli, P. 1971. An analytical model for the appraisal of public irrigation projects. Rpt. EEP-43. Stanford Univ., Stanford, CA. 179 p.

37 Ehrler, W. L., and C. H. M. Van Bavel. 1968. Leaf diffusion resistance, illuminance, and transpiration. Plant Physiol. 43:208-214.

38 English, B. C., and D. Dvoskin. 1977. National and regional water production functions reflecting weather conditions. Iowa State Univ. Misc. Rpt.

39 Erie, L. J., O. F. French and K. Harris. 1965. Consumptive use of water by crops in Arizona. Ariz. Agr. Exp. Sta. Bul. 169, 44 p.

40 Erie, L. J., and D. E. Longenecker. 1968. Irrigation water management. In: Cotton. Iowa Univ. Press. Ames. pp. 21-24.

41 Fischbach, P. E. 1965. Scheduling irrigation by electrical resistance blocks. Nebr. Agr. Exp. Sta. Ext. Cir. 65-752, 13 p.

42 Fischbach, P. E., and A. Martin. 1975. Herbigation—applying herbicide through sprinkler systems. ASAE Paper No. 75-1527, ASAE, St. Joseph, MI 49085. 7 p.

43 Fischbach, P. E., and R. E. Schleusener. 1961. Tensiometers a tool to help control irrigation water. Univ. of Nebr. Agr. Exp. Sta., Lincoln, Ext. Cir. 61-716, 8 p.

44   Fischbach, P. E., and Somerholder. 1971. Efficiencies of an automated surface irrigation system with and without a runoff re-use system. TRANSACTIONS of the ASAE 14(4): 717-719.

45   Fischbach, P. E., and B. R. Somerholder. 1974. Irrigation design requirements for corn. TRANSACTIONS of the ASAE 17(1):162-165, 171.

46   Fischbach, P. E. 1977. Scheduling irrigations by electrical resistance blocks. Univ. of Nebr. Agr. Exp. Sta. Nebguide G77-340, 4 p.

47   Fonken, D. W., J. C. Steele and P. E. Fischbach. 1974. Sprinkler irrigation design criteria for sugar beets. TRANSACTIONS of the ASAE 17(3):889-891.

48   Gallagher, J. N., P. V. Biscoe and R. K. Scott. 1975. J. Appl. Ecol. 12:319-336.

49   Gardner, W. R. 1958. Some steady state solutions of the unsaturated moisture flow equation with application to evaporation from a water table. Soil Sci. 85:222-232.

50   Gardner, W. R. 1964. Relation of root distribution to water uptake and availability. Agron. J. 56:41-45.

51   Gardner, W. R. 1965a. Dynamic aspects of water availability to plants. Ann. Rev. of Plant Physiol. 16:323-342.

52   Gardner, W. R. 1965b. Soil water movement and root absorption. In: Plant Environment and Efficient Water Use, W. H. Pierre et al., (Eds.). Am. Soc. Agron. and Soil Sci. Soc. Am., Madison, WI. pp. 127-149.

56   Gates, D. M. 1968. Transpiration and leaf temperature. Ann. Rev. of Plant Physiology. 19:211-238. Annual Reviews, Inc., Palo Alto, CA.

57   Gerard, J. J., and L. N. Namken. 1966. Influence of soil texture and rainfall on the response of cotton to moisture regime. Agron. J. 58:39-42.

58   Gerber, J. F., and D. S. Harrison. 1964. Sprinkler irrigation for cold protection of citrus. TRANSACTIONS of the ASAE 7(4):404-407.

59   Gerber, J. F., and J. D. Martsolf. 1966. Protecting citrus from cold damage. Univ. of Florida Agr. Ext. Ser., Cir. 287, 29 p.

60   Gilbert, D. E., J. L. Meyer and J. J. Kissler. 1971. Evaporation cooling of vineyards. TRANSACTIONS of the ASAE 14(5):841-843.

61   Gilley, J. R., and D. G. Watts. 1976. Energy reduction through improved irrigation practices. Presented at Energy and Agr. Conf., St. Louis, MO, June 16-19.

62   Grable, A. R. 1966. Soil aeration and plant growth. Adv. in Agron. 18:57-106. Academic Press, Inc., New York.

63   Gregory, J. M. 1978. Xylem pressure prediction equation derived by use of mass theory. ASAE Paper No. MC-78-502. 11 p.

64   Griffin, R. E. 1976. Micro-climate control of deciduous fruit production with overhead sprinklers. In: Reducing Fruit Losses Caused by Low Spring Temperatures. Utah Agr. Exp. Sta., Doc. No. 10550101: Appendix F.

65   Grimes, D. W. 1977. Physiological response of cotton to water and its impact on economical production. Proc. Western Cotton Prod. Conf., Lubbock, TX, Feb. 22-24, 1977.

66   Grimes, D. W., and W. L. Dickens. 1977. Cotton responses to irrigation. Calif. Agr. 31(5):16-17.

67   Haddock, J. L. 1959. Yield, quality and nutrient content of sugar beet as affected by irrigation regime and fertilizers. J. Am. Soc. of Sugar Beet Tech. 10:344-355.

68   Hagan, R. M., and J. I. Stewart. 1972. Water deficits-irrigation design and programming. Proc. of ASCE, Irrig. and Drain. Div. 98(IR2):215-237.

70   Hagood, M. A. 1964. Irrigation scheduling from evaporation reports. Wash. State Univ. Ext. Cir. No. 341.

71   Haise, H. R., and R. M. Hagan. 1967. Predicting irrigation needs. In: Irrigation of Agricultural Lands, R. M. Hagan et al. (Eds.), Am. Soc. of Agron., Madison. Mono. No. 11, pp. 577-604.

72   Hall, W. A., and W. S. Butcher. 1968. Optimal timing of irrigation. Proc. Am. Soc. Civil Engr., J. Irrig. and Drain. Div. 94(IR2):267-275.

73   Hanks, R. J. 1974. Model for predicting plant yield as influenced by water use. Agron. J. 66:660-664.

74   Hanks, R. J., H. R. Gardner and R. L. Florian. 1969. Plant growth-evapotranspiration relations for several crops in Central Great Plains. Agron. J. 61:30-34.

75   Hanson, E. G. 1967. Influence of irrigation practices on alfalfa yield and consumptive use. New Mex. Agr. Exp. Sta. Bul. 514.

76   Hargreaves, G. H. 1968. Consumptive use derived from evaporation pan data. Proc. Am. Soc. Civ. Engr., J. Irrig. and Drain. Div. 94(IR1):97-105.

77   Hay, D. R., and D. L. Pope. 1975. Grain sorghum irrigation Kans. Agr. Exp. Sta., Ext. Cir. 6, 4 p.

78   Hay, D. R., and D. L. Pope. 1975. Irrigating soybeans. Kans. Agr. Exp. Sta., Ext. Cir. 5, 4 p.

79   Heermann, D. F. 1975. Irrigation scheduling for energy and water conservation. Proc. Sprinkler Irrig. Assoc. Tech. Conf., pp. 7-13.
80   Heermann, D. F., H. R. Haise and R. H. Mickelson. 1976. Scheduling center pivot sprinkler irrigation for corn production in eastern Colorado. TRANSACTIONS of the ASAE 19(1):284-287.
81   Heikes, E. 1976. Pre-emergence herbicide applied through center pivot sprinkler: four years weed control performance tests. Presented at Northwestern Herbigation Seminar, Boise, ID, Oct. 14. 14 p.
82   Herpich, R. 1974. Irrigate properly-Grain Sorghum Handbook. Kans. Coop. Ext. Serv. Cir. 494, pp. 14-16.
83   Hiler, E. A., and R. N. Clark. 1971. Stress day index to characterize effects of water stress. TRANSACTIONS of the ASAE 14(4):757-761.
84   Hiler, E. A., T. A. Howell, R. B. Lewis and R. P. Boos. 1974. Irrigation timing by the stress day index method. TRANSACTIONS of the ASAE 17(2):393-398.
85   Hillel, D. 1971. Soil and water physical principles and processes. Academic Press, Inc., New York, 288 p.
86   Hillel, D. 1977. Computer simulation of soil-water dynamics; a compendium of recent work. Internal Develop. and Res. Ctr., Ottawa, Canada, 214 p.
87   Hillel, D., and Y. Guron. 1973. Relation between evapotranspiration rate and maize yield. Water Resources Res. 9:743-749.
88   Hobbs, E. H. 1972. Crop cooling with sprinklers. Can. Agr. Eng. 15(1):6-8.
89   Hobbs, E. H., K. K. Krogman and L. G. Sonmor. 1963. Effects of levels of minimum available soil moisture on crop yields. Can. J. Plant Sci. 43:441-446.
90   Howe, O. W., and H. F. Rhoades. 1955. Irrigation practice for corn production in relation to stage of plant development. Soil Sci. Soc. Am. Proc. 19:94.
91   Howell, T. A., E. A. Hiler and R. D. Reddell. 1975. Optimization of water use efficiency under high frequency irrigation: II system simulation and dynamic programming. TRANSACTIONS of the ASAE 18(5):879-887.
92   Howell, T. A., E. A. Hiler and C. H. M. Van Bavel. 1971. Crop response to mist irrigation. TRANSACTIONS of the ASAE 14(5):906-910
93   Hsiao, T. C. 1973. Plant responses to water stress. Ann. Rev. Plant Physiol. 24:519-570.
94   Hurd, E. A. 1968. Growth of roots of seven varieties of spring wheat at high and low moisture levels. Agron. J. 60:201-205.
95   Hurd, E. A. 1974. Phenotype and drought tolerance in wheat. Agr. Meteorol. 14:39-55.
96   Huszer, P. E., M. D. Skold and R. E. Danielson. 1970. Evaluation of irrigation water and nitrogen fertilizer in corn production. Colorado Agr. Exp. Sta. Tech. Bul. 107, 38 p.
97   Idso, S. B., D. G. Baker and D. M. Gates. 1966. The energy environment of plants. Adv. in Agron. 18:171-218. Academic Press, Inc., New York.
98   Idso, S. B., R. D. Jackson and R. J. Reginato. 1977. Remote sensing of crop yields. Science 196:19-25.
99   Israelson, O. W., and V. E. Hansen. 1962. Irrigation principles and practices. 3rd ed., John Wiley and Sons, Inc., New York, NY, pp. 412.
100 Jackson, R. D., R. J. Reginato and S. B. Idso. 1977. Wheat canopy temperatures: A practical tool for evaluating water requirements. Water Resource Research (in press).
101 Jensen, M. C., and J. E. Middleton. 1970. Scheduling irrigation from pan evaporation. Washington Agr. Exp. Sta. Cir. 527.
102 Jensen, M. E. 1969. Scheduling irrigations with computers. J. Soil and Water Conserv. 24:193-195.
103 Jensen, M. E. 1972. Programming irrigation for greater efficiency. In: Optimizing the Soil Physical Environment Toward Greater Crop Yields. D. Hillel (Ed.). Academic Press, Inc., New York. pp. 133-161.
104 Jensen, M. E. 1975. Scientific irrigation scheduling for salinity control of irrigation return flows. Environ. Prot. Agency Tech. Ser. Rpt. 600/2-75-064, 100 p.
105 Jensen, M. E., and J. T. Musick. 1960. The effect of irrigation treatments on evapotranspiration and production of sorghum and wheat in the Southern Great Plains. Internatl. Soil Sci. Soc. 7th Cong. Trans. 1:386-393.
106 Jensen, M. E., D. C. N. Robb and G. E. Franzoy. 1970. Scheduling irrigations using climate-crop-soil data. Proc. Am. Soc. Civ. Eng., J. Irrig. and Drain. Div. 96(IR1):25-38.
107 Jensen, M. E., and W. H. Sletten. 1965a. Evapotranspiration and soil moisture-fertilizer interrelations with irrigated grain sorghum in the Southern High Plains. USDA Conserv. Res. Rpt. No. 5, 27 p.
109 Jensen, M. E., J. L. Wright and B. J. Pratt. 1971. Estimating soil moisture depletion from climate, crop and soil data. TRANSACTIONS of the ASAE 14(5):954-959.
110 Kanemasu, E. T., A. J. Chen, W. L. Powers and I. D. Teare. 1973. Stomatal resistance as an indicator of water stress. Trans. Kansas Acad. Sci. 76(2):159-166.

111 Kanemasu, E. T., L. R. Stone and W. L. Powers. 1976. Evapotranspiration model tested for soybean and sorghum. Agron. J. 68:569-572.

112 Kanemasu, E. T., G. W. Thurtell and C. B. Tanner. 1969. Design, calibration and field use of a stomatal diffusion parometer. Plant Physiol. 44:881-885.

113 Kassam, A.H., and J. F. Elston. 1974. Ann Bot. (London) 38:419-429.

114 Keller, W., and C. W. Carlson. 1967. Forage crops. In: Irrigation of Agricultural Land, R. M. Hagan et al. (Eds.). Am. Soc. Agron. Monog., 11 p. 607-621.

115 Kmoch, H. G., R. E. Ramig, R. L. Fox and F. E. Kochler. 1957. Root development of winter wheat as influenced by soil moisture and fertilization. Agron. J. 49:20-25.

116 Korven, H. C., and W. L. Pelton. 1967. Advection in southwest Saskatchewan. Can. Soc. Agr. Eng. 9:88-90.

117 Korven, H. C., and J. C. Wilcox. 1965. Correlation between evaporation from Bellani plates and evapotranspiration from orchards. Can. J. Plant Sci. 45:132-139.

118 Kozlowski, T. T. (Ed.). 1968a. Water deficits and plant growth. Vol. 1, Development, Control and Measurement. Academic Press, Inc., New York, 390 p.

119 Kozlowski, T. T. (Ed.). 1968b. Water deficits and plant growth. Vol. 2, Plant Water Consumption and Response. Academic Press, Inc., New York, 333 p.

120 Kozlowski, T. T. (Ed.). 1972. Water deficits and plant growth. Vol. 3, Plant Responses and Control of Water Balance. Academic Press, Inc., New York, 368 p.

121 Kozlowski, T. T. (Ed.). 1976. Water deficits and plant growth. Vol. 4, Soil Water Measurement, Plant Responses, and Breeding for Drought Resistance. Academic Press, Inc., New York. 384 p.

122 Kramer, P. J. 1969. Plant and soil water relationships: A modern synthesis. McGraw Hill Book Co., New York, 482 p.

123 Krogman, K. K., and E. H. Hobbs. 1976. Scheduling irrigation to meet crop demands. Can. Dept. of Agr., Ottawa, Publ. 1590, 18 p.

124 Landers, J. N. and K. White. 1967. Irrigation for frost production. In: Irrigation of Agricultural Land. R. M. Hagan et al. (Eds.). Am. Soc. Agron. Monog. 11, pp. 1037-1057.

125 Larson, W. E., and W. B. Johnston. 1955. The effect of soil moisture level on yield, consumptive use for water and root development by sugarbeets. Soil Sci. Soc. Am. Proc. 19:275-279.

126 Lemon, E. R. 1965. Energy conversion and water use efficiency in plants. In: Plant Environment and Efficient Water Use. W. H. Pierre et al. (Eds.). Am. Soc. Agron. and Soil Sci. Soc. Am., Madison, WI. pp. 28-48.

127 Levin, I. and E. Shmueli. 1964. The response of cotton to various irrigation regimes in the Hula Valley. Isael J. Agr. Res. 14:211-25.

128 Lewis, R. B., E. A. Hiler and W. R. Jodean. 1974. Susceptibility of grain sorghum to water deficit at three growth stages. Agron. J. 66:589-591.

129 Lomas, J. and M. Mandel. 1973. The quantitative effects of two methods of sprinkler irrigation on the microclimate of a mature avocado plantation. Agr. Meteorol. 12:35-48.

130 Lundstrom, D. R. and E. C. Stegman. 1977. Checkbook method of irrigation scheduling. ASAE Paper No. NCR77-1001.

131 Mapp, H. P., Jr., V. R. Eidman, J. F. Stone and J. M. Davidson. 1975. Simulating soil water and atmospheric stress-crop yield relationships for economic analysis. Okla. Agr. Exp. Sta. Tech. Bul. No. T140, 63 p.

132 Mech, S. J. and N. P. Woodruff. 1967. Wind erosion on irrigated land. In: Irrigation of Agricultural Land, R. M. Hagan et al. (Eds.). Am. Soc. Agron. Monog. 11, pp. 964-973.

133 Menzies, J. D. 1967. Plant diseases related to irrigation. In: Irrigation of Agricultural Land, R. H. Hagan et al. (Eds.). Am. Soc. Agron. Monog. 11, pp. 1058-1064.

134 Merriam, J. L. 1960. Field method of approximating soil moisture for irrigation. TRANSACTIONS of the ASAE 3(1):31.

135 Meyer, B. S., D. B. Anderson, R. H. Bohning and D. G. Fratianne. Nostrand Co., New York, 565 p.

136 Meyer, R. F. and J. S. Boyer. 1972. Planta 108:77-87.

137 Miller, D. E. 1976. Deficit high frequency irrigation of sugarbeets, wheat, and beans. USDA-ARS Scientific Paper No. 4785, 6 p.

138 Miller, D. E., and J. S. Aarstad. 1971. Available water as related to evapotranspiration rates and deep drainage. Soil Sci. Soc. Am. Proc. 35:131-134.

139 Miller, D. E., and J. S. Aarstad. 1973. Effective available water and its relation to evapotranspiration rate, depth of wetting, and soil texture. Soil Sci. Soc. Amer. Proc. 37: 763-766.

140 Miller, D. E., and J. S. Aarstad. 1974. Calculations of the drainage component of soil water depletion. Soil Sci. 118:11-15.

141 Miller, D. E., and J. S. Aarstad. 1976. Yields and sugar content of sugarbeets as affected by deficit high-frequency irrigation. Agron. J. 68:231-234.

142 Miller, S. F., L. L. Boersma and E. N. Castle. 1965. Irrigation water values in the Willamette Valley: A study of alternative valuation methods. Ore. Agr. Exp. Sta. Tech. Bul. 85, 34 p.

143 Morey, R. V., J. R. Gilley, F. G. Bergsrud and L. R. Dirkzwager. 1975. Yield response of corn related to soil moisture conditions. ASAE Paper No. 75-2004, St. Joseph, MI 49085.

144 Musick, J. T. 1970. Effect of antecedent soil water on preseason rainfall storage in a slowly permeable irrigated soil. J. Soil and Water Conserv. 25:99-101.

145 Musick, J. T. 1976. Irrigation timing and scheduling to increase water use efficiency. "Conflicts and issues in water quality and use" Seminar of Water Resources Comm., Great Plains Agr. Council, Denver, CO, April 7-8.

146 Musick, J. T., and D. A. Dusek. 1971. Grain sorghum response to number, timing and size of irrigations in the Southern High Plains. TRANSACTIONS of the ASAE 14(3):401-404, 410.

147 Musick, J. T., and D. W. Grimes. 1961. Water management and consumptive use of irrigated grain sorghum in Western Kansas. Kans. Agr. Exp. Sta. Tech. Bul. 113.

148 Musick, J. T., D. W. Grimes and G. M. Herron. 1963. Water management, consumptive use and nitrogen fertilization of irrigated winter wheat in western Kansas. USDA Prod. Res. Rpt. No. 75, 37 p.

149 Musick, J. T., and W. H. Sletten. 1966. Grain sorghum irrigation water management on Richfield and Pullman soils. TRANSACTIONS of the ASAE 9(2):369-371, 373.

150 Musick, J. T., W. H. Sletten and D. A. Dusek. 1971. Preseason irrigation of grain sorghum in the southern high plains. TRANSACTIONS of the ASAE 14(1):93-97.

151 Neghassi, H., D. F. Heermann and D. E. Smika. 1975. Wheat yield models with limited water. TRANSACTIONS of the ASAE 18:549-557.

152 Newman, E. I. 1969. Resistance to water flow in soil and plant. I. Soil resistance in relation to amounts of root: Theoretical estimates. J. Appl. Ecol. 6:261-272.

153 Ogata, G., and L. A. Richards. 1957. Water content changes following irrigation of bare-soil that is protected from evaporation. Soil Sci. Soc. Am. Proc. 21:355-356.

154 Pearson, R. W. 1965. Soil environment and root development. In: Plant Environment and Efficient Water Use, W. H. Pierre et al. (Eds.). Am. Soc. Agron. and Soil Sci. Soc. Am., Madison, WI, pp. 96-126.

155 Pelton, W. L., and H. C. Korven. 1969. Evapotranspiration estimates from atmometers and pans. Can. J. Plant Sci. 49:615-621.

156 Peterson, L. E., and J. L. Weigle. 1970. Varietal response of potatoes to air conditioning irrigation. Am. Potato J. 47:94-98.

157 Philip, J. R. 1957. The physical principles of soil water movement during the irrigation cycle. Proc. 3rd Internl. Cong. Irrig. Drain. pp. 8.125-8.154.

158 Pierre, W. H., D. Kirkham, J. Pesek and R. Shaw (Eds.). 1965. Plant Environment and Efficient Water Use. Am. Soc. Agron. and Soil Sci. Soc. Am., Madison, WI, 295 p.

159 Power, J. F., J. J. Bond, W. A. Sellner and H. M. Olson. 1973. Effect of supplemental water on barley and corn production in a subhumid region. Agron. J. 65:464-467.

160 Pruitt, W. O. 1971. Interactions of microclimate and energy balance under a wide range of surface resistance. Univ. of Calif., Water Resources Ctr. Ann. Rpt. pp. 16-25.

161 Pruitt, W. O., and S. K. Gupta. 1976. The capricious nature of combination evapotranspiration equations. ASAE Paper No. 76-2063. ASAE, St. Joseph, MI 49085.

162 Raney, F. C., and Y. Mihara. 1967. Water and soil temperature. In: Irrigation of Agricultural Land, R. M. Hagan et al. (Eds.). Am. Soc. Agron. Monog. 11, pp. 1024-1036.

163 Rawlins, S. L., and P. A. C. Raats. 1975. Prospects for high frequency irrigation. Science 188:604-610.

164 Reicosky, D. C., and J. T. Ritchie. 1976. Relative importance of soil resistance and plant resistance in root water absorption. Soil Sci. Soc. Am. J. 40:293-297.

165 Reicosky, D. C., and J. R. Lambert. 1978. Field measured and simulated corn leaf water potential. Agron. J. 42(2):221-228.

166 Richardson, E. A., and R. E. Griffin. 1976. Basic concepts of freeze protection. In: Reducing Fruit Losses Caused by Low Spring Temperatures. Utah Agr. Exp. Sta., Logan, UT. Doc. No. 10550101: Appendix G.

167 Ritchie, G. A. and T. M. Hinckley. 1975. Adv. Ecol. Res. 9:165-254.

168 Ritchie, J. T. 1971. Dryland evaporative flux in a subhumid climate. I. Micrometeorological influences. Agron. J. 63:51-55.

169 Ritchie, J. T. 1972. Model for predicting evapotranspiration from a row crop with incomplete cover. Water Resources Res. 8(5):1204-1213.

170 Ritchie, J. T., E. Burnett and R. C. Henderson. 1972. Dryland evaporative flux in a subhumid climate. III. Soil water influence. Agron. J. 64:168-173.

171 Ritchie, J. T. 1973. Influence of soil water status and meteorological conditions on evaporation from a corn canopy. Agron. J. 65:893-897.

172   Ritchie, J. T. 1974. Atmosphere and soil water influence on plant water balances. Agr. Meteorol. 14:183-198.

173   Robins, J. S., and G. E. Domingo. 1953. Some effects of severe soil moisture stress deficits at specific growth stages in corn. Agron. J. 45:618.

174   Robins, J. S., and C. E. Domingo. 1956. Moisture deficits in relation to the growth and development of dry beans. Agron. J. 48:67-70.

175   Robins, J. S., and C. E. Domingo. 1956. Potato yield and tuber shape as affected by severe soil moisture deficits and plant spacing. Agron. J. 48:488.

176   Robins, J. S., and C. E. Domingo. 1962. Moisture and nitrogen effects on irrigated spring wheat. Agron. J. 54:135.

177   Robinson, F. E. 1970. Modifying an arid microclimate with sprinklers. AGRI-CULTURAL ENGINEERING 51(8):465.

178   Rosenberg, H. J. 1969. Advective contribution of energy utilized in evapotranspiration by alfalfa in the east central Great Plains. Agr. Meteorol.. 6:179-184.

179   Ross, Ron. 1974. Herbigation. Irrig. Age 9(3):5-9.

180   Salter, P. J., and J. E. Goode. 1967. Crop responses to water at different stages of growth. Commonwealth Agr. Bur., Farnham Royal, Bucks, England, 246 p.

181   Schearer, M. N. 1963. Electrical resistance gypsum blocks for scheduling irrigation. Oregon State Univ. Ext. Bul. 810.

182   Schneider, A. D., J. T. Musick and D. A. Dusek. 1969. Efficient wheat irrigation with limited water. TRANSACTIONS of the ASAE 12(1):23-26.

183   Scholander, P. F., H. T. Hammel, E. D. Bradstreet and E. A. Hemmingsen. 1965. Sap pressure of vascular plants. Science 148:339-346.

184   Scofield, C. S. 1945. The water requirement of alfalfa. USDA Cir. No. 735.

185   Shalhavet, J., and L. Bernstein. 1968. Effects of vertically heterogeneous soil salinity on plant growth and water uptake. Soil Sci. 106:85-93.

186   Shannon, J. W. 1968. Use of atmometers in estimating evapotranspiration. Proc. Am. Soc. Civ. Eng., J. Irrig. and Drain. Div. 94(IR3)309-320.

187   Shaw, R. H., and D. R. Laing. 1965. Moisture stress and plant response. In: Plant Environment and Efficient Water Use, W. H. Pierre et al. (Eds.). Am. Soc. Agron. and Soil Sci. Soc. Am., Madison, WI, pp. 73-94.

188   Shipley, J., and C. Regier. 1975. Water response in the production of irrigated grain sorghum, High Plains of Texas. Tex. Agr. Exp. Sta. MP-12-2, 8 p.

189   Shipley, J., and C. Regier. 1976. Corn yield response to limited irrigations, high plains of Texas. Tex. Agr. Exp. Sta. Prog. Rpt. PR-3379C. 12 p.

190   Siefert, W. 1976. Weed management tool: Herbigation for irrigated cropland. Crops and Soils, 10-11.

191   Slayter, R. O. 1967. Plant-water relationships. Academic Press, Inc., New York.

192   Smerdon, E. T. 1974. Energy—our most critical resource. Irrig. J., Mar.-April 30-31.

193   Splinter, W. E. 1974. Modeling of plant growth for yield prediction. Agr. Meteorol. 14:243-253.

194   Stanberry, C. O., C. D. Converse, H. R. Haise and O. J. Kelley. 1955. Effect of moisture and phosphate variables on alfalfa hay production on the Yuma Mesa. Soil Sci. Soc. Am. Proc. 19:303.

195   Staple, W. J. 1974. Modified penman equation to provide the upper boundary condition in computing evaporation from soil. Soil. Sci. Soc. Am. Proc. 38(5):837-838.

196   Stapleton, H. N., D. R. Buxton, F. L. Watson, D. J. Nalting and D. N. Baker. 1973. Cotton: A computer simulation of cotton growth. Ariz. Agr. Exp. Sta. Tech. Bul. No. 206, 124 p.

197   Stegman, E. C., and L. D. Ness. 1974. Evaluation of alternative scheduling schemes for center pivot sprinkler systems. North Dakota Agr. Exp. Sta. Res. Rpt. No. 48, 20 p.

198   Stegman, E. C., L. H. Schiele and A. Bauer. 1976. Plant water stress criteria for irrigation scheduling. TRANSACTIONS of the ASAE 19(5):850-855.

199   Stegman, E. C., and H. M. Olson. 1976. Water management relationship for irrigated pinto beans. North Dakota Agr. Exp. Sta. Res. Rpt. No. 61, 6 p.

200   Stegman, E. C., and A. Bauer. 1977. Sugarbeet response to water stress in sandy soils. TRANSACTIONS of the ASAE 20(3):469-473, 477.

201   Stetson, L. E., D. G. Watts, F. C. Corey and I. D. Nelson. 1975. Irrigation system management for reducing peak electrical demands. TRANSACTIONS of the ASAE 18(2):303-306, 311.

202   Stetson, L. E. 1976. Outlook for energy. Irrig. Age. 10(6):20-22.

203   Stewart, J. I., and R. M. Hagan. 1969. Predicting effects of water shortage on crop yield. Proc. Am. Soc. Civ. Eng., J. Irrig. and Drain. Div. 95(IR1):91-104.

204   Stewart, J. I., and R. M. Hagan. 1973. Functions to predict effects of crop water deficits. Proc. Am. Soc. Civ. Eng., J. Irrig. and Drain. Div. 99(IR4):421-439.

205 Stewart, J. I., R. M. Hagan, and W. O. Pruitt. 1974. Functions to predict optimal irrigation programs. Proc. Am. Soc. Civ. Eng., J. Irrig. and Drain. Div. 100(IR2):179-199.

206 Stewart, J. I., R. D. Misra, W. O. Pruitt and R. M. Hagan. 1975. Irrigating corn and grain sorghum with a deficient water supply. TRANSACTIONS of the ASAE 8(2):270-280.

207 Stockton, J. R., J. R. Carreker, and M. Hoover. 1967. Irrigation of cotton and other fiber crops. In: Irrigation of Agricultural Land, R. M. Hagan et al. (Eds.). Am. Soc. Agron. Monog. 11, pp. 661-673.

208 Stone, J. F., R. H. Griffin, II, and B. J. Ott. 1964. Irrigation studies of grain sorghum in the Oklahoma Panhandle, 1958-1962. Okla. Agr. Exp. Sta. Bul. B-619, 35 p.

209 Sutter, R. J., and G. L. Cory. 1970. Consumptive irrigation requirements for crops in Idaho. Idaho Agr. Exp. Sta. Bul. 516, 97 p.

210 Sweeney, M. D. 1970. Soil and water characteristics important in irrigation. North Dakota Agr. Exp. Sta. Cir. S&F-573, 6 p.

211 Tanner, C. B., and W. L. Pelton. 1960. Potential evapotranspiration estimates by the approximate energy balance method. J. Geophys. Res. 65:3391.

212 Tanner, C. B., and W. A. Jury. 1976. Estimating evaporation and transpiration from a row crop during incomplete cover. Agron. J. 68:239-243.

213 Taylor, S. A. 1965. Managing irrigation water on the farm. TRANSACTIONS of the ASAE 8(3):433-435.

214 Teare, I. D., and E. T. Kanemasu. 1972. Stomatal-diffusion resistance and water potential of soybean and sorghum leaves. New Phytol. 71:805-810.

215 Thomas, J. C., K. W. Brown and W. R. Jordan. 1976. Stomatal response to leaf water potential as affected by preconditioning water stress in the field. Agron. J. 68:706-708.

216 Trava, J., D. F. Heermann and J. W. Labadie. 1976. Optimal on-farm allocation of irrigation water. TRANSACTIONS of the ASAE 20(1):85-88.

217 USDA, SCS, Eng. Div. 1967. Irrigation water requirements. Tech. Rel. No. 21, 83 p.

218 USDA, SCS. 1976. Irrigation guide for Minnesota. St. Paul, MN.

219 Unrath, C. R. 1972. The evaporative cooling effects of overtree sprinkler irrigation in red delicious apples. J. Am. Soc. Hort. Sci. 97(1):55-58.

220 Van Schilfgaarde, J., L. Bernstein, J. D. Rhoades and S. L. Rawlins. 1974. Irrigation management for salt control. Proc. Am. Soc. Civ. Engr., J. of the Irrig. and Drain. Div. 100(IR3):321-338.

221 Van den Honert, T. H. 1948. Water transport as a catanery process. Disc. Faraday Soc. 3:146-153.

222 Vitosh, M. L. 1977. Irrigation scheduling guide for field crops and vegetables. Ext. Bul. E-1110. Michigan State Univ.

223 Wardlow, I. F. 1967. Aust. J. Biol. Sci. 20:25-39.

224 Watts, D. G. 1975. Irrigation practices in southwest Nebraska in 1974. Proc. Nebr. Irrig. Short Course. Agr. Eng. Dept., Univ. of Nebraska.

225 Westerman, P. W., B. J. Barfield, O. J. Loewer and J. M. Walker. 1976. Evaporative cooling of a partially-wet and transpiring leaf: Parts I & II. TRANSACTIONS of the ASAE 19(5):881-893.

226 Wheaton, R. Z., and E. H. Kidder. 1966. To control leaf stress in plants. AGRICULTURAL ENGINEERING 47(6):325.

227 Wheaton, R. Z., and R. C. Carolus. 1975. Environmental control. In: Sprinkler Irrigation, 4th ed., C. H. Pair et al. (Eds.). Sprinkler Irrig. Assoc., Silver Springs, MD. pp. 381-386.

228 Willardson, L. S., and R. J. Hanks. 1976. Irrigation management affecting quality and quantity of return flow. Environ. Prot. Agency Tech. Ser. Rpt. 600/2-76-226, 191 p.

229 Wilcox, J. C., and W. K. Sly. 1974. A weather based irrigation scheduling procedure. Can. Dept. Agr. Tech. Bul. 83, 23 p.

230 Williams, D. W., and Chancellor, W. J. 1975. Irrigated agricultural production response to constraints in energy-related inputs. TRANSACTIONS of the ASAE 18:459-466.

231 Woodruff, C. M., M. R. Peterson, D. H. Schnarre and C. F. Cromwell. 1972. Irrigation scheduling with planned soil moisture depletion. ASAE Paper No. 72-722, ASAE, St. Joseph, MI 49085.

232 Wright, J. L., and M. E. Jensen. 1972. Peak water requirements of crops in southern Idaho. Proc. Am. Soc. Civ. Engr., J. Irrig. and Drain. Div. 96:193-201.

233 Yaron, D. 1971. Estimation and use of water production functions in crops. Proc. Am. Soc. Civ. Engr., J. Irrig. and Drain. Div. 97(IR2):291-303.

234 Yaron, D., and G. Strateener. 1973. Wheat response to soil moisture and optimal irrigation policy under conditions of unstable rainfall. Water Resource Res. 9(5):1145-1154.

235 Zentnyer, G. A. 1966. Soil aeration and plant diseases. Proc. Conf. on Drainage for efficient crop production, ASAE, St. Joseph, MI 49085. pp. 15-16.

PARTIAL LIST OF FREQUENTLY USED SYMBOLS AND UNITS*

Symbol	Explanation	Common Units
A	Area	m^2, cm^2, mm^2
C	Coefficient	---
C	Concentration of salts	---
D	Diameter (pipe)	mm, m
E	Efficiency, decimal fraction, subscripts denote specific uses	---
E	Evaporation, depth rate	mm/d
E_t	Evapotranspiration, depth rate	mm/d
EC_e	Electrical conductivity of saturation extracts	dS/m
H	Hydraulic head	m, cm‡
I	Cumulative infiltration	length
K	Hydraulic conductivity	length per unit time‡
K_c	Crop coefficient	---
K_{ca}	Crop coefficient, water not limiting	---
H_f	Head loss due to friction	m/m, m/100 m
L	Liter	$0.001\ m^3$
P	Precipitation	mm
P	Pressure (pressure head)	kPa (m)
P	Precipitation or application rate	mm/h
Q	Discharge or flow	m^3/s, L/s
R	Hydraulic radius	m, cm
R_A	Extraterrestrial solar radiation	$MJ/m^2 \cdot d$
R_n	Net radiation	$MJ/m^2 \cdot d$
R_s	Solar radiation	$MJ/m^2 \cdot d$
S	Slope	m/m
S_o	Bottom slope	m/m
S	Distance	m
T	Temperature	K, °C
T_d	Dewpoint temperature	°C
T_w	Wet bulb temperature	°C
U_e	Water use efficiency, dry matter per unit ET	kg/kg, kg/m^3
V	Mean velocity, Q/A	m/s
V, ∀	Volume	m^3, dm^3, cm^3
W	Watt	J/s
d_a	Vapor pressure deficit, $d_a = e^o - e$	kPa, mb‡
e	Water vapor pressure	kPa, mb‡
e^o	Saturation vapor pressure	kPa, mb‡
e_w^o	Saturation vapor pressure at wetbulb temperature	kPa, mb‡
g	Gram	g
g	Acceleration of gravity	m/s^2
h	Water pressure head	m, cm
m	Meter	m
n	Manning coefficient	$m^{1/6}$
q	Flow per unit border width	m^2/s
t	Time (second, hour, day, year)	s, h, d, a
u_z	Horizontal wind speed at height z	m/s, km/h
x	Horizontal coordinate	m, cm
y	Depth of water (vertical)	m, mm
z	Elevation	m, cm
δ	Difference	see usage
ψ	Water potential (subscripted as needed)	kPa, bar‡
α	Shortwave reflectance coefficient or albedo	---
β	Bowen ratio, $\beta = A/E$	---
θ	Volumetric soil water content	---
π	Pi (3.1416)	---
ρ	Density (subscripted as needed)	kg/m^3, g/cm^3

*Subscripts for special uses and symbols for empirical equations are defined in the text.
†Suggested SI metric units listed first with other acceptable units (Other dimensions are as specified in the text.)
‡Preferred units not established partly because of the wide range of values encountered.

APPENDIX
SELECTED CONVERSION FACTORS

To convert from	to	multiply by
Power		
Btu (Intern'l. Table)/min	watt (W)	17.584
cal (Intern'l. Table)/s	watt (W)	4.1868*
horsepower (550 ft·lbf/s)	kilowatt (kW)	0.745 70
Pressure		
atmosphere (standard)	kilopascal (kPa)	101.325*
bar	kilopascal (kPa)	100.00*
foot of water (39.2 °F)	kilopascal (kPa)	2.9890
inch of mercury (60 °F)	kilopascal (kPa)	3.376 85
meter of water ((4 °C)	kilopascal (kPa)	9.8064
millibar	kilopascal (kPa)	0.100 00*
lbf/ft²	kilopascal (kPa)	0.047 88
lbf/in² (psi)	kilopascal (kPa)	6.8948
Temperature		
degree Celsius	kelvin (K)	$t_k = t_{oc} + 273.15$
degree Fahrenheit	degree Celsius	$t_{oc} = (t_{of} - 32)/1.8$
degree Fahrenheit	kelvin (K)	$t_k = (t_{of} + 459.67)/1.8$
Time		
day (mean solar)	second (s)	86 400.
hour (mean solar)	second (s)	3 600.00
minute (mean solar)	second (s)	60.000
Velocity (Includes Speed)		
ft/min	meter per second (m/s)	0.005 080*
ft/s	meter per second (m/s)	0.304 80*
km/h	meter per second (m/s)	0.277 778
mi/h	meter per second (m/s)	0.447 04*
Viscosity		
centipoise	pascal second (Pa·s)	0.001 000*
poise	pascal second (Pa·s)	0.100 00*
Volume per Unit Time		
ft³/s	meter³ per second (m³/s)	0.028 317
gal (U.S. liquid)/h	liter per second (L/s)	0.227 125
gal (U.S. liquid)/min	meter³ per second (m³/s)	6.309 0 E-5
gal (U.S. liquid)/min	liter per second (L/s)	0.063 090

*An asterisk after the fifth significant digit indicates that the conversion factor is exact and that all subsequent digits are zero.

†Abstracted mainly from: ASTM. 1976. Standard for Metric Practice. E380-76, 37 p. and rounded to 5 significant figures.

To convert from	to	multiply by
Length		
foot	meter (m)	0.304 80*
inch	millimeter (mm)	25.400*
mile (statute)	kilometer (km)	1.6093
yard	meter (m)	0.9144*
Area		
acre (U.S. Survey)	meter^2 (m^2)	4 046.9
acre (U.S. Survey)	hectare (ha)	0.404 69
ft^2	meter^2 (m^2)	0.092 903
in^2	millimeter^2 (mm^2)	645.16*
Volume		
acre-foot (U.S. Survey)	meter^3 (m^3	1 233.5
acre-inch	meter^3 (m^3)	102.79
bushel (U.S.)	meter^3 (m^3)	0.035 239
ft^3	meter^3 (m^3)	0.028 317
hectare-millimeter	meter^3 (m^3)	10.000*
hectare-meter	meter^3 (m^3)	10 000.*
gallon (U.S. liquid)	liter (L)	3.7854
gallon (Canadian liquid)	liter (L)	4.5461
in^3	liter (L)	0.016 387
yd^3	meter^3 (m^3)	0.764 55
Electricity		
mho	siemens (S)	1.0000*
mmho/cm	decisiemens/meter (dS/m)	1.0000*
Energy (and Work)		
British thermal unit (Intern'l. Table)	joule (J)	1 055.06
calorie (Intern'l. Table)	joule (J)	4.1868*
kW·h	megajoule (MJ)	3.6000*
therm	megajoule (MJ)	105.51
Energy per Unit Area·Time		
Btu/ft^2·h	watt per meter^2 (W/m^2)	3.1525
cal (thermochemical)/cm^2·min	watt per meter^2 (W/m^2)	697.33
Mass		
gram	kilogram (kg)	1 000.0*
pound (avoirdupois)	kilogram (kg)	0.453 59
ton (2,000 lb)	kilogram (kg)	907.18
tonne	kilogram (kg)	1 000.0*
Mass per Unit Volume		
g/cm^3	kilograms per meter^3 (kg/m^3)	1 000.0*
lb/ft^3	kilograms per meter^3 (kg/m^3)	16.018
mg/L	kilograms per meter^3 (kg/m^3)	1.0000*

*An asterisk after the fifth significant digit indicates that the conversion factor is exact and that all subsequent digits are zero.

†Abstracted mainly from: ASTM. 1976. Standard for Metric Practice. E380-76, 37 p. and rounded to 5 significant figures.

SUBJECT INDEX